HANDBOOK OF
GENETICS

Volume 2
Plants,
Plant Viruses,
and Protists

HANDBOOK OF GENETICS

Volume 1: Bacteria, Bacteriophages, and Fungi

Volume 2: Plants, Plant Viruses, and Protists

Volume 3: Invertebrates of Genetic Interest

Volume 4: Vertebrates of Genetic Interest

Volume 5: Molecular Genetics

HANDBOOK OF GENETICS

ROBERT C. KING, EDITOR

Professor of Genetics, Department of Biological Sciences
Northwestern University, Evanston, Illinois

Volume 2
Plants,
Plant Viruses,
and Protists

PLENUM PRESS · NEW YORK AND LONDON

Library of Congress Cataloging in Publication Data

King, Robert C.
  Plants, plant viruses, and protists.

  (His Handbook of genetics; v. 2)
  Includes bibliographies and index.
  1. Plant genetics. 2. Plant viruses. 3. Unicellular organisms. I. Title.
  [DNLM: 1. Genetics. QH431 K54h]
  QH433.K56   581.1'5   74-23531
  ISBN 0-306-37612-1

© 1974 Plenum Press, New York
A Division of Plenum Publishing Corporation
227 West 17th Street, New York, N.Y. 10011

United Kingdom edition published by Plenum Press, London
A Division of Plenum Publishing Company, Ltd.
4a Lower John Street, London, W1R 3PD, England

Printed in the United States of America

Preface

The purpose of this and future volumes of the *Handbook of Genetics* is to bring together a collection of relatively short, authoritative essays or annotated compilations of data on topics of significance to geneticists. Many of the essays will deal with various aspects of the biology of certain species selected because they are favorite subjects for genetic investigation in nature or the laboratory. Often there will be an encyclopedic amount of information available on such a species, with new papers appearing daily. Most of these will be written for specialists in a jargon that is bewildering to a novice and sometimes even to a veteran geneticist working with evolutionarily distant organisms. For such readers what is needed is a written introduction to the morphology, life cycle, reproductive behavior, and culture methods for the species in question. What are its particular advantages (and disadvantages) for genetic study, and what have we learned from it? Where are the classic papers, the key bibliographies, and how does one get stocks of wild type or mutant strains? A list giving the symbolism for unknown mutations is helpful, but it need include only those mutants that have been retained and are thus available for future studies. Other data, such as up-to-date genetic and cytological maps, listings of break points for chromosomal aberrations, mitotic karyotypes, and haploid DNA values, will be included when available.

Genetics dates back to the work of the primitive agriculturalists who, thousands of years ago, in the ancient centers of civilization domesticated the various plants and animals we use to this day. However, modern genetics originated with the rediscovery in 1900 of a scientific article originally published in 1866 by the Augustinian monk, Gregor Mendel. About two thirds of this volume is made up of chapters covering plants that are favorites for genetic research, and, of course, Mendel's experimental organism, the garden pea, figures prominently among these.

In cases where a plant genus has been extensively studied there will be data presented for many related species, and here the nomenclature may become cumbersome and bewildering to geneticists working in other areas. In the chapter on cotton, for example, one sees reference to *Gossypium incanum* (Schwartz) Hillcoat and *G. stocksii* Masters ex Hooker. Lyle Phillips has explained to me that the name in parentheses is that of the original author of the specific epithet. The second authority changed the use of the epithet, i.e., moved the species from one genus to another or changed the rank of the taxon. In the *G. incanum* example, Schwartz described *incanum* as a species of *Cienfuegosia* and Hillcoat later transferred *incanum* to *Gossypium*. The "ex" is used when an author of a species name feels he should "share credit". In the *G. stocksii* example, Masters was the collector of the type specimen and had indicated on the herbarium sheet that he thought the specimen to be a representative of a new species. For unknown reasons he never published the species epithet, and when Hooker did subsequently, he gave Masters his due.

The plant chapters are followed by a chapter cataloguing the viruses attacking plants, and the volume ends with four chapters dealing with protists (as defined in the five kingdom classification described by Lynn Margulis in volume 1 of this series). Volume 3 will cover the invertebrates of genetic interest.

I am particularly grateful for the splendid assistance provided by Pamela Khipple and Karen Slusser during the preparation of this volume.

Robert C. King

Evanston
October, 1974

Contributors

Stig Blixt, Weibullsholm Plant Breeding Institute, Landskrona, Sweden

Edward H. Coe, Jr., Division of Biological Sciences and United States Department of Agriculture, University of Missouri, Columbia, Missouri

Heinz Fraenkel-Conrat, Department of Molecular Biology and Virus Laboratory, University of California, Berkeley, California

Edward D. Garber, Barnes Laboratory, University of Chicago, Chicago, Illinois

Ahmed F. Hadidi, Department of Virology and Cell Biology, Litton Bionetics, Inc., Kensington, Maryland

Cornelia Harte, Institüt fur Entwicklungsphysiologie, Universität zu Köln, Köln-Lindenthal, West Germany

Gurdev S. Khush, Plant Breeder and Head, Varietal Improvement Department, International Rice Research Institute, Los Baños, Laguna, Philippines

R. P. Levine, The Biological Laboratories, Harvard University, Cambridge, Massachusetts

Myron G. Neuffer, Division of Biological Sciences and United States Department of Agriculture, University of Missouri, Columbia, Missouri

Robert A. Nilan, Program in Genetics and Department of Agronomy and Soils, Washington State University, Pullman, Washington

Lyle L. Phillips, Department of Crop Science, North Carolina State University, Raleigh, North Carolina

György P. Rédei, Agronomy Department, University of Missouri, Columbia, Missouri

Charles M. Rick, Department of Vegetable Crops, University of California, Davis, California

Richard W. Robinson, New York State Agricultural Experimental Station, Department of Seed and Vegetable Sciences, Hedrick Hall, Geneva, New York

Edward F. Rossomando, School of Dental Medicine, University of Connecticut Health Center, Farmington, Connecticut

Ernest R. Sears, Agricultural Research Service, United States Department of Agriculture, University of Missouri, Columbia, Missouri

Harold H. Smith, Department of Biology, Brookhaven National Laboratory, Upton, New York

Tracy M. Sonneborn, Department of Zoology, Indiana University, Bloomington, Indiana

Erich Steiner, Matthaei Botanical Gardens, University of Michigan, Ann Arbor, Michigan

Maurice Sussman, Department of Biology, Brandeis University, Waltham, Massachusetts

Jack Van't Hof, Biology Department, Brookhaven National Laboratory, Upton, New York

Thomas W. Whitaker, United States Department of Agriculture, Agricultural Research Service, Horticultural Field Station, La Jolla, California

Contents

D. THE PLANTS

Chapter 1
  Corn (Maize) 3
  MYRON G. NEUFFER AND EDWARD H. COE, JR.

Chapter 2
  Rice 31
  GURDEV S. KHUSH

Chapter 3
  The Wheats and Their Relatives 59
  ERNEST R. SEARS

Chapter 4
  Barley (Hordeum vulgare) 93
  ROBERT A. NILAN

Chapter 5
  Cotton (Gossypium) 111
  LYLE L. PHILLIPS

Chapter 6
  Cucurbita 135
  THOMAS W. WHITAKER

Chapter 7
  Cucumis 145
  RICHARD W. ROBINSON AND THOMAS W. WHITAKER

Chapter 8
  Arabidopsis thaliana 151
  GYÖRGY P. RÉDEI

Chapter 9
  The Pea 181
    STIG BLIXT

Chapter 10
  Oenothera 223
    ERICH STEINER

Chapter 11
  The Tomato 247
    CHARLES M. RICK

Chapter 12
  Nicotiana 281
    HAROLD H. SMITH

Chapter 13
  Antirrhinum majus L. 315
    CORNELIA HARTE

Chapter 14
  Collinsia 333
    EDWARD D. GARBER

Chapter 15
  *The Duration of Chromosomal DNA Synthesis, the Mitotic
  Cycle, and Meiosis of Higher Plants* 363
    JACK VAN'T HOF

E.   PLANT VIRUSES

Chapter 16
  Host-Range and Structural Data on Common Plant Viruses .. 381
    AHMED F. HADIDI AND HEINZ FRAENKEL-CONRAT

F.   PROTISTS OF GENETIC INTEREST

Chapter 17
  Chlamydomonas reinhardi 417
    R. P. LEVINE

Chapter 18
  Cellular Slime Molds 427
    MAURICE SUSSMAN AND EDWARD F. ROSSOMONDO

Chapter 19
    Tetrahymena pyriformis 433
        TRACY M. SONNEBORN

Chapter 20
    Paramecium aurelia 469
        TRACY M. SONNEBORN

Author Index 595

Subject Index 609

Contents of Other Volumes 623

PART D
THE PLANTS

1

Corn (Maize)

MYRON G. NEUFFER AND EDWARD H. COE, JR.

Corn, *Zea mays* L., is a robust, monoecious summer annual of the grass family (Gramineae) with broad alternate leaves, an erect stalk terminated by a staminate inflorescence (tassel) composed of a main spike and several branches, and one to several pistillate inflorescences. The pistillate inflorescences are composed of 8–16 or more rows of kernel-bearing spikelets attached to a woody rachis (cob), and the whole (ear) is enclosed in several large foliaceous bracts (husks) and arises in the axils of the leaves at the mid-section of the plant. Long styles (silks) protrude from the tip of the ear through the husks as long silky threads.

Pollination is accomplished by wind dispersion, thus resulting in both self- and cross-fertilization. Cross-fertilization, wide distribution, and selection by man have brought about a high degree of variability. Corn grows only under cultivation, there being no truly wild variants known, although wild teosinte and corn show reciprocal introgression wherever they both occur (Galinat, 1971; Wilkes, 1972).

The international common name "maize" is used in most literature because the term "corn" is ambiguous internationally; however, some ambiguity exists for the term "maize" as there is a type of sorghum which is called "milo maize."

Corn exhibits several characteristics that make it a unique and useful organism for genetic and plant-breeding studies—these include high genetic variability; separate inflorescences, thus simplifying crossing techniques; free crossability; large ears with many kernels, thus providing large easily stored populations; kernels that express a wide array of easily

MYRON G. NEUFFER AND EDWARD H. COE, JR.—Division of Biological Sciences and United States Department of Agriculture, University of Missouri, Columbia, Missouri.

observable phenotypes; adaptability to cultivation practices; and major economic usefulness.

As a result of easy recognition of segregating characteristics and the continuing investigations of many scientists, a significant number of special techniques and phenomena have surfaced involving corn as the investigated organism. These include: demonstration of multiple allelism (Emerson, 1911); discovery and utilization of hybrid vigor (heterosis) (Shull, 1911); genetic control of mutability (Emerson, 1914; Rhoades, 1941; McClintock, 1950); interactions in multiple-factor control of phenotypes (Emerson, 1918, 1921); maternal inheritance (Anderson, 1923); Mendelizing behavior in the gametophyte (Brink and MacGillivray, 1924; Demerec, 1924); inheritance of chromosome aberrations—translocations (Burnham, 1930, 1950), inversions (McClintock, 1933), rings (McClintock, 1932, 1938), deficiencies (McClintock, 1941); genic control of meiosis (Beadle, 1930, 1931, 1932a,b, 1933, 1937; Clark, 1940; Nelson and Clary, 1952; Rhoades and Dempsey 1966b; Sinha, 1962); association of cytological exchange with crossing over (Creighton and McClintock, 1931); pachytene chromosome morphology and its use in chromosome analysis (McClintock, 1931; Longley, 1938, 1939); cytoplasmic inheritance and its relationship to nuclear genotype (Rhoades, 1931, 1933; Josephson and Jenkins, 1948; Duvick, 1965); chromosome pairing, movement, and disjunction (Burnham, 1932, 1950; McClintock, 1933, Randolph, 1941; Roman, 1947; Laughnan, 1961; Rhoades and Dempsey, 1966a,b, 1972; Doyle, 1969; Burnham et al., 1972); genetic effects of x rays and ultraviolet light (Stadler, 1932, 1939; Stadler and Roman, 1948; Nuffer, 1957); systematic linkage mapping (Emerson et al., 1935); nucleolar inheritance and behavior (McClintock, 1934); variation in morphological features of chromosomes among populations (Longley, 1938, 1939; Longley and Kato, 1965); breakage-fusion-bridge-breakage behavior of broken chromosome ends (McClintock, 1939); behavior of accessory (B-type) chromosomes (Randolph, 1941) and the use of A–B translocations to locate genes on chromosomes and to control dosages (Roman, 1947; Beckett, 1972); relating DNA to genetic material by correlation of induced mutation with uv absorption spectrum (Stadler and Uber, 1942); genetic control of cytoplasmic inheritance of chloroplasts (Rhoades, 1943, 1947); cytological demonstration of pseudoallelism (McClintock, 1944); mutation analysis of selected loci (Stadler, 1946); discovery of transposable genetic elements and their control of chromosome breakage and gene expression (McClintock, 1950); compound loci (Stadler and Nuffer, 1953; Stadler and Emmerling, 1956; Laughnan, 1952, 1961); relationships among chlorophyll, carotenoid pigments, and seed dormancy (Robertson, 1955); allelic isoenzymes and the formation of hybrid enzymes (Schwartz, 1960); determination of a gene action se-

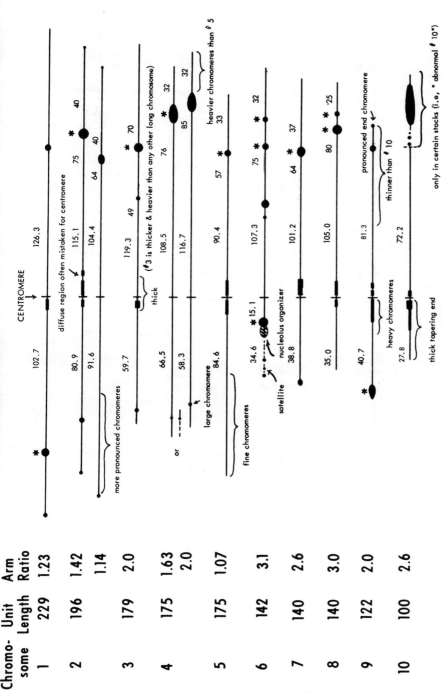

Figure 1. Cytological map of maize chromosomes. The smallest dots represent prominent chromomeres. An asterisk indicates that knobs are found in more than 50 percent of the races. The more common form of chromosomes 2 and 4 is given first. An error in the arm ratio for chromosome 6 printed in earlier versions of this map was called to our attention by Dr. B. G. S. Rao. Therefore, a correction has been made which conforms to the original data. The figure is drawn to scale and is based on published and unpublished data from Longley, McClintock, and Rhoades and Longley and Kato (1965).

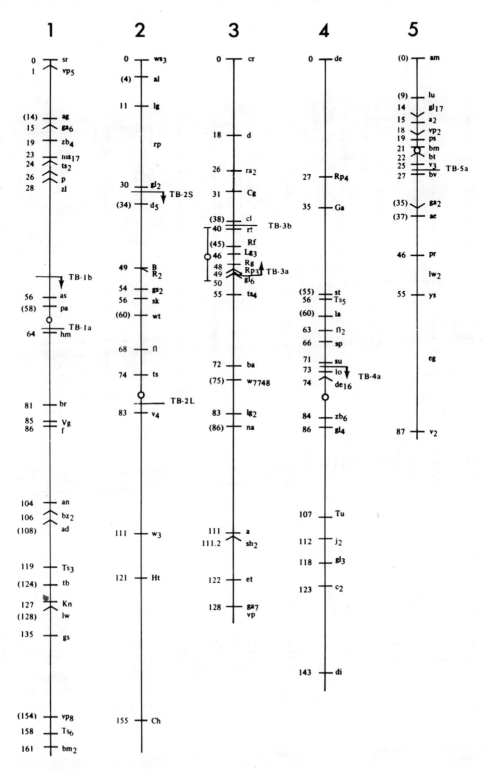

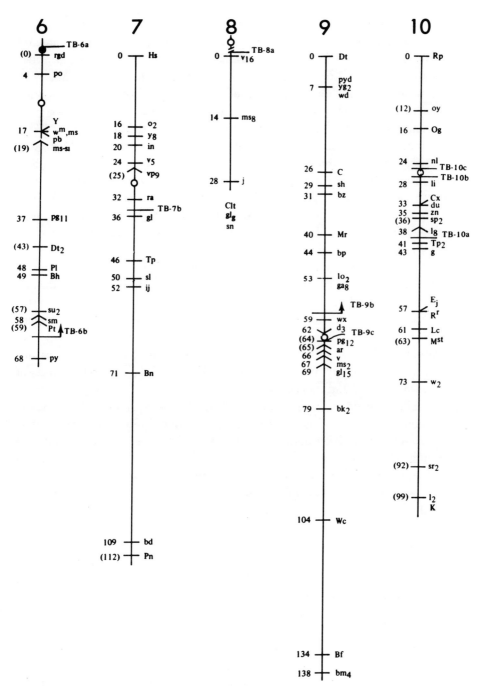

*Figure 2. Linkage map of maize. Parentheses indicate probable position based on insuffi-
cient data, ○ indicates centromere position, and ● indicates organizer. Positions designated
TB identify the genetic location of breakpoints of A-B translocations, which generate terminal deficiencies; ——↘indicates that the TB breakpoint is in that position or is some
distance in the direction indicated. Revised from Neuffer et al. (1968).*

quence for the anthocyanin biosynthetic pathway (Reddy and Coe, 1962); virus mutagenesis (Sprague *et al.*, 1963) and the associated phenomenon of aberrant ratio (Sprague and McKinney, 1971); discovery of nutritionally adequate amino acid balance in certain mutant strains (Mertz *et al.*, 1964); genetic demonstration of endosperm morphogenesis pattern (McClintock, 1965); combining of subunits to form multiple isozymes (Scandalios, 1965); paramutation (Coe, 1966; Brink *et al.*, 1968); preferential segregation and neocentric activity (Rhoades and Dempsey, 1966*a*); effects of ploidy relationships in seed development (Rhoades and Dempsey, 1966*b*; Sarkar and Coe, 1971); genetic demonstration of shoot apex morphogenesis pattern (Steffensen, 1968); isoenzyme variations (summarized by MacDonald and Brewbaker, 1972). The above, somewhat arbitrary, list presents many examples; however, numerous other important contributions could be included.

The biology and utilization of this species is given detailed consideration in *Corn and Corn Improvement,* edited by G. F. Sprague (1955). In addition to sections on the specific botany, culture, diseases, and insect pests, this book contains a thorough section by M. M. Rhoades on cytogenetics. The origin of maize has been reviewed recently by Galinat (1971). Specific chromosomal features in the diverse races are described by Longley and Kato (1965); Wilkes (1972) discusses aspects of the wild relatives and their survival.

The linkage summary prepared by Emerson *et al.* (1935) is the primary source of original descriptions and information on interactions of genetic factors, linkage data, and references. More recent aids include the annual *Maize Genetics Cooperation Newsletter* (the organ of information exchange among contributors, used with the understanding that contributions will not be cited in publications without the consent of the author); a Symbol Index to the *Maize Newsletter* printed in 1962 (available from Coe); a book by Neuffer *et al.* (1968) containing maps, color photographs of most mutant types, descriptions, and interaction information; and an annotated gene list with reference leads to biochemical studies on genetic types by Neuffer and Coe (1970) in the *Handbook of Biochemistry.*

Small quantities (20–50 kernels) of maize genetic strains may be obtained from Dr. R. J. Lambert, Maize Genetics Cooperative, Department of Agronomy, University of Illinois, Urbana, Illinois 61801.

For a cytological map with important chromosome features expressed diagrammatically and to scale, see Figure 1. A linkage map with those genetic markers that have been placed with relation to other markers is shown in Figure 2. A list of all the known genetic markers that have been reported and verified is given in Table 1 in the form specified in the new Recommendations for Nomenclature [*Maize Newsletter* **48**:201–202 (1974)].

TABLE 1. Genetic Markers

Symbol	Name	Location	Phenotype	Reference
a	Anthocyaninless	3–111	Colorless aleurone, green or brown plant, brown pericarp	Emerson et al., 1935
α	Component at A (see β)	3–111	Pale aleurone, red-brown plant, dark brown pericarp	Laughnan, 1952
a2	Anthocyaninless	5–15	Like a but red pericarp with P-rr	Emerson et al., 1935
a3	Anthocyanin	3L	Red pigment in sheath, culm and husks; resembles B, but is recessive	
Ac	Activator	Various	Transposable factor, regulates Ds activity	McClintock, 1950
ad	Adherent	1–(108)	Seedling leaves adhere	Emerson et al., 1935
Adh	Alcohol dehydrogenase	1–near lw	Electrophoretic mobility of enzyme	MacDonald and Brewbaker, 1972
Adh2	Alcohol dehydrogenase	—		
ae	Amylose extender	5–(37)	High amylose content of kernel	Creech, 1965
ag	Grasshopper resistant	1–14		Coe, 1962
al	Albescent	2–4	Erratic development of chlorophyll; pale yellow endosperm	Emerson et al., 1935
am	Ameiotic	5–0	Meiosis fails, sporogenous tissue degenerates	Sinha, 1962
Amy	Amylase	—	Electrophoretic mobility	MacDonald and Brewbaker, 1972
Amy2	Amylase	—		
an	Anther ear	1–104	Dwarf, anthers in ear florets, few tassel branches; responds to gibberellins	Emerson et al., 1935
Ap	Acid phosphatase	—	Electrophoretic mobility	MacDonald and Brewbaker, 1972
Ap2	Acid phosphatase	—		
ar	Argentia	9–65	Virescent seedling, greens rapidly	Emerson et al., 1935
as	Asynaptic	1–56	Synaptic failure of meiotic prophase chromosomes	

TABLE 1. Continued

Symbol	Name	Location	Phenotype	Reference
B	Booster	2–49	Strong anthocyanin pigment in plant	Emerson et al., 1935
β	Component at A (see α)	3–111	Purple or red aleurone and plant color, red pericarp	Laughman, 1952
B chr	B chromosome		Supernumerary chromosome	Randolph, 1941
ba	Barren stalk	3–72	Ear shoot, tassel florets missing	Emerson et al., 1935
ba2	Barren stalk	2–	Like ba, but tassel more normal	
bd	Branched silkless	7–109	Branched ear and tassel, silks absent	
Bf	Blue fluorescent	9–134	Seedlings fluoresce blue under ultraviolet; anthranilic acid present	Teas and Anderson, 1951
bf2	Blue fluorescent	10L	Like Bf, but is recessive	Coe, 1962
Bh	Blotched aleurone	6–49	Colored patches on colorless (c) aleurone	Emerson et al., 1935
bk2	Brittle stalk	9–79	Brittle leaves and stalk	Neuffer et al., 1968; Coe, 1962
bl	Blotched leaf	2–		
bm	Brown midrib	5–21	Brown pigment over vascular bundles of leaf sheath, midrib, and blade	Emerson et al., 1935
bm2	Brown midrib	1–161	Like bm	
bm3	Brown midrib	4–near su	Like bm	
bm4	Brown midrib	9–138	Like bm	Coe, 1962
Bn	Brown aleurone	7–71	Yellowish-brown aleurone color	
bp	Brown pericarp	9–44	Changes red pericarp of P-rr to brown	Emerson et al., 1935
br	Brachytic	1–81	Short internodes, stiff erect leaves	
br2	Brachytic	—	Short plant	
bt	Brittle endosperm	5–22	Mature kernel collapsed, often translucent and brittle	Emerson et al., 1935

bt2	Brittle endosperm	4–near *su*	Like *bt*	Emerson *et al.*, 1935
bv	Brevis	5–27	Shortened internodes	
bz	Bronze	9–31	Modifies purple aleurone and plant color to pale or reddish-brown	Rhoades, 1952
bz2	Bronze	1–106	Like *bz*	Neuffer *et al.*, 1968; Coe, 1962
C	Aleurone color	9–26	Colored aleurone; *c* = colorless, *C-I* = dominant colorless	Emerson *et al.*, 1935
c2	Colorless aleurone	4–123	Near colorless aleurone, reduced plant color	Neuffer *et al.*, 1968; Coe, 1962
Cat	Catalase (was *Ct*)	—	Electrophoretic mobility	MacDonald and Brewbaker, 1972
Cg	Corngrass	3–31	Narrow leaves, extreme tillering	Singleton, 1950
Ch	Chocolate pericarp	2–155	Dark brown pericarp	Emerson *et al.*, 1935
cl	Chlorophyll	3–38	Small white to pale yellow-green seedling, depending on *Clm*; pale yellow endosperm	
Clm	Modifer of *cl* (was *Cl*_m)	—		Robertson, 1966
Clt	Clumped tassel	8–	Variable dwarfing, developmental anomalies	Gelinas *et al.*, 1966
cms-S	Cytoplasmic male sterility		Female-transmitted male sterility, S type	Singh and Laughnan, 1972; Duvick, 1965
cms-T	Cytoplasmic male sterility		Female-transmitted male sterility, Texas type	Duvick, 1965
cp	Collapsed endosperm	7–near *vp9*	Endosperm collapsed, partially defective	
cp2	Collapsed endosperm	7–near *in*	Endosperm rough, collapsed, partially defective; seedling very light green with darker streaks	Neuffer *et al.*, 1968
cr	Crinkly leaf	3–0	Plant short, leaves broad, crinkled	Emerson *et al.*, 1935
ct	Compact	8–	Semi-dwarf plant	Nelson and Ohlrogge, 1961
Cx	Catechol oxidase	10–33	Electrophoretic mobility of enzyme; null allele is known	MacDonald and Brewbaker, 1972

TABLE 1. Continued

Symbol	Name	Location	Phenotype	Reference
d	Dwarf	3–18	Andromonoecious dwarf; plant short, compact; responds to gibberellins	Emerson *et al.*, 1935
d2	Dwarf	3–	Like *d*	
d3	Dwarf	9–62	Like *d*	
d5	Dwarf	2–34	Like *d*	
D8	Dwarf	1–	Dominant dwarf; resembles *d*, not responsive to gibberellins	Neuffer *et al.*, 1968; Coe, 1962
de	Defective endosperm	4–0	Kernels small, distorted	Emerson *et al.*, 1935
de16	Defective endosperm	4–74	Like *de*; semi-dwarf plant	
dp	Distal pale (was *di*)	4–143	Seedling leaf tip virescent	
Ds	Dissociation	Various	Transposable factor, associated with chromosome breakage and/or control of expression of adjacent genes; regulated by *Ac*	McClintock, 1950
Dt	Dotted	9–0	Regulates controlling element at *a* locus; colored dots on colorless *a* kernels, purple sectors on brown *a* plants	Emerson *et al.*, 1935
Dt2	Dotted	6–43	Like *Dt*	Nuffer, 1955
Dt3	Dotted	7L	Like *Dt*	
du	Dull endosperm	10–33	Dull endosperm; with *wx*, endosperm is shrunken	Emerson *et al.*, 1935
dv	Divergent spindle	—	Spindle nonconverging in meiosis	Clark, 1940
dy	Desynaptic	—	Chromosomes unpaired in microsporocytes	Nelson and Clary, 1952
E	Esterase	7–	Electrophoretic mobility	MacDonald and Brewbaker, 1972
E2	Esterase	—		

Symbol	Name	Location	Description	Reference
E3	Esterase	—	Electrophoretic mobility	MacDonald and Brewbaker, 1972
E4	Esterase	3–	Electrophoretic mobility of enzyme; null allele is known	
E5-I	Esterase	—	Presence-absence	
E5-II	Esterase	—	Presence-absence	
E6	Esterase	—	Presence-absence	
E7	Esterase	—	Presence-absence	
E8	Esterase	—	Presence-absence	
E9	Esterase	—	Electrophoretic mobility of enzyme; null allele is known	
E10	Esterase	—	Electrophoretic mobility	Neuffer et al., 1968
eg	Expanded glumes	5L	At anthesis glumes open at right angle	
Ej	Extension of japonica	10–57	Extreme expression of j, sr2, etc.	
el	Elongate chromosomes	—	Chromosomes uncoiled during meiotic metaphase and anaphase; ears with normal triploid and shriveled pentaploid endosperms	Rhoades and Dempsey, 1966b
En	Enhancer	Various	Transposable factor, regulates pg14-m; equivalent to Spm	Peterson, 1960
et	Etched endosperm	3–122	Pitted, scarred endosperm; virescent seedling	Neuffer et al., 1968; Coe, 1962
f	Fine stripe	1–86	Virescent seedling, fine white stripes on base and margin of older leaves	Emerson et al., 1935
fl	Floury endosperm	2–68	Endosperm opaque, soft; dosage effect	
fl2	Floury endosperm	4–63	Like fl; phenotypically dominant	Levings and Stuber, 1971
fv	Flavones	—	Polyphenols in silks absent	Emerson et al., 1935
g	Golden plant	10–43	Seedling and plant with distinct yellow cast	
g2	Golden plant	7–	Like g	Emerson et al., 1935; Coe, 1962

TABLE 1. Continued

Symbol	Name	Location	Phenotype	Reference
Ga	Gametophyte factor	4–35	Only Ga pollen functions on Ga Ga silks	Emerson et al., 1935
ga2	Gametophyte factor	5–35		Neuffer et al., 1968; Coe, 1962
ga6	Gametophyte factor	1–15	ga6 pollen grains nonfunctional on Ga6 silks	Neuffer et al., 1968
ga7	Gametophyte factor	3–128	ga7 pollen from heterozygotes 10–15 percent functional regardless of silk genotype	Neuffer et al., 1968; Coe, 1962
ga8	Gametophyte factor	9–44 to 59		Neuffer et al., 1968; Coe, 1962
ga9	Gametophyte factor	4–		Neuffer et al., 1968
gl	Glossy leaf	7–36	Cuticle wax altered; leaf surface bright, water adheres	
gl2	Glossy leaf	2–30	Like gl	
gl3	Glossy leaf	4–118	Like gl	Emerson et al., 1935
gl4	Glossy leaf	4–86	Like gl; expression poor	
gl5	Glossy leaf	5–	Like gl	
gl6	Glossy leaf	3–50	Like gl	
gl7	Glossy leaf	—	Like gl	
gl8	Glossy leaf	5L	Like gl	
gl9	Glossy leaf (allele gy)			
gl11	Glossy leaf	2–near B	Like gl; abnormal seedling morphology	
gl12	Glossy leaf	—	Like gl	Coe, 1962
gl14	Glossy leaf	2–	Like gl	
gl15	Glossy leaf	9–69	Like gl; expressed after third leaf	
gl16	Glossy leaf (allele gl4)			

gl17	Glossy leaf	5-14	Like gl; semi-dwarf with necrotic crossbands on leaves	Coe, 1962
gl18	Glossy leaf (was gl9)	8-	Like gl; expression poor	Neuffer et al., 1968; Coe, 1962
gm	Germless	—	Embryo fails	
gs	Green stripe	1-135	Grayish green stripes between vascular bundles on leaves; tissue wilts	Emerson et al., 1935
gs2	Green stripe	2-54	Like gs2 but pale green stripes, no wilting	
gt	Grassy tillers	—		Shaver, 1967
h	Soft starch	—		Emerson et al., 1935
hm	Susceptibility to Helminthosporium carbonum	1-64	Disease lesions on leaves, black masses of fruiting bodies on ears	Neuffer et al., 1968; Coe, 1962
hm2	Susceptibility to Helminthosporium carbonum	9-	Like hm	
Hs	Hairy sheath	7-0	Excess hairs on leaf sheath	Emerson et al., 1935
Ht	Resistance to Helminthosporium turcicum	2-121		Neuffer et al., 1968
I	(See C)			
id	Indeterminate growth	1-near an	Requires short day for flowering	Emerson et al., 1935
Idf	Diffuse (allele c2)			Singleton, 1946
ig	Indeterminate gametophyte		Polyembryony, heterofertilization, polyploidy	Kermicle, 1971
ij	Iojap	7-52	Many variable white stripes on leaves; conditions chloroplast defects that are cytoplasmically inherited	Emerson et al., 1935
in	Intensifier	7-20	Intensifies anthocyanin pigments	
is	Cupulate interspace	—		Galinat, 1971
j	Japonica	8-28	White stripes on leaf and sheath (see Ej)	
j2	Japonica	4-112	Extreme white striping of leaves, etc.	Emerson et al., 1935

TABLE 1. Continued

Symbol	Name	Location	Phenotype	Reference
K–10	Abnormal chromosome	10–99	Heterochromatic appendage on long arm of chromosome 10; neocentric activity distorts segregation of linked genes	Rhoades and Dempsey, 1966a
Kn	Knotted leaf	1–127	Scattered proliferation of vascular bundles on leaf	
l	Luteus	10–	Yellow pigment in white tissue of chlorophyll mutants w, j, ij, etc.	Emerson et al., 1935
l2	Luteus	10–99	Lethal yellow seedling	
l3	Luteus	—		
l4	Luteus	—		
l6	Luteus	9–	Like l2	
l7	Luteus	9–	Yellow seedling and plant	
l8	Luteus	10–38	Like l2	Coe, 1962
l10	Luteus	6–near y	Like l2	
l*–4120	Luteus	6–	Like l2	Neuffer et al., 1968
l*–4920	Luteus	6–		
la	Lazy	4–60	Prostrate growth habit	Emerson et al., 1935
Lc	Red leaf stripe	10–61	Red color in leaf surface	Kermicle, 1970
lg	Liguleless	2–11	Ligule and auricle missing; leaves upright, enveloping sheath	Emerson et al., 1935
lg2	Liguleless	3–83	Like lg, less extreme	Emerson et al., 1935
Lg3	Liguleless	3–46	Dominant, no ligule; leaves broad, flat and erect	Neuffer et al., 1968
li	Lineate	10–28	Fine white striations on basal half of mature leaves	Emerson et al., 1935

Symbol	Name	Map position	Description	Reference
ln	Linoleic acid	—	Ratio of oleate to linoleate	de la Roche et al., 1971
lo	Lethal ovule	4-73	Ovules containing lo gametophytes abort	Emerson et al., 1935
lo2	Lethal ovule	9-53	Like lo	Nelson and Clary, 1952; Coe, 1962
Lp	Leucine aminopeptidase	—	Electrophoretic mobility	MacDonald and Brewbaker, 1972
Lp2	Leucine aminopeptidase	—	Electrophoretic mobility	MacDonald and Brewbaker, 1972
lu	Lutescent seedling	5-9	Pale yellow-green leaves; duplicate factor with lu2	Neuffer et al., 1968
lu2	Lutescent seedling		See lu	Neuffer et al., 1968
lw	Lemon white	1-128	White seedling, pale yellow endosperm	Tulpule, 1954; Neuffer et al., 1968
lw2	Lemon white	5-46+	Like lw	Tulpule, 1954
lw3	Lemon white	5-	Like lw; duplicate factor with lw4	Tulpule, 1954
lw4	Lemon white	4-71 to 84	See lw3	Tulpule, 1954
mi	Midget	1-	Small plant	Emerson et al., 1935
mn	Miniature seed	2-	Small, somewhat defective kernel	Neuffer et al., 1968; Coe, 1962
Mp	Modulator of pericarp	1-26, various	Transposable factor affecting P locus; parallel to Ac-Ds	Brink and Nilan, 1952
Mr	Mutator of R-m	9-40	Transposable factor, regulates R-m	Neuffer et al., 1968
ms	Male sterile	6-17	Anthers shriveled, not usually exserted	Emerson et al., 1935
ms2	Male sterile	9-67	Like ms	
ms5	Male sterile (allele po)	5-		
ms6	Male sterile			
ms7	Male sterile	7-near gl		
ms8	Male sterile	8-14	Like ms	
ms9	Male sterile	1-	Like ms	
ms10	Male sterile	10-	Like ms	
ms11	Male sterile	—		
ms12	Male sterile	1-	Like ms	

TABLE 1. Continued

Symbol	Name	Location	Phenotype	Reference
ms13	Male sterile	5–	Like ms	Emerson et al., 1935
ms14	Male sterile	1–	Like ms	
ms17	Male sterile	1–23	Like ms	
ms-si	Male sterile-silky (allele si)	6–19	Excessive silks on ear, silks in tassel florets	
Mst	Modifier of R-st (was M^{st})	10–63	Affects expression of R-st	Ashman, 1960
na	Nana	3–86	Short, erect dwarf; no response to gibberellins	Emerson et al., 1935
na2	Nana	5L	Like na	
ncs	Nonchromosomal stripe	—	Maternally inherited light green leaf striping	Shumway and Bauman, 1967
nl	Narrow leaf	10–24	Leaf blade narrow, some white streaks	Emerson et al., 1935
o	Opaque endosperm	4–near gl3	Endosperm starch soft, opaque	
o2	Opaque endosperm	7–16	Like o, high lysine content	
o4	Opaque endosperm	—	Like o	
o5	Opaque endosperm	7–near gl	Like o	
Og	Old gold	10–16	Variable bright yellow stripes on leaf blade	Emerson et al., 1935
oy	Oil yellow	10–12	Seedling oily greenish-yellow	
P	Pericarp color	1–26	Red pigment in cob and pericarp	Coe, 1962
pa	Pollen abortion	1–58		
pb	Piebald	6–17	Very light, irregular green patches on leaf	Emerson et al., 1935
pb4	Piebald	6–	Like pb	
pd	Paired rows	—	Single vs. paired pistillate spikelets	Galinat, 1971
pe	Perennialism	—		Shaver, 1967

pg2	Pale green	3	Seedling light yellowish-green	Emerson et al., 1935
pg11	Pale green	6–37	Like pg2; duplicate factor with pg12	Neuffer et al., 1968; Coe, 1962
pg12	Pale green	9–64	See pg11	Coe, 1962
pg14	Pale green mutable (was pg^m)	3S	Pale green leaves with normal green sectors, controlled by En (Spm)	Peterson, 1960
Phos4	Phosphatase, alkaline	—	Electrophoretic mobility	MacDonald and Brewbaker, 1972
Phos8	Phosphatase, alkaline	—	Electrophoretic mobility	
Pl	Purple plant	6–48	Sunlight independent purple pigment in plant	Emerson et al., 1935
pm	Pale midrib	3-near ts4	Midrib and adjacent tissue lighter green	Emerson et al., 1935
Pn	Papyrescent glumes	7–112	Long thin papery glumes on ear and tassel	Neuffer et al., 1968; Coe, 1962
po	Polymitotic	6–4	Microspore division without chromosome division	Emerson et al., 1935
pr	Red aleurone	5–46	Changes purple aleurone to red	
ps	Pink scutellum (=vp7)	5–19	Viviparous; coleoptiles and scutellum pink, seedling white with pink flush	Robertson, 1955
Pt	Polytypic	6–59	Proliferation of pistillate tissue to produce amorphous growth on ear	Neuffer et al., 1968; Coe, 1962
Pu	Purple plumule	—	Purple plumule	Emerson et al., 1935
Pu2	Purple plumule	—	Purple plumule	
Px	Peroxidase	—	Electrophoretic mobility of enzyme; null allele is known	MacDonald and Brewbaker, 1972
Px2	Peroxidase	—	Electrophoretic mobility	
Px3	Peroxidase	—	Electrophoretic mobility	
Px4	Peroxidase	—	Electrophoretic mobility	
Px5	Peroxidase	—	Electrophoretic mobility	
Px6	Peroxidase	—	Electrophoretic mobility	

TABLE 1. *Continued*

Symbol	Name	Location	Phenotype	Reference
Px7	Peroxidase	—	Electrophoretic mobility of enzyme; null allele is known	MacDonald and Brewbaker, 1972
py	Pigmy	6–68	Leaves short, pointed; fine white streaks	Emerson *et al.*, 1935
pyd	Pale yellow deficiency	9S	Pale yellow seedling	McClintock, 1944
R	Colored aleurone and plant	10–57	Red or purple color in aleurone and/or anthers, leaf tip, brace roots, etc.	Emerson *et al.*, 1935
R2	Colored aleurone	2–49	Duplicate factor with *R*; either *R* or *R2* is required for aleurone color	Styles, 1970
ra	Ramosa	7–32	Ear branched, tassel conical	Emerson *et al.*, 1935
ra2	Ramosa	3–26	Irregular kernel placement; tassel many branched	
ra3	Ramosa	4–		Neuffer *et al.*, 1968
rd	Reduced	1L	Dwarf plant	Glover, 1970
rd2	Reduced (= *spl*)	6L	Like *rd*	
Rf	Restorer of fertility	3–45	Restores fertility to *cms-T*; complementary to *Rf2*	Snyder and Duvick, 1969; Duvick, 1965
Rf2	Restorer of fertility	9–	See *Rf*	
Rf3	Restorer of fertility	—		Duvick, 1965
Rg	Ragged	3–48	Chlorotic tissue between veins of older leaves, causing holes and torn appearance	Emerson *et al.*, 1935
rgd	Ragged seedling	6–0	Seedling leaves narrow, threadlike, have difficulty in emerging	Neuffer *et al.*, 1968; Coe, 1962
Rp	Rust resistant	10–0	Resistant to *Puccinia spp.*	Emerson *et al.*, 1935

Symbol	Name	Location	Description	Reference
Rp3	Rust resistant	3–49	Resistant to *Puccinia sorghi*	Wilkinson and Hooker, 1968
Rp4	Rust resistant	4–27	Resistant to *Puccinia sorghi*	Wilkinson and Hooker, 1968
rp7	Rust susceptibility (was *rp$_x$*)	2–11+	Susceptible to *Puccinia sorghi*	Coe, 1962
Rs	Rough sheath	—		
rs2	Rough sheath	1-near *as*		
rt	Rootless	3–40	Secondary roots few or absent	Emerson *et al.*, 1935
S	Scutellum color	—		
Sks	Suppressor of sterility (was *S^{Kus}*)	2-near *v4*	In *Ms*ms*Skssks* plants only *Sks* pollen functions	Neuffer *et al.*, 1968; Coe, 1962
Sg	String cob	—	Reduced pedicels	Galinat, 1971
sh	Shrunken endosperm	9–29	Endosperm collapses; smoothly indented kernels	Emerson *et al.*, 1935
sh2	Shrunken endosperm	3–111.2	Large, transparent, sweet kernels that collapse on drying	Neuffer *et al.*, 1968; Coe, 1962
sh4	Shrunken endosperm (= *sh-fl*)	5–	Like *sh2*, but opaque	Tsai and Nelson, 1969
si	Silky (= *ms-si*)	6–19	Excess silks in ear, sterile tassel with silks	Emerson *et al.*, 1935
sk	Silkless	2–56	Pistils abort, no silks	Emerson *et al.*, 1935
sl	Slashed leaf	7–50	Leaves slit longitudinally by necrotic streaks	Emerson *et al.*, 1935
sm	Salmon silk	6–58	Silks salmon colored with *P-rr*, brown with *P-vrw*	Neuffer *et al.*, 1968
sn	Sienna	8–	Seedling light brown	Neuffer *et al.*, 1968
so	Orange scutellum	—		
sp	Small pollen	4–66	Pollen grains small, not competitive with normal	Emerson *et al.*, 1935
sp2	Small pollen	10–36	Like *sp*	Coe, 1962
Spm	Suppressor-mutator	Various	Transposable factor, regulates responsive element at *a-m-1*, *c2-m*, *pg14-m*, etc.	McClintock, 1965

TABLE 1. Continued

Symbol	Name	Location	Phenotype	Reference
sr	Striate leaf	1–0	Many white striations on leaves	Emerson et al., 1935
sr2	Striate leaf	10–92	Like sr	Neuffer et al., 1968; Coe, 1962
sr3	Striate leaf	10S	Like sr	
st	Sticky chromosome	4–55	Small plant, striate leaves, pitted kernels resulting from sticky chromosomes	
su	Sugary endosperm	4–71	Endosperm wrinkled, translucent when dry; sweet at milk stage	Emerson et al., 1935
su2	Sugary endosperm	6–57	Endosperm translucent, sometimes wrinkled	
sy	Yellow scutellum	—		
T	Translocations	Various	Exchange of parts between two nonhomologous chromosomes	
Ta	Transaminase	—	Electrophoretic mobility	MacDonald and Brewbaker, 1972
tb	Teosinte branched	1–124	Many tillers; nodes with slender branches ending in unbranched tassel	Neuffer et al., 1968
td	Thick tassel dwarf	5–		Emerson et al., 1935
tn	Tinged leaf tip	5–	Small plant, pale green leaf tip	Neuffer et al., 1968
tn2	Tinged leaf tip	10–	Like tn	Emerson et al., 1935
Tp	Teopod	7–46	Many tillers, narrow leaves, many small partially podded ears, tassel simple	
Tp2	Teopod	10–41	Like Tp	Neuffer et al., 1968; Coe, 1962
Tr	Two-ranked ear	—	Distichous vs. decussate phyllotaxy	Galinat, 1971
ts	Tassel seed	2–74	Tassel pistillate; if removed, small ear with irregular kernel placement develops	Emerson et al., 1935

ts2	Tassel seed	1–24	Like *ts*	
Ts3	Tassel seed	1–119	Dominant tassel seed with both pistillate and staminate florets	Emerson *et al.*, 1935
ts4	Tassel seed	3–55	Tassel compact with pistillate and staminate florets	
Ts5	Tassel seed	4–56	Like *Ts3*	
Ts6	Tassel seed	1–158	Tassel completely pistillate, compact	Neuffer *et al.*, 1968; Coe, 1962
Tu	Tunicate	4–107	Kernels enclosed in long glumes; tassel large, coarse	Emerson *et al.*, 1935
ub	Unbranched tassel	—	Tassel with one spike	Neuffer *et al.*, 1968
v	Virescent seedling	9–66	Yellowish-white seedling, greens rapidly	
v2	Virescent seedling	5–87	Like *v*, greens slowly	
v3	Virescent seedling	5–25	Light yellow seedling, greens rapidly	
v4	Virescent seedling	2–83	Like *v2*	
v5	Virescent seedling	7–24	Like *v*, older leaves with white stripes	
v8	Virescent seedling	4–near *j2*	Like *v2*	Emerson *et al.*, 1935
v12	Virescent seedling	5–	Like *v3*	
v13	Virescent seedling	—		
v16	Virescent seedling	8–0	Like *v2*	
v17	Virescent seedling	—		
v18	Virescent seedling	10–	Like *v*	
v19	Virescent seedling	1–		
va	Variable sterile	7–	Male sterile with some fertile anthers	
Vg	Vestigial glumes	1–85	Glumes very small, cob and anthers exposed	Neuffer *et al.*, 1968; Coe, 1962
Vm	Virescent mutable (was *V^m1817*)	10–	Pale green, dominant; endosperm etched	Neuffer *et al.*, 1968
vp	Viviparous	3–128	Premature germination, inhibits anthocyanin in endosperm, no effect on chlorophyll or carotenoids	Emerson *et al.*, 1935

TABLE 1. Continued

Symbol	Name	Location	Phenotype	Reference
vp2	Viviparous	5–18	Premature germination, white endosperm, white seedling	Emerson et al., 1935
vp5	Viviparous	1–1	Like vp2	} Robertson, 1955
vp8	Viviparous	1–154	Like vp; no effect on seed color; small, pointed-leaf seedling	
vp9	Viviparous	7–25	Like vp2	} Emerson et al., 1935
w	White seedling	6–	White seedling	
w2	White seedling	10–73	White seedling	
w3	White seedling	2–111	Like vp2	
w11	White seedling	9-near wx		
w*-7748	White seedling	3–75	Like w	Coe, 1962
Wc	White cap	9–104	Kernel with white crown	Emerson et al., 1935
wd	White deficiency	9S	White seedling from homozygous deficiency for distal half of last chromomere on short arm of chromosome 9	McClintock, 1944
wi	Wilted	6–	Chronic wilting, delayed differentiation of metaxylem vessels	Neuffer et al., 1968
wl	White leaf base	6–		Emerson et al., 1935
w*-mut	White mutable (was w^mut)	6–17	White endosperm and white seedling with yellow and green sectors, respectively; temperature variations result in bands of pale green and white on seedling leaves	Neuffer et al., 1968
ws	White sheath	—	Like ws3	} Emerson et al., 1935
ws2	White sheath	—		

ws3	White sheath	2–0	White leaf sheath, culm, husks	Emerson *et al.*, 1935
wt	White tip	2–60	Tips of seedling leaves are white	Neuffer *et al.*, 1968
wx	Waxy endosperm	9–59	Amylopectin replaces amylose in endosperm and pollen. Stained red by iodine.	Emerson *et al.*, 1935
Y	Yellow endosperm	6–17	Carotenoid pigments in endosperm	
y8	Yellow endosperm	7–18	Light yellow endosperm	Coe, 1962
y9	Yellow endosperm	10–near *bf2*	Pale yellow endosperm, slightly viviparous	
yd	Yellow dwarf	6–	Yellow dwarf	
yg	Yellow-green seedling	5–	Like *yg2*	
yg2	Yellow-green seedling	9–7	Yellow-green seedling and plant	Emerson *et al.*, 1935
ys	Yellow stripe	5–55	Yellow tissue between leaf veins, reflects iron-deficiency symptoms	
ys3	Yellow stripe	3–near *Lg3*	Like *ys*	Coe, 1962
zb	Zebra striped	—	Like *zb4*	
zb2	Zebra striped	—	Like *zb4*	
zb3	Zebra striped	5–near *v2*	Like *zb4*	Emerson *et al.*, 1935
zb4	Zebra striped	1–19	Regularly spaced yellowish crossbands on earlier leaves	
zb6	Zebra striped	4–84	Like *zb4*	Coe, 1962
zl	Zygotic lethal	1–28	Sporophyte fails	Emerson *et al.*, 1935
zn	Zebra necrotic	10–35	Necrotic tissue appears between veins in transverse leaf bands on half-grown plants	Neuffer *et al.*, 1968; Coe, 1962
zn2	Zebra necrotic	—	Like *zn*	Giesbrecht, 1965

Acknowledgment

We greatly appreciate information and advice from numerous colleagues, especially J. B. Beckett, R. J. Lambert, A. E. Longley, and M. M. Rhoades.

Literature Cited

Anderson, E. G., 1923   Maternal inheritance of chlorophyll in maize. *Bot. Gaz.* **76:**411–418.

Ashman, R. B., 1960   Stippled aleurone in maize. *Genetics* **45:**19–34.

Beadle, G. W., 1930   Genetical and cytological studies of Mendelian asynapsis in *Zea mays. Cornell Univ. Agric. Exp. Stn. Mem.* **129**.

Beadle, G. W., 1931   A gene in maize for supernumerary cell divisions following meiosis. *Cornell Univ. Agric. Exp. Stn. Mem.* **135**.

Beadle, G. W., 1932*a*   A gene for sticky chromosomes in *Zea mays. Z. Indukt. Abstammungs- Vererbungsl.* **63:**195–217.

Beadle, G. W., 1932*b*   A gene in *Zea mays* for the failure of cytokinesis during meiosis. *Cytologia (Tokyo)* **3:**142–155.

Beadle, G. W., 1933   Further studies of asynaptic maize. *Cytologia (Tokyo)* **4:**269–287.

Beadle, G. W., 1937   Chromosome aberration and gene mutation in sticky chromosome plants of *Zea mays. Cytologia (Tokyo)* **Fujii Jubilee Volume:**43–56.

Beckett, J. B., 1972   An expanded set of B-type translocations in maize. *Genetics* **71:**s3–4.

Brink, R. A. and J. M. MacGillivray, 1924   Segregation for the waxy character in maize pollen and differential development of the male gametophyte. *Am. J. Bot.* **11:**465–469.

Brink, R. A. and R. A. Nilan, 1952   The relation between light variegated and medium variegated pericarp in maize. *Genetics* **37:**519–544.

Brink, R. A., E. D. Styles and J. D. Axtell, 1968   Paramutation: Directed genetic change. *Science (Wash. D.C.)* **159:**161–170.

Burnham, C. R., 1930   Genetical and cytological studies of semisterility and related phenomena in maize. *Proc. Natl. Acad. Sci. USA* **16:**269–277.

Burnham, C. R., 1932   An interchange in maize giving low sterility and chain configurations. *Proc. Natl. Acad. Sci. USA* **18:**434–440.

Burnham, C. R., 1950   Chromosome segregation in translocations involving chromosome 6 in maize. *Genetics* **35:**446–481.

Burnham, C. R., J. T. Stout, W. H. Weinheimer, R. V. Kowles and R. L. Phillips, 1972   Chromosome pairing in maize. *Genetics* **71:**111–126.

Clark, F. J., 1940   Cytogenetic studies of divergent meiotic spindle formation in *Zea mays. Am. J. Bot.* **27:**547–559.

Coe, E. H., Jr., 1962   Symbol Index to *Maize Newsletters* Vol. 12–35. *Maize Genetics Cooperation Newsl.* **36:**appendix.

Coe, E. H., Jr., 1966   The properties, origin, and mechanism of conversion-type inheritance at the *B* locus in maize. *Genetics* **53:**1035–1063.

Creech, R. G., 1965   Genetic control of carbohydrate synthesis in maize endosperm. *Genetics* **52:**1175–1186.

Creighton, H. B. and B. McClintock, 1931   A correlation of cytological and genetical crossing over in *Zea mays. Proc. Natl. Acad. Sci. USA* **21**:148–150.

de la Roche, I. A., D. E. Alexander and E. J. Weber, 1971   Inheritance of oleic and linoleic acids in *Zea mays* L. *Crop Sci.* **11**:856–859.

Demerec, M., 1924   A case of pollen dimorphism in maize. *Am. J. Bot.* **11**:461–464.

Doyle, G. G., 1969   Preferential pairing in trisomics of *Zea mays.* In *Chromosomes Today,* Vol. 2, edited by C. D. Darlington and K. R. Lewis, pp. 12–20, Oliver and Boyd, Edinburgh.

Duvick, D. N., 1965   Cytoplasmic pollen sterility in corn. *Adv. Genet.* **13**:1–56.

Emerson, R. A., 1911   Genetic correlation and spurious allelomorphism in maize. *Neb. Agric. Exp. Stn. Rep.* **24**:59–90.

Emerson, R. A., 1914   Inheritance of a recurring somatic variation in variegated ears of maize. *Am. Nat.* **48**:87–115.

Emerson, R. A., 1918   A fifth pair of factors, *A a,* for aleurone color in maize, and its relation to the *C c* and *R r* pairs. *Cornell Univ. Agric. Exp. Stn. Mem.* **16**.

Emerson, R. A., 1921   The genetic relations of plant colors in maize. *Cornell Univ. Agric. Exp. Stn. Mem.* **39**.

Emerson, R. A., G. W. Beadle and A. C. Fraser, 1935   A summary of linkage studies in maize.*Cornell Univ. Agric. Exp. Stn. Mem.* **180**.

Galinat, W. C., 1971   The origin of maize. *Annu. Rev. Genet.* **5**:447–478.

Gelinas, D. A., S. N. Postlethwait and L. F. Bauman, 1966   Developmental studies in the *Zea mays* mutant clumped tassel (*Ct*). *Am. J. Bot.* **53**:615.

Giesbrecht, J., 1965   A second zebra-necrotic gene in maize. *J. Hered.* **56**:118, 130.

Glover, D. V., 1970   Location of a gene in maize conditioning a reduced plant stature. *Crop Sci.* **10**:611–612.

Josephson, L. M. and M. T. Jenkins, 1948   Male sterility in corn hybrids. *J. Am. Soc. Agron.* **40**:267–274.

Kermicle, J. L., 1970   Somatic and meiotic instability of *R*-stippled, an aleurone spotting factor in maize. *Genetics* **64**:247–258.

Kermicle, J. L., 1971   Pleiotropic effects on seed development of the indeterminate gametophyte gene in maize. *Am. J. Bot.* **58**:1–7.

Laughnan, J. R., 1952   The action of allelic forms of the gene *A* in maize. IV. On the compound nature of A^b and the occurrence and action of its A^d derivatives. *Genetics* **37**:375–395.

Laughnan, J. R., 1961   The nature of mutations in terms of gene and chromosome changes. *Natl. Acad. Sci.-Natl. Res. Counc. Sci. Ser. Rep.* **891**:3–29.

Levings, C. S., III and C. W. Stuber, 1971   A maize gene controlling silk browning in response to wounding. *Genetics* **69**:491–498.

Longley, A. E., 1938   Chromosomes of maize from North American Indians. *J. Agric. Res.* **56**:177–195.

Longley, A. E., 1939   Knob positions on corn chromosomes. *J. Agric. Res.* **59**:475–490.

Longley, A. E. and T. A. Kato Y., 1965   Chromosome morphology of certain races of maize in Latin America, Centro Internacional de Mejoramiento de Maiz y Trigo, Chapingo, Mexico.

McClintock, B., 1931   Cytological observations of deficiencies involving known genes, translocations and an inversion in *Zea mays. Mo. Agric. Exp. Stn. Res. Bull.* **163**.

McClintock, B., 1932   A correlation of ring-shaped chromosomes with variegation in *Zea mays. Proc. Natl. Acad. Sci. USA* **18**:667–681.

McClintock, B., 1933   The association of non-homologous parts of chromosomes in the mid-prophase of meiosis in *Zea mays. Z. Zellforsch. Mikrosk. Anat.* **19**:191–237.

McClintock, B., 1934   The relation of a particular chromosomal element to the development of the nucleoli in *Zea mays. Z. Zellforsch. Mikrosk. Anat.* **21**:294–328.

McClintock, B. 1938   The production of homozygous deficient tissues with mutant characteristics by means of the aberrant mitotic behavior of ring-shaped chromosomes. *Genetics* **23**:315–376.

McClintock, B., 1939   The behavior in successive nuclear divisions of a chromosome broken at meiosis. *Proc. Natl. Acad. Sci. USA* **25**:405–416.

McClintock, B., 1941   The association of mutants with homozygous deficiencies in *Zea mays. Genetics* **26**:542–571.

McClintock, B., 1944   The relation of homozygous deficiencies to mutations and allelic series in maize. *Genetics* **29**:478–502.

McClintock, B., 1950   The origin and behavior of mutable loci in maize. *Proc. Natl. Acad. Sci. USA* **36**:344–355.

McClintock, B., 1965   Components of action of the regulators *Spm* and *Ac. Carnegie Inst. Wash. Year Book* **64**:527–536.

MacDonald, T. and J. L. Brewbaker, 1972   Isoenzyme polymorphism in flowering plants. VIII. Genetic control and dimeric nature of transaminase hybrid maize isoenzymes. *J. Hered.* **63**:11–14.

Mertz, E. T., L. S. Bates and O. E. Nelson, Jr., 1964   Mutant gene that changes protein composition and increases lysine content of maize endosperm. *Science (Wash., D.C.)* **145**:279–280.

Nelson, O. E., Jr. and G. B. Clary, 1952   Genic control of semi-sterility in maize. *J. Hered.* **43**:205–210.

Nelson, O. E., Jr. and A. J. Ohlrogge, 1961   Effect of heterosis on the response of *compact* strains of maize to population pressures. *Agron. J.* **53**:208–209.

Neuffer, M. G. and E. H. Coe, Jr., 1970   Linkage map and annotated list of genetic markers in maize. In *Handbook of Biochemistry,* edited by H. A. Sober, pp. I 88–I95, Chemical Rubber Co., Cleveland, Ohio.

Neuffer, M. G., L. M. Jones and M. Zuber, 1968   *The Mutants of Maize,* Crop Science Society of America, Madison, Wisc.

Nuffer, M. G., 1955   Dosage effect of multiple *Dt* loci on mutation of *a* in the maize endosperm. *Science (Wash., D.C.)* **121**:399–400.

Nuffer, M. G., 1957   Additional evidence on the effect of X-ray and ultraviolet radiation on mutation in maize. *Genetics* **42**:273–282.

Peterson, P. A., 1960   The pale green mutable system in maize. *Genetics* **45**:115–133.

Randolph, L. F., 1941   Genetic characteristics of the B chromosomes in maize. *Genetics* **26**:608–631.

Reddy, G. M. and E. H. Coe, Jr., 1962   Inter-tissue complementation: A simple technique for direct analysis of gene-action sequence. *Science (Wash., D.C.)* **138**:149–150.

Rhoades, M. M., 1931   Cytoplasmic inheritance of male sterility in *Zea mays. Science (Wash., D.C.)* **73**:340–341.

Rhoades, M. M., 1933   The cytoplasmic inheritance of male sterility in *Zea mays. J. Genet.* **27**:71–93.

Rhoades, M. M., 1941   The genetic control of mutability in maize. *Cold Spring Harbor Symp. Quant. Biol.* **9**:138–144.

Rhoades, M. M., 1943   Genic induction of an inherited cytoplasmic difference. *Proc. Natl. Acad. Sci. USA* **29**:327–329.

Rhoades, M. M., 1947   Plastid mutations. *Cold Spring Harbor Symp. Quant. Biol.* **11**:202–207.

Rhoades, M. M., 1952   The effect of the bronze locus on anthocyanin formation in maize. *Am. Nat.* **86:**105–108.

Rhoades, M. M. and E. Dempsey, 1966a   The effect of abnormal chromosome 10 on preferential segregation and crossing over in maize. *Genetics* **53:**989–1020.

Rhoades, M. M. and E. Dempsey, 1966b   Induction of chromosome doubling at meiosis by the elongate gene in maize. *Gentics* **54:**505–522.

Rhoades, M. M. and E. Dempsey, 1972   On the mechanism of chromatin loss induced by the B chromosome of maize. *Genetics* **71:**73–96.

Robertson, D. S., 1955   The genetics of vivipary in maize. *Genetics* **40:**745–760.

Robertson, D. S., 1966   Allelic relationships and phenotypic interactions of four dominant modifiers of the cl_1 locus in maize. *Heredity* **21:**1–7.

Roman, H., 1947   Mitotic nondisjunction in the case of interchanges involving the B-type chromosome in maize. *Genetics* **32:**391–409.

Sarkar, K. R. and E. H. Coe, 1971   Anomalous fertilization in diploid–tetraploid crosses in maize. *Crop Sci.* **11:**539–542.

Scandalios, J. G., 1965   Subunit dissociation and recombination of catalase isozymes. *Proc. Natl. Acad. Sci. USA* **53:**1035–1040.

Schwartz, D., 1960   Genetic studies on mutant enzymes in maize: Synthesis of hybrid enzymes by heterozygotes. *Proc. Natl. Acad. Sci. USA* **46:**1210–1215.

Shaver, D. L., 1967   Perennial maize. *J. Hered.* **58:**270–273.

Shull, G. H., 1911   Experiments with maize. *Bot. Gaz.* **52:**480–485.

Shumway, L. K. and L. F. Bauman, 1967   Nonchromosomal stripe of maize. *Genetics* **55:**33–38.

Singh, A. and J. R. Laughnan, 1972   Instability of S male-sterile cytoplasm in maize. *Genetics* **71:**607–620.

Singleton, W. R., 1946   Inheritance of indeterminate growth in maize. *J. Hered.* **37:**61–64.

Singleton, W. R., 1950   Corn-grass, a dominant monogenic spontaneous mutant, and its possible significance as an ancestral type of corn. *Genetics* **35:**691–692.

Sinha, S. K., 1962   Cytogenetic and biochemical studies of the action of a gene controlling meiosis in maize. *Diss. Abstr.* **21:**2443–2444.

Snyder, R. J. and D. N. Duvick, 1969   Chromosomal location of Rf_2, a restorer gene for cytoplasmic pollen sterile maize. *Crop Sci.* **9:**156–157.

Sprague, G. F., editor, 1955   *Corn and Corn Improvement,* Academic Press, New York.

Sprague, G. F. and H. H. McKinney, 1971   Further evidence on the genetic behavior of AR in maize. *Genetics* **67:**533–542.

Sprague, G. F., H. H. McKinney and L. W. Greeley, 1963   Virus as a mutagenic agent in maize. *Science (Wash. D.C.)* **141:**1052–1053.

Stadler, L. J., 1932   On the genetic nature of induced mutation in plants. *Proc. VI Int. Congr. Genet.* **1:**274–294.

Stadler, L. J., 1939   Genetic studies with ultraviolet radiation. *Proc. VII Int. Congr. Genet.* 269–276.

Stadler, L. J., 1946   Spontaneous mutation at the *R* locus in maize. I. The aleurone-color and plant-color effects. *Genetics* **31:**377–394.

Stadler, L. J. and M. H. Emmerling, 1956   Relation of unequal crossing over to the interdependence of R^r elements (P) and (S). *Genetics* **41:**124–137.

Stadler, L. J. and M. G. Nuffer, 1953   Problems of gene structure. II. Separation of R^r elements (S) and (P) by unequal crossing over. *Science (Wash., D.C.)* **117:**471–472.

Stadler, L. J. and H. Roman, 1948   The effect of X-rays upon mutation of the gene *A* in maize. *Genetics* **33:**273–303.

Stadler, L. J. and F. M. Uber, 1942   Genetic effects of ultraviolet radiation in maize. IV. Comparison of monochromatic radiations. *Genetics* **27**:84–118.

Steffensen, D. M., 1968   A reconstruction of cell development in the shoot apex of maize. *Am. J. Bot.* **55**:354–369.

Styles, E. D., 1970   Functionally duplicate genes conditioning anthocyanin formation in maize. *Can. J. Genet. Cytol.* **12**:397.

Teas, H. J. and E. G. Anderson, 1951   Accumulation of anthranilic acid by a mutant of maize. *Proc. Natl. Acad. Sci. USA* **37**:645–649.

Tsai, C.-Y. and O. E. Nelson, Jr., 1969   Mutations at the shrunken-4 locus in maize that produce three altered phosphorylases. *Genetics* **61**:813–821.

Tulpule, S. H., 1954   A study of pleiotropic genes in maize. *Am. J. Bot.* **41**:294–301.

Wilkes, H. G., 1972 Maize and its wild relatives. *Science (Wash., D.C.)* **177**:1071–1077.

Wilkinson, D. R. and A. L. Hooker, 1968   Genetics of reaction to *Puccinia sorghi* in ten corn inbred lines from Africa and Europe. *Phytopathology* **58**:605–608.

2

Rice

GURDEV S. KHUSH

Introduction

Rice is the world's most important food crop. It is the principal food of more than half of mankind. Ninety percent of all rice is grown and consumed in Asia.

Cultivated rice belongs to the genus *Oryza*. The genus is highly variable and distributed throughout the world. It has about 23 well-defined species, two of which are cultivated. *Oryza glaberrima,* grown on limited acreage in a few African countries, is gradually being replaced by the other cultivated species *Oryza sativa.* This review is confined to studies on the cytogenetics and improvement of *O. sativa.* Very little research of this type has been done on *O. glaberrima.*

From tropical Asia, where it originated, rice has spread from latitude 40° south to 44° north. It grows at sea level as well as at elevations of 2500 meters or even higher. Although typically grown on irrigated or rain-fed, puddled, lowland soils, rice is the only crop that grows in river deltas where water may rise as high as 4 meters. It also grows without standing water, as an upland crop, particularly on rolling land. These diverse cultural, geographic, and climatic conditions, along with the force of natural selection, have brought about the great varietal diversity in rice.

Cultivated varieties of *O. sativa* are grouped into three types: indicas, japonicas, and javanicas. Indica rices are grown throughout the tropics and subtropics. In 1962, the International Rice Research Institute launched the first major effort to improve their yielding ability. Many

GURDEV S. KHUSH—Varietal Improvement Department, International Rice Research Institute, Los Baños, Laguna, Philippines.

31

Asian, African, and Latin American countries are now involved in the improvement of these rices. For the past 5 decades, Indian and American researchers have studied their genetics and cytology.

Japonica rices are limited to temperate zones and the subtropics. The genetics and cytology of japonica rices have been well explored, particularly by Japanese workers, and their yielding ability has been improved.

Javanicas are confined mainly to parts of Indonesia. Very little work has been done on their genetics and cytology. There has been no active program to improve javanica rices, nor have they been used in improving the indicas and japonicas.

Biosystematics

The genus *Oryza* L. belongs to the tribe Oryzeae in the family Gramineae. Different authors have proposed at least 75 specific names for its various entities. Only 20–25 species however, are recognized by modern taxonomists. Roschevicz (1931) published a comprehensive study of 19 species which provided a basis for later taxonomic investigations of the genus. Chatterjee (1948) and Sampath (1962) listed 23 species; Tateoka (1964) listed 21 species.

The genus *Oryza* is distributed throughout tropical Asia, Africa, and Latin America. *O. sativa* has world-wide distribution, being the only cultivated species adapted to temperature climate. *O. glaberrima,* the other cultivated species, is confined to tropical West Africa. Of the wild species, one is endemic to Australia, nine grow in South and Southeast Asia, seven in Africa, and three in Latin America. One wild species (*O. perennis*) grows in Asia, Africa, and the West Indies (Table 1).

Subgroups of Genus *Oryza*

Sampath (1966) divided the genus *Oryza* into four series to show the evolutionary relationships of the species.

1. Series *sativae*. This group includes the *O. sativa* complex, the *O. perennis* complex, and *O. nivara,* as well as *O. glaberrima* and *O. breviligulata*. All species belonging to this series are diploid and have more or less homologous genomes.
2. Series *latifoliae*. This group includes diploid and tetraploid species, such as *O. alta, O. latifolia, O. grandiglumis, O. punctata, O. eichingeri, O. officinalis,* and *O. minuta*.

TABLE 1. The Chromosome Number, Genomic Classification, and Geographical Distribution of Different Species of Genus Oryza

Species	Chromosome No. (n=12)	Genome	Distribution
O. schlechteri	—	—	New Guinea
O. meyeriana	24	—	Indonesia, Philippines, Thailand
O. coarctata	48	—	India, Pakistan, Bangladesh
O. ridleyi	48	—	Malaysia, Thailand, New Guinea
O. longiglumis	—	—	New Guinea
O. granulata	24	—	India, Bangladesh, Ceylon, Burma, Thailand, Indonesia
O. eichingeri	48	BBCC	Tanzania, Uganda
O. punctata	48	—	Northeast tropical Africa
O. minuta	48	BBCC	Malaysia, Philippines, Indonesia
O. officinalis	24	CC	India, Bangladesh, Burma
O. latifolia	48	CCDD	Central and South America, West Indies
O. alta	48	CCDD	Honduras, Brazil, Paraguay
O. grandiglumis	48	CCDD	Brazil
O. angustifolia	24	—	Tropical Africa
O. brachyantha	24	FF	West tropical and central Africa
O. perrieri	24	—	Madagascar
O. tisseranti	24	—	Tropical Africa
O. australiensis	24	EE	Western northern Australia
O. breviligulata	24	A^gA^g	West tropical Africa to Sudan
O. glaberrima	24	A^gA^g	West tropical Africa
O. perennis	24	AA	
Asiatic (subsp. balunga)	24	AA	Asia
American (subsp. cubensis)	24	$A^{cu}A^{cu}$	West Indies
African (subsp. barthii)	24	A^bA^b	Africa
O. nivara	24	AA	India
O. sativa	24	AA	World-wide
var. fatua (or var. spontanea)	24	AA	South and Southeast Asia
var. formasana	24	AA	Taiwan

3. Series *angustifoliae.* This group includes only diploid species such as *O. brachyantha, O. angustifolia, O. tisseranti,* and *O. perrieri.* The genomes of *O. brachyantha* have been designated as FF. Sampath (1966) conjectured on the basis of morphological similarities that the species of this series have the same genome and may be intercrossable.

4. Series *granulatae.* This series includes diploid as well as tetraploid taxa, such as *O. meyeriana, O. ridleyi,* and *O. longiglumis.* Very little is known about the interrelationships of the species of this group.

O. australiensis and *O. schlechteri* have not been placed in any of the series mentioned, although Sampath (1966) believes that *O. australiensis* may belong to series *latifoliae.*

Interrelationships within Series *Sativae*

Since cultivated rice belongs to the series *sativae,* this section reviews the interrelationships within this group. *O. glaberrima, glaberrima* var. *stapfii,* and *breviligulata* are annuals, and are endemic to tropical west Africa. Hybrids between members of this group have normal meiosis and are usually completely fertile, although some may have a slight degree of sterility (Yeh and Henderson, 1962). *O. sativa* and *O. glaberrima* differ in morphology, distribution, and adaptation. Although the interspecific hybrids are sterile, an almost normal synapsis in the microsporocytes of F_1 plants shows that the genomes of these two species are partially homologous (Bouharmont, 1962). Genomes of *O. sativa* and *O. glaberrima* have therefore been designated as AA and A^gA^g to show both their cytological affinity and genetic differentiation.

The relationships between *O. sativa* and closely related taxonomic entities (*perennis, rufipogon, balunga, barthii, fatua, spontanea,* and *formosana*) have been subjects of controversy between taxonomists. *Oryza sativa* var. *fatua,* also called *sativa* var. *spontanea* by Roschevicz (1931), differs only slightly from cultivated varieties of *sativa* and is interfertile with them. It grows in or near fields of cultivated varieties and intercrosses with them to produce hybrid swarms (Sampath and Govindaswami, 1958). It may be an introgressant form that resulted from the hybridization of cultivated varieties of *sativa* with *O. perennis.* No cytogenetic barriers exist between *sativa* and *sativa* var. *formosana,* nor do they differ in key morphological characters. Most taxonomists regard *fatua* or *spontanea* and *formosana* as botanical varieties of *sativa.*

The perennial forms belonging to series *sativae* are found in Africa, Asia, and America. Chevalier (1932) and Chatterjee (1948) grouped them under the specific name *perennis*. Asiatic forms of *perennis* have been variously designated as subsp. *rufipogon* or subsp. *balunga*. The American form has been designated as subsp. *cubensis*; the African form, subsp. *barthii*. The three entities are taxonomically related both to *sativa* and to each other. The relationships between *sativa* and various subspecies of *perennis* were investigated by Yeh and Henderson (1961). Hybrids of *sativa* and subsp. *cubensis* were completely sterile, while hybrids of *sativa* and Asiatic forms of *perennis* were 80–100 percent fertile. Only one hybrid between African *perennis* and American *perennis* was obtained. Its meiosis was essentially regular. Similarly, numerous attempts to cross *sativa* and *perennis* subsp. *barthii* resulted in only one hybrid. The plant was vigorous, its meiosis was essentially normal, and 5 percent of the florets set seed.

According to Tateoka (1963, 1964) the Asiatic forms of *perennis* cannot always be distinguished from the American forms. He suggested merging the two under the name *rufipogon,* previously used for Asiatic forms of *perennis*. Sampath (1964) agreed with this revision, but Henderson (1964) pointed out that the hybrids of *sativa* and Asiatic forms of *perennis* are completely fertile while those between *sativa* and American *perennis* are completely sterile. This observation shows that from the evolutionary standpoint *O. sativa* and the Asiatic *perennis* are much more closely related than *O. sativa* and the American *perennis*. Henderson argued that placing the two wild forms in the same species without including *sativa* would be illogical. He therefore proposed that the Asiatic wild forms be designated as an independent species, *Oryza balunga.*

Tateoka (1964) pointed out that subsp. *barthii* differs from Asiatic forms of *perennis* by its well-developed rhizomes and other morphological traits. Cytogenetic relationships between *barthii* and *cubensis* have not been studied, but probably they are no more closer to each other than they are to *O. sativa*. I think that the phylogenetic relationships of different forms of *perennis* and *sativa* may be best expressed either by merging *balunga, barthii,* and *cubensis* with *sativa* as its subspecies, or by recognizing *sativa, balunga, barthii,* and *cubensis* as independent species (after all, *balunga, barthii,* and *cubensis* are geographically isolated and genetically differentiated). This would eliminate *O. perennis,* a name of doubtful validity (Tateoka, 1964).

Figure 1 shows the relationships of the taxonomic entities of series *sativae*. All these taxa with AA genome probably originated from a common ancestor which may now be extinct. The center of variability of *O.*

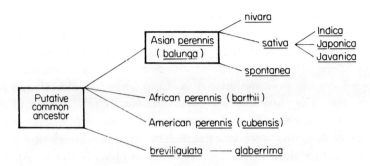

Figure 1. A diagrammatic representation of species relationships in the series sativae. Modified from Oka, 1964.

sativa is South and Southeast Asia. It probably originated from the perennial Asian wild rices (*O. balunga*), to which it is closely related. The gene exchange between *O. sativa* and *O. balunga* is probably still going on contributing to the variability of the cultivated varieties.

O. nivara, an annual weedy form from central India that was recently described by Sharma and Shastry (1965) as an independent species, appears to be another botanical variety or subspecies of *O. sativa.* It is interfertile with *O. sativa* and their F_1 hybrid has regular meiosis (Khush, unpublished).

Cytogenetics

Karyotype

Kuwada (1910) first reported that rice has a somatic number of 24 and a haploid number of 12 chromosomes. His observation has been verified by countless subsequent investigations. Numerous researchers have tried to identify the individual members of the somatic chromosome complement on the basis of length, position of centromere, presence or absence of satellites, and other characteristics. Rau (1929) and Sethi (1937) reported that of the 12 pairs of the complement, five were large, four were medium, and the remaining three were small. From the data on secondary association (Sakai, 1935) and his own observations on karyomorphology, Nandi (1936) hypothesized that *O. sativa* is a secondarily balanced allotetraploid and that two chromosomes, which he designated as A and B, duplicated after hybridization between putative ancestral species, each with the basic chromosome number $X = 5$. This hypothesis, however, received little support from subsequent cytogenetic investigations. Stebbins (1950) pointed out that not only is the basic number $X = 12$ found in *Oryza* and most other genera of the tribe *Oryzeae,* but it is also, by far, the most common number in the primitive genera of family Gramineae. On the other hand, species with five pairs of chromosomes are very uncommon among the Gramineae, and this number is not found in any primitive genus, nor in any genus related to *Oryza.* Recent studies of trisomic phenotypes also repudiated Nandi's hypothesis. All the primary trisomics of rice are morphologically distinct from each other. If, as Nandi postulated, two duplicated chromosomes are present in the complement, then two trisomic pairs should be phenotypically similar.

Hu (1958) examined the mitotic metaphase in a haploid of *O. sativa* and concluded that the complement consists of four median and eight submedian, or subtelocentric, chromosomes. He pointed out that one chromosome has a secondary constriction and the other has a satellite. Bhaduri *et al.* (1959) studied the haploid chromosome complement during pollen mitosis, but they made no attempt to study the karyomorphology.

Yao *et al.* (1958) began studies on karyomorphology at pachynema in some intervarietal hybrids. But, Shastry *et al.* (1960) first analyzed the entire chromosome complement at pachynema. They identified each chromosome on the basis of length, arm ratio, and presence or absence of dark-

Chromosomes

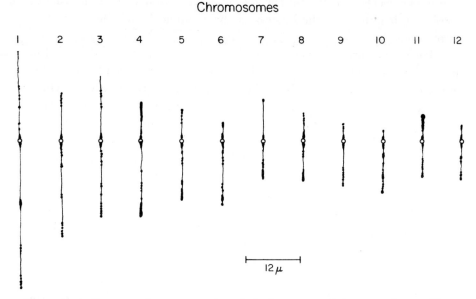

Figure 2. A diagrammatic representation of the karyotype of rice at pachytene. From Shastry, Ranga Rao, and Misra (1960) with the permission of the Indian Journal of Genetics and Plant Breeding.

TABLE 2. *Length, Arm Ratio, and Classification of the Chromosomes at Pachynema in the Strain Norin 6 of Oryza sativa L.*[a]

| | Length (in μ) | | | Arm ratio, | |
Chromosome	Short arm	Long arm	Total	short: long	Classification[b]
1	29.0	50.0	79.0	0.58	SM
2	15.0	32.5	47.5	0.46	SM
3	21.0	26.0	47.0	0.81	M
4	12.5	26.0	38.5	0.48	SM
5	10.0	20.5	30.5	0.49	SM
6	5.5	22.0	27.5	0.25	ST
7	13.0	13.5	26.5	0.96	M
8	8.5	14.5	23.0	0.59	SM
9	5.0	16.0	21.0	0.31	SM
10	3.0	18.0	21.0	0.17	ST
11	8.0	12.5	20.5	0.64	SM
12	4.5	13.5	18.0	0.33	SM

[a] From Shastry, Ranga Rao and Misra (1960) with permission of the Indian Journal of Genetics and Plant Breeding.
[b] M = Median; SM = Sub-median; ST = Subtelocentric chromosomes.

staining knobs (Figure 2). The chromosomes were numbered in the order of decreasing length, with the longest as chromosome 1 and the shortest as chromosome 12.

Pachytene chromosomes vary considerably in length (Table 2) and in arm ratios. Only chromosome 3 and 7 have truly median centromeres; the rest are submedian or subtelocentric. The nucleoli are attached to chromosomes 3 and 4 in most varieties. Small heterochromatic regions flank the centromeres on each side. In addition, dark-staining chromomeres are present in all chromosome arms. Pachytene chromosomes of rice, however, are not strongly differentiated into heterochromatic and euchromatic blocks, as are those of tomato (Brown, 1949; Barton, 1950) and numerous other genera (Rick and Khush, 1966).

Chu (1967) studied the karyotype of haploid plants of japonica varieties at pachynema and observed nonhomologous pairing between different members of the complement. The pairing seemed to occur at random. Thus, chromosome 1 showed pairing with six other chromosomes in different cells. Associations of three and four chromosomes were also observed. These observations refuted the hypothesis that the basic number of chromosomes in rice is five. Chu (1967) also tried to identify the individual chromosomes of the complement at pachynema in the haploids. But since pachytene univalents are more difficult to identify than the bivalents, he was not always certain about the position of the centromere. Therefore, one might expect certain differences between his observation and those of Shastry *et al.* (1960) about the arm ratios of a few chromosomes.

Khan and Rutger (1972) studied three japonica varieties at pachynema and clearly identified chromomeric patterns of each chromosome. Shastry and Mohan Rao (1961) and Shastry (1962, 1964) studied the karyotype of several other species of *Oryza* and certain interspecific hybrids at pachynema.

Chromosome Aberrations

Several workers have observed plants with aberrant karyotypes such as inversions, segmental deficiencies, or translocations. Most of the aberrations resulted from radiation treatments; only a few were of spontaneous origin. Radiation-induced aberrations, mainly translocations, were studied by Parthasarathy (1938), Nishimura (1957), Bora and Rao (1958), Huang and Chang (1958), Soriano (1959, 1961), Shastry and Ramiah (1961), Shah *et al.* (1961), and Katayama (1963a). Reciprocal translocation homozygotes were obtained by Nishimura and Kurakami (1952), Oka *et al.* (1953), and Huang and Chang (1958) from the selfed progenies of translocation heterozygotes. Translocation homozygotes are phenotypically indistinguishable from the original variety (Hsieh *et al.*, 1962).

Nishimura (1961) cytologically examined 224 semisterile lines from irradiated rice cultures and found 153 which carried reciprocal translocations. Translocation homozygotes were established and individual chromosomes involved in these translocations were identified from intercrosses of 34 translocation stocks. The 12 chromosomes were numbered I–XII in the order in which they were identified from intercrosses. This numbering, however, is arbitrary and not related to the numbers assigned to the individual chromosomes on the basis of cytological observations. Nishimura (1961) and Iwata and Omura (1971a,b) used 29 of these translocation stocks for locating marker genes on the chromosomes. Twenty-one marker genes were tested and linkage groups corresponding to 10 of the 12 chromosomes were identified.

Haploids, Polyploids, and Aneuploids

Haploids

Since the first discovery of a haploid rice by Ramiah *et al.* (1933a), several haploids of spontaneous origin and haploids from the progenies of irradiated materials have been studied. Yasui (1941) and Hu (1958, 1964) examined the somatic chromosomes of haploids and compared the morphology of individual members of the complement. Hu (1958) observed that chromosomes tended to pair somatically in the haploid plant. Configurations similar to those observed by Hu may also result from chromosome stickiness or clumping caused by poor fixations. Chu (1967) noted paired chromosomes at pachynema in haploids. Some chromosomes paired with several other chromosomes. At diakinesis and metaphase I, bivalents, trivalents, and quadrivalents were noted. But pairing in haploids is not cogent in throwing light on the presence of duplications in the chromosome complement.

Niizeki and Ono (1968) induced haploids from anther culture. Haploids studied to date are characterized by complete sterility, and by reduction in tiller number, height, number of spikelets, leaf width, and length.

Triploids

In rice it is difficult to get triploids by crossing a tetraploid and a diploid plant (Okura, 1940). However, Morinaga and Kuriyama (1959) obtained a few triploid plants from the $2n \times 4n$ cross, and Nagamatsu *et al.* (1964) from the $4n \times 2n$ cross. Spontaneous triploids are occasionally found in rice fields, however. Nakamori (1932), Ramiah *et al.* (1933*b*), Morinaga and Fukushima (1935), Hu and Ho (1963), and Watanabe *et al.* (1969) studied triploids of spontaneous origin. Morinaga and Fukushima (1935), Ramanujam (1937), and Hu and Ho (1963) studied the cytology, especially meiotic configurations at metaphase I. The modal pairing at metaphase I was observed to be 10 trivalents, two bivalents, and two univalents. Several associations resulting from nonhomologous pairing, such as cells with a few hexavalents, or with 10 III + 3 II or 8 III + 6 II, were also reported.

Triploids are distinguishable from disomic sibs by their broad leaves, large spikelets, and somewhat taller height. Their seed fertility is less than 1 percent upon self-pollination. However, cross-pollination with pollen from the $2n$ parent improves seed setting. Progenies of triploids have been studied by Ramanujam (1937), Katayama (1963*b*), Watanabe *et al.* (1969), and Khush (unpublished). Plants with $2n + 1$ and $2n + 2$ chromosomes are predominant in such progenies (Table 3). The maximum number of extra chromosomes tolerated by rice plants is six. Very few of the rice plants have more than four extra chromosomes, so the tolerance limit for extra chromosomes is narrow. Gametes or zygotes with higher

TABLE 3. Chromosome Numbers in the Progenies of Autotriploid Rice Plants

	Chromosome number													Total	Reference
	24	25	26	27	28	29	30	31	32	33	34	35	36		
No.	6	9	8	9	11	4	3							50	Ramanujam, 1937
%	12.0	18.0	16.0	18.0	22.0	8.0	6.0							100	
No.	3	6	8	1	2	1								21	Katayama, 1963b
%	14.3	28.5	38.1	4.8	9.5	4.8								100	
No.	20	42	61	14	1	0	1							139	Hu, 1968
%	14.4	30.2	43.9	10.1	0.7	0	0.7							100	
No.	6	35	31	12	2	1								87	Watanabe et al., 1969
%	6.9	40.2	35.6	13.8	2.3	1.1								100	
No.	2	20	25	14	8	3								72	Khush, unpublished
%	2.8	27.8	34.7	19.4	11.1	4.2								100	

numbers of extra chromosomes abort because the duplication of extra chromatin causes imbalance. Polyploid species, however, can tolerate a much higher number of extra chromosomes (Khush, 1973). Narrower limits for duplications and the absence of reports on transmissible monosomics indicate that rice is a basic diploid with little duplication in its genome.

Tetraploids

Spontaneous tetraploids were studied by Nakamori (1933) and Morinaga and Fukushima (1937). The plants were somewhat stunted, reduced to about four-fifths the height of the diploid plants. They had incomplete panicle exsertion, longer panicles, awns on the spikelets, thicker culms coarser leaves, and fewer tillers. Their fertility varied from 5 to 35 percent. Tetraploids were induced by colchicine and high-temperature treatments by Beachell and Jones (1945) and Cua (1951). Katayama (1963a) obtained tetraploids by x-ray treatment.

Morinaga and Fukushima (1937) studied meiosis in tetraploids. Tetravalents formed at an average frequency of 8.66 per cell while the rest of the chromosomes were present as bivalents or univalents. Masima (1952) reported that pollen fertility in a tetraploid was 70 percent, while seed set varied from 0 to 27 percent; many shriveled seeds were produced. He also noted that fertility of the line varied considerably from year to year and could not be improved by selection. He calculated that 15- to 20-percent sterility was caused by unbalanced gametes which resulted from irregular segregation of chromosomes from tetravalents and misdivision of univalents. Fifty percent of the sterility was due to zygotic lethality, probably caused by physiological imbalance; 15 percent was attributed to environmental causes.

Oka et al. (1954) examined meiosis in a number of tetraploid rice varieties and in the F_1 hybrids of these tetraploids. The number of tetravalents and univalents was generally lower, and segregation of chromosomes at anaphase I was more regular in intervarietal hybrids than in tetraploid varieties. Hybrids were also generally more fertile. But the tetravalent or univalent frequency and the percentage of fertility among the hybrids or varieties were not correlated.

To breed highly fertile tetraploid rice varieties, Masima and Uchiyamada (1955) induced 41 tetraploids from F_1 hybrids of distant varieties and compared their fertilities with those of autotetraploids of 16 varieties. On the average, the hybrid tetraploids had higher fertility than the tetraploid varieties, but none were fertile enough to be of commercial value. Moreover, all of them were inferior to the disomics in agronomic traits. Oka (1955) also noted the higher fertility of the hybrid tetraploids, but he demonstrated that this could not be attributed to bivalent pairing in such hybrids since several Mendelian genes segregated in a typically tetragenic manner.

Aneuploids

Several authors have obtained primary trisomics of rice from various sources. Triploids have proved to be the best sources of trisomics. Progenies of triploids reported by Morinaga and Fukushima (1935) were grown for several years and investigated by Yunoki and Masuyama (1945). At least six morphologically distinguishable primary trisomics were identified. Primary trisomics from the progenies of triploids were also obtained by Ramanujam (1937), Karibasappa (1961), Katayama (1963a), Jachuck (1963), Sen (1965), Hu (1968), Watanabe et al. (1969), and Iwata et al. (1970); but only Hu (1968) was able to isolate a complete set of 12 primary trisomics in variety Kehtze (an indica rice). All primary trisomics are morphologically distinct from the disomic sibs, as well as from each other. The trisomics generally have short stature, delayed heading, and vary in the size of panicle and spikelets. Seed setting is generally low. Some have longer grains than the disomic controls, but others have shorter grains. All the triosomics are transmitted to the next generation upon self-pollination. The average transmission rate for the 12 trisomics was reported to be 36.7 percent through the female side. Primary trisomics were also isolated from the progenies of asynaptic and partially sterile plants by Jones and Longley (1941), Jones (1952), and Katayama (1966). Primary trisomics have not yet been used to associate linkage groups with respective chromosomes.

Little is known about the secondary or tertiary trisomics of rice. Some of the $2n + 1$ plants found in the progenies of translocation heterozygotes must have been tertiary trisomics.

Only two cases of monosomy have been reported. Chandrasekharan (1952) obtained a plant which had fertile and sterile tillers. The sterile tillers were monosomic. The plant was obviously a chimera. Seshu and Venkataswamy (1958) selected a sterile plant from the F_5 generation of a cross between an indica and a japonica variety. Cytological examination proved it to be a monosomic. Its progeny consisted entirely of disomics. The monosomic condition was not transmitted to its progeny. In this respect rice behaves as a basic diploid; monosomy is tolerated only at the sporophytic level, not at the gametophytic level.

Marker Genes and Linkage Groups

Marker Genes

Although rice is the world's most important food crop, present knowledge of its basic genetics lags far behind that of other genetically well-known species such as maize, barley, and tomato. The progress of genetic studies has been slow because the research done in Japan was reported in Japanese, while that done in India and the U.S.A. was reported in English. Furthermore, the limited contact between scientists in different countries resulted in a confusing lack of uniformity in symbols to designate genes.

Yamaguchi (1927) proposed standard symbols for known genes but did not suggest a guide for devising additional symbols. He proposed the symbols for 43 genes, most of which were derived from German names for the traits. Later, Yamaguchi (1939) listed 130 loci and proposed quite different symbols; these were derived from Latin names for the respective characters. Several of the symbols suggested earlier were changed. Kadam and Ramiah (1943) proposed certain rules for gene symbolization in rice, summarized the literature on rice genetics published so far in India, and listed recommended symbols. The system they proposed was followed in the U.S.A., but in Japan a different set of symbols for the same or similar genes was used. The confusion that resulted from nonuniform gene symbolization is exemplified by the six different symbols used by various authors for one recessive gene which conditions glutinous endosperm (Table 4). Moreover, the symbol *gl* is used for another entirely different locus, which conditions the glabrousness of plant parts.

Realizing the importance of uniformity in gene symbolization in rice, the Sixth Meeting of the FAO International Rice Commission (IRC)

TABLE 4.   *Different Symbols Used for the Gene for Glutinous or Waxy Endosperm*[a]

Name	Symbol	Author
Mochi (glutinous)	*m*	Yamaguchi, 1918
Uruchi (nonglutinous)	*U*	Takahashi, 1923
Amylacea	*am*	Yamaguchi, 1927
Glutinous	*gl*	Chao, 1928
Glutinous	*g*	Enomoto, 1929
Waxy	*wx*[b]	

[a] From Kihara (1964).
[b] Approved by the IRC.

Working Party on Rice Breeding, held in 1955, appointed a committee to standardize rules for gene symbolization and to compile recommendations on the symbols for the already-known genes. The committee adopted, with minor modifications, the Rules for Standardization of Gene Symbols prepared by the Tenth International Congress of Genetics (1959). Mr. N. E. Jodon (U.S.A.) as convener, Dr. M. E. Takahashi (Japan) an associate of Dr. S. Nagao, Dr. R. Seetheraman (India) on behalf of Dr. N. Parthasarathy, and Dr. K. Ramiah prepared tabulations of known genes for the U.S.A., Japan, and India. These tabulations, along with recommended symbols for 90 genes, were published in the 1959 IRC Newsletter. Later, expanded tabulations, gene symbols, linkage maps, and a comprehensive bibliography were published by the committee (Anonymous, 1963).

Authoritative reviews describing mutant traits were published by Yamaguchi (1927, 1939), Jones (1936, 1952), Kadam and Ramiah (1943), Nagao (1951), Ramiah and Rao (1953), Nagai (1958), and Ghose et al. (1960). Table 5 lists some of the important genes and the symbols recommended by the IRC.

There are more dominant mutants in rice than in maize, barley, and tomato. Approximately, half of the mutant traits listed in Table 5 are governed by dominant genes. Moreover, the inheritance of several traits seems to be much more complex in rice than in other crop species. Many complementary loci as well as inhibitors are known. An example of complicated inheritance in rice is the interaction of several loci which govern the pigmentation of various plant parts.

Two complementary series of alleles at the C and A loci determine the formation of anthocyanin. The basic gene for the production of chromogen is C, and the conversion of chromogen into anthocyanin is controlled by A. There are multiple alleles at the two loci; six alleles at the C locus and four at the A (Nagao et al., 1962). Five other loci control the distribution of anthocyanin in plant parts: P controls the distribution of chromogen over the apiculus; Pr controls the color over the glumes and rachilla; Pl, with three alleles, conditions the distribution of pigment over the leaf blade, leaf sheath, pulvinus, auricle, ligule, node, internode, and rachis; Pn governs the color in leaf apex and margin; and Ps controls the localization of pigments in the stigma (Takahashi, 1957; Nagao et al., 1962). An inhibitor gene, $I\text{-}Pl$, suppresses the effect of Pl (Nagao et al., 1962; Kondo, 1963a). Inhibitor genes for colored apiculus ($I\text{-}P$) and colored leaf apex ($I\text{-}Pla$) have also been reported (Kondo, 1963a,b). The interaction of basic genes, spreader genes, and inhibitor genes produces various color distributions and intensities. Color development is further complicated by the fading and leaching of certain colors; and some red

TABLE 5.   Well-Known Genes of Rice[a]

Gene	Phenotype
A, A^d, a	Allelic anthocyanin activator genes (complementary action with C genes produces red or purple pigment in apiculus)
al	Albino
An	Awned
au	Rudimentary auricle
bc	Brittle culm
Bd	Beaked hull (tip of lemma recurved over palea)
Bf	Brown furrow (dark brown color in furrows of lemma and palea)
I–Bf	Inhibitor of dark-brown furrow
Bh	Black hull
bl	Physiologic diseases showing dark brown or blackish mottled discoloration of leaf
bn	Bent node—culm forms angle at node
C, C^B, C^{Bp}, C^{Bt}, C^{Br}, c	Allelic basic genes for anthocyanin color; higher alleles have pleiotropic expression in internode
C (alone)	Tawny-colored apiculus
CA	Red- or purple-colored apiculus
CAP	Completely and fully purple-colored apiculus
Ce	*Cercospora* resistance
chl	Chlorina (chlorophyll deficiency)
Cl	**Clustered spikelets, also super cluster**
cls	Cleistogamous spikelets
clw	Claw-shaped spikelets
d	Dwarf
da	Double awn
Dn	Dense or compact; very close arrangement of spikelets (vs. normal panicle); epistatic to Ur
Dn_2	Dense vs. lax[b]
Dn_3	Normal vs. lax
Dp	Depressed palea and underdeveloped palea
dw	Deep-water paddy, so-called floating rice
Ef	Early flowering (low photosensitivity)
er	Erect growth habit, recessive to spreading or procumbent
Ex	Exerted vs. enclosed panicle
Fgr	Fragrant flower
fs	Fine stripe
g	Long glume exceeding ⅓ length of the spikelet
gh	Gold hull (golden yellow hull, recessive to straw color)
gl	Glabrous (nonhairy) leaf
Gm	Semidominant long glume
He	*Helminthosporium* resistance

TABLE 5. Continued

Gene	Phenotype
hsp	Hull spot
la	Lazy, procumbent growth habit
Lf	Late flowering (highly photosensitive)
lg	Liguleless (auricle and collar also absent)
Lh	Very hairy (long hair), dominant to ordinary pubescence
lk	Grain length (long grain)
lmx	Extra lemma
lu	Lutescent
Lx	Lax vs. normal panicle
Lx_2	Lax vs. compact
me	Multiple embryos
mp	Multiple pistils
nal	Narrow leaf
nl	Neck leaf
nk or I-Nk	Notched kernel
o	Open hull (parted lemma and palea)
P	Completely purple apiculus (complementary action with C and A)
Pau	Purple auricle, basic to Plg
Pg	Purple outerglumes
Ph	Phenol staining of hull and bran
Pi	Piricularia resistance
Pin	Purple internode
Pl	Purple leaf
Pla	Purple leaf apex (complementary action with C and A)
Plg	Purple ligule
I-Pl	Colorless leaf except margin
Pn	Purple node
Pr	Purple hull (lemma and palea) (complementary action with CAP)
Prp	Purple pericarp
Ps	Purple stigma (complementary action with CAP)
Psh	Purple sheath
Pu	Purple pulvinus
Pw	Purplewash (faded purple)
Px	Purple axil
Rc	Brown pericarp (basic to Rd)
Rd	Red pericarp (complementary action with Rc)
ri	Verticillate (whorled) arrangement of rachis
Rk	Round spikelet (kernel)
rl	Rolled leaf
Sk	Scented kernel
sn	Sinuous neck

TABLE 5. Continued

Gene	Phenotype
spr	Spreading panicle branches
th	Shattering or easy threshing recessive to difficult separation
tl	Twisted leaves
tri	Triangular hull (spikelet)
Ur	Undulate rachis vs. normal
v	Virescent, also green and white stripe
wb	White belly (endosperm)
wb$_2$	White core (endosperm)
Wh	White hull, epistatic to gh
wx	Waxy (glutinous) endosperm
y	Yellow leaf
l-y	Lethal yellow, also xantha
z	Zebra stripe

[a] From Anonymous, 1963.
[b] The number of grains per unit length of panicle is greater (*dense*) or less (*lax*) than in wild type.

colors disappear by maturity (Nagao, 1951; Jodon, 1955; Takahashi, 1957). An example of the complexity of pigmentation patterns as conditioned by the interaction of different alleles at different loci is illustrated for apiculus color in Table 6. Several other examples of multiple allelic series and of complex gene action in rice have been reviewed by Seetharaman (1964) and Chang (1964).

The search for genes which control disease resistance and insect resistance has acquired considerable importance as emphasis has shifted to the breeding of varieties resistant to diseases and insect pests. At least seven independent loci with dominant alleles convey resistance to different races of blast (*Pericularia oryzae*) (Kiyosawa, 1972). Jodon *et al.* (1944) ascribed the resistance to cercospora leaf spot (*Cercospora oryzae*) to a dominant gene, *Ce*. The dominant gene *He* controls the resistance to brown spot (*Helminthosporium oryzae*) (Nagai and Hara, 1930); *Xa* governs resistance to bacterial leaf blight (*Xanthomonas oryzae*) (Nishimura and Sakaguchi, 1959). Resistance to grassy stunt virus is also conferred by a single dominant gene (Khush, unpublished). Athwal *et al.* (1971) have identified three independent loci with dominant alleles (*Glh*1, *Glh*2, *Glh*3), for resistance to green leafhopper (*Nephotetix virescens*), and two very closely linked loci (*Bph*1 and *bph*2) for resistance to the brown planthopper (*Nilaparvata lugens*). These useful genes are being incorporated into improved varieties.

Linkage Groups

The first instance of linkage in rice, that between black hull and colored internode, was reported by Parnell *et al.* (1917). Yamaguchi (1921, 1926) reported a linkage between the apiculus color and waxy endosperm. The linkage was confirmed by Chao (1928), Ramiah *et al.* (1931), Jodon, (1940), and Nagao and Takahashi (1942). These reports established the "waxy" linkage group or linkage group I. Morinaga and Nagamatsu (1942) established the second linkage group or "purple-leaf" group when he discovered the linkages between genes for purple leaf, ligulelessness, and phenol staining. Several researchers have summarized data on linkages and advocated provisional linkage groups (Nagao, 1951; Ramiah and Rao, 1953; Jodon, 1955; Nagao and Takahashi, 1960, 1963). Nagao and Takahashi (1963) and Takahashi (1964) have proposed 12 linkage groups and presented extensive segregation data. Some of the linkage groups consist of only two genes each. Forty genes have been mapped. Figure 3 presents the latest chromosome map of rice.

Iwata and Omura (1971*a,b*) established the independence of 10 linkage groups through the use of a set of translocation testers. The chromosomes involved in the translocation testers have not been identified

TABLE 6.   *Apiculus Color as Conditioned by the Interaction of Different Alleles at Three Loci* [a]

Genotype	Color shade at flowering	Color shade at ripening
$C^B\ A\ P$	Blackish red-purple	Faded purple
$C^{Bp}\ A\ P$	Pansy purple	Faded red-purple
$C^{Bt}\ A\ P$	Tyrian rose	Faded pink
$C^{Br}\ A\ P$	Rose red	Straw white
$C^{Bm}\ A^E\ P$	Faint red	White or almost white
$C^+\ A\ P$	White	Straw white
$C^B\ A^d\ P$	Amaranth purple	Tawny
$C^{Bp}\ A^d\ P$	Pomegranate purple	Light tawny
$C^{Bt}\ A^d\ P$ $C^{Br}\ A^d\ P$ }	Seashell pink	Yellow-white or white
$C^+\ A^d\ P$	White	White
$C^B\ A^+\ P$	White	Russet
$C^{Bp}\ A^+\ P$	White	Tawny
$C^{Bt}\ A^+\ P$	White	Ochraceous buff
$C^{Br}\ A^+\ P$	White	Warm buff
$C^+\ A^+\ P$	White	Straw white

[a] Takahaski (1957) and Nagao *et al.* (1962).

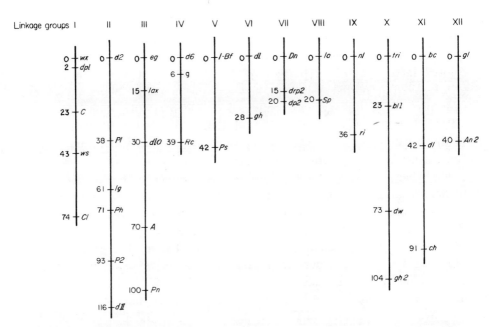

Figure 3. Linkage maps of rice based on linkage data of Nagao and Takahashi (1960, 1963) and Iwata and Omura (1971a,b).

cytologically, but they have been arbitrarily numbered 1–12 in the order of their identification from intercrosses of different translocations.

The primary trisomic series now available should be examined cytologically, and the extra chromosome of each trisomic should be identified according to the nomenclature of Shastry *et al.* (1960). The representative markers of each linkage group should be concurrently tested against each trisomic to associate the linkage groups with their respective chromosomes.

Breeding for Higher Productivity and Wider Adaptability

Average yields are dismally low in the less-developed countries where more than 90 percent of the world's rice is grown. In tropical Asia, for example, the average rice yields in the early 1960's were 1.5 metric tons per hectare—little different from what they had been for centuries. In the more-developed temperate parts of the world, such as Japan, California, Australia, Spain, and Italy, national rice yields averaged more than 5 tons per hectare—three or four times the best tropical yields.

In the late 1950's, alarmed by the widening gap between population growth and food supply, the Ford Foundation and The Rockefeller Foundation in cooperation with the Government of the Philippines decided to establish the International Rice Research Institute (IRRI) to improve rice yields in the less-developed areas of the world. IRRI scientists noted that while high rates of fertilizers were being used in the developed countries, very little was being used in the tropics. Tall indica rice varieties grown in the tropics developed under an alternating monsoon–dry climate and in areas of poor drainage, low solar radiation, high temperatures, infertile soils, and high weed competition. Consequently, the basic tropical plant type is characterized by rapid and vigorous seedling development and initial nutrient uptake, long and drooping leaves, tall and weak stems, profuse tillering, and late maturity. Long droopy leaves form a canopy which intercepts most of the sunlight, thus causing senescence of lower leaves and weak, elongated stems. When grown with moderate to high rates of nitrogen fertilizer, the plant grows even taller and lodges. Since the lodged plants can no longer intercept sunlight efficiently, they stop manufacturing carbohydrates; the grains do not fill properly, and yields are even lower than those in unfertilized fields.

Japonica varieties grown in temperate areas and in subtropical Taiwan, on the other hand, are short, with narrow, erect leaves. They are adapted to intense management practices, such as weed control, good drainage and water management, and fertilizer application. They are responsive to nitrogen fertilizers since they do not lodge as easily as do the tropical varieties, but they are not adapted to the tropics. Most attempts to develop nitrogen-responsive varieties for the tropics by crossing indica and japonica varieties were unsuccessful, primarily because the breeders continued to select for tall indicas in the segregating generations.

Initially IRRI plant breeders realized that they needed short-statured and sturdy-stemmed varieties which would respond to high rates of nitrogen (Jennings, 1964). From the collection of rice germ plasm assembled in 1961, they took three closely related short-statured indica-type varieties from Taiwan. In 1962, their first year of research, IRRI breeders made 38 crosses; a third of which involved such dwarfs as Dee-geo-woo-gen and I-geo-tze. One cross was with Peta, a tall traditional variety from Indonesia adapted to tropical conditions, which had good disease and insect resistance. Three years later, an experimental selection from this cross yielded 6.6 metric tons per hectare. This selection was widely tested in regional, national, and international trials. It gave record yields in most locations. This variety, named IR8 by IRRI in 1966, is the forerunner of the green revolution in rice. It has short stature (about 100 cm, compared with 150 cm for typical indicas), sturdy straw, narrow dark green and upright leaves which use solar energy more efficiently, and high

Figure 4. Traditional tropical variety (Peta) characterized by tall habit, long droopy leaves, and lodging susceptibility (left). Dwarf high-yieldings selection (IR773-36) characterized by short sturdy stems, heavy tillering, lodging resistance, and short, dark green, erect leaves (right).

tillering ability (Figure 4). Highly responsive to nitrogen, it has yielded up to 10 metric tons per hectare in the dry seasons (when solar radiation is high). With the development of this variety, the yield potential of rice in the tropics has been more than doubled.

IR8 sold the concept of short plant type to breeders across the world. Today rice breeders in Asia, Africa, and Latin America are developing dwarf varieties with IR8 or other IRRI-developed short-statured lines as the dwarfing sources. Today, more than 12 million hectares of rice land are planted with four IRRI-named dwarf varieties, 23 improved varieties developed from IRRI-bred experimental lines, and more than 20 improved varieties from national programs. All these high-yielding varieties have the same gene for short stature. This dwarfing source probably originated as a spontaneous mutant in mainland China and was introduced to Taiwan before the Japanese occupation of that country (Chandler, 1968; Athwal, 1971). Probably no other single gene in rice has benefited mankind more than this dwarfing gene.

Several other traits are being incorporated into high-yielding dwarfs to stabilize their yield potential and increase their adaptability (Beachell and Khush, 1969). Early-maturing varieties would permit farmers to grow more than one crop on the same land. Varieties with long growth duration and with photoperiod sensitivity are being developed for areas where heavy rainfall during the monsoon season and standing water due to inadequate drainage hamper rice harvesting, threshing, and drying operations. Cold-tolerant varieties are needed for high-altitude areas. Large areas are planted with rice in Asia where water depth is 1–4 m during the growing season. "Floating-rice" varieties grown in these areas are able to elongate with the rising water level. These varieties need to be improved.

The major thrust of rice breeding effort at IRRI is being directed toward developing disease- and insect-resistant varieties (Khush and Beachell, 1972). Sources of resistance to internationally important diseases such as blast, bacterial blight, tungro, and grassy stunt, and to such insects as stem borers, brown planthoppers, and green leafhoppers have been identified from IRRI's world germ plasm collection (Chang *et al.,* 1973). A set of high-yielding experimental lines with varying growth durations and with multiple disease resistance and insect resistance have been developed and are being evaluated for adaptability to different environmental conditions (IRRI, 1973).

Literature Cited

Anonymous, 1963 Rice gene symbolization and linkage groups. *US Dep. Agric. Res. Serv. ARS (Ser.)* **34–28**:1–56.

Athwal, D. S., 1971 Semidwarf rice and wheat in global food needs. *Q. Rev. Biol.* **46**:1–34.

Athwal, D. S., M. D. Pathak, E. H. Bacalangco and C. D. Pura, 1971 Genetics of resistance to brown planthoppers and green leafhoppers in *Oryza sativa* L. *Crop Sci.* **11**:747–750.

Barton, D. W., 1950 Pachytene morphology of the tomato chromosome complement. *Am. J. Bot.* **37**:639–643.

Beachell, H. M. and J. W. Jones, 1945 Tetraploids induced in rice by temperature and colchicine treatments. *J. Am. Soc. Agron.* **37**:165–175.

Beachell, H. M. and G. S. Khush, 1969 Objectives of the IRRI rice breeding program. *Sabrao Newsl.* **1**:69–80.

Bhaduri, P. N., A. T. Natarajan and R. N. Mohanty, 1959 Pollen mitosis in paddy. *Indian J. Genet. Plant Breed.* **18**:249–251.

Bora, K. C. and N. S. Rao, 1958 Experience with rice (*Oryza sativa*) on the induction of mutation by ionizing radiations. *Proc. 2nd Int. Conf. Peaceful Uses Atomic Energy, UN Geneva* **27**:306–313.

Bouharmont, J., 1962 Recherches cytogenetiques chez quelques hydrides interspecifiques d'*Oryza. Cellule* **63**:51–132.

Brown, S. W., 1949 The structure and meiotic behavior of the differentiated chromosomes of tomato. *Genetics* **34**:437–461.

Chandler, R. F., 1968 Dwarf rice—a giant in tropical Asia. *US Dep. Agric. Yearb. Agric.* **1968**:252–255.

Chandrasekharan, P., 1952   The occurrence of barren tillers in a heterozygous rice. *Madras Agric. J.* **39:**605–610.

Chang, T. T., 1964   Present knowledge of rice genetics and cytogenetics. *Int. Rice Res. Inst. (Los Banos) Tech. Bull.* **1:**1–96.

Chang, T. T., S. H. Ou, M. D. Pathak, K. C. Ling and H. E. Kauffman, 1973   The search for disease and insect resistance in rice. Paper presented at the FAO/IBP technical conference on Crop Genetic Resources, March 12–16, 1973, FAO, Rome.

Chao, L. F., 1928   Linkage studies in rice. *Genetics* **13:**133–169.

Chatterjee, D., 1948   A modified key and enumeration of species of *Oryza* L. *Indian J. Agric. Sci.* **18:**185–192.

Chevalier, A., 1932   Nouvelle contribution a l'étude systematique de *Oryza*. *Rev. Bot. Appl. Agric. Trop.* **12:**1014–1032.

Chu, Y. E., 1967   Pachytene analysis and observations of chromosome association in haploid rice. *Cytologia (Tokyo)* **32:**87–95.

Cua, L. D., 1951   Artificial polyploidy in the oryzeae. II. Fertile tetraploids of Japonica × Indica in rice. *Proc. Jap. Acad.* **27:**43–48.

Enomoto, N., 1929   Mutation on the endosperm character in rice plant. *Jap. J. Genet.* **5:**49–72.

Ghose, R. L. M., M. B. Ghatge and V. Subrahmanyam, 1960   Rice in India, second edition, Indian Council of Agricultural Research, New Delhi.

Henderson, M. T., 1964   Cytogenetic studies at the Luisiana Agricultural Experiment Station of species relationships in *Oryza*. In *Rice Genetics and Cytogenetics*, pp. 103–110, American-Elsevier, New York.

Hsieh, S. C., W. T. Chang and T. M. Chang, 1962   Studies on agronomic characters in reciprocal translocation homozygotes of rice. *Jap. J. Breed.* **12:**45–48.

Hu, C. H., 1958   Karyological studies in haploid rice. II. Analysis of karyotype and somatic pairing. *Jap. J. Genet.* **33:**296–301.

Hu, C. H., 1964   Further studies on the chromosome morphology of *Oryza sativa* L. In *Rice Genetics and Cytogenetics*, pp. 51–61, American-Elsevier, New York.

Hu, C. H., 1968   Studies on the development of twelve types of trisomics in rice with reference to genetic study and breeding program. *J. Agric. Assoc. China* **63:**53–71.

Hu, C. H. and K. M. Ho, 1963   Karyological studies of triploid rice plants. I. Chromosome pairing in autotriploid of *Oryza sativa* L. *Bot. Bull. Acad. Sin. (Taipei)* **4:**30–36.

Huang, C. S. and T. D. Chang, 1958   Cytogenetic studies on X-rayed rice. *J. Agric. Assoc. China* **24:**10–31.

International Rice Research Institute, 1973   Annual Report for 1972. Los Baños, Philippines.

Iwata, N. and T. Omura, 1971a   Linkage analysis by reciprocal translocation method in rice plants (*Oryza sativa* L.). I. Linkage groups corresponding to the chromosome 1, 2, 3, and 4. *Jap. J. Breed.* **21:**19–28.

Iwata, N. and T. Omura, 1971b   Linkage analysis by reciprocal translocation method in rice plants (*Oryza sativa* L.). II. Linkage groups corresponding to the chromosomes 5, 6, 8, 9, 10 and 11. *Sci. Bull. Fac. Agric. Kyushu Univ.* **25:**137–153.

Iwata, N., T. Omura and M. Nakagahara, 1970   Studies on the trisomics in rice plants (*Oryza sativa* L.). I. Morphological classification of trisomics. *Jap. J. Breed.* **20:**230–236.

Jachuck, P. J., 1963   Isolation of chromosome variants in cultivated rice. *Oryza* **1:**131–132.

Jennings, P. R., 1964   Plant type as a rice breeding objective. *Crop Sci.* **4**:13–15.

Jodon, N. E., 1940   Inheritance and linkage relationships of a chlorophyll mutation in rice. *J. Am. Soc. Agron.* **32**:342–346.

Jodon, N. E., 1955   Present status of rice genetics. *J. Agric. Assoc. China* **10**:5–21.

Jodon, N. E., T. C. Ryker and S. J. P. Chilton, 1944   Inheritance of reaction to physiologic races of *Cercospora oryzae* in rice. *J. Am. Soc. Agron.* **36**:497–507.

Jones, J. W., 1936   Improvement in rice. *US Dep. Agric. Yearb. Agric.* **1936**:415–454.

Jones, J. W. 1952   Inheritance of natural and induced mutations in Caloro rice and observations on sterile Caloro types. *J. Hered.* **43**:81–85.

Jones, J. W. and A. F. Longley, 1941   Sterility and aberrant chromosome numbers in Caloro and other varieties of rice. *J. Agric. Res.* **62**:381–399.

Kadam, B. S. and K. Ramiah, 1943   Symbolization of genes in rice. *Indian J. Genet. Plant Breed.* **3**:7–27.

Karibasappa, B. K., 1961   An auto-triploid of rice and its progeny. *Current Sci. (Bangalore)* **30**:432–433.

Katayama, T., 1963a   X-ray induced chromosomal aberrations in rice plants. *Jap. J. Genet.* **38**:21–31.

Katayama, T., 1963b   Study on the progenies of autotriploid and asynaptic rice plants. *Jap. J. Breed.* **13**:83–87.

Katayama, T., 1966   Relationship between seed weight and somatic chromosome number in the progeny of partially asynaptic rice plant. *Jap. J. Breed.* **16**:10–14.

Khan, S. H. and J. N. Rutger, 1972   Revision of the rice karyotype. *Proc. Fourteenth Rice Technical Working Group (USA) Meetings,* p. 20.

Khush, G. S., 1973   *Cytogenetics of Aneuploids,* Academic Press, New York.

Khush, G. S. and H. M. Beachell, 1972   Breeding for disease and insect resistance at IRRI. In *Rice Breeding,* pp. 309–322, International Rice Research Institute, Los Baños, Philippines.

Kihara, H., 1964   Need for standardization of genetic symbols and nomenclature in rice. In *Rice Genetics and Cytogenetics,* pp. 3–11, American-Elsevier, New York.

Kiyosawa, S., 1972   Genetics of blast resistance. In *Rice Breeding,* pp. 203–225, International Rice Research Institute, Los Baños, Philippines.

Kondo, A., 1963a   Fundamental studies on rice breeding through hydridization between Japanese and foreign varieties. VII. Identification of gene system controlling anthocyanin coloration in Japanese and foreign varieties. *Jap. J. Breed.* **13**:92–98.

Kondo, A., 1963b   Fundamental studies on rice breeding through hybridization between Japanese and foreign varieties. VIII. On some genes modifying anthocyanin coloration and their suppressors. *Jap. J. Breed.* **13**:241–245.

Kuwada, Y., 1910   A cytological study of *Oryza sativa* L. *Bot. Mag. (Tokyo)* **24**:267–281.

Masima, I., 1952   Causes of sterility in tetraploid rice. *Jap. J. Breed.* **1**:179–181.

Masima, I. and H. Uchiyamada, 1955   Studies on the breeding of fertile tetraploid plants of rice. *Bull. Natl. Inst. Agric. Sci. Ser. D (Plant. Physiol. Genet. Crops Gen.)* **5**:104–136.

Morinaga, T. and E. Fukushima, 1935   Cyto-genetical studies on *Oryza sativa* L. II. Spontaneous autotriploid mutants in *Oryza sativa* L. *Jap. J. Bot.* **7**:207–225.

Morinaga, T. and E. Fukushima, 1937   Cyto-genetical studies on *Oryza sativa* L. III. Spontaneous autotetraploid mutants in *Oryza sativa* L. *Jap. J. Bot.* **9**:71–94.

Morinaga, T. and H. Kuriyama, 1959   A note on the cross results of diploid and tetraploid rice plants. *Jap. J. Breed.* **9**:187–193.

Morinaga, T. and T. Nagamatsu, 1942   Linkage studies on rice, *Oryza sativa* L. *Jap. J. Genet.* **18**:192–200.

Nagai, I., 1958   *Japonica Rice: Its Breeding and Culture,* Yokendo Ltd., Tokyo.

Nagai, I. and S. Hara, 1930   Inheritance of a kind of leaf spot disease in rice. *Jap. J. Genet.* **5**:140–144.

Nagamatsu, T., T. Omura and Y. Koga, 1964   Studies on the crossing experiments between diploid and auto-tetraploid rice plants, *Oryza sativa* L., with special reference to the development of hydrid seeds. *Sci. Bull. Fac. Agric. Kyushu Univ.* **21**:25–34.

Nagao, S., 1951   Genetic analysis and linkage relationship of characters in rice. *Adv. Genet.* **4**:181–212.

Nagao, S. and M. E. Takahashi, 1942   Genetical studies on rice plant. III. Type and inheritance of awnedness. *J. Sapporo Agric. Forest.* **34**:36–43.

Nagao, S. and M. E. Takahashi, 1960   Genetical studies on rice plant. XXIV. Preliminary report of twelve linkage groups in Japanese rice. *J. Fac. Agric. Hokkaido Univ.* **51**:289–298.

Nagao, S. and M. E. Takahashi, 1963   Genetical studies on rice plant. XXVII. Trial construction of twelve linkage groups of Japanese rice. *J. Fac. Agric. Hokkaido Univ.* **53**:72–130.

Nagao, S., M. E. Takahashi and T. Kinoshita, 1962   Genetical studies on rice plant. XXVI. Mode of inheritance and causal genes for one type of anthocyanin color character in foreign rice varieties. *J. Fac. Agric. Hokkaido Univ.* **52**:20–50.

Nakamori, E., 1932   On the occurrence of the triploid plant of rice, *Oryza sativa* L. *Proc. Imp. Acad.* **8**:528–529.

Nakamori, E., 1933   On the occurrence of the tetraploid plant of rice, *Oryza sativa* L. *Proc. Imp. Acad.* **9**:340–341.

Nandi, H. K., 1936   The chromosome morphology, secondary association and origin of cultivated rice. *J. Genet.* **33**:315–336.

Niizeki, H. and K. Ono, 1968   Induction of haploid rice plant from anther culture. *Proc. Jap. Acad.* **44**:554–557.

Nishimura, Y., 1957   Genetic and cytological studies on the progeny of rice plants exposed to the atomic bomb as well as those irradiated by X-rays. *Proc. Int. Genet. Symp. Tokyo and Kyoto (1956)*, pp. 265–270.

Nishimura, Y., 1961   Studies on the reciprocal translocations in rice and barley. *Bull. Natl. Inst. Agric. Sci. D (Plant. Physiol. Genet. Crops Gen.)* **9**:171–235.

Nishimura, Y. and H. Kurakami, 1952   Mutations in rice induced by X-rays. *Jap. J. Breed.* **2**:65–71.

Nishimura, Y. and S. Sakaguchi, 1959   Inheritance of resistance in rice to bacterial leaf blight, *Bacterium oryzae* (Uyeda *et* Ishiyama) Nakata. *Jap. J. Breed.* **9**:58.

Oka, H. I., 1955   Studies on tetraploid rice. VI. Fertility variation and segregation ratios for several characters in tetraploid hybrids of rice, *Oryza sativa* L. *Cytologia (Tokyo)* **20**:258–266.

Oka, H. I., 1964   Pattern of interspecific relationships and evolutionary dynamics in *Oryza.* In *Rice Genetics and Cytogenetics,* pp. 71–90, American-Elsevier, New York.

Oka, H. I., T. D. Chang and M. S. Hong, 1953   Reciprocal translocations in rice. *Jap. J. Genet.* **28**:87–91.

Oka, H. I., S. C. Hsieh and T. S. Huang, 1954   Studies on tetraploid rice varieties and their hybrids. *Jap. J. Genet.* **29**:205–214.

Okura, E., 1940 Diploid F_1 hybrids produced from the cross between tetraploid and diploid rice in *Oryza sativa* L. *Jap. J. Genet.* **16:**228–233.

Parnell, F. R., G. N. R. Ayyangar and K. Ramiah, 1917 Inheritance of characters in rice. I. *Mem. Dep. Agric. India Bot. Ser.* **9:**75–105.

Parthasarathy, N., 1938 Cytological studies in Oryzeae and Phalarideae. II. Further studies in *Oryza. Cytologia (Tokyo)* **9:**307–318.

Ramanujam, S., 1937 Cytogenetical studies in the Oryzeae. II. Cytogenetical behavior of an autotriploid in rice (*Oryza sativa* L.). *J. Genet.* **35:**183–221.

Ramiah, K. and M. B. V. N. Rao, 1953 Rice Breeding and Genetics, Scientific Monograph No. 19, Indian Council of Agricultural Research, New Delhi.

Ramiah, K., S. Jobitharaj and S. D. Mudaliar, 1931 Inheritance of characters in rice. IV. *Mem. Dep. Agric. India Bot. Ser.* **18:**229–259.

Ramiah, K., N. Parthasarathy and S. Ramanujam, 1933a A triploid plant in rice (*Oryza sativa* L.). *Current Sci. (Bangalore)* **2:**170–171.

Ramiah, K., N. Parthasarathy and S. Ramanujam, 1933b Haploid plant in rice (*Oryza sativa*). *Current Sci. (Bangalore)* **1:**277–278.

Rau, N. S., 1929 Further contributions to the cytology of some crop plants of South India. *J. Indian Bot. Soc.* **8:**201–206.

Rick, C. M. and G. S. Khush, 1966 Chromosome engineering in *Lycopersicon*. In *Chromosome Manipulations and Plant Genetics,* (Suppl. to *Heredity,* Vol. 20) edited by R. Riley and K. R. Lewis, pp. 8–20, Oliver and Boyd, Edinburgh.

Roschevicz, R. J., 1931 A contribution to the knowledge of rice. *Bull. Appl. Bot. Genet. Plant Breed.* **27**(4):3–133.

Sakai, K., 1935 Chromosome studies in *O. sativa*. I. The secondary association of meiotic chromosomes. *Jap. J. Genet.* **11:**145–156.

Sampath, S., 1962 The genus *Oryza:* its taxonomy and species interrelationships. *Oryza* **1:**1–29.

Sampath, S., 1964 Suggestions for a revision of the genus *Oryza*. In *Rice Genetics and Cytogenetics,* pp. 22–23, American-Elsevier, New York.

Sampath, S., 1966 The genus *Oryza:* an evolutionary perspective. *Oryza* **3**(1):30–34.

Sampath, S. and S. Govindaswami, 1958 Wild rices of Orissa—their relationship to cultivated varieties. *Rice News Teller* **6**(3):17–20.

Seetharaman, R., 1964 Certain considerations on genic analysis and linkage groups in rice. In *Rice Genetics and Cytogenetics,* pp. 203–214, American-Elsevier, New York.

Sen, S. K., 1965 Cytogenetics of trisomics in rice. *Cytologia (Tokyo)* **30:**229–238.

Seshu, D. V. and T. Venkataswamy, 1958 A monosome in rice. *Madras Agric. J.* **45:**311–314.

Sethi, B., 1937 Cytological studies in paddy varieties. *Indian J. Agric. Sci.* **7:**687–706.

Shah, H. M., H. M. Beachell and I. M. Atkins, 1961 Morphological and cytological changes in Century Patna 231 and Bluebonnet 50 rice resulting from X-ray and thermal neutron irradiation. *Crop Sci.* **1:**97–102.

Sharma, S. D. and S. V. S. Shastry, 1965 Taxonomic studies in genus *Oryza* L., *O. rufipogon* Griff, *Sensu stricto* and *O. nivara* Sharma *et.* Shastry nom. nov. *Indian J. Genet. Plant Breed.* **25:**157–167.

Shastry, S. V. S., 1962 Pachytene analysis in the genus *Oryza. Bull. Bot. Surv. India* **4:**241–247.

Shastry, S. V. S., 1964 New approach to study of rice karyomorphology. In *Rice Genetics and Cytogenetics,* pp. 62–67, American-Elsevier, New York.

Shastry, S. V. S. and P. K. Mohan Rao, 1961   Pachytene analysis in *Oryza*. IV. Chromosome morphology of *O. australiensis* Dom., *O. glaberrima,* Steud, and *O. stapfii* Rosch. *Proc. Indian Acad. Sci. Sect. B* **54**:100–112.

Shastry, S. V. S. and K. Ramiah, 1961   Cytogenetical effects of X-rays, thermal neutrons and beta-particles in *Oryza sativa* L. *Indian J. Genet. Plant Breed.* **21**:43–51.

Shastry, S. V. S., D. R. Ranga Rao and R. N. Misra, 1960   Pachytene analysis in *Oryza*. I. Chromosome morphology in *Oryza sativa*. *Indian J. Genet. Plant Breed.* **20**:15–21.

Soriano, J. D., 1959   X-ray induced reciprocal translocations and chlorophyll mutations in rice. *Bot. Gaz.* **120**:162–165.

Soriano, J. D., 1961   Mutagenic effects of gamma radiation on rice. *Bot. Gaz.* **123**:57–63.

Stebbins, G. L., 1950   Variation and Evolution in Plants, Columbia University Press, New York.

Takahashi, M. E., 1957   Analysis of apiculus color genes essential to anthocyanin coloration in rice. *J. Fac. Agric. Hokkaido Univ.* **50**:266–362.

Takahashi, M. E., 1964   Linkage groups and gene schemes of some striking morphological characters in Japanese rice. In *Rice Genetics and Cytogenetics,* pp. 213–236, American-Elsevier, New York.

Takahashi, N., 1923   Linkage in rice. *Jap. J. Genet.* **2**(1):23–30.

Tateoka, T., 1963   Taxonomic studies of *Oryza*. III. Key to the species and their enumeration. *Bot. Mag. (Tokyo)* **76**:165–173.

Tateoka, T., 1964   Taxonomic studies of the genus *Oryza*. In *Rice Genetics and Cytogenetics,* pp. 15–21, American-Elsevier, New York.

Watanabe, Y., S. Ono, Y. Mukai and Y. Koga, 1969   Genetic and cytogenetic studies on the trisomic plants of rice, *Oryza sativa* L. I. On the autotriploid plant and its progenies. *Jap. J. Breed.* **19**:12–18.

Yamaguchi, Y., 1918   Beitrag zur Kenntnis der Xenien bei *Oryza sativa* (Vorlaufig Mitteilung). *Bot. Mag. (Tokyo)* **32**:83–90.

Yamaguchi, Y., 1921   Etudes d'heredite sur la couleur des glumes Chez la riz. *Bot. Mag. (Tokyo)* **35**:106–112.

Yamaguchi, Y., 1926   Kreuzungsuntersuchungen an Reispflanzen, I. Genetische Analyse der Spelzenfarbe, und der Endosperm-beschaffeuheit bei einigen Sorten des Reises. *Ber. Ohara Inst. Landwirtsch. Biol. Okayama Univ.* **3**:1–126.

Yamaguchi, Y., 1927   New genetic investigations on the rice plant. *Z. Indukt. Abstrammungs- Vererbungsl.* **45**:105–122.

Yamaguchi, Y., 1939   Genosymbola *Oryzae sativae*—Danswashitsu *Circ. Genet. Soc. Japan (National Institute of Genetics)* **4**:1–4.

Yao, S. Y., M. T. Henderson and N. E. Jodon, 1958   Cryptic structural hydridity as a probable cause of sterility in intervarietal hybrids of cultivated rice, *Oryza sativa* L. *Cytologia (Tokyo)* **23**:46–55.

Yasui, K., 1941   Diploid-bud formation in a haploid *Oryza* with some remarks on the behavior of nucleolus in mitosis. *Cytologia (Tokyo)* **11**:515–525.

Yeh, B. and M. T. Henderson, 1961   Cytogenetic relationship between cultivated rice, *Oryza sativa* L. and five wild diploid forms of *Oryza*. *Crop Sci.* **1**:445–450.

Yeh, B. and M. T. Henderson, 1962   Cytogenetic relationship between African annual diploid species of *Oryza and cultivated rice, O. sativa* L. *Crop Sci.* **2**:463–467.

Yunoki, T. and Y. Masuyama, 1945   Investigations on the later generations of autotriploid rice plants. *Sci. Bull. Fac. Agric. Kyushu Univ.* **11**:182–216.

3

The Wheats and Their Relatives

Ernest R. Sears

Classification

The wheat genus *Triticum* L. belongs to the subtribe Triticinae of the tribe Triticeae of the family Gramineae. Other genera of the Triticinae are *Agropyron* (the wheat grasses), *Secale* (rye), and *Haynaldia*. Until recently, another genus, *Aegilops,* was recognized, but after the discovery that at least one of the three genomes (basic sets of seven pairs of chromosomes) of hexaploid or common wheat had come from *Aegilops,* there was no way to maintain *Aegilops* as a separate genus and still continue to put all the wheat species in the genus *Triticum.* Tetraploid wheat, if it is an amphiploid of diploid *Triticum* and a diploid *Aegilops* species, could not be designated *Triticum;* nor could hexaploid wheat, an amphiploid of tetraploid *Triticum* and another diploid *Aegilops,* be included in the same genus as only one of its parents.

The simplest solution, to relegate the species of *Aegilops* to the genus *Triticum,* as Bowden (1959) has done, is also a reasonable one, for *Aegilops* is a small genus whose most primitive members closely resemble the primitive wheats. Mac Key (1968) prefers to retain *Aegilops,* remove diploid wheat from *Triticum* (calling it *Crithodium*), and designate *Triticum* a hybrid genus comprised of the polyploid wheats and amphiploids involving them. Bowden's more conventional classification was adopted by Morris and Sears (1967) and is followed here (Table 1).

Ernest R. Sears—Agricultural Research Service, United States Department of Agriculture, University of Missouri, Columbia, Missouri.

TABLE 1. The Species of Triticum and Their Genomic Formulae

Species	Formula	Former designations
1. Diploids		
T. monococcum	A	T. boeoticum (T. aegilopoides) + T. thaoudar + T. monococcum
T. speltoides	S	Ae. speltoides (Ae. aucheri + Ae. ligustica)
T. bicorne	S^b	Ae. bicornis
T. longissimum	S^l	Ae. longissima (+ Ae. sharonensis?)[a]
T. tripsacoides	Mt	Ae. mutica
T. tauschii	D	Ae. squarrosa, T. aegilops
T. comosum	M	Ae. comosa + Ae. heldreichii
T. uniaristatum	M^u	Ae. uniaristata
T. dichasians	C	Ae. caudata
T. umbellulatum	C^u	Ae. umbellulata

2. Allopolyploid wheats

T. turgidum	AB	*T. dicoccoides* + *T. dicoccon* (*T. dicoccum*) + *T. durum* + *T. turgidum* + *T. polonicum* + *T. carthlicum* (*T. persicum*)
T. timopheevii		
var. *timopheevii*	AG	*T. timopheevi* + *T. araraticum* (*T. armeniacum*) + *T. dicoccoides* var. *nudiglumis*
var. *zhukovskyi*	AAG	*T. zhukovskyi*
T. aestivum	ABD	*T. aestivum* (*T. vulgare*) + *T. compactum* + *T. sphaerococcum* + *T. spelta* + *T. vavilovi*

3. Other allopolyploids

T. ventricosum	DM^{ub}	*Ae. ventricosa*
T. crassum	DM, DDM	*Ae. crassa*
T. syriacum	DMS^l	*Ae. crassa* ssp. *vavilovi*
T. juvenale	DMC^u	*Ae. juvenalis*
T. kotschyi	C^uS^l	*Ae. variabilis* + *Ae. kotschyi*
T. triaristatum	C^uM, C^uMM^u	*Ae. triaristata*
T. macrochaetum	C^uM	*Ae. biuncialis*
T. columnare	C^uM	*Ae. columnaris*
T. triunciale	C^uC	*Ae. triuncialis*
T. cylindricum	CD	*Ae. cylindrica*

[a] Waines and Johnson (1972) maintain that *sharonensis* deserves specific rank.
[b] Underlining indicates substantial modification of the genome.

The wheats proper, which formerly constituted the whole of the genus *Triticum,* fall into three natural groups: the diploids, the tetraploids, and the hexaploids. At each ploidy level the differences are superficial, except for the relatively rare *T. timopheevii,* which forms nearly sterile, poor-pairing hybrids with the other tetraploids. There is a hexaploid variant, *zhukovskyi,* which resembles *timopheevii.*

The group of species formerly separated as *Aegilops* also includes diploids, tetraploids, and hexaploids. Most of the sections into which the genus was originally divided cannot be retained, because intersectional amphiploids were placed in the same section as one parent species.

Less closely related to the wheats are the genera *Secale* and *Haynaldia.* However, *Secale cereale,* common rye, hybridizes readily with certain wheats, crossability being genetically determined. Wheat–rye amphiploids (*Triticale*), involving tetraploid wheats, are now being commercially grown in limited competition with wheat. *Haynaldia villosa,* a diploid species (the genus includes but one other, a tetraploid) can be crossed with *Triticum* species.

In the large genus *Agropyron,* only a few species are close enough to the wheats to cross with them. The two that have been used to a significant extent are *Agropyron intermedium* ($n = 21$) and *Agropyron elongatum* ($n = 7$ and 35).

Origin and Relationships

One reason for combining *Triticum* and *Aegilops* into a single genus is that the most primitive *Triticum,* the diploid wheat *T. monococcum,* does not differ greatly from the members of the primitive *Sitopsis* group of *Aegilops.* Presumably, *T. monococcum* and the *Sitopsis* species (the diploids *T. speltoides, T. longissimum,* and *T. bicorne*) had a common ancestor, now extinct.

From the Sitopsis diploids or their ancestor, six other diploids evolved in various directions, resulting in a range of variation in spike and spikelet characters at least as great as in all the rest of the *Hordeae* (Eig, 1929). Diploid wheat diversified relatively little on the diploid level; instead, it added the genome of one of its diploid close relatives (or a composite genome derived from two or more of these) to produce *T. turgidum.* Later, after it was taken into cultivation, this tetraploid wheat added still another *Aegilops* genome, that of *T. tauschii* (*Ae. squarrosa*), one of the highly evolved diploids, to form hexaploid wheat, *T. aestivum.*

Besides contributing to hexaploid and presumably to tetraploid wheat, the *Aegilops* diploids combined among themselves to form some 10 tetraploids and 3 hexaploids.

The exact manner of origin of tetraploid (emmer) wheat *T. turgidum* is far from clear, the chief question being the source of the second or B genome. Following the observation by Sarkar and Stebbins (1956) that *T. speltoides* has the proper morphological characters to modify *T. monococcum* strongly in the direction of *T. turgidum, T. speltoides* came to be widely accepted as the sole source of this genome. Unfortunately, the degree of homology of the *speltoides* chromosomes with those of the B genome could not be tested directly, because presence of the *speltoides* genome caused chromosomes to pair that were only homoeologous (related). Eventually Kimber and Athwal (1972) discovered that some strains of *speltoides* did not cause homoeologous pairing, and that the chromosomes of these strains were not homologous with those of the B genome. It now appears probable that B came from a diploid which is now extinct or that it is a composite of chromosomes or parts of chromosomes from two or more species, most likely of the *Sitopsis* group, as suggested by Sarkar and Stebbins (1956) and elaborated by Zohary and Feldman (1962). The latter authors believe that two or more different allotetraploids, each with one A genome, hybridized, and that rearrangement and reassortment of the chromosomes of the various second genomes resulted in the distinctive B genome of tetraploid wheat. Johnson (1972) disagrees, suggesting on the basis of seed-protein studies that a variant of *T. monococcum* supplied the B.

The tetraploid wheat *T. timopheevii* was long thought to have a second genome other than B. Wagenaar (1961), however, proposed that it possesses the same genomes as other tetraploid wheats but has a genetic mechanism which causes asynapsis in hybrids with *turgidum* and *aestivum* wheats. Feldman (1966a) contradicted this by showing that pairing of most chromosomes of the *aestivum* A genome with *timopheevii* chromosomes is much better than when B-genome chromosomes are involved. Shands and Kimber (1973) conclude that *timopheevii* is AS, its second genome coming from *T. speltoides*.

The hexaploid *T. timopheevii* var. *zhukovskyi* is evidently an amphiploid of *T. monococcum* and tetraploid *T. timopheevii*.

When hexaploid wheat is artificially synthesized from *T. turgidum* and *T. tauschii* (McFadden and Sears, 1946), it has a long, lax spike with tough glumes and a somewhat fragile rachis, very similar to present-day spelta wheats. We may, therefore, assume that the original hexaploid was of spelta type. The evolution of the aestivum-type common wheat mainly required the mutation of a gene *q* to *Q*. Another mutation from *c* to *C* (compactum) further shortened the spike. The round-kerneled sphaerococcum wheats of India and Pakistan also differ from aestivum by a single gene, presumably acquired by mutation.

The wheats that feed the world are primarily common wheats, the aestivum group, with club (compactum) and durum wheats also important. Durum is a free-threshing tetraploid which evidently evolved from a cultivated emmer (which differs from the primitive or wild emmer mainly in having a spike with a nonfragile rachis) not by a change of q to Q, but by a series of other gene mutations. It is used for making macaroni and, in some regions, unleavened bread.

For determining relationships, wheat geneticists can go well beyond the usual criteria of phenotypic similarity, crossability, ability of chromosomes to pair in hybrids, and fertility of hybrids. They can determine the homologies, both cytological and genetic, of each individual chromosome.

Chromosome pairing often leads to equivocal results in the wheat group, because the polyploid wheats, and presumably other polyploid Triticinae, have acquired a genetic means of suppressing the pairing of homoeologous (related) chromosomes. Thus, two chromosomes that normally do not pair with each other at all may be very closely related genetically and may, in fact, pair frequently if the pairing inhibition is removed or suppressed. Since this inhibition depends primarily on a single gene, it is only necessary to remove the chromosome concerned (5B) to determine the latent ability of any chromosome to pair with those of wheat. The same result can be obtained by suppressing the pairing gene through addition of the chromosome set of certain biotypes of *T. speltoides* or *T. tripsacoides* (*Ae. mutica*).

For determining the ability of nonhomologous wheat chromosomes to pair with each other when there is no suppression of homoeologous pairing, Riley and Chapman (1964) used telocentric (one-armed) chromosomes. Two of these were combined in the same plant, along with the chromosomes of *T. speltoides* to condition homoeologous pairing. The telocentrics could be identified cytologically and thus scored for frequency of pairing with each other (and with other chromosomes).

The capacity of an alien chromosome to pair with a particular one of its wheat homoeologues can be determined rather easily if the proper substitution line (with an alien pair substituted for a wheat pair) is available. In two or three generations the proper crosses and selection will yield plants that are nulli-5B and monosomic for both the alien chromosome and its wheat homoeologue. The pairing of the two homoeologues with each other can then be scored. If the alien chromosome carries a marker gene, the amount of pairing can be checked by determining the number of offspring with a part-wheat, part-alien chromosome (Sears, 1972a).

If a telocentric chromosome is available for either arm of the alien chromosome in an addition or substitution line, the alien one can be tested for pairing with any desired wheat telocentric one in the presence of the chromosomes of *T. speltoides* which induce homoeologous pairing (Johnson and Kimber, 1967). However, the result may underestimate pairing potential, since pairing is evidently reduced by the telocentric condition (Sears, 1972c).

The genetic relationship of a particular alien chromosome to a wheat chromosome is determined by the degree to which the alien chromosome substitutes for the wheat chromosome. With two wheat chromosomes, the approach is similar; it consists of determining the degree, if any, to which extra dosage of one chromosome compensates for absence of the other. In practice this means synthesizing the nullisomic–tetrasomic combination and observing whether or not it is more vigorous or fertile than the simple nullisomic.

By means of nullisomic–tetrasomic tests, the 21 chromosomes of wheat have been placed in seven homoeologous groups of three chromosomes each, each group comprised of one chromosome from each genome (Sears, 1965). Although this result does not exclude the possibility that minor homologies exist between certain chromosomes of different homoeologous groups, it does establish that the major relationships lie within the groups. On the basis of the seven homoeologous groups and the three genomes they belong to, the 21 chromosomes of wheat are designated 1A–7A, 1B–7B, and 1D–7D.

Of the relatives of wheat, only *S. cereale* has thus far been studied intensively enough to provide a good idea of the degree of its relationship with wheat. Genetically, rye proves to be close to wheat, and its chromosomes have undergone surprisingly little alteration. Six of the seven chromosomes have been identified with particular homoeologous groups of wheat. It appears that one of these six is weakly related to the wheat chromosomes of a second group (Lee *et al.*, 1969). Substitutions of rye chromosomes for their wheat homoeologues are not completely successful, but some substituted rye chromosomes provide substantial compensation. Homologous pairing does not take place between wheat and rye chromosomes, and there is some doubt that homoeologous pairing can be induced with rye chromosomes (Riley and Kimber, 1966; Siddiqui, 1972). Bielig and Driscoll (1970) observed occasional pairing of 5R, however, in nulli-5B material, and Riley *et al.* (1959) believed they had induced some pairing of wheat with rye chromosomes. It is not known to what extent the pairing difficulties are attributable to a general asynaptic effect of the wheat background on the rye chromosomes, but it is true that fully homologous rye chromosomes often do not pair regularly as additions to wheat (see the section Addition Lines).

The few chromosomes of *Agropyron intermedium* and *Agropyron elongatum* that have been tested have been found to substitute well for wheat chromosomes and to be capable of a substantial amount of homoeologous pairing with their wheat homoeologues.

Of the nonwheat *Triticum* (*Aegilops*) species, *T. umbellulatum* is one of the most highly evolved diploids in karyotype as well as morphology. Nevertheless, one of the two chromosomes tested, $5C^u$, compensates reasonably well for its wheat homoeologue 5B (Chapman and Riley, 1970). The other $6C^u$, compensates poorly (Athwal and Kimber, 1972). Chromosome 2M of *T. comosum* compensates very well for 2A, 2B, or 2D (Riley *et al.*, 1966a), and what may be called 1C of *T. dichasians* compensates fairly well for at least 1D (Kihara, 1963).

Aneuploids

As a consequence of the fact that it was formed by the addition of three closely related diploids whose chromosomes had changed only slightly from those of the common parent of the three, hexaploid wheat has a great amount of duplication and triplication of genes. This, coupled with the fact that 21 pairs of chromosomes are too many for easy detection of linkages, makes the genetic analysis of wheat by conventional means exceedingly difficult. However, it also makes possible the establishment and exploitation of aneuploids. More different aneuploids have been obtained in wheat than in any other organism, and their use has resulted in substantial progress in the genetic analysis of wheat.

Monosomics and Nullisomics

Of greatest value have been monosomics and nullisomics, all of which have been available in the variety "Chinese Spring" since 1954 (Sears, 1954), and which have now been transferred to a number of other varieties.

Nullisomics, which lack one entire pair of chromosomes, are potentially the most useful of all aneuploids. All genes on a particular chromosome that have a phenotypic effect are revealed by absence of that chromosome. Similarities between different nullisomics were the basis for the first assignment of chromosomes to homoeologous groups.

The usefulness of nullisomics is severely limited, however, by the fact that most of them are sterile, either as male or female. Thus, they cannot be maintained as nullisomic lines, but must be recovered anew from monosomics in each generation. Those that are fertile are relatively stable chromosomally, although they do have a tendency to acquire a trisome or tetrasome for one of the homoeologous chromosomes. These compensating additions arise as a result of the enhanced transmission of the male gametes that have an extra homoeologue that are produced occasionally as a result of random asynapsis, or as a result of mitotic nondisjunction and subsequent overgrowth of a sector carrying an extra homoeologue.

Monosomics are much more nearly normal than nullisomics, most being difficult to distinguish from normal when grown under favorable conditions. The partially dominant gene Q on chromosome 5A, however, makes identification of this particular monosomic easy, and various others can also be distinguished if carefully studied.

Because they are frequently lost during meiosis, monosomes are distributed to only about 25 percent of gametes, with some variation in this value, depending on the particular chromosome concerned (Tsunewaki, 1964). On the female side, there is no selection against the deficient eggs, but on the male side, pollen with 21 chromosomes is strongly favored over that with only 20. About 1–10 percent of the pollen that functions is the 20-chromosome type. The result of this is that nullisomics are recovered in frequencies of less than 1 percent to more than 7 percent (Sears, 1954). [From certain monosomics in other varieties, the frequency of nullisomics may be substantially higher, e.g., Bhowal (1964).] Disomics comprise 20–25 percent and monosomics about 75 percent of the offspring.

Most of the monosomics of the Chinese Spring variety were originally obtained from two sources: haploids and asynaptics. However, the frequency of spontaneous occurrence of monosomes is high enough in most varieties that analysis of a few hundred or thousand plants might lead to recovery of all 21 different monosomes. Ionizing radiation increases the frequency, but at the same time produces other aberrations.

Monosomic analysis, the use of monosomics for locating genes to chromosomes, is conducted in different ways, depending on the kind of gene concerned—dominant, active recessive, or inactive recessive.

For a simple dominant, advantage is taken of the fact that a gene located on a monosome is present in all but the nullisomics in the next generation, i.e., in 93–99 percent of the offspring. Locating a dominant gene is, therefore, only a matter of crossing to each of the 21 monosomics, identifying the F_1 monosomics, and observing which F_2 population deviates strongly from a 3:1 ratio. A check is available in that the recessives are all nullisomic in the critical F_2 family.

An inactive recessive is one that behaves like a deficiency or null allele, i.e., like the vast majority of the mutants in diploid organisms. Its effect is the same when hemizygous (or even absent) as when homozygous. When plants homozygous for such a gene are crossed to the 21 monosomics, 20 F_1's show only the dominant phenotype, whereas in the

21st F_1 all the monosomics (about 75 percent of the population) are of recessive phenotype. If an F_2 is grown from a monosomic plant of the critical cross, all F_2 individuals are recessive.

Active recessives, while rare in diploids, are common in hexaploid wheat. Also known as hemizygous-ineffective alleles, they are active genes whose effect depends largely or entirely on their dosage. In particular, the hemizygote is like the heterozygote, or at least resembles it more than it does the homozygote. With typical genes of this type, the critical F_1 monosomic differs little or not at all from any of the other F_1's. To determine which is the critical chromosome, it is therefore usually necessary to grow the 21 different F_2's. Since the critical F_2 segregates about 75 percent dominants (monosomics and nullisomics) to 25 percent recessives (disomics), just like the other 20 F_2's, cytological study is necessary. In the critical family the recessives are all disomic, whereas in the other 20 families only about 25 percent of them are disomic.

Whatever the nature of the gene concerned, locating it may sometimes be best accomplished by identification of a few disomics in F_2 and observation or test of their F_3 progeny. Where the critical chromosome is involved, each F_2 disomic is homozygous for the gene being studied and consequently does not segregate in F_3. For the other 20 chromosomes, each F_2 disomic has only one chance in four of being homozygous for the allele concerned. Therefore an F_3 test of only three disomics from each of the 21 F_2 populations has a good chance of revealing which chromosome carries the gene under study.

The foregoing technique may be carried further, with the production of substitution lines. In this procedure, the chromosome monosomic in F_1 is retained through several backcrosses made to the proper monosomic of the recurrent parent, and through selfing, it is finally obtained disomic. Since it has been kept monosomic throughout the backcrossing, the chromosome has had no chance to cross over with any other chromosome, and is, therefore, exactly the same as in the nonrecurrent variety. Provided the number of backcrosses has been sufficient, practically all genes on the other 20 chromosomes of the nonrecurrent parent have been eliminated. Eight backcrosses, for example, should eliminate 99.6 percent of these genes. It is then expected that each distinguishing gene of the nonrecurrent parent will be expressed in one and only one of the 21 substitution lines.

In the production of substitution lines, advantage is taken of the fact that some 90–99 percent of functioning pollen from a monosomic has the euploid complement of 21 chromosomes, whereas about 75 percent of eggs have only 20. Therefore, when a monosomic of the recurrent parent is pollinated by the corresponding F_1 or backcross monosomic, up to 75 percent of the offspring have a monosome that is the desired chromosome from the nonrecurrent parent. However, 1 percent or more, depending on which chromosome is involved, may have a monosome that came from the recurrent (female), rather than the nonrecurrent, parent. It is, therefore, essential that the monosome of the recurrent parent be marked in some way so that when it is recovered monosomic it can be identified and the plant eliminated.

Either cytologically or genetically marked monosomes can be used to ensure that they are not retained, but cytological markers have been used almost exclusively. The marked monosomes are iso- or telocentric chromosomes, which can be readily identified at meiosis. Anderson and Driscoll (1967) and Bielig and Driscoll (1973) suggest using monosomes derived from alien species such as rye or *Agropyron,* with each such chromosome carrying one or more genetic markers. Such alien monosomic lines permit easy cytological detection of univalent shift (see next paragraph), since the alien chromosome is unable to pair with the homoeologous wheat chromosome when this is also present.

A less reliable method of substituting chromosomes involves selfing after each backcross in order to obtain a disomic for use in making the next backcross. This method involves some risk of failure to produce a true substitution, because of the occasional occurrence of "univalent shift" (Person, 1956). This is the production by a monosomic plant of offspring monosomic for the wrong chromosome, because of the less-than-complete regularity of bivalent pairing in hexaploid wheat. The tendency toward asynapsis varies widely among varieties, but exists in all. A monosome may thus be transmitted at the same time that another chromosome is lost, with resultant production of a plant monosomic for a different chromosome than was monosomic in the parent. Unless this plant is tested for authenticity of its monosome, the shift will go undetected.

Many substitution lines exist. Of these, those involving substitutions of all the chromosomes of several different varieties into Chinese Spring, the cultivar in which a complete monosomic series was first available, have existed longest and, consequently, have been most studied (Sears *et al.,* 1957; Sheen and Snyder, 1964, 1965). Many additional lines have been or will be produced, particularly because of their great usefulness in the discovery and study of genes with quantitative effects. Law's (1967, 1968) demonstration of the precision of genetic analysis possible with substitution lines has resulted in the formation of the European Wheat Aneuploids Cooperative, with the primary objective of producing and eventually exploiting large numbers of substitution lines. The essential step in the exploitation, originally suggested by Unrau (1958), is that an F_1 between a substitution line and its nonrecurrent parent (from which it differs by only the one chromosome) is backcrossed to the monosomic for the chromosome concerned. Monosomic plants of the next generation are then monosomic for the critical chromosome from the F_1, and these monosomes cannot undergo any further crossing over. Following a selfing, each can be obtained disomic and, hence, completely homozygous. Populations can then be grown in which there is no genetic variation, and the effects of whatever combination of genes each critical chromosome may have acquired through crossing over can be assessed very precisely.

At the tetraploid level, monosomics can also be obtained, but there has been a serious problem in maintaining them, as they tend to be male sterile and to have poor female transmission of the monosomic condition. Mochizuki (1970), however, discovered that monosomic seeds of most types tended to be shriveled, and he made use of this phenomenon to obtain all 14 possible monosomics in a durum cultivar.

Trisomics and Tetrasomics

Trisomes, like monosomes, were mostly obtained from haploids and nullisomic 3B, but some came from a triploid, and a few first appeared in nullisomics (Sears, 1954), where their ability to compensate for the missing chromosome identified them as homoeologous. Kihara and Wakakuwa (1935) and Matsumura (1952) had previously reported the occurrence of such "gigas" types in their collection of nullisomics for D-genome chromosomes.

Trisomics differ even less from normal than do monosomics. Some, however, can be recognized by the effect of the added dose of a particular gene, such as Q on 5A and the awn promoters on 2A, 2B, and 2D.

Transmission of trisomes is slightly reduced (to about 40 percent) on the female side because of occasional failure of trivalent formation and subsequent frequent loss of the univalent chromosome. On the male side, there is selection in favor of pollen without the extra chromosome. The result of these factors is the production of about 1–10 percent tetrasomics, about 45 percent trisomics, and the rest disomics (Sears, 1954).

Once obtained, tetrasomics are mostly more vigorous and fertile than nullisomics. However, the tetrasomics of group 7 produce fewer seeds per plant than the nullisomics, except 7A, whose nullisomic suffers from pistillody. Tetra-4B is weak and poorly fertile (but still superior to the nullisomic), because it has necrosis of leaves.

Since a tetrasome sometimes forms a trivalent and a univalent at meiosis, and the univalent is subject to loss, some of the gametes produced by tetrasomics have only 21 chromosomes. The frequency of these varies with the chromosome concerned, and the amount of selection in their favor on the male side varies greatly. As a consequence, certain tetra-somics, most notably those of group 3, have a high frequency of trisomics (about 50 percent) among their offspring when selfed. Most others, however, produce 80 percent or more of tetrasomics.

The trisomics and tetrasomics serve chiefly to extend the dosage range for the chromosomes and the genes they carry. The effect of each wheat chromosome in dosages from 0 to 4 can thus be studied.

Tetrasomes have also been very useful in the determination of the relationships of chromosomes. When a particular tetrasome compensates for a missing chromosome, it is established that the two chromosomes are homoeologous. Compensating nullisomic tetrasomics have been used in some cases instead of nullisomics, especially where the latter are sterile. For example, nulli-5B tetra-5D is often used as a source of pollen lacking chromosome 5B, which inhibits homoeologous pairing. Of course, nullisomic tetrasomics are not a satisfactory substitute for nullisomics if, as is often the case, the gene or genes under study are present on both chromosomes. This appears to be true for the genes involved in the production of various isozymes (Brewer *et al.*, 1969).

Secondary Aneuploids

Besides the primary aneuploids, which involve losses or duplications of complete chromosomes, there are what may be called secondary aneuploids, which involve various dosages or combinations of chromosomes that have been derived through misdivision of the centromere—namely, telocentrics and isochromosomes. Since there may be some disagreement as to the essential difference between aneuploidy and euploidy, aneuploids are here defined as individuals with a changed number of chromosomes, chromosome arms, or centromeres, or with arms so arranged (as in double monoisosomics, with an isochromosome for each arm of a chromosome) that normal pairing and disjunction cannot occur at meiosis.

When a chromosome is unpaired at meiosis, it is ordinarily late in coming onto the metaphase plate (if it does at all), and it may or may not divide. In wheat, it does divide if it reaches the plate, but it may divide transversely instead of longitudinally, with the result that a chromosome with two identical arms goes to each pole; or the division may be such that three arms go to one pole and only one to the other. The frequency of such misdivision of univalents varies with the genotype and reportedly with the chromosome. It reaches about 40 percent for chromosome 5A of Chinese Spring (Sears, 1952), and even higher frequencies have been observed for rye chromosomes present in wheat plants.

Telocentric chromosomes are now available for 41 of the 42 arms of the wheat chromosomes (Sears, 1954, and unpublished). Only one arm of 7D is lacking, and there is little reason to doubt that this one can also be obtained. The two telocentrics of a chromosome are designated L(long) and S(short) if their relative lengths are known; otherwise they are α and β (Kimber and Sears, 1968).

One important use of telocentrics is as monosomes that are distinguishable from other chromosomes of the complement. The use of such monosomes eliminates the danger of undetectable univalent shift and recovery of maternal instead of paternal monosomes.

Isochromosomes are useful as monosomes in the same way as are telocentric chromosomes. Some are in fact better, because a monoisosomic plant, with two doses of the arm concerned, is genetically more nearly normal than the corresponding monotelosomic. This may be important where the monotelocentric is of marginal fertility, as in monotelo-4Aα, or is of late maturity, as in monotelo-5DL. Only about one-fourth of the chromosome arms are presently represented by isochromosomes.

Monotelosomics (or monoisosomics) for about three-fourths of the 42 arms are fertile enough for maintenance. Ditelosomics are maintained for all of these. They constitute reasonably stable lines and are used as sources of telocentrics for identifying new or doubtful aneuploids and for use in linkage experiments.

The telocentrics that cannot be maintained in the di- or monotelosomic condition are kept as ditelosomics with the second arm present as a monotelosome. Thus, every female gamete carries the arm concerned, and about three-fourths lack the other arm. Some of these telocentrics are also maintained in company with the homologous complete chromosome.

For some chromosomes, e.g., 2A and 2B, neither monotelosomic is easily maintained. For these, it is convenient to use plants with both telocentrics present in one dose. These double monotelosomics produce only about 9/16 of 20-chromosome gametes instead of 3/4, but otherwise they have all the advantages of monotelosomics. Double monoisosomics would be even more useful, since they are genetically normal, but they are not yet available.

Even better than ditelosomics for checking the identity of aneuploids are double ditelosomics. These have a pair of each of the two telocentrics of the chromosome concerned. Although they have 22 pairs instead of 21, they are genetically normal and, consequently, are fully vigorous and fertile. Their gametes contain both telocentrics.

Various combinations of telocentrics, isochromosomes, and complete chromosomes have been obtained by accident or design, and some have served useful purposes. Muramatsu (1963), for example, selfed a *qqq* 5A trisomic and obtained a tetrasomic offspring in which one member of the tetrasome was an isochromosome for the arm carrying q. The plant was therefore $qqqqq$, and it resembled QQ. Feldman (1966b) obtained plants with three isochromosomes among the offspring of plants with two isochromosomes of 5A, 5B, and 5D, and observed the effect of six doses of genes affecting pairing. Such plants occur because the two isochromosomes often fail to pair with each other and then occasionally go to the same pole.

Telocentric chromosomes have great utility in gene-mapping work (Sears, 1966). From a cross to the two ditelocentric lines for the chromosome concerned, the arm on which a particular gene is located can easily be determined (provided linkage to the centromere is not too tight), since the recovery of segregant telosomes carrying the gene is only possible for one arm. Furthermore, the frequency with which such telosomes are recovered is a measure of the distance of the gene from the centromere.

In determining the linkage with the centromere, advantage can be taken of the fact that most telocentrics are poorly transmitted through pollen in competition with their complete homologue. If there were no transmission of the telocentric, a heterozygote having one allele on a telocentric chromosome could be used as male in a cross to the homozygous recessive, and the amount of crossing over could be read directly from the percentage recovery of the allele carried by the telocentric. Since there is always some transmission, cytological examination of at least a sample of the critical class is necessary to discover how many are carrying the telocentric rather than being crossovers. If a marker is present on the other arm of the complete chromosome, the necessity for cytological analysis is eliminated, and, in fact, it is practicable to use the heterozygote as female, if desired.

There is reason to suspect that crossing over is reduced near the centromere when one chromosome of a pair is telocentric. The evidence to support this theory comes mainly from work of Endrizzi and Kohel (1966) with a chromosome of cotton, but a degree of confirmation has been obtained with wheat (Sears, 1972c).

Genes and Maps

About half the genes available to wheat workers are ones that condition resistance to disease organisms, particularly the rust fungi. Of the slightly more than 100 loci listed by McIntosh (1973) as clearly identified and with suitable stocks available, there are 17 for resistance to stem rust, 15 for resistance to leaf rust, and 8 for resistance to stripe rust (Table 2). On the other hand, mutants have been identified at only 4 chlorophyll

TABLE 2.   Genes Identified in Wheat, Their Symbols, and Their Chromosome Location
(Where Known)

Name of gene(s)[a]	Symbol(s)	Chromosome(s)
Gross morphological characters:		
Squarehead (or aestivum)	Q	5AL
Compactum (or club)	C	2D
Sphaerococcum	s1	3D
Sphaerococcum simulator	S2	
Short culm:		
Dwarf	Sd1, Sd2	
Tom Thumb		4A
Hybrid weakness:		
Hybrid necrosis	Ne1, Ne2	5BL, 2BS
Hybrid chlorosis	Ch1, Ch2	2A, 3Dα
Grass-clump dwarfness	D1, D2, D3, D4	2Dα, 2BL, 4BL, 2D
Corroded	co1, co2	6BS, 6D
Abnormal chlorophyll:		
Neatby's virescent	v1a	3BS
Viridis-508	v2a	3A
Hermsen's virescent	v1b, v2b	3BS, 3A
Chlorina-1	cn-A1a	7AL
Yellow	cn-A1b	7AL
Chlorina-448	cn-A1c	7AL
Chlorina-214	cn-D1	7DL
Hairy:		
Hairy node	Hn	5AL
Hairy neck	Hp, Hp(Tp5B), Hp(Tp6D)	4Aβ, 5BS, 6D
Hairy glume	Hg	1AS
Waxy:		
Waxy-1	W1	2BS
Waxy-1 inhibitor	W1[1]	2BS
Waxy-2	W2	2Dα
Waxy-2 inhibitor	W2[1]	2Dα
Lack of ligules:		
Liguleless	lg1, lg2	2B, 2D

TABLE 2.   Continued

Name of gene(s)[a]	Symbol(s)	Chromosome(s)
Colored:		
Purple coleoptile	*Rc1, Rc2, Rc3*	7A, 7BS, 7DS
Purple culm	*Pc*	7BS
Red glume	*Rg1, Rg2*	1BS, 1DL
Red grain	*R1, R2, R3*	3A, 3B, 3Dα
Black glume	*Bg*	1AS
Psuedo-black-chaff	*Pbc*	3B
Awn characters:		
Tipped-1 inhibitor	*B1*	5AL
Tipped-2 inhibitor	*B2*	6BL
Hooded	*Hd*	4BS
Modifiers of meiosis:		
Homoeologous pairing	*Ph*	5BL
Low-temperature pairing	*ltp*	5D
Physiological characters:		
Early	*e1*	7BS
Vernalization response	*Vrn3*	5D
	Vrn1, Vrn2, Vrn4	
Response to photoperiod	*Pfd1, Pfd2*	
Crossability with rye	*kr1, kr2*	5B, 5A
Pollen killer	*Ki*	6BL
Fertility restoration	*Rf1, Rf2, Rf3, Rf4, Rf5*	1A, 7D, 1B, 6B, 6D
Basal sterility in speltoids	*bs*	5D
Proteins:		
Alcohol dehydrogenase	*Adh-A1, Adh-B1, Adh-D1*	4Aα, 4BL, 4DS
Aminopeptidase	*Amp-A1, Amp-B1, Amp-D1*	6Aα, 6BS, 6Dα
Reaction to Puccinia graminis:		
Stem-rust reaction	*Sr5, Sr6, Sr7a,b, Sr8*	6D, 2Dα, 4BL, 6Aα
	Sr9a,b,d, Sr11, Sr12, Sr13	2BL, 6BL, 3B, 6Aβ
	Sr14, Sr15, Sr16, Sr17	1BL, 7AL, 2BL, 7BL
	Sr18, Sr19, Sr20, Sr21	1D, 2B, 2B, 2A
	Sr22, Sr23	7AL, 4A
	Sr2, Sr10	

TABLE 2.   *Continued*

Name of gene(s)[a]	Symbol(s)	Chromosome(s)
Reaction to Puccinia recondita:		
Leaf-rust reaction	*Lr1, Lr2a,b,c, Lr3, Lr9*	5DL, 2Dα, 6BL, 6BL
	Lr10, Lr11, Lr12,	1AS, 2A, 4A
	Lr14a,b,ab, Lr15, Lr16	7BL, 20D α, 4A
	Lr19, Lr20, Lr21, Lr22	7D, 7AL, 1D, 2Dα
	Lr13, Lr17, Lr18	
Reaction to Puccinia glumarum:		
Yellow-rust reaction	*Yr1, Yr7, Yr8*	2A, 2B, 2D
	Yr2, Yr3a,b,c	
	Yr4a,b, Yr5, Yr6	
Reaction to Erysiphe graminis:		
Powdery-mildew reaction	*Pm1, Pm2, Pm3a,b,c*	7AL, 5DS, 1AS
	Pm4, Pm5, Pm6	2A, 7BL, 2B
Reaction to Tilletia spp.		
Bunt reaction	*Bt1, Bt4, Bt5*	2B, 1B, 1B
	Bt6, Bt7	1B, 2D
	Bt2, Bt3, Bt8, Bt9, Bt10	
Resistance to Mayetiola destructor:		
Hessian fly resistance	*H1, H2, H3, H4, H5, H6*	
Resistance to Toxoptera graminum:		
Greenbug resistance	*gb*	
Glume tenacity:		
Tenacious glumes	*Tg*	2Dα

[a] Only those genes are included for which stocks are known to exist. See McIntosh (1973) for sources and references.

loci, although such loci are very numerous in most diploids that have been analyzed. The large number of resistance genes identified may be attributed to the great interest of breeders in disease resistance, leading to intensive searches for sources of resistance, and to the fact that resistance genes do not ordinarily interact with each other, each being effective against a particular virulence gene in the fungus regardless of the rest of the wheat genotype. The paucity of chlorophyll mutants is presumably due to the general triplication of the genes for chlorophyll production, so deficiency or mutation of any one locus usually has no detectable effect on chlorophyll development.

Of the genes listed, about two-thirds have been located as to chromosome, and about half of these as to chromosome arm. For several chromosomes, linkage maps are available, with centromere positions known (Figure 1).

Alien Additions

Because hexaploid wheat is so genetically well buffered, it easily tolerates the addition of chromosomes from other species and genera. Alien-addition lines, which have an added pair of foreign chromosomes, are available for each of the seven chromosomes of rye, for the seven chromosomes of *T. umbellulatum,* for one of *T. comosum,* and for at least three of *A. intermedium* and two of *A. elongatum.*

Some addition lines have arisen more or less fortuitously after crossing of wheat with a foreign species, usually one or more backcrosses to wheat and then some selfings, with selection for a character, e.g., disease resistance, of the nonwheat parent. Other addition lines have been produced in a systematic manner using O'Mara's (1940) technique. This involves making a wheat–alien amphiploid, crossing this to wheat to make each alien chromosome monosomic, backcrossing to wheat, and selecting plants with a single alien chromosome. Differences in the phenotypes of these plants, or in the morphology of the added chromosome, usually suffice to distinguish between additions involving different chromosomes. In the disomic additions (*i.e.,* addition lines), which arise after selfing of the monosomics, the distinctive phenotypic characteristics are more pronounced.

From selfing of addition monosomics, the frequency of disomics is much less than from monosomics for wheat chromosomes. Female transmission is more or less the same, about 25 percent [although Riley *et al.,* (1966a) found about 50 percent for the *T. comosum* 2M chromosome], but whereas pollen selection strongly favors wheat monosomes, it usually acts against alien-addition monosomes; that is, the 21-chromosome male gametes are favored in both situations. The adverse pollen selection against alien chromosomes varies greatly. It is very little for certain *Agropyron* chromosomes, which are consequently transmitted through as much as 25 percent of pollen; somewhat more for rye chromosomes (Riley, 1960; Evans and Jenkins, 1960; Sears, 1967); and severe for *T. umbellulatum* chromosome $6C^u$ [1.3 percent transmission (Sears, 1956)].

Addition disomics vary in stability, depending on what alien chromosome is involved. If the alien disome pairs and disjoins regularly, there is no opportunity for loss, regardless of how intense the selection may be in favor of 21-chromosome pollen. If there is asynapsis, its rate of occur-

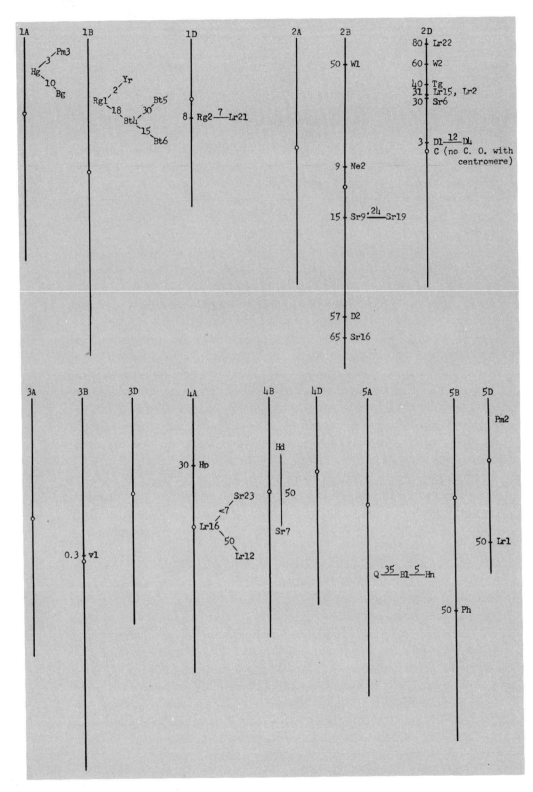

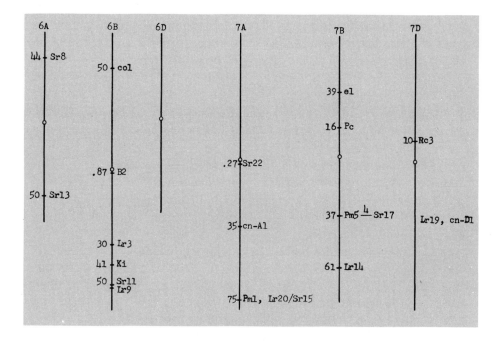

Figure 1. The chromosomes of wheat and the genes they carry for which linkage data are available. Numbers written to the left of a chromosome are distances to the centromere, as determined through the use of telocentrics or as calculated from the distance to another gene. Many of the values are tentative, being based on small populations. The true distances to the centromere of nearby genes are probably substantially greater than those given. Many of the 50-unit distances require substantial revision, particularly in the upward direction, since they are based only on a finding of independence between genes or between gene and centromere. The relative lengths of the 21 chromosomes and the centromere positions are those obtained by Sears (1954) from measurement of monosomes of Chinese Spring at TII of meiosis. See McIntosh (1973) for the linkage data on which the maps are based, and for the sources of these data.

rence, along with the intensity of selection in favor of pollen lacking the alien chromosome, determines the speed with which the line returns to ordinary wheat. The rate of asynapsis can be substantial [possibly as much as 30 percent according to the data of O'Mara (1951)]. In the absence of selection, each asynaptic event, whether on the female or male side, has a probability of about 9/16 of resulting in a monosomic offspring. The selective advantage enjoyed by normal pollen, of course, increases the frequency of monosomics obtained.

Once the monosomic condition is reached, the line very quickly returns to nullisomy for the alien chromosome, whether there is adverse selection or not. Even without selection, an alien monosome is expected to become nullisomic in about 9/16 of the progeny in each generation and disomic in only about 1/16.

Complete series of seven different addition lines have been produced for three different varieties of rye: King II (Riley and Chapman, 1958a; and Riley in Riley and Kimber, 1966), Dakold (Evans and Jenkins, 1960), and Imperial (Driscoll and Sears, 1971).

Kimber (1967) and Chapman and Riley (1970) have obtained all seven possible addition lines from *T. umbellulatum.* Riley *et al.* (1966a) added a disease-controlling chromosome to wheat from *T. comosum.* Cauderon (1966) and Wienhues-Ohlendorf (1960) have produced addition lines involving three different chromosomes from *A. intermedium;* and Schlehuber and Sebesta (1959), Driscoll (1960), and Knott (1961) have added two different chromosomes from polyploid *A. elongatum.* All the *Agropyron* additions carry genes for resistance to rust fungi.

At the tetraploid level, more difficulty is experienced in producing alien-addition lines. Mochizuki (1962), however, succeeded in producing all seven additions of chromosomes of *A. elongatum* ($n = 7$) to durum wheat. The six disomic additions obtained showed 1–58 percent self-fertility.

Alien Substitutions

Chromosomes from several relatives of wheat have been successfully substituted for wheat chromosomes. As a rule, a particular alien chromosome substitutes well for the chromosomes of only the homoeologous group, and such an alien chromosome is considered to belong to that homoeologous group. It is given a letter suffix indicating its source, e.g., 5R for the rye chromosome belonging to group 5.

Several substitution lines have arisen from hybrids which were simply backcrossed to the wheat parent with selection for a particular character, such as resistance to a certain fungus. In these instances, chance asynapsis involving a wheat homoeologue of the alien chromosome concerned presumably occurred and resulted in production of a gamete in which the alien chromosome was present and the wheat homoeologue absent. Depending on the degree of relatedness of the two chromosomes, male gametes of this substitution type may compete quite successfully with normal, whereas those carrying the alien chromosome as an addition usually compete poorly.

An alien chromosome may also be substituted for one of its homoeologues by crossing the alien-addition disomic to the proper monosomic. The F_1, which has $20'' + 2'$, on selfing gives rises to some offspring with $20'' + 1''$ alien, which may be identified by genetic markers on the alien chromosome and confirmed by hybridizing with normal and obtaining $20'' + 2'$ in all offspring. If desired, the F_1 may be backcrossed to the addition line and a daughter selected with $20'' + 1''$ alien $+ 1'$ wheat. Selfing can then eliminate the wheat monosome.

Although alien-substitution lines are generally stable, and several, particularly those involving *Agropyron* chromosomes, are reasonably vigorous and fertile, very few have proven to be useful agronomically. However, Zeller and Fischbeck (1971), Zeller (1972), Zeller and Sastrosumarjo (1972), and Mettin *et al.* (1973) showed that "Zorba", "Salzmünde 14/44", "Weique" and some of their derivatives have a pair of rye chromosomes substituted for 1B of wheat. Perhaps the chief value of substitution lines is as material for use in the induced transfer of a critical part of the alien chromosome to a homoeologous wheat chromosome. When a substitution line is crossed with normal, both the alien chromosome and its wheat homoeologue are monosomic and can be made to pair by either removing chromosome 5B or suppressing its effect through addition of the chromosomes of certain genotypes of *T. speltoides*.

As a rule, a chromosome that substitutes successfully for one member of a homoeologous group will substitute also for the other two members of the group. The degree of compensation for the wheat chromosome will not ordinarily be the same in all three substitutions, however. In most groups, the wheat chromosomes themselves differ significantly in their genetic content, as shown by differences in the ability of tetrasomes to compensate for nullisomes of their homoeologues.

The Genetic Control of Chromosome Pairing

The discovery by Okamoto (1957), Riley and Chapman (1958*b*), and Sears and Okamoto (1958) that the long arm of chromosome 5B carries a gene which prevents homoeologous pairing, *Ph*, provided an immediate answer to the question of why chromosomes so closely related genetically fail to pair. It was now only necessary to remove chromosome 5B or suppress its effect in order to ascertain true pairing affinities. Under these circumstances, the chromosomes of wheat were found to pair not only with their homoeologues of the other two wheat genomes (Riley and Chapman, 1964, 1966; Riley and Kempanna, 1963; Kempanna and Riley, 1964), but also with chromosomes of related species and genera (Riley and Chapman, 1958a; Riley *et al.*, 1959; Riley, 1966; Riley and Law, 1965).

Feldman (1966b) and Riley *et al.* (1966b) showed that chromosomes 5D and 5A also carry genes affecting pairing, but promoting rather than suppressing it. According to Riley and Chapman (1967) and Feldman and Mello-Sampayo (1967), the short arm of 5B also promotes pairing, and the latter authors conclude that the amount of homoeologous pairing that can occur depends on a balance between the promoters and *Ph*.

Riley *et al.* (1961) and Riley (1966) further showed that the genomes of *T. speltoides* and *T. tripsacoides* (*Ae. mutica*) prevent *Ph* from functioning, thus allowing homoeologous pairing to occur. The other diploid species tested had no detectable effect on pairing; in particular, there were apparently none that reduced pairing below what could reasonably be expected. Therefore, it could be concluded that the suppressor of homoeologous pairing had arisen as a mutation in the newly produced AABB amphidiploid, immediately diploidizing its meiosis and thereby improving its fertility and stability.

The manner in which *Ph* controls homoeologous pairing was declared by Feldman (1966b) and Feldman *et al.* (1966) to be through an effect on somatic association. It was held that *Ph* reduces the forces causing somatic association enough to prevent homoeologues from associating, but not enough to keep homologues apart. As a result, homologues come into meiosis already effectively paired, and homoeologues are excluded from pairing. Darvey and Driscoll (1972) have been unable to confirm the effect of *Ph* on somatic association [but their data are from root tips (as are those of Feldman *et al.*)], whereas the degree of association immediately preceding meiosis is the critical consideration.

Chromosomes other than those of group 5 are known to affect chromosome pairing. Absence of the long arm of chromosome 3B results in partial pairing failure, but there is evidence (Li *et al.*, 1945; Kempanna and Riley, 1962) that desynapsis rather than asynapsis is involved. Mello-Sampayo (1971) finds a minor effect of the β arm of the 3D chromosome on homoeologous pairing, and Driscoll (1972) not only confirms this but finds that 3Dα and 3A also affect the pairing. The genes on 3Dβ and 3A are suppressors, like *Ph,* but weaker. According to Mello-Sampayo and Canas (1973), the 3A suppressor is located on the β arm, and 3Aβ and 3Dβ have a combined effect approximately as great as that of 5B.

Transfer of Characters from Alien Species to Wheat

Since many of its relatives have characters that would be useful if present in wheat, and since wheat can be crossed with nearly all these relatives, there has long been interest in transferring some of these characters to wheat. Early efforts, however, were almost entirely unsuccessful. Some transfers could be made through a simple crossing and backcrossing procedure, but these transfers nearly always proved to involve addition or substitution of an entire alien chromosome, and almost always there were agronomically undesirable effects. Because the foreign chromosome did not pair with any wheat chromosome, there was no easy way to reduce the alien contribution.

Exchange of parts between chromosomes can of course be induced by ionizing radiation, and in recent years radiation has been used to transfer desired alien-chromosome segments to wheat chromosomes (Sears, 1956; Knott, 1971). Obviously the alien segment cannot replace a random segment of wheat chromosome, because any wheat segment of substantial size is very likely to contain essential genes. In practice, the satisfactory transfers strongly tend to involve substitution of the alien segment for a homoeologous wheat segment. Although such transfers might be expected to be of almost prohibitively low frequency, there is good evidence that they can be obtained in experiments of modest size. As Knott (1968) points out, the tendency of homoeologues to associate somatically must increase the frequency with which transfers to homoeologues are obtained.

With the discovery of the genetic control of pairing in wheat, it became obvious that transfers from alien chromosomes might be effected through induced homoeologous pairing. Certain *Agropyron* chromosomes that seldom, if ever, pair with wheat chromosomes under ordinary conditions, show up to 30 percent pairing with particular homoeologues when chromosome 5B is absent (Johnson and Kimber, 1967; Sears, 1972a). This not only makes it possible to get high frequencies of transfers, but also ensures that each transfer will be made to the most favorable wheat chromosome. High frequencies of transfer will certainly not be obtainable with all alien chromosomes, but there is good reason to think that a useful frequency of homoeologous pairing and resultant transfer can be induced with almost every chromosome of the wheat relatives, except possibly rye.

Various techniques for using homoeologous pairing are possible, from simply producing an interspecific or -generic hybrid lacking chromosome 5B, to synthesizing plants monosomic for a single alien chromosome, monosomic for a homoeologous wheat chromosome, and nullisomic for 5B (Sears, 1972*a*). Riley *et al.* (1968) have successfully exploited the ability of *T. speltoides* to produce homoeologous pairing in hybrids; they transferred disease resistance from *T. comosum* by crossing *speltoides* to a wheat line with an added pair of *comosum* chromosomes.

Translocations

It is clear from the essential integrity of the homoeologous groups that translocation between chromosomes was not an important factor in the evolution of the diploids that supplied the three sets of seven chromosomes of hexaploid wheat. Furthermore, at the tetraploid and hexaploid levels, few translocations can have occurred unless they involved homoeologues. Riley *et al.* (1967) studied several hexaploid cultivars (including Chinese Spring) that had neither translocation differences from primitive tetraploid wheats nor from the D-genome donor, *T. tauschii.*

A few translocations do exist between cultivars, however, at both the tetraploid and hexaploid levels (Riley *et al.*, 1967). When heterozygous, these frequently do not form a figure-eight pattern, but instead make an open ring, two bivalents or a trivalent and a univalent, with the result that duplication-deficiency gametes frequently occur. Unlike in diploid species, most of these pass rather freely through the pollen, and homozygous duplication-deficiency variants can be obtained.

Translocations have caused some concern in the transfer of monosomy from one cultivar to another. The problem is whether to retain the chromosome arrangement of the nonrecurrent parent through the necessary series of backcrosses, thus ending up with a monosomic equivalent that has 19˝ + chain-of-3; or whether to attempt to recover, after irregular behavior of the three chromosomes concerned, a true monosomic of the recurrent parent. The latter result has the disadvantage that the new monosome is not identical with any of those of the standard cultivar; for example, with a 4A–6B translocation, either a 4AL–6BS or a 4AS–6BL monosome may be obtained from an original cross with mono-4A. Since the same two monosomes may be derived from the transfer series starting with mono-6B, the monosome selected from the 6B series of backcrosses may be identical with that from the 4A series.

In the substitution of a chromosome involved in a translocation with respect to the recipient cultivar, it is essential that the chromosome architecture of the recipient be retained, i.e., that each backcross generation have the same 19˝ + 1‴ that distinguishes the F_1. Otherwise, less than the equivalent of an entire chromosome will come unchanged from the donor cultivar.

Mutation

Mutants are easily induced by ionizing radiation in diploid wheat (Smith, 1939) and appear with considerable diversity and reasonably high frequency in tetraploid wheat (Mac Key, 1960; D'Amato *et al.*, 1964), but in hexaploids there is quite a different response. In particular, hexaploids produce virtually no chlorophyll mutations when irradiated. They do, however, give rise to very high frequencies of such "peripheral" mutants (Mac Key, 1954) as speltoids, late-maturing types, and awn variants, attributable in all or nearly all cases to simple deficiency for the loci concerned. These results agree with the assumption that the predominant effect of ionizing radiation is to delete chromosome segments. At the hexaploid level, where the vast amount of duplication and triplication results in greatly reduced screening against such losses, deficiencies and duplications are practically the only mutations recovered. As is then obvious from the fact that none of the 21 nullisomics or tetrasomics is chlorophyll defective, no simply inherited chlorophyll mutations are to be expected from ionizing radiation. A speltoid, on the other hand, will appear every time chromosome 5A, or any portion of it that includes the locus of Q, is deleted.

At the tetraploid level, there has evidently been some diploidization with respect to the genes involved in chlorophyll production, i.e., some of these are no longer duplicated. These nonduplicated loci can, of course, be mutated with neutrons and x rays, and because of the relatively high tolerance of the tetraploid to deficiency, at a high rate compared to what is obtainable at the diploid level. Presumably, many fewer loci mutate in the tetraploid, but their higher rate of mutation leads to an overall rate of chlorophyll mutation which, according to Mac Key (1960), is not much lower than that of the diploid.

Even at the hexaploid level, some cultivars are evidently diploidized with respect to one or more series of chlorophyll loci. A high rate of chlorophyll mutation can be obtained with only a single locus diploidized, i.e., mutable (Natarajan *et al.*, 1958).

In spite of the high level of triplication of chlorophyll genes in hexaploid wheat, simply inherited chlorophyll-deficient mutants have occurred spontaneously (Neatby, 1933; Pettigrew *et al.*, 1969) and have been induced by the chemical mutagen ethyl methanesulfonate (Shama Rao and Sears, 1964; Prabhakara Rao and Washington, 1969). As expected from the nondefectiveness of all 21 nullisomics, all the mutant genes actively interfere with the production of chlorophyll (Washington and Sears, 1970; Sears, 1972*b*; Pettigrew, 1972).

All six mutants thus far analyzed belong to only two series of homoeoalleles: a *virescens* series on chromosomes 3A, 3B, and 3D, and a *chlorina* series on 7A, 7B, and 7D. In each series, mutants have occurred at only two of the three loci, but the presence of a nonmutated allele on the third homoeologue has been demonstrated by its effect on mutant expression. The mutant genes compete with their nonmutant alleles and homoeoalleles in such a way that the degree of mutant expression goes up with the number of mutant alleles and down with the number of normal alleles. Mutant alleles at different loci reinforce each other. Three different *chlorina* alleles, one substantially more effective than the other two, occur at one locus (Pettigrew, 1972).

The hypothesis put forward (Sears, 1969, 1972*b*; Pettigrew, 1972) concerning the nature and origin of the chlorophyll mutations is based on the following assumptions: the polypeptides (monomers) produced by the three different members of a series are very similar; they combine more or less at random to form a multimeric enzyme; the mutant alleles produce defective monomers; and enzyme molecules containing one or more defective monomers are partly or wholly inactive.

Although it seems likely that defective-monomer mutations have occurred in wheat, affecting other functions than chlorophyll production and including some of evolutionary importance, little or no concrete evidence yet exists for this idea. The mutants *Q*1 and *B*1 have been shown (Muramatsu, 1963; Sears, 1944) to be probable duplications, in that each has the same effect as its normal allele but is more extreme.

Literature Cited

Anderson, L. M. and C. J. Driscoll, 1967   The production and breeding behaviour of a monosomic alien substitution line. *Can. J. Genet. Cytol.* **9**:399–403.

Athwal, R. S. and G. Kimber, 1972   The pairing of an alien chromosome with homoeologous chromosomes of wheat. *Can. J. Genet. Cytol.* **14**:325–333.

Bhowal, J. G., 1964   An unusual transmission rate of the deficient male gamete in a substitution monosomic of chromosome 3D in wheat. *Can .J. Bot.* **42**:1321–1328.

Bielig, L. M. and C. J. Driscoll, 1970   Substitution of rye chromosome 5R^L for chromosome 5B and its effect on chromosome pairing. *Genetics* **65**:241–247.

Bielig, L. M. and C. J. Driscoll, 1973   Release of a series of *MAS* lines. In *Proceedings of the Fourth International Wheat Genetics Symposium,* pp. 893–937, Missouri Agricultural Experimental Station, Columbia, Missouri.

Bowden, W. M., 1959 The taxonomy and nomenclature of the wheats, barleys, and ryes and their wild relatives. *Can. J. Bot.* **37**:657–684.

Brewer, G., C. F. Sing and E. R. Sears, 1969 Studies of isozyme patterns in nullisomic–tetrasomic combinations of hexaploid wheat. *Proc. Natl. Acad. Sci. USA* **64**:1224–1229.

Cauderon, Y., 1966 Étude cytogénétique de l'évolution du material issu de croisement entre *Triticum aestivum* et *Agropyron intermedium*. I. Création de types d'addition stable. *Ann. Amelior. Plant. (Paris)* **16**:43–70.

Chapman, V. and R. Riley, 1970 Homoeologous meiotic chromosome pairing in *Triticum aestivum* in which chromosome 5B is replaced by an alien homoeologue. *Nature (Lond.)* **226**:376–377.

D'Amato, F., G. T. Scarascia Mugnozza and A. Bozzini, 1964 Mutanti vitali di frumento duro ottenuti del laboratorio per le applicazioni in agricoltura de C.N.E.N. con impiego di radiazioni ionizzanti e mutageni chimici. *Genet. Agrar.* **18**:132–141.

Darvey, N. L. and C. J. Driscoll, 1972 Evidence against somatic association in wheat. *Chromosoma* **36**:140–149.

Driscoll, C. J., 1960 Cytogenetical studies of wheat in relation to monosomic analyses and disease resistance derived from *Agropyron elongatum*. *J. Austral. Inst. Agric. Sci.* **26**:372.

Driscoll, C. J., 1972 Genetic suppression of homoeologous pairing in hexaploid wheat. *Can. J. Genet. Cytol.* **14**:39–42.

Driscoll, C. J. and E. R. Sears, 1971 Individual addition of the chromosomes of 'Imperial' rye to wheat. *Agron. Abs.* **1971**:6.

Eig, A., 1929 Monographisch-Kritische Übersicht der Gattung *Aegilops*. *Repert. Spec. Nov. Reg. Veg. Beih.* **55**:1–228.

Endrizzi, J. E. and R. J. Kohel, 1966 Use of telosomes in mapping three chromosomes in cotton. *Genetics* **54**:535–550.

Evans, L. E. and B. C. Jenkins, 1960 Individual *Secale cereale* chromosome additions to *Triticum aestivum*. I. The addition of individual "Dakold" fall rye chromosomes to "Kharkov" winter wheat and their subsequent identification. *Can. J. Genet. Cytol.* **2**:205–215.

Feldman, M., 1966a Identification of unpaired chromosomes in F_1 hybrids involving *Triticum aestivum* and *T. timopheevii*. *Can. J. Genet. Cytol.* **8**:144–151.

Feldman, M., 1966b The effect of chromosomes 5B, 5D, and 5A on chromosomal pairing in *Triticum aestivum*. *Proc. Natl. Acad. Sci. USA* **55**:1447–1453.

Feldman, M. and T. Mello-Sampayo, 1967 Suppression of homoeologous pairing in hybrids of polyploid wheats × *Triticum speltoides*. *Can. J. Genet. Cytol.* **9**:307–313.

Feldman, M., T. Mello-Sampayo and E. R. Sears, 1966 Somatic association in *Triticum aestivum*. *Proc. Natl. Acad. Sci. USA* **56**:1192–1199.

Johnson, B. L., 1972 Protein electrophoretic profiles and the origin of the B genome of wheat. *Proc. Natl. Acad. Sci. USA* **69**:1398–1402.

Johnson, R. and G. Kimber, 1967 Homoeologous pairing of a chromosome from *Agropyron elongatum* with those of *Triticum aestivum* and *Aegilops speltoides*. *Genet. Res.* **10**:63–71.

Kempanna, C. and R. Riley, 1962 Relationships between the genetic effects of deficiencies for chromosomes III and V on meiotic pairing in *Triticum aestivum*. *Nature (Lond.)* **195**:1270–1273.

Kempanna, C. and R. Riley, 1964 Secondary association between genetically equivalent bivalents. *Heredity* **18**:287–306.

Kihara, H., 1963   Nucleus and chromosome substitution in wheat and *Aegilops. II.* Chromosome substitution. *Seiken Ziho* **15**:13–23.

Kihara, H. and S. Wakakuwa, 1935   Veränderung von Wuchs, Fertilität und Chromosomenzahl in den Folgegenerationen der 40-chromosomigen Zwerge bei Weizen. *Jap. J. Genet.* **11**:102–108.

Kimber, G., 1967   The addition of the chromosomes of *Aegilops umbellulata* to *Triticum aestivum* (var. Chinese Spring). *Genet. Res.* **9**:111–114.

Kimber, G. and R. S. Athwal, 1972   A reassessment of the course of evolution of wheat. *Proc. Natl. Acad. Sci. USA* **69**:912–915.

Kimber, G. and E. R. Sears, 1968   Nomenclature for the description of aneuploids in the Triticinae. In *Proceedings of the Third International Wheat Genetics Symposium,* pp. 468–473, Australian Academy of Science, Canberra.

Knott, D. R., 1961   The inheritance of rust resistance. VI. The transfer of stem rust resistance from *Agropyron elongatum* to common wheat. *Can. J. Plant Sci.* **41**:109–123.

Knott, D. R., 1968   Translocations involving *Triticum* chromosomes carrying rust resistance. *Can. J. Genet. Cytol.* **10**:695–696.

Knott, D. R., 1971   The transfer of genes for disease resistance from alien species to wheat by induced translocations. In *Mutation Breeding for Disease Resistance,* pp. 67–77, International Atomic Energy Agency, Vienna.

Law, C. N., 1967   The location of genetic factors controlling a number of quantitative characters in wheat. *Genetics* **56**:445–461.

Law, C. N., 1968   Genetic analysis using intervarietal chromosome substitutions. In *Proceedings of the Third International Wheat Genetics Symposium,* pp. 331–342, Australian Academy of Science, Canberra.

Lee, Y. H., E. N. Larter and L. E. Evans, 1969   Homoeologous relationship of rye chromosome VI with two homoeologous groups from wheat. *Can. J. Genet. Cytol.* **11**:803–809.

Li, H. W., W. K. Pao and C. H. Li, 1945   Desynapsis in common wheat. *Am. J. Bot.* **32**:92–101.

McFadden, E. S. and E. R. Sears, 1946   The origin of *Triticum spelta* and its free-threshing hexaploid relatives. *J. Hered.* **37**:81–89, 107–116.

McIntosh, R. A., 1973   Gene symbols in wheat. In *Proceedings of the Fourth International Wheat Genetics Symposium,* pp. 893–937, Missouri Agricultural Experimental Station, Columbia, Missouri.

Mac Key, J., 1954   Neutron and X-ray experiments in wheat and a revision of the speltoid problem. *Hereditas* **40**:65–180.

Mac Key, J., 1960   Radiogenetics in *Triticum. Genet. Agrar.* **12**:201–230.

Mac Key, J., 1968   Relationships in the *Triticinae.* In *Proceedings of the Third International Wheat Genetics Symposium,* pp. 39–50, Australian Academy Science, Canberra.

Matsumura, S., 1952   Chromosome analysis of the dinkel genome in the offspring of a pentaploid wheat hybrid. I. Nullisomics deficient for a pair of D-chromosomes. *Cytologia (Tokyo)* **4**:265–287.

Mello-Sampayo, T., 1971   Genetic regulation of meiotic chromosome pairing by chromosome 3D of *Triticum aestivum. Nat. New Biol.* **230**:22–23.

Mello-Sampayo, T. and A. P. Canas, 1973   Suppressors of meiotic chromosome pairing in common wheat. In *Proceedings of the Fourth International Wheat Genetics Symposium,* pp. 709–713, Missouri Agricultural Experimental Station., Columbia, Missouri.

Mettin, D., W. D. Bluethner and G. Schlegel, 1973   Additional evidence on spontaneous 1B/1R wheat-rye substitutions and translocations. In *Proceedings of the Fourth International Wheat Genetics Symposium,* pp. 179–184, Missouri Agricultural Experimental Station., Columbia, Missouri.

Mochizuki, A., 1962   *Agropyron* addition lines of *durum* wheat. *Seiken Ziho* **13:**133–138.

Mochizuki, A., 1970   Production of three monosomic series in emmer and common wheat. *Seiken Ziho* **22:**39–49.

Morris, R. and E. R. Sears, 1967   The cytogenetics of wheat and its relatives. In *Wheat and Wheat Improvement,* edited by L. P. Reitz and K. S. Quisenberry, pp. 19–87, American Society of Agronomy, Madison, Wisconsin.

Muramatsu, M., 1963   Dosage effect of the spelta gene *q* of hexaploid wheat. Genetics **48:**469–482.

Natarajan, A. T., S. M. Sikka and M. S. Swaminathan, 1958   Polyploidy, radiosensitivity, and mutation frequency in wheats. *Proc. 2nd UN Int. Conf. Peaceful Uses Atom. Energy. Geneva* **27:**321–331.

Neatby, K. W., 1933   A chlorophyll mutation in wheat. *J. Hered.* **24:**159–162.

Okamoto, M., 1957   Asynaptic effect of chromosome V. *Wheat Info. Serv.* **5:**6.

O'Mara, J. G., 1940   Cytogenetic studies on Triticale. I. A method for determining the effects of individual Secale chromosomes on Triticum. *Genetics* **25:**401–408.

O'Mara, J. G., 1951   Cytogenetic studies on Triticale. II. The kinds of intergeneric chromosome addition. *Cytologia (Tokyo)* **16:**225–232.

Person, C. L., 1956   Some aspects of monosomic wheat breeding. *Can. J. Bot.* **34:**60–70.

Pettigrew, R., 1972   Studies on a group of chlorophyll mutants in hexaploid wheat. Ph.D. Thesis, University of New South Wales, Sydney, Australia.

Pettigrew, R., C. J. Driscoll and K. G. Rienits, 1969   A spontaneous chlorophyll mutant in hexaploid wheat. *Heredity* **24:**481–487.

Prabhakara Rao, M. V. and W. J. Washington, 1969   EMS-induced chlorophyll defective sectors in hexaploid wheat. *Proc. Symp. Radiations Radiomimetic Subst. Mut. Breed. (Bombay),* pp. 228–233.

Riley, R., 1960   The meiotic behaviour, fertility and stability of wheat-rye chromosome addition lines. *Heredity* **14:**89–100.

Riley, R., 1966   The genetic regulation of meiotic behaviour in wheat and its relatives. *Proceedings of the Second International Wheat Genetics Symposium,* Lund, 1963, *Hereditas Suppl. Vol.* **2:**395–408.

Riley, R. and V. Chapman, 1958*a*   The production and phenotypes of wheat-rye chromosome addition lines. *Heredity* **12:**301–315.

Riley, R. and V. Chapman, 1958*b*   Genetic control of the cytologically diploid behaviour of hexaploid wheat. *Nature (Lond.)* **182:**713–715.

Riley, R. and V. Chapman, 1964   Cytological determination of the homoeology of chromosomes of *Triticum aestivum. Nature (Lond.)* **203:**156–158.

Riley, R. and V. Chapman, 1966   Estimates of the homoeology of wheat chromosomes by measurements of differential affinity at meiosis. In *Chromosome Manipulations and Plant Genetics,* edited by R. Riley and K. R. Lewis pp. 46–58, Oliver and Boyd, Edinburgh.

Riley, R. and V. Chapman, 1967   Effect of $5B^{s}$ in suppressing the expression of altered dosage of $5B^{L}$ on meiotic chromosome pairing in *Triticum aestivum. Nature (Lond.)* **216:**60–62.

Riley, R. and C. Kempanna, 1963   The homoeologous nature of the non-homologous meiotic pairing in *Triticum aestivum* deficient for chromosome V (5B). *Heredity* **18**:287–306.

Riley, R. and G. Kimber, 1966   The transfer of alien genetic variation to wheat. *Rep. Plant Breed. Inst. (Cambridge)* **1964–1965**:6–36.

Riley, R. and C. N. Law, 1965   Genetic variation in chromosome pairing. *Adv. Genet.* **13**:57–114.

Riley, R., V. Chapman and G. Kimber, 1959   Genetic control of chromosome pairing in intergeneric hybrids with wheat. *Nature (Lond.)* **183**:1244–1246.

Riley, R., G. Kimber and V. Chapman, 1961   Origin of genetic control of diploid-like behavior of polyploid wheat. *J. Hered.* **52**:22–25.

Riley, R., V. Chapman and R. C. F. Macer, 1966a   The homoeology of an *Aegilops* chromosome causing stripe rust resistance. *Can. J. Genet. Cytol.* **88**:616–630.

Riley, R., V. Chapman, R. M. Young and A. M. Belfied, 1966b   The control of meiotic chromosome pairing by the chromosomes of homoeologous group 5 of *Triticum aestivum. Nature (Lond.)* **212**:1475–1477.

Riley, R., H. Coucoli and V. Chapman, 1967   Chromosomal interchanges and the phylogeny of wheat. *Heredity* **22**:233–248.

Riley, R., V. Chapman and R. Johnson, 1968   The incorporation of alien disease resistance in wheat by genetic interference with the regulation of meiotic chromosome synapsis. *Genet. Res.,* **12**:199–219.

Sarkar, P. and G. L. Stebbins, 1956   Morphological evidence concerning the origin of the B genome in wheat. *Am. J. Bot.* **43**:297–304.

Schlehuber, A. M. and E. E. Sebesta, 1959   Progress in wheat-grass breeding. *Proc. Okla. Acad. Sci.* **39**:6–16.

Sears, E. R., 1944   Cytogenetic studies with polyploid species of wheat. II. Additional chromosomal aberrations in *Triticum vulgare. Genetics* **29**:232–246.

Sears, E. R., 1952   Misdivision of univalents in common wheat. *Chromosoma* **4**:535–550.

Sears, E. R., 1954   The aneuploids of common wheat. *Mo. Agric. Exp. Stn. Res. Bull.* **572**:1–59.

Sears, E. R., 1956   The transfer of leaf-rust resistance from *Aegilops umbellulata* to wheat. *Brookhaven Sym. Biol.* **9**:1–22.

Sears, E. R., 1965   Nullisomic–tetrasomic combinations in hexaploid wheats. In *Chromosome Manipulations and Plant Genetics.* edited by R. Riley and K. R. Lewis, Suppl. to *Heredity* **20**:29–45.

Sears, E. R., 1966   Chromosome mapping with the aid of telocentrics. *Proceedings of the Second International Wheat Genetics Symposium, Lund, 1963 Hereditas Suppl. Vol.* **2**:370–381.

Sears, E. R., 1967   Induced transfer of hairy neck from rye to wheat. *Z. Pflanzenzuecht.* **57**:4–25.

Sears, E. R., 1969   Wheat cytogenetics. *Annu. Rev. Genet.* **3**:451–468.

Sears, E. R., 1972a   Chromosome engineering in wheat. *Stadler Genet. Symp.* **4**:23–38.

Sears, E. R., 1972b   The nature of mutation in hexaploid wheat. *Symp. Biol. Hung.* **12**:73–82.

Sears, E. R., 1972c   Reduced proximal crossing-over in telocentric chromosomes of wheat. *Genêt. Ibér.,* **24**:233–239.

Sears, E. R. and M. Okamoto, 1958   Intergenomic chromosome relationships in hexaploid wheat. *Proc. X Int. Congr. Genet.* **2**:258–259.

Sears, E. R., W. Q. Loegering and H. A. Rodenhiser, 1957   Identification of chromosomes carrying genes for stem rust resistance in four varieties of wheat. *Agron. J.* **49**:208–212.

Shama Rao, H. K. and E. R. Sears, 1964   Chemical mutagenesis in *Triticum aestivum. Mutat. Res.* **1**:387–399.

Shands, H. and G. Kimber, 1973   Reallocation of the genomes of *Triticum timopheevii* Zhuk. In *Proceedings of the Fourth International Wheat Genetics Symposium,* pp. 101–108, Missouri Agricultural Experimental Station, Columbia, Missouri.

Sheen, S. J. and L. A. Synder, 1964   Studies on the inheritance of resistance to six stem rust cultures using chromosome substitution lines of a Marquis wheat selection. *Can. J. Genet. Cytol.* **6**:74–82.

Sheen, S. J. and L. A. Synder, 1965   Studies on the inheritance of resistance to six stem rust cultures using chromosome substitution lines of a Kenya wheat. *Can. J. Genet. Cytol.* **7**:374–387.

Siddiqui, K. A., 1972   The influence of the B genome on chromosome pairing in trigeneric *Aegilops* × *Triticum* × *Secale* hybrids. *Hereditas* **70**:97–104.

Smith, L., 1939   Mutants and linkage studies in *Triticum monococcum* and *T. aegilopoides. Mo. Agric. Exp. Stn. Res. Bull.* **298**:26.

Tsunewaki, K., 1964   The transmission of the monosomic condition in a wheat variety, Chinese Spring. II. A critical analysis of nine year records. *Jap. J. Genet.* **38**:270–281.

Unrau, J., 1958   Genetic analysis of wheat chromosomes. I. Description of proposed methods. *Can. J. Plant Sci.* **38**:415–418.

Wagenaar, E. B., 1961   Studies on the genome constitution of *Triticum timopheevi* Zhuk. I. Evidence for genetic control of meiotic irregularities in tetraploid hybrids. *Can. J. Genet. Cytol.* **3**:204–225.

Waines, J. G. and B. L. Johnson, 1972   Genetic differences between *Aegilops longissima, A. sharonensis,* and *A. bicornis. Can. J. Genet. Cytol.* **14**:411–416.

Washington, W. J. and E. R. Sears, 1970   Ethyl methanesulfonate-induced chlorophyll mutations in *Triticum aestivum. Can. J. Genet. Cytol.* **12**:851–859.

Wienhues-Ohlendorf, A., 1960   Die Ertragsleistung rostresistenter 44- und 42-chromoso-miger Weizenquecken-Bastarde. *Züchter* **30**:194–202.

Zeller, F. J., 1972   Cytologischer Naschweis einer Chromosomensubstitution in dem Weizenstamm Salzmünde 14/44 (*T. aestivum* L.). *Z. Pflanzenzuecht.* **67**:90–94.

Zeller, F. J. and G. Fischbeck, 1971   Cytologische Untersuchungen zur Identifizierung des Fremdchromosoms in der Weizensorte Zorba (W565). *Z. Pflanzenzuecht.* **66**:189–202.

Zeller, F. J. and S. Sastrosumarjo, 1972   Zur Cytologie der Weizensorte Weique (*T. aestivum* L.). *Z. Pflanzenzuecht.* **68**:312–321.

Zohary, D. and M. Feldman, 1962   Hybridization between amphidiploids and the evolution of polyploids in the wheat (*Aegilops-Triticum*) group. *Evolution* **16**:44–61.

4

Barley (*Hordeum vulgare*)

ROBERT A. NILAN

Introduction

Cultivated barley, *Hordeum vulgare* L., is one of the leading experimental organisms in genetic studies of flowering plants. The wide use of this important agricultural crop plant in genetic studies may be attributed to its diploid nature, low chromosome number ($2n = 14$), world-wide distribution, high degree of self-fertility, ease of hybridization, relatively large chromosomes which allow detection of several kinds of chromosome aberrations, and numerous easily classified hereditary characters. Many of these characters have occurred as mutants after physical- or chemical-mutagen treatments. Since barley is one of the chief models among higher plants for induced-mutation studies, new characters are being added at a considerable rate.

The genetic studies, as briefly reviewed here, are aiding in improving this plant for the production of grain for livestock feed and human food and of malt. They are also basic to the extensive use of barley as an experimental or model plant for numerous studies in cytogenetics; population and biochemical genetics; radiation and chemical mutagenesis; breeding systems; isogenic analysis; physiology, including photoperiodism; malting and brewing chemistry; enzymology; radiobiology and biophysics; pathology, including host-pathogen interactions; and virology.

Literally all of the world literature about barley genetics and cytology has been very well listed and summarized in several key publications. Smith

ROBERT A. NILAN—Program in Genetics and Department of Agronomy and Soils, Washington State University, Pullman, Washington.

(1951) summarized over 1000 references; he included most of the barley genetic literature published from the beginning of genetic investigations of this plant through 1950. Nilan (1964) provided an additional summary involving 1200 references published from 1951 to 1962. Much of the information about barley genes and linkage and recombination data was summarized by Robertson (1971) and Robertson *et al.* (1941, 1947, 1955, 1965). The proceedings of the two International Barley Genetics Symposia held in 1963 (Lamberts, 1964) and in 1969 (Nilan, 1971) constitute two other major publications on barley genetics. Detailed new information about barley genetics and cytology appears annually in the *Barley Genetics Newsletter* (Vol. 1, 1971; Vol. 2, 1972; Vol. 3, 1973).

This brief review will only highlight some of the major developments in barley genetics as reported in these publications.

Chromosomes

Cultivated barley has 7 pairs of chromosomes, while wild species occur with 7, 14, and 21 pairs of chromosomes. The 7 somatic chromosomes of barley are relatively large, measuring 6–8 μ in length, and each chro-

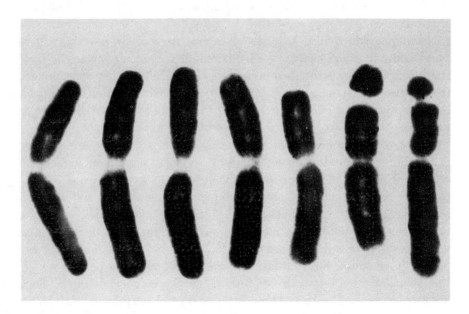

Figure 1. Standard haploid karyotype of barley; left to right, chromosomes 1–7. Prepared by Tsuchiya.

TABLE 1. *Relative Length of Barley Chromosomes and Arm-Length Ratios*

Chromosome number	Relative length[a]	Short arm/ long arm ratio
1	1.00	0.75
2	0.96	0.86
3	0.89	0.92
4	0.87	0.77
5	0.77	0.73
6	0.73[b] (0.88)[c]	0.61[b] (0.94)[c]
7	0.81[b] (0.92)[c]	0.41[b] (0.60)[c]

[a] Relative to chromosome 1.
[b] Without satellite.
[c] With satellite.

mosome can be fairly easily distinguished at metaphase of mitosis (Figure 1). There are two satellite chromosomes, 6 and 7, with chromosome 6 having the longer satellite and chromosome 7 the shorter. The other chromosomes are best characterized by length and arm ratios. Chromosome 1 is the longest and chromosome 5 is the shortest. A detailed description of each chromosome appears in Nilan (1964), and the relative lengths and the short arm to long arm ratios are summarized in Table 1.

Meiotic chromosomes are best seen at diakinesis (Figure 2) or metaphase. The analysis of barley chromosomes at the pachytene stage is quite difficult because of the clumped and knotted chromosome figures and the difficulty in distinguishing centromeres. Nevertheless, certain modified staining techniques have permitted the identification, description, and mapping of the 7 chromosomes at early and late pachynema. The relative measurements of the pachytene chromosomes agree very well with those published for the somatic chromosomes. However, because of the lack of distinct landmarks, particularly heterochromatic segments, the practical value of these chromosomes for cytogenetic studies appears limited.

Alterations in Chromosome Number and Structure

Because of the low number of relatively large chromosomes in barley, alterations in chromosome number and structure are quite easy to detect, isolate, maintain, and utilize in various genetic investigations, e.g., in gene mapping, in assigning linkage groups to specific chromosomes, and in certain breeding techniques. Thus, relatively extensive collections of translocations, tetraploids, and trisomics, and lesser collections of telo-

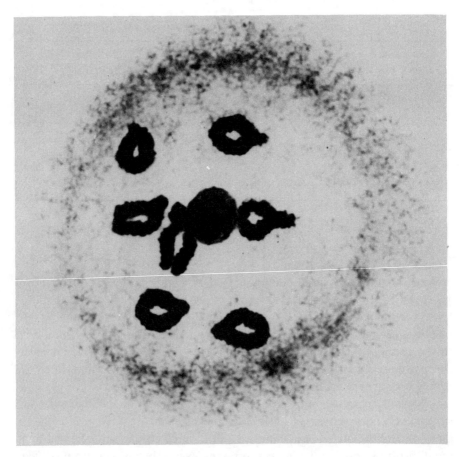

Figure 2. Barley chromosomes at diakinesis of meiosis. Note the two satellited chromosomes (6 and 7) attached to the nucleolus.

centrics, haploids, triploids, and inversions are maintained for genetic and breeding research.

Haploid ($n = 7$) plants are weaker and shorter, have narrower leaves and smaller stomates compared to the diploid, and are almost completely sterile. Until recently they were rare, but now can be produced in relatively high frequencies following crosses between *Hordeum vulgare* and diploid *Hordeum bulbosum* (Kasha and Kao, 1970). Viable embryos from such a cross produce only *vulgare*-type haploids because the *bulbosum* chromosomes are eliminated during formation of the haploid embryos. Homozygous diploid lines for plant bleeding and genetic research are produced by colchicine treatment.

Triploid plants are rare and are seldom found in nature. They are obtained chiefly among progeny of crosses between autotetraploids and diploids. Some heterosis exists in triploid hybrid plants; stomatal size is intermediate between tetraploid and diploid parents and the average frequency of trivalents is about 4.5 while that of univalents about 2.5 per microsporocyte. Pollen viability and seed set are about 50 and 25 percent of the diploid progenitor, respectively.

Tetraploids occur in nature, but most have been induced, chiefly by colchicine. In the world collection of autotetraploids, there are 15 six-row winter, 32 six-row spring, and 90 two-row spring types.* The fertility of most of these strains ranges from 50 to 70 percent seed set. Compared to the diploid parents, they have shorter and thicker stems, larger kernels, 3–5 more kernels per spike, 1–2 more culms per plant, and are 3–5 days later in flowering.

The irregular chromosome pairing at meiosis of tetraploid barley yields univalents, trivalents, quadrivalents, and associations above four chromosomes. In addition, other abnormalities such as lagging chromosomes, bridges, and micronuclei occur. Tetraploids have been tested for economic value, but none have been found agronomically superior to the diploid progenitor.

Primary trisomics have been obtained, usually among progeny of triploids as well as translocation heterozygotes. Three complete sets of almost fertile primary trisomics have been developed. The phenotype of each trisomic is altered in the characteristic manner depending upon which primary chromosome is present as the extra chromosome. Thus, the seven different trisomics within each set maintained can be identified. In general, trisomic plants differ in characteristics such as general plant vigor, height, time of maturity, tiller number, fertility, leaf color, leaf width, and leaf length. The primary trisomics have been very useful in associating linkage groups with specific chromosomes and also for associating specific genes with their respective chromosomes.

Tertiary trisomics and balanced tertiary trisomics have also been induced. The latter, which, to date, total 13, are widely utilized in the production of hybrid barley.

Telocentric chromosomes have been obtained spontaneously in the progeny of triploids and primary trisomics induced by radiation and chemical mutagens. Seven telocentric trisomic lines have been identified cytogentically. Because these lines exhibit less sterility than the primary trisomics, they are more useful in genetic studies. They are used for locating genes on a particular arm of a chromosome and thus aid in locating the cen-

---

* Here the "two" and "six" refer to the number of rows of kernels along spike, and "winter" and "spring" refer to the season when the seeds are sown.

tromere position of each linkage group and chromosome map (Tsuchiya, 1972).

Reciprocal translocations or interchanges occur spontaneously, but they have been induced in high numbers by ionizing-radiation and chemical-mutagen treatments. Thus, there is now available a collection of over 260 translocations which have been classified as to the chromosomes involved. For many of these, the breakpoints have been located at specific sites on the chromosome arms. Most homozygous translocations produce no specific effect on the phenotype, although about 30 percent of the translocations exhibit reduced vigor and yield. Some translocations appear to be connected with single gene changes. There also has been some indication of "position effects" in translocation stocks, but no real proof of this phenomenon exists.

As a rule, there are no classification problems in distinguishing fertile from partially sterile plants. These usually show 30–50 percent sterility. Thus, translocations have been useful in a variety of genetic and cytogenetic studies, such as locating genes on specific chromosomes, positioning genes in relation to the centromeres, determining crossover and chiasmata positions, moving genes into desirable linkage relationships, separating them from undesirable linkage situations, and creating duplicated segments.

Spontaneous inversions are rare, but a few have been isolated following mutagen treatment. Because of relatively low sterility, inversions are not as useful as translocations for cytogenetic and linkage studies.

Mutations

The natural variation within barley is extensive. Many of the spontaneous mutations which contribute to this variation can be found in the collection containing over 10,000 entries maintained by the U.S. Department of Agriculture. This collection and others in the U.S.S.R., Canada, and West Germany have been analyzed frequently to determine the extent of variation and distribution of a number of important characteristics such as disease resistance, early maturity, lysine content, and insect resistance.

Mutations have been induced more frequently in barley than in probably any other flowering plant. Starting with the pioneering work of Nilsson-Ehle and followed by the extensive investigations of Gustafsson and colleagues in Sweden, this plant has been a model for studying the genetic and cytological effects of a very wide variety of mutagens, particularly ionizing radiations and chemicals. More recently, it has been used extensively in the investigations of the experimental control of the induced-mutation process and of the mechanisms of action of the mutagens (see Nilan, 1964). Barley has also been an important test plant for the induction of beneficial mutations for plant breeding. To date, 19 improved commercial varieties have been created in several countries with the aid of induced mutations (information provided by Dr. B. Sigurbjörnsson, I.A.E.A., Vienna).

Barley is ideal for studying the genetic effects of mutagens because of its characteristic and frequent chlorophyll-deficient mutations and its small number of relatively large chromosomes. Also, barley has particular use in mutagen investigations since it can be handily cultured in both field and greenhouse and since it has easily measured growth patterns.

In general, frequencies of chlorophyll-deficient mutations as measured on an M_1-spike* basis can be increased to 15–17 percent by x and gamma rays and neutrons. Chemicals, such as ethyl methanesulfonate, diethyl sulfate, nitroso urea, and sodium azide induce chlorophyll-deficient mutations in frequencies as high as 45–50 percent on an M_1 spike basis. Frequencies up to 80 percent have recently been obtained following sodium azide treatment of germinating seeds.

Extensive collections of induced mutations in barley are maintained in the German Democratic Republic and in Sweden. Lesser collections are maintained in West Germany, Belgium, and Japan.

The nature of events leading to mutations in barley has not been determined. There is evidence that many mutations are associated with or due to chromosome alterations. Chemical mutagens such as the alkylating compounds and sodium azide, which induce relatively few chromosome aberrations when compared with x rays, may induce changes at the DNA level. It will be necessary, however, to improve the resolving power of mutant analysis in barley before the true nature of mutations events can be determined.

Genetic Characters and Linkage Maps

About 100 genetic characters (phenotypes) involving about 250 loci have now been associated with the seven linkage groups of the seven chromosomes, and of these, about 100 loci have been mapped to specific locations within the linkage groups. The official collection of genetic stocks is cataloged and maintained by Dr. T. Tsuchiya, Department of Agronomy, Colorado State University.

* An M_1 spike is the head of the barley plant raised from a mutagen-treated seed.

Table 2 presents the genetic characters that have been assigned to a specific linkage group and chromosome. This list is by no means complete since there are many new characters that have not yet been assigned to a chromosome. Those genes that have been assigned to a specific site within the linkage groups are shown in the linkage and chromosome maps (Figure 3). Centromeres for chromosomes 1–5 are shown in these maps. The localities of these centromeres have been recently determined by analyses of telosomic trisomic lines (Fedak *et al.*, 1972).

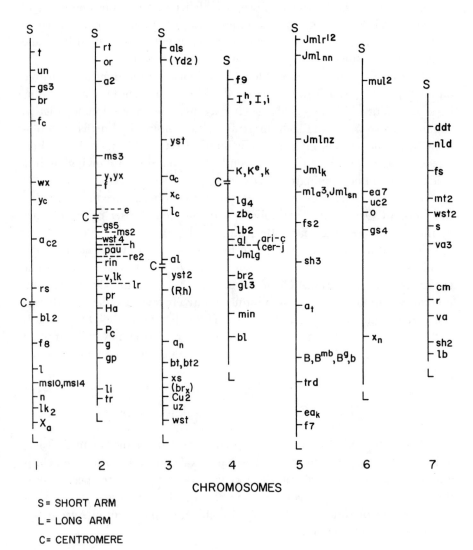

CHROMOSOMES

S = SHORT ARM

L = LONG ARM

C = CENTROMERE

Figure 3. Linkage and chromosome maps of barley (1973). Prepared with the help of T. Tsuchiya and T. E. Haus.

TABLE 2.   List of Barley Genes that Have Been Located on or Associated with a Chromosome

Name	Gene symbol[a]	Chromosome
Albino seedling	$a2$—2, a_c—3, a_c2—1, a_c a_n—3, a_t—5	
Albino lemma	al	3
Absent lower laterals	als	3
Alpha amylase	Amy	6
Black, gray, medium black, and white lemma and pericarp	B, B^g, B^{mb}, b	5
Broad leaf (Brodblädig)	bb	7
Blue aleurone (complementary factor)	Bl	4
Blue aleurone (complementary factor)	$Bl2$	1
Brachytic	br	1
Brachytic 2	$br2$	4
Brachytic	br_x	3
Brittle rachis (complementary factor)	$Bt1$, $Bt2$	3
Eceriferum (waxless)	cer-b—7, cer-f—1, cer-g—2, cer-i—7, cer-j—4, cer-n—2, cer-s—2, cer-t—7, cer-v—2, cer-x—7, cer-zd—3, cer-zj—7, cer-zm—2, cer-zn—3, cer-zp—7	
Cream seedling	cm	7
Cream seedling	$cm2$	5
Corn stalk	cs	7
Curved (curled) stem	$cu2$	3
Resistance to DDT	ddt	7
High diastatic power	Dip	2
Desynaptic chromosome	$des2b$	3
Long-glume awn	e	2
Early heading	Ea—1, $Ea2$—4, $Ea5$—7, $ea7$—6, ea_k—5	
Erectoides	ert-a—1, ert-d—1, ert-e—6, ert-g—7, ert-i—4, ert-k—6, ert-m—1, ert-n—7, ert-r—7, ert-ii—3	
Esterase isozyme activity	Est-1	3

TABLE 2. Continued

Name	Gene symbol[a]	Chromosome
Chlorina seedling	f—2, f2—3, f4—1, f5—1, f6—6, f7—5, f8—1, f9—4, fc—1	
Fragile stem	fs	7
Fragile stem	fs2	5
Teeth on lemma	G	2
Gametophyte factor	ga	1
Glossy leaf	gl (cer)	4
Glossy leaf 2	gl2 (cer)	4
Glossy leaf 3	gl3 (cer)	4
Glossy leaf 4	gl4 (cer)	6
Grandpa	gp (=gp2)	2
Glossy sheath	gs1—4, gs2—3, gs3—1, gs4—6, gs5—2, gs6—2	
Resistance to root nematode	Ha (Rha)	2
Resistance to *Helminthosporium sativum* (spot blotch)	hl (rhs)	2
Hairy lemma nerves	Hn	4
Hairy rachis	Hr	2
Hairy sheath	Hs	4
Infertility of lateral florets	I, I^h, i	4
Resistance to race IX of *Erysiphe graminis hordei* in *H. spontaneum nigrum* (mildew)[b]	JML-sn	5
		5
Medium resistance to race IX of *E. graminis hordei* in Nigrinudum (mildew)	JMl-nn	5
Resistance to race IV of *E. graminis hordei* in A222 (mildew)	al Jml-sn	5
Resistance to race IX of *E. graminis hordei* in H.E.S. 4 (mildew)	h4 Jml-sn	5
Resistance to race IX of *E. graminis hordei* in Monte Cristo (mildew)	mc JMl-sn	

TABLE 2.   *Continued*

Name	Gene symbol[a]	Chromosome
Resistance to race IV of *E. graminis hordei* in Nigrate (mildew)	*n* *JMl-sn*	5
Medium resistance to race IX of *E. graminis hordei* in Kwan (mildew)	*JMl-k*	5
Resistance to race IV of *E. graminis hordei* in Nakaizumi-zairai (mildew)	*Jml-nz*	5
Resistance to race IV of *E. graminis hordei* in Russian No. 12 (mildew)	*Jml-r$_{12}$*	5
Resistance to race IX of *E. graminis hordei* in Goldfoil (mildew)	*JMl-g*	4
Hooded, elevated hood, awned	*K, K^e, k*	4
Lax spike	*L (Lax)*—1, *L2 (Lax)*—1, *L3 (Lax)*—7, *L5 (Lax)*—2, *L7 (Lax)*—1, *L$_a$*—1, *L$_c$*—3, *lax-a*—7	
Long basal rachis internode	*Lb*	7
Long weak basal rachis, internode 2	*lb2*	4
Long weak basal rachis, internode 3	*lb3*	1
Light green seedling	*lg4*	4
Liguleless	*li*	2
Awnless	*Lk*	2
Short awn	*lk2*	1
Short awn	*lk6*	4
Awnless laterals	*lr*	2
Lysine content	*lys*	7
Many-noded dwarf 2	*m2*	2
Minute plant	*min*	4
Male sterile	*ms*—5, *ms2*—2, *ms3*—2, *ms4*—5, *ms5*—3, *ms6*—6, *ms10*—1, *ms14*—1, *ms19*—7, *ms20*—5, *ms22*—1, *ms23*—1, *ms24*—4	
Midseason stripe	*mss*	2

TABLE 2.   Continued

Name	Gene symbol[a]	Chromosome
Mottled leaves 2	*mt2*	7
Multiflorous	*mul2*	6
Naked caryopsis	*n*	1
Narrow-leafed dwarf	*nld*	7
Orange lemma	*o*	6
Opposite spikelets	*op*	1
Orange seedling	*or*	2
Purple lemma	*P*	2
Purple-veined lemma	P_c	2
Resistance to *Puccinia hordei* (leaf rust)	*Pa (Rph)*	2
Purple auricle	*Pau*	2
Pubescence on outer glume	*Pbg*	1
Purple node	*Pn*	2
Purple stem	*Pr*	2
Smooth awn	*r—7, r2—7*	
Short rachilla	*ra*	2
Purple lemma and pericarp (complementary gene)	*Re*	5
Purple lemma and pericarp (complementary gene)	*Re2*	2
Resistant to *Rhynchosporium secalis* (scald)	*Rh3 (Rrs3)*	3
Resistant to *R. secalis* (scald) in Trebi	*Rh4 (Rrs4)*	3
Low number of rachis internodes	*rin*	2
Red stem	*Rs*	1
Rattail	*rt*	2
Short rachilla hairs	*s*	7
Shrunken endosperm	*se1—1,se2—1, se3—3, se4—1, se5—1, se6—6*	
Spring habit of growth (recessive)	*sh*	4
Spring habit of growth	*Sh2*	7
Spring habit of growth	*Sh3*	5
Subjacent hood	*sk*	2
Resistance to *Puccinia graminis tritici* (Stem rust)	*T (Rpg)*	1

TABLE 2. *Continued*

Name	Gene symbol[a]	Chromosome
Triple-awned lemma	*tr*	2
Third outer glume	*trd*	5
Unbranched style	*u4*	1
Uniculm	*uc2*	6
Resistance to *Ustilago nuda* (Loose smut)	*Un (Run)*	1
Resistance to *U. nuda* (Loose smut)	*un7 (run7)*	1
"Uzu" (Semibrachytic)	*uz*	3
Row number (two row and six row)	V, V^t, V^{M20}, v	2
Six row	*v2*	7
Variegated	*va, va2, va3*	7
Low number of vascular bundles	*vbn*	1
White streak	*wst, wst3*	3
White streak 2	*wst2*	7
White streak 4	*wst4*	2
Waxy endosperm	*wx*	1
Xantha (Incomplete dominance)	X_a	1
Xantha seedling (Colsess)	x_c	3
Xantha seedling	x_n	6
Xantha seedling (Smyrna)	x_s	3
Virescent seedling	y, y_x	2
Virescent seedling (Coast III)	y_c	1
Resistance to yellow dwarf	*Yd2 (Ryd2)*	3
Yellow head	*yh*	4
Yellow streak	*yst, yst2*	3
Zebra stripe	*zb*	3
Zebra stripe	zb_c	4

[a] Proposed new symbols are in parentheses. Characters controlled by several genes are designated with gene symbols followed by chromosone designation, e.g., a2—2; a_b—7.
[b] The terminology here indicates resistance to a particular race of the pathogen, e.g., race IX. The gene for this resistance was first found in the 2-rowed species *H. Spontaneum*. Other designations are varieties, e.g., Nigrinudum or A222.

While the numerous translocation breakpoints which have been assigned to specific locations in the linkage group are not shown in Figure 3, many of these have been published (Ramage *et al.,* 1961).

The genc symbols used are those that have appeared in the literature and that have been checked with the Committee on Nomenclature of Barley Genes, T. Tsuchiya, Chairman. At the present time, an extensive revision of the nomenclature of genetic characters and of the appropriate symbols is under way by the Committee on Genetic Marker Stocks, Nomenclature, and Symbolization of the International Barley Genetics Symposium. Since this revision is not yet completed only a few of the proposed new names and gene symbols for genetic characters will be introduced in this review.

Most of the phenotypes reported here have been described as to biochemistry, physiology, morphology, and anatomy in Smith (1951) and Nilan (1964). Subsequent descriptions of more recently discovered ones have appeared in Nilan (1971) and in *Barley Genetics Newsletters,* Vols. 1, 2, and 3.

Barley characters can be grouped in various ways, and certainly no single classification system is satisfactory. The reader is referred to the reviews of Smith (1951) and Nilan (1964) for a grouping that has been more commonly used.

The most frequently detected and studied genetic characters in barley are those that have been readily classified by visual inspection. Most of these concern characteristics of the spike, possibly because of the practical interest in the reproductive parts of the plant and because the spike exhibits easily measured botanical traits. Some of the better-known spike characters are two-row vs. six-row spike, hulled vs. naked kernels, hoods vs. awns, black vs. white kernels, rough vs. smooth awns, and short vs. long rachilla hairs. Currently, chemical tests and biochemical analyses are uncovering additional loci, and, of course, an increasing number of genes controlling disease responses are being detected as a result of improved techniques in phytopathology and biochemistry. Certain other characters are intensively studied because of their application to barley breeding and improvement.

Most of the characters are controlled by a single gene, and usually a 3:1 ratio is obtained. Multiple-gene action is indicated for several characters such as *erectoides* (erect plant and spike—26 loci), *eceriferum* (waxless—44 loci), plant height, straw strength, and yield. Complementary gene action exists for some characters such as blue aleurone color. Examples of xenia expression are waxy endosperm and blue aleurone, while cytoplasmic inheritance has been detected for certain chlorophyll-deficient mutations.

Linkage between characters is common, and the literature on the subject is extensive. Most of the actual linkage data can be found in Smith

(1951), Nilan (1964), Robertson (1971), and Robertson *et al.* (1941, 1947, 1955, 1965).

Examples of Genetic Characters

Only a few characters will be discussed here. These have been chosen because they have been intensively studied and/or because they provide examples of the different types of characters found in barley.

Erectoides (ert). Among the spike and plant characters, none have been more intensively studied than the *erectoides* mutant (dense spikes and stiff, short stems). It is one of the most common morphological characters, induced by both physical and chemical mutagens.

Well over 200 mutants located at 26 different loci have been induced. For some loci, as many as 20 different mutant alleles are known. Nine of the loci have been assigned to definite chromosomes. Chromosome 1 has 3, 3 has 2, 4 has 1, 5 has 1, and 7 has 2 (Persson and Hagberg, 1969).

All but three of the genes determining *erectoides* are recessive, and at least twelve are associated with structural changes in the chromosomes. Certain of the loci exhibit interlocus specificity in relation to their response to physical or chemical mutagens. For instance, of 34 mutant alleles at locus *ert-c*, 16 were induced by neutrons, 7 by chemicals, and the rest by x or gamma rays. On the other hand, of 32 mutant alleles at locus *ert-a*, only 1 was induced by neutrons, 17 by chemicals and the rest by x or gamma rays.

Chlorophyll Variants. Numerous types of chlorophyll variants occur in barley, and most are detected at the seedling stage. They are described as albina (*a*), viridis (*y*), xantha (*x*), striata (*st*), chlorina (*f*), etc., depending upon pigment content and placement, and are utilized extensively in measuring the effects of physical and chemical mutagens.

Many of the chlorophyll-deficient seedlings are recessive, and most are inherited monofactorially. Seven cytoplasmically inherited chlorina mutants are known. Aberrant ratios of chlorophyll-deficient seedlings occur, and some of these are due to transmission problems caused by associated chromosome aberrations.

Work by (Wettstein *et al.* (1971) has shown that a great many loci are involved in chlorophyll synthesis, as revealed by the already-detected numerous loci involved in *albina, xantha, striata,* and *chlorina* mutants. They have also shown how plastid ultrastructure and development are controlled by certain of these loci.

Eceriferum (cer). Eceriferum (reduction or absence of wax) is another character in barley that occurs very often following mutagen treatment. To date, over 660 induced and 4 spontaneous mutants have been analyzed (Lundqvist *et al.,* 1968). Many of these mutants have been localized to 46 loci which control the synthesis and deposition of the epidermal wax coating. Twenty of these are randomly distributed throughout the genome. Some phenotypes exhibit a reduction of wax over the entire plant, while others have a reduction of wax on only a portion of the plant, i.e., the leaf, stem, spike, leaf and stem, leaf and spike, spike and stem, etc.

Interlocus mutagen specificity is apparent for certain of the eceriferum loci. Mutations at some loci are induced only by x or gamma rays, whereas at other loci most are induced by neutrons or certain chemical mutagens.

Ultrastructural and chemical analyses of several of eceriferum mutants have revealed striking differences in the physical structure and the chemical composition of the wax (Wettstein-Knowles, 1971).

Commercial Characteristics. Important agronomic characteristics such as yield, winter-hardiness, and quality characteristics for malting, brewing, and feed, have received relatively little genetic attention. Progress in genetic analyses of these important characteristics has been slow because the inheritance of these characters is quite complex. Genetic studies of the components of these characters are beginning.

The need for a more nutritious barley grain for both human and livestock consumption has initiated many genetic studies of feed quality. The most advanced aspect of this development involves the high-protein, high-lysine gene (*lys*) (Munck *et al.*, 1971).

Habit of Growth (sh). Spring vs. winter habit of growth is an important characteristic that has been studied extensively by Japanese workers (Takahashi and Yasuda, 1971). At least three pairs of genes, *Shsh*, *Sh_2sh_2*, and *Sh_3sh_3*, located on chromosome 4, 7, and 5, respectively, affect growth habit. One recessive and two dominant genes, *sh, Sh_2* and *Sh_3*, control spring habit; their respective alleles control winter habit.

DDT Resistance (ddt). Surprisingly, barley has a gene that confers susceptibility to DDT. This character has no apparent associated morphological characters, and the resistance allele occurs in about 10 percent of the world barley varieties. It has been used in one of the genetic techniques for the production of hybrid barley seed.

Enzyme Mutants (Est). Electrophoretic techniques have permitted mass screening for enzyme mutants in barley. Those most extensively studied are isozymes of esterase. The inheritance and distribution of these isozymes among barley varieties have been described by Allard *et al.* (1971) and by Nielsen and Frydenberg (1971). Genetic analyses have shown two isozymes which are controlled by very closely linked genes, *Est-1* and *Est-2*. Three different alleles have been found at each locus. The third esterase isozyme is controlled by a third locus, *Est-3*, and here three alleles have also been found. The genetic control of polymorphism in endosperm proteins has also been analyzed (Solari and Favret, 1971).

Susceptibility to Pests. Genetics of resistance to powdery mildew (*Ml*) has been the most thoroughly studied disease reaction. There appear to be at least 20 genes which determine the reaction to the pathogen *Erysiphe graminis hordei,* and most of these genes have been located on chromosome 5. For each of several host genes, there is a complementary gene for virulence in the pathogen. This gene-for-gene concept has been described in detail in numerous papers (see Moseman, 1971).

The genetics of reaction to the pathogens causing leaf rust (*Pa*), yellow dwarf (*Yd*), loose smut (*Un*), and scald (*Rh*) have received considerable attention. Genetics of reaction to green bug (*Grb*), cereal leaf beetle, and root nematode (*Ha*) have also been studied.

Meiotic and Sterility Mutants. Numerous meiotic and/or sterility mutants have been investigated. For instance, eight loci controlling the desynaptic chromosome character (*des*) have been identified.

With the advent of hybrid barley, male sterile genes have become important. Since the search for them was initiated a number of years ago, over 30 different mutants at no fewer than 24 loci have been obtained.

Literature Cited

Allard, R. W., A. L. Kahler and B. S. Weir, 1971   Isozyme polymorphisms in barley populations. In *Barley Genetics II,* edited by R. A. Nilan, pp. 1–13, Washington State University Press, Pullman, Wash.

Fedak, G., T. Tsuchiya and S. B. Helgason, 1972   Use of monotelotrisomics for linkage mapping in barley. *Can. J. Genet. Cytol.* **14**:949–957.

Kasha, K. J. and K. N. Kao, 1970   High-frequency haploid production in barley (*Hordeum vulgare* L.). *Nature (Lond.)* **225**:874–875.

Lamberts, H., editor, 1964   *Barley Genetics I.* Pudoc, Centre for Agricultural Publications and Documentation, Wageningen, The Netherlands.

Lundqvist, U., P. von Wettstein-Knowles and D. von Wettstein, 1968   Induction of *eceriferum* mutants in barley by ionizing radiations and chemical mutagens. II. *Hereditas* **59**:473–504.

Moseman, J. G., 1971   Co-evolution of host resistance and pathogen virulence. In *Barley Genetics II,* edited by R. A. Nilan, pp. 450–456, Washington State University Press, Pullman, Wash.

Munck, L., K-E. Karlsson and A. Hagberg, 1971   Selection and characterization of a high-protein, high-lysine variety from the World Barley Collection. In *Barley Genetics II,* edited by R. A. Nilan, pp. 544–558, Washington State University Press, Pullman, Wash.

Nielsen, G. and O. Frydenberg, 1971   The inheritance and distribution of esterase isozymes in barley. In *Barley Genetics II,* edited by R. A. Nilan, pp. 14–22, Washington State University Press, Pullman, Wash.

Nilan, R. A., 1964   *The Cytology and Genetics of Barley, 1951–1962,* Monograph Supplement 3, Research Studies Vol. 32, No. 1, Washington State University Press, Pullman, Wash.

Nilan, R. A., editor, 1971   *Barley Genetics II,* Washington State University Press, Pullman, Wash.

Persson, G. and A. Hagberg, 1969   Induced variations in a quantitative character in barley. Morphology and cytogenetics of erectoides mutants. *Hereditas* **61:** 145–178.

Ramage, R. T., C. R. Burnham and A. Hagberg, 1961   A summary of translocation studies in barley. *Crop Sci.* **1:** 277–279.

Robertson, D. W., 1971   Recent information of linkage and chromosome mapping. In *Barley Genetics II,* edited by R. A. Nilan, pp. 220–242, Washington State University Press, Pullman, Wash.

Robertson, D. W., G. A. Wiebe and F. R. Immer, 1941   A summary of linkage studies in barley, *J. Am. Soc. Agron.* **33:** 47–64.

Robertson, D. W., G. A. Wiebe and R. G. Shands, 1947   A summary of linkage studies in barley, Supplement I, 1940–1946. *J. Am. Soc. Agron.* **39:** 464–473.

Robertson, D. W., G. A. Wiebe and R. G. Shands, 1955   A summary of linkage studies in barley, Supplement II, 1947–1953. *Agron. J.* **47:** 418–425.

Robertson, D. W., G. A. Wiebe, R. G. Shands and A. Hagberg, 1965   A summary of linkage studies in cultivated barley, *Hordeum* species, Supplement III, 1954–1963. *Crop Sci.* **5:** 33–43.

Smith, L., 1951   Cytology and genetics of barley, *Bot. Rev.* **17:** 1–51, 133–202, 285–355.

Solari, R. M. and E. A. Favret, 1971   Polymorphism in endosperm proteins of barley and its genetic control. In *Barley Genetics II,* edited by R. A. Nilan, pp. 23–31, Washington State University Press, Pullman, Wash.

Takahashi, R. and S. Yasuda, 1971   Genetics of earliness and growth habit in barley. In *Barley Genetics II,* edited by R. A. Nilan, pp. 388–408, Washington State University Press, Pullman, Wash.

Tsuchiya, T., 1972   Revision of linkage map of chromosome 5 in barley by means of telotrisomic analysis. *J. Hered.* **63:** 373–375.

Wettstein, D. von, K. W. Henningsen, J. E. Boynton, G. C. Kannangara, and O. F. Nielsen, 1971   The genic control of chloroplast development in barley. In *Autonomy and Biogenesis of Mitochondria and Chloroplasts,* edited by N. K. Boardman, A. W. Linnane and R. M. Smillie, pp. 205–233, Amsterdam.

Wettstein-Knowles, P. von, 1971   The molecular phenotypes of the *eceriferum* mutants. In *Barley Genetics II,* edited by R. A. Nilan, pp. 146–193, Washington State University Press, Pullman, Wash.

5

Cotton (*Gossypium*)*

Lyle L. Phillips

Introduction

The genus *Gossypium* is comprised of about 35 diploid ($n = 13$) and tetraploid ($n = 26$) species which occur as elements of the native flora, or as cultigens, throughout the tropics and subtropics. The tetraploid taxa, two of which are cultivated, are the result of hybridization of two diploid species and subsequent amphidiploidy; indeed, the New World tetraploid cottons, with *Nicotiana tabacum* and *Triticum aestivum,* are classic examples of crop plant evolution through hybridization and natural amphidiploidy.

Cytotaxonomy and Distribution of *Gossypium* Diploids

The 31 diploid taxa have been categorized into genome groups on the basis of the cytology of interspecific hybrids (Beasley, 1942; Phillips and Strickland, 1966). Thus, interspecific, intragenomic hybrids have an average of 12.50–13.00 bivalents per cell and a chiasma frequency of 1.50–2.00 per bivalent at meiotic metaphase I (MI). Interspecific, intergenomic hybrids range from less than one to eleven bivalents per cell at MI, and the chiasma frequency ranges from 1.0 to 1.4 per bivalent. The species of *Gossypium,* their geographical distributions, and their genome symbols are listed in Table 1.

* This chapter has been published as paper no. 3756 of the Journal Series of the North Carolina State University Agricultural Experiment Station, Raleigh, North Carolina.

Lyle L. Phillips—Department of Crop Science, North Carolina State University, Raleigh, North Carolina.

TABLE 1. *The Species of Gossypium Linnaeus*

Species	Distribution	Genome designation[a]
G. *herbaceum* Linnaeus	Old World cultigen	A_1
G. *arboreum* Linnaeus	Old World cultigen	A_2
G. *anomalum* Wawra and Peyritch	Northern and southern Africa	B_1
G. *triphyllum* Hockreutener	Southern Africa	B_2
G. *barbosanum* Phillips and Clement	Cape Verde Islands	B_3
G. *capitis-viridis* Mauer	Cape Verde Islands	*
G. *sturtianum* Willis	Central to eastern Australia	C_1
var. *nandewarense* (Derera) Fryxell	Eastern Australia	C_{1-n}
G. *robinsonii* F. Mueller	Western Australia	C_2
G. *australe* F. Mueller	Northern Australia	*
G. *bickii* Prokhanov	North-central Australia	*
G. *costulatum* Todaro	Northwestern Australia	*
G. *populifolium* (Bentham) Todaro	Northwestern Australia	*
G. *cunninghamii* Todaro	Northern Australia	*
G. *pulchellum* (Gardner) Fryxell	Northwestern Australia	*
G. *thurberi* Todaro	Sonora, Mexico; Arizona	D_1
G. *armourianum* Kearney	Baja California, Mexico	D_{2-1}
G. *harknessii* Brandegee	Baja California, Mexico	D_{2-2}
G. *klotzschianum* Andersson	Galapagos Islands	D_{3-k}
var. *davidsonii* (Kellogg) Hutchinson	Sonora and Baja Calif., Mexico	D_{3-d}
G. *aridum* (Rose and Standley) Skovsted	West-central Mexico	D_4
G. *raimondii* Ulbrich	West-central Peru	D_5
G. *gossypioides* (Ulbrich) Standley	Oaxaca, Mexico	D_6
G. *lobatum* Gentry	Michoacan, Mexico	D_7
G. *trilobum* (Mocino and Sesse) Skovsted	West-central Mexico	D_8
G. *laxum* Phillips	Guerrero, Mexico	D_9
G. *stocksii* Masters ex Hooker	Arabia; Pakistan	E_1
G. *somalense* (Gurke) Hutchinson	Northeastern Africa	E_2
G. *areysinanum* (Deflers) Hutchinson	Arabian Penninsula	E_3
G. *incanum* (Schwartz) Hillcoat	Arabian Penninsula	E_4
G. *longicalyx* Hutchinson and Lee	Northeastern Africa	F_1

TABLE 1.   Continued

Species	Distribution	Genome designation[a]
G. hirsutum Linnaeus	New World cultigen	(AD)$_1$
G. barbadense Linnaeus	New World cultigen	(AD)$_2$
G. tomentosum Nuttall ex Seemann	Hawaiian Islands	(AD)$_3$
G. caicoense Aranha, Filho, and Gridi-Papp	Northeastern Brazil	(AD)$_4$

[a] An asterisk indicates that the genome designation has not been determined cytologically.

Four of the six genome groups (A,B,E, and F) are found in Africa. The two A-geonome species, *G. herbaceum* and *G. arboreum,* have in years past been commercially important cultigens* in the Mediterranean area and in Asia. At the present time, *G. herbaceum* exists in the wild in southern Africa; whether it is truly wild or feral is a matter of some controversy. *G. arboreum* is unknown in the wild and is assumed to have arisen from *G. herbaceum* under cultivation. The two A-genome species differ cytologically by a major reciprocal chromosome translocation, the *G. herbaceum* × *G. arboreum* hybrid typically showing a quadrivalent at meiotic MI (Gerstel, 1953).

The B-genome group includes three species. *G. anomalum* is a wide-ranging species with a large distributional area in sub-Sahara Africa and a smaller range in southern Africa. *G. triphyllum* is distributed in southwestern Africa, and *G. barbosanum* is found on one island of the Cape Verde archipelago. *G. capitis-viridis* will be an enigma until it can be introduced into garden culture and its hybrids studied cytologically. This species is sympatric with B-genome *G. barbosanum,* but the type description of *G. capitis-viridis* is more suggestive of *G. herbaceum* (yellow corolla, no petal spot, 5-loculate capsule) of the A-genome than of any of the B-genome species (which have cream petals, large petal spots, and 3-loculate capsules).

The species of the E genome are distributed from extreme northeastern Africa (*G. somalense*) through the Arabian peninsula (*G. incanum, G. areysianum,* and *G. stocksii*) to Pakistan (*G. stocksii*). The northern African species, *G. longicalyx,* was tentatively placed by its authors in the E-genome group. Phillips and Strickland (1966), on the basis of cytological studies of the *G. hirsutum* × *G. longicalyx* hybrid,

* A cultigen is a cultivated species, strain, form, stock, variety, etc.

concluded that *G. longicalyx* represented a novel cytotype not referable to any established genome group; *G. longicalyx* is thus the only species of the F genome.

The two non-African genome groups are found in Australia (C genome) and in the Americas (D genome). When Beasley (1942) established the genome groupings of *Gossypium*, there were two closely related species representing the C genome, *G. sturtianum* and *G. robinsonii.* Fryxell's (1965) monograph on Australian *Gossypium* lists six additional species, only two of which, the closely related *G. australe* and *G. bickii*, are in garden culture. The cytology of a hybrid between *G. australe* and *G. robinsonii*, which exhibits 3.84 univalents and 11.08 bivalents per cell and shows a chiasma frequency of 1.48 per bivalent (Phillips, unpublished), indicates that *G. australe* (with *G. bickii*) represents a new cytotype (genome grouping). However, a formal proposal to this effect must await a comprehensive cytological study of the Australian species.

The D-genome group is the largest (11 taxa) of the *Gossypium* cytotypes. The constituent species of this group are basically of Mexican distribution, but taxa of the D genome are also represented in Arizona (*G. thurberi*), Peru (*G. raimondii*), and the Galápagos archipelago (*G. klotzschianum*).

Cytological studies of the interspecific hybrids of the diploid species of *Gossypium* have been in progress for 35 years and now include crosses representing all of the possible intergenomic combinations (see Phillips, 1966, for summary). Table 2 contains a summary of phylogenetic relationships among *Gossypium* genome groups based on average univalent frequencies for intergenomic F_1 hybrids. The theory that phyletic affinity is directly correlated to reduction of chromosome pairing in hybrids rests on the assumptions that reduced pairing is the result of chromosome structural divergence and that such structural evolution is progressive and cumulative over time. Genetic asynapsis or desynapsis has not been reported for the hybrids summarized in Table 2. The F-genome species, *G. longicalyx*, has been successfully crossed to only one other diploid, this being *G. raimondii* of the D-genome group. However, Phillips and Strickland (1966), on the basis of cytological studies of the hybrid between *G. hirsutum* and *G. longicalyx*, concluded that the F genome is more closely related to the A genome than to other cytotypes.

Gossypium is one of many plant groups, including such other Malvaceous genera as *Cienfugosia* and *Hibiscus*, that have distributions on two or more continents (Fryxell, 1969; Good, 1947; Menzel and Martin, 1971). It is thus apparent that the early evolution of *Gossypium* antedated the late Triassic–early Jurassic geotectonic fracturing of the su-

TABLE 2.   *Summary of Average Univalent Frequency for Diploid Intergenomic Hybrids of Gossypium*

Hybrid type	Univalents per cell	Chiasmata per bivalent
A × B	2.82	1.44
A × C	8.50	1.19
A × D	11.70	1.10
A × E	17.13	1.08
B × C	11.17	1.11
B × D	18.19	1.07
B × E	22.35	—
C × D	11.28	1.16
C × E	24.68	1.01
D × E	25.15	1.00
D × F	21.60	1.02

percontinent of Pangaea (Dietz and Sproll, 1970; Dietz, 1971), and that continental drift resulted in the present distribution of the *Gossypium* species to all continents that extend into tropical or subtropical areas.

Origin and Distribution of *Gossypium* Tetraploids

The *Gossypium* tetraploids, a group of extreme taxonomic and nomenclatural complexity, originated in the New World tropics and are presently distributed, as both wild and cultivated biotypes, throughout the arid, frostfree regions of the world. Most *Gossypium* specialists recognize four taxa at the specific level: (1) *G. hirsutum* includes the Upland cotton cultivars (providing 90 percent of commercial cotton fibers) and several wild and/or feral entities. Fifty to sixty published epithets (Smith, 1964) have been applied to types referable to *G. hirsutum*. (2) *G. barbadense* includes cultivars grown in the Nile river drainage area (Egyptian cotton), in Peru (Tanguis cotton), and in restricted areas of the southwestern United States (Pima cotton). Together, these account for about 10 percent of world commercial cotton production. In addition, there are reasonably distinct wild *G. barbadense* biotypes in the Galápagos Islands (var. *darwinii*) and near the Gulf of Guayaquil in Peru and Ecuador. (3) The wild *G. tomentosum* grows on several islands of the Hawaiian archipelago. (4) Wild *G. caicoense* found in Rio Grande do Norte, Brazil.

It has long been known that the New World cottons are natural amphidiploids which contain a set of chromosomes from one of the two A-genome species and a chromosome set from a species of the D-genome group (Beasley, 1940; Harland, 1940; Skovsted, 1934, 1937). Gerstel (1953) has shown that the A genome of the natural amphidiploid differs from *G. herbaceum* by two reciprocal translocations and from *G. arboreum* by three translocations, thus indicating that *G. herbaceum* contributed the A genome to New World cotton (and indicating also that two additional reciprocal translocations have occurred in the amphidiploid since its origin).

Cytogenetic studies (Phillips, 1962, 1963) and seed-protein electrophoresis studies (Cherry *et al.,* 1970; Johson and Thein, 1970) have indicated that *G. raimondii* is the most closely related of the known D species to the contributor of the New World D genome. This species is found on the Pacific foothills of the Andean Cordillera in Peru.

With the identification of the species (or their close relatives) that contributed the A and D genomes to the New World amphidiploid, the "hard facts" relative to the origin of the amphidiploid are exhausted. Phillips (1962, 1964) reported that hexaploids synthesized of hybrids between the New World tetraploids and species of the A-genome group gave near autoploid segregation ratios for marker genes located on the A genome and showed chromosome multivalent frequencies characteristic of autoploids. Such data, indicating little or no chromosome structural differentiation (other than major segmental interchanges) between the A genome of the New World tetraploid and its donor species, augur for a relatively recent origin of the natural amphidiploid, i.e., since the start of Pleistocene.

The "where" of amphidiploid origin is even more conjectural than the "when." A necessary precondition for the origin of the amphidiploid is a New World distribution of the A-genome donor, and this assumes oceanic drift dispersal of an A-genome propagule from Africa. Extrapolating on the basis of present-day African *Gossypium* distributions and ocean currents, migrule drift* would have to have been *G. herbaceum* var. *africanum* from southern Africa via the South Atlantic Equatorial Current or *G. capitis-viridis* from the Cape Verde Islands via the North Atlantic Equatorial Current. (The possibility that *G. capitis-viridis* is allied to *G. herbaceum* is based on the type description and observations of the type-specimen photograph; *G. capitis-viridis* is not in culture and its cytotype is unknown.) Both of the Atlantic Equatorial Currents flow along the North Atlantic and Caribbean coasts of South and Central America, and a

---

* Migrule drift is the over-water dispersal (or migration) of seeds or rootstocks.

pre-Pleistocene distribution of an A-genome taxon in this area seems probable. Sympatry of both the A- and D-genome contributors of the New World amphidiploid, assuming an A-genome donor distribution near the Caribbean coast of South or Central America, requires a distribution of the D-genome donor some 1200–1500 miles north of the present range of *G. raimondii*. A consideration of all available evidence indicates that the origin of the New World amphidiploid probably took place in northwestern South America or Mesoamerica.

There have been proposals that man introduced an A-genome cultigen via the Pacific (Hutchinson *et al.*, 1947) or the Atlantic (Schwerin, 1970) to the New World where, under cultivation, it hybridized with one or more D species. Since *G. arboreum* is the only A-genome species ever cultivated in Asia, the possibility of the Pacific introduction of the A-genome species to the New World has not been seriously-entertained since Gerstel (1953) indicated that *G. herbaceum,* and not *G. arboreum,* was involved in the origin of the New World tetraploid. The arguments against a recent introduction of a *G. herbaceum* cultigen across the Atlantic by man are based on the chronologies involved. The earliest known date for archeological New World cottons is 4000–3000 B.C. (Bird and Mahler, 1951; Smith and MacNeish, 1964); this antedates the earliest known archeological cotton in Africa by 2000 years, the earliest date for West African agriculture by 1000 years, and the earliest date for long-distance ocean voyages in the African area by 3000 years (Schwerin, 1970). Another "time-interval" aspect of any recent-origin hypothesis is the improbability that a few thousand years is sufficient time for the post-origin evolution of at least four species, each exhibiting considerable genetic and cytological diploidization (Beasley, 1942; Endrizzi, 1962; Rhyne, 1962c, 1965), and the dispersal of one of them to the Hawaiian Islands (Stephens, 1964).

Both oceanic drift and man are responsible for the pan-tropic distribution of the New World amphidiploid species. *G. tomentosum* is unique among the New World species in that its seed fibers are undifferentiated (other New World cottons have seed fibers differented into lint and fuzz layers), a primitive feature which it shares with the *Gossypium* wild diploids, and, perhaps, also a characteristic of the ancestral amphidiploid. Stephens (1966) has shown that the seeds of *G. tomentosum* will retain viability for up to 9 months in salt water, and he explains the Hawaiian endemism of this species on the basis of oceanic drift.

The center of variability for *G. hirsutum* is in Mesoamerica and the Caribbean, with wild forms distributed throughout Polynesia and Micronesia. The seeds of some wild *G. hirsutum* collections have sufficient salt-water tolerance (Stephens, 1966, 1967) to enable them to "island-hop" in the Caribbean, but Stephens (1963) suggests that Spanish trans-Pacific shipping during the 16th and 17th centuries had a major role in the Pacific distribution of *G. hirsutum.*

The *G. barbadense* center of variability is in northern South America. Connate-seeded forms (*G. barbadense* var. *brasiliense*) predominate east of the Andes, free-seeded forms to the west. The Galápagos Island form of *G. barbadense* (var. *darwinii*) is probably the result of oceanic drift (Stephens, 1966) from the Peru–Ecuador coast, where similar forms are found.

Cotton has a long history of cultivation in the New World. Archeological cotton, probably *G. barbadense* and dated to 2500 B.C., has been found at Hauca Prieta in Peru (Bird and Mahler, 1951). Another Peruvian site has yielded cotton fiber, seeds, and capsule fragments (2500 B.C.) that have been identified as *G. barbadense* (Stephens and Moseley, 1973); the primitive nature of these plant remains (small seeds and capsules and brown lint) suggest that they represent an early stage in the domestication of cotton in South America. Cotton capsules and fiber, probably *G. hirsutum,* have been recovered from the Tehuacan Valley of Mexico and dated from 3500–2300 B.C. (Smith and MacNeish, 1964). A later record, this positively identified as *G. hirsutum,* has been found in the Oaxaca Valley of Mexico (Smith and Stephens, 1971). Even the oldest of these Mexican archeological records probably do not represent original sites of cotton domestication, however, since they appear to be fully cultivated types and were recovered with other domesticated crop plants.

Gossypium Cytodifferentiation

The cytology and genetics of *Gossypium* have been intensively studied during the last half-century, especially in the last 35 years. The number of "journal-type" articles on *Gossypium* cytogenetics and qualitative genetics approximates 800; this literature, and also that concerned with cotton breeding and quantitative genetics, has been bibliographed for the years 1900–1950 by Knight (1954), and for 1950–1964 by Murray and Verhalen (1965).

Cytology of Interspecific Hybrids: Diploid × Diploid

The cytogenetics of interspecific hybrids of the diploid species of *Gossypium* has been persued rather intensively for 30 years. In that time, about one third of the possible interspecific combinations have been synthesized and studied. Synthesis of another third of the possible hybrids has been attempted, but success has been blocked by embryonic abortion, (Brown, 1951; Oakes, 1966a,b; Skovsted, 1935a; Weaver, 1957, 1958) or postgerminative lethality (Brown and Menzel, 1952b; Phillips and Merritt, 1972). Attempts to obtain the remaining third of the possible combinations have not been reported, in part because some diploid species have only recently been brought under culture. Studies of interspecific hybrids have primarily involved assessment of chromosome pairing and chiasma frequency at the diploid level (Beasley, 1942; Brown and Menzel, 1952a; Douwes, 1951, and 1953; Endrizzi and Phillips, 1960; Iyengar, 1944; Phillips, 1961, 1963, 1966; Phillips and Clement, 1963; Saunders, 1961; Skovsted, 1934, 1935a,b, 1937), but some studies have also included multivalent frequency and marker gene segregation of colchicine-induced allotetraploids (Gerstel, 1956; Gerstel and Phillips, 1957; Sarvella, 1958). Table 2 contains a summary of cytogenetical data of intergenomic hybrids of *Gossypium* diploids.

Cytology of Tetraploid × Diploid Interspecific Hybrids

The New World tetraploids, *G. hirsutum* or *G. barbadense*, have been successfully crossed with all but four of the diploids, including species from all genome groups (Beasley, 1942; Boza Barducci and Madoo, 1941; Brown and Menzel, 1952a; Endrizzi, 1957; Gerstel, 1955; Phillips, 1962, 1963, 1964, 1966; Phillips and Gerstel, 1959; Phillips and Strickland, 1966; Sarvella, 1958; Skovsted, 1934; Stephens, 1945). Most of the studies cited above have treated the hybrids at both the triploid level (chromosome pairing and chiasma frequency) and the synthetic hexaploid level (multivalent frequency and marker-gene segregation). Table 3 is a summary of cytogenetical data for triploids and synthetic hexaploids of New World × diploid hybrids.

Comparative Genetics

Before colchicine came into general use, studies on *Gossypium* comparative genetics were confined to the fertile F_1 hybrids (and derivatives) of *G. hirsutum* × *G. barbadense* (Harland, 1929, 1932, 1933, 1934; Hutchinson, 1946; Knight, 1944, 1945), *G. herbaceum* × *G. arboreum* (Hutchinson, 1931; Silow, 1941, 1944; Skovsted, 1935b), and *G. herbaceum* or *G. arboreum* × *G. amonalum* (Silow, 1941, 1944). The F_2 and later generations of *G. hirsutum* × *G. barbadense* hybrids exhibit "hybrid breakdown," and only those individuals which resemble one or the other of the parental species can be stabilized by selection. Based on his studies with *G. hirsutum* × *G. barbadense*, Harland (1933, 1936a) proposed that speciation in *Gossypium* consisted of a step-by-step accumulation of major-gene modifier complexes; according to this theory, hybrid break-

TABLE 3. *Summary of Cytogenetical Data for Triploids and Synthetic Hexaploids of New World* × *Diploid Hybrids*

	Triploid		Hexaploid	
Hybrid	*Univalents per cell*	*Chiasmata per bivalent*	*Multivalents per cell*	*Segregation, percent recessives[a]*
AD × A	13.20	1.89	7.15	15.4
AD × B	26.91	1.07	0.91	0.4
AD × C	26.81	1.03	0.75	0.3
AD × D$_1$	13.56	1.83	3.61	3.0
AD × D$_{2-1}$	13.94	1.76	3.96	5.5
AD × D$_{2-2}$	13.55	1.75	3.65	5.5
AD × D$_3$	13.44	1.79	3.96	4.8
AD × D$_4$	13.35	1.86	4.48	4.5
AD × D$_5$	13.07	1.95	6.16	9.6
AD × D$_6$[b]	20.95	1.40	1.13	1.4
AD × D$_7$	13.21	1.89	3.66	4.0
AD × D$_8$	13.44	1.76	—	—
AD × D$_9$	13.40	1.82	—	—
AD × E	37.90	1.00	0.19	—
AD × F	22.25	1.17	2.00	—

[a] Expected percent recessives from autoploid testcross is 16.6.
[b] Partially asynaptic.

down resulted from the disruption of coadapted modifier complexes. Silow (1941, 1944) invoked a similar scheme to account for speciation of *G. herbaceum*, *G. arboreum*, and *G. anomalum*.

Stephens (1950*b*), in a general review and reexamination of comparative genetics research data, concluded that cryptic chromosomal structural differentiation constituted a more important factor in species evolution than multiple gene substitution, and suggested several genetic tests for cryptic cytodifferentiation: (1) skewed ratios for marker genes in backcross progenies, (2) tighter linkages for marker genes in F_2 and testcross progenies of inter- as compared to intraspecific crosses, and (3) fewer recessives in testcrosses of synthetic allopolyploids as compared to autopolyploids. These genetic procedures have been applied to *G. herbaceum* × *G. arboreum* hybrids (Gerstel and Phillips, 1957, 1958; Phillips, 1961) and to *G. hirsutum* × *G. barbadense* hybrids (Lewis and MacFarland, 1952; Loden, 1951; Stephens, 1949). Collectively, these studies demonstrated significant reduction in linkage values in hybrid testcrosses. The linkage studies, however, usually involved relatively short segments of chromosome length; in no case was more than a single chromosome arm involved. Stephens (1961) has shown, using a five-point linkage group (involving most of the chromosome map length) in a *G. hirsutum* × *G. barbadense* hybrid, that reduction in crossing over one segment of a chromosome is balanced by a compensatory increase in crossing-over in other segments of the mapped chromosome.

The utilization of chochicine-induced "gene-bridges" (commonly the synthetic amphidiploid of A-genome × D-genome hybrids) have permitted studies on interspecific gene homologies (Giles, 1962; Harland, 1935; Harland and Atteck, 1941; Silow, 1946; Stephens, 1954a,b) and linkage-group homologies (Giles, 1961; Rhyne, 1958, 1960, 1962a) in New World × A-genome and New World × D-genome hybrids of much the same scope and implication as those studies described above that involved hybrids within the New World and Asiatic groups.

Qualitative Genetics

The genetics of mutant and marker loci of *Gossypium* has been extensively studied for 60 years, and the number of research papers on this subject exceeds 200 (Knight, 1954; Murray and Verhalen, 1965). A major share of this literature have been concerned with *G. hirsutum*, in which 10 linkage groups are now recognized (Kohel, 1972; Kohel *et al.*, 1965; Stephens, 1955); six of these linkage groups have been associated with specific chromosomes on the basis of translocation or aneuploid analysis, and two additional linkage groups have been assigned to either the A or D genome (Endrizzi, 1963; Endrizzi and Brown, 1962; Poisson, 1968; White and Endrizzi, 1965). The genetics of mutant and marker genes in *G. barbadense* has received less study than in *G. hirsutum*. Some 20–25 mutants and markers, including two linkages, have been studied and symbolized in *G. barbadense*. Table 4 contains a selected list of mutant and marker loci, and linkage data, for *G. hirsutum* and *G. barbadense*.

Developmental Genetics

Haploidy

The association of haploidy and polyembryony in *G. hirsutum* and *G. barbadense* has been known and studied for a number of years (Blank and Allison, 1963; Endrizzi, 1959; Harland, 1936b; Kimber, 1958; Owings *et al.*, 1964; Silow and Stephens, 1944; Turcotte and Feaster, 1963; Webber, 1938). In twin embryo seeds, one embryo is characteristically diploid and the other is haploid. The frequency of polyembryony is

TABLE 4.   *Selected List of Marker and Mutant Genes in G.* hirsutum *and G.* barbadense

Symbol[a]	Character	Reference	Linkage group [b]
bw_1	Withering bract ⎱	Knight, 1952	V
bw_2	Withering bract ⎰		IX
cl_1	Cluster	Hutchinson and Silow, 1939	III
cl_2	Cluster	Silow, 1946	I
cr	Crinkle leaf	Knight, 1952	II
cu	Cup leaf	Lewis, 1954	—
Dw	Dirty white	Rhyne, 1957	III
fg	Narrow bract	Green, 1955	VI
gl_1	Glandless stem	McMichael, 1954	—
gl_2	Glandless leaf ⎱	Rhyne, 1962a, b	V
gl_3	Glandless leaf ⎰		IX
H_1	Hairy stem ⎱	Knight, 1952	—
H_2	Pilose stem ⎰		IV
ia	Accessory involucre	Kohel, 1965	VI
L	Narrow leaf ⎱	Hutchinson and Silow, 1939	II
Lc_1	Brown lint ⎰		I
Lc_2	Brown lint	Ware, 1932	IV
Lg	Green lint	Hutchinson and Silow, 1939	II
Li	Ligon lintless	Kohel, 1972	—
lp_1	Abnormal palisade ⎱	Kohel, 1964	VII
lp_2	Abnormal palisade ⎰		II
ml	Mosaic leaf	Lewis, 1958	VIII
ms_3	Male sterile	Justus et al., 1963	III
mt	Mottled leaf	Lewis, 1960	—
N	Naked seed	Hutchinson and Silow, 1939	—
ne_1	Nectariless leaf ⎱	Meyer and Meyer, 1961	V
ne_2	Nectariless leaf ⎰		IX
P_1	Yellow pollen	Hutchinson and Silow, 1939	—
P_2	Orange pollen	Turcotte and Feaster, 1966	—
R_1	Red plant body ⎱		III
R_2	Petal spot ⎬	Hutchinson and Silow, 1939	I
Rd	Red dwarf ⎰		—
Rg	Ragged leaf	Kohel and Lewis, 1962	X
rl	Round leaf	Kohel, 1972	X
rs	Rudimentary style	Turcotte and Feaster, 1964	—
Ru	Rugate leaf	Turcotte and Feaster, 1965	—
Sm	Smooth leaf	Meyer, 1957	—
st_1	Stippled leaf	Kohel, 1972	VIII
v	Virescent yellow ⎱	Hutchinson and Silow, 1939	—
Y	Yellow petal ⎰		—
yg_1	Yellow green ⎱	Rhyne, 1954	III
yg_2	Yellow green ⎰		I

[a] The mutant form of the locus, as contrasted with the "normal" condition, is listed in each entry.
[b] Kohel, 1972.

1:10000–1:25000 in *G. hirsutum* (Blank and Allison, 1963), and as high as 1:300–1:500 in some lines of *G. barbadense* (Owings *et al.*, 1964). It has been shown (Blank and Allison, 1963) that the selfed progeny of the diploid member of a twin embryo exhibits a significantly higher frequency of twin-embryo seeds (from 1:13490 in the original seed to 1:2639 for selfed diploid twin). Haploidy has been reported in monoembryonic seeds in a doubled haploid of *G. barbadense* at the very high frequency of about 60 percent (Turcotte and Feaster, 1963), and breeding tests have shown that the haploid-producing tendency is simply inherited. Numerous colchicine-induced diploids have been derived from *G. hirsutum* and *G. barbadense* haploids, including a large number in commercial cultivars.

Semigamy

Semigamy, a form of apomixis in which the sperm nucleus enters the egg but fails to fuse with the egg nucleus, has been demonstrated in *G. barbadense* (Turcotte and Feaster, 1967). In this semigametic system, the seed and the plant derived from it consist of maternal and paternal chimeras derived independently from the sperm and egg nuclei. In about 75 percent of the chimeral plants both maternal and paternal tissues are haploid, but in the remaining 25 percent each sector is diploid, this presumably resulting from the formation of restitution nuclei at the initial division in both male and female cell lines.

Interspecific Incompatibility

Developmental, abnormalities, affecting both pre- and postgermination stages of plant differentiation, are rather common among interspecific hybrids of *Gossypium*. The extreme variability of expression among interspecific combinations does not readily submit to a rational categorization, but for convenience two general types will be recognized: (1) abnormal development conditioned by the complementary effect of simply inherited genetic factors and (2) physiological–biochemical interaction of maternal and zygotic tissues resulting in embryo abortion.

Simply inherited interspecific incompatibility has been demonstrated in the hybrids of *G. hirsutum* with *G. arboreum* var. *sanguineum* (Gerstel, 1954), *G. klotzschianum* (Phillips, 1963), *G. gossypioides* (Phillips, 1962), and *G. barbadense* (Stephens, 1946, 1950a). It has been shown that hybrid necrosis in the *G. hirsutum* × *G. gossypioides* combination results from stem-tumor formation and ultimate inactivation of the vascular cambium (Phillips and Merritt, 1972). The *G. hirsutum* × *G. barbadense* example above involves the "corky" alleles ck^x (known only in *G. hirsutum*) and *ck^y* (known only in *G. barbadense*); when heterozygous, these alleles condition a semilethal syndrome characterized by corky stems and petioles, leaf mottling, and partial sterility. The *ck^y* allele is widespread in populations of *G. barbadense*, but the *ck^x* allele is restricted to a few West Indian islands and apparently originated in mixed plantings of *G. hirsutum* and *G. barbadense* that were common in the Caribbean in the 18th century (Stephens and Phillips, 1972).

In addition to the postgermination semilethals described above, there is a large number of interspecific hybrids that abort during embryogenesis. Most of the hybrids involving *G. klotzschianum* and *G. gossypioides* fall into this category; these are assumed to be genetic lethals, in part because postgermination semilethals of these species are known to have a genetic basis, and also because their embryos cannot be cultured successfully on artificial medium.

The only definitive studies on the etiology of the type of arrested embryogenesis that I have classed as "physiological–biochemical" are those of Weaver (1957, 1958) on the histology of embryo development in the hybrid of *G. hirsutum* × *G. arboreum,* and the reciprocal cross. When *G. hirsutum* is the pistillate parent of this hybrid, the endosperm becomes cellular precociously and ceases normal development within 5 or 6 days following fertilization; the embryo degenerates soon thereafter. Normal embryogenesis and germination is readily achieved on a culture medium if the embryos of this hybrid are removed from their ovular environment before degeneration commences. When diploid *G. arboreum* is the pistillate parent of this hybrid, the endosperm develops rapidly without cell-wall formation, and though the embryo does not differentiate normally, its persists for 15–20 days following fertilization and achieves one-fifth normal size. The maternal tissues of the "seed" grow at a normal rate and often reach full size. Embryos of this hybrid are rarely culturable on an artificial medium. The embryogenesis of the *G. arboreum* × *G. hirsutum* hybrid is typical of all heteroploid crosses in which the pistillate parent is of lower ploidy level than the staminate parent ($2X \times 4X$, $4X \times 6X$, etc.).

Chromosome Interchanges in *Gossypium*

The cytological investigations of several researchers (Beasley, 1942; Iyengar, 1944; Silow, 1946) showed that hybrids of the New World tetraploids and the A-genome species exhibited mutivalents at meiotic metaphase I (2 quadrivalents for the hybrid with *G. herbaceum*, 1 quadrivalent and 1 hexavalent for the *G. arboreum* hybrid), but it was Gerstel (1953) and Gerstel and Savella (1956) that first concluded that these multivalents were indicative of translocation heterozygosity. They accounted for the segmental interchanges by assuming that one reciprocal translocation occurred and became established during the differentiation of *G. arboreum* (from *G. herbaceum*), and that two reciprocal translocations occurred in the A genome of the New World amphidiploid; one of the latter interchanges must have involved a chromosome homologous with a chromosome of the *G. arboreum* translocation to account for the hexavalent in hybrids of New World × *G. arboreum*. Menzel and Brown (1954b) and Menzel (1955) studied these segmental interchanges in three-species hybrids [e.g., *G. hirsutum* × (*G. arboreum* × *G. thurberi*)], and in serially numbering the five chromosomes involved in the three translocations, initiated the system of nomenclature that has been consistently used for *G. hirsutum* chromosome identification. Chromosome numbers 1–13 were reserved for the A genome, chromosomes 14–26 being allocated to the D genome. Several hundred segmental interchanges, derived both from hybrid and irradiated material, have been added to this base of naturally occurring translocations (Brown, 1949; Endrizzi and Brown, 1962; Menzel and Brown, 1952, 1954b; Tiranti and Endrizzi, 1963; Wilson and Kohel, 1970). These stocks, which are maintained at the Texas Agricultural Experiment Station, include translocations involving all except one or two of the chromosomes of the *G. hirsutum* complement.

Aneuploidy in *Gossypium*

New World species, being tolerant of chromosome deficiencies and duplications because of their amphidiploid origin, are very favorable material for the recovery and study of aneuploids. Also, the allohexaploids which can be synthesized from New World × diploid species hybrids provide a rich source of aneuploids in addition to those occurring within the New World species. The origin and study of aneuploids have been largely restricted to *G. hirsutum* and its hybrids (Beasley and Brown, 1943; Brown, 1949, 1966; Menzel and Brown, 1954a), although in recent years, aneuploids (primarily monosomics) in *G. barbadense* have drawn some attention (Endrizzi, 1966).

Monosomy

The type of aneuploidy that has received by far the most research attention in *Gossypium* is monosomy (Brown, 1966; Brown and Endrizzi, 1964; Douglas and Brown, 1971; Endrizzi, 1963; Endrizzi and Brown, 1964; Endrizzi and Taylor, 1968; Endrizzi *et al.*, 1963; Kammacher,

1968; Kammacher and Schwendiman, 1967; Kammacher *et al.*, 1957; Poisson, 1968). Approximately 150 monosomic lines have been isolated from inter- and intraspecific hybrids, commercial varieties, wild strains, irradiated material, and various types of aneuploid stocks. About one half of these monosomic lines have been identified as to the chromosome involved; identified chromosomes are 1, 2, 4, 6, and 7 of the A genome and 16, 17, and 18 of the D genome. As can be discerned, there is a tendency for A chromosomes to be involved in monosomy more frequently then D chromosomes; also, two A chromosomes (2 and 4) are recovered as monosomes at greater than random frequency (Brown and Endrizzi, 1964). The frequency of transmission of the monosomic condition from one generation to the next is 20–35 percent through the pistillate parent; nil through the staminate parent. The monosomic lines now available are being exploited to produce interspecific chromosome substitution lines (White *et al.*, 1967). Thus, a *G. hirsutum* monosomic is crossed, as female, to a standard *G. barbadense* stock. Then, five or six generations of backcrossing to the monosomic parent, selecting at each generation of monosomy, produces nearly pure *G. hirsutum* monosomic for a *G. barbadense* chromosome, which, when selfed, results in a true-breeding substitution line. Lines with the *G. hirsutum* background genotype are being developed at Texas Agricultural Experiment Station, whereas the *G. barbadense* counterpart lines are being synthesized at the University of Arizona.

Trisomy

Trisomics occur spontaneously in *G. hirsutum* and *G. barbadense* (Endrizzi *et al.*, 1963), but are recovered more frequently in the backcrosses of the synthetic auto-octoploid of *G. hirsutum* and synthetic allohexaploid hybrids of *G. hirsutum* and the wild diploids (Brown, 1966). Intra-*hirsutum* trisomics exhibit trivalents in about 50 percent of the cells at meiosis; the trivalent frequency for trisomic derivatives of hybrids ranges from 50 percent (when the extra chromosome is contributed by the A-genome species) to nil, depending on the residual homology of the diploid-species chromosome involved and its *G. hirsutum* homeologue. Tetrasomics are rather easily derived by selfing trisomics, and their cytology reflects the pairing observed in the parental trisomic, ranging from 25 II 1 IV to 27 II (Brown, 1966).

Telocentrics

G. hirsutum lines with telocentric chromosomes derived from both the long and short arms of chromosome 6, from the long arm of chromosome 15, and with an isochromosome from the long arm of chromosome 6 have been reported (Endrizzi and Kohel, 1966). All of these lines were isolated from monosomic or trisoimic material, apparently by misdivision of the unpaired univalent (Brown, 1958). These lines have been useful in associating marker genes with the chromosomes involved.

Literature Cited

Beasley, J. O., 1940   The origin of American tetraploid *Gossypium* species. *Am. Nat.* **64**:285–286.

Beasley, J. O., 1942   Meiotic chromosome behavior in species, species hybrids, haploids and induced polyploids of *Gossypium*. *Genetics* **27**:25–54.

Beasley, J. O. and M. S. Brown, 1943   The production of plants having an extra pair of chromosomes from species hybrids of cotton. *Rec. Gen. Soc. Am.* **12**:43.

Bird, J. and J. Mahler, 1951   America's oldest cotton fabrics. *Am. Fabrics* **20**:73–78.

Blank, L. M. and D. C. Allison, 1963   Frequency of polyembryony in certain strains of *Gossypium hirsutum* L. *Crop Sci.* **3**:97–98.

Boza Barducci, T. and R. M. Madoo, 1941   Investigaciones acerca del parentiesco de las especie peruana de algodonero *Gossypium raimondii* Ulbrich. *Min. Fom. Direct Agric. Ganaderia Bol.* **22**:1–29.

Brown, M. S., 1949   Polyploids and aneuploids derived from species hybrids in *Gossypium*. Proceedings of the Eighth International Congress of Genetics, *Hereditas*, Suppl. Vol., 543–545.

Brown, M. S., 1951   The spontaneous occurrence of amphidiploidy in species of *Gossypium*. *Evolution* **5**:25–41.

Brown, M. S., 1958   The division of univalent chromosomes in *Gossypium*. *Am. J. Bot.* **45**:24–32.

Brown, M. S., 1966   Attributes of intra- and interspecific aneuploidy in *Gossypium*. In *Chromosome Manipulations and Plant Genetics*, edited by R. Riley and C. F. Lewis, pp. 98–112, Oliver and Boyd, Edinburgh, Scotland.

Brown, M. S. and J. E. Endrizzi, 1964   The origin, fertility and transmission of monosomics in *Gossypium*. *Am. J. Bot.* **51**:108–115.

Brown, M. S. and M. Y. Menzel, 1952a   Polygenomic hybrids in *Gossypium*. I. Cytology of hexaploids, pentaploids, and hexaploid combinations. *Genetics* **37**:242–263.

Brown, M. S. and M. Y. Menzel, 1952b   The cytology and crossing behavior of *Gossypium gossypioides*. *Bull. Torrey Bot. Club* **79**:110–125.

Cherry, J. P., F. R. H. Katterman and J. E. Endrizzi, 1970   Comparative studies of seed proteins of species of *Gossypium* by gel electrophoresis. *Evolution* **24**:431–447.

Dietz, R. S., 1971   Plate tectonic evolution of Caribbean-Gulf of Mexico region. *Nature (Lond.)* **232**:20–23.

Dietz, R. S. and W. P. Sproll, 1970   Fit between Africa and Antarctica: a continental drift reconstruction. *Science (Wash., D.C.)* **167**:1612–1614.

Douglas, C. R. and M. S. Brown, 1971   A study of triploid and $3x - 1$ aneuploid plants in the genus *Gossypium*. *Am. J. Bot.* **58**:65–71.

Douwes, H., 1951   The cytological relationships of *Gossypium somalense* Gurke. *J. Genet.* **50**:179–191.

Douwes, H., 1953   The cytological relationships of *Gossypium areysianum* Deflers. *J. Genet.* **53:**611–624.

Endrizzi, J. E., 1957   The cytology of two hybrids of *Gossypium*. *J. Hered.* **48:**221–226.

Endrizzi, J. E., 1959   Cytogenetics for four sets of twins in cotton. *J. Hered.* **50:**222–226.

Endrizzi, J. E., 1962   The diploid-like cytological behavior of tetraploid cotton. *Evolution* **16:**325–329.

Endrizzi, J. E., 1963   Genetic analysis of six primary monosomes and one tertiary monosome in *Gossypium hirsutum*. *Genetics* **48:**1625–1633.

Endrizzi, J. E., 1966   Use of haploids in *Gossypium barbadense* L. as a source of aneuploids. *Curr. Sci. (Bangalore)* **35:**34–35.

Endrizzi, J. E. and M. S. Brown, 1962   Identification of a ring of four and two chains of three chromosomes from the *Gossypium arboreum-hirsutum* ring of six. *Can. J. Genet. Cytol.* **4:**458–468.

Endrizzi, J. E. and M. S. Brown, 1964   Identification of monosomes for six chromosomes in *Gossypium hirsutum*. *Am. J. Bot.* **51:**117–120.

Endrizzi, J. E. and R. J. Kohel, 1966   Use of telosomes in mapping three chromosomes in cotton. *Genetics* **54:**535–550.

Endrizzi, J. E. and L. L. Phillips, 1960   A hybrid between *Gossypium arboreum* L. and *G. raimondii* Ulbr. *Can. J. Genet. Cytol.* **2:**311–319.

Endrizzi, J. E. and T. Taylor, 1968   Cytogenetic studies of *N*, Lc_1, yg_2, and R_2 marker genes and chromosome deficiencies in cotton. *Genet. Res.* **12:**295–304.

Endrizzi, J. E., S. C. McMichael and M. S. Brown, 1963   Chromosomal constitution of "stag" plants of *Gossypium hirsutum* 'Acala 4-42'. *Crop Sci.* **3:**1–3.

Fryxell, P. A., 1965   A revision of the Australian species of *Gossypium* with observations on the occurrence of *Thespesia* in Australia (Malvaceae). *Aust. J. Bot.* **13:**71–102.

Fryxell, P. A., 1969   The genus *Cienfugosia* Cav. (Malvaceae) *Ann. Mo. Bot. Gard.* **56:**179–250.

Gerstel, D. U., 1953   Chromosomal translocations in interspecific hybrids of the genus *Gossypium*. *Evolution* **7:**234–245.

Gerstel, D. U., 1954   A new lethal combination in interspecific cotton hybrids. *Genetics* **39:**628–639.

Gerstel, D. U., 1955   Segregation in new allopolyploids of *Gossypium*. I. The *R* locus in certain New World × wild American hexaploids. *Genetics* **41:**31–44.

Gerstel, D. U., 1956   The use of segregation ratios of synthetic allopolyploids as a taxonomic tool. *J. Elisha Mitchell Sci. Soc.* **72:**193.

Gerstel, D. U. and L. L. Phillips, 1957   Segregation in new allopolyploids of *Gossypium*. II. Tetraploid combinations. *Genetics* **42:**783–797.

Gerstel, D. U. and L. L. Phillips, 1958   Segregation in synthetic amphidiploids in *Gossypium* and *Nicotiana*. *Cold Spring Harbor Symp. Quant. Biol.* **23:**225–237.

Gerstel, D. U. and P. A. Sarvella, 1956   Additional observation on chromosomal translocations in cotton hybrids. *Evolution* **10:**408–414.

Giles, J. A., 1961   A third case of compensatory recombination in interspecific hybrids of *Gossypium*. *Genetics* **46:**1381–1384.

Giles, J. A., 1962   The comparative genetics of *Gossypium hirsutum* L. and the synthetic amphiploid *Gossypium arboreum* × *Gossypium thurberi* Tod. *Genetics* **47:**45–59.

Good, R., 1947   *The Geography of the Flowering Plants,* Longmans, Green and Co., New York.

Green, J. M., 1955   *Frego bract,* a genetic marker in Upland cotton. *J. Hered.* **46:**232.

Harland, S. C., 1929   The work of the Genetics Department of the Cotton Research Station, Trinidad. *Emp. Cotton Grow. Rev.* **6:**304–314.

Harland, S. C., 1932   The genetics of cotton. V. Reversal of dominance in the interspecific cross *G. barbadense* L. × *G. hirsutum* L., and its bearing on Fisher's theory of dominance. *J. Genet.* **25**:261–270.

Harland, S. C., 1933   The genetical conception of the species. *Mem. Acad. Sci. U.S.S.R.* **4**. (Reprinted in *Trop. Agri.* **11**:51–53.)

Harland, S. C., 1934   The genetics of cotton. XI. Further experiments on the inheritance of chlorophyll deficiency in the New World Cottons. *J. Genet.* **29**:181–195.

Harland, S. C., 1935   The genetics of cotton. XII. Homologous genes for anthocyanin pigmentation in New World and Old World cottons. *J. Genet.* **30**:465–477.

Harland, S. C., 1936*a*   The genetical conception of the species. *Biol. Rev. (Camb.)* **11**:83–112.

Harland, S. C., 1936*b*   Haploids in polyembryonic seed of Sea Island Cotton. *J. Hered.* **27**:229–231.

Harland, S. C., 1940   New polyploids in cotton by the use of colchicine. *Trop. Agric.* **17**:53–55.

Harland, S. C. and O. M. Atteck, 1941   The genetics of cotton. XVIII. Transference of genes from diploid North American wild cotton (*Gossypium thurberi, G. armourianum,* and *G. aridum*) to tetraploid New World cotton (*G. barbadense* and *G. hirsutum*). *J. Genet.* **42**:1–19.

Hutchinson, J. B., 1931   The genetics of cotton. IV. The inheritance of corolla colour and petal size in Asiatic cotton. *J. Genet.* **24**:325–353.

Hutchinson, J. B., 1946   The crinkle dwarf allelomorph series in New World cotton. *J. Genet.* **47**:178–207.

Hutchinson, J. B. and R. A. Silow, 1939   Gene symbols for use in cotton genetics. *J. Hered.* **30**:461–464.

Hutchinson, J. B., R. A. Silow and S. G. Stephens, 1947   *The Evolution of Gossypium and the Differentiation of the Cultivated Cottons,* Oxford University Press, London.

Iyengar, N. K., 1944   Cytological investigations on auto- and allotetraploid Asiatic cottons. *Indian J. Agric. Sci.* **14**:30–40.

Johnson, B. L. and M. M. Thein, 1970   Assessment of evolutionary affinities in *Gossypium* by protein electrophoresis. *Am. J. Bot.* **57**:1081–1091.

Justus, N., J. R. Meyer and J. B. Roux, 1963   A partially male sterile character in Upland cotton. *Crop Sci.* **3**:428–429.

Kammacher, P., 1968   New investigations of linkage group I of *Gossypium hirsutum.* *Coton Fibres Trop.* **23**:179–183.

Kammacher, P. and S. Schwendiman, 1967   Etude de la localization chromosomique du gine ms_3 de sterilite pollinique du contonnier. *Coton Fibres Trop.* **22**:417–420.

Kammacher, P. A., M. S. Brown and J. S. Newman, 1957   A quadruple monosomic in cotton. *J. Hered.* **48**:135–138.

Kimber, G., 1958   Cryptic twin plants in New World cotton. *Emp. Cotton Grow. Rev.* **35**:24–25.

Knight, R. L., 1944   The genetics of blackarm resistance. IV. *Gossypium punctatum* (Sch. and Thon.) crosses. *J. Genet.* **46**:1–27.

Knight, R. L., 1945   The theory and application of the backcross technique in cotton breeding. *J. Genet.* **47**:76–86.

Knight, R. L., 1952   The genetics of Jassid resistance in cotton. I. The genes H_1 and H_2. *J. Genet.* **51**:47–66.

Knight, R. L., 1954   Abstract bibliography of cotton breeding and genetics, 1900–1950. Commonwealth Bureau of Plant Breeding and Genetics, Technical Communication No. 17, London.

Kohel, R. J., 1964   Inheritance of abnormal palisade mutant in American Upland cotton, *Gossypium hirsutum* L. *Crop. Sci.* **4**:112–113.

Kohel, R. J., 1965   Inheritance of accessory involucre mutant in American Upland cotton, *Gossypium hirsutum* I. *Crop Sci.* **5**:119–120.

Kohel, R. J., 1972   Linkage studies in Upland cotton, *Gossypium hirsutum* L. II. *Crop Sci.* **12**:66–69.

Kohel, R. J. and C. F. Lewis, 1962   Inheritance of ragged leaf mutant in American Upland cotton, *Gossypium hirsutum* L. *Crop Sci.* **2**:61–62.

Kohel, R. J., C. F. Lewis and T. R. Richmond, 1965   Linkage tests in Upland cotton, *Gossypium hirsutum* L. *Crop Sci.* **5**:582–585.

Lewis, C. F., 1954   The inheritance of cup leaf in cotton. *J. Hered.* **45**:127–128.

Lewis, C. F., 1958   Genetic studies of a mosaic leaf mutant. *J. Hered.* **49**:267–271.

Lewis, C. F., 1960   The inheritance of mottled leaf in cotton. *J. Hered.* **51**:209–212.

Lewis, C. F. and E. F. McFarland, 1952   The transmission of marker genes in intraspecific backcrosses of *Gossypium hirsutum* L. *Genetics* **37**:353–358.

Loden, H. D., 1951   Genetic evaluation of the role of cryptic structural differences as a mechanism of speciation in *Gossypium hirsutum* and *G. barbadense.* Ph.D. Dissertation, A & M College of Texas, College Station, Texas.

McMichael, S. C., 1954   Glandless boll in Upland cotton and its use in the study of natural crossing. *Agron. J.* **46**:527–528.

Menzel, M. Y., 1955   A cytological method for genome analysis in *Gossypium. Genetics* **40**:214–223.

Menzel, M. Y. and M. S. Brown, 1952   Viable deficiency-duplications from a translocation in *Gossypium hirsutum. Genetics* **37**:678–692.

Menzel, M. Y. and M. S. Brown, 1954a   The tolerance of *Gossypium hirsutum* for deficiencies and duplications. *Am. Nat.* **88**:407–418.

Menzel, M. Y. and M. S. Brown, 1954b   The significance of multivalent formation in three-species *Gossypium* hybrids. *Genetics* **39**:546–577.

Menzel, M. Y. and D. W. Martin, 1971   Chromosome homology in some intercontinental hybrids in *Hibiscus* sect. *furcaria. Am. J. Bot.* **58**:191–202.

Meyer, J. R., 1957   Origin and inheritance of D_2 smoothness in Upland cotton. *J. Hered.* **48**:249–250.

Meyer, J. R. and V. G. Meyer, 1961   Origin and inheritance of nectariless cotton. *Crop Sci.* **1**:167–169.

Murray, J. C. and L. M. Verhalen, 1965   A selected annotated bibliography on cotton breeding and genetics. 1950–1964. *Okla. Agric. Exp. Stn. Processed Ser.* **P**:508.

Oakes, A. J., 1966a   Sterility in certain *Gossypium* hybrids. I. Prefertilization phenomena. *Can. J. Genet. Cytol.* **8**:818–829.

Oakes, A. J., 1966b   Sterility in certain *Gossypium* hybrids. II. Postfertilization phenomena. *Can. J. Genet. Cytol.* **8**:830–845.

Owings, A., P. Sarvella and J. R. Meyer, 1964   Twinning and haploidy in a strain of *Gossypium barbadense. Crop Sci.* **4**:652–653.

Phillips, L. L., 1961   The cytogenetics of speciation in Asiatic cotton. *Genetics* **46**:77–83.

Phillips, L. L., 1962   Segregation in new allopolyploids of *Gossypium.* IV. Segregation in New World × Asiatic and New World × wild American hexaploids. *Am. J. Bot.* **49**:51–57.

Phillips, L. L., 1963   The cytogenetics of *Gossypium* and the origin of New World cottons. *Evolution* **17**:460–469.

Phillips, L. L., 1964   Segregation in new allopolyploids of *Gossypium.* V. Multivalent formation in New World × Wild American hexaploids. *Am. J. Bot.* **51**:324–329.

Phillips, L. L., 1966 The cytology and phylogenetics of the diploid species of *Gossypium. Am. J. Bot.* **53:**328–335.

Phillips, L. L. and D. Clement, 1963 The cytological affinities of *Gossypium barbosanum. Can. J. Genet. Cytol.* **5:**459–461.

Phillips, L. L. and D. U. Gerstel, 1959 Segregation in new allopolyploids of *Gossypium.* III. Leaf shape segregation in hexaploid hybrids of New World cottons. *J. Hered.* **50:**103–108.

Phillips, L. L. and J. F. Merritt, 1972 Interspecific incompatibility in *Gossypium.* I. Stem histogenesis of *G. hirsutum* × *G. gossypioides. Am. J. Bot.* **59:**118–124.

Phillips, L. L. and M. A. Strickland, 1966 The cytology of a hybrid between *Gossypium hirsutum* and *G. longicalyx. Can. J. Genet. Cytol.* **8:**91–95.

Poisson, C., 1968 Preliminary note on a monosomic in *Gossypium hirsutum* corresponding to linkage group I. *Coton Fibres Trop.* **23:**183–185.

Rhyne, C. L., 1954 The inheritance of yellow green, a possible mutation in cotton. *Genetics* **40:**235–245.

Rhyne, C. L., 1957 Duplicated linkage groups in cotton. *J. Hered.* **48:**59–62.

Rhyne, C. L., 1958 Linkage studies in *Gossypium.* I. Altered recombination in allotetraploid *G. hirsutum* L. following linkage group transference from related diploid species. *Genetics* **43:**822–834.

Rhyne, C. L., 1960 Linkage studies in *Gossypium.* II. Altered recombination values in a linkage group of allotetraploid *G. hirsutum* L. as a result of transferred diploid species genes. *Genetics* **45:**673–681.

Rhyne, C. L., 1962a Enhancing linkage-block breakup following interspecific hybridization and backcross transference of genes in *Gossypium hirsutum* L. *Genetics* **47:**61–69.

Rhyne, C. L., 1962b Inheritance of the glandless leaf phenotype in Upland cotton. *J. Hered.* **53:**115–123.

Rhyne, C. L., 1962c Diploidization in *Gossypium hirsutum* as indicated by glandless stem and boll inheritance. *Am. Nat.* **96:**265–276.

Rhyne, C. L., 1965 Duplicate linkage blocks in glandless leaf cotton. *J. Hered.* **56:**247–252.

Sarvella, P., 1958 Multivalent formation and genetic segregation in some allopolyploid *Gossypium* hybrids. *Genetics* **43:**601–619.

Saunders, J. H., 1961 *The Wild Species of Gossypium,* Oxford University Press, London.

Schwerin, K. H., 1970 Winds across the Atlantic—possible African origins for some pre-Columbian New World cultures. *Res. Rec. Univ. Mus. S. Ill. Univ. Mesoamerican Studies* **6:**1–33.

Silow, R. A., 1941 The comparative genetics of *Gossypium anomalum* and the cultivated Asiatic cottons. *J. Genet.* **42:**259–358.

Silow, R. A., 1944 The genetics of species development in Old World cottons. *J. Genet.* **46:**62–77.

Silow, R. A., 1946 Evidence on chromosome homology and gene homology in the amphidiploid New World cottons. *J. Genet.* **47:**213–221.

Silow, R. A. and S. G. Stephens, 1944 Twinning in cotton. *J. Hered.* **35:**76–78.

Skovsted, A., 1934 Cytological studies in cotton. II. Two interspecific hybrids between Asiatic and New World cottons. *J. Genet.* **28:**407–424.

Skovsted, A., 1935a Cytological studies in cotton. III. A hybrid between *Gossypium davidsonii* and *G. sturtii. J. Genet.* **30:**397–405.

Skovsted, A., 1935*b*   Some interspecific hybrids in the genus *Gossypium. J. Genet.* **30**:447–463.

Skovsted, A., 1937   Cytological studies in cotton. IV. Chromosome conjugation in interspecific hybrids. *J. Genet.* **34**:97–134.

Smith, C. E., Jr., 1964   *Gossypium:* Names available for specific and subspecific taxa. *Taxon* **13**:211–217.

Smith, C. E., Jr. and R. S. MacNeish, 1964   Antiquity of American polyploid cottons. *Science (Wash., D.C.)* **143**:675–676.

Smith, C. E., Jr. and S. G. Stephens, 1971   Critical identification of Mexican archaeological cotton remains. *Econ. Bot.* **25**:160–168.

Stephens, S. G., 1945   Colchicine produced polyploids in *Gossypium.* II. Old World triploid hybrids. *J. Genet.* **46**:303–312.

Stephens, S. G., 1946   The genetics of *"corky."* I. *J. Genet.* **47**:150–161.

Stephens, S. G., 1949   The cytogenetics of speciation in *Gossypium.* I. Selective elimination of the donor parent genotype in interspecific backcrosses. *Genetics* **34**:627–637.

Stephens, S. G., 1950*a*   The genetics of *"corky."* II. *J. Genet.* **50**:9–20.

Stephens, S. G., 1950*b*   The internal mechanism of speciation in Gossypium. *Bot. Rev. (Camb.)* **26**:115–149.

Stephens, S. G., 1954*a*   Interspecific homologies between gene loci in *Gossypium.* I. Pollen color. *Genetics* **39**:701–711.

Stephens, S. G., 1954*b*   Interspecific homologies between gene loci in *Gossypium.* II. Corolla color. *Genetics* **39**:712–723.

Stephens, S. G., 1955   Linkage in Upland cotton. *Genetics* **40**:903–917.

Stephens, S. G., 1961   Recombination between supposedly homologous chromosomes of *Gossypium barbadense* L. and *G. hirsutum* L. *Genetics* **46**:1483–1500.

Stephens, S. G., 1963   Polynesian Cottons. *Ann. Mo. Bot. Gard.* **50**:1–22.

Stephens, S. G., 1964   Native Hawaiian cotton. *Pac. Sci.* **18**:385–398.

Stephens, S. G., 1966   The potentiality for long-range dispersal of cotton seeds. *Am. Nat.* **100**:199–210.

Stephens, S. G., 1967   Evolution under domestication of the New World cottons (*Gossypium* spp.). *Cienc. Cult. (Sao Paulo)* **19**:118–134.

Stephens, S. G. and M. E. Mosely, 1973   Cotton remains from archeological sites in central coastal Peru. *Science* (Wash., D.C.) **180**:186–188.

Stephens, S. G. and L. L. Phillips, 1972   The history and geographical distribution of a polymorphic system in New World cottons. *Biotropica* **4**:49–60.

Tiranti, I. N. and J. E. Endrizzi, 1963   Identification of three rings of four from a ring of six chromosomes in cotton. *Can. J. Genet. Cytol.* **5**:374–379.

Turcotte, E. L. and C. V. Feaster, 1963   Haploids: High-frequency production from single-embryo seeds in a line of Pima cotton. *Science (Wash., D.C.)* **140**:1407–1408.

Turcotte, E. L. and C. V. Feaster, 1964   Inheritance of a mutant with a rudimentary stigma and style in Pima cotton, *Gossypium barbadense* L. *Crop Sci.* **4**:377–378.

Turcotte, E. L. and C. V. Feaster, 1965   The inheritance of rugate leaf in Pima cotton. *J. Hered.* **56**:81–83.

Turcotte, E. L. and C. V. Feaster, 1966   A second locus for pollen color in Pima cotton, *Gossypium barbadense* L. *Crop. Sci.* **6**:117–119.

Turcotte, E. L. and C. V. Feaster, 1967   Semigamy in Pima cotton. *J. Hered.* **58**:55–57.

Ware, J. C., 1932   Inheritance of lint colors in Upland cotton. *J. Agron.* **24**:550–562.

Weaver, J. B., Jr., 1957   Embryological studies following interspecific crosses in *Gossypium.* I. *G. hirsutum* × *G. arboreum. Am. J. Bot.* **44**:209–214.

Weaver, J. B., Jr., 1958   Embryological studies following interspecific crosses in *Gossypium*. II. *G. arboreum* × *G. hirsutum*. *Am. J. Bot.* **45**:10–16.

Webber, J. M., 1938   Cytology of twin cotton plants. *J. Agric. Res.* **57**:155–160.

White, T. G. and J. E. Endrizzi, 1965   Tests for the association of marker loci with chromosomes in *Gossypium hirsutum* by the use of aneuploids. *Genetics* **51**:605–613.

White, T. G., T. R. Richmond and C. F. Lewis, 1967   Use of cotton monosomics in developing interspecific substitution lines. *U.S. Department of Agriculture Bulletin ARS-34-91*. U.S. Department of Agriculture, Washington.

Wilson, F. D. and R. J. Kohel, 1970   Linkage of green-lint and okra-leaf genes in a reciprocal translocation stock of Upland cotton. *Can. J. Genet. Cytol.* **12**:100–104.

6

Cucurbita

Thomas W. Whitaker

Introduction

Cucurbita is a genus of about 27 species, and is indigenous to the Americas (Bailey, 1943; Castetter and Erwin, 1927). Most of the species are concentrated in Mexico and in the southwestern United States. Of the five cultivated species, all except *Cucurbita maxima* are found in Mexico and Central American. *C. maxima* was confined to Argentina, Bolivia, Chile, and Peru in pre-Columbian times (see Table 1).

The plants are annual or perennial herbs; long-running and climbing, or short and bushy. The stems are normally angled or furrowed, often rooting at the nodes. The tendrils are large and branched. The leaves are simple, and often deeply lobed. The flowers are monoecious, and numerous, with conspicuous creamy white to deep yellow corollas. The fruits come in many sizes and shapes. Some of them are among the largest in the plant kingdom. The seeds are numerous; flat, colored white, buff, brown, or black; usually ovate to oblong; and normally with a thickened margin.

Genetics and Cytology

All species of *Cucurbita* thus far examined have 20 pairs of small, rod-shaped chromosomes (Weiling, 1959). The chromosomes of species in this genus are diffcult to work with because they do not respond to

Thomas W. Whitaker—United States Department of Agriculture Agricultural Research Service, Horticultural Field Station, La Jolla, California.

TABLE 1. *The Cultivated Species of Cucurbita, Their Common Names, and Geographic Distribution in Pre-Columbian Times*

Species	Common names	Distribution in pre-Columbian times
C. pepo	Summer squash	North American from about Mexico City northward to northeastern U.S. and Southern Canada
	Some varieties of winter or baking squash, pumpkins (jack-o-lantern, sugar), marrows, ornamental gourds	Southwestern U.S. and northwestern Mexico
C. mixta	Squash, cymlins, cushaws	From southern Mexico to northwestern Mexico and southwestern U.S.
C. moschata	Winter or baking squash, field pumpkins	From southwestern U.S. through Mexico, Central America to northern portion of South America (Peru, Columbia)
C. maxima	Winter or baking squash (Hubbard and Banana groups), large pumpkins, turban squash	In South America (Peru, Bolivia, Chile, northern Argentina)
C. ficifolia	Fig-leaf gourd, Malabar gourd	At high altitudes from southern Mexico to Bolivia

conventional cytological methods. This problem has led some authors to suggest that the cell chemistry of *Curcurbita,* and perhaps of the entire family Cucurbitaceae, is different from that of other plant material. This suggestion has not been supported by adequate investigation as yet.

Inbreeding and Heterosis

Numerous experiments have failed to demonstrate marked inbreeding depression in *Cucurbita,* even after prolonged selfing (Cummings and Jenkins, 1928). There is no experimental record of naturally occurring self-incompatibility in the genus. Surprisingly, there is considerable heterosis, and in one species, *Cucurbita pepo,* F_1 hybrids are being used for commercial crops (Curtis, 1941). There are reports (H. M. Munger, private communication) that Japanese seed companies have successfully harnessed the heterotic effects common to F_1 species hybrids by solving the

technology of producing seed of an F_1 hybrid of *Cucurbita maxima* ×
Cucurbita moschata. This is one of the few examples, perhaps the only
one, of an F_1 species hybrid being used to produce the commercial crop.

Interspecific Crossing

Experimental genetic work in *Cucurbita* has been devoted largely to
determining the parameters of species crossing in the genus (Whitaker and
Bemis, 1965). The general purpose of this experimentation was to shift a
character or group of characters from a donor to a recipient species to
combat disease or insect pests (see Pearson *et al.,* 1951; Rhodes, 1959,
1964).

The only report of recombination in species crosses in *Cucurbita* is
that of Wall (1961). He found free recombination in the F_2 of the cross *C.
pepo* × *C. moschata* for the genes determining internode length, sepal
width, and androecium length. Based on fertility relationships of the F_1
and the recombination generations, Wall has suggested that some of the
differences between these species are determined by chromosome
structural differences, as well as by accumulated gene differences.

The studies involving species crossing are fairly complete, and are
reported in a series of papers by several investigators. From these studies,
the relationship between species has been worked out, and the course of
evolution in the genus is clearly discernible (Hurd, *et al.,* 1971).

The essential information may be summarized as follows:

1. There are two groups of xerophytic species, and they are incom-
    patible with each other. Both groups are, for the most part,
    confined to the deserts of northern Mexico, Baja California, and
    southwestern United States. The first group contains the poly-
    morphic species *Cucurbita foetidissima*. The second group
    contains *Cucurbita palmata, C. digitata, Cucurbita cordata,* and
    Cucurbita cylindrata. These xerophytic types are evidently deriva-
    tives of a mesophytic ancestor. The xerophytic species are more
    closely related to *C. moschata* than to other species, either wild
    or cultivated.

2. The mesophytic species mostly occur from slightly south of
    Mexico City to the Guatemalan border. Within this group are
    two subgroups. The first subgroup contains *Cucurbita martinezii,
    Cucurbita lundelliana,* and *Cucurbita okeechobeensis,* which are
    mostly perennials. The second subgroup centers around *Cucurbita
    sororia* and is probably closely related to *C. pepo* in the cultivated

group; species in this second subgroup are annuals. *C. lundelliana* is the key species in the mesophytic group. Used as the female parent, it is compatible with nearly all of the cultivated species and many species of the wild group. The compatibility pattern of the cultivated group suggests that *C. moschata* is the axis through which the cultivated species are related to each other.

3. Data from numerical taxonomy (Bemis *et al.*, 1970) and isozyme studies (Wall and Whitaker, 1971) support the pattern of species relationships derived from conventional sources.
4. The evidence suggests that the cultivated species were domesticated in at least three different areas, and probably at different periods in the evolutionary time scale (Cutler and Whitaker, 1961).

Interspecific Polyploid Hybrids

Generally, interspecific crosses between the five cultivated species of *Cucurbita*, if successful, are more or less sterile. To restore an acceptable degree of fertility to the F_1 hybrids, which may carry valuable genes for resistance to various diseases and insect pests, doubling the chromosome number of the sterile hybrids by means of colchicine treatments has been used with some success. Whitaker and Bohn (1950) have reported on the behavior of amphidiploids from the following interspecific crosses: *C. maxima* × *C. pepo*; *C. maxima* × *C. mixta*; and *C. maxima* × *C. moschata*. Amphidiploids derived from the cross *C. maxima* × *C. pepo* are self-sterile or nearly so. Those from the cross *C. maxima* × *C. mixta* are slightly self-fertile, and display pronounced segregation. Amphidiploids from the cross *C. maxima* × *C. moschata* are self-fertile, and cross-sterile with the parent species. These amphidiploids display some segregation, but they are stable enough to be regarded as a new species. Pearson, *et al.* (1951) have made further studies of the vines and fruit of the *C. maxima* × *C. moschata* amphidiploids. They found that some amphidiploid lines from this cross produced fruits that compared favorably in baking quality with those of several varieties of baking squash.

Bemis (1970) has studied polyploid hybrids from the cross *Cucurbita moschata* × *Cucurbita foetidissima*. The latter is a wild xerophytic species. Using embryo culture techniques, several F_1 hybrid plants were obtained, but they were sterile. Amphidiploids were obtained by means of colchicine treatments of the F_1 plants. The amphidiploid was female fertile but genic male sterile.

The practical use of the amphidiploids from the *C. moschata* × *C.*

foetidissima cross is limited by the poor quality of the fruits introduced by the *C. foetidissima* parent. Bemis has indicated that there must be intergenomic transfer within the amphidiploid in order to select against the undesirable fruit quality genes introduced by *C. foetidissima*. Such intergenomic transfer does not occur, as shown by cytological examination of the amphidiploid and unbalanced derivatives of this cross, where there is little if any intergenomic association.

Genetics of *C. pepo* and *C. maxima*

Studies on specific characters in *Cucurbita* are sketchy and confined largely to the cultivated species, specifically *C. pepo* and *C. maxima*. Lack of genetic information about the group stems mainly from their space requirements; the plants are extremely large. To develop conventional genetic ratios, much space would be required, especially for the study of mature vine and fruit characters.

Sinnott (1927) and his co-workers have investigated the genetic control of developmental relationships in cucurbits, mostly in *C. pepo*. They found, for example, that shape determination involves primarily a control of cell number rather than cell shape. Differences in the shape of the organ result from the relative number of cells along certain axes compared to others. But fruit-shape determination is not as simple as the above statement suggests. There are at least three other distinct types of shape determination, all under genetic control.

Shifriss (1947) and later Denna and Munger (1963) have studied the bush habit in *C. pepo* and *C. maxima*. The latter investigators have found that major genes for the bush vs. vine habit in both species are probably located at the same locus. The action of these genes is peculiar in that they undergo a developmental reversal of dominance. In *C. pepo* the bush gene is almost completely dominant to the vine gene during early growth, but incompletely dominant during later growth. In *C. maxima* the bush gene is completely dominant during early growth, but completely recessive in the later stages of growth.

There is much confusion in the genetic work surrounding these two species, particularly *C. pepo*. The confusion originates mainly from the fact that the earlier work done by Sinnott and his students was not coordinated with work done by later investigators. Since different materials were used, it is not possible to tell whether the investigators were working with the same gene, different genes, or alleles of the same gene.

The genetic information for *C. pepo* is summarized in Table 2 and for *C. maxima* in Table 3.

TABLE 2. The Inheritance of Several Characters in Cucurbita pepo L.

Character	Dominance	Number of genes involved	Gene action	Reference
Resistance to squash bug (*Anasa tristis*)	Resistance partly dominant over susceptibility	3 pairs	Additive	Benepal and Hall, 1967
Resistance to striped cucumber beetle	Resistance partially dominant over susceptibility	3 pairs	Additive	Nath and Hall, 1963
Male sterility	Fertility	1 pair (ms_2ms_2)	—	Eisa and Munger, 1968
Mottled vs. uniform green leaf	Mottled	1 pair (MM)	—	Scarchuk, 1954
Rosette vs. normal leaf	Normal	1 pair (ro)	—	Mains, 1952
Dark green vs. light green stem	Dark green	1 pair (DD)	—	Globerson, 1968
Yellow vs. green seedling	Green	1 pair (ys)	Lacks chlorophyll, lethal	Mains, 1952
Bush vs. vine	Bush	1 major gene + modifiers	Dominance changes during development	Denna and Munger, 1963
Green vs. white fruit	Green	2 pairs (CCRR)	(C) dominant with epistatic control	Globerson, 1968
Nonpersistent, faint green to white color	Green	2 pairs (ccRR)	—	Globerson, 1968
Green-striped vs. solid white fruit	Green striped	1 pair (StSt)	—	
Yellow vs. green fruit	Yellow	1 pair (BB)	Uniformly yellow at maturity	Shifriss, 1947

Character	Dominant	Gene pairs	Notes	Reference
Yellow *vs.* (pigment control) green fruit	Yellow	1 pair (YY)	—	Shifriss, 1947
White *vs.* yellow or green fruit	White	1 pair (WW)	Epistatic to Y	Sinnott and Durham, 1922
Striped *vs.* uniform fruit color	Striped	1 pair (MM)	—	Shifriss, 1955
Uniform heavy pigmentation *vs.* light pigmentation of fruit	Uniform heavy	1 pair (LL)	—	Scarchuk, 1954
Warty *vs.* smooth fruit surface	Warty	1 pair	—	Sinnott and Durham, 1922
Hard *vs.* soft rind of fruit	Hard	1 gene pair + modifiers	—	Mains, 1952
Fruit shape	Disc	2 pairs: (AB)—disc, (Ab or aB)—sphere, (ab) — elongate	—	Sinnott, 1927
Bitter *vs.* mild flesh	Bitter	1 pair (Bi)	—	Grebenscikov, 1954
White *vs.* cream flesh	White	1 pair	—	Sinnott and Durham, 1922
"Naked" (nonlignified) *vs.* lignified seed coat	Lignified	1 major gene + modifiers	—	Grebenscikov, 1954
Bush *vs.* vine	Bush	1 major gene + modifiers	Dominance changes during development	Denna and Munger, 1963
Yellow *vs.* green seedlings	Green	1 pair	—	—
Rosette *vs.* normal leaf	Normal	1 pair	—	Mains, 1952
Hard *vs.* smooth rind of fruit	Hard	1 pair	—	Mains, 1952

TABLE 3. The Inheritance of Several Characters in C. maxima

Character	Dominance	Number of genes involved	Gene action	Reference
♂ sterility	Complete, fertile over sterile	1 pair	—	Francis and Bemis, 1970; Scott and Riner, 1946; Singh and Rhodes, 1961
♀ and ♂ sterility	Completely sterile	1 pair	—	Hutchins, 1944
Mottle leaf	Mottle leaf over normal	1 pair	—	Scott and Riner, 1946
Growth habit (vine vs. bush)	Intermediate	2 pairs	—	
Number of days from seeding to anthesis of first staminate flower	Complete, fewer days over more days	3 pairs	Additive	
Number of days from seeding to anthesis of first pistillate flower	Complete, fewer days over more days	3 pairs	Additive	Singh, 1949
Number of days between opening of first staminate flower and first pistillate flower	Complete, shorter duration over longer duration	2 pairs	Partly complementary	
Color of fruit (green vs. orange)	Partly dominant, green	2 pairs	—	
Precocious fruit pigmentation	Precocious	1 pair	—	Shifriss, 1966
Weight of fruit	Partly phenotypic, for large fruit size	3 pairs	Complex intra- and inter-allelic interaction	Singh, 1949
Total solids	Intermediate	2 pairs	Additive	

Literature Cited

Bailey, L. H., 1943   Species of *Cucurbita*. *Gentes Herb.* **6**:267–322.

Bemis, W. P., 1970   Polyploid hybrids from the cross *Cucurbita moschata* Poir × *C. foetidissima* HBK. *J. Am. Soc. Hortic. Sci.* **95**:529–531.

Bemis, W. P., A. M. Rhodes, T. W. Whitaker and S. G. Carmer, 1970   Numerical taxonomy applied to *Cucurbita* relationships. *Am. J. Bot.* **57**:404–412.

Benepal, P. S. and C. V. Hall, 1967   Biochemical studies of plants of *Cucurbita pepo* L. varieties as related to feeding response of squash bug, *Anasa tristis* DeGeer. *Proc. Am. Soc. Hortic. Sci.* **91**:361–365.

Castetter, E. F. and A. T. Erwin, 1927   A systematic study of squashes and pumpkins. *Bull. Iowa Agric. Exp. Stn.* **244**:107–135.

Cummings, M. B. and E. W. Jenkins, 1928   Pure-line studies with 10 generations of Hubbard squash. *Bull. Vt. Agric. Exp. Stn.* **270**:29 p.

Curtis, L. C., 1941   Comparative earliness and productiveness of first-and second-generation summer squash (*Cucurbita pepo*), and the possibilities of using second-generation seed for commercial planting. *Proc. Am. Soc. Hortic. Sci.* **38**:596–598.

Cutler, H. C. and T. W. Whitaker, 1961   History and distribution of the cultivated cucurbits in the Americas. *Am. Antiquity* **26**:469–485.

Denna, D. W. and H. M. Munger, 1963   Morphology of the bush and vine habits and the allelism of the bush genes in *Cucurbita maxima* and *C. pepo* squash. *Proc. Am. Soc. Hortic. Sci.* **82**:370–377.

Eisa, H. M. and H. M. Munger, 1968   Male sterility in *Cucurbita pepo*. *Proc. Am. Soc. Hortic. Sci.* **92**:473–479.

Francis, R. R. and W. P. Bemis, 1970   A cytomorphological study of male sterility in a mutant of *Cucurbita maxima* Duch. *Econ. Bot.* **24**:325–332.

Globerson, D., 1968   The inheritance of white fruit and stem color in summer squash, *Cucurbita pepo* L. *Euphytica* **18**:249–255.

Grebenscikov, I., 1954   Notulae cucurbitologicae I. Zur vererbung der Bitterheit and Kurztriebizkeit bei *Cucurbita pepo* L. *Kulturpflanze* **2**:145–154.

Hurd, P. D., Jr., E. G. Linsley and T. W. Whitaker, 1971   Squash and gourd bees (*Peponapis, Xenoglossa*) and the origin of the cultivated *Cucurbita*. *Evolution* **25**:218–234.

Hutchins, A. E., 1944   A male and female sterile variant in squash, *Cucurbita maxima* Duch. *Proc. Am. Soc. Hortic. Sci.* **44**:494–496.

Mains, E. B., 1952   Inheritance in *Cucurbita pepo*. *Mich. Acad. Sci. Arts Letters* **36**:27–30.

Nath, P. and C. V. Hall, 1963   Inheritance of fruit characteristics in *Cucurbita pepo* L. *Indian J. Hortic.* **20**:215–221.

Pearson, O. H., R. Hopp and G. W. Bohn, 1951   Notes on species crosses in *Cucurbita*. *Proc. Am. Soc. Hortic. Sci.* **57**:310–322.

Rhodes, A. M., 1959   Species hybridization and interspecific gene transfer in the genus *Cucurbita*. *Proc. Am. Soc. Hortic. Sci.* **74**:546–551.

Rhodes, A. M., 1964   Inheritance of powdery mildew resistance in the genus *Cucurbita*. *Plant Dis. Rep.* **48**:54–55.

Scarchuk, J., 1954   Fruit and leaf characters in summer squash. *J. Hered.* **45**:295–297.

Scott, D. H. and M. E. Riner, 1946   Inheritance of male sterility in winter squash. *Proc. Am. Soc. Hortic. Sci.* **47**:375–377.

Shifriss, O., 1947   Developmental reversal of dominance in *Cucurbita pepo*. *Proc. Am. Soc. Hortic. Sci.* **50**:330–346.

Shifriss, O., 1955   Genetics and the origin of bicolor gourds. *J. Hered.* **46**:213–222.

Shifriss, O., 1966   Behavior of gene *B* in *Cucurbita*. *Veg. Improvement Newsl.* **8**:7–8.

Singh, D., 1949   The inheritance of certain economic characters in squash, *Cucurbita maxima* Duch. *Tech. Bull. Minn. Agric. Exp. Stn.* **186**:30 p.

Singh, S. P. and A. M. Rhodes, 1961   A morphological and cytological study of male sterility in *Cucurbita maxima*. *Proc. Am. Soc. Hortic. Sci.* **78**:375–378.

Sinnott, E. W., 1927   A factorial analysis of certain shape characters in squash fruits. *Am. Nat.* **61**:333–344.

Sinnott, E. W. and G. B. Durham, 1922   Inheritance in the summer squash. *J. Hered.* **13**:177–186.

Wall, J. R., 1961   Recombination in the genus *Cucurbita*. *Genetics* **46**:1677–1685.

Wall, J. R. and T. W. Whitaker, 1971   Genetic control of leucine aminopeptidase and esterase isozymes in the interspecific cross, *Cucurbita ecuadorensis* × *C. maxima*. *Biochem. Genet.* **5**:223–229.

Weiling, F., 1959   Genomanalytische Untersuchungen bei Kurbis (*Cucurbita* L.). *Zuchter* **29**:161–179.

Whitaker, T. W. and W. P. Bemis, 1965   Evolution in the genus *Cucurbita*. *Evolution* **18**:553–559.

Whitaker, T. W. and G. W. Bohn, 1950   The taxonomy, genetics, production and uses of the cultivated species of *Cucurbita*. *Econ. Bot.* **4**:52–81.

7

Cucumis

Richard W. Robinson and Thomas W. Whitaker

Introduction

Cucumis comprises a genus of nearly forty species (Whitaker and Davis, 1962), including several of considerable economic importance such as cucumber (*Cucumis sativus* L.), the muskmelon (*Cucumis melo* L.), and the West Indian gherkin (*Cucumis anguria* L.). All species are indigenous to Africa except for *C. anguria,* which apparently was introduced to the West Indies from Africa, and *C. sativus* and *Cucumis hardwickii,* which are natives of Asia.

The plants are long, trailing, annual or perennial herbs, with angular or lobed leaves, simple tendrils, and branched, hirsute stems. They are mostly monoecious, but dioecious and andromonoecious forms also occur. The yellow flowers are fascicled or solitary and usually trimerous, rarely pentamerous, and often borne at every node. Fruits are usually fleshy, indehiscent, and contain many tan or white seeds.

Genetics and Cytology

The cucumber (*C. sativus*) and the closely related *C. hardwickii* have seven pairs of chromosomes. *Cucumis heptadactylus* and *Cucumis ficifolius* have 24 pairs of chromosomes, and are evidently tetraploids. All other

Richard W. Robinson—New York State Agricultural Experiment Station, Department of Seed and Vegetable Sciences, Hedrick Hall, Geneva, New York. Thomas W. Whitaker— United States Department of Agriculture, Agricultural Research Service, Horticultural Field Station, La Jolla, California.

species of *Cucumis* studied have twelve pairs (Heimlich, 1927; Whitaker, 1930; Shimotsuma, 1965).

Polyploidy and Monoploidy

Tetraploidy is readily induced in *Cucumis* by application of colchicine. Tetraploid cucumbers are distinctive, but have not played an important role in plant breeding (Shifriss, 1942). Tetraploid muskmelons have been reported to have better fruit quality than diploids (Batra, 1952), but they are not used commercially.

Aalders (1958) devised a simple method of detecting spontaneous monoploids of the cucumber. He found that diploid seeds usually sink in water, but some of the seeds that float have the haploid number of chromosomes. The monoploids are reduced in vigor and fertility.

Inbreeding and Heterosis

Inbred lines of cucumber can be developed without loss of vigor (Tiedjens, 1928; Hawthorn and Wellington, 1930; Jenkins, 1942). Despite the lack of depression of vigor by inbreeding, considerable heterosis was noted by Hayes and Jones (1917), Porter (1931), Cochran (1937), and later workers. F_1 hybrids are commercially important at present, especially for pickling cucumbers.

Interspecific Crossing

Deakin *et al.* (1971) have recently reviewed their research and that of other investigators concerned with compatibility relationships of *Cucumis* species. Many of the noncultivated species can be crossed with each other. *C. hardwickii* is fully fertile with the cucumber. Therefore, it is not a distinct species, but more likely a botanical variety of *C. sativus*. Deakin *et al.* concluded that *Cucumis* species can be classified into four cross-sterile groups. They are:

1. The spiny-fruited species of African origin that can be crossed with the West India gherkin, *Cucumis anguria*. There are at least 8 of these.
2. *Cucumis metuliferus*, the African horned cucumber.
3. *C. sativus*, cucumbers.
4. African species with young fruit having hairs, but not operculate spines. This group includes *C. melo, Cucumis humifructus*, and *Cucumis sagittatus*.

Kishi and Fujishita (1969) have shown that while pollen germinates in incompatible interspecific *Cucumis* crosses, pollen tube development is abnormal.

Genetics of *C. sativus* and *C. melo*

Although more is known about the genetics of the cucumber than of other cucurbits, only 55 genes of this species have been identified. Genes of economic importance are listed for the cucumber in Table 1 and for the muskmelon in Table 2.

The genetics of sex expression in *Cucumis* has received considerable attention, and is the basis for a method of producing hybrid seed. Two major genes, plus modifiers, interact to determine whether plants will be monoecious (male flowers and female flowers on the same plant), gynoecious (only female flowers), gynomonoecious (mostly female flowers, few male flowers), andromonoecious (bisexual flowers and male flowers), or hermaphroditic (only bisexual or perfect flowers). A single recessive gene in each species is responsible for the andromonoecious sex expression. Another gene, dominant in the cucumber but recessive in the muskmelon, governs gynoecious or gynomonoecious sex expression. Plants homozygous for both genes have perfect flowers, and plants with the normal allele of each gene are monoecious. When hybrids are desired, a gynoecious line is preferable as the female parent since it can not self-pollinate, and thus all the seed it produces will be hybrid. A gynoecious line may be perpetuated by treatment with a growth regulator to induce it to produce staminate flowers, or by crossing it with an isogenic line having perfect flowers.

Another interesting gene in the cucumber enables the fruit to develop without pollination (parthenocarpy). This gene apparently originated as a spontaneous mutation in the last century, and was perpetuated by European horticulturists because of its beneficial effect on yield in greenhouses, where insects are scarce and are inefficient pollinators. There has been increased interest in this gene recently because of development of gynoecious cultivars that otherwise need to be grown with monoecious plants for pollination. The current practice of blending seed of a monoecious pollinator with seed of a gynoecious hybrid can be replaced by breeding the parthenocarpic gene into the gynoecious cultivar.

Cucurbitacins, terpene compounds that impart a bitter flavor to the foliage and occasionally to the fruit, are under genetic control. Andeweg (1959) found a spontaneous mutant lacking cucurbitacins, and breeders have used this gene to breed nonbitter cucumbers. The lack of cucurbitacins makes the plants resistant to some insects, but more attractive to others (Costa and Jones, 1971).

Genes of muskmelon that have been important in plant breeding include those which condition resistance to powdery mildew. This

TABLE 1.   *The Inheritance of Several Economically Important Characters in*
Cucumis sativus

Character	Gene	Phenotype
Plant habit	*dw*	Dwarf; short internodes
	de	Determinate; short vines terminating in flowers
	In (de)	Intensifier of *de*; modifies expression of *de*
Sex expression	*F*	Female; high degree of female sex expression
	m	Male; high degree of male sex expression; *m* plants are andro-monoecious if recessive and her-maphroditic if dominant for *F*
	a	Androecious; producing only male flowers
	Tr	Trimonoecious; producing male, fe-male, and bisexual flowers
Fertility	*ms, ms*-2	Male sterile; male flowers abort be-fore anthesis
	co	Green corolla; female sterile
Fruit set	*Pc*	Parthenocarpic; sets fruit without pollination
Fruit type	*B, B*-2	Black spines on fruit; dominant to white
	s	Size and number of fruit spines; many fine spines recessive to few, coarse
	u	Uniform color of immature fruit; lack of mottling or stippling
	R, c	Mature fruit color; *R*+ is red, *Rc* orange, ++ yellow, and +*c* cream
	te	Tender skin
Flavor	*bi*	Bitter free; lacking cucurbitacins
	Bt	Very bitter
Disease resistance	*Ar*	One of several genes for anthracnose resistance
	Bw	Bacterial wilt resistance
	Cm	One of several genes for resistance to cucumber mosaic virus
	pm-1, *pm*-2, *pm*-3	Resistance to powdery mildew

TABLE 2.   *The Inheritance of Several Economically Important Characters in
Cucumis melo*

Character	Gene	Phenotype
Plant habit	*dw*	Dwarf; short internodes
	ab	Abrachiate; lacking lateral branches
Sex expression	*a*	Andromonoecious
	g	Gynomonoecious
Fertility	*ms*-1, *ms*-2	Male sterile; male flowers abort after anthesis
Fruit type	*gf*	Green flesh; recessive to salmon color
	Y	Yellow skin color; dominant to white color of mature fruit
	Wi	White color of immature fruit; dominant to green
	st	Striped fruit
	s	Prominent sutures or vein tracts
	sp	Spherical fruit shape
Flavor	*so*	Sour taste
Disease resistance	*Mc, Mc*-2	Resistance to gummy stem blight
	Pm-1, *Pm*-2, *Pm*-3, *Pm*-4, *Pm*-5	Resistance to powdery mildew

devastating disease threatened the muskmelon industry of California until
California and USDA breeders found a resistance trait in plants introduced
from India and developed a resistant cultivar. Later, mildew occurred on
plants having this gene, thus suggesting that a second race of the fungus was
operating. Resistance to the second race was also found (Bohn and
Whitaker, 1964) and incorporated into useful cultivars. Three additional
genes for powdery mildew resistance have subsequently been identified
(Harwood and Markarian, 1968).

Literature Cited

Aalders, L. E., 1958   Monoploidy in cucumbers. *J. Hered.* **49**:41–44.
Andeweg, J. M., 1959   The breeding of non-bitter cucumbers. *Euphytica* **8**:13–20.
Batra, S., 1952   Induced tetraploidy in muskmelons. *J. Hered.* **43**:141–148.
Bohn, G. W. and T. W. Whitaker, 1964   Genetics of resistance to powdery mildew Race 2
   in muskmelon. *Phytopathology* **54**:589–591.

Cochran, F. D., 1937   Breeding cucumbers for resistance to downy mildew. *Proc. Am. Soc. Hortic. Sci.* **35**:541–543.

DaCosta, C. P.   and C. M. Jones, 1971   Cucumber beetle resistance and mite susceptibility controlled by the bitter gene in *Cucumis sativus* L. *Science (Wash., D.C.)* **172**:1145–1146.

Deakin, J. R., G. W. Bohn and T. W. Whitaker, 1971   Interspecific hybridization in *Cucumis. Econ. Bot.* **25**:195–211.

Harwood, R. R. and D. Markarian, 1968   A genetic survey of resistance to powdery mildew in muskmelon. *J. Hered.* **59**:213–217.

Hawthorn, L. R. and R. Wellington, 1930   Geneva, a greenhouse cucumber that develops fruit without pollination. *N.Y. Agric. Exp. Stn. Geneva Bull.* **580**:1–11.

Hayes, H. K. and D. R. Jones, 1917   First generation crosses in cucumbers. *Conn. Storrs Agric. Exp. Stn. Res. Rpt.* **40**:319–322.

Heimlich, L. F., 1927   Microsporogenesis in the cucumber. *Proc. Natl. Acad. Sci. USA* **13**:113–115.

Jenkins, J. M., Jr., 1942   Natural self-pollination in cucumbers. *Proc. Am. Soc. Hortic. Sci.* **40**:411–412.

Kishi, Y. and N. Fujishita, 1969   Studies on interspecific hybridization in the genus *Cucumis.* I. Pollen germination and pollen tube growth in selfings and incompatible crossings. *J. Jap. Soc. Hortic. Sci.* **38**:329–334.

Porter, R. H., 1931   The reaction of cucumber to types of mosaic. *Iowa State J. Sci.* **6**:95–130.

Shifriss, O., 1942   Polyploids in the genus *Cucumis,* preliminary account. *J. Hered.* **33**:144–152.

Shifriss, O., 1945   Male sterilities and albino seedlings in cucurbits. *J. Hered.* **36**:47–52.

Shimotsuma, M., 1965   Chromosome studies of some *Cucumis* species. *Seiken Ziho* **17**:11–16.

Tiedjens, V. A., 1928   Sex ratios in cucumber flowers as affected by different conditions of soil and light. *J. Agric. Res.* **36**:721–746.

Whitaker, T. W., 1930   Chromosome numbers in cultivated cucurbits. *Am. J. Bot.* **17**:1033–1040.

Whitaker, T. W. and G. N. Davis, 1962   *Cucurbits,* Interscience, New York.

8

Arabidopsis thaliana

György P. Rédei

History

Botanical information on *Arabidopsis thaliana* (L.) Heynh. has been available since the 16th century. The species name honors Johannes Thal, a German physician, who in 1577 described the plant in his *Harzflora*. During the ensuing centuries, nearly three dozen botanical synonyms have been used, and a multitude of common names are known, including the "mouse-ear-cress" or "wall cress." In 1935 a journal "transliterated" the name "Drosophila végétative" from which R. O. Whyte (1946) anglicized the nickname "botanical Drosophila."

The chromosome number ($n = 5$) was determined by Friedrich Laibach in 1907, then a student in the Bonn laboratory of Eduard Strasburger. Genetic investigations were begun in the 1940's by Laibach, then professor of botany at the University of Frankfurt am Main, who maintained a research laboratory with his personal funds in his home town, Limburg. *Arabidopsis* was started on its genetic career by the doctoral thesis of a student of Laibach, Erna Reinholz, which was published in 1947 in *Field Information Agency Technical Report* (the Report of the Military Command of Germany), and by a letter in *Nature* by John Langridge (1955), then a student of D. G. Catcheside.

A comprehensive review of the literature up to 1968 has appeared recently (Rédei, 1970). The *Arabidopsis Information Service*, a newsletter, is published annually by G. Röbellen of the University of Göttingen, West Germany.

György P. Rédei—Agronomy Department, University of Missouri, Columbia, Missouri.

Merits of *Arabidopsis* for Genetic Studies

This angiosperm exhibits some characteristics of special interest for the geneticists: (1) the chromosome number is low; (2) autopolyploid series are remarkably fertile; (3) all primary trisomics are available and morphologically distinguishable; (4) the life cycle may be completed within a month; (5) autogamy facilitates the establishment of true isogenic lines; (6) single plants can produce over 50,000 seeds; (7) the size of the plant is small, and therefore populations of millions can be studied under laboratory conditions; (8) the plant is adaptable to a wide variety of environmental conditions; (9) it can be raised to maturity aseptically in standard-size test tubes on simple, defined nutrient media; (10) a wide range of mutants, including certain auxotrophs, can be easily induced; (11) ecotypes adapted to diverse ecological niches are available, and (12) it can be hybridized with a number of species of different chromosome number.

Research conducted with this plant has contributed basic information primarily to physiological and developmental genetics, and the organism is useful for testing pharmacological and mutagenic effects of various agents. Geneticists have been attracted to population and quantitative genetics of this plant, and even college students are taking advantage of this plant in one-semester laboratory classes (Rédei, 1968).

Culture Techniques

Large quantities of seed suspended in a cold, viscose, agar solution can be uniformly distributed into pots with the aid of a separatory funnel (Rédei, 1973a), and cloning is feasible (Napp-Zinn and Berset, 1966; Reinholz, 1972).

Seeds can be disinfected in calcium hypochlorite (Rédei, 1962a) or in alcohol–peroxide solutions (Langridge, 1957b). Intact plants can be grown on mineral–agar media (Langridge, 1957b; Rédei, 1965b), on silica gel (Langridge, 1957b), or perlite-supported media (Feenstra, 1964). Shaken liquid cultures are also successful (Rédei and Perry, 1971; Rédei, 1972a).

There are suitable techniques for isolating embryos (Rijven, 1956), roots (Neales, 1968a,b), and hypocotyl sections (Nitsch, 1967). Callus tissue grows slowly and may display redifferentiation (Ziebur, 1965; Yokoyama and Jones, 1965; Shen-Miller and Sharp, 1966; Anand, 1966; Loewenberg and Thompson, 1967). By means of anther culture, haploid callus and plants can be produced (Gresshoff and Doy, 1972).

Cytology

The nuclear volume (30 μ^3) is the smallest known among the angiosperms. The C value is 2×10^9 nucleotides (Sparrow *et al.*, 1972). The chromosome number is low ($n = 5$), and the chromosome size is small, in the range of 0.8–3.7 μ at telophase I (Steinitz-Sears, 1963). There is one major and probably one secondary nucleolus. The chromosomes often display mitotic association (Laibach, 1907; Steinitz-Sears, 1963). The number of "prochromosomes" (prominent heterochromatic bodies in the interphase nucleus) coincides with the chromosome number (Laibach, 1907; Röbbelen, 1957a; Steinitz-Sears, 1963). The microtechnical method developed by Steinitz-Sears (1963) has made possible the analysis of the diplotene stage and the detection of pachynema (Sears and Lee-Chen, 1970). In spite of the small number of chromosomes, *Arabidopsis* is not a favorable material for cytological studies with the light microscope because of the small size of its chromosomes. An idiogram of the somatic chromosomes has been published by Měsíček (1967).

Polyploidy had been induced with colchicine (Bouharmont, 1964, 1969) and by x rays (Sosna, 1965). Contrary to expectations, polyploids display greater sensitivity to ionizing radiation (Bouharmont, 1964).

Studies of haploids appear promising with *Arabidopsis* (Melchers, 1972). Androgenesis has been observed (3/19,600) under conditions where both the male and female genotype and the female cytoplasm were genetically marked (Barabás and Rédei, 1971). Attempts to differentiate pollen grains directly into seedlings have not yet been successful [Barabás (1970), private communication]. However, haploid callus can redifferentiate into flowering plantlets (Gresshoff and Doy, 1972). Aneuploidy of an unexpected type (8 somatic chromosomes) has also been reported (Arnold and Cruse, 1965).

All five primary trisomics have been identified, and one tertiary, telotrisomic type has been verified cytologically and genetically. The trisomics have been correlated with linkage information, including the position of the centromere (Lee-Chen and Burger, 1967; Lee-Chen and Steinitz-Sears, 1967; Lee-Chen and Sears, 1969; Sears and Lee-Chen, 1970). All five primary trisomics of *Arabidopsis* have distinct phenotypes, thus facilitating gene assignment to chromosomes.

Chromosomal rearrangements have been described by Langridge (1958a) and Steinitz-Sears (cf. Fujii, 1964b). *A. thaliana* apparently shares chromosomes, as judged by meiotic pairing, with *Hylandra suecica*, *Cardaminopsis arenosa* and *Arabidopsis pumila* (Berger, 1968).

Recombination

Linkage is estimated in F_2 by the product method (Immer, 1930). The small size of the plants and the large seed yield make possible the detection of chromatid segregation (double reduction) in triplex tetraploids or duplex trisomics (Lee-Chen and Sears, 1967; Bouharmont, 1969), and the study of centromere location becomes feasible by such genetic methods.

Maps of very primitive type are known (Rédei and Hirono, 1964; Rédei, 1965b; McKelvie, 1965). Linkage groups can be assigned to the five chromosomes by trisomic analysis (Lee-Chen and Steinitz-Sears, 1967; Sears and Lee-Chen, 1970).

A novel method of gene localization was worked out by Hirono (1964), utilizing the mathematical correlations among marker transmission rates, segregation ratios, and map distances. A simplified graphical solution which considers either male or female gametophyte factors separately has been developed by Li (1968).

Gametophyte factors affecting male (Rédei, 1964; Hirono and Rédei, 1965a,b; Li, 1967; Van der Veen and Wirtz, 1968) or female transmission (Rédei, 1965a) are known. A female-gametophyte factor is known which modifies recombinationation frequency in its vicinity (Rédei, 1965a).

Premeiotic recombination has been shown to occur between appropriately marked chromosomes following x-irradiation (Hirono and Rédei, 1965a,b); somatic crossing over has been claimed to cause mosaicism after chemical mutagenesis (Demchenko *et al.*, 1972; Ahnström *et al.*, 1972). Somatic juxtaposition of chromosomes has been actually observed (Steinitz-Sears, 1963), and it provides a plausible physical basis for the phenomenon.

Mutation

Few other higher plants have been used as extensively for testing radiobiological, pharmacological, and mutagenic agents as *Arabidopsis* (Ehrenberg, 1971). This autogamous plant lends itself favorably to fast evaluation of various mutagens because of its short life cycle and small size. The normally great genetic stability is combined with favorable response to mutagens and the exhibition of a wide spectrum of induced morphological and physiological variations. The genetic effect of about 100 physical and chemical agents on *Arabidopsis* has been tabulated recently (Rédei, 1970).

Detection methods are concerned primarily with forward mutation.

Müller's embryo test (1963, 1965*d*) detects lethal or chlorophyll-defective embryos in the 11–13-day-old siliques of the plants which develop from mutagen-treated seeds. Thus, the planting of the second generation can be avoided. Although this method may provide information on mutagenicity within a month, direct physiological effects and differences in transmission are not separated, and therefore the genetic information provided is limited.

An alternative procedure is to dispense, with the aid of a separatory funnel, approximately 50 mutagen-treated seeds in an agar suspension into a five-inch pot. The survivors are harvested in bulk, and their seed is spread onto pots of the same size; mutants can then be isolated during several development stages (Li and Rédei, 1969*a*; Rédei, 1973*a*). By this technique, thousands of mutants can be isolated for further use. To improve the efficiency of isolation of mutations, Harle (1972) has recommended the use of relatively small M_1 populations and testing in M_2 all the offspring of each inflorescence. The advantages of maximal M_1 and minimal M_2 populations are more economical, however (Rédei, 1974*a,b*).

Somatic mutations, deletions or recombination involving specific loci, can be detected by the treatment of plants heterozygous for suitable marker genes (Hirono and Rédei, 1965*a*; Gichner and Velemínshý, 1965; Fujii, 1964*a*; Fujii *et al.*, 1966; Hirono *et al.*, 1970). Although the majority of genes display sectoring if treated with mutagenic agents in heterozygous condition (Rédei, 1967*e*), only a few of the available markers can be safely used for such tests (Rédei and Li, 1969*a*) because chemical mutagens induce a wide variety of types of "dominant" sectoring which are impossible to identify on a genotypic basis. Rapid and effective chemical (Rédei, 1967*a*) or progeny tests, if the sector develops into fruit-bearing structures (Hirono and Rédei, 1965*a*; Röbbelen, 1972), are necessary. Röbbelen (1972) has developed a somatic reversion test for chlorophyll-deficiency mutants.

Calculation of mutation rates are simple if haploid germ cells are treated with the mutagen (Röbbelen, 1960, 1962*b*; Müller, 1965*c*). Caution is necessary in the interpretation of the data (Röbbelen, 1962*b*) if treatment of the gametophytic stage is compared with treatments given in multicellular diploid stages (Rédei, 1970). A theoretically satisfactory general method for the estimation of mutation rate has been provided (Li and Rédei, 1969*c*).

Spontaneous mutation rate is extremely difficult to estimate in all organisms because contaminations can rarely be ruled out in an entirely satisfactory manner, even in multiply marked material. In various laboratories spontaneous mutation rates have been estimated by different procedures and found to range between 0.01 and 1–3 percent (Rédei, 1970). Mutator factors affecting the mutability of one (Röbbelen, 1966b) or several (Rédei, 1973a; Rédei and Plurad, 1973) specific sites in the plastome are known.

The dependence of the mutation spectrum on the genotype, the stage of treatment, or the agent used has been discussed by Röbbelen (1960, 1962b, 1965c), Jacobs (1964, 1969c), and Müller (1964, 1967a). McKelvie (1962b) has warned that "claims for mutation specificity in higher plants must be regarded with caution," although differences between chemical and x-ray mutations were observed. The frequency of mutation for thiamine auxotrophy was remarkably close in *Arabidopsis* to that in microbes, yet no other obligate auxotrophs could be obtained, though thousands were expected (Rédei, 1965b; Li *et al.*, 1967). A hypothesis of developmental selection has been proposed by Langridge (1958a) to explain the absence of certain types of mutations in multi-cellular organisms.

Reverse mutation has been studied in somatic tissues of leaf-color mutants, and followed by progeny analysis. In the overwhelming majority of the cases, supressor mutations have been involved (Röbbelen, 1972).

Only a very small fraction of the available mutants have been characterized or reported. Certain morphological mutants have been listed and described by Reinholz (1947) and by McKelvie (1962a). Many chlorophyll mutants have been characterized by Röbbelen (1957a), and thiamine auxotrophs at four loci have been listed by Li and Rédei (1969e). A large collection of natural races and ecotypes have been described by Röbbelen (1965a,b).

Radiation Biology

The radiosensitivity of *Arabidopsis* is one of the lowest among species of higher plants (Sparrow *et al.*, 1972; Gomez-Campo and Delgado, 1964). This high tolerance of radiation may be related to the extremely small volume and the very low DNA content of the nucleus. The effect of radiation upon seed germination has been studied by Ivanov *et al.* (1967) and by Reinholz (1947, 1967, 1968). For the induction of useful mutants, doses not exceeding 13 kr for imbibed seeds and below 60 kr for dry seeds seem desirable (Reinholz, 1947; Rédei, 1962c; Mesken and Van der Veen, 1968). The irradiation of the seed, may lead to tumorous growth (Hirono *et al.*, 1968). The deleterious effects upon seed viability of ex-posures to fast neutrons relative to x and gamma rays have been studied by Fujii (1967) and by Timofeeff-Ressovsky *et al.* (1971).

The small seed of *Arabidopsis* (2×10^{-3} mm^3) have been used for studying the biological effects of ionizing radiations of widely different linear energy transfers (Fujii *et al.*, 1966; Hirono *et al.*, 1968, 1970). The absolute radiation sensitivities of *Arabidopsis* and *Vicia faba* are widely different, yet the RBE's (relative biological effectiveness) of gamma rays and fission neutrons in both species are remarkably similar, if expressed on the basis of nuclear volume (Donini *et al.*, 1967; Yamakawa and Sparrow, 1966).

The mutagenic effects of ultraviolet irradiation and photoreactivation were studied by Fujii (1965). High mutagenicity of modulated-pulse ruby laser beams was reported by Usmanov *et al.* (1970*b*).

Modification of the radiation sensitivity has been accomplished by temperature treatments (Reinholz, 1954*a*, 1965*a*; Rajewsky, 1953; Nikolov and Ivanov, 1969; Daly, 1971), and various chemicals have been tested for possible protection effects against radiation injury (Reinholz and Aurand, 1954; Kucera, 1966; Van der Veen *et al.*, 1969; Reinholz, 1968) or for enhancing x-ray sensitivity (McKelvie, 1962*b*; Koo, 1969). According to Reinholz and Aurand (1954) N$_2$ atmosphere provides effective radiation protection, but Müller (1966*b*) failed to observe differences between O$_2$ and N$_2$ atmosphere. Radiation applied in different developmental stages was studied by Reinholz (1959), Röbbelen (1962*b*), Müller (1965*c*), and Usmanov and Müller (1970).

Chemical mutagenesis in *Arabidopsis* has been reviewed recently in detail (Rédei, 1970; Ehrenberg, 1971; Müller, 1972). Methanesulfonic esters have been shown to be highly mutagenic (Röbbelen, 1962*c*; McKelvie, 1963; Müller, 1966; Gichner *et al.*, 1968). The imbibition of the seed prior to ethyl methanesulfonate administration increases the effectiveness of the treatment by facilitating the uptake of the mutagen (Röbbelen, 1965*d*). The degree of labeling of *Arabidopsis* is proportional to the mutation frequency when the seed is treated with ^{3}H-ethyl-methanesulfonate (Walles and Ahnström, 1965). Dimethyl sulfoxide facilitates the absorption of ethyl methanesulfonate (Bhatia, 1967) and ZnSO$_4$ and CuSO$_4$ increase the effectiveness of this mutagen in a pH-dependent fashion (Bhatia and Narayanan, 1965). Chelating does not appear to be influential on mutagenicity of this chemical (Gichner and Veleminský, 1965). The nitrosoureas are also strong mutagens (Müller, 1964, 1965*a,b*; Gichner and Veleminský, 1967; Veleminský *et al.*, 1967, 1970).

The nitrosoguanidines and nitrosoamines are highly effective mutagens (Müller and Gichner, 1964; Gichner and Veleminský, 1967; Veleminský and Gichner, 1968, 1971). The effectiveness of nitrosoureas is reduced by enzyme inhibitors, but the nitrosoguanidine effect is not. This

suggests that enzymes convert the former group into active compounds
(Müller, 1965e). Nitroso-imidazolidine deserves special mention on account of its superior efficiency of inducing back mutation (Röbbelen,
1972). Nitrous acid is a poor mutagen (Brock, 1971); nucleoside analogs
are weak mutagens in *Arabidopsis* (Hirono and Smith, 1969; Jacobs,
1967).

Physiological Genetics

In the field of physiological genetics, *Arabidopsis* has a greater
potential than any other higher plant because of the availability of nutritional mutants.

Auxotrophy is known for at least four loci in the thiamine pathway
(Feenstra, 1964; Rédei, 1965b; Li and Rédei, 1969e). The majority of the
mutants require the pyrimidine moiety (Rédei, 1960), several are blocked
in the synthesis of the thiazole molecule (Rédei, 1962a), and there are two
loci that respond to the intact thiamine molecule (Li and Rédei, 1969e;
Langridge, 1958a, 1965). The majority of the mutants are obligate
auxotrophs. They die at the cotyledon stage in the absence of the required
compound, but in its presence their growth is not distinguishable from the
wild type. Three alleles at the pyrimidine locus are temperature sensitive
(Li and Rédei, 1969e). Part of the thiamine pathway appears to be subject
to catabolite repression (Li and Rédei, 1969d).

A favorable response to leucine supplementation has been observed in
certain viable plastid mutants (Kasyanenko *et al.*, 1971). This finding requires the type of scrutiny that has been applied to the "leucine mutants"
of barley (Land and Norton, 1970). Coconut milk, carbohydrates, choline
(Langridge, 1958a), biotine, cytidine (Langridge and Griffing, 1959) cy-
tosine, uracil, arginine, cysteine, methionine, histidine, tryptophan, nicotinic acid, paraaminobenzoic acid, ascorbic acid, and inositol requirements
(Jacobs, 1965) have been reported, but not confirmed (e.g., Neales,
1968b). A mutant is known that responds favorably to elevated osmotic
pressure of the medium provided by salts or carbohydrates (Langridge,
1958b). Some correction of the phenotype of dwarf mutants was accomplished by gibberellin (Napp-Zinn and Bonzi, 1970). Chlorate-resistant
mutants with an elevated level of nitrate reductase activity have been observed by Oostindier-Braaksma and Feenstra (1973).

Except for those mutants involved in the thiamine pathway, the most
extensive studies on auxotrophy in *Arabidopsis* have failed to reveal
mutants with verified requirements for specific metabolites. [In the case of
the thiamine pathway, over 60 independent mutations are available (Li

and Rédei, 1969a).] An analysis of the auxotrophic mutation spectra of a wide range of organisms has indicated unexpected limitations in all chloroplast-containing organisms (Li *et al.*, 1967). Late-flowering mutants (Rédei, 1962c) responding to nucleoside analogs with early flowering have been studied (Hirono and Rédei, 1966a,b; Rédei, 1969).

Another "antimetabolite-requiring" locus, *im*, is involved in the control of plastid differentiation. Some mutant plants display a tenfold improvement in plastid differentiation when grown on azauracil-, azauridine-, or azacytidine-containing media (Rédei, 1965d, 1967d). The *im* locus appears to contain regulatory information; the synthesis of a cytosol-localized ribonuclease (Rédei, 1967b) and an orotidylic acid pyrophosphorylase and decarboxylase seem to be under its control (Rédei *et al.*, 1973, Chung and Rédei, 1974).

Arabidopsis has been used more extensively than any other higher plant for the study of uptake of foreign DNA (Ledoux *et al.*, 1971a,b). Seed exposed for four days to bacterial DNA's of distinct buoyant density take up these macromolecules without substantial degradation. The labeled foreign DNA stays mainly in the cotyledons until flowering, when it moves to the developing embryo and is transmitted to subsequent generations of nuclei; the genetic information of the bacterial DNA seems to correct the genetic defects of auxotraphic mutants of the plant.

Doy and associates (1973a,b) have reported that λ and φ80 phages confer galactose and lactose resistance on haploid cells of the plant. The information transfer, termed transgenosis, requires the functional bacterial operon; the basic nature of the process is not understood.

Isozyme studies have revealed variations in acid phosphatase, leucine amino peptidase, transaminase, and peroxidase in different populations of *Arabidopsis* (Jacobs and Schwind, 1972; Grover and Byrne, 1972), and Lavrinetskaya and Strekhalov (1970) have discussed the importance of such protein analyses for taxonomic purposes. Fructose-1, 6-diphosphate aldolase was found to display three identical bands in *A. thaliana*, *Cardaminopsis arenosa*, and *Hylandra suecica* and also a fourth band in the two latter species (Rédei, 1973b).

Studies of chlorophyll mutants by Röbbelen (1957a) have shown that plants can survive only if the total leaf-pigment content stays above 6 percent of the wild-type value. One recessive mutant produces only protochlorophyll, and this can be transformed into chlorophyll by extracts of the wild type (Röbbelen, 1956). Another mutant which lacks chlorophyll b fails to convert sugar into starch (Röbbelen, 1957c). Three different alleles of the *ch* locus in linkage group 4 are characterized by the absence or by reduced levels of chlorophyll b. They can still synthesize starch (Hirono and Rédei, 1963), although their photosynthetic activity is reduced (Kranz, 1968). The ultrastructure of the chloroplast in the chlorophyll-b-free mutant ch^1 appears normal (Veleminský and Röbbelen, 1966). Illumination with yellow light (400 nm) reduces the chlorophyll-a and carotenoid content in the wild type and in the mutants, but it does not affect the chlorophyll-b level. Extracts of the wild-type plants appear to promote chlorophyll formation in the mutants (Kranz, 1971).

Mutants with increased content of total leaf pigment, modified spectra of pigments, and variable photosynthetic ability have been found (Kasyanenko and Timofeeff-Ressovsky, 1967). An accumulation of glutamine, asparagine, and total nitrogen has been reported for certain chlorophyll mutants (Svachulová, 1967), and an abnormal cathodic structural protein has been detected in two lethal pigment mutants (Svachulová and Hadacova, 1972). A reduction of leaf-pigment synthesis during thiamine deficiency (Rédei, 1965b) and an interference with chlorophyll and carotenoid synthesis in the presence of the degradation products of fructose were observed (Rédei, 1973b,c, 1974c).

Developmental Genetics

The genetic control of chloroplast differentiation has been studied by Röbbelen (1959). The fine structure and development of lamellae and grana were revealed by electron microscopic analyses (Röbbelen and Wehrmeyer 1965; Wehrmeyer and Röbbelen, 1965; Röbbelen, 1966a), and environmental and genetic interactions during plastid differentiation have been analyzed (Röbbelen, 1965b). The dispensable nature of chlorophyll b as opposed to chlorophyll a in plastid differentiation has been shown (Veleminský and Röbbelen, 1966). Several plastid mutants have been characterized (Röbbelen, 1957a; Kasyanenko et al., 1971). A quantitative method using a linear integrator has been developed for the estimation of plastid defects (Abdullaev et al., 1972).

A locus, im, in linkage group 4 controls plastid differentiation by regulating the synthesis of specific enzymes involved in nucleic acid metabolism (Rédei, 1967c). Chloroplast differentiation in im mutants can be controlled by light of different wave lengths (Rédei, 1965d; Röbbelen, 1968), and can be promoted by azapyrimidines, kinetin (Rédei, 1965d, 1967d), cysteine (Rédei, 1963), and heavy doses of x rays (Rédei, 1967b). On 6-azauracil media, an unusual curved but highly functional lamellar structure develops in the mutant plastid. In the absence of the analog, at high intensity illumination, there is no lamella differentiation in the mutant (Chung *et al.*, 1974, Rédei *et al.*, 1973). The *im* locus has the largest number of alleles, nearly 100, in Arabidopsis (Röbbelen, 1968; Rédei, unpublished).

In x-rayed material, the mutation rate for the plastome increases to several times the natural value (Röbbelen, 1962*a*). Nuclear-mutator factors for various plastome sites have been analyzed genetically and structurally (Röbbelen, 1966*b*; Rédei, 1973*a*; Rédei and Plurad, 1973).

Pollen morphology is influenced by the level of ploidy (Bronckers, 1963). The growth rates for pollen tubes on the stigma and in the pistillar tissues were studied in species crosses by Berger (1968). Several male-gametophyte factors have been described (Rédei, 1964; Hirono and Rédei, 1965*a*; Van der Veen and Wirtz, 1968), and genetic methods for mapping male-gametophyte factors have been developed by Hirono (1964) and Li (1968). Mutations were induced in pollen by Müller (1965*c*), Röbbelen (1960, 1962*b*, 1965*c*, 1966*c*), and Usmanov and Müller (1970). Androgenesis has been observed (Arnold and Cruse, 1968; Barabás and Rédei, 1971); semigamy has been induced by irradiation, and it has been detected by microscopic genetic methods (Gerlach-Cruse, 1970).

Under normal conditions, megasporogenesis follows the pattern previously described for *Polygonum* (Vandendries, 1909; Misra, 1962; Polyakova, 1964; Rédei, 1965*a*). A rare mutation induces the formation of double tetrads and thus permits megaspore selection and the exclusively male transmission of the trait. This pleiotropic, female-gametophyte factor acts as a segregation distorter for genes on one arm of the chromosome and suppresses crossing over in its vicinity (Rédei, 1965*a,c*).

Embryogenesis normally represents the *Capsella* variation of the Onagrad type (Vendendries, 1909; Reinholz, 1959). If the pistil is irradiated, di- and trispermic fertilization and abnormal development are seen (Gerlach-Cruse, 1969; Reinholz, 1954*b*, 1959). Irradiation of the female gametophyte occasionally prevents the fusion of the gametes. This results in the separate division of egg and sperm nuclei (semigamy) and in other cytological disturbances (Gerlach-Cruse, 1970). Thalidomide (0.1 percent) applied before the first division blocks the development of two-thirds of the zygotes (Arnold and Cruse, 1967). The embryogenesis is sometimes abnormal in polyploids (Bouharmont, 1965), and various anomalies in embryo differentiation in species hybrids have been described (Berger, 1968). Embryonic lethality has been classified into six major types (Múller, 1963). The trifolia mutant of Reinholz (1965*b*) displays an unusually high penetrance of tricotyledony, a character that fascinated, but eluded, De Vries at the turn of the century. Other variants changing cotyledon number are described by Napp-Zinn (1955).

Germination in *Arabidopsis* is facilitated by exposure to red light (660 nm) and inhibited by far-red (720-740 nm) illumination (Shropshire *et al.*, 1961). Germination in the dark is determined by the quality of light exposed to in the preceding period of light (McCullogh and Shropshire, 1970). The degree of dormancy and dark germination is genetically determined (Niethammer, 1926; Kugler, 1951; Ratcliffe, 1961; Baskin and Baskin, 1972), and it is also temperature dependent (Laibach, 1956; Corcos, 1969). Some filter papers inhibit germination (Rehwaldt, 1968). X-irradiation can break the requirement for light in the light-requiring genotypes (Reinholz, 1968). Gibberellic acid (0.02 percent) treatment mitigates the radiation-caused germination failure (Reinholz, 1967).

Lists and descriptions of morphological mutants have been published by Reinholz (1947), McKelvie (1962a), and Rédei (1970). Descriptions of the wild-type plant are given by Müller (1961), Röbbelen (1957b), and Napp-Zinn (1957a, 1963a).

Flowering in *Arabidopsis* is controlled by vernalization and/or photoperiodic requirements (Laibach, 1940, 1943, 1951; Laibach and Zenker, 1954; Zenker, 1955). The genetic basis of the flowering in crosses with vernalization-requiring species has been investigated by Napp-Zinn (1957c, 1962). The genetic studies on vernalization were greatly hampered by the multigenic control of this physiological phenomenon and by the lack of monogenic mutants sharing identical genetic backgrounds. Biochemical and physiological characterization of the vernalization reaction has been attempted by Napp-Zinn (1954, 1957b,c,d, 1960a,b, 1962, 1963a,b, 1971) and Bonzi and Napp-Zinn (1967). A theoretical pathway for flower induction in cold-requiring genotypes has been proposed (Napp-Zinn, 1965).

The development of the vegetative and reproductive apices are well characterized in *Arabidopsis* (Vaughan, 1955; Miksche and Brown, 1965). The effect of vernalization on the transition from the vegetative to generative phase of the apex has been studied with autoradiographic techniques (Wibaut, 1966).

The photoperiodic response is controlled by one major and by a few minor genes in natural ecotypes of this long-day plant (Härer, 1951). Single recessive mutations with major qualitative effects on photoperiodism have been induced (Rédei, 1962c; Reinholz, 1947; McKelvie, 1962a; Brown, 1964). The photoperiodic response can be modified by various chemical treatments (Rédei, 1962c, 1969; Brown, 1962, 1968, 1972; Hirono and Rédei, 1966a,b; James, 1969; Michneiwicz and Kamiénska, 1965; Tsuboi and Yanagishima, 1971; Langridge, 1957a; Brown *et al.*, 1964; Wibaut, 1966; Besnard-Wibaut, 1966, 1968, 1970).

Extra-Chromosomal Inheritance

The spontaneous rate of plastome mutations is increased by x-irradiation (Röbbelen, 1962a). Nuclear genes are also capable of inducing mutations in the plastome, which then are transmitted maternally and autonomously (Röbbelen, 1966b; Rédei, 1973a; Rédei and Plurad, 1973). Homoplastidic mutant lines have been isolated and maintained (Rédei, 1973a; Rédei and Plurad, 1973). Halogenated deoxyribonucleoside analogs produce some odd types of laggard plants (Brown and Smith, 1964) which appear to be controlled cytoplasmically (Jacobs, 1969a).

Quantitative Inheritance

In *Arabidopsis* a variety of polygenic traits has been subjected to study (Barthelmess, 1964, 1965; Dobrovolná, 1969; Van der Veen, 1965; Westerman, 1971; Westerman and Lawrence, 1970; Pederson and Matzinger, 1972). Quantitative variations induced by physical and chemical mutagens have been the subject of several extensive studies (Bhatia and Van der Veen, 1965; Brock, 1965, 1967a,b, 1968; Brock and Shaw, 1969; Daly, 1960a,b, 1961, 1973; Hussein, 1968; Kučera, 1965, Lawrence, 1965, 1968; Dragavtsev et al., 1970).

Heterosis appears to be the result of interactions of dominant genes (Dierks, 1958; Van der Veen, 1965; Hussein, 1968). Superdominance in combinations of two monogenic mutants on verified isogenic background has been demonstrated (Rédei, 1962b). The superiority of three doses of the wild-type allele and one dose of the *chlorina* mutation over four wild-type genes has been established (Wricke, 1955). Monogenic heterosis in chlorophyll-deficient mutants has been reported by Röbbelen (1957a). Allelic complementation by various temperature-sensitive pyrimidine auxotrophs provides the best genetic evidence so far for the existance of overdominance and for the improved phenotypic stability of the hybrids in higher plants (Li and Rédei, 1969a,b; Rédei and Li, 1969b).

A mechanism for heterosis has been proposed in terms of the favorable interaction of temperature-sensitive alleles, and manifested at elevated temperatures in inbreeders and also at low temperatures in outbreeders (Langridge, 1962, 1963; Griffing and Langridge, 1962). A dependence of heterosis upon environmental factors has also been shown (Griffing and Zsiros, 1971; Pederson, 1968).

Population Genetics

The geographical distribution of *A. thaliana* in the Northern Hemisphere is very wide. In the tropical areas it is limited to the high elevations, and the species is very uncommon in the southern half of the globe (Rédei, 1970). The genecenter of *Arabidopsis* is probably in Western Europe.

The distribution of certain genes in natural populations of *Arabidopsis* is determined by the ecological conditions (Ratcliffe, 1961, 1965; Brodführer, 1955; Lockhart and Brodführer-Franzgrote, 1961). In relatively small geographical areas, considerable genetic diversity can be found (Laibach, 1951; Cetl, 1965; Usmanov *et al.*, 1970a; Napp-Zinn, 1957c, 1963b, 1964; Härer, 1951; Kugler, 1951; Jones, 1971; Langridge and Griffing, 1959; Griffing and Langridge, 1962; Ashraf, 1970). Under laboratory conditions, several induced mutants display an extremely high selective advantage in competition with the wild type (Rédei, 1962c). A selective advantage of the autotetraploids has been noted (Bouharmont and Mace, 1972).

In the laboratory most of the genotypes are obligate selfers. In nature, however, outcrossing may approach 2 percent (Snape and Lawrence, 1971) because small insects may frequent the plants, which are characterized by protogyny and often display protruding stigmae (Jones, 1971).

Taxonomy

The taxonomical position of *Arabidopsis* is not entirely settled (Berger, 1965; Rédei, 1970). None of the numerous *Arabidopsis* species have the same basic chromosome number as *A. thaliana*. Only 6 of the 194 interspecific combinations attempted produced viable seed, although fertilization took place in several other cases. Chromosomal homologies between *A. thaliana* and *A. pumila*, *Cardaminopsis arenosa*, and *Hylandra suecica* have been detected (Berger, 1968).

Hybrids have been made between *A. thaliana* and *Hylandra suecica* ($n = 13$), *Hylandra* and *Cardaminopsis arenosa* ($n = 16$) (Laibach, 1958; Kribben, 1965; Berger, 1968), and *Cardaminopsis petraea* ($n = 8$) and *A. thaliana* ($n = 5$) (Měsíček, 1967).

Hylander (1957) proposed that *Hylandra suecica* is an amphiploid of a diploid *Cardaminopsis arenosa* ($n = 8$) and a diploid *A. thaliana*. Löve (1961) has suggested that the original hybridization took place at the tetraploid level. Crossing of these two species remained unsuccessful, however, until recently (Rédei, 1972b), but the hybrid is different in many characters from Hylandra, though they can form fertile offspring when crossed (Rédei, 1974d).

Literature Cited

Abdullaev, H. A., P. D. Usmanov, S. V. Tageeva and Y. S. Nasyrov, 1972   Quantitative estimation of plastid microstructure in *Arabidopsis thaliana.* Arabidopsis Inf. Serv. **9**:26–27.

Ahnström, G., A. T. Natarjan and J. Veleminský, 1972   Chemically induced somatic mutations in *Arabidopsis. Hereditas* **72**:319–22.

Anand, R., 1966   Preliminary studies on callus culture of *Arabidopsis thaliana. Arabidopsis Inf. Serv.* **3**:15.

Arnold, C. G. and D. Cruse, 1965   Eine 8-chromosomige Mutante von *Arabidopsis thaliana. Flora (Jena)* **155**:474–476.

Arnold, C. G. and D. Cruse, 1967   Die Wirkung von Thalidomid auf die Embryoentwicklung von *Arabidopsis thaliana. Z. Pflanzenphysiol.* **56**:292–294.

Arnold, C. G. and D. Cruse, 1968   Pflanzen mit vaterlichen Eigenschaften nach Röntgenbestrahlung. *Arabidopsis Inf. Serv.* **5**:39.

Ashraf, J., 1970   Variability of cell-temperature resistance in ecological races of *Arabidopsis thaliana* (L.) Heynh. *Arabidopsis Inf. Serv.* **7**:11–12.

Barabás, Z. and G. P. Rédei, 1971   Frequency of androgenesis. *Arabidopsis Inf. Serv.* **8**:9–10.

Barthelmess, I., 1964   Merkmalskorrelationen und Selektion bei *Arabiodopsis thaliana* (L.) Heyhn. *Z. Pflanzenzücht.* **52**:273–332.

Barthelmess, I., 1965   Vom Blühalter-abhängige positive oder negative Merkmalskorrelationen. In *Arabidopsis Research, Report of the International Symposium, Göttingen,* edited by G. Röbbelen, pp. 72–78, University of Göttingen, Göttingen.

Baskin, J. M. and C. C. Baskin, 1972   Ecological life cycle and physiological ecology of seed germination of *Arabidopsis thaliana. Can. J. Bot.* **50**:353–360.

Berger, B., 1965   The taxonomic confusion within Arabidopsis and allied genera. In *Arabidopsis Research, Report of the International Symposium,* Göttingen, edited by G. Röbbelen, p. 19–25, University of Göttingen, Göttingen.

Berger, G., 1968   Entwicklungsgeschichtliche und chromosomale Ursachen dar verschiedenen Kreuzungverträglichkeit zwischen Arten des Verwandtschaftkreises Arabidopsis. *Beitr. Biol. Pflanz.* **45**:171–212.

Besnard-Wibaut, C., 1966   Étude historadioautographique des modifications du fonctionnement apical de l'*Arabidopsis thaliana* (L.) Heynh. en photoperiode défavorable à la floraison. *C. R. Hebd. Séances Acad. Sci. Ser. D Sci. Nat.* **263**:1582–1585.

Besnard-Wibaut, C., 1968   Modification des synthèse nucléiques dans l'apex de l'*Arabidopsis thaliana* (L.) Heynh. Lors du passage à l'état reproducteur. Analyse histoautoradiographique de l'action de la photopériode et de l'acide gibbérellique sur le fonctionnement apical. In *Cellular and Molecular Aspects of Floral Induction, International Symposium,* edited by G. Bernier p. 7–9, Longman's Green and Co., London.

Besnard-Wibaut, C., 1970   Action comparée de la photopériode et de l'acide gibbérellique sur le fonctionnement apical de l'*Arabidopsis thaliana.* Étude histoautoradiographique des syntheses nucléiques et de leur modification lors du passage à l'état reproducteur. *Rev. Cytol. Biol. Vèg.* **33**:265–280.

Besnard-Wibaut, C., 1972   Analyse cytochimique et histoautoradiographique des processus de floraison chez l'*Arabidopsis thaliana,* race Stockholm vernalisé à l'état de graines. *C. R. Hebd. Séances Acad. Sci. Ser. D. Sci. Nat.* **274**:1161–1164.

Bhatia, C. R., 1967   Increased mutagenic effect of ethyl methanesulfonate when dissolved in dimethylsulfoxide. *Mutat. Res.* **4**:375–376.

Bhatia, C. R. and K. R. Narayanan, 1965   Genetic effects of ethyl methanesulfonate in combination with copper and zinc ions on *Arabidopsis thaliana. Genetics* **52**:577–581.

Bhatia, C. R. and J. H. Van der Veen, 1965   The use of induced mutations in plant breeding. Two-way selection for EMS-induced micro mutations in *Arabidopsis thaliana* (L.) Heynh. *Radiat. Bot.* **5**:*Suppl.* 497–503.

Bonzi, G. and K. Napp-Zinn, 1967   Accélération de la mise à fleur par des géranyl-valérates chez la race "Zürich" d'*Arabidopsis thaliana* (L.) Heynh. *C. R. Hebd. Séance Acad. Sci Ser. D. Sci. Nat.* **265**:962–964.

Bouharmont, J., 1964   Action des rayons x sur une sérine auto-polyploide d'Arabidopsis. In *The Use of Induced Mutations in Plant Breeding. Radiat. Bot. Suppl.* **5**:649–657.

Bouharmont, J., 1965   Fertility studies in polyploid *Arabidopsis thaliana.* In *Arabidopsis Research, Report of the International Symposium, Göttingen,* edited by G. Röbbelen, pp. 31–36, University of Göttingen, Göttingen,

Bouharmont, J., 1969   Étude de l'hérédité tétrasomique des mutations induites chez *Arabidopsis thaliana.* In *International Atomic Energy Agency/Food and Agricultural Organization of the United Nations Symposium: Induced Mutations in Plants,* pp. 603–610, IAEA, Vienna.

Bouharmont, J. and F. Mace, 1972   Valeur compétitive des plantes autotétraploides d'*Arabidopsis thaliana. Can. J. Genet. Cytol.* **14**:257–263.

Brock, R. D., 1965   The use of induced mutations in plant breeding. Induced mutations affecting quantitative characters. *Radiat. Bot. Suppl.* **5**:451–464.

Brock, R. D., 1967a   Quantitative variation in *Arabidopsis thaliana* induced by ionizing radiations. *Radiat. Bot.* **7**:193–203.

Brock, R. D., 1967b   Induced quantitative variation in *Arabidopsis thaliana. Abh. Dtsch. Akad. Wiss. Berl. Kl. Med.* **2**:263–267.

Brock, R. D., 1968   Induced quantitatively inherited variation in *Arabidopsis thaliana.* In *Mutation in Plant Breeding. II. Int. At. Energy Agency Panel Proc. Ser.* pp. 57–58.

Brock, R. D., 1971   Nitrous acid as a mutagenic agent for *Arabidopsis thaliana. Radiat. Bot.* **11**:309–311.

Brock, R. D. and H. F. Shaw, 1969   Response to a second-cycle mutagenic treatment in *Arabidopsis thaliana.* In *Proceedings of the International Atomic Energy Agency/Food and Agricultural Organization of the United Nations Symposium: Induced Mutations in Plants,* pp. 457–467, IAEA, Vienna.

Brodführer, U., 1955   Der Einfluss einer abgestuften Dosierung von ultravioletter Sonnenstahlung auf das Wachstum der Pflanzen. *Planta (Berl.)* **45**:1–56.

Brodführer, U., 1957   The effect of temperature on the reaction of plants to ultraviolet radiation. *Carnegie Inst. Wash. Year Book* **56**:288–291.

Bronckers, F., 1963   Variations pollinique dans une série d'autopolyploides artificiels d'*Arabidopsis thaliana* (L.) Heynh. *Pollen Spores* **5**:233–238.

Brown, J. A. M., 1962   Effect of thymidine analogues on reproductive morphogenesis in *Arabidopsis thaliana* (L.) Heynh. *Nature (Lond.)* **196**:51–53.

Brown, J. A. M., 1964   A temperature-sensitive late-flowering mutant of Arabidopsis. *Genetics* **50**:237–238.

Brown, J. A. M., 1968   The role of competitive halogen analogues of thymidine in the induction of floral morphogenesis. In *Cellular and Molecular Aspects of Floral Induc-*

tion, International Symposium, edited by G. Bernier, p. 22–25, Longman's Green and Co., London.

Brown, J. A. M., 1972 Distribution of ³H-thymidine in *Arabidopsis* vegetative meristems after 5-iododeoxyuridine treatment. *Am. J. Bot.* **59**:228–232.

Brown, J. A. M. and H. H. Smith, 1964 Incorporation and effects of thymidine analogues in gametophytic tissue of *Arabidopsis thaliana. Mutat. Res.* **1**:45–53.

Brown, J. A. M., J. P. Miksche and H. H. Smith, 1964 An analysis of H³-thymidine distribution throughout the vegetative meristem of *Arabidopsis thaliana* (L.) Heynh. *Radiat. Bot.* **4**:107–113.

Brown, J. A. M., C. R. Bhatia and H. H. Smith, 1965 Lethal, laggard, and flowering-time variants in generations following thymidine analogue treatment of *Arabidopsis* buds. In *Arabidopsis Research, Report of the International Symposium, Göttingen,* edited by G. Röbbelen, p. 171–183, University of Göttingen, Göttingen.

Cetl, I., 1965 Some developmental features in natural populations of *Arabidopsis thaliana* (L.) Heynh. In *Arabidopsis Research, Report of the International Symposium, Göttingen,* edited by G. Röbbelen, p. 46–52, University of Göttingen, Göttingen.

Chung, S. C. and G. P. Rédei, 1974 An anomaly of the genetic regulation of the *de novo* pyrimidine pathway in the plant *Arabidopsis. Biochem. Genet.,* in press.

Chung, S. C., G. P. Rédei and J. A. White, 1974 Plastid differentiation on 6-azauracil media. *Experientia (Basel),* **30**:92–94.

Corcos, A., 1969 Isolation of germination temperature-sensitive mutants in *Arabidopsis thaliana. Arabidopsis Inf. Serv.* **6**:15.

Daly, K., 1960*a* The induction of quantitative variability by γ-radiation in *Arabidopsis thaliana. Genetics* **45**:983.

Daly, K., 1960*b* Effect of temperature on survival of gamma-irradiated Arabidopsis seed. *Radiat. Res.* **12**:430.

Daly, K., 1961 The effect of fast neutrons on quantitative variability in *Arabidopsis thaliana. Genetics* **46**:861.

Daly, K. R., 1971 Effect of temperature on survival of γ-irradiated seeds of *Arabidopsis thaliana. Experientia (Basel)* **27**:81–83.

Daly, K., 1973 Quantitative variation induced by gamma rays and fast neutrons in *Arabidopsis thaliana. Radiat. Bot.* **13**:149–154.

Demchenko, S. I., R. Ya. Serova and V. A. Avetisov, 1972 The effect of concentration and exposure on leaf spotting in M_1 in *Arabidopsis* after HMU treatment. *Arabidopsis Inf. Serv.* **9**:31–33.

Dierks, W., 1958 Untersuchungen zum Heterosisproblem. *Z. Pflanzenzuecht.* **40**:67–102.

Dobrovolná, J., 1969 The estimation of heritability of some developmental characters in natural populations of *Arabidopsis thaliana* (L.) Heynh. *Biol. Plant. (Prague)* **11**:310–318.

Donini, B., A. H. Sparrow, L. A. Schairer and R. C. Sparrow, 1967 The relative biological efficiency of gamma rays and fission neutrons in plant species with different nuclear volumes. *Radiat. Res.* **32**:692–705.

Doy, C. H., P. M. Gresshoff and B. Rolfe, 1973 Transgenosis of bacterial genes from Escherichia coli to cultures of haploid *Lycopersicon esculentum* and haploid *Arabidopsis thaliana* plant cells. In *The Biochemistry of Gene Expression in Higher Organisms,* pp. 21–38, Australia and New Zealand Book Co., Artarmon, New South Wales, Australia.

Doy, C. H., P. M. Gresshoff and B. G. Rolfe, 1973   Biological and molecular evidence for the transgenosis of genes from bacteria to plant cells. *Proc. Natl. Acad. Sci. USA* **70**:723-726.

Dragavtsev, V. A., G. F. Privalov and A. G. Babkishev, 1970   Primenenie statisticheskogo effekta bufernosti k mutatsiyam pri poiske fonovykh priznakov u rastenii. *Genetika* **6**:38-42.

Ehrenberg, L., 1971   Higher plants. In *Chemical Mutagens*. edited by A. Hollaender, pp. 365-386, Plenum Press, New York.

Feenstra, W. J., 1964   Isolation of nutritional mutants in *Arabidopsis thaliana*. *Genetica (The Hague)* **35**:259-269.

Feenstra, W. J., 1967   Complementatie van thiamineless mutanten bij *Arabidopsis thaliana*. *Genen Phaenen* **11**:46.

Fujii, T., 1964*a*   Radiation effects on *Arabidopsis thaliana* I. Comparative efficiencies of γ-rays, fission and 14 MeV neutrons in somatic mutation. *Jap. J. Genet.* **39**:91-101.

Fujii, T., 1964*b*   *Arabidopsis* an experimental organism in genetics. *Idengaku Zasshi* **18**:22-26 (in Japanese).

Fujii, T., 1965   Effects of UV-rays on *Arabidopsis* seedlings. In *Arabidopsis Research, Report of the International Symposium, Göttingen*, edited by G. Röbbelen, pp. 147-151, University of Göttingen, Göttingen.

Fujii, T., 1967   Comparison of the killing effect of gamma rays and thermal neutrons. *Arabidopsis Inf. Serv.* **4**:59.

Fujii, T., M. Ikenaga and J. T. Lyman, 1966   Radiation effects on *Arabidopsis thaliana* II. Killing and mutagenic efficiencies of heavy ionizing particles. *Radiat. Bot.* **6**:297-306.

Gerlach-Cruse, D., 1969   Embryo- und Endospermentwicklung nach einer Röntgenbestrahlung der Fruchtknoten von *Arabidopsis thaliana* (L.) Heynh. *Radiat. Bot.* **9**:433-442.

Gerlach-Cruse, D., 1970   Experimentelle Auslösung von Semigamie bei *Arabidopsis thaliana* (L.) Heynh. *Biol. Zentralbl.* **89**:435-456.

Gichner, T. and J. Velemínský, 1965   Induction of chlorophyll chimeras by x-rays and ethyl methanesulphonate (EMS) in different heterozygous strains of *Arabidopsis thaliana*. In *Proceedings of the Symposium: Induction of Mutations and the Mutation Process, Prague, Sept. 27-29, 1963*, pp. 54-56, Publishing House Czechoslovak Academy of Science, Prague.

Gichner, T. and J. Velemínský, 1967   The mutagenic activity of 1-alkyl-1-nitrosoureas and 1-alkyl-3-nitro-1-nitrosoguanidines. *Mutat. Res.* **4**:207-212.

Gichner, T., L. Ehrenberg and C. A. Wachtmeister, 1968   The mutagenic activity of β-hydroxyethyl methanesulfonate, β-methoxyethyl methanesulfonate and diethyl 1,3-propanedisulfonate. *Hereditas* **59**:253-262.

Glubrecht, H., 1965   Mode of action of incorporated nuclides. In *The Use of Induced Mutations in Plant Breeding. Radiat. Bot.* **5**:*Suppl.* 91-99.

Gómez-Campo, C. and L. Delgado, 1964   Radioresistance in crucifers. *Radiat. Bot.* **4**:479-483.

Gresshoff, P. M. and C. H. Doy, 1972   Haploid *Arabidopsis thaliana* callus and plants from anther culture. *Aust. J. Biol. Sci.* **25**:259-264.

Griffing, B. and J. Langridge, 1962   Phenotypic stability of growth in the self-fertilized species, *Arabidopsis thaliana*. In *Statistical Genetics and Plant Breeding. Natl. Acad. Sci. Natl. Res. Council. Publ.* **982**:368-394.

Griffing, B. and E. Zsiros, 1971   Heterosis associated with genotype environment interactions. *Genetics* **68**:443–455.

Grover, N. S. and O. R. Byrne, 1972   Isozyme studies in *Arabidopsis thaliana*. *Arabidopsis Inf. Serv.* **9**:10–11.

Härer, L., 1951   Die Vererbung des Blühalters früher und später sommereinjähriger Rassen von *Arabidopsis thaliana* (L.) Heynh. *Beitr. Biol. Pflanz.* **28**:1–35.

Harle, J. R., 1972   A revision of mutation breeding procedures in *Arabidopsis* based on a fresh analysis of the mutant sector problem. *Can. J. Genet. Cytol.* **14**:559–572.

Hirono, Y., 1964   A genetic method for the localization of chromosome defects. *Genetics* **50**:255.

Hirono, Y. and G. P. Rédei, 1963   Multiple allelic control of chlorophyll b level in *Arabidopsis thaliana*. *Nature (Lond.)* **197**:1324–1325.

Hirono, Y. and G. P. Rédei, 1965a   Induced premeiotic exchange of linked markers in the angiosperm *Arabidopsis*. *Genetics* **51**:519–526.

Hirono, Y. and G. P. Rédei, 1965b   Concurrent products of premeiotic recombination. In *Arabidopsis Research, Report of the International Symposium, Göttingen*, edited by G. Röbbelen, pp. 85–90, University of Göttingen, Göttingen.

Hirono, Y. and G. P. Rédei, 1966a   Acceleration of flowering of the long-day plant *Arabidopsis* by 8-azaadenine. *Planta (Berl.)* **68**:88–93.

Hirono, Y. and G. P. Rédei, 1966b   Early flowering in *Arabidopsis* induced by DNA base analogs. *Planta (Berl.)* **71**:107–112.

Hirono, Y. and H. H. Smith, 1969   Mutation induced in *Arabidopsis* by DNA nucleoside analogs. *Genetics* **61**:191–199.

Hirono, Y., H. H. Smith and J. T. Lyman, 1968   Tumor induction by heavy ionizing particles and X-rays in *Arabidopsis*. *Radiat. Bot.* **8**:449–456.

Hirono, Y., H. H. Smith, J. T. Lyman, K. H. Thompson and J. W. Baum, 1970   Relative biological effectiveness of heavy ions in producing mutations, tumors, and growth inhibition in the crucifer plant, *Arabidopsis*. *Radiat. Res.* **44**:204–223.

Hussein, H. A. S., 1968   Genetic analysis of mutagen-induced flowering time variation in *Arabidopsis thaliana* (L.) Heynh. Doctoral Dissertation, Agricultural University, Wageningen, Netherlands.

Hylander, N., 1957   *Cardaminopsis suecica* (Fr.) Hiit., a northern amphidiploid species. *Bull. Jard. Bot. Natl. Brux.* **27**:591–604.

Immer, F. R., 1930   Formulae and tables for calculating linkage intensities. *Genetics* **15**:81–98.

Ivanov, V. I., A. V. Sanina and H. A. Timofeeff-Ressovsky, 1967   Studies in the radiation genetics of *Arabidopsis thaliana* (L.) Heynh. II. Survival and fertility of G_1 plants after γ-irradiation of dormant seeds. *Genetika* **3**(5):16–23 (Russian with English summary).

Ivanov, V. I., A. B. Sanina and H. A. Timofeeff-Ressovsky, 1968   Effect of γ-irradiation of the seed on the survival, growth, development and fertility of *Arabidopsis thaliana* (L.) Heynh. *Radiobiologiya* **8**:118–123 (in Russian).

Jacobs, M., 1964   Étude comparative des fréquences et types de mutations induites par le méthane sulfonate d'éthyle et le méthane sulfonate de méthyle chez *Arabidopsis thaliana* (L.) Heynh. *Bull. Soc. Bot. Belg.* **98**:43–66.

Jacobs, M., 1965   Isolement de mutants biochimiques chez *Arabidopsis thaliana*. Méthods et résultats préliminaires. *Bull. Acad. R. Sci. Belg. Cl. Sci. Ser. 5* **51**:735–747.

Jacobs, M., 1967   Effets induits chez *Arabidopsis thaliana* (L.) Heynh. par un analogue de la thymidine incorporé dans le milieu de culture. *Bull. Soc. Bot. Belg.* **100**:259–281.

Jacobs, M., 1969a   Studies on the genetic activity of thymidine base analogues in *Arabidopsis thaliana. Mutat. Res.* **7**:51–62.

Jacobs, M., 1969b   An attempt to sensitize *Arabidopsis* seeds to gamma irradiation after BUdR-treatment. *Arabidopsis Inf. Serv.* **6**:24.

Jacobs, M., 1969c   Comparaison de l'action mutagénique d'agents alkylants et des radiations gamma chez l'*Arabidopsis thaliana. Radiat. Bot.* **9**:251–268.

Jacobs, M. and F. Schwind, 1972   Isozyme variability in *Arabidopsis thaliana*. Genetic basis of the acid phosphatase and leucine aminopeptidase variation. *Arabidopsis Inf. Serv.* **9**:11–12.

James, T., 1969   Effects of 8-azaguanine and p-fluorophenylalanine on the development of *Arabidopsis. Arabidopsis Inf. Serv.* **6**:5.

Jones, M. E., 1971   The population genetics of *Arabidopsis thaliana. Heredity* **27**:39–50, 51–58, 59–72.

Kasyanenko, A. G. and N. V. Timofeeff-Ressovsky, 1967   On some interesting "chlorophyll" mutations in *Arabidopsis thaliana* (L.) Heynh. *Bjull. Mosk. Obsh. Icp. Prir. Otdel Biol.* **72**:100–105. (in Russian).

Kasyanenko, A. G., Y. S. Nasyrov and E. A. Smolina, 1971   Leucine mutations of *Arabidopsis thaliana*. In *Genetical Aspects of Photosynthesis*, edited by Y. S. Nasyrov, pp. 56–76, Academy Sciences Tadzh. SSR., Institute of Plant Physiology and Biophysics, Dushanbe.

Koo, F. K. S., 1969   Potential use of target atom irradiation in control of mutation induction. In *Proceedings of the International Atomic Energy Agency/Food and Agricultural Organization of the United Nations Symposium: Induced Mutations in Plant Breeding*, pp. 305–312, IAEA, Vienna.

Kranz, A. R., 1968   Endogene und exogene Beeinflussung der apparenten Strahlungsenergienutzung annueller Pflanzen. *Angew. Bot.* **41**:271–278.

Kranz, A. R., 1971   Genphysiologie quantitativer Merkmale bei *Arabidopsis thaliana* (L.) Heynh. *Theor. Appl. Genet.* **41**:45–51, 91–99, 191–196.

Kribben, F. J., 1957   Die Abkürzung der Samenruhe bei Arabidopsis durch Gibberellinsäure. *Naturwissenschaften* **44**:313.

Kribben, F. J. 1965   Interspecific hybridization with Arabidopsis. In *Arabidopsis Research, Report of the International Symposium, Göttingen*, edited by G. Röbbelen, pp. 26–30, University of Göttingen, Göttingen.

Kučera, J., 1965   Induction of developmental mutations by N-methyl-N-nitrosourea in *Arabidopsis thaliana*. In *Mechanism of Mutation and Inducing Factors, Proceedings of the Symposium, Prague, Aug., 1965*, pp. 313–316, Publishing House Czechoslovak Academy of Science, Prague.

Kučera, J., 1966   The effect of temperature, cysteine, and gibberellin on radiation damage of seed germination. *Arabidopsis Inf. Serv.* **3**:32.

Kugler, I., 1951   Untersuchungen über das Keimverhalten einiger Rassen von *Arabidopsis thaliana* (L.) Heynh.—Ein Beitrag zum Problem der Lichtkeimung. *Beitr. Biol. Pflanz.* **28**:211–243.

Laibach, F., 1907   Zur Frage nach der Individualität der Chromosomen im Pflanzenreich. *Beih. Bot. Cbl. 1 Abt.* **22**:191–210.

Laibach, F., 1940   Die Ursachen der Blütenbildung und das Blühhormon. *Natur Volk (Frankfurt)* **70**:55–65.

Laibach, F., 1943   *Arabidopsis thaliana* (L.) Heynh. als Objekt für genetische und entiwichlungsphysiologische Untersuchungen. *Bot. Arch.* **44**:439–455.

Laibach, F., 1951   Über sommer-und winterannuelle Rassen von *Arabidopsis thaliana* (L.) Heynh. Ein Beitrag zur Ätiologie der Blütenbildung. *Beitr. Biol. Pflanz.* **28**:173–210.

Laibach, F., 1956   Über die Brechung der Samenruhe bei *Arabidopsis thaliana* (L.) Heynh. *Naturwissenschaften* **43**:164–166.

Laibach, F., 1958   Über den Artbastard *Arabidopsis suecica* (Fr.) Norrl. × *A. thaliana* (L.) Heynh. und die Bezeihungen zwischen den Gattungen Arabidopsis Heynh. und Cardaminopsis (C. A. Meyer) Hay. *Planta (Berl.)* **51**:148–166.

Laibach, F. and A. M. Zenker, 1954   Zur Kältebeeinflussung der Blütenbildung bei Langtagspflanzen. *Planta (Berl.)* **43**:250–252.

Land, J. B. and G. Norton, 1970   The nature of the leucine requirement of the barley mutant Xan-b[61]. *Genet. Res.* **15**:135–137.

Langridge, J., 1955   Biochemical mutations in the curcifer *Arabidopsis thaliana* (L.) Heynh. *Nature (Lond.)* **176**:260–261.

Langridge, J., 1957a   Effect of day-length and gibberellic acid on the flowering of *Arabidopsis*. *Nature (Lond.)* **180**:36–37.

Langridge, J., 1957b   The aseptic culture of *Arabidopsis thaliana* (L.) Heynh. *Aust. J. Biol. Sci.* **10**:243–252.

Langridge, J., 1958a   A hypothesis of developmental selection exemplified by lethal and semilethal mutants of *Arabidopsis*. *Aust. J. Biol. Sci.* **11**:58–68.

Langridge, J., 1958b   An osmotic mutant of *Arabidopsis thaliana*. *Aust. J. Biol. Sci.* **11**:457–470.

Langridge, J., 1962   A genetic and molecular basis for heterosis in *Arabidopsis* and *Drosophila*. *Am. Nat.* **96**:5–27.

Langridge, J., 1963   The genetic basis of climatic response. In *Environmental Control of Plant Growth; Symposium in Canberra, Aug. 25–31, 1962*, pp. 367–379, Academic Press, London.

Langridge, J., 1965   Temperature-sensitive, vitamin-requiring mutants of *Arabidopsis thaliana*. *Aust. J. Biol. Sci.* **18**:311–321.

Langridge, J. and B. Griffing, 1959   A study of high-temperature lesions in *Arabidopsis thaliana*. *Aust. J. Biol. Sci.* **12**:117–135.

Lavrinetskaya, T. E. and A. A. Strekhalov, 1970   Elektroforez rastvorimykh belkov listyev *Arabidopsis* v poliakrilamidnom gele. *Fiziol. Rast.* **17**:843–846.

Lawrence, C. W., 1965   Radiation-induced polygenic mutation. In *The Use of Induced Mutations in Plant Breeding. Rad. Bot.* **5**: *Suppl.* 491–496.

Lawrence, C. W., 1968   Radiation-induced polygenic mutation in *Arabidopsis thaliana*. *Heredity* **23**:321–337, 573–589.

Ledoux, L., R. Huart and M. Jacobs, 1971a   Fate of exogenous DNA in *Arabidopsis thaliana*. Translocation and integration. *Eur. J. Biochem.* **23**:96–108.

Ledoux, L., R. Huart and M. Jacobs, 1971b   Fate of exogenous DNA in *Arabidopsis thaliana*. II. Evidence for replication and preliminary results at the biological level. In *Informative Molecules in Biological Systems*, edited by L. Ledoux, pp. 159–175, North-Holland, Amsterdam.

Lee-Chen, S. and D. Burger, 1967   The location of linkage groups on the chromosomes of *Arabidopsis* by the trisomic method. *Arabidopsis Inf. Serv.* **4**:4–5.

Lee-Chen, S. and L. M. S. Sears, 1969   A telotrisomic in *Arabidopsis thaliana*. *Arabidopsis Inf. Serv.* **6**:22.

Lee-Chen, S. and L. M. Steinitz-Sears, 1967   The location of linkage groups in *Arabidopsis thaliana. Can. J. Genet. Cytol.* **9**:381–384.

Li, S. L., 1967   A new segregation distorter factor in *Arabidopsis. Arabidopsis Inf. Serv.* **4**:5–6.

Li, S. L., 1968   Genetics of thiamine metabolism in *Arabidopsis*. Doctoral Dissertation. University of Missouri, Columbia, Missouri.

Li, S. L. and G. P. Rédei, 1969*a*   Direct evidence for models of heterosis provided by mutants of *Arabidopsis* blocked in the thiamine pathway. *Theor. Appl. Genet.* **39**:68–72.

Li, S. L. and G. P. Rédei, 1969*b*   Allelic complementation at the pyrimidine (*py*) locus of the crucifer *Arabidopsis. Genetics* **62**:281–288.

Li, S. L. and G. P. Rédei, 1969*c*   Estimation of mutation rates in autogamous diploids. *Radiat. Bot.* **9**:125–131.

Li, S. L. and G. P. Rédei, 1969*d*   Gene locus specificity of the glucose effect in the thiamine pathway of the angiosperm, *Arabidopsis. Plant Physiol.* **44**:225–229.

Li, S. L. and G. P. Rédei, 1969*e*   Thiamine mutants of the crucifer, *Arabidopsis. Biochem. Genet.* **3**:163–170.

Li, S. L., G. P. Rédei and C. S. Gowans, 1967   A phylogenetic comparison of mutation spectra. *Mol. Gen. Genet.* **100**:77–83.

Lockhart, J. and U. Brodführer-Franzgrote, 1961   The effects of ultra-violet radiation on plants. In *Handbuch der Pflanzenphysiologie,* Vol. 16, edited by W. Ruhland, pp. 532–554, Springer Verlag, Berlin.

Loewenberg, J. R. and P. J. Thompson, 1967   Nutritional requirements of callus from *Ei-6. Arabidopsis Inf. Serv.* **4**:68–70.

Löve, Å., 1961   Hylandra–A new genus of the cruciferae. *Sven. Bot. Tidskr.* **55**:211–217.

McCullogh, J. M. and W. Shropshire, Jr., 1970   Physiological predetermination of germination responses in *Arabidopsis thaliana* (L.) Heynh. *Plant Cell Physiol.* **11**:139–148.

McKelvie, A. D., 1962*a*   A list of mutant genes in *Arabidopsis thaliana* (L.) Heynh. *Radiat. Bot.* **1**:233–241.

McKelvie, A. D., 1962*b*   Differential response to mutagens in *Arabidopsis thaliana. Nature (Lond.)* **195**:409–410.

McKelvie, A. D., 1963   Studies in the induction of mutations in *Arabidopsis thaliana* (L.) Heynh. *Radiat. Bot.* **3**:105–123.

McKelvie, A. D., 1965   Preliminary data on linkage groups in Arabidopsis. In *Arabidopsis Research, Report of the International Symposium,* Göttingen, edited by G. Röbbelen, pp. 79–84, University of Göttingen, Gottingen.

Melchers, G., 1972   Haploid higher plants for plant breeding. *Z. Pflanzenzuecht.* **67**:19–32.

Mĕsíček, J., 1967   The chromosome morphology of *Arabidopsis thaliana* (L.) Heynh. and some remarks on the problem of *Hylandra suecica* (Fr.) Löve. *Folia Geobot. Phytotaxon.* **2**:433–436.

Mesken, M. and J. H. Van der Veen, 1968   The problem of induced sterility. A comparison between EMS and X-rays in *Arabidopsis thaliana. Euphytica* **17**:363–370.

Michniewicz, M. and A. Kamieńska, 1965   Flower formation induced by kinetin and vitamin E treatment in long-day plant (*Arabidopsis thaliana*) grown in short day. *Naturwissenschaften.* **52**:623.

Miksche, J. P. and J. A. M. Brown, 1965 Development of vegetative and floral meristems of *Arabidopsis thaliana*. *Am. J. Bot.* **52**:533–537.

Misra, R. C., 1962 Contribution to the embryology of *Arabidopsis thaliana* (Gay and Monn.) *Agric. Univ. J. Res. Sci.* **11**:191–199.

Müller, A. J., 1961 Zur Charakterisierung der Blüten und Infloreszenzen von *Arabidopsis thaliana* (L.) Heyhn. *Kulturpflanze* **9**:364–393.

Müller, A. J., 1963 Embryonenstest zum Nachweis rezessiver Letalfaktoren bei *Arabidopsis thaliana*. *Biol. Zentralbl.* **83**:133–163.

Müller, A. J., 1964 Mutationsauslösung durch Nitrosomethyl-Harnstoff bei Arabidopsis. *Züchter* **34**:102–120.

Müller, A. J., 1965a Reparation chemisch induzierter pramutativer Läsionen durch Rücktrocknung der behandelten Samen? *Biol. Zentralbl.* **84**:759–762.

Müller, A. J., 1965b Über den Zeitpunkt der Mutationsauslösung nach Einwirkung von N-nitroso-N-methylharnstoff auf quellende Samen von Arabidopsis. *Mutat. Res.* **2**:426–437.

Müller, A. J., 1965c Durch Röntgenbestrahlung des Pollens von Arabidopsis induzierte diplophasische und haplophasische Letalmutationen. *Kulturpflanze* **13**:163–171.

Müller, A. J. 1965d The chimerical structure of M_1 plants and its bearing on the determination of mutation frequencies in Arabidopsis. In *Induction of Mutations and the Mutation Process, Proceedings of the Symposium: Prague, Sept. 27–29, 1963*, pp. 46–52, Publishing House Czechoslovak Academy of Science, Prague.

Müller, A. J., 1965e Beeinflussung der radiominetischen Wirksamkeit von Nitrosamiden durch Stoffwechselinhibitoren. *Naturwissenschaften* **52**:213–214.

Müller, A. J. 1966a Die Induktion von rezessiven Letal-Mutationen durch Äthylmethansulfonat bei Arabidopsis. I. Dosis-Effekt-Beziehungen und deren Beeinflussung durch die Behandlungsbedingungen. *Züchter* **36**:201–220.

Müller, A. J. 1966b Induction of recessive lethals by X-rays. *Arabidopsis Inf. Serv.* **3**:22.

Müller, A. J., 1967a Die Induktion von rezessiven Letal-Mutationen durch Äthylmethansulfonat bei Arabidopsis II. Sensibilitätsänderungen in quellenden und keimenden Samen. *Biol. Zentralbl. (Suppl.)* **86**:89–106.

Müller, A. J., 1967b Genetic analysis of sterility induced by highly efficient mutagens in Arabidopsis. *Abh. Dtsch. Akad. Wiss. Ber. Kl. Med.* **2**:89–97.

Müller, A. J., 1968 Genic male sterility in *Arabidopsis*. *Arabidopsis Inf. Serv.* **5**:53–54.

Müller, A. J., 1972 Mutagenitätsprüfung von Chemikalien bei *Arabidopsis thaliana*. *Biol. Zentralbl.* **91**:31–48.

Müller, A. J. and T. Gichner, 1964 Mutagenic activity of 1-methyl-3-nitrosoguanidine on *Arabidopsis*. *Nature (Lond.)* **201**:1149–1150.

Napp-Zinn, K., 1954 Vergleichende Atmungsmessungen an Sommer und Winterannuellen. Untersuchungen an Caryopsen und Embryonen von *Secale cereale* und an Samen von *Arabidopsis thaliana*. *Z. Naturforsch. Teil B* **9b**:218–229.

Napp-Zinn, K., 1955 Spontanes Auftreten von Kotylvarianten bei *Arabidopsis thaliana* (L.) Heyhn. *Ber. Dtsch. Bot. Ges.* **68**:369–373.

Napp-Zinn, K., 1957a Untersuchungen uber den Aufbau der Infloreszenz bei *Arabidopsis thaliana*. *Beitr. Biol. Pflanz.* **34**:113–128.

Napp-Zinn, K., 1957b Die Abhängigkeit des Vernalisationseffektes bei *Arabidopsis thaliana* vom Quellungsgrad der Samen und vom Lichtgenuss der Pflanzen nach der Kältebehandlung. *Flora (Jena)* **144**:403–419.

Napp-Zinn, K., 1957*c* Untersuchungen zur Genetik des Kältebedürfnisses bei *Arabidopsis thaliana. Z. Indukt. Abstammungs.-Vererbungsl.* **88**:253–285.

Napp-Zinn, K., 1957*d* Die Abhangigkeit des Vernalisationseffektes bei *Arabidopsis thaliana* von der Dauer der Vorquellung der Samen sowic wom Alter der Pflanzen bei Beginn der Vernalisation. *Z. Bot.* **45**:379–394.

Napp-Zinn, K., 1960*a* Vernalisation, Licht und Alter bei *Arabidopsis thaliana* (L.) Heynh. I. Mitteilung. Licht und Dunkelheit während Kälte-und Warmebehandlung. *Planta (Berl.)* **54**:409–444.

Napp-Zinn, K., 1960*b* Vernalisation, Licht und Alter bei *Arabidopsis thaliana* (L.) Heynh. II. Mitteilung. Die Rolle der vor und nach der Kältebehandlung herrschenden Lichtintensität. *Planta (Berl.)* **54**:445–452.

Napp-Zinn, K., 1962 Über die genetischen Grundlagen des Vernalisationsbedürfnisses bei *Arabidopsis thaliana* I. Mitt. Die Zahl der beteiligten Faktoren. *Z. Vererbungsl.* **93**:154–163.

Napp-Zinn, K., 1963*a* Zur Genetik der Wuchsformen. *Beitr. Biol. Pflanz.* **38**:161–177.

Napp-Zinn, K., 1963*b* Über den Einfluss von Genen und Gibberellinen auf die Blütenbildung von *Arabidopsis thaliana. Ber. Dtsch. Bot. Ges.* **76**:77–89.

Napp-Zinn, K., 1964 On the genetic and developmental physiological basis of seasonal aspects of plant communities. *Arb. Landwirt. Hochsch. Hohenheim* **30**:33–49.

Napp-Zinn, K., 1965 Theory of vernalization—New experiments with *Arabidopsis*. In *Arabidopsis Research, Report of the International Symposium, Göttingen*, edited by G. Röbbelen, pp. 56–61, University of Göttingen, Göttingen.

Napp-Zinn, K., 1971 Gibberellinartige "Antigibberellin"—Wirkungen. *Z. Pflanzenphysiol.* **65**:351–358.

Napp-Zinn, K. and D. Berset, 1966 Kultur von *Arabidopsis-Blattstecklingen. Arabidopsis Inf. Serv.* **3**:37.

Napp-Zinn, K. and G. Bonzi, 1970 Gibberellin effects in dwarf mutants of *Arabidopsis thaliana. Arabidopsis Inf. Serv.* **7**:8–9.

Neales, T. F., 1968*a* The nutritional requirements of excised roots of three genotypes of *Arabidopsis thaliana* (L.) Heynh. *New Phytol.* **67**:159–165.

Neales, T. F., 1968*b* Effects of high temperature and genotype on the growth of excised roots of *Arabidopsis thaliana. Aust. J. Biol. Sci.* **21**:217–223.

Niemann, E. G., 1962 Wirkung eines Künstlichen ^{90}Sr-Fallout auf Pflanzen. II. Strahlenbiologische Wirkungen. *Atompraxis* **8**:51–56.

Niethammer, A., 1926 Der Einfluss von den Reizchemikalien auf die Samenkeimung. II. *Mitt. Jahbr. Wiss. Bot.* **67**:223–241.

Nikolov, Ch. V. and V. I. Ivanov, 1969 Vliyanie teplovykh shokov i gamma-oblucheniya semyan *Arabidopsis thaliana* (L.) Heynh. na chastotu mutatsii v M_2. *Genetika* **5**(1):168–170.

Nitsch. J. P., 1967 Towards a biochemistry of flowering and fruiting: contributions of the *"in vitro"* technique. *Proc. XVII Int. Hortic. Congr.* **3**:291–308.

Oostindier-Braaksma, F. and W. J. Feenstra, 1973 Isolation and characterization of chlorate-resistant mutants of *Arabidopsis thaliana. Mutat. Res.* **19**:175–85.

Pederson, D. G., 1968 Environmental stress, heterozygote advantage and genotype–environment interaction in *Arabidopsis. Heredity* **23**:127–138.

Pederson, D. G. and D. F. Matzinger, 1972 Selection for fresh weight in *Arabidopsis thaliana* under two mating systems. *Theor. Appl. Genet.* **42**(2):75–80.

Polyakova, T. F., 1964 The development of the male and female gametophytes of

Arabidopsis thaliana (L.) Heynh. *Issledov. Genet. SSR* **2**:125–133 (Russian with English summary).

Rajewsky, B., 1953 The limits of the target theory of the biological action of radiation. *J. Radiol.* **25**:550–552.

Ratcliffe, D., 1961 Adaptation to habitat in a group of annual plants. *J. Ecol.* **49**:187–203.

Ratcliffe, D., 1965 The geographical and ecological distribution of Arabidopsis and comments on physiological variation. In *Arabidopsis Research, Report of the International Symposium, Göttingen,* edited by G. Röbbelen, pp. 37–45, University of Göttingen, Göttingen.

Rédei, G. P., 1960 Genetic control of 2,5-dimethyl-4-amino-pyrimidine requirement in *Arabidopsis thaliana. Genetics* **45**:1007.

Rédei, G. P., 1962*a* Genetic block of "vitamin thiazole" synthesis in *Arabidopsis. Genetics* **47**:979.

Rédei, G. P., 1962*b* Single locus heterosis. *Z. Vererbungsl.* **93**:164–170.

Rédei, G. P., 1962*c* Supervital mutants of *Arabidopsis. Genetics* **47**:443–460.

Rédei, G. P., 1963 Somatic instability caused by a cysteine sensitive gene in *Arabidopsis. Science (Wash., D. C.)* **139**:767–769.

Rédei, G. P., 1964 A pollen abortion factor. *Arabidopsis Inf. Serv.* **1**:10–11.

Rédei, G. P., 1965*a* Non-mendelian megagametogenesis in *Arabidopsis. Genetics* **51**:857–872.

Rédei, G. P., 1965*b* Genetic blocks in the thiamine synthesis of the angiosperm *Arabidopsis. Am. J. Bot.* **52**:834–841.

Rédei, G. P., 1965*c* Genetic basis of an abnormal segregation in *Arabidopsis.* In *Arabidopsis Research, Report of the International Symposium, Göttingen,* edited by G. Röbbelen, pp. 91–99, University of Göttingen, Göttingen.

Rédei, G. P., 1965*d* Genetic control of subcellular differentiation. In *Arabidopsis Research, Report of the International Symposium, Göttingen,* edited by G. Röbbelen, pp. 119–127, University of Göttingen, Göttingen.

Rédei, G. P., 1967*a* Improved method of leaf pigment chromotography. *Arabidopsis Inf. Serv.* **4**:64.

Rédei, G. P., 1967*b* X-ray induced phenotypic reversions in *Arabidopsis. Radiat. Bot.* **7**:401–407.

Rédei, G. P., 1967*c* Biochemical aspects of a genetically determined variegation in *Arabidopsis. Genetics* **56**:431–443.

Rédei, G. P., 1967*d* Suppression of a genetic variegation by 6-azapyrimidines. *J. Hered.* **58**:229–235.

Rédei, G. P., 1967*e* Genetic estimate of cellular autarky. *Experientia (Basel)* **23**:584.

Rédei, G. P., 1968 *Arabidopsis* for the classroom. *Arabidopsis Inf. Serv.* **5**:5–7.

Rédei, G. P., 1969 Gene function and phenotypic repair. In *Proceedings of the International Atomic Energy Agency/Food and Agricultural Organization of the United Nations Sympsoium: Induced Mutations in Plants,* pp. 51–60, IAEA, Vienna.

Rédei, G. P., 1970 *Arabidopsis thaliana* (L.) Heynh. A review of the genetics and biology. *Bibliogr. Genet.* **20**:1–151.

Rédei, G. P., 1972*a* Obligate photoorganotrophy in submerged culture. *Arabidopsis Inf. Serv.* **9**:41.

Rédei, G. P., 1972*b* Compatibility of *Arabidopsis thaliana* with *Cardaminopsis arenosa. Arabidopsis Inf. Serv.* **9**:42.

Rédei, G. P., 1973a   Extra-chromosomal mutability determined by a nuclear gene locus in *Arabidopsis. Mutat. Res.* **18**:149–162.

Rédei, G. P., 1973b   Effects of the degradation products of fructose on the glycolytic pathway. *Z. Pflanzenphysiol.* **70**:97–106.

Rédei, G. P., 1973c   Effect of autoclaved fructose media on metabolites in three cruciferous plants. *Z. Pflanzenphysiol.* **70**:107–114.

Rédei, G. P., 1974a   Analysis of the diploid germline of plants by mutational techniques. *Can. J. Genet. Cytol.,* in press.

Rédei, G. P., 1974b   Economy in mutation experiments, Zeitschr. Pflanzenzücht, in press.

Rédei, G. P., 1974c   "Fructose effect" in higher plants. *Ann. Bot.,* in press.

Rédei, G. P., 1974d   The origin of Hylandra suecica (Fr.) Löve In *International Symposium on the Biology and Chemistry of the Cruciferae, London, Jan. 7–9, 1974,* Phytochem. Soc. and Linnean Soc. London, p. 21.

Rédei, G. P. and Y. Hirono, 1964   Linkage studies. *Arabidopsis Inf. Serv.* **1**:9–10.

Rédei, G. P. and S. L. Li, 1969a   Effects of X-rays and ethyl methanesulfonate on the chlorophyll b locus in the soma and on the thiamine loci in the germline of *Arabidopsis. Genetics* **61**:453–459.

Rédei, G. P. and S. L. Li, 1969b   Physiological resolution of the *py* locus of *Arabidopsis* by means of allelic complementation. *Proc. XI. Int. Bot. Congr. Abst.,* p. 178.

Rédei, G. P. and C. M. Perry, 1971   Submerged culture of intact plants in liquid medium. *Arabidopsis Inf. Serv.* **8**:34.

Rédei, G. P. and S. B. Plurad, 1973   Hereditary structural alterations of plastids induced by a nuclear mutator gene in *Arabidopsis. Protoplasma* **77**:361–380.

Rédei, G. P., S. C. Chung and S. B. Plurad, 1974   Mutants, antimetabolites and differentiation. *Brookhaven Symp. Biol.* **25** in press.

Rehwaldt, C. A., 1968   Filter paper effect on seed germination of *Arabidopsis thaliana. Plant Cell. Physiol.* **9**:609–611.

Reinholz, E., 1947   Auslösung von Röntgenmutationen bei *Arabidopsis thaliana* (L.) Heynh. und ihre Bedeutung für die Pflanzenzüchtung und Evolutionstheorie. *Field Information Agency Technical Report* **1006**:1–70.

Reinholz, E., 1954a   Beiträge zur indirekten Strahlenwirkung I. Röntgenbestrahlung biologischer Objekte in fester Phase. *Strahlentherapie* **95**:131–147.

Reinholz, E., 1954b   Weitere Untersuchungen zur Induktion von Keimblattveränderungen durch Röntgenstrahlen. *Experientia (Basel)* **10**:486–488.

Reinholz, E., 1959   Beeinflussung der Morphogenese embryonaler Organe durch ionisierende Strahlunge. I. Keimlingsanomalien durch Röntgenbestrahlung von *Arabidopsis thaliana*-Embryonen in verschiedenen Entwicklungsstadien. *Strahlentherapie* **109**:537–553.

Reinholz, E., 1965a   Biologische Strahlenwirkung bei tiefen Temperaturen. In *25 Jahre Max-Planck-Institut für Biophysik (1937–1962),* pp. 88–99, Max-Planck Institut, Frankfurt am Main.

Reinholz, E., 1965b   X-ray induction of embryonic malformations. In *Arabidopsis Research, Report of the International Symposium, Göttingen,* edited by G. Röbbelen, pp. 142–146, University of Göttingen, Göttingen.

Reinholz, E., 1967   The influence of gibberellic acid on the germination of irradiated and non-irradiated *Arabidopsis* seeds. *Arabidopsis Inf. Serv.* **4**:16–17.

Reinholz, E., 1968   Germination of light requiring races of *Arabidopsis thaliana* in the dark after x-irradiation. *Arabidopsis Inf. Serv.* **5**:18.

Reinholz, E., 1972   Vegetative reproduction. *Arabidopsis Inf. Serv.* **9**:37.

Reinholz, E. and K. Aurand, 1954   Untersuchungen über die Wirksamkeit von Strahlenschutzsubstanzen bei Pflanzen. *Strahlentherapie* **94**:646–656.

Rijven, A. H. G. C., 1956   Glutamine and asparagine as nitrogen sources for the growth of plant embryos *in vitro*: A comparative study of 12 species. *Aust. J. Biol. Sci.* **9**:511–527.

Röbbelen, G., 1956   Über die Protochlorophyllreduktion in einer Mutante von *Arabidopsis thaliana* (L.) Heynh. *Planta (Berl.)* **47**:532–546.

Röbbelen, G., 1957*a*   Untersuchungen an strahieninduzierten Blattfarbmutaten von *Arabidopsis thaliana* (L.) Heynh. *Z. Indukt. Abstammungs-Vererbungsl.* **88**:189–252.

Röbbelen, G., 1957*b*   Über heterophyllie bei *Arabidopsis thaliana* (L.) Heynh. *Ber. Dtsch. Bot. Ges.* **70**:39–44.

Röbbelen, G., 1957*c*   Eine Balttfarbmutante ohne Chlorophyll b von *Arabidopsis thaliana* (L.) Heynh. *Naturwissenschaften* **44**:288–289.

Röbbelen, G., 1959   Untersuchungen über die Entwicklung der submikroskopischen Chloroplastenstruktur in Blattfarbmutanten von *Arabidopsis thaliana*. *Z. Vererbungsl.* **90**:503–506.

Röbbelen, G., 1960   Über die unterschideliche genetische Reaktion von *Arabidopsis thaliana* auf eine Röntgenbestrahlung in verschiedenen Entwicklungsstadien. *Ber. Dtsch. Bot. Ges.* **73**:41–42.

Röbbelen, G., 1962*a*   Plastommutationen nach Röntgenbestrahlung von *Arabidopsis thaliana* (L.) Heynh. *Z. Vererbungsl.* **93**:25–34.

Röbbelen, G., 1962*b*   Über Unterschiede in den genetischen Folgen einer Röntgenbestrahlung Verschiedenartiger Pflanzenzellen. Untersuchungen an *Arabidopsis thaliana* (L.) Heynh. *Z. Vererbungsl.* **93**:127–153.

Röbbelen, G., 1962*c*   Wirkungsvergleich zwischen Äthylmethansulfonat und Röntgenstrahlen in Mutationsversuch mit *Arabidopsis thaliana*. *Naturwissenschaften* **49**:65.

Röbbelen, G., 1965*a*   The Laibach standard collection of natural races. *Arabidopsis Inf. Serv.* **2**:36–47.

Röbbelen, G., 1965*b*   Submicroscopic structure of mutant chloroplasts. In *Arabidopsis Research, Report of the International Symposium, Göttingen,* edited by G. Röbbelen, pp. 138–141. University of Göttingen, Göttingen.

Röbbelen, G., 1965*c*   The effects of two endogenous factors on artificial mutagenesis in *Arabidopsis thaliana*. In *Proceedings of the Symposium Induction of Mutations and the Mutation Process, Prague, Sept. 27–29, 1963,* pp. 42–45, Publishing House Czechoslovak Academy of Science, Prague.

Röbbelen, G., 1965*d*   Wirkung von Äthylmethansulfonate nach verschiden langer Vorquellung Der Samen in Mutations-Versuch mit *Arabidopsis thaliana* (L.) Heynh. *Beitr. Biol. Pflanz.* **41**:323–335.

Röbbelen, G., 1966*a*   Gestörte Thylakoidbildung in Chloroplasten einer xantha-Mutante von *Arabidopsis thaliana* (L.) Heynh. *Planta (Berl.)* **69**:1–26.

Röbbelen, G., 1966*b*   Chloroplastendifferenzierung nach geninduzierter Plastommutation bei *Arabidopsis thaliana* (L.) Heynh. *Z. Pflanzenphysiol.* **55**:387–403.

Röbbelen, G., 1966*c*   EMS efficiency in seeds and gametes of *Arabidopsis thaliana*. In *Mechanism of Mutation and Inducing Factors, Proceedings of the Symposium Prague, Aug., 1965,* pp. 309–312, Publishing House Czechoslovak Academy of Science, Prague.

Röbbelen, G., 1968 Genbedingte Rotlicht-Empfindlichkeit der Chloroplastendifferenzierung bei Arabidopsis. *Planta (Berl.)* **80**:237–254.

Röbbelen, G., 1972 Untersuchungen zur genetischen Characterisierung von induzierten phänotypischen Reversionen bei Arabidopsis Mutanten. *Z. Pflanzenzuecht.* **67**:177–196.

Röbbelen, G. and W. Wehrmeyer, 1965 Gestörte Granabildung in Chloroplasten einer chlorina-Mutante von *Arabidopsis thaliana* (L.) Heynh. *Planta (Berl.)* **65**:105–128.

Sears, L. M. S. and S. Lee-Chen, 1970 Cytogenetic studies in *Arabidopsis thaliana. Can. J. Genet. Cytol.* **12**:217–223.

Seyffert, W., 1960 Untersuchungen über die Vererbung quantitativer Charaktere an *Arabidopsis thaliana* (L.) Heynh. *Z. Pflanzenzuecht.* **42**:356–401.

Shen-Miller, J. and W. R. Sharp, 1966 An improved medium for rapid initiation of Arabidopsis tissue culture from seed. *Bull. Torrey Bot. Club* **93**:68–69.

Shropshire, W., Jr., W. H. Klein and V. B. Elstad, 1961 Action spectra of photomorphogenetic induction and photoinactivation of germination in *Arabidopsis thaliana. Plant Cell Physiol.* **2**:63–69.

Sosna, M., 1965 Induction of polyploidy by means of X-rays in Arabidopsis thaliana. In *Induction of Mutations and the Mutation Process,* edited by J. Veleminský and T. Gichner, 53, Publishing House, Czechosolovak Academy of Science, Prague.

Snape, J. W. and M. J. Lawrence, 1971 The breeding system of *Arabidopsis thaliana. Heredity* **27**:299–302.

Sparrow, A. H., H. J. Price and A. G. Underbrink, 1972 A survey of DNA content per cell and per chromosome of prokaryotic and eukaryotic organisms: some evolutionary considerations. *Brookhaven Symp. Biol.* **23**:451–494.

Steinitz-Sears, L. M., 1963 Chromosome studies in *Arabidopsis thaliana. Genetics* **48**:483–490.

Svachulová, J., 1967 Content of total chlorophyll and free amino acids of the chlorina mutant of *Arabidopsis thaliana* in artificial glucose nutrition. *Biol. Plant. (Prague)* **9**:34–40.

Svachulová, J. and V. Hadacova, 1972 Strukturproteine und lösliche Proteine in zwei letalen, auf Saccharosemedium kultivierten Chlorophyllmutanten von *Arabidopsis thaliana. Biol. Plant (Prague)* **14**:297–301.

Timofeeff-Ressovsky, N. V., E. K. Ginter, N. V. Glotov and V. I. Ivanov, 1971 Geneticheskie i somaticheskie effekty rentgenovykh luchei i vystrykh neutronov (opyty na Arabidopsise i Drozofile). *Genetika* **7**:42–52.

Tsuboi, M. and M. Yanagashima, 1971 Effects of antimetabolites on flowering *Arabidopsis. Plant Cell Physiol.* **12**:813–816.

Usmanov, P. D. and A. Müller, 1970 Primenenie embriontesta dlya analiza embrinalynikh letalei indutsirovannikh oblucheniem pylytsevykh zeren *Arabidopsis thaliana* (L.) Heynh. *Genetika* **6**:(7)50–60.

Usmanov, P. D., M. A. Loginov, U. Israfilova, A. Ya. Achmedov and S. Yu. Yunosov, 1970a Physiological peculiarities of species and ecotypes of the genus *Arabidopsis* from the Pamir-Alay region. *Arabidopsis Inf. Serv.* **7**:34.

Usmanov, P. D., G. A. Startsev, V. V. Shebalov and Yu. S. Nasyrov, 1970b On the mutagenic effect of laser irradiation on the seeds of *Arabidopsis thaliana* (L.) Heynh. *Dokl. Akad. Nauk. SSSR Ser. Biol.* **193**:455–457.

Vandendries, R., 1909 Contribution à l'histoire du développement des crucifères. *La Cellule* **25**:414–459.

Van der Veen, J. H., 1965 Genes for late flowering in *Arabidopsis thaliana*. *Arabidopsis Research, Report of the International Symposium, Göttingen,* edited by G. Röbbelen, pp. 62–71, University of Göttingen, Göttingen.

Van der Veen, J. H. and P. Wirtz, 1968 EMS-induced genic male sterility in *Arabidopsis thaliana* (L.) Heynh.: A model selection experiment. *Euphytica* **17**:371–377.

Van der Veen, J. H., G. H. M. Van Brederode and M. F. Vis, 1969 Sulfhydryl protection against X-rays. *Arabidopsis Inf. Serv.* **6**:23.

Vaughn, J. G., 1955 The morphology and growth of the vegetative and reproductive apices of *Arabidopsis thaliana* (L.) Heynh., *Capsella bursa-pastoris* (L.) Medic., and *Anagallis arvensis* (L.). *J. Linn. Soc. Lond. Bot.* **55**:279–301.

Veleminský, J. and T. Gichner, 1968 The mutagenic activity of nitrosoamines in *Arabidopsis thaliana*. *Mutat. Res.* **5**:417–428.

Veleminský, J. and T. Gichner, 1970 The influence of pH on the mutagenic effectiveness of nitroso compounds in *Arabidopsis*. *Mutat. Res.* **10**:43–52.

Veleminský, J. and T. Gichner, 1971 Two types of dose response curves in *Arabidopsis thaliana* after the action of nitrosoamines. *Mutat. Res.* **12**:65–70.

Veleminský, J. and G. Röbbelen, 1966 Beziehungen zwischen Chlorophyllgehalt und Chloroplastenstruktur in einer Chlorina-Mutante von *Arabidopsis thaliana* (L.) Heynh. *Planta (Berl.)* **68**:15–35.

Veleminský, J., T. Gichner and V. Pokorný, 1965 The effect of inhibitors on the mutagenic activity of N-methyl-N-nitrosourea and N-ethyl-N-nitrosourea in *Arabidopsis thaliana*. *Biol. Plant. (Prague)* **7**:325–329.

Veleminský, J., T. Gichner and V. Pokorný, 1967 The action of 1-aklyl-1-nitrosoureas and 1-alkyl-3-nitro-1-nitrosoguanidines on the M_1 generation of barley and *Arabidopsis thaliana* (L.) Heynh. *Biol. Plant. (Prague)* **9**:249–262.

Veleminský, J., V. Pokorný and T. Gichner, 1968 The effect of visible light on mutagenic activity of 1-methyl-1-nitrosourea and 1-methyl-3-nitro-1-nitrosoguanidine. *Biol. Plant. (Prague)* **10**:85–94.

Veleminský, J., S. Osterman-Golkar and L. Ehrenberg, 1970 Reaction rates and biological action of N-methyl- and N-ethyl-N-nitrosourea. *Mutat. Res.* **10**:169–174.

Viviand-Morel, M., 1877–1878 Exemple de nanisme chez un pied d'*Arabis thaliana*. *Ann. Soc. Bot. Lyon* **6**(2):98–99.

Walles, S. and G. Ahnström, 1965 Correlation between the mutation frequency and the alkylation of deoxyribonucleic acid upon treatment of seeds of *Arabidopsis thaliana* with ethyl methanesulfonate. In *Arabidopsis Research, Report of the International Symposium, Göttingen,* edited by G. Röbbelen, pp. 165–170, University of Göttingen, Göttingen.

Wehrmeyer, W. and G. Röbbelen, 1965 Raumliche Aspekte zur Membranschichtung in den Chloroplasten einer *Arabidopsis*-Mutant unter Auswertung von Serienschnitten. III. Mitt. Über Membranbildungsprozessen im Chloroplasten. *Planta (Berl.)* **64**:312–329.

Westerman, J. M., 1971 Genotypic-environmental interaction in developmental regulation in *Arabidopsis thaliana*. *Heredity* **26**:93–106, 373–382, 383–395.

Westerman, J. M. and M. J. Lawrence, 1970 Genotype–environment interaction and developmental regulation in *Arabidopsis thaliana*. *Heredity* **25**:609–627.

Whyte, R. O., 1946 *Crop Production and Environment,* Faber and Faber, London.

Wibaut, C., 1966 Analyse quantitative des synthèses nucléiques dans l'apex de l'*Arabidopsis thaliana* (L.) Heynh. au cours des diverses phases de son développement, en photopériode inductive. *C. R. Hebd. Séances Acad. Sci. Ser. D. Sci. Nat.* **263**:343–346.

Wricke, G., 1955 Ein Fall von Superdominanz bei einer Experimentell hergestellten autotetraploiden von *Arabidopsis thaliana*. *Z. Indukt. Abstammungs-Vererbungsl.* **87**:47–64.

Yamakawa, K. and A. H. Sparrow, 1966 The correlation of interphase chromosome volume with pollen abortion induced by chronic gamma irradiation. *Radiat. Bot.* **6**:21–38.

Yokoyama, K. and W. H. Jones, 1965 Tissue cultures of *Arabidopsis thaliana*. *Plant Physiol.* **40**:LXXVII.

Zenker, A. M., 1955 Jarowisationsuntersuchungen an sommerannuellen Arabidopsis-Rassen. *Beitr. Biol. Pflanz.* **32**:135–170.

Ziebur, M. K., 1965 Tissue culture. *Arabidopsis Inf. Serv.* **2**:34–35.

9

The Pea[*]

STIG BLIXT

Introduction

Experimental evidence has shown that all forms of peas previously described as species have a diploid chromosome number of 14, that no sterility barriers exist, and that gene exchange is complete. The genus *Pisum* is therefore best regarded as monospecific in accordance with Lamprecht's (1966) view. He classified the different forms as ecotypes included under *Pisum arvense* Linné, the wild-growing form of the two described by Linné. The ecotypes *abyssinicum* Braun; *arvense* (Linné) Lamprecht (including *elatius* Steven, *jomardi* Schrank, and *transcaucasicum* Stankov); *fulvum* Sibthorp and Smith; and *humile* Boissier (including *syriacum*/Berger/Lehmann) occur as wild-growing populations. All man-made genetic variations were collected together under the name *sativum,* the domesticated race. This system of classification is practical and workable, though perhaps not taxonomically orthodox. *Pisum formosum* Steven, which is a tuber-forming perennial, was separated to form the genus *Alophotropsis* (Boissier) Lamprecht.

Further details and fuller literature references for this, as well as for the following sections, are found in Blixt (1972).

Ontogeny

The ovary is one carpel, united above and containing up to ten ovules. In temperate climates, the pea may be regarded as cleistogamous, fertiliza-

---

* Part of the information on which this work is based was the result of investigations sponsored by Swedish Research Councils.

STIG BLIXT—Weibullsholm Plant Breeding Institute, Landskrona, Sweden

tion taking place approximately 24 hours before the opening of the flowers. Under subtropic to tropic conditions cross-fertilization occurs (Govorov, 1928; Harland, 1948). The two male nuclei are in the pollen tube surrounded by cytoplasm, and transfer of male cytoplasm to the zygote is possible. One nucleus fertilizes the egg; the other fuses with the two polar bodies to form a triploid endosperm which is later resorbed. The mature pea seed thus contains no endosperm; the nutritive reserves are stored in the two cotyledons (Håkansson, 1931; Monti and Devreux, 1964; Kapoor, 1966–1967).

The developing embryo contains two distinct tunica layers. The outer is continuous with the epidermis; the generative tissue originates from the inner layer (Reeve, 1948). Cells from these different layers can, however, replace each other (Sachs, 1969; Balkema, 1971).

In the mature seed, the seedling is partly developed. Five or six leaves are discernible; the first, not counting the primary scales, is almost fully developed, while the fifth to sixth are only present as a meristem. Also, primordia of basal branches, called stem-branches, are formed in certain genotypes. Lateral branches result from primordia formed later during ontogenesis in the axils of higher leaves. In the mature seed, the seedling embryo constitutes only about 1–3 percent of the total seed weight, about 90 percent comes from the cotyledons and the rest from the seed coat (Blixt and Gelin, 1965; Highkin and Lang, 1966). The seed coat shows an abundance of hereditary variation in color and shape. The determination of sterility from counts of undeveloped and developed seeds in the mature pods is a valuable aid in finding chromosome rearrangements because of its simplicity and reliability in relation to pollen fertility scoring.

Germination in the sense of visible root elongation is initiated after about 25 hours at about 20–25°C, and lateral roots appear after about 75 hours. Epicotyl growth is apparent on the seed after 3–4 days, and above the soil after about 6–7 days. The first three leaves are developed within 2–3 weeks (Klasova et al., 1971). During the first days, the seeds are very sensitive to low oxygen supply, and prolonged soaking in water will cause death (Larsson and Lwanga, 1969).

Because of the storage capacity of the cotyledons, a large number of "lethal" mutants, including those with defective chlorophylls, survive at least to the five-leaf stage, or as long as organs initiated during embryological development remain. As a considerable number of good or reasonably workable marker genes can also be observed in this stage, gene mapping based on cotyledon and seedling character observation of F_2 generation is possible (Blixt, 1972).

The pea leaf is paripinnate with about 0–12 pairs of leaflets and none too many tendrils, in different combinations. The leaf goes through a series of metamorphoses during the ontogenesis. The two primary scales are followed by about three simpler leaves with one pair of leaflets and one terminal tendril. Full complexity, with the number and combination of leaflets and tendrils which is characteristic of the particular genotype, is reached at about the seventh leaf. At the end of ontogenesis, a tendency to revert to greater simplicity is again noticed, and one (pseudo-) stipule might be the equivalent of a full compound leaf (Gottschalk, 1970; Blixt, 1972).

Flower initiation may take place as early as the stage at which the seedlings have but two leaves, i.e., already in the embryo. The number of sterile nodes before the first flower is known to vary from two to at least 34 (Marx, 1969). The flower initiation seems to be under very strict genetic control, the environment playing only a minor part (Haupt, 1952). The removal of the cotyledons seems to postpone flower initiation in early varieties by 1–3 nodes, indicating the presence of a flower-inhibiting substance in late flower-initiating varieties (Barber and Paton, 1952; Johnston and Crowden, 1967).

The typical inflorescence is a raceme with one too many flowers which are unilaterally inserted. The flowers are pentamerous, with the sepals largely united (the odd one anterior). The corolla is also pentamerous; it is descending, with a big posterior petal, the vexillum (also called the standard or the sail). Two lateral petals, the alae or the wings, are loosely joined with the anterior ones by a conjunction apparatus. These are, in turn, accreted, forming the carina or the keel. Nine stamens are joined in a staminal tube around the carpel, and the tenth one, above, is free. The carpel is joined above with one line of ovules on each side of the suture. The carpel grows out to a pod. The pod contains, on an average, from four to ten seeds, depending on genotype. In wild forms, dehiscing occurs by a rather sudden rolling-up movement.

For almost every step of the ontogenesis, including the generative stages of meiosis and fertilization, a varying number of controlling genes are already known. The largest number of mutants recorded control readily observed characters like flower and seed color, while the smallest number regulate root formation.

Spontaneous and Induced Mutation

Pisum seems to mutate comparatively easily to yield mutations of the most varied kind. Out of 1988 mutations recorded from different sources, 468 were spontaneous, 812 were induced by different radiations, and 708 were induced by chemicals. Different mutagenic agents, as well as different genotypes of peas treated with the same agent, seem to induce different spectra of mutation types, at least quantitatively. The possibilities of in-

ducing a wide spectrum of genetic variations in peas by means of spontaneous and induced mutation, therefore, seem very large. In addition, it is
possible to obtain very high mutation rates, particularly with such chemical
mutagens as ethyl methanesulfonate, ethyleneimine, or diethylsulfate. By
the application of an appropriate selection method on the treated
generation, M_1, it is also possible to increase the yield of mutants from
radiation experiments. A mutation yield of over five cases of different mutations from every treated and selected plant is possible, and has been
repeatedly obtained in my experiments (Blixt, 1972).

The vast majority of isolated mutants segregate, upon crossing with
the original form, into 1 dominant : 2 heterozygous : 1 recessive mutant.
For most of these mutant loca* the actual cause (i.e., chromosome deficiency or deletion or base exchange or deletion) is unknown. Such mutants
are simply referred to as "Mendelian." Exceptions would be those cases
where there is evidence for chromosome mutations, as shown by sterility or
meiotic behavior, to be due to translocations or other kinds of gross chromosome rearrangements.

As only a small number of the mutants have, as yet, been worked with,
it is very difficult to estimate the total number of loca available now in living
material. Certainly 500 are readily available, but probably the number
3000, given by Gottschalk (Gottschalk and Blixt, 1973) is closer to the real
figure.

Radiation studies have shown that individual genes can affect gamma-
ray sensitivity. Dominance at the a locus (A), gives a high level of resistance
(Blixt, 1970).

Cytology

The somatic chromosome number of all known peas is 14. The somatic complement has been analyzed, and all seven chromosomes
characterized qualitatively and quantitatively (Blixt, 1958). Translocations
and other chromosome rearrangements are common in Pisum, and by
agreement the normal chromosome structure is represented by the one of
line 110 in the Weibullsholm collection. The cytological analysis of Pisum
has been hampered by the unsuitability of the early meiotic stages (Lamm

---

* According to linguistic experts (for instance, Prof. G. Bendz, University of Lund,
Sweden), the correct plural form of locus in this context is loca. The use of loci is restricted
to a few particular cases, of which one is the reference to a place in a book (locus citatus). It
seems odd to banish the correct form, but loci has been in long and universal use, and,
therefore, maybe both forms should be accepted.

and Miravalle, 1959; Morrison and Lin, 1955; Gottschalk, 1968*a*). Nevertheless, Lamm has succeeded in completing a translocation tester set (Lamm, 1951; Lamm and Miravalle, 1959; Snoad, 1966). It has also been possible to coordinate the cytological chromosomes with the linkage groups (Blixt, 1959) by utilizing translocation lines genetically analyzed by Lamprecht (1953*a*; 1954*a*; 1955*a*,*b*; 1957*a*).

The position of the centromeres on the genetic map is still not known. However, tentative localizations for some of the chromosomes (Figure 1) have been made by utilizing translocations and gamma-ray-sensitivity data (Blixt, 1972). It should be pointed out, that some uncertainty with respect to the coordination of the chromsomes and the linkage groups still remains, as Lamm (1960) found the cytological results of a study of meiosis in the cross between the lines 21 and 379 to be at variance with the expectations from genetical data. The chromosomes showed a switch of direction with respect to the chromosomes 1 and 5 when comparing the cytologic map (Figure 1) with the genetic map (Figure 2), the latter being the traditional

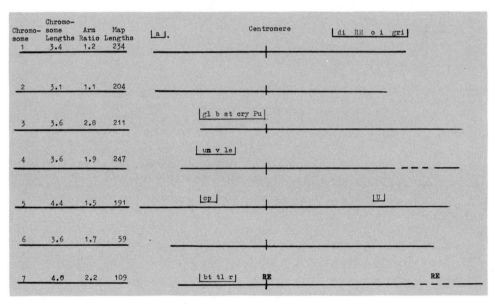

Figure 1. The chromosomes of the pea. The left side of the figure gives the chromosome lengths in microns, the arm-ratio values obtained for a given chromosome by dividing the length of the long arm by the length of the short arm, and the genetic map lengths in crossover units. The right side of the figure shows the relative lengths of the chromosome arms and the tentative locations of certain genes and radiation sensitive regions (RE) with respect to the centromeres. The nucleolus organizers are most probably located on chromosome 7. The chromosomes are from root-tip metaphase preparations from line 110.

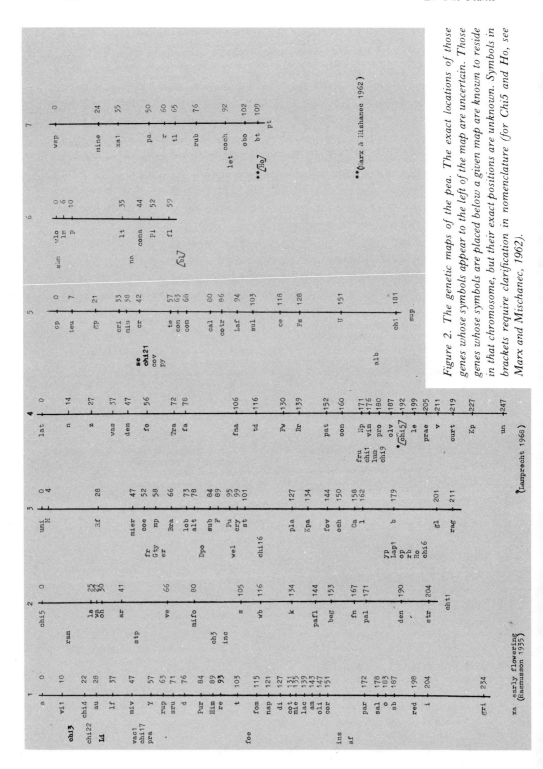

Figure 2. The genetic maps of the pea. The exact locations of those genes whose symbols appear to the left of the map are uncertain. Those genes whose symbols are placed below a given map are known to reside in that chromosome, but their exact positions are unknown. Symbols in brackets require clarification in nomenclature (for Chi5 and Ho, see Marx and Mischanec, 1962).

presentation of the chromosomes. Obviously, there is a need for a revision of the genetic map when more conclusive evidence is at hand.

The work on chromosome rearrangements has not been particularly extensive in peas. Despite this, a number of different structural types have been recorded. Translocations, mainly of spontaneous origin, are now available in known combinations as follows: T(1–2), T(1–3), T(1–5), T(1–7), T(2–7), T(3–4), T(3–5), T(3–7), T(4–6), and T(5–7). Besides these simple translocations, a number of more complicated ones have been analyzed: T(1–7, 4–6), T(1–4–7), T(2–3–5), T(3–5–7), and T(5–3–4).

Peas of ecotype *fulvum* seem to be characterized by a series of translocations involving at least six of the seven chromosomes. Five duplications are recorded, of which, two are preserved—one involving chromosome 4, and another involving chromosome 5. Inversions seem to be present in ecotype *humile* and some other lines.

The main bulk of material containing structural changes remains, however, unanalyzed. In the Weibullsholm collection, at least 20 lines with rearrangements of spontaneous origin are available, and Gottschalk (1968*b*) alone has isolated more than 80 translocated mutants after x-ray irradiation.

Trisomics can be produced from translocation heterozygotes and, possibly, also from certain induced mutants. A trisomic tester set is being developed at the John Innes Institute (Marks, 1971).

Gene Mapping

Genetic analyses in *Pisum* have a tradition even longer than that of *Drosophila*. Mendel (1866) studied the following seven characters or "character pairs":

1. The shape of mature seeds, either smooth or wrinkled. These characters are determined by the alleles *R* and *r*, respectively, in chromosome 7.
2. Seed color, yellow or green. These characters are due to the alleles *I* and *i*, respectively, in chromosome 1.
3. Seed coat colored and flowers purple-violet, or seed coat and flowers white. The alleles *A* and *a*, respectively, in chromosome 1 control these phenotypes.
4. Mature pods smooth and expanded or wrinkled and indented. The typically smooth and expanded pod is determined by the simultaneous presence of genes *P* and *V*; the strongly wrinkled and brittle pod by recessivity (*p,v*). *P, v* gives pods with small patches of

sclerenchyma on the pod wall, a phenotype that fits Mendel's description. He therefore probably studied the effect of the alleles *V* and *v* in chromosome 4.

5. Color of the unripe pods green or yellow. This characteristic is due to the alleles *Gp* and *gp,* respectively, in chromosome 5.

6. Inflorescences from the leaf axils or umbellate from the top of the plant. The alleles *Fa* or *fa* in chromosome 4 control these characters.

7. Plant height over one meter or around half a meter. This characteristic is due to alleles *Le* and *le* in chromosome 4.

Thus, Mendel probably dealt with the genes *a* and *i* in chromosome 1; *le, fa,* and *v* in chromosome 4; *gp* in 5; and *r* in 7. One might ask why he did not run into the complication of linkage. The answer is, that with respect to *a* and *i* in chromosome 1 and *fa* in relation to *le* and *v* in 4, these genes are so distant that linkage is never found. The only two genes studied between which linkage could have been found were *le* and *v*. As far as is known, Mendel never studied the simultaneous segregation in these two (Lamprecht, 1968*a*).

The first mention of linkage in peas was the one by Vilmorin and Bateson in 1912, and the first coherent presentation of the seven linkage groups was then given by Lamprecht in 1948. He put forward a genetic map comprising 37 genes, and few changes have had to be made subsequently. Presently, 169 genes are more or less accurately mapped (Figure 2), most of them the result of Lamprecht's work. From about 1931 to 1968, he published no less than 155 papers concerning the genetics of *Pisum* (Blixt, 1969*a*). It should be pointed out, that almost the entire bulk of the data is based on dihybrid combinations derived from analyses of F_2 generations.

The gene loca presently established are presented in Table 1. Further details with particular regard to synonyms and existing type lines are found in Blixt (1972). The characters that have a — in the Location column have been investigated to the point of ascertation of their nature as distinct loca, and, therefore, are symbolized; their exact location however, has, not yet been ascertained. A question mark after the chromosome number in the Location column indicates only weak supporting evidence for the given localization; the reason for including this information is, again, to point out the particular need for further investigation. Capitalized gene symbols indicate that the mutant in question is dominant in relation to the as-yet vaguely defined *Pisum* standard type, i.e., the "commonly grown type in northwestern Europe." Conversely, lower-case symbols refer to genes that are recessive to the standard type.

TABLE 1. *The Marker Genes of the Pea*

Symbol	Name	Type of mutant	Location	Phenotype	Reference
a	Anthocyan inhibition	Flower or generative apparatus	1	Uncolored flowers, seeds, axils	Lamprecht, 1948; Tschermak, 1912
acu	Acutifolius	Foliage	—	Pointed apex of leaflets	Lamprecht, 1964a, 1967a
af	Afila	Foliage	1	Leaflets transformed into tendrils	Gottschalk, 1965; Khangildin, 1966; Solovjeva, 1958
age	Ageotropum	Root and shoot	—	Roots growing upward	Blixt et al., 1958; Ekelund and Hemberg, 1966; Grout, 1904
alb	Albina	Chlorophyll	5?	White seedling, colorless cotyledons, lethal	Blixt, 1961, 1966a,b
alt	Albina-terminalis	Chlorophyll	3	Plant green below, white top; lethal	Blixt, 1961; Lamprecht, 1955d, 1959a
alte	Alternilateralis	Inflorescence	—	Flower insertion on inflorescence all-sided	Lamprecht, 1967b
am	Albicans	Flower or generative apparatus	1	Flowers white-pinkish stained	Haan, 1930; Lamprecht, 1957b
Amp1	Aminopeptidase	Physiological character	—	Electrophoretic mobility of enzyme	Scandalios and Espiritu, 1969
Amp2	Aminopeptidase	Physiological character	—	As Amp1	
ang	Angustifolia	Pleiotropic-complex	—	Leaflets linear, very narrow; sterile	Lamprecht, 1939a
ar	Caerulicans	Flower or generative apparatus	2	Flowers bluish violet	Lamprecht, 1948; Tedin et al., 1925
as	Asynapsis	Physiological character	—	Asynaptic; male sterile	Gottschalk, 1968a; Klein, 1970

TABLE 1. Continued

Symbol	Name	Type of mutant	Location	Phenotype	Reference
Asc	Ascendens	Root and shoot	—	Stem branches growing semiprostrate	Lamprecht, 1951
asre	Asymmetrically reduced	Foliage	—	Leaflets on one side of petiol smaller	Lamprecht, 1962a
Astr	Astriatus	Pod	—	Pods longitudinally violet striped	Lamprecht, 1961a
au	Xantha	Chlorophyll	1	Seedlings reddish yellow; lethal	Lamprecht, 1952a
auv	Aurea-terminalis	Chlorophyll	—	Plant green below, reddish yellow top	Lamprecht, 1957c
b	Clarroseus	Flower or generative apparatus	3	Flowers deep rose pink	Lamprecht, 1948; Tschermak, 1912
beg	Begoniaerubrum	Flower or generative apparatus	2	Modifier of b	Lamprecht, 1962b
bila	Bilateralis	Inflorescence	—	Flowers bilaterally inserted on inflorescence	Lamprecht, 1967c
bip	Bipartitus	Flower or generative apparatus	—	Wings lengthwise two-colored	Lamprecht, 1962b
bl	Incerata	Foliage	6	Plant waxless all over	Wellensiek, 1971b; White, 1917
Br	Bracteatus	Inflorescence	4	Flowers with bracts	Lamprecht, 1963a; Lamprecht and Mrkos, 1950
Bra	Bracteatus	Inflorescence	3	Flowers with bracts	Lamprecht, 1953b, 1963a
brac	Frondosus	Inflorescence	—	Flowers with large bracts	Gottschalk, 1961

brev	Breviflamentosus	Flower or generative apparatus	Anther filaments shortened	—	Lamprecht, 1935
bri	Breviramosus	Pleiotropic-complex	Inflorescences on short shoots; sterile	—	Lamprecht, 1953c
bt	Acutilegumen	Pod	Pod-apex pointed	7	Lamprecht, 1948; White, 1917
Ca	Caneo	Seed	Gray color of part-colored seeds	3	Lamprecht, 1957d, 1960a
cal	Calvitium	Seed	Part-coloring of seed coat	5	Lamprecht, 1956a, Lamprecht and Åkerberg, 1939
cat	Calvitium	Seed	As *cal*	—	Lamprecht, 1960a
ce	Roseus	Flower or generative apparatus	Flowers cerise colored	5	Wellensiek, 1951
ceo	Centrobscurum	Flower or generative apparatus	Vexillum with darker center	—	Lamprecht, 1962c
ch	Chlorina	Chlorophyll	Symbol for greenish yellow seedlings; lethal–semilethal		Blixt, 1972
ch1	Chlorina	Chlorophyll	See *ch*	5	Lamprecht, 1960b
ch2	Chlorina	Chlorophyll	See *ch*	—	Blixt, 1969b
ch3	Chlorina	Chlorophyll	See *ch*	2	Blixt, 1966c
chi	Chlorotica	Chlorophyll	Symbol for light green to yellowish green plants		Blixt, 1972
chi1	Chlorotica	Chlorophyll	See *chi*	4	Lamprecht, 1960c; Wells, 1951a
chi2	Chlorotica	Chlorophyll	See *chi*	—	Blixt, 1968a
chi3	Chlorotica	Chlorophyll	See *chi*	1	Blixt, 1965, 1968a,b
chi4	Chlorotica	Chlorophyll	See *chi*	1	Blixt, 1965, 1968a
chi5	Chlorotica	Chlorophyll	See *chi*	2	Blixt, 1968a, c
chi6	Chlorotica	Chlorophyll	See *chi*	3	Blixt, 1968a, 1969c

TABLE 1. *Continued*

Symbol	Name	Type of mutant	Location	Phenotype	Reference
chi7	Chlorotica	Chlorophyll	—	See *chi*	Blixt, 1968a
chi8	Chlorotica	Chlorophyll	—	See *chi*	
chi9	Chlorotica	Chlorophyll	4?	See *chi*	Blixt, 1968a, 1969c
chi10	Chlorotica	Chlorophyll	—	See *chi*	
chi11	Chlorotica	Chlorophyll	—	See *chi*	
chi12	Chlorotica	Chlorophyll	—	See *chi*	Blixt, 1968a
chi13	Chlorotica	Chlorophyll	—	See *chi*	
chi14	Chlorotica	Chlorophyll	—	See *chi*	
chi15	Chlorotica	Chlorophyll	—	See *chi*	
chi16	Chlorotica	Chlorophyll	3	See *chi*	Blixt, 1968a, d
chi17	Chlorotica	Chlorophyll	1	See *chi*	Blixt, 1969c
chi18	Chlorotica	Chlorophyll	—	See *chi*	
chi19	Chlorotica	Chlorophyll	—	See *chi*	Blixt, 1972
chi20	Chlorotica	Chlorophyll	—	See *chi*	
chi21	Chlorotica	Chlorophyll	5	See *chi*	
chi22	Chlorotica	Chlorophyll	1?	See *chi*	Blixt, 1968a, 1972
chi23	Chlorotica	Chlorophyll	—	See *chi*	Blixt, 1972; Lamprecht, 1957e; Wells, 1951b
chrw	Marginata	Chlorophyll	—	Leaflets with center and margin of different color	Rosen, 1942
chtl	Terminalis	Chlorophyll	2?	Plant green below, top yellowish	Lamprecht, 1952a
chves1	Chlorotica	Chlorophyll	—	See *chi*	Lamprecht, 1960b
chves2	Chlorotica	Chlorophyll	—	See *chi*	Blixt, 1969b
Cit	Clariluteus	Flower or generative apparatus	—	Flowers citrus yellow to cream colored	Lamprecht, 1961b

Symbol	Name	Category	No.	Description	References
Cm	Cereus	Flower or generative apparatus	—	Flowers cream colored	Fedotov, 1930; Lamprecht, 1961b
co	Convexum	Pod	—	Pods convexley curved	Lamprecht, 1953b
coch	Cochleata	Foliage	7	Stipules with lamina on long petiole	Wellensiek, 1959, 1962
coe	Internodes slightly longer	Root and shoot	3	Increasing internode length	Lamprecht, 1962d
coh	Brevi-internodium	Root and shoot	5	Reducing internode length	
com	Comprimere	Seed	5	Seeds compressed, almost cubic	Lamprecht, 1960c
con	Convexum	Pod	4	As *co*	Lamprecht, 1936, 1955c
coma	Brevi-internodium	Root and shoot	6	As *coh*	Lamprecht, 1962b
cont	Contrahere	Foliage	—	Modifier, reducing *st* stipules	Lamprecht, 1963b
cor	Corona	Seed	1	Hilum-region colored *ochraceous*	Lamprecht, 1947a, 1954b
cot	Brevi-internodium	Root and shoot	1	As *coh*	Lamprecht, 1962b
cotr	Contractivus	Pod	5	Reducing pod length	Lamprecht, 1963b
cov	Coeruleovirens	Chlorophyll	5	Bluish green seedlings	Lamprecht, 1957f; Marx, 1971a
cp	Concavum	Pod	5	Concavely curved pods	Lamprecht, 1948; Wellensiek, 1925a
cpa	Concavum	Pod	—	Concavely curved pods	Lamprecht, 1956a, 1963c; Rasmusson, 1927–1928; Rosen, 1944
cr	Fuscopurpureus	Flower or generative apparatus	5	Flowers dull red purple	Fedotov, 1930; Lamprecht, 1950a
cri	Crispa	Foliage	5	Tissues folded	Lamm, 1949
crif	Crispifolius fertilis	Foliage	—	Tissues folded	
cris	Crispifolius sterilis	Foliage	—	Tissues folded; sterile	Gottschalk, 1964a

TABLE 1. Continued

Symbol	Name	Type of mutant	Location	Phenotype	Reference
crpt	Crumpled petals	Flower or generative apparatus	—	Anthers and petals folded; flowers not opening	Sharma and Aravindian, 1971a
cry	Internodes much longer	Root and shoot	3	cry le la with very long internodes	Lamm, 1937; Rasmusson, 1927–1928
curt	Contractivus	Pod	4	Reducing pod length	Lamprecht, 1968a, b
Cv	Intensifier	Flower or generative apparatus	—	Intensifies flower color	Fedotov, 1930
cvit	Chlorotica	Chlorophyll	—	See chi	Lamprecht, 1960b
d	Maculum	Foliage	1	Absence of maculum ring	Lamprecht, 1961c; Tschermak, 1912
dem	Deminutio	Seed	4	Part-colored seed coat with fork and large moon	Lamprecht, 1957g; Lamprecht and Åkerberg, 1939
den	Deminutio	Seed	2	As dem	Lamprecht, 1962e
di	Lacunosus	Seed	1	Seeds with ± deep depressions	Lamprecht, 1969a; Wellensiek, 1943
dim	Determine epicotyle growth	Root and shoot	—	Epicotyle not developing; stem branches ± sterile	Gottschalk, 1964a
dip	Dipetala	Pleiotropic-complex	—	Number of petals and anthers reduced	Gottschalk, 1961
disp	Dispergere	Seed	—	Part-colored seed coat, with decolored patches	Lamprecht, 1963d
dn	Photoperiodism	Physiological character	—	Not responding to day length	Bremer and Weiseth, 1961
dp	Dark pod	Pod	—	Darkens pod color	Marx, 1970

Symbol	Name	Character		Description	Reference
Dpo	Dehisching	Pod	3	Pods tough and leathery, "wild-type" dehiscing	Blixt, 1972; Marx, 1971b
ds	Desynapsis	Physiological character		Symbol for genes causing more or less severe desynapsis	
ds 82A	Desynapsis	Physiological character	—	See ds	Gottschalk and Baquar, 1971
ds 94	Desynapsis	Physiological character	—	See ds	
ds 232	Desynapsis	Physiological character	—	See ds	
ds 239	Desynapsis	Physiological character	—	See ds	
ds 242	Desynapsis	Physiological character	—	See ds	
dt	Distans	Inflorescence	—	Shortening distance from axil to first flower	Lamprecht, 1949a
E	Earliness	Physiological character	—	Early flowering	Keeble and Pellew, 1910–1911; Murfet, 1971a, b, 1974
Ed	Rough testa	Seed	—	Seed coat surface rough	Govorov, 1937
Ef	Earliness	Physiological character	—	Dominant early flowering	White, 1917
elo	Elongata	Foliage	—	Reducing foliage width	Kellenbarger, 1952
em1	Emergences	Foliage	—	Stem with 1–2 mm long tendril-like outgrowths	Haan, 1932
em2	Emergences	Foliage	—	As em1	
En	PEMV resistance	Physiological character	—	Resistance to pea enation mosaic virus	Schroeder and Barton, 1958
ep	Testa thickness	Seed	—	Reducing thickness of seed coat	Kaznowski, 1926
ep1	Testa thickness	Seed	3	As ep	
er	Mildew resistance	Physiological character	3	Resistance to Erysiphe polygoni	Harland, 1948; Marx, 1971b
er1	Mildew resistance	Physiological character	—	As er	Keringa et al., 1969
er2	Mildew resistance	Physiological character	—	As er	

TABLE 1. *Continued*

Symbol	Name	Type of mutant	Location	Phenotype	Reference
F	Violaceopunctata	Seed	3	Seed coat with violet spots	Lamprecht, 1961c; Tschermak, 1912; White, 1917
fa	Fasciata	Root and shoot	4	Stem fasciated	Lamprecht, 1952b; White, 1917
fas	Fasciata	Root and shoot	—	As fa	Lamprecht, 1952b
fil	Filiformis	Pleiotropic-complex	—	Leaves and stipules thread narrow, sterile	Monti, 1970a
fl	Aeromaculata	Foliage	6	Leaves and stipules without air pockets under epidermis	Lamprecht, 1948; Tedin and Tedin, 1925–1926
fn	Flower number	Inflorescence	2	Increasing flower number	Lamprecht, 1952c; White, 1917
fna	Flower number	Inflorescence	4	As fn	Lamprecht, 1947b
Fnw	Fusarium resistance	Physiological character	—	Resistance to Fusarium oxysporum race 2	Hare et al., 1949
fo	Folia oblonga	Foliage	4	Decreasing foliage width	Härstedt, 1950
fob	Folia oblonga	Foliage	—	As fo	Härstedt, 1950
foe	Corrugatus	Seed	1?	Seeds with irregular, small; ± wrinkled depressions	Kaznowski, 1926; Lamprecht, 1969b
fol	Folia oblonga	Foliage	—	As fo	Härstedt, 1950
fom	Folia oblonga	Foliage	1	As fo	Lamprecht, 1960d
fov	Foveatus	Seed	3	Seed over radicula with notch	Lamprecht, 1959b
fr	Fruticosa	Root and shoot	3	Increasing number of stem branches	Blixt, 1968e; Lamprecht, 1950b

fru	Fruticosa	Root and shoot	4	As *fr*	Blixt, 1968*e*; Lamprecht, 1950*b*
Fs	Violaceopunctata	Seed	5	As *F*	Lamprecht, 1942*a*; Winge, 1936
Fw	Fusarium resistance	Physiological character	4	Resistance to *Fusarium oxysporum* race 1	Govorov, 1937; Wade, 1929
gl	Radicula ochracea	Seed	3	Seed coat over radicula ochre colored	Lamprecht, 1946; Tedin and Tedin, 1928
gla	Glaucescens	Seed	—	Seed coat greenish blue	Lamprecht, 1959*c*
gp	Luteo-legumina	Pod	5	Pods yellow	Lamprecht, 1948; White, 1917
gri	Griseostriata	Seed	1	Seed coat with a gray longitudinal stripe	Lamprecht, 1944*a*, 1957*h*
Gty	Gritty	Seed	3	Seed-coat surface gritty	Marx, 1971*b*
Him	Hilum major	Seed	1	Hilum size one half of seed diameter	Lamprecht, 1963*e*
ho	Horizontalis	Root and shoot	—	Lateral branches grow horizontally	Lamprecht, 1958*a*
i	Cotyledons green	Seed	1	Green cotyledons	Lamprecht, 1948; White, 1916
ib	Sterile nodes 2–5	Physiological character	—	2–5 sterile nodes before first flower	Lamprecht, 1956*b*
iba	Sterile nodes 2–5	Physiological character	—	As *ib*	Lamprecht, 1958*b*; Rasmusson, 1938
if	Infundibulum	Pleiotropic-complex	—	Leaflets funnel-shaped; sterile	
inc	Inflorescentia-conversa	Inflorescence	2?	Inflorescences on short axillary shoots	Lamprecht, 1958*b*
Inci	Incisus	Foliage	—	Leaflets and stipules deeply incised	Lamprecht, 1962*f*
ins	Insecatus	Foliage	1?	Leaflet apex deeply incised	Lamprecht, 1959*d*
Int	Incrementum	Foliage	—	Enhances *Td* dentation	Lamprecht, 1962*f*

TABLE 1. Continued

Symbol	Name	Type of mutant	Location	Phenotype	Reference
k	Alae keel-like	Flower or generative apparatus	2	Wings reduced, keellike	Lamprecht, 1948; Pellew and Sverdrup, 1923
kl	Cross-over regulator	Physiological character	—	Controlling CrO intensity	Håkansson, 1929; Matsuura, 1933
Kp	Keel colored	Flower or generative apparatus	4	Keel sides colored	Fedotov, 1935; Lamprecht, 1963f
Kpa	Keel colored	Flower or generative apparatus	3	As Kp	Lamprecht, 1963f
l	Impremere	Seed	3	Seed coat with few, larger depressions	Tedin et al., 1925; Wellensiek, 1950
la	Internodes much longer	Root and shoot	2	Increasing internode length	Haan, 1927; Lamprecht, 1961c
lac	Laciniata	Pleiotropic-complex	1	Leaflets funnel-shaped, lacinated; tendrils over-developed; sterile	Lamprecht, 1945a, 1956c
Laf	Latior	Pod	5	Broad pods	Lamprecht, 1954c
Lap1	Aminopeptidase	Physiological character	3	Electrophoretic mobility of aminopeptidase	Almgard and Öhlund, 1970
lat	Latum	Foliage	4	Increasing foliar area	Lamm, 1957
lath	Lathyroides	Pleiotropic-complex	—	Leaflets and stipules pointed, narrow; sterile	Lamprecht, 1959e
lc	Internode increase	Root and shoot	—	Slight increase of length and number of internodes	Haan, 1930
ld	Internode increase	Root and shoot	—	As lc	
le	Brevi-internodium	Root and shoot	4	Shortening internodes	Lamprecht, 1948; White, 1917

let	Lethality	Physiological character	7	Lethality in second half of ontogenesis	Gottschalk, 1968c
lf	Sterile nodes 4–11	Physiological character	1	4–11 sterile nodes to first flower	Hoshino, 1915; Murfet, 1971b; White, 1917
Li	Flowering inhibitor	Physiological character	1	± absence of vegetative-growth-inhibiting substance	Barber and Paton, 1952; Blixt, 1969b
lm	Brevi-internodium	Root and shoot	6	Short internodes	Lindquist, 1951; Rasmusson, 1938
lo	Semicompactum	Root and shoot	—	Very short internodes; female sterile	Rasmusson, 1938
lob	Lobata	Seed	3	Seed part-colored; legs of fork elongated, bending	Lamprecht, 1947a, 1957g; Lamprecht and Åkerberg, 1939
los1	Seed length	Seed	—	Reduces seed length	Blixt, 1972; Kaznowski, 1926
los2	Seed length	Seed	—	As los1	
los3	Seed length	Seed	—	As los1	
los4	Seed length	Seed	—	As los1	
lr	Resistance to PLV	Physiological characters	—	Resistance to pea leaf-roll virus	Drijfhout, 1968
lt	Latus	Pod	6	Broad pods	Lamprecht, 1957c
lum	Costata	Chlorophyll	4	Foliage with veins and lamina of different color	Monti, 1970b
M	Marmoreus	Seed	3	Seed coat marbled in lighter and darker brown	Lamprecht, 1948; Lock, 1907; White, 1917
ma	Seed size and weight	Seed	—	Affects seed size and weight	Lamprecht, 1957f
mal	Malehabitus	Seed	—	Uncolored pattern on the seed	Blixt, 1969b; Pohjakallio, 1941

TABLE 1. Continued

Symbol	Name	Type of mutant	Location	Phenotype	Reference
mare	Maximo-reductus	Foliage	—	Foliage threadlike, reduced	Lamprecht, 1967a
mex	Alterno-marmorata	Seed	—	Absence of M marbling; plants sterile, dwarfs	Lamprecht, 1960e
mie	Minuere	Root and shoot	1	Decreasing number of internodes	Lamprecht, 1962d
mier	Minuere	Root and shoot	3	As mie	Lamprecht, 1968a
mifo	Minute-foveatus	Seed	2	Seeds with close standing ± regular small and shallow despressions	Lamprecht, 1962e
min	Minulus	Pleiotropic-complex	—	Plants with an over-all size reduction	Lamprecht, 1957f
mine	Minuere	Root and shoot	7	As mie	Lamprecht, 1962d, 1968a
mis	Microsurculus	Pleiotropic-complex	—	Short axillary shoots instead of inflorescences; sterile	Lamprecht, 1964b
miu	Minuere	Root and shoot	5	As mie	Lamprecht, 1962d
miv	Minor intervallum	Flower or generative apparatus	1	Affects distance between seeds in pod	Lamprecht, 1952c
mo	Resistance to BV2 and PV2	Physiological character	—	Resistance to bean yellow mosaic virus and pea mosaic virus	Yen and Fry, 1956
mp	Furca	Seed	3	Part-color; zone around hilum, the "fork", decolored	Lamprecht, 1948; Tedin and Tedin, 1928
msl	Male sterility	Physiological character	—	Male sterility from breakdown of microsporogenesis	Gottschalk and Jahn, 1964

ms2	Male sterility	Physiological character	—	As ms1	Gottschalk and Jahn, 1964
mut	Mutagenic	Physiological character	—	A mutagenic gene	Wellensiek, 1959
n	Pod-wall thick	Pod	4	Thick and fleshy pod wall	Lamprecht, 1961c; Wellensiek, 1925a
na	Nana	Root and shoot	6	Extremely shortened and fewer internodes	Wellensiek, 1971b
nap	Navicula apertus	Flower or generative apparatus	1	Keel leaves not accreted above	Lamprecht, 1953c
no	Sterile nodes 4–11	Physiological character	4	4–11 sterile nodes to first inflorescence	Wellensiek, 1964, 1972
nod	Nodulum	Root and shoot	—	Affects number of bacterial root nodules	Gelin and Blixt, 1964
noda	Nodulum	Root and shoot	—	As nod	Blixt, 1972; Gelin and Blixt, 1964
nol	Notched leaflets	Foliage	—	Leaflet apex incised	Sharma, 1972a
Np	Neoplastic pods	Pod	4	Neoplastic outgrowths on pods	Dodds and Matthews, 1966; Nuttal and Lyall, 1964
nr	Narrow rogue	Foliage	—	Foliage narrow with pointed apex	Matsuura, 1933; Pellew, 1927
o	Flavoviridis	Chlorophyll	1	Yellowish green foliage color	Lamprecht, 1948; White, 1917
obo	Obovatus	Pleiotropic-complex	7	Leaflets obovate; sterile	Lamprecht, 1945b, 1958c
Obs	Obscuratum	Seed	—	Light intensity-sensitive increase of F, F_S, and U_{st} color pattern	Lamprecht, 1956d, 1958d
och	Ochraceus	Seed	3	Seed coat ochraceous colored	Lamprecht, 1959f
oh	Testa reddish brown	Seed	2	Seed-coat color reddish brown	Lamprecht, 1948; Tschermak, 1912; Wellensiek, 1925b

TABLE 1. Continued

Symbol	Name	Type of mutant	Location	Phenotype	Reference
oli	Griseolivaceus	Seed	1	Seed-coat color grayish olive	Lamprecht, 1969a
olv	Pallidolivaceus	Seed	4	Seed-coat color bleached greenish olive gray	Lamprecht, 1968a
op	Ovula pistilloida	Pleiotropic-complex	3	Ovules developing rudimentary pistils; sterile	Monti and Saccardo, 1971
p	Pod-wall sclerenchyma	Pod	6	Small stripes of sclerenchyma on inside of pod wall	Lamprecht, 1948; White, 1917
pa	Viridis	Chlorophyll	7	Dark green foliage color	Lamprecht, 1948; White, 1917; Winge, 1936
Paf	Bluish pink	Flower or generative apparatus	—	Flower color bluish pink	Fedotov, 1930; Lamprecht, 1957b
pafl	Parviflora	Flower or generative apparatus	2	Flowers small, vexillum about 18 × 17 mm	Lamprecht, 1963g
pal	Pallens	Seed	2	Part-color, seed coat almost uncolored	Lamprecht, 1961c; Lamprecht and Åkerberg, 1939
par	Thousand-grain weight	Seed	1	Reduced 1000-grain weight	Lamprecht, 1969a
pat	Patelliformis	Foliage	4	Leaflets irregularly bowl shaped	Lamprecht, 1968a
pe	Petalosus	Flower or generative apparatus	—	Anthers ± petaloid	Nilsson, 1932
ph	Photoperiodism	Physiological character	—	Not responding to day length	Barton et al., 1964

Symbol	Character	Organ/category	No.	Phenotype	References
Pl	Hilum black	Seed	6	Black hilum	Lamprecht, 1948; Lock, 1907; White, 1917
pla	Planatus	Seed	3	Seeds flattened sideways	Lamprecht, 1960c
pn	Pod number	Inflorescence	—	Affects flower shedding	Omar, 1960
pr	Praecidere	Inflorescence	—	Shortens inflorescences	Lamprecht, 1949a
pra	Sterile nodes 4–11	Physiological character	1	As *no*	Monti, 1970b; Monti and Scarascia-Mugnozza, 1969
prae	Internodiolum	Root and shoot	4	Increases number of internodes	Lamprecht, 1968a
pre	Praecidere	Inflorescence	—	As *pr*	Lamprecht, 1949a
pro	Procumbens	Root and shoot	4	Stem branches growing out horizontally, then rising at 45° angle	Lamprecht, 1963h
pt	Pollen-tube growth	Flower or generative apparatus	7?	Slow-growing pollen tubes	Wellensiek, 1925a
Pu	Purple pods	Pod	3	Pod color purple	Lamprecht, 1953b, 1961b; Sverdrup, 1927; White, 1917
Pur	Purple pods	Pod	1	As *Pu*	Lamprecht, 1938, 1948; White, 1917
py	Precocious yellowing	Chlorophyll	5	Plant suddenly yellowing shortly before maturity	Marx, 1971a
Q	Seed abortion	Physiological character	—	Seed early-aborting	Wellensiek and Keyser, 1928
qua	Comprimere	Seed	—	As *com*	Lamprecht, 1960c
r	Rugosus	Seed	7	Cotyledons wrinkled; starch phenotypically compound	Lamprecht, 1948; White, 1917
rag	Radicula grisea	Seed	3	Radicula region colored gray	Lamprecht, 1961d

TABLE 1. Continued

Symbol	Name	Type of mutant	Location	Phenotype	Reference
ram	Ramosus	Root and shoot	2	Larger number of branches	Monti, 1970b; Monti and Scarascia-Mugnozza, 1967
rb	Rugosus	Seed	3?	Cotyledons wrinkled; starch phenotypically simple	Gritton, 1971; Kooistra, 1962
re	Infecunda	Flower or generative apparatus	1	Anthers shortened; flowers budlike; self-sterile	Lamprecht, 1960f; Nilsson, 1932
red	Reductus	Foliage	1	Leaflets very narrow	Lamprecht, 1942b, 1948
Ro	Planatus	Seed	3?	As pla	Wellensiek, 1943
Rf	Radicula fusca	Seed	3	Radicula region colored brownish	Lamprecht, 1953d; Tedin and Tedin, 1928
rpv1	Resistance to Peronospora	Physiological character	—	Resistance to Peronospora viciae	Blixt, 1972; Matthews and Dow, 1969
rpv2	Resistance to Peronospora	Physiological character	—	As rpv1	
ru	Ruby	Seed	—	Immature seed coat dark red	Dickerson, 1931
rub	Rubicundus	Flower or generative apparatus	7	Modifies b flowers to brick-red color	Lamprecht, 1962g
rup	Rubropulvis	Pod	1	Pods with minute anthocyan-colored spots	Lamprecht, 1963i
rups	Rubropulvis	Pod	—	As rup	
s	Chenille	Seed	2	Tragacanth excretion on seed coat outside; seeds stick together	Lamprecht, 1948; White, 1917

sa1	Stomata number	Foliage	—	Reducing number of stomata	Matsuura, 1933; Tavcar, 1926
sa2	Stomata number	Foliage	—	As sa1	
sa3	Stomata number	Foliage	—	As sa1	
sal	Salmoneus	Seed	1	Seed coat salmon colored	Lamprecht, 1961e
sat	Sativoides	Pleiotropic-complex	—	Pleiotropic change to sativum character in ect. abyssinicum	Sharma and Aravindan, 1971b
sb	Screwball	Foliage	1	Crimpled, rigid leaves with whitish wax cover	Kellenbarger, 1952
sbm	Resistance to PSBMV	Physiological character	6	Resistance to pea seed-borne mosaic virus (fizzletop virus)	Hagedorn and Gritton, 1971
sc	Stipulae connatae	Foliage	5	Stipule base connected and accreted to stem	Wellensiek, 1971b
Ser	Serratus	Foliage	—	Deep serrated dentation of foliage	Lamprecht, 1962f
Sg1	Seed size and weight	Seed	—	Affects seed size and weight	Wellensiek, 1925b
Sg2	Seed size and weight	Seed	—	As Sg1	
Sg3	Seed size and weight	Seed	—	As Sg1	
Sg4	Seed size and weight	Seed	—	As Sg1	
sifl	Sine-floribus	Inflorescence	—	Inflorescence with big bracts, but no flowers	Lamprecht, 1958a
sin	Sine-fili	Pod	—	Stringless pods	Lamprecht, 1938
siv	Sine-vexillum	Flower or generative apparatus	—	Flowers of first four nodes without vexillum and sterile	Lamprecht, 1957c,f
sn	Sterile nodes 4–11	Physiological character	—	As lf	Murfet, 1971b; Tedin and Tedin, 1923

TABLE 1. Continued

Symbol	Name	Type of mutant	Location	Phenotype	Reference
sob	Superobscurum	Flower or generative apparatus	—	Vexillum upwards gradually with increasing strength of color	Lamprecht, 1962b
sre	Serenum	Flower or generative apparatus	—	Center of vexillum darker, more reddish	Lamprecht, 1962c
sru	Stria rubra	Pod	1	Pod along upper suture with anthocyanin stripe	Lamprecht, 1963i
srub	Stria rubra	Pod	—	As sru	
st	Stipules reduced	Foliage	3	Stipules reduced	Pellew and Sverdrup, 1923; Wellensiek, 1925b
ster	Female sterility	Physiological character	—	Female sterile	Gottschalk, 1964b
sti	Stipuloides	Pleiotropic-complex	—	All elements of flower stipulelike	Lamprecht, 1953c
stim	Stipula imminuata	Foliage	—	Stipules narrow, diamond shaped	Lamprecht, 1960g
stp	Stamina pistilloida	Flower or generative apparatus	2	Two foremost anthers carpellike	Monti and Devreux, 1969
str	Brunneostriata	Seed	2	A brownish stripe running from caruncula around the seed	Lamprecht, 1947a, 1960h
sub	Minulus	Pleiotropic-complex	3	As min	Lamprecht, 1968a, 1969b
sul	Foveatus	Seed	5	As fov	Lamprecht, 1960c
sup	Superpetaloidum	Flower or generative apparatus	5?	Two anthers petaloid	Lamprecht, 1939a
t	Stem thickness	Root and shoot	1	Thicker stem	Keeble and Pellew, 1910–1911

Symbol	Name	Category	No.	Description	Reference
tac	Tendrilled acacia	Foliage	—	Leaf with tendrils and apical leaflet	Sharma, 1972*b*
td	Scalaris forma	Foliage	4	No dentation of foliage	Lamprecht, 1945*c*; Wellensiek, 1925*b*
te	Tenuis	Pod	5	Narrow pod	Lamprecht, 1953*e*
ten	Tenuifolius	Foliage	—	Narrow leaflets	Lamprecht, 1949*b*
teu	Tenuis	Pod	5	As *te*	Lamprecht, 1960*i*
tl	Clavicula	Foliage	7	Leaflets in the place of tendrils	Lamprecht, 1948; White, 1917
Tra	Tragacanthum	Seed	4	Seed coat with "oily" spots from tragacanth excretion on inside	Lamprecht, 1944*a*, 1956*e*
trip	Tripistillum	Pleiotropic-complex	—	Three pistils; foliage very narrow; sterile	Lamprecht, 1944*b*
trp	Triangular pod	Pod	—	Pods triangular in shape	Sharma, 1972*c*
U	Violet testa	Seed	5	Seed coat violet colored	Lamprecht, 1948; White, 1917
Umb	Umbra	Seed	—	Seed coat umbra colored	Lamprecht, 1961*b*
un	Undulatifolius	Foliage	4	Leaflet margin undulated	Lamprecht, 1958*e*
uni	Unifoliata	Pleiotropic-complex	3	Leaf a single lamina; sterile	Lamprecht, 1933–1934, 1964*b*
up	Unipetiolle	Foliage	—	Leaves with one pair of leaflets only	Rosen, 1944
v	Pod-wall sclerenchyma	Pod	4	Pod-wall inside with patches of sclerenchyma	Lamprecht, 1948; White, 1917
vac	Vario-maculata	Chlorophyll	—	Symbol for vario-maculata; foliage with sectors of different color and structure	Blixt, 1972
vac1	Vario-maculata	Chlorophyll	1	See *vac*	Lamprecht, 1958*e*; Wells, 1951*b*

TABLE 1. Continued

Symbol	Name	Type of mutant	Location	Phenotype	Reference
ve	Ventriosus	Seed	2	Part-color; fork very broad, moon pattern missing	Lamprecht, 1954d, 1957j
vil	Chlorotica	Chlorophyll	1	See *chi*	Kellenbarger, 1952; Lamprecht, 1953b
vim	Viridis	Chlorophyll	4	As *pa*	Lamprecht, 1955e, 1957j
Vl	Bleached violet testa	Seed	—	Seed coat bleached violet	Fedotov, 1935
wa	Vix-cerata	Foliage	2	Upper surface of stipules and lower of leaflets, stems, and pods with very reduced wax layer	Lamprecht, 1955f; Wellensiek, 1928
was	Vix-cerata	Foliage	4	As *wa*	Lamm, 1957
wb	Vix-cerata	Foliage	2	As *wa*	Lamprecht, 1948; Wellensiek, 1928; White, 1917
wel	Incerata	Foliage	3	As *bl*	Marx, 1969
wex	Cerosa	Foliage	—	Excess of wax	Lamprecht, 1948; Nilsson, 1933
wlo	Supra-incerata	Foliage	6	Upper surface of leaflets waxless	Marx, 1971a
wp	Waxless pods	Pod	—	Pods waxless	Lamprecht, 1939b, 1954e
wsp	Subtus-incerata	Foliage	7	Upper surface of leaflets normally waxy, other parts waxless	
X	Rogue	Foliage	—	Foliage and pods narrow with pointed apexes	Brotherton, 1923; Matthews, 1969

Symbol	Name	Category		Description	References
xa	Xantha	Chlorophyll	—	Symbol for reddish yellow to yellow to yellowish white seedlings, lethal at 5-7-leave stage	Lamprecht, 1952*a*
xa1	Xantha	Chlorophyll	7	See *xa*	Lamprecht, 1952*a*, 1955*g*
xa2	Xantha	Chlorophyll	—	See *xa*	Lamprecht, 1952*a*; Rasmusson, 1938
xa3	Xantha	Chlorophyll	—	See *xa*	Blixt, 1961; Lamprecht, 1952*a*; Rasmusson, 1938
xa4	Xantha	Chlorophyll	—	See *xa*	
xa5	Xantha	Chlorophyll	—	See *xa*	Blixt, 1961; Lamprecht and Åkerberg, 1939
xat	Xantha-terminalis	Chlorophyll	—	Plant green below, top turns yellow	Lamprecht, 1952*a*
Y	Foliage width reduced	Foliage	1	Reduces width of leaflets, sepals, etc.	Brotherton, 1923; Lamprecht, 1967*a*
yp	Yellow pollen	Flower or generative apparatus	3	Pollen grains yellow	Murfet, 1967
z	Furca	Seed	4	As *mp*	Kajanus, 1923; Lamprecht, 1948

Most mutants have now been assigned to a so-called type line, which is normally the line in which the mutant was first isolated or analyzed. A locus is then defined by its position and effect in that line. The type-line concept has two main advantages. First, it allows a limitation to be put on the number of lines which have to be preserved by gene banks. Instead of maintaining an unlimited number of unique, unanalyzed genotypes, the maintenance can be restricted to the individual loca and chromosomal structural types as defined by and carried in the type lines, plus a more limited number of genotypes of practical or otherwise particular interest. This procedure thus aids in minimizing costs and losses of genetic material. Second, a new mutation may be crossed with plants from existing type lines with similar phenotypes which are thought to contain an allelic gene before a new gene symbol is assigned, an "identity test-cross." The confusion resulting from the creation of synonymous gene symbols can thus largely be avoided. The preservation and distribution of type lines is considered one of the main tasks and responsibilities of the PGA (*Pisum* Genetics Association).

The finding of the appropriate type lines for identity test-crosses would undoubtedly be facilitated by a common system for classifying the different mutations obtained. One system, including details on about 2000 spontaneous and induced mutants recorded in peas, was presented by Blixt (1972). According to phenotypic or other similarities, the individual *mutation cases* were assigned to *mutation types,* and these were arranged in the following groups.

1. Chromosome mutants
2. Physiological character mutations
3. Seed mutants
4. Root and shoot mutants
5. Foliage mutants
6. Chlorophyll mutants
7. Inflorescence mutants
8. Mutations in flowers or generative apparatus
9. Pod mutants
10. Mutations with extensively pleiotropic effects and complex mutants

A fairly large number of mutants, type lines, and test lines are, at the moment, available in different collections. Most of these can be obtained from the *Pisum* Genetics Association (Chairman, Prof. G. A. Marx, Cornell University, Geneva, New York; Secretary, S. Blixt, Weibullsholm Plant Breeding Institute, Landskrona, Sweden). The PGA issues the *Pisum Newsletter* annually to its members.

Literature Cited

Almgård, G. and K. Öhlund, 1970   Inheritance and location of a biochemical character in *Pisum*. *Pisum News.* **2**:9.

Balkema, G. H., 1971   *Chimerism and Diplontic Selection,* A. A. Balkema, Rotterdam.

Barber, H. N. and D. M. Paton, 1952   A gene-controlled flowering inhibitor in *Pisum*. *Nature (Lond.)* **169**:592.

Barton, D. W., W. T. Schroeder, R. Provvidenti and W. Mishanec, 1964   Clones from segregating progenies of garden pea demonstrate that resistance to BV2 and PV2 is conditioned by the same genotype. *Plant Dis. Rep.* **48**:353–355.

Blixt, S., 1958   Cytology of *Pisum*. II. The normal karyotype. *Agri Hort. Genet.* **16**:221–237.

Blixt, S., 1959   Cytology of *Pisum*. III. Investigation of five interchange lines and coordination of linkage groups with chromosomes. *Agri Hort. Genet.* **17**:47–75.

Blixt, S., 1961   Quantitative studies of induced mutations in peas. V. Chlorophyll mutations.*Agri Hort. Genet.* **19**:402–447.

Blixt, S., 1965   Linkage studies in Pisum. I. Linkages of the genes *Chi2, Chi3,* and *Chi4,* causing chlorophyll deficiency. *Agri Hort. Genet.* **23**:26–42.

Blixt, S., 1966*a*   Studies of induced mutation in peas. XIII. Segregation of an *albina* mutant. *Agri Hort. Genet.* **24**:48–55.

Blixt, S., 1966*b*   Linkage studies in *Pisum*. V. The mutant *albina*$_{R2}$. *Agri Hort. Genet.* **24**:168–172.

Blixt, S., 1966*c*   Linkage studies in *Pisum*. IV. The mutant *Chlorina*$_{EO1}$ and the linkage of the determining gene, *Ch3*. *Agri Hort. Genet.* **24**:164–167.

Blixt, S., 1968*a*   Linkage studies in *Pisum*. VIII. Gene-symbolization of 15 *chlorotica*-mutants. *Agri Hort. Genet.* **26**:82–87.

Blixt, S., 1968*b*   Linkage studies in *Pisum*. X. Linkage of the gene *Chi3*. *Agri Hort. Genet.* **26**:100–106.

Blixt, S., 1968*c*   Linkage studies in *Pisum*. IX. Linkage of the gene *Chi5* in chromosome II. *Agri Hort. Genet.* **26**:88–99.

Blixt, S., 1968*d*   Linkage studies in *Pisum*. XI. Linkage relations of the gene *chi16* determining chlorophyll-deficiency *chlorotica*. *Agri Hort. Genet.* **26**:107–110.

Blixt, S., 1968*e*   Linkage studies in *Pisum*. XII. Linkage relations of the genes *Fr* and *Fru,* determining ramification. *Agri Hort. Genet.* **26**:136–141.

Blixt, S., 1969*a*   Publications by Herbert Lamprecht. *Agri Hort. Genet.* **27**:7–18.

Blixt, S., 1969*b*   Gene list with citations. *Pisum Newsl.* **1**:23–60.

Blixt, S., 1969*c*   Linkage relations in *Pisum*. XIV. The chlorophyll deficiency genes *chi6, chi9* and *chi17,* determining *chlorotica*-type, and *ch4* and *ch5,* determining *chlorina*. *Agri Hort. Genet.* **27**:36–42.

Blixt, S., 1970   Studies of induced mutations in peas. XXVI. Genetically conditioned differences in radiation sensitivity. 4. *Agri Hort. Genet.* **28**:55–116.

Blixt, S., 1972   Mutation genetics in *Pisum*. *Agri Hort. Genet.* **30**:1–293.

Blixt, S. and O. Gelin, 1965   The relationship between leaf spotting (A-sectors) and mutation rate in *Pisum*. *Radiat. Bot.* **5**:Suppl. 251–262.

Blixt, S., L. Ehrenberg and O. Gelin, 1958   Quantitative studies of induced mutations in peas. I. Methodological investigations. *Agri Hort. Genet.* **16**:238–250.

Bremer, A. H. and G. Weiseth, 1961   Problems of vegetable production in relation to day-length on latitutdes, 58–70°N. I. Limitation of day-length in vegetable culture.

II. The genetics of long-day and day-neutral peas. (*Proc. XVth Int. Hort. Congr., Nice, 1958*) *Adv. Hortic. Sci. Appl.* **I**:426–435.

Brotherton, J. W., 1923   Further studies of the inheritance of "Rogue" types in garden peas. (*Pisum sativum* L.) *J. Agric. Res.* **24**:815–852.

Dickerson, L. M., 1931   The inheritance of Ruby seed coat color in peas. *J. Hered.* **22**:319–321.

Dodds, K. S. and P. Matthews, 1966   Neoplastic pod in the pea. *J. Hered.* **57**:83–85.

Drijfhout, E., 1968   Testing for pea leaf-roll virus and inheritance of resistance in peas. *Euphytica* **17**:224–235.

Ekelund, R. and T. Hemberg, 1966   A comparison between geotropism and geoelectric effect in *Pisum sativum* and its mutant *ageotropum*. *Physiol. Plant.* **19**:1120–1124.

Fedotov, V. S., 1930   On the hereditary factors of flower colour and of some other characters in the pea. *Proc. USSR Congr. Genet. Plant Animal Breed., 1930* **2**:523–537.

Fedotov, V. S., 1935   The genetics of anthocyanin pigmentation in peas. *Tr. Prikl. Bot. Genet. Sel.* 1935(1936) *Ser. 2.* **9**:163–274.

Gelin, O. and S. Blixt, 1964   Root nodulation in peas. *Agri Hort. Genet.* **22**:149–159.

Goldenberg, J. B., 1965   "Afila," a new mutation in pea (*Pisum sativum* L.). *Bol. Genet.* **1**:27–31.

Gottschalk, W., 1961   Über die mutative Abänderung Pflanzlicher Organizationsmerkmale. *Planta (Berl.)* **57**:313–330.

Gottschalk, W., 1964*a*   Eine Gruppe pleiotroper Gene mit weitgehend übereinstimmenden Wirkungsspektren. Ein neuer Fall von Polymerie bei *Pisum. Radiat. Bot.* **4**:267–274.

Gottschalk, W., 1964*b*   Untersuchungen über die Abgrenzung von pleiotropie und absoluter Koppelung. *Z. Vererbungsl.* **95**:17–24.

Gottschalk, W., 1968*a*   Investigation on the genetic control of meiosis. *Nucleus* (Calcutta) **11**:346–361.

Gottschalk, W., 1968*b*   Origin and behavior of rings of 4 and 6 chromosomes during meiosis of *Pisum. Nucleus* **11**:61–74.

Gottschalk, W., 1968*c*   Simultaneous mutation of closely linked genes. In *Mutations in Plant Breeding*, Vol. II, pp. 97–109, IAEA, Vienna.

Gottschalk, W., 1970   Möglichkeiten der Blattevolution durch Mutation und Rekombination. Ein Modell für die Entwicklung und Weiterentwicklung des Leguminosenblattes. *Z. Pflanzenphysiol.* **63**:44–54.

Gottschalk, W. and S. R. Baquar, 1971   Desynapsis in *Pisum sativum* induced through gene mutation. *Can. J. Genet. Cytol.* **13**:138–143.

Gottschalk, W. and S. Blixt, 1975   Legumes. In *Mutations and Crop Improvement* (in press).

Gottschalk, W. and A. Jahn, 1964   Cytogenetische Untersuchungen an desynaptischen und männlich-sterilen Mutanten von *Pisum. Z. Vererbungsl.* **95**:150–167.

Govorov, L. I., 1928   The peas of Afghanistan. *Trudy Prikl. Bot. Gene. Sel.* **19**:517–522.

Govorov, L. I., 1937   Erbsen. In *Flora of Cultivated Plants,* Vol. IV, edited by N. L. Vavilov and E. V. Wulff, pp. 231–336. State Agricultural Publishing Company, Moscow.

Gritton, E. T., 1971   A case of apparent linkage between the gene *rb* for wrinkled seed and *st* for reduced stipules. *Pisum Newsl.* **3**:15.

Grout, A. J., 1904   A peculiar pea seedling. *Torreya (New York)* **4**:171.

Haan, H. de, 1927   Length-factors in *Pisum. Genetica (The Hague)* **9**:481–498.

Haan, H. de, 1930   Contributions to the genetics of *Pisum. Genetica (The Hague)* **12**:321–439.

Haan, H. de, 1932   The heredity of emergences in *Pisum sativum. Genetica (The Hague)* **14**:319–320.

Hagedorn, D. J. and E. T. Gritton, 1971   Inheritance and linkage of resistance to the pea seed borne mosaic virus. *Pisum Newsl.* **3**:16.

Håkansson, A., 1929   Chromosomenringe in *Pisum* und ihre Mutmassliche Genetische Bedeutung. *Hereditas* **12**:1–10.

Håkansson, A., 1931   Uber Chromosomenverkettung in *Pisum. Hereditas* **15**:17–61.

Hare, W. W., J. C.Walker and E. H. Delwiche, 1949   Inheritance of a gene for near-wilt resistance in the garden pea. *J. Agric. Res.* **78**:239–250.

Harland, S. C., 1948   Inheritance of immunity to mildew in Peruvian forms of *Pisum sativum. Heredity* **2**:263–269.

Härstedt, E., 1950   Über die Vererbung der Form von Laub- und Kelchblättern von *Pisum sativum. Agri Hort. Genet.* **8**:7–32.

Haupt, W., 1952   Untersuchungen über den Determinationsvorgang der Blütenbildung bei *Pisum sativum. Z. Bot.* **40**:1–32.

Heringa, R. J., A. Van Norel and M. F. Tazelaar, 1969   Resistenz gegen Echten Mehltau (*Erisyphe polygoni* D.C.) bei Erbsen (*Pisum sativum* L.). *Euphytica* **18**:163–169.

Highkin, H. R. and A. Lang, 1966   Residual effect of germination temperature on the growth of peas. *Planta (Berl.)* **68**:94–98.

Hoshino, Y., 1915   On the inheritance of the flowering time in peas and rice. *J. Coll. Agric. (Sapporo)* **6**:229–288.

Johnston, M. J. and R. K. Crowden, 1967   Cotyledon excision and flowering in *Pisum sativum. Aust. J. Biol. Sci.* **20**:461–463.

Kajanus, B., 1923   Genetische Studie an *Pisum. Z. Pflanzenzuecht.* **9**:1–22.

Kapoor, B. M., 1966–1967   Contributions to the cytology of endosperm in angiosperms. XII. *Pisum sativum* L. *Genetica (The Hague)* **37**:557–568.

Kaznowski, L., 1926   Recherches sur le pois (*Pisum*). *Mem. Inst. Natl. Pol. Econ. Rurale Pulawy* **7**:1–91.

Keeble, F. W. and C. Pellew, 1910–1911.   The mode of inheritance of stature and of time of flowering in peas. *J. Genet.* **1**:47–56.

Kellenbarger, Sh., 1952   Inheritance and linkage data of some characters in peas (*Pisum sativum*). *J. Genet.* **51**:41–46.

Khangildin, V. V., 1966   A new gene *leaf* inducing the absence of leaflets in peas. Interaction between genes *leaf* and *Tl*^w. *Genetika* **6**:88–96.

Klasova, A., J. Kolek and J. Klas, 1971   A statistical study of the formation of lateral roots in *Pisum sativum* L. under constant conditions. *Biol. Plant. (Prague)* **13**:209–215.

Klein, H. D., 1970   Asynapsis and extensive chromosome breakage in *Pisum. Caryologia* **23**:251–257.

Kooistra, E., 1962   On the differences between smooth and three types of wrinkled peas. *Euphytica* **11**:357–373.

Lamm, R., 1937   Length factors in dwarf peas. *Hereditas* **23**:38–48.

Lamm, R., 1949   Contributions to the genetics of the *Gp*-chromosome of *Pisum. Hereditas* **35**:203–214.

Lamm, R., 1951 Cytogenetical studies on translocations in *Pisum. Hereditas* **37**:356–372.

Lamm, R., 1957 Three new genes in *Pisum. Hereditas* **43**:541–548.

Lamm, R., 1960 Studies on chromosome 1 in *Pisum. Hereditas* **46**:737–744.

Lamm, R. and R. J. Miravalle, 1959 A translocation tester set in *Pisum. Hereditas* **45**:417–440.

Lamprecht, H., 1933–1934 Ein *unifoliata*-Typus von *Pisum* mit gleichzeitiger Pistilloidie. *Hereditas* **18**:56–64.

Lamprecht, H., 1935 Eine *Pisum*-Form mit compactum-Verzweigung und verkürzten Staubfäden. *Hereditas* **20**:94–102.

Lamprecht, H., 1936 Genstudien an *Pisum sativum*. I. Über den Effekt der Genpaare *Con-con* und *S-s. Hereditas* **22**:336–360.

Lamprecht, H., 1938 Über Hülseneigensachaften bei *Pisum,* ihre Vererbung und ihr züchterischer Wert. *Züchter* **10**:150–157.

Lamprecht, H., 1939*a* Über Blüten- und Komplexmutationen bei *Pisum. Z. Indukt. Abstammungs.- Vererbungsl.* **77**:177–185.

Lamprecht, H., 1939*b* Translokation, Genspaltung und Mutation bei *Pisum. Hereditas* **25**:431–458.

Lamprecht, H., 1942*a* Genstudien an *Pisum sativum*. V. Multiple Allele für Punktierung der Testa: Fs_{ex}-*Fs-fs. Hereditas* **28**:157–164.

Lamprecht, H., 1942*b* Die Koppelungsgruppe *Gp-Cp-Fs-Ast* von *Pisum. Hereditas* **28**:143–156.

Lamprecht, H., 1944*a* Genstudien and *Pisum sativum,* VI-VIII. VIII. Das Testamerkmal *griseostriata* und seine Vererbung. *Hereditas* **30**:613–630.

Lamprecht, H., 1944*b* Die Genisch-Plasmatische Grundlage der Artbarriere. *Agri. Hort. Genet.* **2**:75–142.

Lamprecht, H., 1945*a* Durch Komplexmutation bedingte Sterilität und ihre Vererbung. *Arch. Julius Klaus- Stift. Vererbungsforsch. Sozialanthropol. Rassenhyg.* **20**: Suppl. 126–141.

Lamprecht, H., 1945*b* Intra- and interspecific genes. *Agri Hort. Genet.* **3**:45–60.

Lamprecht, H., 1945*c* Die Koppelungsgruppe *N-Z-Fa-Td* von *Pisum. Hereditas* **31**:347–382.

Lamprecht, H., 1946 Die Koppelungsgruppe *Uni-M-Mp-F-St-B-Gl* von *Pisum. Agri Hort. Genet.* **4**:15–42.

Lamprecht, H., 1947*a* Die Testafärbung von *Pisum*-Samen und ihre Vererbung. *Agri Hort. Genet.* **5**:85–105.

Lamprecht, H., 1947*b* The inheritance of the number of flowers per inflorescence and the origin of *Pisum,* illustrated by polymeric genes. *Agri Hort. Genet.* **5**:16–25.

Lamprecht, H., 1948 The variation of linkage and the course of crossingover. *Agri Hort. Genet.* **6**:10–48.

Lamprecht, H., 1949*a* Die Vererbung verschiedener Inflorescenztypen bei *Pisum. Agri Hort. Genet.* **7**:112–133.

Lamprecht, H., 1949*b* Über Entstehung und Vererbung von schmalblättrigen Typen bei *Pisum. Agri Hort. Genet.* **7**:134–153.

Lamprecht, H., 1950*a* Koppelungsstudien in Chromosom V von *Pisum. Agri Hort. Genet.* **8**:163–184.

Lamprecht, H., 1950*b* The degree of ramification in *Pisum* caused by polymeric genes. *Agri Hort. Genet.* **8**:1–6.

Lamprecht, H., 1951   Genanalytische Studien zur Artberechtigung von *Pisum humile* Boiss. *et* Noë. *Agri Hort. Genet.* **9**:107–134.

Lamprecht, H., 1952*a*   Über Chlorophyllmutanten bei *Pisum* und die Vererbung einer neuen, goldgelben Mutante. *Agri Hort. Genet.* **10**:1–18.

Lamprecht, H., 1952*b*   Polymere Gene und Chromosomenstruktur bei *Pisum. Agri Hort. Genet.* **10**:158–168.

Lamprecht, H., 1952*c*   Weitere Koppelungsstudien an *Pisum sativum,* insbesondere im Chromosom II (*Ar*). *Agri Hort. Genet.* **10**:51–74.

Lamprecht, H., 1953*a*   Further studies of the interchange between the chromosomes III and V of *Pisum. Agri Hort. Genet.* **11**:141–148.

Lamprecht, H., 1953*b*   New and hitherto known polymeric genes of *Pisum. Agri Hort. Genet.* **11**:40–54.

Lamprecht, H., 1953*c*   Bisher bekannte und neue Gene für die Morphologie der *Pisum*-Blüte. *Agri Hort. Genet.* **11**:122–132.

Lamprecht, H., 1953*d*   Ein neues Gen *Rf* und die Genenkarte des Chromosoms III von *Pisum. Agri Hort. Genet.* **11**:55–65.

Lamprecht, H., 1953*e*   Ein Gen für schmale Hülsen bei *Pisum* und seine Koppelung. *Agri Hort. Genet.* **11**:15–27.

Lamprecht, H., 1953*a*   Zur Kenntnis der Chromosomenstruktur von *Pisum.* Eine Übersicht und ein neuer Fall, mit chromosomal bedingter Ausspaltung von sterile Zwergen. *Agri Hort. Genet.* **12**:121–149.

Lamprecht, H., 1954*b*   The inheritance of the orange-coloured radicula and corona of *Pisum. Agri Hort. Genet.* **12**:50–57.

Lamprecht, H., 1954*c*   Weitere Studien über die Vererbung der Hülsenbreite von *Pisum. Agri Hort. Genet.* **12**:202–210.

Lamprecht, H., 1954*d*   Ein neues Gen für Teilfarbigkeit von *Pisum*-Samen und seine Koppelung. *Agri Hort. Genet.* **12**:58–64.

Lamprecht, H., 1954*e*   Die Koppelung des Gens *Wsp* und die Genenkarte von Chromosom VII von *Pisum. Agri Hort. Genet.* **12**:115–120.

Lamprecht. H., 1955*a*   Ein Interchange zwischen den Chromosomen I und VII von *Pisum. Agri Hort. Genet.* **13**:173–182.

Lamprecht, H., 1955*b*   Ein neuer Strukturtyp von *Pisum.* Ein Fall mit einer einfachen Translokation und einer Duplikation. *Agri Hort. Genet.* **13**:85–94.

Lamprecht, H., 1955*c*   Über die Wirkung der Gene *Con* und *Co* bei *Pisum* mit einer Übersicht von bisher über die Vererbung der Hülsenform bekanntem. *Agri Hort. Genet.* **13**:19–36.

Lamprecht, H., 1955*d*   Die Vererbung der Chlorophyllmutante *albinaterminalis* von *Pisum* sowie Allgemeines zum Verhalten von Chlorophyll- und andere Genen. *Agri Hort. Genet.* **13**:103–114.

Lamprecht, H., 1955*e*   Zur Kenntnis der Genenkarte von Chromosom VII von *Pisum* sowie die Wirkung der Gene *Tram* und *Vim. Agri Hort. Genet.* **13**:214–229.

Lamprecht, H., 1955*f*   Studien zur Genenkarte von Chromosom II von *Pisum.* Die Koppelung des Gens *Wa* und die Wirkung der übrigen Wachsgene. *Agri Hort. Genet.* **13**:154–172.

Lamprecht, H., 1955*g*   Die Koppelung des Chlorophyllgens *Xa1* von *Pisum. Agri Hort. Genet.* **13**:115–120.

Lamprecht, H., 1956*a*   Die Koppelung des Gens *Cal* für Teilfarbigkeit der *Pisum*-Samen. *Agri Hort. Genet.* **14**:34–44.

Lamprecht, H., 1956*b*   Ein *Pisum*-Typ mit grundständigen Infloreszenzen. *Agri Hort. Genet.* **14**:195–202.

Lamprecht, H., 1956*c*   Studien zur Genenkarte von Chromosom I von *Pisum*. *Agri. Hort. Genet.* **14**:66–106.

Lamprecht, H., 1956*d*   Zum Auftreten von *obscuratum*-Samen bei *Pisum*. *Agri. Hort. Genet.* **14**:19–33.

Lamprecht, H., 1956*e*   Über Wirkung und Koppelung des Gens *Tram* von *Pisum*. *Agri Hort. Genet.* **14**:45–53.

Lamprecht, H., 1957*a*   Die Genenkarte von Chromosom VI und das Interchange der Chromosomen IV/VI von *Pisum*. *Agri Hort. Genet.* **15**:115–141.

Lamprecht, H., 1957*b*   Eine neue Blütenfarbe von *Pisum* mit einer Übersicht über bisher bekannte Blütenfarben und deren genische Bedingtheit. *Agri Hort. Genet.* **15**:155–168.

Lamprecht, H., 1957*c*   Röntgeninduzierte spezifische Mutationen bei *Pisum* in ihrer Abhängigkeit von der Genotypischen Konstitution. *Agri. Hort. Genet.* **15**:169–193.

Lamprecht, H., 1957*d*   Dominantes Drabgrau, eine neue genbedingte Testafarbe bei *Pisum*. *Agri Hort. Genet.* **15**:207–213.

Lamprecht, H., 1957*e*   Zur Genenkarte von Chromosom II von *Pisum*. *Agri Hort. Genet.* **15**:12–47.

Lamprecht, H., 1957*f*   Durch Röntgenbestrahlung von *Pisum*-Samen erhaltene neue und bekannte Genmutationen. *Agri Hort. Genet.* **15**:142–154.

Lamprecht, H., 1957*g*   Die Koppelung der Teilfarbigkeitsgene *Dem* und *Lob* von *Pisum*. Weitere Beiträge zu den Genenkarten der *Pisum*-Chromosomen. *Agri Hort. Genet.* **15**:48–57.

Lamprecht, H. 1957*h*   Die Koppelung des Gens *Gri* von *Pisum*. Ein Beitrag zur Genenkarte von Chromosom I. *Agri Hort. Genet.* **15**:90–97.

Lamprecht, H., 1957*i*   Über die Vererbung der Hülsenbreite bei *Pisum*. *Agri Hort. Genet.* **15**:105–114.

Lamprecht, H., 1957*j*   Die Lage der Gene *Ve* und *Vim* in den Chromosomen II besw. IV von *Pisum* sowie weitere Koppelungsstudien. *Agri Hort. Genet.* **15**:1–11.

Lamprecht, H., 1958*a*   Über grundlegende Gene für die Gestaltung höherer Pflanzen sowie über neue und bekannte Röntgenmutanten. *Agri Hort. Genet.* **16**:145–192.

Lamprecht, H., 1958*b*   Eine *Pisum*-Mutante mit in diminutive Stammverzweigungen umgewandelte Infloreszenzen und ihre Vererbung. *Agri Hort. Genet.* **16**:112–128.

Lamprecht, H., 1958*c*   Die Koppelung des Gens *Obo* mit *R* im Chromosom VII von *Pisum*, sowie über Lamm's genenleeres Chromosom und genenreiche Chromosomensegmente von *Pisum*. *Agri Hort. Genet.* **16**:38–48.

Lamprecht, H., 1958*d*   Zur Genbedingtheit des *obscuratum*-Merkmals von *Pisum*. *Agri Hort. Genet.* **16**:49–53.

Lamprecht, H., 1958*e*   Gekräuselte Blättchen bei *Pisum* und ihre Vererbung. *Agri Hort. Genet.* **16**:1–8.

Lamprecht, H., 1959*a*   Über Wirkung und Koppelung des Gens *Alt* von *Pisum*. *Agri Hort. Genet.* **17**:15–25.

Lamprecht, H., 1959*b*   Ein Gen für eingesenkte Radicula und seine Lage im Chromosom III von *Pisum*. *Agri Hort. Genet.* **17**:37–46.

Lamprecht, H., 1959*c*   Die Vererbung der Farben von *a*-samen von *Pisum*. *Agri Hort. Genet.* **17**:1–8.

Lamprecht, H., 1959*d*   Das Merkmal *insecatus* von *Pisum* und seine Vererbung sowie einige Koppelungsstudien. *Agri Hort. Genet.* **17**:26–36.

Lamprecht, H., 1959*e* Der Artbegriff, seine Entwicklung und experimentelle Klarlegung. *Agri Hort. Genet.* **17**:105–264.

Lamprecht, H., 1959*f* Ein Gen für ockergelbe Färbung der Testa und seine Lage im Chromosome III von *Pisum. Agri Hort. Genet.* **17**:9–14.

Lamprecht, H., 1960*a* Zur Wirkung und Koppelung des Gens *Ca,* ein neues Gen für Teilfarbigkeit sowie weitere Koppelungsergebnisse bei *Pisum. Agri Hort. Genet.* **18**:74–85.

Lamprecht, H., 1960*b* Zwei neue Chlorophyllmutanten von *Pisum, chlorinavirescens* und *chlorina-virescens-chlorotica-terminalis,* sowie zur Koppelung des *Ch*-Gens. *Agri Hort. Genet.* **18**:169–180.

Lamprecht, H., 1960*c* Zur Vererbung der Samenformen bei *Pisum* sowie über zwei neue, diese beeinflussende Gene. *Agri Hort. Genet.* **18**:1–22.

Lamprecht, H., 1960*d* Zur Wirkung und Koppelung des Gens *Fom* für die Blattform von *Pisum. Agri Hort. Genet.* **18**:62–73.

Lamprecht, H., 1960*e* Zwei bemerkenswerte genbedingte Chimären von *Pisum. Agri Hort. Genet.* **18**:125–134.

Lamprecht, H., 1960*f* Studien zur Manifestation und Koppelung des Sterilität bedingenden Gens *Re* von *Pisum. Agri Hort. Genet.* **18**:181–204.

Lamprecht, H., 1960*g* Eine neue Stipel-Mutante bei *Pisum. Agri Hort. Genet.* **18**:209–213.

Lamprecht, H., 1960*h* Zur Manifestation und Koppelung der Gene *Pal* und *Str* von *Pisum.* Ein Beitrag zur Genenkarte von Chromosom II. *Agri Hort. Genet.* **18**:86–96.

Lamprecht, H., 1960*i* Weitere Studien zur Genenkarte von Chromosom V von *Pisum. Agri Hort. Genet.* **18**:23–56.

Lamprecht, H., 1961*a* Ein neues monogen bedingtes Merkmal bei *Pisum*: Purpurviolette Streifung der Hülsen. *Agri Hort. Genet.* **19**:241–244.

Lamprecht, H., 1961*b* *Pisum fulvum* Sibth. and Sm. Genanalytical studies with regard to the classification as species. *Agri Hort. Genet.* **19**:269–297.

Lamprecht, H., 1961*c* Die Genenkarte von *Pisum* bei normaler Struktur der Chromosomen. *Agri Hort. Genet.* **19**:360–401.

Lamprecht, H., 1961*d* Ein neues Gen für lokale Färbung der Testa von *Pisum*-Samen und seine Koppelung sowie weitere Koppelungsstudien insbesondere im Chromosom V. *Agri Hort. Genet.* **19**:197–212.

Lamprecht, H., 1961*e* Eine neue Testafarbe von *Pisum*-Samen: *salmoneus. Agri Hort. Genet.* **19**:213–222.

Lamprecht, H., 1962*a* Über Symmetrie und Asymmetrie von Stipeln und Blättchen bei *Pisum* sowie ihre Vererbung. *Agri Hort. Genet.* **20**:214–219.

Lamprecht, H., 1962*b* Weitere neue Blütenfarbe von *Pisum* sowie zur Kenntnis der Koppelungsgruppe von Chromosom II. *Agri Hort. Genet.* **20**:156–166.

Lamprecht, H., 1962*c* Über Gene für die Farbenverteilung auf den Blumenblättern von *Pisum. Agri Hort. Genet.* **20**:11–16.

Lamprecht, H., 1962*d* Studien zur Vererbung des Höhenwachstums bei *Pisum* sowie Koppelungsstudien. *Agri Hort. Genet.* **20**:23–62.

Lamprecht, H., 1962*e* Über ein neues, die Form der *Pisum*-Samen beeinflussendes Gen sowie ein neues Gen für Teilfarbigkeit. *Agri Hort. Genet.* **20**:137–155.

Lamprecht, H., 1962*f* Zur Vererbung der Blättchenzähnung bei *Pisum. Agri Hort. Genet.* **20**:63–74.

Lamprecht, H., 1962*g* Gleichzeitiges Mutieren in mehreren Genen bei *Pisum* nach Behandlung mit Äthylenimin. *Agri Hort. Genet.* **20**:167–179.

Lamprecht, H., 1963a   Über Wirkung und Koppelung der polymeren Gene *Br* und *Bra* von *Pisum*. *Agri Hort. Genet.* **21**:159–165.

Lamprecht, H., 1963b   Zur Vererbung der Hülsenlänge bei *Pisum* sowie über ein neues Gen für die Reduktion der Stipel. *Agri Hort. Genet.* **21**:25–34.

Lamprecht, H., 1963c   Zur Kenntnis von *Pisum arvense* L. Oect. *Abyssinicum* Braun, mit genetischen und zytologischen Ergebnissen. *Agri Hort. Genet.* **21**:35–55.

Lamprecht, H., 1963d   Ein neuer Typ von Teilfarbigkeit der Samenschale von *Pisum*. *Agri Hort. Genet.* **21**:174–177.

Lamprecht, H., 1963e   Die Vererbung des Samentyps einer Erbse aus dem Jordantal in Israel. Allgemeines über Grösse und Form des Hilums sowie Koppelungs- und Strukturstudien. *Agri Hort. Genet.* **21**:111–136.

Lamprecht, H., 1963f   Ein zweites Gen für Anthocyanfärbung des Schiffchens von *Pisum* und seine Koppelung. *Agri Hort. Genet.* **21**:166–173.

Lamprecht, H., 1963g   Ein Gen für einen bedeutenden Unterschied in der Blütengrösse von *Pisum* und seine Koppelung. *Agri Hort. Genet.* **21**:137–148.

Lamprecht, H., 1963h   Die Vererbung des ascendens-Merkmals von *Pisum*. Genmanifestation und Chromosomenzugehörigkeit. *Agri Hort. Genet.* **21**:87–110.

Lamprecht, H., 1963i   Die Merkmale anthocyanfarbiger Streifen längs der Naht sowie anthocyanfarbige Bestäubung von *Pisum*-Hülsen und ihre Vererbung. *Agri Hort. Genet.* **21**:149–158.

Lamprecht, H., 1964a   Genanalytische Studien zur Artberechtigung von *Pisum transcaucasicum* (Govorov) Stankov. *Agri Hort. Genet.* **22**:243–255.

Lamprecht, H., 1964b   Der Effekt interspezifischer Gene sowie neue Ergebnisse mit der Exmutante microsurculus u.a. Zur Kenntnis der naturbedingten Spezies. *Agri Hort. Genet.* **22**:1–55.

Lamprecht, H., 1966   Die Entstehung der Arten und höheren Kategorien, Springer-Verlag, Wien.

Lamprecht, H., 1967a   Die maximo-reductus-Mutante von *Pisum*. *Phyton (Austria)* **12**:252–265.

Lamprecht, H., 1967b   Die alternilateralis- Mutante und die übrigen Infloreszenztypen von *Pisum*. *Phyton (Austria)* **12**:266–277.

Lamprecht, H., 1967c   Eine bilateral entwickelte *Pisum*-Infloreszens und ihre Vererbung. *Phyton (Austria)* **12**:1–5.

Lamprecht, H., 1968a   Die neue Genenkarte von *Pisum* und warum Mendel in seinen Erbsen-kreuzungen keine Genenkoppelung gefunden hat. *Arb Steiermärkischen Landesbibliothek Joanneum Graz* **10**:1–29.

Lamprecht, H., 1968b   Neue Beiträge zur Genenkarte von Chromosom IV von *Pisum*. Cited in *Arb. Steiermarkischer. Landesbibliothek Joanneum Graz* **10**:1–29.

Lamprecht, H., 1969a   Die Beziehungen zwischen Crossover-Werten und relativer Länge der Chromosomen. Eine Studie in Chromosom I von *Pisum*. *Sitzungsber. Oesterr. Akad. Wiss. Math.-Naturwiss. Kl.* **98**:195–204.

Lamprecht, H., 1969b   Uber ein Gen für einen auffallend grazilen Wuchs von *Pisum*. *Phyton (Austria)* **13**:161–167.

Lamprecht, H. and E. Åkerberg, 1939   Über neue Gene für die Ausbildung von Testafarbe bei *Pisum*. *Hereditas* **25**:323–348.

Lamprecht, H. and H. Mrkos, 1950   Die Vererbung des Vorblattes bei *Pisum* sowie die Koppelung des Gens *Br*. *Agri Hort. Genet.* **8**:153–162.

Larson, L. A., and K. Lwanga, 1969   The effect of prolonged seed soaking on seedling growth of *Pisum sativum*. *Can. J. Bot.* **47**:707–709.

Lindqvist, K. 1951   The mutant "micro" in *Pisum. Hereditas* **37**:389–420.

Lock, R. H., 1907   The present state of knowledge of heredity in *Pisum. Ann. R. Bot. Gard. Peradeniya* **4**:93–111.

Marks, G. E., 1971   Trisomy in peas. *John Innes Inst. Annu. Rep.* **62**:39.

Marx, G. A., 1969   Two additional genes conditioning wax formation. *Pisum Newsl.* **1**:10–11.

Marx, G. A., 1970   An apparent case of pleiotropism involving chlorophyll and anthocyanin development. *Pisum Newsl.* **2**:19.

Marx, G. A., 1971*a*   New linkage relations for chromosome V of *Pisum. Pisum Newsl.* **3**:20–21.

Marx, G. A. 1971*b*   New linkage relations for chromosome III of *Pisum. Pisum Newsl.* **3**:18–19.

Marx, G. A. and W. Mishanec, 1962   Inheritance of ovule number in *Pisum sativum. Proc. Am. Soc. Hortic. Sci.* **80**:462–467.

Matsuura, H., 1933   A bibliographical monograph on plant genetics 1900–1929, Sapporo. Hokkaido Imperial University.

Matthews, P., 1969   Rogue peas. *John Innes Inst. Annu. Rep.* **60**:19:

Matthews, P. and K. P. Dow, 1969   Pea pathogenes: Wilt (*Fusarium oxysporum f. pisi*), Downy Mildew (*Peronospora viciae*), Pod spot of pea (*Aschocyta pisi*). *John Innes Inst. Annu. Rep.* **60**:18.

Mendel, G., 1866   Versuche über Pflanzen-Hybriden. *Verh. Naturf. Vereins Brünn* **IV**:3–47. (Reprinted 1951 in *J. Hered.* **42**:3–47.)

Monti, L. M., 1970*a*   The *filiformis* leaf mutant of peas. *Pisum Newsl.* **2**:21.

Monti, L. M., 1970*b*   Linkage studies on four induced mutants of peas. *Pisum Newsl.* **2**:21.

Monti, L. M. and M. Devreux, 1964   Action des rayons gamma sur les gametes, le zygote et le proembryon d'un pois fourrager. *Caryologia* **17**:433–441.

Monti, L. M. and M. Devreux, 1969   *Stamina pistilloida:* a new mutation induced in pea. *Theor. Appl. Genet.* **39**:17–20.

Monti, L. M. and F. Saccardo, 1971   *Ovula-pistilloida* (*op*): a mutation without sporogeneous tissue in pea. *Pisum Newsl.* **3**:28.

Monti, L. M. and G. T. Scarascia-Mugnozza, 1967   Mutatzioni per precocitá e ramositá indotte in pisello. *Genet. Agrar.* **21**:301–312.

Monti, L. M. and G. T. Scarascia-Mugnozza, 1969   The use of some induced and spontaneous traits of agronomic value in pea breeding. *Pisum Newsl.* **1**:16–17.

Morrison, J. W. and Shu-Chang Lin, 1955   Chromosomes and nucleoli in *Pisum sativum. Nature (Lond.)* **175**:343–344.

Murfet, I. C., 1967   Yellow pollen—a new gene in *Pisum. Heredity* **22**:602–607.

Murfet, I. C., 1971*a*   Flowering in *Pisum*. Three distinct phenotypic classes determined by the interaction of a dominant early and a dominant late gene. *Heredity* **26**:243–257.

Murfet, I. C., 1971*b*   Flowering in *Pisum*. A three-gene system. *Heredity* **27**:93–110.

Nilsson, E., 1932   Erblichkeitsversuche mit *Pisum*. III-V. III. Ein reproduktionsletaler Biotypus und seine Spaltungsverhältnisse. IV. Ein Fall von monohybrider Petaloidie. V. Eine Monohybride Spaltung mit drei Phenotypen. *Hereditas* **17**:71–99.

Nilsson, E., 1933   Erblichkeitsversuche mit *Pisum*. VIII. Ein Lokalisationsgen für Wachs und sein Verhalten zu den Genen *Wa* und *Wb. Hereditas* **17**:216–222.

Nuttal, V. W. and L. H. Lyall, 1964   Inheritance of neoplastic pod in the pea. *J. Hered.* **55**:184–186.

Omar, A., 1960   A study on flower and pod setting in *Pisum. Ann. Agric. Sci. (Cairo)* **5**:111–123.

Pellew, C., 1927   Further data on the genetics of "rogues" among culinary peas (*Pisum sativum*). *Z. Indukt. Abstammungs.- Vererbungsl. Suppl.* **2**:1157–1181.

Pellew, C. and A. Sverdrup, 1923   New observations on the genetics of peas. *J. Genet.* **13**:125–131.

Pohjakallio, O., 1941   Untersuchungen über Auftreten und Vererbung weisser Flecken an Erbsensamen. *J. Sci. Agric. Soc. Finl.* **13**:165–171.

Rasmusson, J., 1927–1928   Genetically changed linkage-values in *Pisum. Hereditas* **10**:1–152.

Rasmusson, J., 1935   Studies on the inheritance of quantitative characters in *Pisum.* I. Preliminary note on the genetics of time of flowering. *Hereditas* **20**:162–180.

Rasmusson, J., 1938   Notes on some mutants in *Pisum. Hereditas* **24**:231–257.

Reeve, R. M., 1948   Late embryogeny and histogenesis in *Pisum. Am. J. Bot.* **35**:591–602.

Rosen, G. von, 1942   Röntgeninduzierte Mutationen bei *Pisum sativum. Hereditas* **28**:313–338.

Rosen, G. Von, 1944   Artkreuzung in der Gattung *Pisum,* insbesondere zwischen *P. sativum* L. und *P. abyssinicum* Braun. *Hereditas* **30**:261–400.

Sachs, T., 1969   Regeneration experiments on the determination of the form of leaves. *Israel J. Bot.* **18**:21–30.

Scandalios, J. G. and L. G. Espiritu, 1969   Mutant aminopeptidases of *Pisum sativum.* I. Developmental genetics and chemical characteristics. *Mol. Gen. Genet.* **105**:101–112.

Schroeder, W. T. and D. W. Barton, 1958   The nature and inheritance of resistance to the pea enation mosaic virus in garden pea, *Pisum sativum* L. *Phytopathology* **48**:628–632.

Sharma, B., 1972*a*   "Notched leaf", a new mutation controlling shape of leaflet in *Pisum sativum. Pisum Newsl.* **4**:53.

Sharma, B., 1972*b*   "Tendrilled acacia", a new mutation controlling tendril formation in *Pisum sativum.* Pisum Newsl. **4**:50.

Sharma, B., 1972*c*   "Triangular pod", a new mutation controlling pod shape and size in *Pisum sativum. Pisum Newsl.* **4**:55.

Sharma, B. and K. V. Aravindan, 1971*a*   "Crumpled petal", a new mutation in *Pisum sativum* causing mechanical sterility. *Pisum Newsl.* **3**:50–51.

Sharma, B. and K. V. Aravindan, 1971*b*   Inductions of sativoid mutation in *Pisum abyssinicum. Pisum Newsl.* **3**:49.

Snoad, B., 1966   The chromosomes and linkage groups of *Pisum. Genetica (The Hague)* **37**:247–254.

Solovjeva, V. K., 1958   A new form of garden-shelling-peas. *Agribiologija* **1**:124–126.

Sverdrup, A., 1927   Linkage and independent inheritance in *Pisum sativum. J. Genet.* **17**:221–243.

Tavcar, A., 1926   Die Vererbung der Anzahl von Spaltoffenungen bei *Pisum sativum* L. *Z. Pflanzenzuecht.* **11**:241–259.

Tedin, H. and O. Tedin, 1923   Contributions to the genetics of *Pisum* III. Internode length, stem thickness and place of the first flower. *Hereditas* **4**:351–362.

Tedin, H. and O. Tedin, 1925–1926   Contributions to the genetics of *Pisum* IV. Leaf axil colour and grey spotting on the leaves. *Hereditas* **7**:102–160.

Tedin, H. and O. Tedin, 1928 Contributions to the genetics of *Pisum* V. Seed coat colour, linkage and free combination. *Hereditas* **11**:1–62.

Tedin, H., O. Tedin and S. J. Wellensiek, 1925 Note on the symbolization of flower-colour factors in *Pisum*. *Genetica (The Hague)* **7**:533–535.

Tschermak, E., 1912 Bastardierungsversuche an Levkojen, Erbsen und Bohnen mit Rücksicht auf die Faktorenlehre. *Z. Indukt. Abstammungs.- Vererbungsl.* **7**:81–234.

Wade, B. L., 1929 The inheritance of *Fusarium* wilt resistance in canning peas. *Wis. Agric. Exp. Stn. Res. Bull.* **97**:1–32.

Wellensiek, S. J., 1925a *Pisum*-crosses I. *Genetica (The Hague)* **7**:3–64.

Wellensiek, S. J., 1925b Genetic monograph on *Pisum*. *Bibliogr. Genet.* **2**:343–476.

Wellensiek, S. J., 1928 Preliminary note on the genetics of wax in *Pisum*. *Am. Nat.* **62**:94–96.

Wellensiek, S. J., 1943 *Pisum*-crosses VI: Seed-surface. *Genetica (The Hague)* **23**:77–92.

Wellensiek, S. J., 1950 New linkages in *Pisum, Cr-Gp* and *B-L*. *Genetica (The Hague)* **25**:183–187.

Wellensiek, S. J., 1951 *Pisum*-crosses IX. The new flower colour "cerise." *Genetica (The Hague)* **25**:525–529.

Wellensiek, S. J., 1959 Neutronic mutations in peas. *Euphytica* **8**:209–215.

Wellensiek, S. J., 1960 Mutagenic genes. *Proc. K. Ned. Akad. Wet. Ser. C Biol. Mea. Sci.* **63**(1):38–42.

Wellensiek, S. J., 1962 The linkage relations of the *cochleata*-mutant in *Pisum*. *Genetica (The Hague)* **33**:145–153.

Wellensiek, S. J., 1964 The origin of early-flowering neutronic mutants in peas. *Radiat. Bot.* **5**: Suppl. 393–397.

Wellensiek, S. J., 1971a Wax genes, *bl* or *wsp*. *Pisum Newsl.* **3**:47.

Wellensiek, S. J., 1971b The localization of some new mutants. *Pisum Newsl.* **3**:46.

Wellensiek, S. J. and J. S. Keyser, 1928 *Pisum*-crosses V. Inherited abortion and its linkage-relations. *Genetica (The Hague)* **11**:329–334.

Wells, D. G., 1951a Inheritance and linkage relations of some foliage colour mutants in peas. *J. Genet.* **50**:215–220.

Wells, D. G., 1951b Inheritance and linkage relations of a crinkled variant in peas. *J. Genet.* **50**:230–234.

White, O., 1916 Inheritance studies in *Pisum*. I. Inheritance of cotyledon color. *Am. Nat.* **50**:530–547.

White, O., 1917 Studies of inheritance in *Pisum* II. The present state of knowledge of heredity and variation in peas. *Proc. Am. Philos. Sco.* **56**:487–588.

Winge, Ö., 1936 Linkage in *Pisum*. *C. R. Trav. Lab. Carlsberg Ser. Physiol.* **21**:271–393.

Yen, D. E. and P. R. Fry, 1956 The inheritance of immunity to pea mosaic virus. *Aust. J. Agric. Res.* **7**:272–280.

10

Oenothera

ERICH STEINER

Introduction

Oenothera first became an object of genetic investigation in 1886 when Hugo De Vries introduced into his garden plants of *Oenothera lamarckiana* from a nearby abandoned potato field for an experimental study of the origin of species. De Vries had occasionally observed abberrant individuals in the wild population of this species which he felt might have evolutionary significance. Subsequently, he found among his garden cultures a series of variants, appearing in low frequencies, which differed sufficiently from the original strain that he considered them to be incipient new species. His experiments, carried out over the succeeding years, led to the publication in 1901 of *Die Mutationstheorie* in which he proposed the concept of mutation as the explanation for the origin of new species. A controversy soon arose as to whether the variants were true mutants or simply segregants from a heterozygous stock, thus casting doubt on the validity of De Vries' theory. The question stimulated a great deal of research which over a period of some thirty years finally resulted in an understanding of the main features of the unique genetic behavior in *Oenothera*.

Oenothera is a New World genus consisting of 15 subgenera according to the treatment of Munz (1965), or 10 subgenera if one accepts the views of Raven (1964). Although it is believed to have first been introduced into Europe in the 17th century, in the years following it must have been brought repeatedly to the European ports with the ballast of

ERICH STEINER—Matthaei Botanical Gardens, University of Michigan, Ann Arbor, Michigan.

ships from many parts of America. Since *Oenothera* is characteristic of disturbed areas, it spread easily from the ballast heaps and has now become a common component of the European flora. Most of the genetic work has involved taxa belonging to the subgenus *Oenothera** (referred to as *Euoenothera* or *Onagra* in the earlier genetic literature). Munz lists ten species for the subgenus; these correspond in general to the phylogenetic groups established by Cleland (1972) in his cytogenetic studies (see p. 239–240). De Vries and the workers of his period referred to the numerous variants as species, so that mastery of this nomenclature is helpful to an understanding of the literature.

Oenotheras are generally considered to be biennials, although many undoubtedly grow as winter annuals, forming a rosette in the fall and producing seed the following summer. The tendency toward the biennial habit is strongest in the northern forms. For genetic studies, seeds have usually been placed on filter paper in Petri dishes in January and February and then planted in flats upon germination. The seedlings are later potted and finally planted in the experimental garden in the spring. Approximately six weeks are required for the seed to mature after the date of pollination. A percentage of the seed will germinate immediately upon harvest; germination, particularly of fresh seed, is enhanced by light.

The *Oenotheras* of North America can be classified into two major groups according to their breeding behavior. The first group, those which occur in California, the Southwest, and Mexico along with certain restricted populations in eastern and southern United States, behave genetically as most other organisms. Their 14 chromosomes form 7 bivalents at meiosis; they tend to have large flowers which are normally cross-pollinated. In contrast, the second group, those *Oenotheras* which are collected in most of North America, have small flowers which are naturally self-pollinated, i.e., the anthers develop in close contact with the stigma and are shed before the flower opens. The progeny produced from self-pollination is identical to the parent. If two different strains are crossed, however, they behave as if they are hybrids, i.e., the offspring may include up to four distinct types. Further, such plants have their chromosomes arranged in a ring at meiosis, rather than in pairs.

Belling (1927) first suggested that interchanges between nonhomologous chromosomes can account for formation of rings of chromosomes at meiosis. Figure 1 shows a diagram of the meiotic behavior which results from a single interchange or reciprocal translocation. If a sufficient number of interchanges occurs such that each set of chromosomes has a completely different arrangement of chromosome ends, all of the chro-

* This review is restricted to the work on the subgenus *Oenothera*.

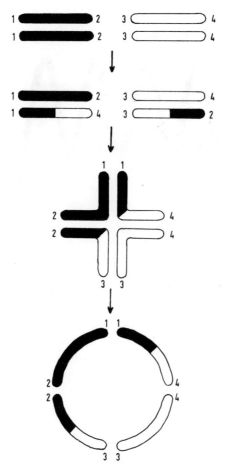

*Figure 1. The meiotic behavior which results from
a single interchange or reciprocal translocation.*

mosomes will become arranged in a single circle during meiosis, as shown in Figure 2. In *Oenothera,* the adjacent chromosomes in the ring regularly go to opposite poles at anaphase I so that both meiotic products have a complete set of chromosome segments, but their arrangement in each set is entirely different. Thus, as a result of circle formation, only two kinds of spores and gametes are produced. If we designate these as *alpha* and *beta,* then from a self-pollination three kinds of progeny, alpha · alpha, alpha · beta, and beta · beta, should be expected. However, as stated above, this type of *Oenothera* breeds true upon self-pollination; the progeny consists of only the ring-forming type like the parent, i.e., the alpha · beta combination. The absence of the homozygous combinations can be explained through balanced lethals; each genome carries a different

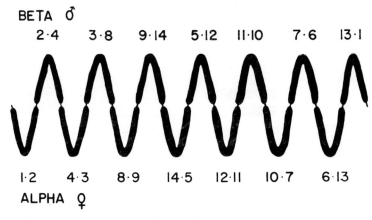

BETA ♂
2·4   3·8   9·14   5·12   11·10   7·6   13·1

1·2   4·3   8·9   14·5   12·11   10·7   6·13
ALPHA ♀

Figure 2. Meiosis in a complex heterozygote, resulting in two types of spores, alpha and beta. On self-pollination only alpha · beta progeny are obtained. In outcrosses, the alpha complex is transmitted predominantly through the egg; the beta complex is transmitted predominantly through the pollen.

recessive lethal so that only the heterozygous combination, in which each lethal is masked by its normal allele, survives.

The crossing behavior of these *Oenotheras* becomes clear with the realization that they are complete translocation heterozygotes. Two races which are crossed may produce as many as four distinct phenotypes among the progeny. This supports the hypothesis that each parent produces two kinds of gametes, each carrying a set of chromosomes with a distinctive segmental arrangement as well as a characteristic phenotype. Because each set of seven chromosomes moves to the pole as a unit at meiosis, the genes behave as if they are linked. The term "complex" was introduced by Renner (1917) to describe the group of characters which are transmitted as a single unit. A *complex-heterozygote* is thus characterized as a plant with two sets of chromosomes, each with a distinctive segmental arrangement and producing a characteristic phenotype. It maintains its identity through successive generations because each complex carries a lethal which is nonallelic to that of the opposing complex. Therefore, the lethals are balanced in the heterozygote, but prevent the occurrence of homozygotes.

Renner (1917) named each of the complexes of a strain according to the phenotype which it produced. Thus *Oenothera lamarckiana* is composed of the complexes *velans* and *gaudens; Oenothera muricata* carries *rigens* and *curvans;* and *Oenothera suaveolens, flavens* and *albicans.* When *Oe. lamarckiana* is crossed either as male or female parent, both *velans* and *gaudens* generally appear among the progeny. In crosses with a great many strains, however, reciprocal differences are the rule. *Oe. muricata,* for example, is strictly heterogamous, transmitting *rigens* only through the egg and *curvans* only through the pollen. In such a race the lethals are gametophytic, rather than as in *Oe. lamarckiana* where they are zygotic. Other races may transmit both complexes through the egg and only one through the pollen or both through the pollen and only one through the egg. Even when both complexes are transmitted, one usually occurs more frequently than the other; the frequency also varies with the particular cross. In order to simplify the nomenclature when dealing with a large number of different strains, Cleland adopted the convention of designating the two complexes of a complex-heterozygote as *alpha* and *beta* according to their transmission behavior, i.e., the alpha is applied to the complex coming predominantly through the egg, the beta to the complex transmitted most frequently through the pollen.

History of *Oenothera* Research

The atypical cytogenetic behavior of *Oenothera,* outlined above, remained a puzzle to geneticists for many years. Mention of the most significant advances leading to the understanding of the *Oenothera* cytogenetic mechanism may be helpful in pointing out some of the key papers in a voluminous literature. A detailed and excellent account of the historical development of the ideas concerning the genetic biology of *Oenothera* can be found in Cleland's book, *Oenothera—Cytogenetics and Evolution* (1972).

De Vries' mutation theory was vulnerable to attack because the contradictory breeding behavior of *Oenothera* suggested that the plant was in some way hybrid; the variants, it was therefore reasoned, must be segregates and not mutants at all. *Oenothera* quickly became the object of numerous investigations, particularly since the mutation concept was a significant one for biological theory. Although several of De Vries' mutants were shown to be chromosomal variants, i.e., tetraploid and trisomic in nature, by Lutz as early as 1907 and 1908, little progress was made toward an understanding of the breeding behavior of *Oenothera* until publication of Renner's classic paper (Renner, 1917). In this study Renner recognized that the transmission of characters occurred in groups or complexes, which could exist only in the heterozygous condition because of balanced lethals. He also found that certain strains, such as *Oenothera hookeri,* exhibited normal genetic behavior and were complex-homozygotes.

Not long after Renner's work was published, Cleland (1922) discovered in a cytological study of *Oenothera franciscana* that a circle of four chromosomes was a constant feature of meiosis in this strain. Subsequently, a survey of other strains revealed chromosome configurations ranging from 7 pairs to a circle of 14. This led Cleland (1923) and Oehlkers (1924) independently to the hypothesis that the peculiar genetic and chromosomal behavior are correlated: If the chromosomes form a single circle, segregation of individual genes will not occur. On the other hand, if smaller circles or pairs are evident at meiosis, the progeny will include segregants and corresponding linkage groups.

This hypothesis was tested and confirmed in studies carried out by Cleland and Oehlkers (1929, 1930) and Renner and Cleland (1933). Belling's idea (1927) that circle formation at meiosis is the result of reciprocal translocation further clarified the picture. Cleland and Blakeslee (1930, 1931) and Emerson and Sturtevant (1931) independently devised an experimental procedure to test the Belling hypothesis. They reasoned that if circles are the result of segmental interchange, then, in the course of evolution of the different *Oenothera* strains, a considerable diversity of chromosomal end arrangements has arisen. If the end arrangements of two complexes could be determined, then according to the hypothesis, it should be possible to predict the chromosome configuration that would result when the two complexes are combined in a hybrid.

In order to be able to determine the chromosomal end arrangement of specific complexes, a system for designating end arrangement was established. The type of analysis carried out by these workers is illustrated below.

The complex *^{h}hookeri** from the complex-homozygote *Oe. hookeri* was arbitrarily assigned the standard arrangement of chromosome ends:

$$^h hookeri = 1 \cdot 2 \quad 3 \cdot 4 \quad 5 \cdot 6 \quad 7 \cdot 8 \quad 9 \cdot 10 \quad 11 \cdot 12 \quad 13 \cdot 14 \dagger$$

When *^{h}hookeri* was combined with the complex *flavens* (*Oenothera suaveolens*), the chromosome configuration of the hybrid proved to be ⊙ 4‡ and 5 pairs. Therefore, *flavens* must differ from *^{h}hookeri* by a single interchange and could be assigned the end arrangement:

$$flavens = 1 \cdot 4 \quad 3 \cdot 2 \quad 5 \cdot 6 \quad 7 \cdot 8 \quad 9 \cdot 10 \quad 11 \cdot 12 \quad 13 \cdot 14$$

---

* The form *^{h}hookeri* denotes *haplo-hookeri;* the *haplo* designation is applied to those complexes which are found in structural homozygotes.

† Each chromosome is designated by two numbered ends.

‡ The symbol ⊙ denotes "circle of."

Combining *velans* with each of these two complexes gave the following configurations:

velans · *ʰhookeri*   ⊙4, 5 pairs
velans · *flavens*   ⊙4, ⊙4, 3 pairs

Velans differs by one interchange from *ʰhookeri*, but it cannot be the same interchange that differentiates *flavens* from *ʰhookeri*. Therefore, the arrangement 1·2 3·4 5·8 7·6 9·10 11·12 13·14 was adopted for *velans*. In this manner, it was possible to assign specific end arrangements to a series of complexes. This made it possible to predict the chromosome configuration of a particular complex combination, produce the hybrid, and examine it cytologically in order to test the interchange hypothesis. The system of end arrangements proved to be entirely consistent, and the validity of the hypothesis was established without question. Further, the set of complexes whose end arrangements had been determined could be used to deduce the arrangement of any unknown complex by combining the latter in hybrids with the known complexes and determining their chromosome configurations at meiosis. A simple example of such an analysis is outlined below:

The complex *ʰTolucca* occurs in a Mexican race belonging to *Oenothera elata*. *ʰTolucca* gives ⊙4, 5 pairs with *ʰblandina*.

ʰblandina = 1·2   3·4   5·6   7·10   9·14   11·12   13·8

It gives ⊙4, ⊙4, 3 pairs with *ʰJohansen*.

ʰJohansen = 1·2   3·4   5·6   7·10   9·8   11·12   13·14

Since *ʰblandina* differs from *ʰJohansen* only in that it possesses the 9 · 14 and 13 · 8, the extra two pairs *ʰTolucca* gives with *ʰblandina* must result from the presence of these two chromosomes.

ʰTolucca gives ⊙6, 4 pairs with *neo-acuens*.

neo-acuens = 1·13   3·2   5·6   7·10   9·8   11·12   4·14

Since *ʰTolucca* has the 13·8 and 9·14, these chromosomes must form part of the ⊙6 in the combination with *neo-acuens;* thus the circle must consist of:

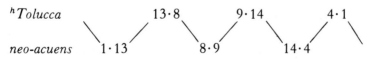

ʰTolucca   13·8   9·14   4·1

neo-acuens   1·13   8·9   14·4

ʰTolucca must have the 1 · 4 to complete the circle. Since the remaining *neo-acuens* chromosomes form pairs, ^{h}*Tolucca* must have them. Thus the end arrangement of ^{h}*Tolucca* must be:

$$^h Tolucca = 1 \cdot 4 \quad 3 \cdot 2 \quad 5 \cdot 6 \quad 7 \cdot 10 \quad 9 \cdot 14 \quad 11 \cdot 12 \quad 13 \cdot 8$$

Although the determination of the end arrangements of the complexes of different *Oenothera* strains was begun for the purpose of testing Belling's interchange hypothesis, it became apparent that the results not only confirmed the hypothesis, but they also presented a unique opportunity for studying the evolutionary history of the group if one assumed that two complexes with identical or similar arrangements tended to be more closely related than those whose arrangements are widely divergent. Cleland (1935) thus began studies which led to the analysis of over 400 complexes from throughout North America. This, along with genetic information, was used to reconstruct the evolutionary history of the group. A discussion of the evolution of the subgenus follows in a later section.

Chromosome Structure and Behavior

In spite of the extensive work on *Oenothera*, a number of cytological problems remain to be clarified. The small size of the chromosomes, which average about 4 microns in length at mitotic prophase, hardly makes them favorable material for the study of structure and morphology. The size range within the chromosome complement has been a matter of interest because the success of circle formation in the complex-heterozygote requires regular disjunction; wide variation in chromosome lengths would likely lead to irregular disjunction and a breakdown of the system. Earlier investigators have described differences in length and location of the centromere, but the relatively recent work of Gardella (1953) supports the view that differences in size and morphology are not significant either within the genome nor between genomes of different races; thus the various translocated chromosomes which have survived through the evolution of the group are not essentially different from those of the ancestral complex-homozygotes.

The generally accepted view is that the chromosomes consist of a median segment which does not pair and is heterochromatic [see for example, Seitz (1935), Marquardt (1937), or Japha (1939)]. The median segments probably correspond to the prochromosomes that are readily visible in the metabolic nucleus at interphase. These segments are also believed to be the location of the genes which characterize the complexes. However, no serious attempt has been made to reconcile the apparent contradiction that in this instance heterochromatin has a significant gene content, although Cleland (1972) has suggested that the heterochromatin may be located close to the centromere with enough euchromatin in the distal portions of the central segment to accomodate the genes characterizing the complexes.

For a detailed account of meiosis in *Oenothera,* the reader is referred to Cleland (1972). The early stages are particularly difficult to study, so many aspects of the division, e.g., the exact nature of the pairing, are still to an extent conjecture.

Supernumerary chromosomes are found in certain races of *Oe. hookeri* (Cleland, 1951). Their characteristics are further described by Cleland and Hyde (1963) and Cleland (1967).

Genetic Analysis in *Oenothera*

The identification and mapping of single loci in *Oenothera* is complicated not only by the ring formation and the effective linkage of the genes into a single complex, but also by the occurrence of both megaspore and pollen-tube competition, each of which may lead to deviations in expected Mendelian segregations.

The occurrence of megaspore competition was first described by Renner (1921) in *Oe. muricata. Oe. muricata* is a strongly heterogamous strain, the *rigens* complex coming through the egg, the *curvans* through the pollen. Ordinarily in *Oenothera* the micropylar megaspore gives rise to the embryo sac. However, in *Oe. muricata* if the *rigens* complex happens to arrive in the chalazal megaspore, the latter may succeed in developing into the embryo sac. This competition between megaspores, often called the "Renner effect," depends upon the genetic constitution of the complexes associated in a particular strain (Renner, 1940*a*).

Similarly, the rate of pollen-tube development may be influenced by the complex it carries, as shown by Renner (1917). Thus, if pollination is sparse, both complexes represented in the pollen may appear among the progeny. On the other hand, if an excess of pollen is applied to the stigma, most or all of the eggs are likely to be fertilized by the tubes which grow most rapidly. The general review of spore competition in the higher plants by Harte (1967) summarizes the cases which have been studied in *Oenothera.*

As long as a complex remains intact, the individual genes cannot be identified as separate entities. Genetic analysis must, therefore, involve the production of hybrids in which the chromosome configurations consist of pairs and/or small circles of chromosomes. For example, if a hybrid with

⊙12, 1 pair shows segregation of a character, one can conclude that the gene for the character is located in the pair. A knowledge of the end arrangements of the complexes permits the.gene to be located on a specific chromosome. In hybrids with small circles, a particular gene can only be assigned to a group of chromosomes. The situation may also be complicated by the presence of a lethal on one or more of the chromosomes. The low frequency of crossing over (see below) limits the possibilities for fine-structure analysis.

In spite of these difficulties, a number of genes have been identified and associated with specific chromosomes. In many cases, however, the degree of genetic resolution is limited. Cleland (1972) has summarized in tabular form a considerable number of the genes which have been identified and assigned to specific chromosome ends by Renner, Harte, as well as other investigators.

Crossing Over

One of the interesting features of the complex-heterozygote is that each complex retains its identity through successive generations. The complexes of many races produce strikingly different phenotypes; they show no evidence of modification from one generation to the next. Since chiasmata are necessary for the circle of chromosomes to remain intact at meiosis, one would expect crossing over to occur between the complexes associated in a plant. However, exchanges between complexes have been detected only infrequently. These facts have led to the view that pairing of the chromosomes is limited to the end segments, and that the genes which characterize a complex reside in the median portion of the chromosome, which ordinarily does not pair. Further, if the plant is heterozygous for a gene located in a pairing segment, crossing over will produce a proportion of offspring which are homozygotes. Thus, it is possible that in most strains the genes in the pairing segments have become largely homozygous. Renner (1942) has shown that a gene may be included in the pairing segment in combination with one complex, but not with another. He also found a higher frequency of crossing over when a hybrid is first produced than after the complexes have been associated for a number of generations. This is interpreted to be a result of an extension of the chromosomal segments which pair so that additional genes are included; the reduction in crossing over in later generations probably represents an increase in the number of genes which have become homozygous.

Although, as stated above, crossing over tends to be rare, a number of instances have been described by various workers; some representative cases have been reviewed by Cleland (1972) to illustrate the type of analysis and the results which have been obtained.

Position Effect

The extensive history of segmental interchange which has characterized the genus *Oenothera* suggests that it might be material particularly suited for the study of the position effect. However, spontaneous translocations appear to be rare, at least in experimental cultures (Cleland, 1972), and the difficulties of genetic analysis in the complex-heterozygotes do not make it easy to obtain rigorous proof of this phenomenon. Nevertheless, it is interesting that one of the well-known cases of the position effect has been demonstrated in *Oenothera* (Catcheside, 1939, 1947).

Catcheside obtained through x irradiation of *Oe. blandina** an interchange between the 3 · 4 and the 11 · 12 chromosomes. The irradiated complex *^{h}blandina* carried the *P^s* allele, located on the 3 end of the chromosome. (This allele produces a bud with longitudinal red stripes, among other effects.) When the complex with the induced interchange (designated *^{h}blandina A*) was combined with the normal *^{h}blandina*, the plant showed buds with very narrow red stripes interrupted by irregular patches of green, giving a mottled effect. In further crossing, the altered complex with its *P^s* allele was combined with an *^{h}blandina* complex carrying the *P^r* allele. (The *P^r* allele produces a fully red cone.) This combination was then backcrossed to normal *^{h}blandina* with *P^s*: *^{h}blandina A P^s · ^{h}blandina P^r* × *^{h}blandina P^s · ^{h}blandina P^s*.

This cross should yield *^{h}blandina A P^s · ^{h}blandina P^s* (narrow stripes with mottling) and *^{h}blandina P^r · ^{h}blandina P^s* (fully red cone). The offspring were as expected, with the exception of one plant which showed the normal *P^s* phenotype. Catcheside interpreted this plant to represent a crossover in which the *P^s* had been transferred from the translocated to the normal 3 end. In the latter position, the allele produced its original effect of broad red stripes on the bud. Since this interpretation was based on only one plant, Catcheside carried out additional experiments in which he was able to confirm the position effect, not only for *P^s* and *P^r*, but also for the gene *s* (sulfur-colored flowers), which is located eight units distant from the *P* locus (Catcheside, 1947).

---

* A complex-homozygote derived from *Oe. lamarckiana*.

The Cruciate and Missing-Petal Characters

The cruciate character is of particular interest because its study led Renner (1937) to interpret its genetic behavior as evidence for gene conversion, i.e., the change of one member of an allelic pair to the state of its associated allele. The condition of missing petals shows certain similarities to the cruciate character in its genetic behavior, and it deserves mention in connection with a discussion of the latter.

Cruciate petals are straplike, with irregular margins, and are often green to greenish yellow in color. This character has long been known and has appeared in different strains; its expression varies according to the strain. The number of cruciate flowers on a plant may be few to many; in some crosses it may behave as a dominant, in others as a recessive.

Oehlkers (1935, 1938) studied the genetic behavior of the character in considerable detail. Its inconstancy led him to a hypothesis which postulated a series of multiple alleles differing in the degree of expression of the character. He further believed that the cruciate locus was highly mutable and, therefore, a large variety of alleles had arisen.

Renner (1937, 1959) obtained the same results as Oehlkers in his studies of the character, but chose to adopt a modified version of the Winkler gene-conversion hypothesis (Winkler, 1930) as their most likely explanation. Cleland (1972), after a thorough review of the earlier work, proposed a third hypothesis, one which views the cruciate character as a mutation of a gene which plays a key role in the biochemical sequence controlling sepal formation. The mutant, in contrast to its normal allele, may also become active in the cells of the petal primordia so that sepaloid rather than petaloid tissue is produced. In such a case, the nonallelic locus initiating petal development fails to act. Thus the appearance of the cruciate character depends upon a competition between two nonallelic genes which control sepal and petal development. Unfortunately, all three hypotheses are difficult to test because the nature of the complex-heterozygote limits the resolving power of genetic analysis.

The genetic behavior of the missing-petal character is described by Cleland (1972). The condition has arisen in different races as well as hybrids. It, like the cruciate character, is mosaic in its expression, i.e., flowers with 0, 1, 2, 3, or 4 petals occur on the same plant. It may be dominant in some crosses and recessive in others. Cleland believes that the genetic behavior of the missing-petal character is best explained by the same hypothesis which he proposed for the cruciate character.

Mutations in *Oenothera*

The appearance of variants among the populations of *Oe. lamarckiana* attracted De Vries' attention to the genus. As it turned out, however, De Vries' interpretation of his mutants proved to be largely in error, even though the concept of mutation in a modified sense has become a central feature of genetic theory. Later analysis has shown that his mutants do conform to the modern definition of the term. As early as 1907 Lutz showed that the mutant *Oe. gigas* was a tetraploid form of *Oe. lamarckiana*. She also reported a chromosome number of 15 for several other mutants (Lutz, 1908). Later, after the essential features of the cytogenetic mechanism of *Oenothera* had been elucidated, studies were carried out, particularly by Catcheside (1933, 1936), Emerson (1935, 1936), and Renner (1940b, 1941), to determine the nature and origin of many of the mutants described by De Vries and other early workers. Although some mutants undoubtedly represented changes in single loci, most of them proved to be trisomics; this was not unexpected, since plants showing a large circle of chromosomes at meiosis can theoretically yield a considerable variety of such types through nondisjunction. For example, if three successive chromosomes in the circle all go to the same pole instead of the normal disjunction of adjacent members to opposite poles, a gamete with one complex plus an extra chromosome from the other complex will result. The chromosome ends composing the extra chromosome in a number of the trisomic mutants have been identified largely through the efforts of Renner (1954, 1949).

In addition to the well-known *Oe. gigas*, recognized as a tetraploid by Lutz, triploid mutants are known. Another class of variants arising from *Oe. lamarckiana* known as "half-mutants" deserves mention. These all possess a normal *velans* complex and a lethalfree complex composed of three *velans* chromosomes, a *gaudens* chromosome, and two interchanged chromosomes requiring more than a single interchange from the *Oe. lamarckiana* complement. This is noteworthy, considering that such half-mutants have arisen repeatedly in the cultures of several investigators. A more detailed summary of the nature and origin of *Oe. lamarckiana* mutants is presented by Cleland (1972). He has compiled a list in which the mutants are classified according to type.

Attempts to induce mutations in *Oenothera* have not been as extensive as one might expect, even though the material has disadvantages for the analysis of point mutations as well as chromosomal changes. Mutations interpreted as genic were obtained by Rudloff and Stubbe (1935)

following x irradiation of *Oe. hookeri* pollen. Lewis (1951) was able to induce mutations in the incompatibility locus of *Oe. organensis* (see
below). Most of the work on the induction of mutations has dealt with effects on chromosome structure. Chromosomal interchanges were obtained
by Catcheside (1935) and by Marquardt (1948). Oehlkers reported in
1943 that the inorganic salts KCl and KNO_3 significantly increased the
number of chromosomal alterations. Certain organic compounds, notably
ethyl urethane, also proved effective, particularly when combined with the
inorganic substances mentioned above. One interesting result was that the
combination of urethane and KCl could be as much as twice as effective on
the chromosomes, depending upon the strain of cytoplasm in which they
were treated (Oehlkers and Linnert, 1951). Oehlkers (1949) tested a variety of chemicals and found such compounds as acetophenone, glycol, acet-
analide, and morphine to be effective chromosomal mutagens.

Chromosomal aberrations have also been obtained through treatment
with electromagnetic radiation of radio frequencies by Harte (1949, 1972).

Self-Incompatibility

That self-incompatibility occurs in the genus *Oenothera* is well
known through the work of Emerson (1938, 1939) and, later, Lewis
(1948, 1949, 1951, 1960) on *Oenothera organensis*. Cleland (1962)
reported that self-incompatibility has been found in nine of the ten
subgenera of *Oenothera*. Emerson (1938) showed that *Oe. organensis*
possesses a gametophytic incompatibility system; this species is known
only from the Organ Mountains of southern New Mexico and occurs in a
population estimated by Emerson to be under 500 in number. In a sample
of plants from this population Emerson (1939) identified 37 different incompatibility alleles.

Lewis chose *Oe. organensis* for a study of the mutability of the incompatibility (*S*) allele. He found that the rate of spontaneous mutation is
surprisingly low, particularly in light of the great diversity of such alleles
in natural populations. Through x irradiation he obtained two kinds of
mutants, permanent and revertible. No spontaneous nor induced mutations involved a change from one functional *S* allele to another. All
permanent mutations involved only the loss of pollen activity; the stylar
function was retained. The results of Lewis and of Emerson have raised
some interesting questions regarding the population genetics of *Oe.
organensis* (e.g., Mayo, 1966).

Incompatibility alleles have also been demonstrated in certain complex-heterozygotes where they function as part of the balanced lethal

system, rather than as a device to enforce outbreeding (Steiner, 1956). When races belonging to the *biennis* group I (see p. 239) are crossed, hybrids consisting of the alpha complex from each parent may be obtained. These alpha · alpha combinations are self-incompatible; the pollen tubes fail to grow in the stigmas and styles of plants of the same genetic constitution. Alpha · alphas derived from different races are in most cases cross-compatible. These results are consistent with the hypothesis that the alpha complexes of the *biennis* group I carry a gametophytic incompatibility system, i.e., the alpha complex of each race carries a self-incompatibility or *SI* allele. The beta, or pollen complex, of *biennis* I, however, has a self-compatibility (*SC*) allele at the same locus. In self-pollinations, pollen bearing the alpha complex does not grow in the style because the stylar tissue bears the same allele. Only pollen tubes carrying the beta complex, which lacks the *SI* allele, reach the ovules. Incompatibility thus acts as a lethal for pollen carrying the alpha complex in self-pollinations.

Evidence of incompatibility alleles in the alpha complexes of phylogenetic groupings other than the *biennis* I (see p. 239–249) has been obtained (Steiner, 1961). The situation in these groups is complicated by the nature of the chromosomal end arrangements which for the most part give circles of varying sizes with the tester complex. As a result, if a pollen lethal is present, it remains linked to the incompatibility locus, and the pollen fails to grow in outcrosses as well as in self-pollinations. Thus the presence of the incompatibility allele can not be detected.

Studies of the distribution of specific incompatibility alleles in *biennis* I populations have revealed a considerable diversity of alleles throughout the geographic range of the group, although no single population, i.e., local stand, has been found with more than two different alleles (Steiner, 1964; Khafaji and Steiner, 1970). These findings make it likely that the ancestral complex-homozygous self-incompatible population from which the alpha *biennis* I complexes have been derived, hybridized repeatedly to give rise to the complex-heterozygotes. Further, the uniformity that single populations show with regard to *SI* alleles is evidence that populations are usually established from one or two colonizers. The implications of self-incompatibility for the evolution of the complex-heterozygotes will be discussed in the section on evolution.

Plastid–Genome Interactions

While the appearance of chlorotic plants or plants with chlorotic sectors among the progeny of certain crosses was noted by De Vries in *Die Mutationstheorie*, Renner (1922) was the first to show that such chlorosis is a result of an incompatibility between the plastids and the genes in specific hybrid combinations. He demonstrated that different species are characterized by distinct plastid types. In combination with certain genomes, a plastid type may be unable to develop and function normally. Plastid–genome harmony depends upon the specific combination brought together in a hybrid.

Stubbe (1959, 1960) has analyzed the plastid–genome interaction of over 400 combinations, involving both European and North American races. He identified five distinct classes of plastids (I–V) and three types of genomes (A, B, and C). For example, genomes AA give normal green plastids with types I and II, periodically pale (virescent) with III, green to gray-green with IV, and lethal (albino) with V. In terms of the groupings established by Cleland for the North American oenotheras (p. 239), the A type includes the *hookeri, elata,* and *strigosa* complexes; B type includes the *biennis* and *grandiflora* complexes; and C type includes the *argillicola* complex. The plastid–genome disharmony does not seem to be simply an incompatibility between the plasmone and the plastids, but actually between the plastid and the genome, regardless of the cytoplasm in which the plastid is found. This was shown by a series of crosses in which cytoplasm, plastids, and the complexes were all controlled (Renner, 1936).

Plastid mutations in *Oenothera* have been noted by Renner (1936), Stubbe (1953), and Schötz (1954).

The behavior of the plastids and the unusual cytogenetic mechanism in *Oenothera* provides a unique opportunity to study a variety of problems relating to plastid structure and function. These have been investigated particularly by Schötz, by Stubbe, and their associates. Their studies have dealt with plastid function in the photosynthetic cycle (e.g., Fork and Heber, 1968), the biochemistry of pigment development (e.g., Schötz *et al.,* 1966), plastid multiplication and competition (e.g. Schötz, 1968), and structure and ultrastructure (e.g., Diers and Schötz, 1969).

Structure and Evolution of North American *Oenothera* Populations

The extensive and often intergrading variability found in the subgenus *Oenothera* presented a difficult taxonomic problem until the detailed cytogenetic analysis of population samples was carried out by Cleland and his associates. As a result of these studies the general population structure and evolution is reasonably well understood.

The phylogenetic implications of the early genetic work were recognized by Hoeppener and Renner (1928), who published a diagram showing the relationships of various complexes which had been analyzed genetically. Cleland, who at the time was engaged in determining the chromosome configurations of different complex combinations, found that chromosome configuration could also serve as a criterion of relationship and correlated well with that based on genetic analysis (Cleland, 1931). With the establishment of the system in which the chromosomal end arrangement of any complex could be determined, the different complexes could be readily compared with regard to end arrangement. Further, in many cases it has been possible to postulate the interchange history of a complex. Adopting the working hypothesis that similarity in segmental arranagement is an indication of degree of relationship between complexes, Cleland and his co-workers analyzed a total of 438 complexes for end arrangement and the phenotypic characters they produced. On this basis, he established ten phylogenetic groupings; these correspond essentially to the species recognized by Munz (1965) in his treatment of the subgenus. The groups are briefly characterized as follows:

> *elata*—Mexico and Central America; large, open-pollinated flowers; chromosomes paired at meiosis. All races studied are two interchanges removed from the ancestral arrangement, although three different arrangements have been found among the ten races analyzed.
>
> *hookeri*—California to Mexico, east to Utah and New Mexico; large, open-pollinated flowers; usually paired chromosomes, a few have small circles. The segmental arrangement most frequently encountered is $1 \cdot 2 \ 3 \cdot 4 \ 5 \cdot 6 \ 7 \cdot 10 \ 9 \cdot 8 \ 11 \cdot 12 \ 13 \cdot 14$; this arrangement is believed to be ancestral for the subgenus.
>
> *strigosa*—Rocky Mountains eastward to the Mississippi; $\odot 14$ chromosomes at meiosis; small, self-pollinating flowers; almost entirely heterogamous. The alpha and beta complexes are quite similar in phenotype and produce thick, relatively narrow, gray-green leaves, spreading bracts, short, thick buds, and spreading sepal tips.
>
> *biennis I*—Midwest to East Coast and the Southeast; $\odot 14$, small, self-pollinating flowers; range from strictly heterogamous to isogamous. Alpha complexes produce a *biennis* phenotype, i.e., dark green, broad, thin, crinkly leaves, limited pubescence, appressed bracts and sepal tips, and thin, tapered buds. Most alpha complexes show an end arrangement one interchange removed from the ancestral arrangement. The *biennis*

phenotype of the alpha complex to a large extent masks the *strigosa* phenotype produced by the beta complex.

biennis II—Canada and northern United States from the Great Lakes to the East, along the Alleghenies south to North Carolina; phenotypically difficult to distinguish from *biennis I.* The alpha complex produces a *strigosa,* the beta a *biennis* phenotype.

biennis III—Virginia, North Carolina, western Pennsylvania. Both alpha and beta complexes produce the *biennis* phenotype. Segmentally, the alpha complexes are close to those of alpha *biennis I* and the beta complexes very similar to the beta *biennis II.*

gradiflora—Restricted to the environs of Mobile, Alabama; large, open-pollinated flowers and paired chromosomes. Phenotypically similar to *biennis,* it possesses the ancestral segmental arrangment.

parviflora I—Range essentially the same as *biennis II;* self-pollinators with a ⊙14; heterogamous. The alpha complex is of the *biennis* type; the beta characterizes it as a *parviflora* (narrow leaves, bent stem tips, subterminal sepal tips).

parviflora II—Range the same as *parviflora I* from which it differs only in having an alpha complex producing the *strigosa* phenotype.

argillicola—A small group restricted to the shale barrens of Appalachia; large, open-pollinated flowers; pairs of chromosomes, on occasion, small circles. Phenotypically, the *argillicolas* have certain similarities to the beta *parvifloras* such as bent stem tips and subterminal sepal tips.

The complex-heterozygote is an evolutionary development unique to the genus *Oenothera.* Its evolution undoubtedly required a combination of features in the ancestral forms not found in other groups; these include isobrachial chromosomes of equal length, heterochromatic median segments probably favoring interchanges, and chiasmata restricted to the terminal segments. The extensive segmental interchange led to differentiation of populations with regard to chromosomal end arrangement. While the originally structurally homozygous populations may have carried a proportion of small circles resulting from an occasional interchange, it is improbable that forms with large circles arose by a gradual accumulation of interchanges within the same population. More likely, isolated, structurally homozygous populations underwent different histories of interchange. Some of these populations possessed a gametophytic incompatibility system. If such a self-incompatible population subsequently extended its range, it may well have come into contact with another population which had previously been isolated. Let us assume that the latter population was self-compatible and had undergone segmental differentiation. If the two populations differed sufficiently in end arrangement, hybrids between them would exhibit a ⊙14 at meiosis. In a self-pollination of this hybrid, the pollen carrying the complex with the *SI* allele would fail to function; only the complex contributed by the self-compatible parent would be transmitted through the pollen. In the formation of the embryo sac from the megaspore, it is possible that the megaspores carrying the different complexes would not be equally capable of giving rise to the embryo sac, so that the one complex would predominate in the eggs. Thus, a complex-heterozygote could have arisen simply through hybridization. With the presence of a large circle at meiosis, any lethals arising subsequently would accumulate and reinforce the effects of the *SI* allele and megaspore competition in eliminating homozygous offspring. The hybrid vigor of the complex-heterozygote in all likelihood provided a selective advantage which enabled the newly arisen form to compete successfully with the structural homozygotes of the parental populations.

It is also possible that ⊙14 forms could have arisen through hybridization between two self-compatible populations which had become segmentally differentiated. If pollen tubes carrying one of the complexes had a markedly slower growth rate, homozygotes of this type would tend to occur at a reduced frequency. Megaspore competition could play a role in eliminating the other homozygous type. The selective advantage of the complex-heterozygote assured not only its survival, but the gradual displacement of the structurally homozygous populations.

The probable evolutionary history of each of the phylogenetic groups comprising the North American *Oenotheras* has been worked out by Cleland [see Cleland (1972) for a detailed review of this topic]. He has postulated the successive evolution of four distinct types at the center of origin of the group, presumed to be in Mexico or Central America. As each of these differentiated, its population underwent a major expansion throughout large areas of North America. The *argillicola* type is a relict of the first population to spread over the continent. This was followed by the evolution and migration of the second type, a *biennis* form which hybridized with type I to give the *parviflora I* group. This *biennis* population is also represented as a relict in the *grandiflora* type. The third type, possessing *strigosa* characters, subsequently spread to the East and Midwest where it crossed with the earlier colonizers to give the *parviflora II* and the *biennis I* and *II* groups. The latter groups later crossed to give the *biennis III* group. The final event in the development of the populations to complex-heterozygotes was the appearance and expansion of a second *strigosa* type, which, as it moved into the western United States, hybridized with type III, yielding the present-day *strigosa* population.

It is not certain whether the *hookeri* group represents a fifth and later-spreading type which retained the ancestral chromosome arrangement, or whether it is a primitive group which early invaded its present range and has remained essentially unchanged. The *elata* group seems to be one which has not spread from the center of origin.

Relatively little attention has been given to the dynamics of single populations or local stands. More recently, the work of Levin *et al.*, (1972) shows promise in arriving at a more precise understanding of the genetic structure of local stands. Single populations appear to be quite uniform genetically, probably arising from only one or two colonizers. Further, individual plants show a fairly high level of heterozygosity in the loci studied, thus supporting the view that the complex-heterozygotes owe their success to hybrid vigor.

Literature Cited

Belling, J., 1927   The attachment of chromosomes at the reduction division in flowering plants. *J. Genet.* **18**:177–205.

Catcheside, D. G., 1933   Chromosome configurations in trisomic oenotheras. *Genetica (The Hague)* **15**:177–201.

Catcheside, D. G., 1935   X-ray treatment of *Oenothera* chromosomes. *Genetica (The Hague)* **15**:313–341.

Catcheside, D. G., 1936   Origin, nature, and breeding behavior of *Oenothera lamarckiana* trisomics. *J. Genet.* **33**:1–23.

Catcheside, D. G., 1939   A position effect in *Oenothera*. *J. Genet.* **38**:345–352.

Catcheside, D. G., 1947   The *P* locus position effect in *Oenothera*. *J. Genet.* **48**:31–42.

Cleland, R. E., 1922   The reduction divisions in the pollen mother cells of *Oenothera franciscana*. *Am. J. Bot.* **9**:391–413.

Cleland, R. E., 1923   Chromosome arrangements during meiosis in certain oenotheras. *Am. Nat.* **57**:562–566.

Cleland, R. E., 1931   Cytological evidence of genetical relationships in *Oenothera*. *Am. J. Bot.* **18**:629–640.

Cleland, R. E., 1935   Cytotaxonomic studies on certain *Oenotheras* from California. *Proc. Am. Philos. Soc.* **75**:339–429.

Cleland, R. E., 1951   Extra diminutive chromosomes in *Oenothera*. *Evolution* **5**:165–176.

Cleland, R. E., 1962   Cytogenetics of *Oenothera*. *Adv. Genet.* **11**:147–229.

Cleland, R. E., 1967   Further evidence bearing upon the origin of extra diminutive chromosomes in *Oenothera hookeri*. *Evolution* **21**:341–344.

Cleland, R. E., 1972   *Oenothera—Cytogenetics and Evolution*, Academic Press, New York.

Cleland, R. E. and A. F. Blakeslee, 1930   Interaction between complexes as evidence for segmental interchange in *Oenothera*. *Proc. Natl. Acad. Sci. USA* **16**:183–189.

Cleland, R. E. and A. F. Blakeslee, 1931   Segmental interchange, the basis of chromosomal attachments in *Oenothera. Cytologia (Tokyo)* **2:**175–233.

Cleland, R. E. and B. B. Hyde, 1963   Evidence of relationship between extra diminutive chromosomes in geographically remote races of *Oenothera. Am. J. Bot.* **50:**179–185.

Cleland, R. E. and F. Oehlkers, 1929   New evidence bearing on the problem of the cytological basis for genetical peculiarities in the oenotheras. *Am. Nat.* **63:**497–510.

Cleland, R. E. and F. Oehlkers, 1930   Erblichkeit und Zytologie verschiedener Oenotheren und ihrer Kreuzungen. *Jahrb. Wiss. Bot.* **73:**1–124.

De Vries, H., 1901–1910   *Die Mutationstheorie,* Vol. I (1901), Vol. II (1903), Von Veit, Leipzig; *The Mutation Theory* (English translation), Vol. I (1909), Vol. II (1910), Open Court, Chicago.

Diers, L. and F. Schötz, 1969   Über ring und schalenförmige Thykaloidbildungen in den Plastiden. *Z. Pflanzenphysiol.* **60:**187–210.

Emerson, S. H., 1935   The genetic nature of De Vries' mutations in *Oenothera lamarckiana. Am. Nat.* **69:**545–559.

Emerson, S. H., 1936   The trisomic derivatives of *Oenothera lamarckiana. Genetics* **21:**200–224.

Emerson, S. H., 1938   The genetics of incompatibility in *Oenothera organensis. Genetics* **23:**190–202.

Emerson, S. H., 1939   A preliminary survey of the *Oenothera organensis* population. *Genetics* **24:**524–537.

Emerson, S. H. and A. H. Sturtevant, 1931   Genetical and cytological studies in *Oenothera.* III. The translocation hypothesis. *Z. Indukt. Abstammungs. -Vererbungsl.* **59:**395–419.

Fork, D. C. and U. Heber, 1968   Studies on electron transport reactions of photosynthesis in plastome mutants of *Oenothera. Plant Physiol.* **43:**606–612.

Gardella, C. 1953   Studies on the chromosome structure of *Oenothera. Diss. Abstr.* **13:**957.

Harte, C., 1949   Mutationsauslösung durch Ultrakurzwellen. *Chromosoma* **3:**440–447.

Harte,C., 1967   Gonenkonkurrenz. *Hand. Pflanzenphysiol.* **18:**447–478.

Harte, C., 1972   Auslösung von Chromosomenmutationen durch Meterwellen in Pollenmutterzellen von *Oenothera. Chromosoma* **36:**329–337.

Hoeppener, E. and O. Renner, 1928   Genetische und Zytologische Oenotherenstudien I. Zur Kenntnis der *Oenothera ammophila* Focke. *Z. Indukt. Abstammungs.-Vererbungsl.* **49:**1–25.

Japha, B., 1939   Die Meiosis von *Oenothera.* II. *Z. Bot.* **34:**321–369.

Khafaji, S. D. Al and E. Steiner, 1970   Further analysis of *Oenothera biennis* populations for incompatibility alleles. *Am. J. Bot.* **57:**183–189.

Levin, D., G. Howland, and E. Steiner, 1972   Protein polymorphism and genic heterozygosity in a population of permanent translocation heterozygotes of *Oenothera biennis. Proc. Natl. Acad. Sci. USA* **69:**1475–1477.

Lewis, D., 1948   Structure of the incompatibility gene. I. Spontaneous mutation rate. *Heredity* **2:**219–236.

Lewis, D., 1949   Structure of the incompatibility gene. II. Induced mutation rate. *Heredity* **3:**339–355.

Lewis, D., 1951   Structure of the incompatibility gene. III. Types of spontaneous and induced mutation. *Heredity* **5:**399–414.

Lewis, D., 1960   Genetic control of specificity and activity of the S-antigen in plants. *Proc. R. Soc. Lond. Ser. B Biol. Sci.* **151:**468–477.

Lutz, A., 1907   Preliminary note on the chromosomes of *Oenothera* and one of its mutants, *Oe. gigas. Science (Wash., D.C.)* **26**:151–152.

Lutz, A., 1908   Chromosomes of somatic cells of the oenotheras. *Science (Wash., D.C.)* **27**:335.

Marquardt, H., 1937   Die Meiosis von *Oenothera*. I. *Z. Zellforsch. Mikrosk. Anat.* **27**:159–210.

Marquardt, H., 1948   Das Verhalten röntgeninduzierte Viererringe mit grossen interstitialen Segmenten bei *Oenothera hookeri. Z. Indukt. Abstammungs.-Vererbungsl.* **82**:415–429.

Mayo, O., 1966   On the problem of self-incompatibility alleles in *Oenothera organensis. Biometrics* **22**:111–120.

Munz, P. A., 1965   Onagraceae. *North Am. Flora Ser.* II. **5**:1–278. N.Y. Bot. Garden.

Oehlkers, F., 1924   Sammelreferat über neuere experimentelle Oenotheren-arbeiten. *Z. Indukt. Abstammungs.- Vererbungsl.* **34**:259–283.

Oehlkers, F., 1935   Studien zum Problem der Polymerie und des multiple Allelomorphismus. III. Die Erblichkeit der Sepaloidie bei *Oenothera* und *Epilobium. Z. Bot.* **28**:161–222.

Oehlkers, F., 1938   Über die Erblichkeit des *cruciata* Merkmals bei den Oenotheren; eine Erwiderung. *Z. Induk. Abstammungs.- Vererbungsl.* **75**:277–297.

Oehlkers, F., 1943   Die Auslösung von Chromosomenmutationen in der Meiosis durch Einwirkung von Chemikalien. *Z. Indukt. Abstammungs.- Vererbungsl.* **81**:313–341.

Oehlkers, F., 1949   Mutationsauslösung durch Chemikalien. *Sitzungsber. Heidelberg Akad. Wiss. Math.-Naturwiss. Kl.* **9**:3–40.

Oehlkers, F. and G. Linnert, 1951   Weitere Untersuchungen über die Wirkungsweise von Chemikalien bei der Auslösung von Chromosomenmutationen. *Z. Indukt. Abstammungs.- Vererbungsl.* **83**:429–438.

Raven, P. H., 1964   The generic subdivision of *Onagraceae*, tribe *Onagreae. Brittonia* **16**:276–288.

Renner, O., 1917   Versuche über die gametische Konstitution der Oenotheren. *Z. Indukt. Abstammungs.- Vererbungsl.* **18**:121–294.

Renner, O., 1921   Heterogamie im weiblichen Geschlecht und Embryosackentwicklung bei den Oenotheren. *Z. Bot.* **13**:609–621.

Renner, O., 1922   Eiplasma und Pollenschlauchplasma als Vererbungsträger bei den Oenotheren. *Z. Indukt. Abstammungs.- Vererbungsl.* **27**:235–237.

Renner, O., 1936   Zur Kenntnis der nichtmendelenden Buntheit der Laubblätter. *Flora (Jena)* **30**:218–290.

Renner, O., 1937   Über *Oenothera atrovirens* Sh. et Bart. und über somatische Konversion im Erbgang des *cruciata*-Merkmals der Oenotheren. *Z. Indukt. Abstammungs.- Vererbsl.* **74**:91–124.

Renner, O., 1940*a*   Kurze Mitteilungen über *Oenothera*. IV. Über die Beziehung zwischen Heterogamie und Embryosackentwicklung und über diplarrhene Verbindungen. *Flora (Jena)* **134**:145–158.

Renner, O., 1940*b*   Zur Kenntnis der 15-chromosomigen Mutanten von *Oenothera lamarckiana. Flora (Jena)* **134**:257–310.

Renner, O., 1941   Über die Entstehung homozygotischer Formen aus Komplexheterozygotischen Oenotheren. *Flora (Jena)* **135**:201–238.

Renner, O., 1942   Über das Crossing-over bei *Oenothera. Flora (Jena)* **136**:117–214.

Renner, O., 1943   Kurze Mitteilungen über *Oenothera*. VI. Über die 15-chromosomigen Mutanten *dependens, incana, scintillans, glossa, tripus. Flora (Jena)* **137**:216–229.

Renner, O., 1949   Die 15-chromosomigen Mutanten der *Oenothera lamarckiana* und ihrer Verwandten. *Z. Indukt. Abstammungs.- Vererbungsl.* **83**:1–25.

Renner, O., 1959   Somatic conversion in the heredity of the *cruciata* character in *Oenothera. Heredity* **13**:283–288.

Renner, O. and R. E. Cleland, 1933   Zur Genetik und Zytologie der *Oenothera chicaginensis* und ihrer Abkömmlinge. *Z. Indukt. Abstammungs.-Vererbungsl.* **66**:275–318.

Rudloff, C. F. and H. Stubbe, 1935   Mutationsversuche mit *Oenothera hookeri. Flora (Jena)* **129**:347–362.

Schötz, F., 1954   Über Plastidenkonkurrenz bei *Oenothera. Planta (Berl.)* **43**:182–240.

Schötz, F., 1968   Über Plastidenkonkurrenz bei *Oenothera.* II. *Biol. Zentrabl.* **87**:33–61.

Schötz, F., F., Senser, and H. Bathelt, 1966   Untersuchungen über die Plastidenentwicklung und Pigmentausstattung der Oenotheren. II. Mutierte *biennis*-Plastiden. *Planta (Berl.)* **70**:125–154.

Seitz, F. W., 1935   Zytologische Untersuchungen an tetraploiden Oenotheren. *Z. Bot.* **28**:481–542.

Steiner, E., 1956   New aspects of the balanced lethal mechanism in *Oenothera. Genetics* **41**:486–500.

Steiner, E., 1961   Incompatibility studies in *Oenothera. Z. Induk. Abstammungs.- Vererbungsl.* **92**:205–212.

Steiner, E., 1964   Incompatibility studies in *Oenothera:* The distribution of S_I alleles in *biennis I* populations. *Evolution* **18**:370–378.

Stubbe, W., 1953   Genetische und Zytologische Untersuchungen an verschiedenen Sippen von *Oenothera suaveolens. Z. Indukt. Abstammungs.- Vererbungsl.* **85**:180–209.

Stubbe, W., 1959   Genetische Analyse des Zusammenwirkens von Genom und Plastom bei *Oenothera. Z. Indukt. Abstammungs.- Vererbungsl.* **90**:288–298.

Stubbe, W., 1960   Untersuchungen zur genetischen Analyse des Plastoms von *Oenothera. Z. Bot.* **48**:191–218.

Winkler, H., 1930   *Die Konversion der Gene,* G. Fischer, Jena.

11

The Tomato

CHARLES M. RICK

Introduction

The cultivated tomato (*Lycopersicon esculentum* Miller) is a genetically well-endowed species of the potato family (Solanaceae). Although normally grown as an annual plant, it can be asexually propagated by several techniques as a facultative perennial. It is easily and widely cultivated, its growth and reproduction not being restricted by day length or any other special requirement. The tomato phenotype, particularly that of leaf shape, texture, and color, is of such a nature that a great variety of heritable modifications can be readily recognized (Kruse, 1968). While the tomato is normally automatically self-pollinated, controlled pollinations can be made readily to produce hybrid seed, even on a large scale. A full-grown plant produces from 10,000 to as much as 25,000 seeds. The requirements for long-term storage of its seed and pollen are known. The tomato possesses a haploid set of 12 chromosomes, each of which can be identified at pachynema by relative arm lengths, distribution of the highly distinguishable heterochromatin and euchromatin, and other cytological landmarks.

The subject of tomato genetics was last reviewed by Rick and Butler (1956). Assuming that the number of known genes provides a rough index of the status of the field, tomato genetics has experienced an expansion of roughly 10-fold during the intervening 17 years: in 1956 the list numbered 118 genes, while at the present time published accounts of at least 900 genes are known, and I am aware of the existence of additional

CHARLES M. RICK—Department of Vegetable Crops, University of California, Davis, California.

247

mutants that would increase the list to well over 1,100. This vast wealth of genetic variation is further augmented by the availability of many primitive cultivars as well as numerous collections of wild forms of *L. esculentum* and intercompatible, closely related species. Additional advantages of the species for genetic research will be elaborated in other parts of the text.

In keeping with the purposes of this volume, I am not attempting a complete review of the literature; instead, my aim is to incorporate recent developments of the greatest scientific value and of the widest interest.

Regenerative Plasticity

A feature in the biology of the tomato that renders it unique among the genetically most extensively investigated angiosperms is its propensity to regenerate in various ways in addition to seed reproduction: propagation by cuttings, graftage, and microspore culture.

Tomato cuttings root rapidly and dependably, thus constituting a method of vegetative propagation that is highly useful for such purposes as increasing the number of propagules of desired genotypes, maintenance of such sterile genotypes as asynaptics, male steriles, species hybrids, etc., and cloning individuals whose root systems must be sacrificed for such requirements as allozyme determinations.

It is in the ease of grafting that the tomato offers particular advantages for a wide range of experimental objectives. For a summary of the early, fairly extensive literature, see the review by Rick and Butler (1956). Attention is called there to applications in (1) large-scale grafting of tomato scions on root stocks of resistant genotypes or species as a means of combating various soil-borne parasites; (2) testing stock–scion influences, in which the bulk of studies, particularly Stubbe's (1954) very extensive research, report no heritable modifications; and (3) production of chimeras in which tissues of both symbionts of a graft are present in the regenerated shoots.

In recent research, tomato grafts have been used most intensively for investigating the physiology of various mutants. By measuring the responses of reciprocal grafts between mutant and normal plants in comparison with appropriate control grafts, much has been learned about the localization of metabolic defects, idiosyncrasies of transport, etc. Specific references are given in the section on physiological genetics.

Interspecific hybridization has been expedited in several ways by grafting. Thus, via grafts, Günther (1961) synthesized a chimera having inner germ layers of *L. peruvianum* surrounded by a sheath of *L. esculentum*, which thereby lost the characteristic self-incompatibility of the former and enjoyed the wider compatibility relations of the latter (Günther, 1964). Majid (1966) contends that, by grafting seedlings, interspecific incompatibilities can be overcome. As an example of additional studies on interspecific chimeras, Brabec (1960) investigated the tissue relationships in an unstable combination of *L. pimpinellifolium* and *Solanum nigrum*.

Another interesting application of grafts was made by Andersen (1965) to test for possible transmission between *L. esculentum* and *Solanum pennellii* of factors that determine the cytoplasmic male sterility in their hybrids.

A noteworthy recent development in tomato regeneration is the culturing of microspores as a means of producing intact haploid plants. Sharp *et al.* (1971) devised techniques that permit development of a haploid callus and stimulation of adventitious roots from the callus. Sharp *et al.* (1972) have reported varying degrees of success in plating mature pollen in nurse cultures. By manipulating media and various external factors, Gresshoff and Doy (1972) succeeded in regenerating plantlets from haploid callus of specific genotypes. A highly promising recent development is a method incorporating chilling of the anthers and use of certain growth supplements to culture haploid tomato plants directly from microspores, thereby bypassing the callus stage (Debergh and Nitsch, 1973). Microspore culturing, thus demonstrated to be feasible for the tomato, adds appreciably to the long list of advantages offered by this species for genetic research. This technique should essentially permit genetic analysis in the haploid stage, facilitate quick development of pure lines, and possibly lead the way to a system of cell genetics for the tomato.

Induced Mutation

The tomato has been utilized to a considerable extent for mutagenic studies. Certain aforementioned features, including its high reproductivity and ease of making controlled matings, account for this popularity. Although not an exceptional subject for general mutation studies, it offers advantages for the following specialized purposes. Many investigations have been justified for the purpose of acquiring mutants for particular purposes. Thus the list of mutants available as genetic markers has been enriched by 300 induced by x rays in *L. esculentum* (Stubbe, 1957, 1958*a*, 1959, 1963, 1964, 1972*a*) and by 200 in the very closely related *L. pimpinellifolium* (Stubbe, 1960, 1961, 1965, 1972*b*). Additional useful mutations have been induced by neutrons (Verkerk, 1959; Yu and Yeager, 1960).

The effectiveness of various chemical mutagens has been demonstrated, and appreciable numbers of new mutants have thereby been acquired. Such reports have been made by Buiatti and Ragazzini (1966) for acridine orange, Emery (1960) for diepoxybutane, Hildering (1963) for ethylene imine, Hildering and van der Veen (1966) for ethyl methanesulfonate, and by Jain *et al.* (1968) for the latter as well as base-specific chemicals. The latter study is noteworthy for its report of anomalously high yields of hydrazine-induced, recessive, mutant homozygotes. The frequencies for the same mutant homozygotes were nearly the same in M_1 as in M_2; whereas they would have been expected to be nearly absent in the former and vastly higher in the latter.

The utilization of induced mutants for various studies that have improved understanding of the physiology of higher plants is outlined in the section Physiological Genetics. Programs have been organized to systematically screen tomato mutants for such types as the overwilting types (Tal, 1966), those with defective root systems (Zobel, 1972*a*), and others.

The normal development of the tomato plant into a large highly branched unit with many multifruited inflorescences renders it particularly useful for the study of chimerism of induced mutations. In such studies, Hildering and Verkerk (1965) discovered that chimeras are much more abundant following EMS than x-ray treatment, possibly because the latter tends to kill cell initials in the embryonic meristems of treated seeds. Their conclusions might also explain the results of Bianchi *et al.* (1963) that intrasomatic selection is of minor importance in x-ray treated tomato material.

The tomato has also proved amenable for studying the effects of such concomitant factors of mutagenesis as desiccation, storage, and radiation intensity (Brock and Franklin, 1966).

The research work of Stubbe has also provided important applications of mutagenesis in studies of evolution of the tomato. Thus, 13 parallel mutations in *L. esculentum* and *L. pimpinellifolium* constitute an interesting confirmation of Vavilov's Law of the Homologous Series (Stubbe, 1970). In further studies (Stubbe, 1971) he demonstrated that, by means of a program of stepwise induced mutation and selection, it is possible to increase fruit size genetically in the latter species to approximate that of the former and, by the same methods, diminish fruit size of the former to nearly the level of the latter. Additionally, the research of Stubbe (1958*b*) and his colleagues has revealed applications of mutants affecting tomatine content in tomato breeding and biosynthesis of other glycosidal alkaloids.

Mutant Genes

Information concerning the name, symbol, phenotype, chromosome locus, and literature reference for 211 mutant genes is given in Table 1. This group includes all located markers and an additional group selected because they determine characters of special interest. Although not intended to be a complete enumeration of genes, this list is representative of the total array known for the tomato.

The genes are named and symbolized according to rules formulated by the Tomato Genetics Cooperative (for latest revision see Clayberg *et al.,* 1970), a self-financed group operating for the exchange of information and stocks for the promotion of research in tomato genetics. The normal or standard (+) alleles are defined as those present in cv. Marglobe. Symbols are designated in capital or small case according to whether the mutant allele is, respectively, dominant or recessive to the allele in Marglobe.

Seed stocks of the mutant genes listed are available in small quantities from stock cooperators of the Tomato Genetics Cooperative. Inquiries concerning these as well as other stocks can be addressed to the writer. A large collection of tomato germplasm is maintained by the North Central Regional Plant Introduction Station at Ames, Iowa.

Linkage

Linkage maps of the tomato were developed by a combination of genetic and cytogenetic techniques. Several kinds of aneuploids have been used extensively to determine the chromosome on which each linkage group resides and, further, for assigning parts of the linkage groups to their respective chromosome arms (Rick and Khush, 1969). Various kinds of trisomics have played a leading role. A complete collection of the primary trisomics (with an extra normal chromosome) has been assembled and identified cytologically and genetically (Rick and Barton, 1954; Rick *et al.,* 1964). As in other basic diploid species, each primary can be identified by its highly distinctive phenotype. The unique morphology of each chromosome in pachynema permits cytological identification of the trisomics. Genetic identification of the linkage groups with their corresponding trisomic is accomplished by the usual trisomic ratio method. This system was pursued for chromosome–linkage-group associations of all of the tomato complement except chromosome 11, for which the first association was accomplished by use of induced deficiencies and the monosomic haplo-11 (Rick and Khush, 1961). Subsequently, monosomics

TABLE 1. *Essential Features of the Linkage Markers and Other Noteworthy Mutants*

Symbol	Name	Locus[a]	Phenotype	Reference[b]
a	anthocyaninless	11–57	Complete absence of anthocyanin in all parts	Rick and Butler, 1956
aa	anthocyanin absent	2–32	Complete absence of anthocyanin in all parts	Clayberg et al., 1970
afl	albifolium	4–3	Leaves whitish, later becoming normal	Clayberg et al., 1966
ag	anthocyanin gainer	10–127	Anthocyanin gained only in later development	Rick and Butler, 1956
ah	anthocyaninless of Hoffman	9–24	Complete absence of anthocyanin in all parts	Clayberg et al., 1960
al	anthocyanin loser	8–67	Anthocyanin faint at first, later eliminated	Rick and Butler, 1956
alb	albescent	12–0	Bold white and light green variegation	Clayberg et al., 1966
apn	albo-punctata	11–21	Fine white speckling on foliage	Clayberg et al., 1973
as-1 to as-5	asynaptic-1 to -5	—	Little or no pairing of chromosomes at MI	Rick and Butler, 1956
atv	atroviolacea	7	Intense anthocyanin pigmentation of herbage	Clayberg et al., 1966
au	aurea	1–32	Foliage bright yellow; corolla and stigma pale yellow	Clayberg et al., 1960
aut	aureata	3L–?	Virescent for bright golden yellow	Clayberg et al., 1970
aw	without anthocyanin	2–59	Anthocyanin completely absent in all parts	Rick and Butler, 1956
B	Beta-carotene	6–94	Beta-carotene the dominant fruit pigment	Rick and Butler, 1956
bi	bifurcate	11–86	Extreme fasciation in combination with f and j	Rick and Butler, 1956

Symbol	Name	Locus	Description	Reference
bip	bipinnata	2–68	Leaves highly divided with small acute segments	Stubbe, 1959
bk	beaked	2–32	Pointed process at stylar end of fruits	Rick and Butler, 1956
bl	blind	11–64	Stem terminating in first inflorescence	Clayberg et al., 1966
bls	baby lea syndrome	3–74	Anthocyanin lacking, compact habit	Rick and Butler, 1956
br	brachytic	1–0	Internodes shortened	Clayberg et al., 1970
brt	bushy root	12	Roots excessively branched; dwarf habit	Clayberg et al., 1967
bs	brown seed	1–16	Brown endosperm color	Clayberg, et al., 1970
bs-2	brown seed-2	7–34	Brown endosperm color	Clayberg et al., 1960
btl	brittle	—	Stems brittle when boron is deficient	Rick and Butler, 1956
bu	bushy	8–18	Internodes, inflorescences shortened; petioles elongated	Rick and Butler, 1956
c	potato leaf	6–93	Number of leaf segments reduced	Rick and Butler, 1956
Cf	Cladosporium fulvum resistance	1–92?	Resistance to specific races of *Cladosporium fulvum*	
Cf-2	Cladosporium fulvum resistance-2	6–32		
Cf-3	Cladosporium fulvum resistance-3	11–0		
ch	chartreuse	8–28	Corolla segments greenish yellow	Clayberg et al., 1966
chln	chloronerva	—	Upper leaves small, chlorotic interveinally; strong graft response	Clayberg et al., 1966
cl-2	cleistogamous-2	6–102	Flowers open only slightly	Rick and Butler, 1956
clau	clausa	4–0	Partly cleistogamous; leaves rugose, excessively divided	Stubbe, 1958a
cm	curly mottled	4–6	Leaves mottled and distorted; environment-sensitive	Clayberg et al., 1960
cn	cana	3–24	Leaves gray-green, tiny unbranched plants	Stubbe, 1958a
co	cochlearis	1–76	Plants small; leaves reduced, gray-green; pinnae concave, spoon-shaped	Stubbe, 1958a

TABLE 1. Continued

Symbol	Name	Locus[a]	Phenotype	Reference[b]
coa	corrotundata	6–64	Plant small; pinnae fewer, broad, rounded, darker; flower parts short, broad	Stubbe, 1964
com	complicata	1–63	Leaves excessively subdivided; plant small, erect	Stubbe, 1959
con	convalescens	3L–?	Yellow-green, especially at growing point	Stubbe, 1957
cpt	compact	8–16	Plant highly branched, compact; stems lax; leaves small, pale	Clayberg *et al.*, 1966
Cu	Curl	2–49	Leaf veins and petiole strongly foreshortened; homozygous viable	Clayberg *et al.*, 1960
d	dwarf	2–70	All parts foreshortened; leaves dark, rugose	Rick and Butler, 1956
d-2	dwarf-2	6–93	Plant slow-growing; semi-sterile	Clayberg *et al.*, 1960
dd	double dwarf	2?	Extreme dwarf syndrome	Clayberg *et al.*, 1967
deb	debilis	7–52	Leaves yellowish, becoming necrotic	Stubbe, 1957
def	deformis	6–57	Flowers deformed; fertility reduced; leaves progreesively reduced	Stubbe, 1957
depa	depauperata	8–18	Plants very small; leaves variably yellowish	Stubbe, 1958a
dgt	diageotropica	1–152	Diageotropic growth of stems and roots; roots unbranched; plant small	Clayberg *et al.*, 1973
di	divergens	4–89	Rachis short; progressively reduced stem, whitish streaked	Stubbe, 1958a
dil	diluta	2–53	Leaves dull light green, small, roughened	Stubbe, 1957

Symbol	Name	Location	Description	Reference
div	divaricata	3–111	Plant small, compact; leaves yellowish intercostally	Stubbe, 1959
dl	dialytic	8–29	Stamens not united; all hairs suppressed	Rick and Butler, 1956
dmt	diminutiva	4–79	Internodes and leaves foreshortened	Clayberg et al., 1973
dpy	dumpy	2–59	Leaves greatly condensed, rugose, dark green; internodes somewhat foreshortened	Clayberg et al., 1967
dt	dilatata	1–161	Plant small; pinnae yellowish; veins darker	Stubbe, 1963
dv	dwarf virescent	2–74	Growing point pale green; plant stunted	Rick and Butler, 1956
e	entire	4–66	Fewer leaf segments; midvein distorted	
ele	elegans	11–19	Leaves reduced, yellow-green; plants tiny	Stubbe, 1957
f	fasciated	11–84	Fruits many-loculed	Rick and Butler, 1956
fa	falsiflora	—	Inflorescence extremely ramified; flowers grossly modified, functionless	Stubbe, 1963
fd	flecked dwarf	12–31	Leaves flecked with light green; plant retarded	Clayberg et al., 1967
fer	fe inefficient	—	Leaves chlorotic due to faulty iron transport	Clayberg et al., 1973
fla	flavescens	1–145	Leaves light green with few segments	Stubbe, 1957
flc	flacca	7–59	Leaves small, dorsally recurved, tending to overwilt	Stubbe, 1959
fms	female sterile	—	Gynoecium distorted, functionless; pollen normal	Clayberg et al., 1966
ful	fulgens	4–24	Leaves bright yellow, turning greener	Stubbe, 1957
Ge	Gamete eliminator	4–27	Ge^c male and female gametes eliminated in Ge^c/Ge^p heterozygotes	Clayberg et al., 1967
gf	green flesh	8–44	Fruit with persistent chlorophyll	Clayberg et al., 1960
gh	ghost	11–32	Plant starts green, later breaks to white	

TABLE 1. Continued

Symbol	Name	Locus[a]	Phenotype	Reference[b]
glg	galápagos light green	8–43	Leaves pale gray-green, darker veins; plant small	Clayberg et al., 1973
glo	globosa	4–72	Internodes short; leaves short, pale green; incompletely dominant	Stubbe, 1957
Gp	Gamete promoter	9–0	Promotes gamete competition in fertilization	Clayberg et al., 1970
gs	green stripe	7–5	Green stripes in epidermis of unripe fruit; golden in ripe fruit	Rick and Butler, 1956
h	hairs absent	10–48	Large hairs absent except on hypocotyl; incompletely dominant	Rick and Butler, 1956
Hero	Heterodera rostochiensis resistance	—	Resistant to pathotype A of Heterodera rostochiensis	Clayberg et al., 1973
hi	hilara	9–42	Pinnae light green, rugose and irregularly crenate	Stubbe, 1964
hl	hairless	11–37	All herbage hairless; stems brittle	Rick and Butler, 1956
hp	high pigment	—	Chlorophyll, carotenoids, ascorbic acid content of fruit intensified	Clayberg et al., 1960
Hr	Hirsute	8–46	Long hairs on upper leaf surface	Clayberg et al., 1966
hy	homogeneous yellow	10–17	All vegetative parts yellow in all stages	Clayberg et al., 1970
I	Immunity	11	Resistance to Fusarium oxysporum f. lycopersici	Rick and Butler, 1956
icn	incana	10–22	Leaves emerge whitish margined; plant retarded	Clayberg et al., 1970
ig	ignava	7–27	Habit erect, little branched; leaves relatively large, light green	Stubbe, 1959
imi	inquieta	11–57	Immature pinnae lighter, involuted; plant reduced	Stubbe, 1964

inv	invalida	1–140	Leaves with fine yellowish flecks; plant reduced	Stubbe, 1957
j	jointless	11–17	Pedicels jointless; inflorescences leafy	Rick and Butler, 1956
l	lutescent	8–0	Leaves prematurely yellowing; unripe fruits yellowish	
l-2	lutescent-2	10–98	Same phenotype as *l*	Clayberg et al., 1960
La	Lanceolate	7–48	Leaves small, simple, entire; stems slender; fruits small; homozygous inviable	
lg-5	light-green-5	7–17	Foliage uniformly light green	Clayberg et al., 1966
Ln	Lanata	3–53	Excessively hairy; all genotypes viable and distinct	Clayberg et al., 1973
Lpg	Lapageria	1–16	Leaves small, dark green, glossy; flowers campanulate	Clayberg et al., 1966
ls	lateral suppressor	4–132	Few or no axillary branches; corolla suppressed; partially male-sterile	Clayberg et al., 1960
ltf	latifolia	7–0	Cotyledons and pinnae extremely broad proportioned	Clayberg et al., 1973
lut	lutea	9–56	Leaves yellow-green, darker veins; growing point much yellower	Stubbe, 1957
Lx	Lax	2–56	Leaves pendent, elongate, acute segmented	
lyr	lyrata	5–31	First leaves entire; later leaves fan-shaped; female-sterile	Clayberg et al., 1966
m	mottled	2–71	Leaves flecked pale green, distorted	Rick and Butler, 1956
m-2	mottled-2	6–69	Leaves with many fine chlorotic spots; temperature sensitive	
marm	marmorata	9–62	White-light-green marbling of leaves environmentally sensitive	Clayberg et al., 1960
mc	macrocalyx	5–0	Sepals leafy; inflorescence leafy	Rick and Butler, 1956

TABLE 1. Continued

Symbol	Name	Locus[a]	Phenotype	Reference[b]
Me	Mouse ears	2–48	Leaves 3–4 pinnately compound with clavate segments; homozygous viable; heterozygote intermediate	Clayberg et al., 1960
Mi	Meloidogyne incognita resistance	6–35	Resistance to the nematode M. incognita	
mnt	miniature	11–40	Small plant; branching and fruit set reduced	Clayberg et al., 1967
mps	miniature phosphorus syndrome	11–74	Extremely reduced; chlorotic and purplish	Clayberg et al., 1970
ms-2	male-sterile-2	2–69	Anthers pale, shrunken; no pollen	
ms-7	male-sterile-7	11–35	Anthers slightly shrunken; aborted pollen in tetrads	
ms-8	male-sterile-8	8–11	Anthers pale, shrunken; no pollen; flowers small	
ms-9	male-sterile-9	3–8	Anthers nearly normal; no pollen	
ms-10	male-sterile-10	2–42	Anthers small, very pale; no pollen; flowers very small	Rick and Butler, 1956
ms-12	male-sterile-12	11–47	Anthers shrunken; no pollen; flowers very small	
ms-14	male-sterile-14	11–77	Anthers shrunken; very pale; no pollen	
ms-15	male-sterile-15	2–62	Anthers dwarfed, very pale; no pollen; flowers small; stigmas exserted	
ms-32	male-sterile-32	1–47	Anthers very small, shrunken, very pale, often brown; no pollen; stigmas exposed	Clayberg et al., 1966
ms-33	male-sterile-33	6L–?	Anthers variable, yellow-green; no pollen	

mta	mutata	9–17	Small broad bush habit; internodes short; leaves long; juvenile yellow-green virescence	Stubbe, 1964
mts	mortalis	1–94	No inflorescences even if grafted on normal stock; small bush habit	Stubbe, 1963
mua	multifurcata	12L–?	First inflorescence multibranched; olive-green interveinal chlorosis; plant small	Stubbe, 1963
nd	netted	10–105	Primary leaves attenuated and chlorotic; darker veins in older leaves	Clayberg et al., 1960
ne	necrosis	2–prox.	With *Cf-2* causes progressive necrosis of leaves	Rick and Butler, 1956
neg	neglecta	11–29	Plants small, weakly branched; leaves pale, darker veined, becoming necrotic	Stubbe, 1957
ni	nitida	8–45	Leaves long petioled with dainty, deeply cut segments	Stubbe, 1957
not	notabilis	7–40	Leaves tiny, delicate, overwilting under stress	Stubbe, 1958a
Nr	Never ripe	9	Fruits ripen to dirty orange color; homozygous viable	Clayberg et al., 1960
nv	netted virescent	9–20	Leaves with fimbriate margins and pale interveinal chlorosis	Clayberg et al., 1960
o	ovate	2–55	Fruit ovate or pyriform	Rick and Butler, 1956
oc	ochroleuca	4–24	Leaves acuminate, chlorotic, becoming white variegated	Stubbe, 1959
og	old gold	6–94	Corolla tawny orange	Clayberg et al., 1966
oli	olivacea	10–71	Leaves blotched pale olive green, crinkled	Stubbe, 1959
op	opaca	2–63	Leaf color light green, yellow at growing point	Stubbe, 1958a

TABLE 1. Continued

Symbol	Name	Locus[a]	Phenotype	Reference[b]
p	peach	2–67	Fruits with dull surface and increased hairiness	Rick and Butler, 1956
pau	pauper	3–0	Plants tiny, weak, unbranched; marked reciprocal graft influence with normal	Stubbe, 1958a
pc	precocious centromere	—	Centromeres divide prematurely in meiosis; highly sterile	Clayberg et al., 1960
pcv	polychrome variegated	11–25	Leaves distorted with fine, striated variegation of white and several shades of green	Clayberg et al., 1973
pds	phosphorus-deficiency syndrome	6–0	Retarded; leaves irregular, blotched dull yellow green and purple	Clayberg et al., 1970
pen	pendens	2–54	Leaves broad, convex, pendant, glossy	Stubbe, 1959
Ph (Pi)	Phytophthora infestans resistance	7	Resistance to race To of P. infestans	Clayberg et al., 1960
pl	perlucida	7–40	Pinnae narrow, light green, yellowing prematurely	Stubbe, 1963
pla	plana	9–33	Plant small, rigid; leaves yellowish to light green, darker veined	Stubbe, 1959
pro	procera	—	Growth-rate accelerated; pinnae entire	Stubbe, 1957
prun	prunoidea	2–56	All parts, particularly fruit, elongate; traits weakly dominant	Stubbe, 1957
ps	positional sterile	2–61	Corolla remains unfurled; pollen not shed	Rick and Butler, 1956
pst	persistent style	7–5	Style persistently adnate to fruit, producing pronounced beak on latter	Clayberg et al., 1967
pum	pumila	9–24	Plant tiny, dainty with few or no branches	Stubbe, 1957

Py	Pyrenochaeta lycopersici resistance	—	Resistance to *Pyrenochaeta lycopersici*; heterozygote intermediate	Clayberg et al., 1973
r	yellow flesh	3–29	Yellow color of ripe fruit flesh	Rick and Butler, 1956
ra	rava	4–31	Leaves downcurled, gray-green; hairs dense, elongate	Stubbe, 1957
re	reptans	8–12	Plant elongate, less branched, inclined, becoming recumbent	Stubbe, 1958a
res	restricta	10–105	Plant small, compact; pinnae boat-shaped, yellow-green, purplish ventrally	Stubbe, 1963
ria	rigidula	7–70	Plant small; leaves stiff, yellowish, turning dark green	
rig	rigida	8–35	Plant small, rigidly erect when young; leaves yellowish, turning green	Stubbe, 1959
rin	ripening inhibitor	5–0	Fruits green at maturity, turning bright yellow	Clayberg et al., 1970
ro	rosette	2–52	Internodes extremely short; no flowers	Rick and Butler, 1956
rot	rotundifolia	7–17	Leaves short, broad, blistered, internodes short	Stubbe, 1959
ru	ruptilis	3–63	Pinnae narrow, keeled, dull light green, darker veined; heterozygote temporarily intermediate	Stubbe, 1958a
rust	rustica	8–58	Plants dwarf; leaves broad, blunt, fewer segmented	Stubbe, 1957
rv	reticulate virescent	3–76	New leaves pale green, darker veined, turning to normal color	Rick and Butler, 1956
s	compound inflorescence	2–30	Strong proliferation of inflorescence	
scf	scurfy	1–76	Cotyledons scurfy, striated	Clayberg et al., 1966
scl	seasonal chlorotic lethal	8–36	Cotyledons pale yellow; lethal in early stages except at high light intensity	Clayberg et al., 1970

TABLE 1. Continued

Symbol	Name	Locus[a]	Phenotype	Reference[b]
ses	semisterilis	1–25	Plant small, erect, nearly unbranched; leaves thick, yellowish, light gray-green; reduced fertility	Stubbe, 1963
sf	solanifolia	3–111	Primary leaves entire; segments of later leaves entire, concave; flower parts filiform	Clayberg et al., 1960
sfa	sufflaminata	1–152	Plant smaller, weakly branched; pinnae concave, yellow to yellow-green, purplish ventrally	Stubbe, 1963
si	sinuata	4–36	Reduced growth; leaves small, wavy margined, yellow-green	Stubbe, 1959
sit	sitiens	1–32	Small, weak plant; leaves short, overwilting under stress	Stubbe, 1957
sl	stamenless	4–89	Stamens usually absent; fruit form modified	Rick and Butler, 1956
sms	small seed	—	Endosperm almost absent; testa thin, transparent; germination reduced	Clayberg et al., 1973
sp	self-pruning	6–94	Plant-habit determinate	Rick and Butler, 1956
spa	sparsa	8–6	Plant size reduced; leaves emerge yellow-green, becoming blotched whitish green; responds to thiamine applications	Stubbe, 1959
spl	splendens	4–38	Leaves shiny yellow-green, dark veined	Stubbe, 1963
sub	subtilis	11–78	Plant tiny with short internodes, fastigiate habit; pinnae narrow, acute	Stubbe, 1957
suf	sufflava	2–66	Foliage uniform light green color	

sulf	sulfurea	2-prox.	Cotyledons greenish yellow, turning pale yellow; true leaves yellow; lethal unless grafted on normal stock; frequent somatic conversion[c]	Clayberg *et al.*, 1960
sy	sunny	3-46	Cotyledons bleached whitish; true leaves emerge yellow, becoming green	
syv	spotted yellow-virescent	4-8	Bright yellow growing point; fine whitish speckling on laminae may turn necrotic	Clayberg *et al.*, 1973
t	tangerine	10-102	Fruit flesh and stamens orange colored	Rick and Butler, 1956
tab	tabescens	11-0	Plant stunted; leaves irregular, yellow-green, violet-veined with necrotic speckling	Stubbe, 1958a
ten	tenuis	—	Plant very retarded; leaves light green, reticulated with whitish veins; responds to thiamine	Stubbe, 1957
tf	trifoliate	5-31	Leaf usually with only three segments, petiole elongate	Rick and Butler, 1956
tl	thiaminless	6-27	Leaves blotched yellow; lethal; viable and normal if fed thiamin	Clayberg *et al.*, 1966
Tm	Tobacco mosaic resistance	5	Moderate resistance to certain strains of tobacco mosaic virus	Clayberg *et al.*, 1960
Tm-2	Tobacco mosaic resistance-2	9-22	High-level resistance to many strains of tobacco mosaic virus	
tp	tripinnate	8-22	Leaves tripinnately compound; plant retarded	Clayberg *et al.*, 1966
u	uniform	10-3	Unripe fruits of uniform light green color lacking normal darker shoulder	Rick and Butler, 1956
um	umbrosa	1-119	Mature leaves darker green, wilted appearance; later growth stunted	Stubbe, 1958a

TABLE 1. Continued

Symbol	Name	Locus[a]	Phenotype	Reference[b]
v-2	virescent-2	2-67	Emerging leaves pale under greenhouse conditions often indistinguishable in field	Clayberg et al., 1966
va	varia	8-28	Leaves emerge yellow-green, turning normal	Stubbe, 1957
var	variabilis	7-0	Leaves emerge yellow-green, turning light green; plant slightly smaller	
ven	venosa	4-40	Weak growth; leaves small, folded, whitish yellow, green veined	
ver	versicolor	4-17	Immature leaves finely mottled yellow, green veined	Stubbe, 1958a
vg	vegetative	4-89	Flowers highly deformed, usually functionless	Rick and Butler, 1956
vga	virgulta	8-36	Plant small with short internodes; leaves emerge light yellow-green, turning dull green	Stubbe, 1963
vit	vitiosa	3-111	Leaves become progressively more deformed with twisted, filiform pinnae	Stubbe, 1958a
vms	variable male-sterile	8-34	Anthers abortive under high temperatures	Clayberg et al., 1966
vrd	viroid	1-109	Leaves highly distorted, white-speckled chlorosis; very environment sensitive	Clayberg et al., 1970
w	wiry	4-20	Leaves progressively reduced to midvein filaments in the upper positions	Rick and Butler, 1956
w-4	wiry-4	4-28	Leaves as in w, although expression less extreme	Clayberg et al., 1966

Symbol	Name	Position	Description	Reference
wd	wilty dwarf	9-20	Plants stunted; leaves grayish green and drooping under xerophytic conditions; graft responsive	Rick and Butler, 1956
wf	white flower	3-44	Corolla white to buff colored	
Wo	Woolly	2-46	All parts densely pubescent; homozygous inviable	
wt	wilty	5-46	Leaf margins curl adaxially	
wv	white virescent	2-44	Cotyledons whitish yellow; leaves emerge white speckled, turning normal; plant retarded	Clayberg *et al.*, 1966
wv-2	white virescent-2	9-48	Leaves emerge white, green veined, turning normal	Clayberg *et al.*, 1973
x	gametophytic factor	11-12	With *I* renders microgametes inactive	
Xa	Xanthophyllic	10-88	Leaves yellow; growth retarded; homozygous inviable	Rick and Butler, 1956
Xa-2	Xanthophyllic-2	10-0	Same phenotype as *Xa*	Clayberg *et al.*, 1966
Xa-3	Xanthophyllic-3	10-25	As in *Xa* except leaves more greenish	
y	pigmentless fruit epidermis	1-30	Fruit epidermis lacks pigmentation	Rick and Butler, 1956
yg-6	yellow-green-6	11-50	Seedling etiolated; leaves bright yellow, blotched white	Clayberg *et al.*, 1966
yv	yellow virescent	6-34	Leaves emerge yellow-green, turning normal green	Rick and Butler, 1956

[a] Numbers represent chromosome—position; a single number designates the chromosome for a gene whose position has not been determined.

[b] Serial compilations, which cite original references.

[c] In heterozygotes, one allele of a gene does not normally affect the genetic structure of the other, so they segregate again in the following generations. However, it has been shown that somatic conversion occurs in a few genes of some flowering plants, i.e., one allele undergoes mutational changes in a certain percentage of vegetative cells if a conversion-active allele is present in the same cell nucleus. Somatic conversion is comparable in many respects to Brink's paramutation at the *R* locus in *Zea*.

were found to be viable for chromosomes 12 (Khush and Rick, 1966) and 5 (Ecochard and Merkx, 1972). According to good circumstantial evidence, the tomato sporophyte probably cannot tolerate the monosomic condition for the other chromosomes.

Arm identifications can be made with the use of several types of deficiencies and modified trisomic types. Modified ratios accomplish this purpose in the following kinds of trisomics: secondary (extra isochromosome; Khush and Rick, 1969); tertiary (extra translocated chromosome; Khush and Rick, 1967a); telo (extra telocentric chromosome; Khush and Rick, 1968a); and double-iso compensating (normal chromosome replaced by two isochromosomes, one for each arm; Khush and Rick, 1967b). Collections of these trisomic types are not complete, but they are adequate for arm assignments for each of the chromosomes. With secondary and telo trisomics, such determinations can be made if the marker in question has its locus on the extra arm. If it is located on the other arm, the test is not conclusive unless independent data have ascertained that the gene is located on that specific chromosome. Simultaneous, conclusive tests of chromosome and arm locations are permitted by tertiary and double-iso compensating types.

Aside from these purposes, the tomato trisomics have other applications in testing the distance of genes from the centromeres by means of double reduction and in various experiments concerning gene-dosage.

Translocated or tertiary monosomics (Khush and Rick, 1966; Ramanna, 1969) and haplo-triplo disomics (Khush and Rick, 1967c) have also assisted in arm identifications. The former possesses a translocated chromosome and only one each of the normal set whose arms are represented in the interchange; in the latter, a normal chromosome is replaced by one of its two corresponding isochromosomes. Although useful for pseudodominance tests of recessive alleles in the generation of induction of the aberration, these two cytological deviants cannot be reproduced sexually and, hence, are more limited in their cytogenetic applications.

Since exchanges are centromeric in all of the aforementioned translocated aneuploid types, such identifications have been of great value in positioning the centromere in the genetic linkage maps and have been responsible for most of the centromere locations indicated in Figure 1.

Interstitial deficiencies provide the best opportunity for making more precise cytological determinations of gene loci (Khush and Rick, 1968a; Khush et al., 1964; Rick and Khush, 1961). A homozygous stock of the recessive marker is mated with the homozygous dominant whose pollen has been heavily treated with ionizing radiation. Offspring with recessive (pseudo-dominant) phenotype are selected and studied for signs of deficiency in their pachytene chromosomes. With a group of such deficiencies,

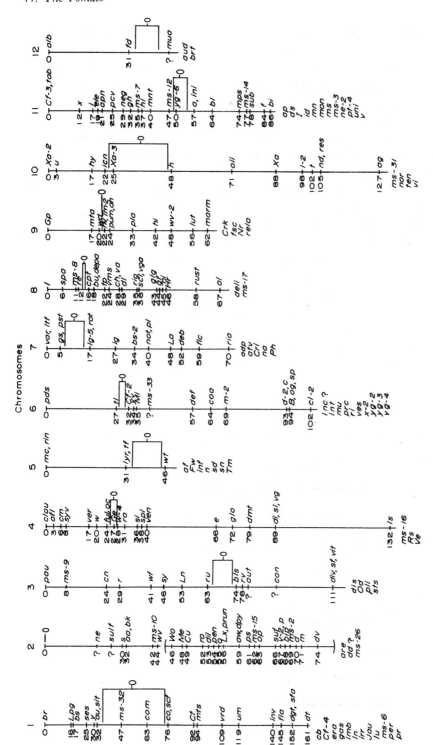

Figure 1. *Revised linkage map of the tomato, based on the summary in Report of the Tomato Genetics Cooperative No. 23 (1973) with several corrections and amendments. The "O" symbols designate centromeres.*

the locus for a gene can be narrowed to that segment of the chromosome that is common to all of the deficiencies. Deficiencies thus induced in mature pollen are transmitted to progeny. If they are located in euchromatin, however, they do not transmit through gametogenesis to the next generation. The cytological localizations thus achieved are summarized in Figure 2, which presents a schematic representation of the chromosome map based on morphology of pachytene chromosomes. The positions of key marker genes, as determined by means of induced deficiencies, are also indicated in this figure.

Cytologically detectable differences in size of the satellite terminating the entirely heterochromatic short arm of chromosome 2 permit this heteromorphy to be used as a marker for linkage tests against genes on the long arm. Exploiting this technique, Moens and Butler (1963) established distances of several genic loci from the centromere that were concordant with values obtained by Snoad (1962) from translocation linkages.

With the framework thus established by various cytogenetic techniques, the tomato genome is in a uniquely favorable position for further genetic analysis. By utilizing this knowledge, it has been possible to synthesize linkage tester stocks in which are compounded useful alleles with seedling-character manifestation and marking each chromosome arm.

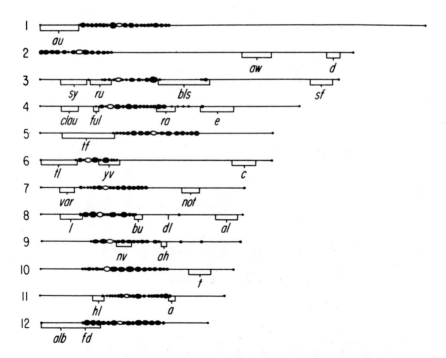

Figure 2. Chromosome maps of the tomato. Loci of key marker genes are largely approximated by means of induced deficiencies (Khush and Rick, 1968b).

A set of six such stocks, covering nearly all of the tomato genome, has been assembled and is available to interested investigators (Rick and Zobel, 1972). Further analysis of the tomato genome has been accomplished mostly by standard genetic linkage tests against such stocks, and the current status of such research has been summarized in the linkage maps of Figure 1. We resort to the aforementioned, more laborious, cytogenetic techniques only for such special situations as further delimitations of the centromeres in the linkage groups.

The current product of these efforts is the linkage map illustrated in Figure 1 which summarizes the status of the map as in Report No. 23 (1973) of the Tomato Genetics Cooperative. A total of 258 genes have been assigned to their respective groups; 190 have been sufficiently tested to provide an approximation of their loci; the remaining 68 have been assigned only by trisomic or two-point tests. Figure 2 presents a schematic representation of the chromosome map based on morphology of the pachytene chromosomes. The positions of key marker genes, as determined by means of induced deficiencies, are also indicated.

Gene Distribution

The linkage map thus constructed affords some interesting comparisons of gene distribution in the tomato genome. An initial analysis of the genome (Rick, 1959) revealed limited evidence for nonrandom distribution of genes between the chromosomes. Similar tendencies were also detected in a survey of the genome by means of primary trisomics (Rick *et al.*, 1964). The most recent summary (Rick, 1971*a*) revealed the same trends and also detected certain asymmetrical aspects of gene distribution within chromosomes. The following conclusions are excerpted from the latter reference.

In respect to numbers of genes per chromosome in comparison with numbers expected for relative lengths of euchromatic portions, the distribution deviates with high significance ($X^2 = 35.61$, 11 df). A large share of this discrepancy is contributed by an excess of markers on chromosome 11 and deficiency on chromosomes 3 and 12. Significant deviations of the same kinds were found among the spontaneous mutations in this sample; a good fit to expected values was obtained for x-ray-induced mutations.

Within chromosomes, several types of nonrandom features were encountered. In respect to quantitative anomalies, markers tend to be clustered in proximal regions and are sparsely distributed distally except for the nucleolar chromosome (No. 2), in which the proximal region has few markers in contrast to the distal regions. Qualitative anomalies include a marked concentration of spontaneous dominants on chromosome 2 and a highly significant restriction of loci for single kinds of mimic genes (genes with the same phenotype) to single chromosomes.

Another conclusion that is evident from these summaries is that the proximal dark-staining regions, present in every arm of the tomato complement, are remarkably free of genetic activity, thus confirming an earlier assumption that these regions are indeed heterochromatic. Extremely few of the active gene loci have been proven to exist in such heterochromatic regions (Khush and Rick, 1968*b*; Khush *et al.,* 1964), and, for chromosomes that can be tested adequately, rates of crossing over are greatly reduced there (Khush and Rick, 1967*d*).

Finally, collating all of the genetic and cytogenetic evidence, the tomato clearly behaves as a basic diploid species. Despite suggestions to the contrary from silent regions and from the nonrandom distribution of mimic loci, the degree of such duplication evidently is not sufficient to buffer monogenic segregation, chromosomal unbalance, etc., which are the true indices of diploidy vs. polyploidy.

Unstable Mutants

The tomato has been the source of numerous interesting unstable mutants. Studies of these mutants have been one of the specialties of the Zentralinstitut für Kulturpflanzenforschung in Gatersleben, East Germany; examples of the work done there are presented in the following paragraphs.

An example of chlorophyll variegation depending upon the presence or absence of a mitotically unstable fragment chromosome is yv^{ms}. Certain homozygous derivatives of yv^{ms} are uniformly yellow colored; others are variegated with the same yellow color on a normal green background. According to the research of Hagemann (1967) and Herrmann and Hagemann (1967), the uniformly yellow-colored individuals have the normal diploid chromosome number; the variegated ones have the normal diploid chromosome number plus varying numbers of tiny chromosome fragments. Intensity of variegation is inversely correlated with the number of extra chromosome fragments, which evidently carry a determiner for normalizing chlorophyll synthesis.

Xa-3, another fragment-dependent instability for intensity of chlorophyll pigmentation, was induced by means of x rays (Gröber, 1963). In further studies Gröber (1967, 1968, 1969) discovered that chlorosis or normal color of cells is determined by the *Xa-3*/+ dosage balance, that the variegation depends upon the presence of a centric fragment carrying the + allele, and that activity levels vary according to karyotype.

A totally different situation exists in the ever-sporting phenomenon of *sulf* alleles (Hagemann, 1958, 1961*a,b,c,* 1966, 1969*a,b*). Recessive

mutant heterozygotes are yellow and lethal in the seedling stage. The
sulf/+ heterozygotes are subject to varying intensities of somatic variegation for the yellow color, the intensity depending on the particular *sulf*
allele. Progeny tests prove that in yellow areas the + allele has been
mutated to a *sulf* allele, not necessarily identical to its homolog. Het-
erochromatization of *sulf*⁺ is believed to be responsible for this
"conversion" or "paramutation" process.

Physiological and Developmental Genetics

Research in this area of tomato genetics continues at an accelerated
pace. Its popularity for such studies stems from several factors. Thanks to
its ease of culture and to readily discerned diagnostic responses to environmental stimuli, the tomato has always been a favorite subject for general
physiological studies. Its adaptation to grafting (see section on
regenerative plasticity) confers on it a very special advantage. In addition,
it is well known genetically, with a wealth of mutants useful for tackling
various physiological problems.

Early research in this general area tended to concentrate on the
physiological genetics of the carotenoid pigments of the fruit, and an appreciable body of knowledge was acquired at the time of the preceding
review (Rick and Butler, 1956). Although research in this field has
continued, it has not received as much attention as others reviewed below.
The incompletely dominant gene *Del* codes for the production of large
quantities of *delta*-carotene at the expense of other carotenoid pigments in
otherwise normal genotypes (Tomes, 1969). This study, as well as many
preceding ones, suggests that the various tomato carotenoids are
synthesized in parallel fashion from a common precursor, rather than in
sequential steps of a single chain.

Recently other constituents of the ripe fruit have received attention.
Stevens studied the relationships between components that determine fruit
quality (1972*a*) and analyzed the inheritance of concentrations of the key
components—2-isobutylthiazole, methyl salicylate, eugenol (1970),
citrate, and malate (1972*b*). In each instance he could demonstrate the effects of a major gene.

The tomato has provided useful and interesting mutants for the study
of mineral transport. Thus, *btl* was found to be deficient in boron content
of stems (Wall and Andrus, 1962)—a situation traced via grafting experiments to inefficient transport (not absorption) of boron by *btl* roots
(Brown and Jones, 1971). The same mutant led to the discovery (well
nigh impossible with the normal genotype) that germanium can occupy

the site normally possessed by boron, but cannot substitute for it meta-bolically (Brown and Jones, 1972). Another mutant aberrant in mineral nutrition, *fer*, suffers inefficient transport of iron, a defect also localized in the root system by means of graft responses (Brown *et al.*, 1971).

The physiology of the wilty mutants *flc*, *not*, and *sit* was studied, and their defect found to be manifest in abnormal stomatal behavior (Tal, 1966). Further studies of *flc* (Tal and Imber, 1970; Tal *et al.*, 1970) revealed that resistance to water flow in roots also contributes to the wilting and that the mutant suffers a hormonal imbalance. Recent investi-gations (Tal and Nevo, 1973) have established that all three mutants have increased root resistance and a deficiency of abscisic acid; treatment with abscisic acid results in their reversion to normal phenotype in respect to wilting and the defects of roots and stomata. Knowledge of the control of the water economy in higher plants is thereby enhanced.

Mutants with modified anthocyanin synthesis were investigated by Wettstein-Knowles (1968). The anthocyanin of the normal type was identified as petanin. Specific alleles of *ag*, *al*, and *Pn* permit synthesis, but regulate anthocyanin distribution in characteristic time–spatial pat-terns. Intermediate stages of synthesis are regulated in *af*, *ah*, *aw*, and *bls*. Two additional mutants are distinguished from the others in the structure of their principal anthocyanins: *a* having peonidin as the aglycone component, and *ai* with a petunidin-based pigment.

The novel mutant *dgt* suffers highly modified geotropic responses of both stems and roots (Zobel, 1972*b*). Reciprocal grafts localized the meta-bolic disorder in the shoots. The response of *dgt* to low-level concentra-tions of ethylene suggests a defect in the auxin–ethylene regulation of growth; this mutant should be useful for further analysis of this phenomenon (Zobel, 1973).

A clear example of temperature sensitivity is provided by the variable male-sterile mutant *vms* (Rick and Boynton, 1967). Flowering normally at lower temperatures, its anthers become highly reduced and sterile at 30–32°C and higher. The sensitive stage is remarkably early—about 10 days prior to meiosis in the anthers.

The incompletely dominant, laceolate-leaved *La* has been a favorite mutant form developmental and physiological investigations (Mathan and Jenkins, 1960, 1962). *La/+* is viable and intermediate, but oddly, *La/La*, even in highly inbred lines, always comprises three morphologically dis-tinct classes: one entirely leafless, another terminating growth in a single cotyledon, and the third capable of continued growth but flowerless. Addi-tional evidence for the lack of genetic determination of the three classes is the fact that their proportions can be drastically modified by age of seed

(Mathan and Stettler, 1962). Among many details investigated in the morphogenesis of the leafless class of *La/La* (Caruso, 1968; Caruso and Cutter, 1970), graft experiments revealed the control by a normal shoot tip of the development of a vascular cambium. Dosage effects of *La* vs. + on leaf shape at $2n$ and $4n$ levels were analyzed (Stettler, 1964). Mathan and Cole (1964) demonstrated an increase in content of four oxidative enzymes in proportion to the dosage of *La*; Haccius and Garrecht (1963) and Mathan (1965*a,b*) reported that, in response to applications of phenylboric acid, normal tomato seedlings develop *La*-like leaves, an effect that could be reversed with treatment by actinomycin-D (Mathan, 1967).

The chlorophyll mutants, owing partly to the great diversity of their phenotypes, amenability to quantification, and responsiveness to environmental variables, are popular subjects for physiological studies.

The first known thiamine-requiring mutant (*tl*) was reported by Langridge and Brock (1961), who also determined that the synthesis is blocked in the assembly of the pyrimidine rather than the thiazole component. Subsequently, Boynton (1966*a*) found thiamine synthesis to be defective but not completely blocked in *spa* and *ten*, the former also suffering a pyrimidine problem, the block in the latter not entirely clear, but supposedly related to the conversion of thiamine to TPP. The synthetic blocks and ultrastructural defects of the chloroplast are developmentally interrelated with the characteristic progressive chlorosis of these mutants (Boynton, 1966*b*).

Among the many other chlorotic mutants investigated, *xan-5* has been studied intensively. Machold (1966) reported a strong response to feeding with NH_4, but not to nitrogen in the NO_3 form—a response not related to the pH of leaf tissues, Fe uptake or transport, or protein metabolism. Also noteworthy is the finding that chloronerva (*chln*) can be normalized by exogenous applications of aqueous extracts of alfalfa foliage, the responsible agent probably being a peptide with glycine, serine, and glutamic acid (Böhme and Scholz, 1960; Scholz, 1964).

Interspecific Relationships

The tomato is particularly well suited to investigations of interspecific relationships. Although this review is not intended to cover the genetics of all tomato species, their contribution of useful germ plasm to the cultivated tomato deserves mention; furthermore, the scientific merit of certain studies on the tomato species hybrids *per se* constitutes one of the genetic attributes of the tomato. The genus *Lycopersicon* consists of eight species, and *Lycopersicon esculentum* can be hybridized with all others, albeit at the expense of much effort in certain combinations. Ample collections exist, and all species can be grown and reproduced in culture. All have 12 pairs of chromosomes. Nearly perfect homology is indicated by the conformity in chromosome morphology and pairing behavior in F_1 hybrids. Speciation within the genus has therefore proceded mainly by gene mutation rather than chromosomal alterations.

Special purposes for which the tomato species have served well are: (1) behavior of the synaptinemal complex in an intergeneric hybrid (Menzel and Price, 1966); (2) natural hybridization between species (Rick, 1950, 1958); (3) unilateral compatibility barriers in relation to self-incompatibility (Chmielewski, 1966; Hardon, 1967; Hogenboom, 1972*a,b*); and (4) experimental introgression between species. Thanks to its many attributes, the combination of *L. esculentum* and *S. pennellii* (a species that by all experimental criteria behaves like a *Lycopersicon* yet has the taxonomic key characters of *Solanum*) qualifies well for the introgression studies (Rick, 1960). Backcrosses between these species have furnished material useful for studies on the nature of genetic differentiation and stability of certain characters that distinguish the parental species (Tal, 1967), on cytoplasmically determined male sterility (Andersen, 1963), and on the effect on segregation and recombination of substituted single chromosomes (Rick, 1969, 1971*b*). Finally, attention should be called to the great economic value of certain traits, including resistance to many diseases, that have been bred from wild species into improved cultivars. Some of the literature pertinent to this subject has been summarized recently (Rick, 1973).

Acknowledgments

Partial support for the bibliographical search by USPHS Grant GM 20492 and expert assistance in editing the manuscript by Dora G. Hunt are gratefully acknowledged.

Literature Cited

Andersen, W. R., 1963   Cytoplasmic sterility in hybrids of *Lycopersicon esculentum* and *Solanum pennellii. Rep. Tomato Genet. Coop.* **13**:7–8.

Andersen, W. R., 1965   Cytoplasmic male sterility and intergrafts between *Lycopersicon esculentum* and *Solanum pennellii* Corr. *J. Minn. Acad. Sci.* **32**:93–94.

Bianchi, A., G. Marchesi and G. P. Soressi, 1963   Some results in radiogenetical experiments with tomato varieties. *Radiat. Bot.* **3**:333–343.

Böhme, H. and G. Scholz, 1960   Versuche zur Normalisierung des Phaenotyps der Mutante Chloronerva von *Lysopersicon esculentum* Mill. *Kulturpflanze* **8**:93–109.

Boynton, J. E., 1966a   Chlorophyll-deficient mutants in tomato requiring vitamin B_1. I. Genetics and physiology. *Hereditas* **56:**171–199.

Boynton, J. E., 1966b   Chlorophyll-deficient mutants in tomato requiring vitamin B_1. II. Abnormalities in chloroplast ultrastructure. *Hereditas* **56:**238–254.

Brabec, F., 1960   Über eine Mesochimäre aus *Solanum nigrum* L. und *Lycopersicon pimpinellifolium* Mill. *Planta Berl.* **55:**687–707.

Brock, R. D. and I. R. Franklin, 1966   The effect of dessication, storage and radiation intensity on mutation rate in tomato pollen. *Radiat. Bot.* **6:**171–179.

Brown, J. C. and W. E. Jones, 1971   Differential transport of boron in tomato (*Lycopersicon esculentum* Mill.). *Physiol. Plant.* **25:**279–282.

Brown, J. C. and W. E. Jones, 1972   Effect of germanium on the utilization of boron in tomato (*Lycopersicon esculentum* Mill.). *Plant Physiol.* **49:**651–653.

Brown, J. C., R. L. Chaney and J. E. Ambler, 1971   A new tomato mutant inefficient in the transport of iron. *Physiol. Plant.* **25:**48–53.

Buiatti, M. and R. Ragazzini, 1966   The mutagenic effect of acridine orange in tomato (*Lycopersicon esculentum*). *Mutat. Res.* **3:**360–361.

Caruso, J. L., 1968   Morphogenetic aspects of a leafless mutant in tomato. I. General patterns in development. *Am. J. Bot.* **55:**1169–1176.

Caruso, J. L. and E. G. Cutter, 1970   Morphogenetic aspects of a leafless mutant in tomato. II. Induction of a vascular cambium. *Am. J. Bot.* **57:**420–429.

Chmielewski, T., 1966   An exception to the unidirectional crossibility pattern in the genus *Lycopersicon*. *Genet. Pol.* **7:**31–39.

Clayberg, C. D., L. Butler, C. M. Rick and P. A. Young, 1960   Second list of known genes in the tomato. *J. Hered.* **51:**167–174.

Clayberg, C. D., L. Butler, E. A. Kerr, C. M. Rick and R. W. Robinson, 1966   Third list of known genes in the tomato. *J. Hered.* **57:**189–196.

Clayberg, C. D., L. Butler, E. A. Kerr, C. M. Rick and R. W. Robinson, 1967   Supplementary list of tomato genes as of January, 1967. *Rep. Tomato Genet. Coop.* **17:**2–11.

Clayberg, C. D., L. Butler. E. A. Kerr, C. M. Rick and R. W. Robinson, 1970   Additions to the list of genes, January, 1967 to January, 1970. *Rep. Tomato Genet. Coop.* **20:**6–11.

Clayberg, C. D., L. Butler, E. A. Kerr, C. M. Rick and R. W. Robinson, 1973   Additions to the list of genes. *Rep. Tomato Genet. Coop.* **23:**3–7.

Debergh, P. and C. Nitsch, 1973   Premiers résultats sur la culture *in vitro* de grains de pollen isolés chez la tomate. *C. R. Hebd. Seances Acad. Sci. Ser. D Sci. Nat.* **276:**1281–1284.

Ecochard, R. and G. Merkx, 1972   A primary monosomic for chromosome 5 in the tomato. *Caryologia* **25:**531–536.

Emery, G., 1960   Biological effects of a chemical mutagen, diepoxybutane, on tomato. *Science (Wash., D.C.)* **131:**1732–1733.

Gresshoff, P. M. and C. H. Doy, 1972   Development and differentiation of haploid *Lycopersicon esculentum* (Tomato). *Planta (Berl.)* **107:**161–170.

Gröber, K., 1963   Somatische Spaltung in dem heterozygotem Genotyp der dominanten Chlorophyllmutation Xanthophyllic₃ von *Lycopersicon esculentum* Mill. I. Erste genetische und cytologische Beobachtungen in der X_1-Generation. *Kulturpflanze* **11:**583–602.

Gröber, K., 1967   Somatische Spaltung in dem heterozygoten Genotyp der Chlorophyll-mutation *Xanthophyllic₃* von *Lycopersicon esculentum* Mill. II. Cytogenetische

Analyse der verschiedenen Scheckungstypen in der X_2- und X_3-Generation. *Kulturpflanze* **15**:351–365.

Gröber, K., 1968   Somatische Spaltung in dem heterozygoten Genotyp der Chlorophyllmutation *Xanthopyllic* von *Lycopersicon esculentum* Mill. III. Weitere genetische und cytologische Untersuchungen an einigen Scheckungstypen. *Kulturpflanze* **16**:189–201.

Gröber, K., 1969   Somatische Spaltung in dem heterozygoten Genotyp der Chlorophyllmutation *Xanthophyllic* von *Lycopersicon esculentum* Mill. IV. Differenzierte Aktivität des *Xa-3* Locus in verschiedenen Karyotypen. *Kulturpflanze* **17**:179–185.

Günther, E., 1961   Durch Chimärenbildung verursachte Aufhebung der Selbstinkompatibilität von *Lycopersicon peruvianum* (L.) Mill. *Ber. Dtsch. Bot. Ges.* **74**:333–336.

Günther, E., 1964   Reziproke Bastarde zwischen *Lycopersicon esculentum* Mill. und *Lycopersicon peruvianum* (L.) Mill. *Naturwissenschaften* **51**:443–444.

Haccius, B. and M. Garrecht, 1963   Durch Phenylborsäure induzierte "Lanzettblättrikeit" bei *Solanum lycopersicum*. *Naturwissenschaften* **50**:133–134.

Hagemann, R., 1958   Somatische Konversion bei *Lycopersicon esculentum* Mill. *Z. Vererbungsl.* **89**:587–613.

Hagemann, R., 1961a   Mitteilungen über somatische Konversion. 1. Ausschluss des Vorliegens von somatische Austausch. *Biol. Zentralbl.* **80**:477–478.

Hagemann, R., 1961b   Mitteilungen über somatische Konversion. 2. In welchem Ausmass ist die somatische Konversion gerichtet? *Biol. Zentralbl.* **80**:549–550.

Hagemann, R., 1961c   Mitteilungen über somatische Konversion. 3. Die Konversionshäufigkeit in Bastarden zwischen *sulfurea* Homozygoten und verschiedenen Sippen des Subgenus *Eulycopersicon*. *Biol. Zentralbl.* **80**:717–719.

Hagemann, R., 1966   Somatische Konversion am *sulfurea* Locus von *Lycopersicon esculentum* Mill. II. Weitere Beweise für die somatische Konversion. *Kulturpflanze* **14**:171–200.

Hagemann, R., 1967   Über eine immerspaltende *yv*-Mutantenlinie von *Lycopersicon esculentum* Mill. I. Genetische Untersuchungen. *Biol. Zentralbl.* **86**: (Suppl.) 181–209.

Hagemann, R., 1969a   Somatic conversion (paramutation) at the *sulfurea* locus of *Lycopersicon esculentum* Mill. III. Studies with trisomics. *Can. J. Genet. Cytol.* **11**:346–358.

Hagemann, R., 1969b   Somatische Konversion (Paramutation) am *sulfurea* Locus von *Lycopersicon esculentum* Mill. IV. Die genotypische Bestimmung der Konversionshäufigkeit. *Theor. Appl. Genet.* **39**:295–305.

Hardon, J. J., 1967   Unilateral incompatibility between *Solanum pennellii* and *Lycopersicon esculentum*. *Genetics* **57**:795–808.

Herrmann, F. and R. Hagemann, 1967   Über eine immerspaltende *yv*-Mutantenlinie von *Lycopersicon esculentum* Mill. 2. Cytogenetische Untersuchungen an gescheckten Pflanzen. *Biol. Zentralbl.* **86**: (Suppl.) 163–180.

Hildering, G. J., 1963   The mutagenic effect of ethylene imine (EI) on the tomato. *Euphytica* **12**:113–119.

Hildering, G. J. and J. H. van der Veen, 1966   Mutual independence of M_1- fertility and mutant yield in EMS-treated tomatoes. *Euphytica* **15**:412–424.

Hildering, G. J. and K. Verkerk, 1965   Chimeric structure of the tomato plant after seed treatment with EMS and X-rays. In *The Use of Induced Mutations in Plant Breeding*, pp. 317–320, Food and Agricultural Organization of The United Nations, Rome.

Hogenboom, N. G., 1972*a*   Breaking breeding barriers in Lycopersicon. 4. Breakdown of unilateral incompatibility between *L. peruvianum* (L.) Mill. and *L. esculentum* Mill. *Euphytica* **21**:397–404.

Hogenboom, N. G., 1972*b*   Breaking breeding barriers in Lycopersicon. 5. The inheritance of the unilateral incompatibility between *L. peruvianum* (L.) Mill. and *L. esculentum* Mill. and the genetics of its breakdown. *Euphytica* **21**:405–414.

Jain, H. K., R. N. Raut and Y. G. Khamankar, 1968   Base specific chemicals and mutation analysis in *Lycopersicon* [*esculentum*]. *Heredity* **23**:247–256.

Khush, G. S. and C. M. Rick, 1966   The origin, identification, and cytogenetic behavior of tomato monosomics. *Chromosoma* **18**:407–420.

Khush, G. S. and C. M. Rick, 1967*a*   Tomato tertiary trisomics: origin, identification, morphology and use in determining position of centromeres and arm location of markers. *Can. J. Genet. Cytol.* **9**:610–631.

Khush, G. S. and C. M. Rick, 1967*b*   Novel compensating trisomics of the tomato: cytogenetics, monosomic analysis, and other applications. *Genetics* **56**:297–307.

Khush, G. S. and C. M. Rick, 1967*c*   Haplo-triplo-disomics of the tomato: origin, cytogenetics, and utilization as a source of secondary trisomics. *Biol. Zentralbl.* **86:** (Suppl.) 257–265.

Khush, G. S. and C. M. Rick, 1967*d*   Studies on the linkage map of chromosome 4 of the tomato and on the transmission of induced deficiencies. *Genetica (The Hague)* **38**:74–94.

Khush, G. S. and C. M. Rick, 1968*a*   Tomato telotrisomics: origin, identification, and use in linkage mapping. *Cytologia (Tokyo)* **33**:137–148.

Khush, G. S. and C. M. Rick, 1968*b*   Cytogenetic analysis of the tomato genome by means of induced deficiencies. *Chromosoma* **23**:452–484.

Khush, G. S. and C. M. Rick, 1969   Tomato secondary trisomics: origin, identification, morphology, and use in cytogenetic analysis of the genome. *Heredity* **24**:129–146.

Khush, G. S., C. M. Rick and R. W. Robinson, 1964   Genetic activity in a heterochromatic chromosome segment of the tomato. *Science (Wash., D.C.)* **145**:1432–1434.

Kruse, J., 1968   Merkmalsanalyse und Gruppenbildung bei Mutanten von *Lycopersicon esculentum* Mill. *Kulturpflanze* **5:** (Suppl.) 1–227.

Langridge, J. and R. D. Brock, 1961   A thiamine-requiring mutant of the tomato. *Aust. J. Biol. Sci.* **14**:66–69.

Machold, O., 1966   Untersuchungen an stoffwechseldefekten Mutanten der Kulturtomate. III. Die Wirkung von Ammonium- und Nitratstickstoff auf den Chlorophyllgehalt. *Flora (Jena)* **157A**:536–551.

Majid, R., 1966   Efficacy of seedling grafting for overcoming interspecific incompatibility in *Lycopersicon*. *Curr. Sci. (Bangalore)* **35**:420.

Mathan, D. S., 1965*a*   Morphogenetic effect of phenylboric acid on various leaf-shape mutants in the tomato, duplicating the effect of the *lanceolate* gene. *Z. Vererbungsl.* **97**:157–165.

Mathan, D. S., 1965*b*   Phenylboric acid, a chemical agent simulating the effect of the *lanceolate* gene in the tomato. *Am. J. Bot.* **52**:185–192.

Mathan, D. S., 1967   Reversing the morphogenetic effect of phenylboric acid and of the lanceolate gene with actinomycin D in the tomato. *Genetics* **57**:15–23.

Mathan, D. S. and R. D. Cole, 1964   Comparative biochemical study of two allelic forms of a gene affecting leaf-shape in the tomato. *Am. J. Bot.* **51**:560–566.

Mathan, D. S. and J. A. Jenkins, 1960   Chemically induced phenocopy of a tomato mutant. *Science (Wash., D.C.)* **131**:36–37.

Mathan, D. S. and J. A. Jenkins, 1962   A morphogenetic study of lanceolate, a leaf-shape mutant in the tomato. *Am. J. Bot.* **49:**504–514.

Mathan, D. S. and R. F. Stettler, 1962   Age of seed effect in homozygous lanceolate seedlings. *Rep. Tomato Genet. Coop.* **12:**32–33.

Menzel, M. Y. and J. M. Price, 1966   Fine structure of synapsed chromosomes in F₁ *Lycopersicon esculentum-Solanum lycopersicoides* and its parents. *Am. J. Bot.* **53:**1079–1086.

Moens, P. and L. Butler, 1963   The genetic location of the centromere of chromosome #2 in the tomato (*Lycopersicon esculentum*). *Can. J. Genet. Cytol.* **5:**364–370.

Ramanna, M. S., 1969   The origin of tertiary monosomics in the tomato. *Genetica (The Hague)* **40:**279–288.

Rick, C. M., 1950   Pollination relations of *Lycopersicon esculentum* in native and foreign regions. *Evolution* **4:**110–122.

Rick, C. M., 1958   The role of natural hybridization in the derivation of cultivated tomatoes of western South America. *Econ. Bot.* **12:**346–367.

Rick, C. M., 1959   Non-random gene distribution among tomato chromosomes. *Proc. Natl. Acad. Sci. USA* **45:**1515–1519.

Rick, C. M., 1960   Hybridization between *Lycopersicon esculentum* and *Solanum pennellii:* Phylogenetic and cytogenetic significance. *Proc. Natl. Acad. Sci. USA* **46:**78–82.

Rick, C. M., 1969   Controlled introgression of chromosomes of *Solanum pennellii* into *Lycopersicon esculentum:* segregation and recombination. *Genetics* **62:**753–768.

Rick, C. M., 1971a   Some cytogenetic features of the genome in diploid plant species. *Stadler Genet. Symp.* **1 & 2:**153–174.

Rick, C. M., 1971b   Further studies on segregation and recombination in backcross derivatives of a tomato species hybrid. *Biol. Zentralbl.* **91:**209–220.

Rick, C. M., 1973   Potential genetic resources in tomato species: clues from observations in native habitats, *In* Srb, A. M. *Genes, Enzymes, and Populations.* pp. 255–269, Plenum, New York.

Rick, C. M. and D. W. Barton, 1954   Cytological and genetical identification of the primary trisomics of the tomato. *Genetics* **39:**640–666.

Rick, C. M. and J. E. Boynton, 1967   A temperature-sensitive male-sterile mutant of the tomato. *Am. J. Bot.* **54:**601–611.

Rick, C. M. and L. Butler, 1956   Cytogenetics of the tomato. *Adv. Genet.* **8:**267–382.

Rick, C. M. and G. S. Khush, 1961   X-ray induced deficiencies of chromosome 11 in the tomato. *Genetics* **46:**1389–1393.

Rick, C. M. and G. S. Khush, 1969   Cytogenetic explorations in the tomato genome. *Genet. Lect.* **1:**45–68.

Rick, C. M. and R. W. Zobel, 1972   New linkage testers. *Rep. Tomato Genet. Coop.* **22:**24.

Rick, C. M., W. H. Dempsey and G. S. Khush, 1964   Further studies on the primary trisomics of the tomato. *Can. J. Genet. Cytol.* **6:**93–108.

Scholz, G., 1964   Versuche zur Normalisierung des Phänotyps der Mutante *chloronerva* von *Lycopersicon esculentum* Mill. 3. Über Isolierung und chemische Charakterisierung des "normalisierenden Faktors." *Flora (Jena)* **154:**589–597.

Sharp, W. R., D. K. Dougall and E. F. Paddock, 1971   Haploid plantlets and callus from immature pollen grains of *Nicotiana* and *Lycopersicon*. *Bull. Torrey Bot. Club* **98:**219–222.

Sharp, W. R., R. S. Raskin and H. E. Sommer, 1972   Use of nurse culture in the development of haploid clones in tomato. *Planta (Berl.)* **104:**357–361.

Snoad, B., 1962 Pachytene chromosomes and the linkage maps. *Rep. Tomato Genet. Coop.* **12**:44–45.

Stettler, R. F., 1964 Dosage effects of the *lanceolate* gene in tomato. *Am. J. Bot.* **51**:253–264.

Stevens, M. A., 1970 Inheritance and flavor contribution of 2-isobutylthiazole, methyl salicylate, and eugenol in tomatoes. *J. Am. Soc. Hort. Sci.* **95**:9–13.

Stevens, M. A., 1972a Relationships between components contributing to quality variation among tomato lines. *J. Am. Soc. Hort. Sci.* **97**:70–73.

Stevens, M. A., 1972b Citrate and malate concentrations in tomato fruits: genetic control and maturational effects. *J. Am. Soc. Hort. Sci.* **97**:655–658.

Stubbe, H., 1954 Über die vegetative Hybridisierung von Pflanzen. *Kulturpflanze* **2**:185–236.

Stubbe, H., 1957 Mutanten der Kulturtomate *Lycopersicon esculentum* Miller. I. *Kulturpflanze* **5**:190–220.

Stubbe, H., 1958a Mutanten der Kulturtomate *Lycopersicon esculentum* Miller. II. *Kulturpflanze* **6**:89–115.

Stubbe, H., 1958b Advances and problems of research in mutations in the applied field. *Proc. X Int. Congr. Genet.* **I**:247–260.

Stubbe, H., 1959 Mutanten der Kulturtomate *Lycopersicon esculentum* Miller. III. *Kulturpflanze* **7**:82–112.

Stubbe, H., 1960 Mutanten der Wildtomate *Lycopersicon pimpinellifolium* (Jusl.) Mill. I. *Kulturpflanze* **8**:110–137.

Stubbe, H., 1961 Mutanten der Wildtomate *Lycopersicon pimpinellifolium* (Jusl.) Mill. II. *Kulturpflanze* **9**:58–87.

Stubbe, H., 1963 Mutanten der Kulturtomate *Lycopersicon esculentum* Miller. IV. *Kulturpflanze* **11**:603–644.

Stubbe, H., 1964 Mutanten der Kulturtomate *Lycopersicon esculentum* Miller. V. *Kulturpflanze* **12**:121–152.

Stubbe, H., 1965 Mutanten der Wildtomate *Lycopersicon pimpinellifolium* (Jusl.) Mill. III. *Kulturpflanze* **13**:517–544.

Stubbe, H., 1970 Parallelmutationen in der Gattung Lycopersicon. *Kulturpflanze* **18**:209–220.

Stubbe, H., 1971 Weitere evolutionsgenetische Untersuchengen in der Gattung Lycopersicon. *Biol. Zentralbl.* **90**:545–559.

Stubbe, H., 1972a Mutanten der Kulturtomate *Lycopersicon esculentum* Miller. VI. *Kulturpflanze* **19**:185–230.

Stubbe, H., 1972b Mutanten der Wildtomate *Lycopersicon pimpinellifolium* (Jusl.) Mill. IV. *Kulturpflanze* **19**:231–263.

Tal, M., 1966 Abnormal stomatal behavior in wilty mutants of tomato. *Plant Physiol.* **41**:1387–1391.

Tal, M., 1967 Genetic differentiation and stability of some characters that distinguish *Lycopersicon esculentum* Mill. from *Solanum pennellii* Cor. *Evolution* **21**:316–333.

Tal, M. and D. Imber, 1970 Abnormal stomatal behavior and hormonal imbalance in *flacca,* a wilty mutant of tomato. II. Auxin- and abscisic acid-like activity. *Plant Physiol.* **46**:373–376.

Tal, M. and Y. Nevo, 1973 Abnormal stomatal behavior and root resistance, and hormonal imbalance in three wilty mutants of tomato. *Biochem. Genet.* **8**:291–300.

Tal, M., D. Imber and C. Itai, 1970 Abnormal stomatal behavior and hormonal imbalance in *flacca,* a wilty mutant of tomato. I. Root effect and kinetin-like activity. *Plant Physiol.* **46**:367–372.

Tomes, M. L., 1969   Delta-carotene in the tomato. *Genetics* **62**:769–780.

Verkerk, K., 1959   Neutronic mutations in tomatoes. *Euphytica* **8**:216–222.

Wall, R. J. and C. F. Andrus, 1962   The inheritance and physiology of boron response in the tomato. *Am. J. Bot.* **49**:758–762.

Wettstcin-Knowles, P. von, 1968   Mutations affecting anthocyanin synthesis in the tomato. I. Genetics, histology, and biochemistry. *Hereditas* **60**:317–346.

Yu, S. and A. F. Yeager, 1960   Ten heritable mutations found in the tomato following ir-radiation with X-rays and thermal neutrons. *Proc. Am. Soc. Hortic. Sci.* **76**:538–542.

Zobel, R. W., 1972*a*   Genetics and physiology of two root mutants in tomato, *Lycopersicon esculentum* Mill. Ph.D. Thesis, Department of Vegetable Crops, University of California, Davis, Calif.

Zobel, R. W., 1972*b*   Genetics of the diageotropica mutant in the tomato. *J. Hered.* **63**:94–97.

Zobel, R. W., 1973   Some physiological characteristics of the ethylene-requiring tomato mutant *diageotropica*. *Plant Physiol.* **52**:385–389.

12

Nicotiana

HAROLD H. SMITH

Introduction

The genus *Nicotiana* has been a favored material for studies on inheritance and evolution in higher plants since the days of the pre-Mendelian hybridizers (Olby, 1966). Two features of the genus have had a profound influence on the type of investigations undertaken: (1) The species have evolved into a broad spectrum of different degrees of divergence in cytogenetic and morphological systems and therefore provide rich material for studying general problems on the origins and interrelationships among plant species. (2) The genus contains the cultivated species *N. tabacum* which because of its commercial value has been the object of extensive studies of cytogenetic relationships with progenitor species, introgressive hybridization to incorporate disease resistance, and studies on the inheritance of alkaloids, quantitative traits, and cytoplasmic effects. More recently, the genus has been prominent in providing model systems for studying somatic cell genetics (Smith, 1974*a,b*). Reviews on the cytogenetics of the genus *Nicotiana* have been published by East (1928), Kostoff (1943), Goodspeed (1954), and Smith (1968).

Phylogeny

The genus *Nicotiana* is classified into three subgenera, fourteen sections, and sixty-four species. These are listed in Table 1 with the chromosome number of each species (Smith, 1968). The origins and evolution

HAROLD H. SMITH—Department of Biology, Brookhaven National Laboratory, Upton, New York.

TABLE 1.   Classification of the Genus Nicotiana[a]

Subgenus	Section	Species	Authority	Somatic chromosome number
Rustica	Paniculatae	*glauca*	Graham	24
		paniculata	Linnaeus	24
		knightiana	Goodspeed	24
		solanifolia	Walpers	24
		benavidesii	Goodspeed	24
		cordifolia	Philippi	24
		raimondii	Macbride	24
	Thyrsiflorae	*thyrsiflora*	Bitter ex Goodspeed	24
	Rusticae	*rustica*	Linnaeus	48
Tabacum	Tomentosae	*tomentosa*	Ruiz and Pavon	24
		tomentosiformis	Goodspeed	24
		otophora	Grisebach	24
		setchellii	Goodspeed	24
		glutinosa	Linnaeus	24
	Genuinae	*tabacum*	Linnaeus	48
Petunioides	Undulatae	*undulata*	Ruiz and Pavon	24
		arentsii	Goodspeed	48
		wigandioides	Koch and Fintelman	24
	Trigonophyllae	*trigonophylla*	Donal	24
	Alatae	*sylvestris*	Spegazzini and Comes	24
		langsdorffii	Weinmann	18
		alata	Link and Otto	18
		forgetiana	Hort. ex Hemsley	18
		bonariensis	Lehmann	18
		longiflora	Cavanilles	20
		plumbaginifolia	Viviani	20
	Repandae	*repanda*	Willdenow ex Lehmann	48
		stocktonii	Brandegee	48
		nesophila	Johnston	48

TABLE 1.   *Continued*

Subgenus	Section	Species	Authority	Somatic chromosome number
	Noctiflorae	*noctiflora*	Hooker	24
		petunioides	(Grisebach) Millán	24
		acaulis	Spegazzini	24
		ameghinoi	Spegazzini	?
	Acuminatae	*acuminata*	(Graham) Hooker	24
		pauciflora	Remy	24
		attenuata	Torrey ex Watson	24
		longibracteata	Philippi	?
		miersii	Remy	24
		corymbosa	Remy	24
		linearis	Philippi	24
		spegazzini	Millán	24
	Bigelovianae	*bigelovii*	(Torrey) Watson	48
		clevelandii	Gray	48
	Nudicaules	*nudicaulis*	Watson	48
	Suaveolentes	*benthamiana*	Domin	38
		umbratica	Burbidge	46
		cavicola	Burbidge	46
		debneyi	Domin	48
		gossei	Domin	36
		amplexicaulis	Burbidge	36
		maritima	Wheeler	32
		velutina	Wheeler	32
		hesperis	Burbidge	42
		occidentalis	Wheeler	42
		simulans	Burbidge	40
		megalosiphon	Heurck and Mueller	40
		rotundifolia	Lindley	44
		excelsior	J. M. Black	38
		suaveolens	Lehmann	32
		ingulba	J. M. Black	40
		exigua	Wheeler	32
		goodspeedii	Wheeler	40
		rosulata	(S. Moore) Domin	40
		fragrans	Hooker	48

[a] From Smith (1968); after Goodspeed (1954).

of species of the genus have been presented in detail by Goodspeed (1954) and Goodspeed and Thompson (1959). These relationships have been summarized diagrammatically in the form of three phylogenetic arcs. In the first two arcs the genus is envisaged as derived from a pregeneric reservoir of related forms with six pairs of chromosomes which evolved into three complexes, at the 12-paired level, that are hypothetical precursors of the three modern subgenera. The third arc contains the present-day species, with their various degrees of genetic interconnection, at the 12- and 24-paired chromosome level. The evidence shows that interspecific hybridization with subsequent amphiploidy, as well as genetic recombination, has played an important role in the evolution of the genus *Nicotiana*.

Electrophoretic separation of proteins and subsequent staining for different enzyme systems has recently been used to give biochemical markers for phylogenetic studies among species of *Nicotiana* (Sheen, 1970; Smith *et al.*, 1970; Reddy and Garber, 1971). At least 61 species have been so analyzed; leaf, root, or seed extracts have been used and stained most commonly for peroxidases or esterases. With few exceptions, each species has had a unique isozyme band pattern. On the assumption that matching band mobility indicated genetic equivalency, the isozyme band patterns of species *within* taxonomic sections were compared with those *between* sections. The results showed significantly more frequent matching of bands among species *within* a section than among those of different sections, and thus were in broad general agreement with the relationships established by the more conventional methods of systematics.

Interspecific Hybrids and Allopolyploids

More than 300 interspecific hybrids have been reported in the genus *Nicotiana* (East, 1928; Kostoff, 1943; Goodspeed, 1954). Those that develop normally show, with few exceptions and regardless of the parental chromosome numbers, intermediate characteristics of leaf and flower (East, 1935; Kehr and Smith, 1952). Of these, 215 were analyzed cytologically (Goodspeed, 1954) for meiotic behavior in the F_1. The pairing of chromosomes in interspecific hybrids is of interest in studies on the evolution of the genus as well as in problems of transferring genes or chromosome segments from one species to another. In the former, the degree of pairing of chromosomes can be considered a measure of chromosome homology, and hence, as evidence of the degree of species relationships and their ancestral origin. In the latter, the success of transferring desirable hereditary traits from wild to cultivated species depends largely on the extent of pairing of their chromosomes.

Allopolyploidy, and particularly amphiploidy, has played an important role in the evolution of the genus *Nicotiana*. Eleven species with 24 pairs of chromosomes, and hence of probable amphiploid origin, are known today. A large number of amphiploids have been produced in experiments with *Nicotiana*, either spontaneously following interspecific hybridization or by artificial means, notably with colchicine. The frequency of multi- vs. bivalent chromosome associations in amphiploids varies with the degree of homology between parental chromosomes, and is reflected in the fertility and variability among progeny.

The main tobacco species of commerce, *N. tabacum*, is of amphiploid origin ($n = 24$), and much interest and research has centered on the derivation of the present-day tobacco from its putative wild progenitors. The original evidence of Goodspeed and Clausen (1928) was interpreted to indicate that *N. tabacum* arose from chromosome doubling following hybridization between a progenitor of *N. sylvestris* (S′ genome, $n = 12$) and a member of the Tomentosae section, either *N. otophora, N. tomentosiformis* or, more likely, an ancestral type similar to, but not identical with, either of these present-day species (T′ genome, $n = 12$). In order to test genetically whether the chromosomes of *N. tomentosiformis* or those of *N. otophora* are more nearly homologous with the T genome of *N. tabacum*, Gerstel (1960) performed experiments, which, based on segregation frequencies, indicated that *N. tomentosiformis* is the more closely related.

Isozyme band patterns of synthesized amphiploids were, for the most part, a summation of the mobility sites found in each of the diploid parental species (Sheen, 1970; Smith *et al.*, 1970; Reddy and Garber, 1971). The matching of band positions could, therefore, be used as a means of assessing the probable diploid progenitor species among putative parents of an established species of presumed amphiploid origin. Using polyacrylamide gel electrophoresis of leaf extracts and staining for eight enzyme systems, Sheen (1972) found that the mobilities of *N. sylvestris* × *N. tomentosiformis* were more similar than *N. sylvestris* × *N. otophora* to *N. tabacum*. This result supports the hypothesis that ancestors of *N. sylvestris* and *N. tomentosiformis* were the likely progenitors of *N. tabacum*.

Aneuploidy and Multiple Genomes

The monosomics of *N. tabacum* are aneuploid types of particular interest since they provide material for a rapid method of locating genes on specific chromosomes. These monosomics have arisen spontaneously

(Clausen and Goodspeed, 1926a), as derivatives from hybridization of N. tabacum and N. sylvestris (Clausen and Goodspeed, 1926b), and by use of a genetically controlled asynaptic condition (Clausen and Cameron, 1944). The 24 monosomic lines, which were assembled largely through the efforts of the late Professor R. E. Clausen, have been characterized on the basis of their most readily identifiable features by Cameron (1959). They are listed in Table 2 and described according to their appearance on a common genetic background of so-called Red Russian tobacco. The listing of genes associated with the monosomic chromosomes was kindly supplied by Dr. D. R. Cameron. All primary trisomic types have been identified in only one species of Nicotiana, N. sylvestris, $n = 12$ (Goodspeed and Avery, 1939, 1941). Eight of the nine possible trisomics have been identified in N. langsdorffii (Abraham, 1947; Lee, 1950; Smith, 1943a). In both of these species the trisomic types are readily distinguishable from the diploid types in morphological features of plant, leaf, and flower.

Aneuploidy, as well as amphiploidy, has played a part in the evolution of the genus Nicotiana, as shown by the occurrence of 9- and 10-paired species in the Alatae section and 16- and 23-paired species in the Suaveolentes section (Table 1). The former are considered to have resulted from chromosomal loss at the 12-paired level, and the latter from loss at the 24-paired level.

Trispecies combinations in Nicotiana have been made by crossing an amphiploid with a third species. Over 40 different hybrids in which three species were combined have been reported in the literature (Kehr and Smith, 1952; Krishnamurty et al., 1960; Smith and Abashian, 1963). Hybrids in which the complete genomes of four different species are combined have also been reported (Kehr and Smith, 1952; Appa Rao and Krishnamurty, 1963; Smith and Abashian, 1963).

In an effort to explore the limits and consequences of multiple allopolyploidy, a hybrid was produced (Kehr and Smith, 1952) which combined the genomes of three distantly related amphiploid species: N. bigelovii ($n = 24$, North America), N. debneyi ($n = 24$, Australia), and N. tabacum ($n = 24$, South America). In an individual which contained the doubled (by colchicine) complement of 144 chromosomes only bivalents were formed, but laggards were observed in metaphase and anaphase stages of meiosis. Inbreeding and selection were practiced for 10 generations, which established three morphologically distinct races, each involving a loss of different chromosomes from the original 144 to give 108 $\pm$ 6 (Smith et al., 1958). The phenomenon observed in this multiple allopolyploid would seem to offer opportunities for exploitation in the natural evolution of some plant groups in that there is wide variability in

TABLE 2. The Monosomic Types of Nicotiana tabacum[a]

Designation	Distinguishing characteristics					Associated genes
	Plant height	Leaves	Flowers,[b] length in mm and characteristics	Pollen	Monosome	
Haplo-A	Somewhat below normal	Smaller; basal constriction more pronounced	48.6–43.1; somewhat paler in color, fading earlier	Essentially normal	Medium small	hf_1, hairy filaments; pa, asynaptic; au', aurea
Haplo-B	Subnormal; sparsely branched	Smaller; narrow; basal constriction less abrupt; auricles strongly reduced	53.6–41.0; more strongly bent; color darker	Essentially normal	Very small	Ml, many leaves; Pp, purple plant; yb_1, yellow burley; Pb, purple buds; N. otophora[c]
Haplo-C	Often taller than normal; longer internodes	Narrow; basal constriction less abrupt	58.6–43.7; longer and broader; color paler in tube and throat	Marked abortion	Medium small	cd, crinkled dwarf; lf, light filaments; wh, white flower; bf, bent flower; Wh-P, pale; wc, white center; N. otophora[c]
Haplo-D	Normal but maturity delayed	Brighter green in young plants; leaf base semibroad	50.6–39.7; slightly reduced in size	Essentially normal	Incorporated in a trivalent in about 50 percent PMC	fs_1, fasciated

TABLE 2. *Continued*

Designation	Plant height	Leaves	Flowers,[b] length in mm and characteristics	Pollen	Monosome	Associated gene
			Distinguishing characteristics			
Haplo-E	Subnormal	Smaller; constriction less abrupt	51.3–41.1; calyx inflated	Essentially normal	Very small	—
Haplo-F	Subnormal; shorter internodes	Small; more erect	44.5–37.5, distinctly shorter limb, fluted	Moderate abortion	Large with characteristic median construction	co, coral flower; mm_1, mammoth; sn_1, spontaneous necrosis
Haplo-G	Subnormal; meager inflorescence; maturity delayed	Small with rounded tips; basal constriction pronounced	55.6–42.3, tapering gradually to limb; style short; capsules small and poorly filled	Variable as to cytoplasmic content but few grains completely aborted	Large	tg, tinged; vb, veinbanding; ws_2, white seedling; vp', variegated plant
Haplo-H	Normal but stems and branches slender; reduced branching	Small; narrow; basal constriction less pronounced	51.8–38.3; narrow tube; limb reduced; calyx lobes pointed	High abortion but variable as to contents	Medium large	Nc, necrotic, *N. glutinosa*,[c] td, toadskin
Haplo-I	Normal; slender branches; delayed maturity	Small; more sharply pointed	53.1–41.2, corolla lobes pointed; capsules long, narrow, poorly filled; calyx inflated	Low abortion but dimorphic	Very small	cc_1, catacorolla,[a] rd, red modifier

Haplo-J	Subnormal; maturity delayed; leaves small, narrow	Small and narrow	52.3–42.6, limb characteristically wavy at maturity; color less intense; capsules small, poorly filled	High abortion, sharp distinction between stainable and aborted grains	Medium	*cy,* calycine[a] *lc,* lacerate[a], *vi A'*, virescent
Haplo-K	Subnormal; maturity delayed	Semibroad at base	48.3–36.7; tube short; infundibulum proportionately longer; anthers small with delayed dehiscence	Low abortion; dimorphic	Very large, medianly constricted (cf. Clausen and Cameron, 1944)	—
Haplo-L	Above normal; stem heavy; maturity somewhat delayed	—	48.6–37.1; tube shorter and broader; color distinctly paler	High abortion; variable in size	Large with prominent constriction	*at,* Ambalema tall; *gb,* green buds; *Tr,* tube retarder, *N. setchellii*[c]
Haplo-M	Subnormal; branching at the base	Large; basal constriction less pronounced	53.9–40.7; color fades to a purplish hue at maturity; calyx conspicuously longer	High abortion; variable in content	Medium large; characteristically ovoid	*Ap,* apetalous; *Rf,* ruffled; *pvb,* progressive vein banding

TABLE 2. Continued

Designation	Distinguishing characteristics					Associated genes
	Plant height	Leaves	Flowers,[b] length in mm and characteristics	Pollen	Monosome	
Haplo-N	Distinctly subnormal; short internodes; compact inflorescence	Small; erect	43.9–34.0; visibly smaller; color darker red	Low abortion; dimorphic	Large with median constriction (cf. Haplo-F)	mm_2, mammoth; sn_2, spontaneous necrosis
Haplo-O	Close to normal	Slightly smaller; basal constriction more pronounced	49.2–39.5; size reduced; paler in color; stamens and pistils slightly exserted; pollen shedding delayed; capsule small and poorly filled	Low abortion	Medium large	hf_2, hairy filaments; yb_2, yellow burley; au'', aurea
Haplo-P	Normal; maturity delayed	Small; tips rounded; semibroad at base	49.3–37.5; limb narrow; corolla lobes less pronounced; capsules small and poorly filled	Marked abortion; subnormal grains variable in size	Medium large with characteristic subterminal constriction	Br, broad; Fs_2, fasciated; pk, pink flower; sg, stigmatoid; $vi\,A''$, virescent; yc, yellow crittenden
Haplo-Q	Reduced; little branching; maturity delayed	Narrow; basal constriction pronounced; auricles strongly reduced; ruffled	54.8–38.8; tube longer; limb spread reduced; capsules pointed, small, and poorly filled	Very high abortion; sharply divided into two classes	Medium	—

Haplo-R	Subnormal; thick stems; profusely branched	Small; darker green; auricles reduced	49.2–41.9; enlarged infundibulum, wide throat; color paler	High abortion, but completely empty grains rare	Very large	mt_2, mosaic tolerant; *Pd*, petioloid; *Su*, sulfur
Haplo-S	Normal; maturity usually retarded	Lighter green; surface smooth	47.3–41.7; color more vivid; stamens and pistils exserted; pollen shedding delayed	Low abortion; grains variable in size	Large, frequently associated with a bivalent	*cl*, chimeral; *yg*, yellowish green
Haplo-T	Subnormal, maturity delayed	Small; darker green; basal constriction elongated	56.0–41.7; tube longer, merging gradually into the infundibulum; stamens and pistils relatively short; capsules small, poorly filled	High abortion	Large, usually with a well-defined constriction	ws_2, white seedling; vp'', variegated plant
Haplo-U	Subnormal; bushy	Large; frequently with a pronounced petiole	48.5–40.9; corolla lobes acutely pointed; tube pale; limb and throat strongly colored	High abortion; aborted grains variable in size	Medium large	—

TABLE 2. *Continued*

| | | Distinguishing characteristics | | | |
| | | | | | *Associated* |
Designation	*Plant height*	*Leaves*	*Flowers,*[b] *length in mm and characteristics*	*Pollen*	*Monosome*	*genes*
Haplo-V	Subnormal	Small; basal constriction less abrupt; auricles reduced	47.0–39.8; tube stout	High abortion, visibly so in freshly opened flowers	Medium large	—
Haplo-W	Subnormal; elongated internodes; sparsely branched; maturity delayed	Long; narrow; sharply pointed; auricles reduced	52.8–40.8; color lighter; pollen scanty, sometimes lacking in early flowers	High abortion; aborted grains small	Large, but PMC frequently unobtainable during early flowering	—
Haplo-Z	Normal; maturity conspicuously delayed	Small; basal constriction less pronounced; auricles less ruffled	52.9–40.3; style tends to be curved; limb frequently fails to open fully	Abortion very high; sharp distinction between normal and aborted grains	Large	—

[a] After Cameron (1959).
[b] The flower measurements are averages of 10 representative flowers and show tube length–limb spread, both measured in millimeters. These are to be compared with normal values of about 53–43.
[c] Transfers to *N. tabacum* from other species.
[d] Not clear-cut distinct characters.

early generations without the serious loss in fertility usually associated with species hybridization at the diploid level.

Interspecific Recombination and Instability

Interspecific recombination has resulted from the transfer of genes between species, with subsequent establishment of new types having a parental chromosome number and containing genes introgressed from an alien species. The urgency to develop disease-resistant commercial tobacco by utilizing resistance found mainly or exclusively in wild species of the genus has been an impetus to studies on introgression. The difficulty of effecting recombination varies directly with the degree of genetic and cytological divergence of the parental species.

Surmounting barriers to crossability and fertility are initial steps to gene transfer. Sterility barriers may be chromosomal, genic (Ar-Rushdi, 1956; Cameron and Moav, 1957), cytoplasmic (Clayton, 1950; Burk, 1960; Smith 1962b; Cameron, 1965; Hart, 1965), or due to a combination of causes. The species that have provided sources of disease resistance are sufficiently remote from *N. tabacum* so that the parental genomes show partial or complete failure of chromosome pairing and thus present serious difficulties for controlled introgression (Chaplin and Mann, 1961). Such hybrids are highly sterile as a result of the various unbalanced chromosomal products of meiosis. Doubling of chromosome complements, before or after hybridization, can be achieved with colchicine, or may occur spontaneously, and a fertile amphiploid or sesquidiploid is produced. This is then used as the starting material for a series of backcrosses to *N. tabacum*, accompanied by selection for disease resistance. Ultimate success of such a program of controlled introgression requires that a chromosomal exchange eventually takes place, and that the gene block governing the inheritance of the desired characteristic of the nonrecurrent parent becomes incorporated in an *N. tabacum* chromosome (Mann *et al.*, 1963). The first successful use of this method was in the transfer of a gene that confers resistance to tobacco mosaic virus from *N. glutinosa* to *N. tabacum* (Holmes, 1938). Backcrosses to *N. tabacum* followed by continued self-fertilization of resistant types led to formation of a type of tobacco that bred true for the local-lesion type of mosaic resistance. It was later shown (Gerstel, 1943, 1945) that an entire *N. glutinosa* chromosome had been substituted for the H chromosome of *N. tabacum* to produce an "alien-substitution" race. With the continued backcrossing and self-fertilization of such strains, the resistance factor was eventually transferred from the *N. glutinosa* chromosome to a chromosome that now contained a sector from *N. tabacum* of sufficient length to permit conjugation in a majority of cells with the homolog; this chromosome was shown to be the *N. tabacum* H chromosome (Gerstel, 1948). The technique that was so successful in transferring the *N. glutinosa* mosaic resistance to commercial tobacco has now been utilized for controlling other diseases whose source of resistance is also found in species of *Nicotiana* remotely related to *N. tabacum* (Burk and Heggestad, 1966). Most successes in experimental introgression of disease resistance have been with the transfer of a single dominant gene or individual chromosome segment accompanied by restricted intrachromosomal recombination.

Hybrid derivatives of a cross between *N. langsdorffii* and *N. sanderae* have been used to study the effect of selection, following interspecific recombination, on morphological divergence and genetic isolation as a problem in evolution under controlled experimental conditions (Smith and Daly, 1959). Analyses of lines selected on the basis of small, intermediate, and large* corolla size showed that by the F_6 generation they were not only morphologically distinct, but as uniform as the parent species in flower size. Investigation of reproductive isolation based on crossability, meiotic aberration frequency, and pollen abortion indicated that each selected F_6 line was partially isolated from its parents on the basis of one or more of these criteria. Populations satisfying two of the formal criteria of speciation, that is, morphological divergence and genetic isolation, were developed, therefore, from one hybridization followed by single plant selection and self-pollination, within six generations. Novel or transgressive morphologies may result from interspecific hybridization and recombination, thus emphasizing its potential innovative role in evolution (Smith, 1974c).

Various manifestations of instability occur in hybrids among species of *Nicotiana*. These include variations in pigment of flower or leaf, variations in morphology and habit during plant development, and variability in growth among hybrid plants (McCray, 1932; East, 1935; Kostoff, 1935, 1943; Kehr and Smith, 1952). Plants with variegated anthocyanin pigmentation in the flower have appeared sporadically in the F_2 and subsequent generations of the cross *N. langsdorffii* × *N. sanderae* (Smith and Sand, 1957). One of these variegated types has been analyzed in detail and is governed by a mutable locus, *v*. Two alleles at this locus are necessary and sufficient to account genetically for the different modal breeding behavior of the three major variegated phenotypes: speckled (v_s/v_s), sectorial (v_S/v_s), and rare sectorial (v_S/v_S). Both the alleles are unstable and somatic mutations occur in both directions, so that a chromosomal loss is apparently not involved. Differences in frequency and developmental timing of these reversible mutations have both heritable and environmental components (Sand, 1957). One striking differential effect of the stable vs. unstable genes is evident under exposure to low levels of gamma irradiation. The slope of the somatic response curve for the unstable gene, v, is about 10 times greater than for a stable gene, R, at dose levels below 12 r/day (Sand *et al.*, 1960). Radiation operating to extinguish a buffering system against final mutation can account for the dose and dose-rate effects observed (Sand and Smith, 1973).

All hybrid combinations of *N. tabacum* with *N. plumbaginifolia* that were studied by Ar-Rushdi (1957), Moav and Cameron (1960), and Moav (1961) showed somatic variegation of the dominant characters carried on the *plumbaginifolia* genome. Mitotic bridge-like structures in the hybrids have been observed; however, the exact cytological nature of the chromosomal elimination that causes variegation is still to be found. Genetic instability has also appeared repeatedly in hybrids and hybrid derivatives between *N. tabacum* and diploid species which as present-day representatives of putative ancestral forms, are closely related to the cultivated amphiploid. In hexaploids synthesized from *N. tabacum* × *N. otophora*, abnormal segregation ratios, variegation, and chromosomal aberrations were encountered (Gerstel, 1960; Gerstel and Burns, 1966*b*). Chromosomes of extraordinary size, which were called "megachromosomes" (Gerstel and Burns, 1966*a*), were found in scattered cells of derivatives from *N. tabacum* × *N. otophora* which showed variegation. The patterns of distribution of heterochromatin in *N. tabacum* and *N. otophora* are greatly different and the total amount of heterochromatin appears to be much larger in *N. otophora* (Burns, 1966). It is the heterchromatic blocks from *N. otophora* and also from *N. tomentosiformis* (Burns and Gerstel, 1973), that undergo spontaneous breakage and great enlargement when transferred into *N. tabacum* (Gerstel and Burns, 1970). When intact genomes of *N. otophora* or *N. tomentosiformis* are introduced into plants capable of these aberrations, both chromosome breakage and formation of megachromosomes are greatly inhibited (Burns and Gerstel, 1971).

In the unstable hybrid between *N. glutinosa* and *N. suaveolens*, cytological and morphological evidence was found for preferential somatic elimination of *N. glutinosa* chromosomes (Gupta and Gupta, 1973).

Inheritance of Alkaloids

Twelve or more different alkaloids have been identified in the cultivated species *N. tabacum*, mainly by Späth and his students, using classical extraction and isolation procedures (for summaries see Henry, 1949; Marion, 1950, 1960). Little is known about the inheritance of these alkaloids. Alkaloid composition of at least 52 species of *Nicotiana* have been determined (Marion, 1950; Jeffrey, 1959; Smith and Abashian, 1963), more recently by paper partition chromotography. Most of the species contain predominantly one of three identified alkaloids: nicotine, nornicotine, and anabasine. In addition at least six other alkaloids, which are separable by chromatography but have not been identified chemically, are found in the genus and are present in varied but characteristic patterns in each species. Numerous studies on the loci of alkaloid formation (Dawson and Solt, 1959; Mothes, 1955; Tso, 1972) have indicated that nicotine is formed in the tobacco root and translocated to the top through xylem; nornicotine is a demethylation product of nicotine in the shoot; and anabasine can be formed in both root and shoot. No clear-cut relationships were observed between types of alkaloid and phylogenetic position (Imai, 1959), geographical distribution, habitat, or habit of growth. However, since *all* species of the genus contain one or more alkaloids, their presence may have had an adaptive significance early in the evolution of the genus and the biochemistry of their formation been fixed with a significant role in metabolism at the cellular level in present day *Nicotiana* plants.

At least 35 two-species combinations, 14 three-species combinations, and 2 four-species combinations have been analyzed for alkaloid composition (Smith and Abashian, 1963). When a predominently "anabasine species" is crossed with an anabasine, nicotine, or nornicotine species, the main alkaloid in the combination is most frequently anabasine (Smith and Smith, 1942). The biosynthesis of this alkaloid is an essentially dominant genetic characteristic. The main alkaloid produced in crosses between predominantly nicotine and nornicotine species is most frequently nornicotine. The genetic factors controlling nornicotine formation are usually partly dominant over those producing nicotine, but the relationship is clearly not simple (Mann and Weybrew, 1958; Burk and Jeffrey, 1958; Smith, 1965a).

Most commercial varieties of *N. tabacum* produce and retain nicotine as their primary alkaloid. However, related species, including the modern descendents of the probable progentiors of *N. tabacum* and certain strains of tobacco, have both the capacity to produce nicotine as well as to convert nicotine to nornicotine (Mann and Weybrew, 1958; Gerstel and Mann, 1964; Mann *et al.*, 1972). There are two dominant loci for nicotine conversion, only one of which, C_1 in the Tomentosae genome, accounts for alkaloid conversion in the nornicotine-containing types of *N. tabacum* (Mann *et al.*, 1964). Most tobacco types are c_1c_1, and produce mainly nicotine; but this recessive locus apparently may mutate to the nicotine-converting allele at a rather high frequency (Wernsman and Matzinger, 1970). The inheritance of the total alkaloid content in certain strains of burley tobacco has been reported to be governed by two independent pairs of genes (Legg *et al.,* 1969; Legg and Collins, 1971; Mann *et al.,* 1972).

Biometrical Studies

Biometrical methods have been used in genetic studies on *Nicotiana* in order to analyze the inheritance of continuous variation in quantitative characters in terms of the nature of the polygenic systems responsible. An analysis of means and variance components applied to data from a series of generations following prescribed breeding programs has made possible interpretations about kinds of gene action, i.e., additive, dominance, and various nonallelic interactions. The total phenotypic variance for quantitative characters can be partitioned by appropriate methods into genetic and environmental components, and the genetic component further partitioned into the proportionate contribution due to additive, dominance, and various epistatic gene effects. Information gained from such genetical analyses can be utilized in designing breeding methods to give maximum expectations for achieving desired practical goals in plant improvement.

Early studies on plant height and other quantitative characters in *N. rustica* and *N. tabacum* (Mather, 1949; Smith, 1952; Mather and Vines, 1952; Robinson *et al.,* 1954) showed that, in general, additive gene effects contributed more to the total variance than dominance effects (allelic interaction). Consequently, heritability was high and the expected efficacy of selection great. Estimates of additive × additive and other epistatic variances were obtained by Matzinger *et al.* (1960) and Hayman (1960). A diallel analysis of F_1 and F_2 generations from crosses among eight flue-cured varieties of *N. tabacum* (Matzinger *et al.,* 1962) again showed the predominance of additive genetic variance and only small amounts of heterosis and inbreeding depression. Most workers have reported, for most characters in crosses among varieties of *N. tabacum,* a preponderance of additive gene effects and high heritability (Oka, 1959; Wittmer and Scossiroli, 1961; Matzinger *et al.,* 1972; Luthra, 1964); however, large genotype × year interactions may exist (Matzinger *et al.,* 1971).

The dimensions of flower parts were used in early experiments on the inheritance of quantitative characters in crosses between species of *Nicotiana* (East, 1913); these studies have been continued, utilizing methods of biometrical analysis to assess gene effects (Smith and Robson, 1959). Measurements of corolla tube length were made on self and backcross generations obtained from crossing a transgressive hybrid derivative, then in the F_{10} generation, to each of its parents, *N. langsdorffii* and *N. sanderae*. In order to gain information on the nature of gene action controlling the transgressive phenotype, an analysis of the means and variance components was carried out by Daly and Robson (1969). The additive genic effect exceeded the dominance effect for each flower dimension measured. Variance-component estimates were consistent with results for gene effects in indicating the presence of nonallelic gene interactions. The major portion of the increase in size of the selected line over its large parent could be attributed to additive gene action, and the remaining smaller portion to additive × additive interaction.

Cytoplasmic Inheritance

Evidence for extrachromosomal or cytoplasmic inheritance has been found for two characters in *Nicotiana*, male sterility and chlorophyll variegation.

East (1932) was the first to report experimental results in *Nicotiana* showing that a factor in the cytoplasm, combined with specific nuclear genes, governed the expression of male sterility. In hybrid generations segregating from the cross *N. langsdorffii* × *N. sanderae*, male-sterile plants appeared only in progeny in which the cytoplasm came from *N. langsdorffii*, indicating that this species has "male-sterile cytoplasm," designated (ms) (Smith, 1962*b*, 1971). *Nicotiana sanderae*, on the other hand, contains heritable cytoplasmic determinant(s) for male fertility (MF). Each species also has a single, independent, dominant, nuclear gene, designated Rf_1 from *N. langsdorffii* and Rf_2 from *N. sanderae*, for restoring male fertility in the presence of (ms) cytoplasm. Genetic information for producing fertile pollen is therefore carried at three sites, and only (ms) $rf_1rf_1rf_2rf_2$ individuals, which lack information in all sites, are male sterile. A number of other examples of cytoplasmic and genic–cytoplasmic control of male sterility have appeared in the *Nicotiana* literature (Clayton, 1950, 1958; Clayton *et al.*, 1967). In fact, it now seems to be a widespread phenomenon in the genus, i.e., the cytoplasm of one species (A) combined with the partial or complete genome of another species (B) will often produce male sterility, and furthermore, genetic restorers to pollen fertility will be found in certain chromosomes of species A. In repeated backcrosses of *N. debneyi* × *N. tabacum*, as female, to *N. tabacum*, both male sterility and morphological aberrations of corolla and stamens appears. Certain alien *debneyi* chromosomes that remain in a restored tobacco genome modify the morphology to more normal types (Sand and Christoff, 1973).

Cytoplasmic inheritance of chlorophyll variegation in tobacco has been reported by Dermen (1960), Wolf (1959), Burk and Grosso (1963), Burk et al. (1964), Burk (1965), and Edwardson (1965). Defective plastid inheritance is transmitted only through the female, and the progeny of variegated plants have varied from all seedlings either white or green (Dermen, 1960; Burk and Grosso, 1963) to segregation into white, variegated, and green seedlings (Wolf, 1959). An explanation for somatic patterns observed and for differences in heritabilities of plastid-controlled chlorophyll variegation in tobacco has been advanced by Burk et al. (1964). Patterns of variegation in which the sporogenous tissue was derived from the second histogenic layer (L-II), containing only one type of plastid, produced only green or only white offspring. When L-II is a mosaic of both normal and deficient mutant plastids, the inheritance pattern depends on the inclusion or loss of the mutant type in mature cells of sporogenous tissue and successive cell generations in sexually produced seedlings. The genetic determiner of the defective plastid type studied by Burk et al. (1964) is located in the plastid itself. It is dominant in the sense that when present in sufficient numbers it suppresses chlorophyll formation by genetically normal plastids in the same cell.

Genetic Tumors

Genetic tumors are neoplastic growths that arise, without any apparent external cause, in organisms of a certain genotype. The occurrence of these spontaneous tumors in interspecific hybrids of Nicotiana was first reported by Kostoff (1930). Over 30 hybrid combinations are now known that produce tumors regularly (Kehr and Smith, 1954; Näf, 1958; Smith and Stevenson, 1961; Takenaka and Yoneda, 1962; Kehr, 1965); and, in addition, at least 22 others develop similar, but restricted or irregular, growth abnormalities (Kehr and Smith, 1954).

Most sections of the genus are represented among the species that contribute to tumor-forming hybrids (Ahuja, 1965), and each tumor-prone cross is an intersectional species combination, i.e., it combines genetically distant genomes. Näf (1958) proposed that species involved in tumorous combinations may be divided into two groups: one consists largely of species of the section Alatae, whereas the other includes a variety of different species. The contribution of these groups is envisaged as differing in some biochemical or physiological way under genetic control (Ahuja, 1968) so that products conducive to abnormal growth are formed in the intergroup hybrids (Hagen, 1969).

Physiological differences between the Nicotiana species and the tumorous hybrids can be demonstrated by their distinctive requirements for growth in tissue culture (Skoog, 1944; Schaeffer and Smith, 1963). The tumor-prone types usually do not need auxins and cytokinins in the media to sustain tissue growth; whereas these substances are required for culture of the species. Furthermore, nontumorous mutants, or segregants of the tumorous hybrids *N. langsdorffii* × *N. glauca* (Schaeffer *et al.*, 1963) and *N. debneyi-tabacum* × *N. langsdorffii* (Ahuja and Hagen, 1966) have phytohormone requirements like those of parental species tissues. The tumorous hybrid plants have been shown to have higher levels of endogenous auxin than their parental species or nontumorous sibs (Bayer, 1965, 1967; Bayer and Ahuja, 1968). In cultures of tumorous segregants from tetraploid *N. rustica* × *N. glauca,* which do not require the addition of either kinetin or auxin, separation into subcultures with different organ-forming capability (shoots vs. unorganized callus) was obtained (Kovacs, 1967).

These results are consistent with the conclusion that a significant property conferred by the genotype of the genetically tumor-prone hybrids is an inherent tendency to synthesize and/or accumulate greater-than-regulatory amounts of growth-promoting substances, as auxins and cytokinins (Smith, 1965*b*). Cheng (1972) has recently demonstrated that indoleacetic acid (IAA) synthetases are inducible in pith explants of a tumor-prone hybrid by IAA and also by the nonhormonal compounds indole and tryptophane. This biochemical characteristic is apparently governed by a genetic condition not present in parent species. Certain other differences in chemical composition of tumorous compared to nontumorous plants have been reported, notably: in high levels of free amino acids (Vester and Anders, 1960), in the presence of hydroxyproline (Steward *et al.*, 1958), in the level of scopoletin and scopolin (Tso *et al.*, 1964), and in different electrophoretically separated proteins and enzymes (Bhatia *et al.*, 1967).

The following lines of evidence have been marshaled to support the conclusion that the basic cause of the occurrence of spontaneous tumors in the interspecific *Nicotiana* hybrids rests in particulate genes located in the chromosomes in the nucleus of cells (Smith 1968, 1972):

1. The restriction of the occurrence of tumors to only about 30 out of more than 300 different interspecific *Nicotiana* hybrids (Kostoff, 1943; Kehr and Smith, 1952) can be considered as general evidence of their genetic basis.
2. No pathogenic causative agent has been isolated from *Nicotiana* hybrid tumors in spite of attempts over a period of years and by many different workers to do so (Kehr, 1951). These efforts in-

clude grafting experiments which have shown that no causative agent is transmitted across the graft union (Kostoff, 1943; Kehr and Smith, 1954; Smith and Stevenson, 1961; Ahuja, 1962; Steitz, 1963).

3. Tumor formation is the same in reciprocal interspecific hybrids, thus indicating that the tumors are caused, not by cytoplasmic elements present in only one of the parental species, but rather, by nuclear elements contributed from each parent.

4. With the hybrid *N. glauca* × *N. langsdorffii*, various genomic combinations were obtained, ranging from two *N. langsdorffii* genomes with one of *N. glauca* (GLL), to one of *N. langsdorffii* and three of *N. glauca* (GGGL). All these combinations are tumorous, indicating that the tumor-forming potential of the hybrid *N. glauca* × *N. langsdorffii* remains qualitatively the same regardless of the ratio of genomes (Kehr and Smith, 1954).

5. By a program of repeated backcrossing, Ahuja (1962) obtained, on the background of the amphiploid *N. debneyi-tabacum,* a single *N. longiflora* chromosome that was associated with, and thus postulated to be causal to, tumor formation.

6. Segregation for tumors was demonstrated in the cross of the F_1 hybrid, *N. langsdorffii* (a tumor-enhancing genotype) × *N. sanderae* (a related tumor-inhibiting species), with a third species (as *N. glauca* or *N. suaveolens*) selected as conducive to tumor formation (Smith, 1958).

7. Correlations, interpretable as evidence of genetic linkage, have been demonstrated in segregating populations: (a) between *N. langsdorffii* genes governing small corolla size and tumor enhancement, and (b) between *N. sanderae* genes governing large corolla size and tumor inhibition (Smith and Stevenson, 1961).

8. Progeny populations, characterized by marked differences in tumor expression, attributable to genetic recombination, were obtained by crossing recombinant inbreds derived from *N. langsdorffii* × *N. sanderae* with a third species conducive to tumor formation (Smith, 1962a).

9. Evidence that the heritable factors controlling tumor formation are subject to mutation was obtained by Izard (1957), who produced by x irradiation, a nontumorous mutant of the amphiploid *N. glauca* × *N. langsdorffii.*

Although these lines of evidence firmly establish a strong genetic component to the tumor-forming phenomenon in *Nicotiana* hybrids, it is equally evident that a variety of environmental factors will affect tumor expression. These include various types of ionizing irradiations (Sparrow *et al.,* 1956; Smith and Stevenson, 1961; Hagen *et al.,* 1961; Ahuja and Cameron, 1963; Conklin and Smith, 1968; 1969) and various chemical compounds, e.g., turpentine and white lead (Kehr and Smith, 1954), indoleacetic acid (Schaeffer, 1962), certain nucleic acid precursors (Tso and Burk, 1962), the nucleosides uridine and thymidine (Conklin and Smith, 1968), and mercaptoethanol (Ames and Smith, 1969). Furthermore, the tumor-forming response may be induced by more subtle environmental perturbations, e.g., higher temperatures, 24–27°C (Schaeffer *et al.,* 1966), or merely crowding seedlings during growth (Smith, 1962*a*).

The general picture that emerges from the response of *Nicotiana* tumor-forming hybrids to irradiation, chemical applications, and other environmental deterrants to normal growth, is that these genotypes maintain a precarious balance between differentiated growth and neoplasia. Without external treatment the plants will eventually reach a metabolic condition conducive to tumor formation. With a variety of treatments, the balance can be tipped early in development from normal to abnormal growth.

The essential feature of the genetically conditioned tumor-forming *Nicotiana* hybrids appears to be that when two evolutionarily diverse genomes are brought together in a single hybrid cell or plant, metabolic products form and accumulate which trigger abnormal growth. One possible explanation, in harmony with the experimental results, is that the initiation of tumorous growth is due to renewed activity of genes, normally repressed in differentiated tissues, to synthesize products essential for cell divisions and undifferentiated growth (Smith, 1958, 1962*a*, 1965*b*, 1968, 1972). In support of this theory is evidence that plant tumor cells frequently revert to the normal nontumorous state (White, 1939*b*; Braun, 1959, 1965, 1969; Smith, 1965*b*; Sacristan and Melchers, 1969).

Another line of evidence, favoring the gene-regulation explanation of genetic tumors, is that nonmutagenic stress conditions such as merely crowding or wounding at leaf scars, trigger tumor formation. A localized change in gene activation in response to a local alteration in level of metabolites (notably auxins) offers a reasonable explanation. Although evidence is not now available to support any detailed model of tumor-yielding gene activation, the bare elements of such a model based on batteries of regulator–operator-structural gene systems can be postulated (Smith, 1965*b*). A key to the problem may be that different repressor substances, synthesized in the same cell by two nonintegrated genetic systems of diverse phylogeny, interact so as to be ineffective, and thus fail to exercise normal controls on gene activity and cell division.

Model Systems for Somatic Cell Genetics

Improved techniques for cell and tissue culture have opened up new possibilities for conducting genetic studies in higher plants at the cellular level. The genus *Nicotiana* has often furnished the material for initial successes in carrying out the necessary procedures. The hybrid *N. glauca* × *N. langsdorffii* was one of the first, if not the first, plant tissue to be cultured continuously on a chemically defined medium (White, 1939a). Callus of *N. tabacum* has been a favorite material for research on the influence of the chemical composition of the media on *in vitro* growth (Hildebrandt *et al.*, 1946; Linsmaier and Skoog, 1965) and the chemical regulation of organ formation in tissue culture, mainly through interactions of cytokinin and auxin (Skoog and Miller, 1957). A further step was the demonstration by Vasil and Hildebrandt (1965) that single isolated cells of the hybrid *N. tabacum* × *N. glutinosa* could be grown in the absence of neighboring cells, and that the derived callus could be made to differentiate into roots, shoots, and eventually fully mature, flowering plants.

The plating of large numbers of suspended, single cells on agar and their growth into colonies was achieved by Bergman (1960) and further developed by Gibbs and Dougall (1965), utilizing cells obtained from *N. tabacum* callus tissue. High plating efficiency was reported.

Another technical advance in tissue cultue, important in establishing somatic cell genetics of higher plants, was the growing of haploid plants from pollen by asceptic culture of excised anthers. Soon after initial successes in *Datura* (Guha and Maheshwari, 1966), Bourgin and Nitsch (1967) reported the production of haploid plants by anther culture in *N. tabacum* and *N. sylvestris*. The list has now been extended to include *N. alata*, *N. glutinosa* (4n), and *N. rustica* (Nitsch, 1969); *N. otophora* (Collins *et al.*, 1972; Collins and Sadasivaiah, 1972); and *N. suaveolens* and *N. glauca* × *N. langsdorffii* (Smith, 1974a,b). Thus, the genus *Nicotiana* is exceptional in furnishing by far the most examples (Sunderland, 1971) of species yielding haploids by the anther-culture technique. It should be noted that haploid plants have occurred spontaneously, though rarely, and were known in a number of species of *Nicotiana* before the advent of another culture (Clausen and Mann, 1924; Ivanov, 1938; Goodspeed, 1954; de Nettancourt and Stokes, 1960; Burk, 1962; Kimber and Riley, 1963; Magoon and Khanna, 1963). Furthermore, haploid *Nicotianas* have been produced experimentally (Ivanov, 1938), e.g., by treating with colchicine (Smith, 1943b) or exposure to radiation (Goodspeed and Avery, 1929; Dulieu, 1964). Homozygous diploids can be reconsitituted from derived tobacco haploids (Tanaka and Nakata, 1969; Kasperbauer and Collins, 1971).

Utilization of haploid cells of *N. tabacum,* as a tool in the cellular genetics of higher plants, was demonstrated in mutation studies by Carlson (1970). Auxotrophic mutants were selected by plating single haploid cells, previously exposed to a chemical mutagen, onto a minimal medium containing 5-bromodeoxyuridine. Normal cells grew on the medium, incorporated the analog into their DNA, and were subsequently killed by exposure to visible light. Auxotrophic mutants, on the other hand, failed to divide or incorporate, were not killed, and were able to be recovered by culturing on a nutritionally supplemented medium. *N. tabacum* is not completely satisfactory for this kind of genetic study since the induced auxotrophic mutants proved to be "leaky," i.e., they continued to grow slowly on an unsupplemented medium (Carlson, 1970), presumably because this species, owing to its amphiploid origin, has more than two copies of many essential genes. Experiments are currently underway to utilize these same techniques with low-chromosome-number species of *Nicotiana* and with *N. glauca* × *N. langsdorffii* in order to obtain biochemical mutants involving genetic control of substances (as auxins and cytokinins) unique to the growth and differentiation requirements of higher plants (Smith, 1974*a,b*).

The obvious advantages of working with cultured haploid cells over multicellular whole plants are: (1) the large numbers of homogeneous individuals available in a small space; (2) the immediate expression of a mutant characteristic due to the haploid condition combined with the small pool of metabolites in a single cell; and (3) the application of biochemical selection techniques based on plating of cells on chemically defined media. Adding to these the capability of regenerating whole plants from single cells, and also their diploidization, provides a significant new array of methods for genetic analysis in higher plants (Melchers, 1972).

Another important technical advance has been the recent success in growing leaf protoplasts of *N. tabacum* (Nagata and Takebe, 1970, 1971; Takebe *et al.,* 1971). The protoplasts were prepared by digesting tissues and cell walls with the enzymes macerozyme and cellulase. The wall-less cells were then plated on an agar nutrient medium; the cell walls regrew, colonies of cells developed, and these were eventually regenerated into whole plants. Protoplast culture enhances the possibilities of altering the genetic constitution of plant cells by more readily introducing foreign genetic materials (DNA molecules), or fusing with other genotypes (Power *et al.,* 1970).

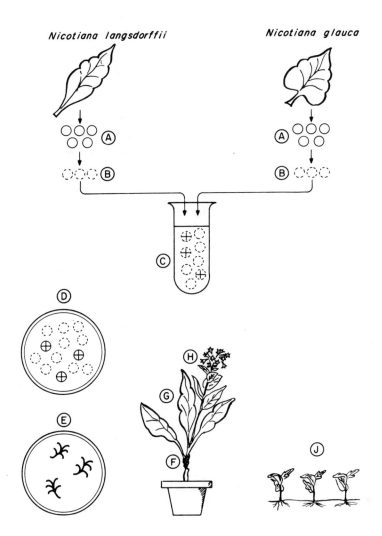

Figure 1. Simplified schematic representation of procedures for parasexual hybridization. Leaf mesophyll cells (A) of the two parental species are treated with enzymes to digest away the cell wall, leaving protoplasts (B). These are suspended in liquid medium containing $NaNO_3$ (C), centrifuged together, then plated on agar culture medium (D). Only fused hybrid cells grow. These then differentiate (E) into leaves and stem, which are grated (F) onto a parent plant. The hybrid scion matures into a plant (G) which produces fertile flowers (H) and seed. The seed germinates to produce seedlings that are essentially similar (J) to a sexually produced amphiploid. (From Smith 1974b) with the permission of BioScience.

It has recently been possible to utilize the various techniques described above to produce a parasexual interspecific hybrid by fusing leaf mesophyll protoplasts of *N. glauca* and *N. langsdorffii*, differentially selecting out the products of fusion on a chemically defined growth medium, and regenerating mature plants from the hybrid calli (Carlson *et al.*, 1972). The plants produced by somatic cell fusion were similar to the amphiploid obtained by doubling the chromosomes of the sexually produced interspecific hybrid, *N. glauca* × *N. langsdorffii*. The method is outlined diagrammatically in Figure 1. The potential offered by somatic cell hybridization may be expected to exceed the limitations imposed by sexual processes, and so extend the possibilities of combining widely divergent genotypes of higher plants.

Literature Cited

Abraham, A., 1947   A cytogenetical study of trisomic types in *Nicotiana langsdorffii*. Ph.D. Thesis, Cornell University, Ithaca, N.Y.

Ahuja, M. R., 1962   A cytogenetic study of heritable tumors in *Nicotiana* species hybrids. *Genetics* **47**:865–880.

Ahuja, M. R., 1965   Genetic control of tumor formation in higher plants. *Q. Rev. Biol.* **40**:329–340.

Ahuja, M. R., 1968   An hypothesis and evidence concerning the genetic components controlling tumor formation in *Nicotiana*. *Mol. Gen. Genet.* **103**:176–184.

Ahuja, M. R., and D. R. Cameron, 1963   The effects of x-irradiation on seedling tumor production in *Nicotiana* species and hybrids. *Radiat. Bot.* **3**:55–57.

Ahjua, M. R. and G. L. Hagen, 1966   Morphogenesis in *Nicotiana debneyi-tabacum, N. longiflora* and their tumor-forming hybrid derivatives *in vitro*. *Dev. Biol.* **13**:408–423.

Ames, I. H. and H. H. Smith, 1969   Effects of mercaptoethanol on tumor induction in a *Nicotiana* amphiploid. *Can. J. Bot.* **47**:921–924.

Appa Rao, K. and K. V. Krishnamurty, 1963   Studies on multiple polyploids in *Nicotiana*. *Genetica* (*The Hague*) **34**:66–78.

Ar-Rushdi, A. H., 1956   Inheritance in *Nicotiana tabacum*. XXVI. Sterility genes from *Tomentosae* species. *J. Genet.* **54**:9–22.

Ar-Rushdi, A. H., 1957   The cytogenetics of variegation in a species hybrid in *Nicotiana*. *Genetics* **42**:312–325.

Bayer, M. H., 1965   Paper chromatography of auxins and their inhibitors in two *Nicotiana* species and their hybrid. *Am. J. Bot.* **52**:883–890.

Bayer, M. H., 1967   Thin-layer chromatography of auxin and inhibitors in *Nicotiana glauca, N. langsdorffii* and three of their tumor-forming hybrids. *Planta* (*Berl.*) **72**:329–337.

Bayer, M. H. and M. R. Ahuja, 1968   Tumor formation in *Nicotiana*: auxin levels and auxin inhibitors in normal and tumor-prone genotypes. *Planta* (*Berl.*) **79**:292–298.

Bergman, L., 1960   Growth and division of single cells of higher plants *in vitro*. *J. Gen. Physiol.* **43**:841–851.

Bhatia, C. R., M. Buiatti and H. H. Smith, 1967   Electrophoretic variation in proteins and enzymes of the tumor-forming hybrid *Nicotiana glauca* × *N. langsdorffii* and its parent species. *Am. J. Bot.* **54:**1237–1241.

Bourgin, J. P., and J. P. Nitsch, 1967   Obtention de *Nicotiana* haploides à partir d'étamines cultivées *in vitro*. *Ann. Physiol. Veg. (Paris)*, **9**:377–382.

Braun, A. C., 1959   A demonstration of the recovery of the crown-gall tumor cell with the use of complex tumors of single cell origin. *Proc. Natl. Acad. Sci. USA* **45:**932–938.

Braun, A. C., 1965   The reversal of tumor growth. *Sci. Am.* **213:**75–83.

Braun, A. C., 1969   *The Cancer Problem. A Critical Analysis and Modern Synthesis*, Columbia University Press, New York.

Burk, L. G., 1960   Male-sterile flower anomalies in interspecific tobacco hybrids. *J. Hered.* **51:**27–31.

Burk, L. G., 1962   Haploids in genetically marked progenies of tobacco. *J. Hered.* 53:222–226.

Burk, L. G., 1965   Inter-histogenic cell migration measured by a plastid defect of tobacco. *Am. J. Bot.* **52:**616–617.

Burk, L. G. and J. J. Grosso, 1963   Plasmagenes in variegated tobacco. *J. Hered.* **54:**23–25.

Burk, L. G. and H. E. Heggestad, 1966   The genus *Nicotiana*: a source of resistance to diseases of cultivated tobacco. *Econ. Bot.* **20:**76–78.

Burk, L. G. and R. N. Jeffrey, 1958   A study of the inheritance of alkaloid quality in tobacco. *Tob. Sci.* **2:**139–141.

Burk, L. G., R. N. Stewart, and H. Dermen, 1964   Histogenesis and genetics of a plastid-controlled chlorophyll variegation in tobacco. *Am. J. Bot.* **51:**713–724.

Burns, J. A., 1966   The heterochromatin of two species of *Nicotiana*: cytological observations. *J. Hered.* 57:43–47.

Burns, J. A. and D. U. Gerstel, 1971   Inhibition of chromosome breakage and of megachromosomes by intact genomes in *Nicotiana*. *Genetics* **69:**211–220.

Burns, J. A., and D. H., Gerstel, 1973   Formation of megachromosomes from heterochromatic blocks of *Nicotiana tomentosiformis*. *Genetics* **75:**497–502.

Cameron, D. R., 1959   The monosomics of *Nicotiana tabacum*. *Tob. Sci.* **3:**164–166.

Cameron, D. R., 1965   Cytoplasmic effects in *Nicotiana*. *Proc. 11th Int. Congr. Genet. (The Hague, 1963)* **1:**203–204.

Cameron, D. R. and R. Moav, 1957   Inheritance in *Nicotiana tabacum*. XXVII. Pollen killer, an alien genetic locus inducing abortion of microspores not carrying it. *Genetics* **42:**326–335.

Carlson, P. S., 1970   Induction and isolation of auxotrophic mutants in somatic cell cultures of *Nicotiana tabacum*. *Science (Wash., D.C.)* **168:** 487–489.

Carlson, P. S., H. H. Smith, and R. D. Dearing, 1972   Parasexual interspecific plant hybridization. *Proc. Natl. Acad. Sci. USA* **69:**2292–2294.

Chaplin, J. F. and T. J. Mann, 1961   Interspecific hybridization, gene transfer and chromosome substitution in *Nicotiana*. *N. C. Agric. Exp. Stn. Tech. Bull.* **145:**1–31.

Cheng, T.-Y., 1972   Induction of indoleacetic acid synthetases in tobacco pith explants. *Plant Physiol.* **50:**723–727.

Clausen, R. E. and D. R. Cameron, 1944   Inheritance in *Nicotiana tabacum* XVIII. Monosomic analysis. *Genetics* **29:**447–477.

Clausen, R. E. and T. H. Goodspeed, 1926a   Inheritance in *Nicotiana tabacum*. VII. The monosomic character "fluted." *Univ. Calif. Publ. Bot.* **11:**61–82.

Clausen, R. E. and T. H. Goodspeed, 1926*b*   Interspecific hybridization in *Nicotiana*. III. The monosomic *Tabacum* derivative, "corrugated," from the *sylvestris–tabacum* hybrid. *Univ. Calif. Publ. Bot.* **11**:83–101.

Clausen, R. E. and M. C. Mann, 1924   Inheritance in *Nicotiana tabacum*. V. The occurrence of haploid plants in interspecific progenies. *Proc. Natl. Acad. Sci. USA* **10**:121–124.

Clayton, E. E., 1950   Male-sterile tobacco. *J. Heredity* **41**:171–175.

Clayton, E. E., 1958   The genetics and breeding progress in tobacco during the last 50 years. *Agron. J.* **50**:352–356.

Clayton, E. E., H. E. Heggestad, J. J. Grosso, and L. G. Burk, 1967   The transfer of blue-mold resistance to tobacco from *Nicotiana debneyi*. I and II. *Tob. Sci.* **165**:91–106.

Collins, G. B. and R. S. Sadasivaiah, 1972   Meiotic analysis of haploid and doubled haploid forms of *Nicotiana otophora* and *N. tabacum*. *Chromosoma* **38**:387–404.

Collins, G. B., P. D. Legg, and M. J. Kasperbauer, 1972   Chromosome numbers in anther-derived haploids of two *Nicotiana* species. *J. Hered.* **63**:113–118.

Conklin, M. E. and H. H. Smith, 1968   Endogenous beta irradiation and tumor production in an amphiploid *Nicotiana* hybrid. *Am. J. Bot.* **55**:473–476.

Conklin, M. E. and H. H. Smith, 1969   Effects of fast neutron versus x- irradiation on development, differentiation and peroxidase isozymes in a genetically tumorous *Nicotiana* amphiploid and its parents. *Int. J. Radiat. Biol.* **16**:311–321.

Daly, K. and D. S. Robson, 1969   Estimates of genetic parameters from a hybrid derivative in *Nicotiana*. *Genetics* **62**:201–213.

Dawson, R. F. and M. L. Solt, 1959   Estimated contributions of root and shoot to the nicotine content of the tobacco plant. *Plant Physiol.* **34**:656–661.

de Nettancourt, D. and G. W. Stokes, 1960   Haploidy in tobacco. *J. Hered.* **51**:102–104.

Dermen, H., 1960   Nature of plant sports. *Am. Hortic. Mag.* **39**:123–173.

Dulieu, H., 1964   Detection of haploid plants among the progeny *Nicotiana tabacum* L. × *N. sanderae* Hort. after pollen irradiation. *C. R. Hebd. Seances Acad. Sci. Ser. D Sci. Nat.* **259**:4126–4129.

East, E. M., 1913   Inheritance of flower size in crosses between species of *Nicotiana*. *Bot. Gaz.* **55**:177–188.

East, E. M., 1928   The genetics of the genus *Nicotiana*. *Bibliogr. Genet.* **4**:243–318.

East, E. M., 1932   Studies on self-sterility. IX. The behavior of crosses between self-sterile and self-fertile plants. *Genetics* **17**:175–202.

East, E. M., 1935   Genetic reactions in *Nicotiana*. II. Phenotypic reaction patterns. *Genetics* **20**:414–442.

Edwardson, J. R., 1965   Gene control of non-Mendelian variegation in *Nicotiana tabacum*. *Genetics* **52**:365–370.

Gerstel, D. U., 1943   Inheritance in *Nicotiana tabacum*. XVII. *Cytogenetical analysis of glutinosa*-type resistance to mosaic disease. *Genetics* **28**:533–536.

Gerstel, D. U., 1945   Inheritance in *Nicotiana tabacum*. XX. The addition of *Nicotiana glutinosa* chromosomes to tobacco. *J. Hered.* **36**:197–206.

Gerstel, D. U., 1948   Transfer of the mosaic-resistance factor between H chromosomes of *Nicotiana glutinosa* and *N. tabacum*. *J. Agric. Res.* **76**:219–223.

Gerstel, D. U., 1960   Segregation in new allopolyploids of *Nicotiana*. I. Comparison of 6 × (*N. tabacum* × *tomentosiformis*) and 6 × (*N. tabacum* × *otophora*). *Genetics* **45**:1723–1734.

Gerstel, D. U. and J. A. Burns, 1966a Chromosomes of unusual length in hybrids between two species of *Nicotiana*. In *Chromosomes Today*, Vol. 1, edited by C. D. Darlington and K. R. Lewis, pp. 41–56, Plenum Press, New York.

Gerstel, D. U. and J. A. Burns, 1966b Flower variegation in hybrids between *Nicotiana tabacum* and *N. otophora*. *Genetics* **53**:551–567.

Gerstel, D. U. and J. A. Burns, 1970 The effect of the *Nicotiana otophora* genome on chromosome breakage and megachromosomes in *N. tabacum* × *N. otophora* derivatives. *Genetics* **66**:331–338.

Gerstel, D. U. and T. J. Mann, 1964 Segregation in new allopolyploids in *Nicotiana*. III. Nicotine-converter genes in allopolyploids from *N. tomentosiformis, N. sylvestris* and *N. tabacum. Crop. Sci.* **4**:387–388.

Gibbs, J. L. and D. K. Dougall, 1965 The growth of single cells from *Nicotiana tabacum* callus tissue in nutrient medium containing agar. *Exptl. Cell. Res.* **40**:85–95.

Goodspeed, T. H., 1954 *The Genus Nicotiana*, Chronica Botanica, Waltham, Mass.

Goodspeed, T. H. and P. Avery, 1929 The occurrence of a *Nicotiana glutinosa* haploid. *Proc. Natl. Acad. Sci. USA* **15**:502–504.

Goodspeed, T. H. and P. Avery, 1939 Trisomic and other types in *Nicotiana sylvestris. J. Genet.* **38**:381–458.

Goodspeed, T. H. and P. Avery, 1941 The twelfth primary trisomic type in *Nicotiana sylvestris. Proc. Natl. Acad. Sci. USA* **27**:13–14.

Goodspeed, T. H. and R. E. Clausen, 1928 Interspecific hybridization in *Nicotiana*. VIII. The *sylvestris–tomentosa–tabacum* hybrid triangle and its bearing on the origin of *tabacum. Univ. Calif. Publ. Bot.* **11**:245–256.

Goodspeed, T. H. and M. C. Thompson, 1959 Cytotaxonomy of *Nicotiana*. II. *Bot. Rev.* **25**:385–415.

Guha, S. and S. C. Maheshwari, 1966 Cell division and differentiation of embryos in pollen grains of *Datura in vitro. Nature (Lond.)* **212**:97–98.

Gupta, S. B., and P. Gupta, 1973 Selective somatic elimination of *Nicotiana glutinosa* chromosomes in the F_1 hybrids of *N. suaveolens* and *N. glutinosa. Genetics* **73**:605–612.

Hagen, G. L., 1969 Tumor growth in hybrid tobacco: the parental contributions. *Abst. 11th Int. Bot. Congr. (Seattle)* pg. 82.

Hagen, G. L., J. E. Gunckel, and A. H. Sparrow, 1961 Morphology and histology of tumor types induced by X, gamma, and beta irradiation of a tobacco hybrid. *Am. J. Bot.* **48**:691–699.

Hart, G. E., 1965 Studies on extrachromosomal male sterility in *Nicotiana*. Ph.D. Thesis, University of California, Berkeley, Calif.

Hayman, B. I., 1960 Separation of epistatic from additive and dominance variation in generation means. II. *Genetica (The Hague)* **31**:133–146.

Henry, T. A., 1949 *The Plant Alkaloids*, fourth edition, Blakiston, Philadelphia, Pa.

Hildebrandt, A. C., A. J. Riker and B. M. Duggar, 1946 The influence of the composition of the medium on growth *in vitro* of excised tobacco and sunflower tissue cultures. *Am. J. Bot.* **33**:591–597.

Holmes, F. O., 1938 Inheritance of resistance to tobacco mosaic disease in tobacco. *Phytopathology* **28**:553–561.

Imai, S., 1959 Alkaloids in *Nicotiana. Hatano Tob. Exp. Stn. Bull.* **44**:129–136 (in Japanese with English summary).

Ivanov, M. A., 1938   Experimental production of haploids of *Nicotiana rustica* L. (and a discussion of haploidity in flowering plants). *Genetica (The Hague)* **20**:295–381.

Izard, C., 1957   Obtention et fixation de lignées tumorales et nontumorales a partir de mutations expérimentales de l'hybride *N. glauca* × *N. langsdorffii. C. R. Seances Acad. Agric. Fr.* **43**:325–327.

Jeffrey, R. N., 1959   Alkaloid composition of species of *Nicotiana. Tob. Sci.* **3**:89–93.

Kasperbauer, M. J. and G. B. Collins, 1971   Reconstitution of diploids from anther-derived haploids in tobacco. *Crop. Sci.* **12**:98–101.

Kehr, A. E., 1951   Genetic tumors in *Nicotiana. Am. Nat.* **85**:51–64.

Kehr, A. E., 1965   The growth and development of spontaneous plant tumors. *Encycl. Plant Physiol.* **XV**/2: 184–196.

Kehr, A. E. and H. H. Smith, 1952   Multiple genome relationships in *Nicotiana. Cornell Univ. Agric. Exp. Stn. Mem.* **311**:1–19.

Kehr, A. E. and H. H. Smith, 1954   Genetic tumors in *Nicotiana* hybrids. *Brookhaven Symp. Biol.* **6**:55–78.

Kimber, G. and R. Riley, 1963   Haploid angiosperms. *Bot. Rev.* **29**:490–531.

Kostoff, D., 1930   Tumors and other malformations on certain *Nicotiana* hybrids. *Zentralbl. Bakteriol. Parasitenkd Infektionskr. Hyg. Abt. 2* **81**:244–260.

Kostoff, D., 1935   On the increase of mutation frequency following interspecific hybridization. *Curr. Sci. (Bangalore)* **3**:302–304.

Kostoff, D., 1943   *Cytogenetics of the Genus Nicotiana*, State Printing House, Sofia, Bulgaria.

Kovacs, E. I., 1967   Organ formation and protein synthesis in instable tissue cultures of the interspecific tumour-forming hybrid of *Nicotiana. Acta Agron. Acad. Sci. Hung.* **16**:41–48.

Krishnamurty, K. V., G. S. Murty and K. Appa Rao, 1960   Cytogenetics of the trispecific hybrid *Nicotiana tabacum* × (*N. glutinosa* × *N. trigonophylla*) and its reciprocal. *Euphytica* **9**:111–121.

Lee, R. E., 1950   A cytogenetic study of extra chromosomes in *Nicotiana langsdorffii* and in crosses with *N. sanderae. Ph.D. Thesis, Cornell University,* Ithaca, N.Y.

Legg, P. D. and G. B. Collins, 1971   Inheritance of percent total alkaloids in *Nicotiana tabacum* L. II. Genetic effects of two loci in Burley 21 × LA Burley 21 populations. *Can. J. Genet. Cytol.* **13**:287–291.

Legg, P. D., J. F. Chaplin and G. B. Collins, 1969   Inheritance of percent total alkaloids in *Nicotiana tabacum* L. *Hered.* **60**:213–217.

Linsmaier, E. M. and F. Skoog, 1965   Organic growth factor requirements of tobacco tissue cultures. *Physiol. Plant.* **18**:100–127.

Luthra, J. K., 1964   Inheritance of quantitative characters in *Nicotiana tabacum. Indian J. Genet. Plant Breed.* **24**:275–279.

McCray, F. A., 1932   Compatibility of certain *Nicotiana* species. *Genetics* **17**:621–636.

Magoon, M. L. and K. R. Khanna, 1963   Haploids. *Caryologia* **16**:191–234.

Mann, T. J. and J. A. Weybrew, 1958   Inheritance of alkaloids in hybrids between flue-cured tobacco and related amphidiploids. *Tob. Sci.* **2**:29–34.

Mann, T. J., D. U. Gerstel and J. L. Apple, 1963   The role of interspecific hybridization in tobacco disease control. In *Proceedings of the third World Tobacco Science, Congress Salisbury, S. Africa, 1963*, pp. 201–205, Mardon Printers, Salisbury, Southern Rhodesia.

Mann, T. J., J. A. Weybrew, D. F. Matzinger and J. L. Hall, 1964   Inheritance of the conversion of nicotine to nornicotine in varieties of *Nicotiana tabacum* L. and related amphiploids. *Crop Sci.* **4**:349–353.

Mann, T. J., D. F. Matzinger and E. A. Wernsman, 1972   Genetic control of tobacco constitutents. In *Coresta/TCRC Symposium*, pp. 77–85, Williamsburg, Virginia.

Marion, L., 1950   The pyridine alkaloids. In *The Alkaloids*, Vol. 1, edited by R. H. F. Manske and H. L. Holmes, pp. 228–253, Academic Press, New York.

Marion, L., 1960   The pyridine alkaloids. In *The Alkaloids*, Vol. 6, edited by R. H. F. Manske, pp. 128–132, Academic Press, New York.

Mather, K., 1949   The genetical theory of continuous variation. In *Proceedings of the Ninth International Congress of Genetics, Stockholm, 1949, Hereditas Suppl.*, pp. 376–401, Mendelian Society, Lund, Sweden.

Mather, K. and A. Vines, 1952   The inheritance of height and flowering time in a cross of *Nicotiana rustica*. In *Quantiative Inheritance (Agricultural Research Council Colloquim)*, pp. 49–79, H. M. Stationery Office, London.

Matzinger, D. F., T. J. Mann and H. F. Robinson, 1960   Genetic variability in flue-cured varieties of *Nicotiana tabacum*. I. Hicks Broadleaf × Coker 139. *Agron J.* **52**:8–11.

Matzinger, D. F., T. J. Mann and C. C. Cockerham, 1962   Diallel crosses in *Nicotiana tabacum*. *Crop. Sci.* **2**:383–386.

Matzinger, D. F., E. A. Wernsman and H. F. Ross, 1971   Diallel crosses among burley varieties of *Nicotiana tabacum* L. in the F_1 and F_2 generations. *Crop. Sci.* **11**:275–279.

Matzinger, D. F., E. A. Wernsman and C. C. Cockerham, 1972   Recurrent family selection and correlated response in *Nicotiana tabacum* L. I. 'Dixie Bright 244' × 'Coker 139'. *Crop Sci.* **12**:40–43.

Melchers, G., 1972   Haploid higher plants for plant breeding. *Z. Pflanzenzucht.* **67**:19–32.

Moav, R., 1961   Genetic instability in *Nicotiana* hybrids. II. Studies of the *Ws (pbg)* locus of *N. plumbaginifolia* in *N. tabacum* nuclei. *Genetics* **46**:1069–1087.

Moav, R. and D. R. Cameron, 1960   Genetic instability in *Nicotiana* hybrids. I. The expression of instability in *N. tabacum* × *N. plumbaginifolia*. *Am. J. Bot.* **47**:87–93.

Mothes, K., 1955   Physiology of alkaloids. *Annu. Rev. Plant Physiol.* **6**:393–432.

Näf, U., 1958   Studies on tumor formation in *Nicotiana* hybrids. I. The classification of the parents into two etiologically significant groups. *Growth* **22**:167–180.

Nagata, T. and I. Takebe, 1970   Cell wall regeneration and cell division in isolated tobacco mesophyll protoplasts. *Planta (Berl.)* **92**:301–308.

Nagata, T. and I. Takebe, 1971   Plating of isolated tobacco mesophyll protoplasts on agar medium. *Planta (Berl.)* **99**:12–20.

Nitsch, J. P., 1969   Experimental androgenesis in *Nicotiana*. *Phytomorphology* **19**:389–404.

Oka, M., 1959   The analysis of inheritance of quantitative characters with flue-cured tobacco varieties in diallel cross. *Jap. J. Breed.* **9**:87–92 (in Japanese).

Olby, R. C., 1966   *Origins of Mendelism*, Schocken Books, New York.

Power, J. B., S. E. Cummings and E. C. Cocking, 1970   Fusion of isolated plant protoplasts. *Nature (Lond.)* **225**:1016–1018.

Reddy, M. M. and E. D. Garber, 1971   Genetic studies of variant enzymes. III. Comparative electrophoretic studies of esterases and peroxidases for species, hybrids and amphiploids in the genus *Nicotiana*. *Bot. Gaz.* **132**:158–166.

Robinson, H. F., T. J. Mann and R. E. Comstock, 1954   An analysis of quantitative variability in *Nicotiana tabacum. Heredity* **8**:365–376.

Sacristan, M. D. and G. Melchers, 1969   The caryological analysis of plants regenerated from tumorous and other callus cultures of tobacco. *Mol. Gen. Genet.* **105**:317 333.

Sand, S. A., 1957   Phenotypic variability and the influence of temperature on somatic instability in cultures derived from hybrids between *Nicotiana langsdorffii* and *N. sanderae. Genetics* **42**:685–703.

Sand, S. A. and G. T. Christoff, 1973   Cytoplasmic–chromosomal interactions and altered differentiation in tobacco. *J. Heredity* **64**:24–30.

Sand, S. A. and H. H. Smith, 1973   Somatic mutational transients. III. Response by two genes in a clone of *Nicotiana* to 24-roentgens of gamma radiation applied at various intensities. *Genetics* **75**:93–111.

Sand, S. A., A. H. Sparrow and H. H. Smith, 1960   Chronic gamma irradiation effects on the mutable *V* and stable *R* loci in a clone of *Nicotiana. Genetics* **45**:289–308.

Schaeffer, G. W., 1962   Tumour induction by an indolyl-3-acetic acid–kinetin interaction in a *Nicotiana* hybrid. *Nature (Lond.)* **196**:1326–1327.

Schaeffer, G. W. and H. H. Smith, 1963   Auxin–kinetin interaction in tissue cultures of *Nicotiana* species and tumor-conditioned hybrids. *Plant Physiol.* **38**:291–297.

Schaeffer, G. W., H. H. Smith and M. P. Perkus, 1963   Growth-factor interactions in the tissue culture of tumorous and nontumorous *Nicotiana glauca-langsdorffii. Am. J. Bot.* **50**:766–771.

Schaeffer, G. W., L. G. Burk, and T. C. Tso, 1966   Tumors of interspecific *Nicotiana* hybrids. I. Effect of temperature and photoperiod upon flowering and tumor formation. *Am. J. Bot.* **53**:928–932.

Sheen, S. J., 1970   Perioxidases in the genus *Nicotiana. Theor. Appl. Genet.* **40**:18–25.

Sheen, S. J., 1972   Isozymic evidence bearing on the origin of *Nicotiana tabacum* L. *Evolution* **26**:143–154.

Skoog, F., 1944   Growth and organ formation in tobacco tissue cultures. *Am. J. Bot.* **31**:19–24.

Skoog, F. and C. O. Miller, 1957   Chemical regulation of growth and organ formation in plant tissues cultured *in vitro. Symp. Soc. Exp. Biol.* **11**:118–131.

Smith, H. H., 1943a   Effects of genome balance, polyploidy, and single extra chromosomes on size in *Nicotiana. Genetics* **28**:227–236.

Smith, H. H., 1943b   Studies on induced heteroploids of *Nicotiana. Am. J. Bot.* **30**:121–130.

Smith, H. H., 1952   Fixing transgressive vigor in *Nicotiana rustica.* In *Heterosis,* edited by J. Gowen, pp. 161–174, Iowa State University Press, Ames, Iowa.

Smith, H. H., 1958   Genetic plant tumors in *Nicotiana. Ann. N.Y. Acad. Sci.* **71**:1163–1177.

Smith, H. H., 1962a   Genetic control of *Nicotiana* plant tumors. *Trans. N.Y. Acad. Sci. Ser. II* **24**:741–746.

Smith, H. H., 1962b   Studies on the origin, inheritance and mutation of genic-cytoplasmic male sterility in *Nicotiana. Genetics* **47**:985–986.

Smith, H. H., 1965a   Inheritance of alkaloids in introgressive hybrids of *Nicotiana. Am. Nat.* **99**:73–79.

Smith, H. H., 1965b   Genetic tobacco tumors and the problem of differentiation. *Brookhaven Lect. Series (BNL 967 T-405)* **52**:1–8.

Smith, H. H., 1968   Recent cytogenetic studies in the genus *Nicotiana. Adv. Genet.* **14**:1–54.

Smith, H. H., 1971   Broadening the base of genetic variability in plants. *J. Hered.* **62**:265–276.

Smith, H. H., 1972   Plant genetic tumors. *Progr. Exp. Tumor Res.* **15**:138–164.

Smith, H. H. 1974a   Model genetic systems for studying mutation, differentiation, and somatic cell hybridization in plants. In *Polyploidy and Induced Mutations in Plant Breeding*, IACA, Vienna.

Smith, H. H. 1974b   Model systems for somatic cell plant genetics. *BioScience* **24**:269–276.

Smith, H. H. 1974c   Interspecific plant hybridization and the genetics of morphogenesis. *Brookhaven Symp. Biol.* **25**:309–328.

Smith, H. H. and D. V. Abashian, 1963   Chromoatographic investigations on the alkaloid content of *Nicotiana* species and interspecific combinations. *Am. J. Bot.* **50**:435–447.

Smith, H. H. and K. Daly, 1959   Discrete populations derived by interspecific hybridization and selection in *Nicotiana*. *Evolution* **13**:476–487.

Smith, H. H. and D. S. Robson, 1959   A quantitative inheritance study of dimensions of flower parts in tobacco. *Biometrics* **15**:147.

Smith, H. H. and S. A. Sand, 1957   Genetic studies on somatic instability in cultures derived from hybrids between *Nicotiana langsdorffii* and *N. sanderae*. *Genetics* **42**:560–582.

Smith, H. H. and C. R. Smith, 1942   Alkaloids in certain species and interspecific hybrids of *Nicotiana*. *J. Agric. Res.* **65**:347–359.

Smith, H. and H. and H. Q., Stevenson, 1961   Genetic control and radiation effects in *Nicotiana* tumors. *Z. Vererbungsl.* **92**:100–118.

Smith, H. H., H. Q. Stevenson and A. E. Kehr, 1958   Limits and consequences of multiple allopolyploidy in *Nicotiana*. *Nucleus (Calcutta)* **1**:205–222.

Smith, H. H., D. E. Hamill, E. A. Weaver, and K. H. Thompson, 1970   Multiple molecular forms of peroxidases and esterases among species and amphiploids in the genus *Nicotiana*. *J. Hered.* **61**:203–212.

Sparrow, A. H., J. E. Gunckel, L. A., Schairer and G. L. Hagen, 1956   Tumor formation and other morphogenetic responses in an amphiploid tobacco hybrid exposed to chronic gamma irradiation. *Am. J. Bot.* **43**:377–388.

Steitz, E., 1963   Untersuchungen über die Tumorbildung bei Bastarden von *Nicotiana glauca* und *N. langsdorffii*. Ph.D. Thesis, University of Saarland, Saarbrüken, Germany.

Steward, F. C., J. F. Thompson and J. K. Pollard, 1958   Contrasts in the nitrogenous composition of rapidly growing and nongrowing plant tissues. *J. Exp. Bot.* **9**:1–10.

Sunderland, N., 1971   Anther culture: a progress report. *Sci. Progr.* **59**:527–549.

Takebe, I., G. Labib and G. Melchers, 1971   Regeneration of whole plants from isolated mesophyll protoplasts of tobacco. *Naturwissenschaften* **6**:318–320.

Takenaka, Y. and Y. Yoneda, 1962   Tumorous hybrids in *Nicotiana*. *Natl. Inst. Genet. Mishima. Annu. Rep.* **13**:63–64.

Tanaka, M. and K. Nakata, 1969   Tobacco plants obtained by anther culture and the experiment to get diploid seeds from haploids. *Jap. J. Genet.* **44**:47–54.

Tso, T. C., 1972   *Physiology and Biochemistry of Tobacco Plants*, Dowden, Hutchinson and Ross, Stroudsburg, Pa.

Tso, T. C. and L. G., Burk, 1962   Effects of certain anti-tumor chemicals on a tumorous *Nicotiana* hybrid. *Nature (Lond.)* **193**:1204–1205.

Tso, T. C., L. G. Burk, L. J. Dieterman and S. H. Wender, 1964   Scopoletin, scopolin and chlorogenic acid in tumours of interspecific *Nicotiana* hybrids. *Nature (Lond.)* **204:**779–780.

Vasil, V. and A. C. Hildebrandt, 1965   Differentiation of tobacco plants from single, isolated cells in microcultures. *Science (Wash., D.C.)* **150:**889–892.

Vester, F. and F. Anders, 1960   Der Gehalt an freien Aminosäuren des spontan tumor-bildenden Artbastards von *Nicotiana glauca* und *N. langsdorffii. Biochem. Z.* **332:**396–402.

Wernsman, E. A. and D. F. Matzinger, 1970   Relative stability of alleles at the nicotine conversion locus of tobacco. *Tob. Sci.* **14:**34–36.

White, P. R., 1939a   Potentially unlimited growth of excised plant callus in an artifical nutrient. *Am. J. Bot.* **26:**59–64.

White, P. R., 1939b   Controlled differentiation in a plant tissue culture. *Bull. Torrey Bot. Club* **66:**507–513.

Wittmer, G. and R. E. Scossiroli, 1961   Estimation of genetic variability for several quantitative traits in two tobacco varieties. *Genet. Agrar.* **14:**223–233.

Wolf, F. A., 1959   Cytoplasmic inheritance of albinism in tobacco. *Tob. Sci.* **3:**39–43.

13

Antirrhinum majus L.

CORNELIA HARTE

Introduction

The snapdragon, *Antirrhinum majus,* is one of the classical species for genetic study. The first publication is that of Godron (1863), who reported the results of crosses between *Antirrhinun majus* and *Antirrhinum barrelieri.* Mendel mentions in letters that he made crosses with snapdragons, but he never published the details. Darwin (1868) crossed plants with normal and with peloric flowers, selfed the F_1 generation, and found in the F_2 generation a ratio of 88 normal to 37 peloric plants. However, he was unable to propose a theoretical explanation for his results. *A. majus* was one of the species used by De Vries in his rediscovery of the Mendelian laws, and further studies in the snapdragon also provided him with insights into the mutation phenomenon. In 1907 Baur in Germany and Wheldale in Great Britain published the first papers on the inheritance of flower colors, and these served as the forerunners of the large and varied literature that now exists for this species [see Stubbe's (1966) monograph].

Advantages for Genetic Study

Antirrhinum majus possesses many characteristics that make it a good subject for genetic research. It is an annual with a short vegetative period. It is relatively small, easy to cultivate, and is self-fertile. The size and structure of the flower make emasculation easy, so that fertilizations by either selfing or outcrossing are facilitated. The time between flowering

CORNELIA HARTE—Institute für Entwicklungsphysiologie, Universität zu Köln, Köln-Lindenthal, West Germany.

315

and the ripening of the seeds is short, fertility is high, and there are many seeds per capsule (usually 50–100, and sometimes as many as 200). Germination is rapid, and more than 95 percent of the seeds are viable. With additional illumination (up to 16 hr/day) and a greenhouse temperature of 20–25 °C during the winter, two generations can be grown per year.

Mendelian Genetics

The laboratory race "Sippe 50" serves as the standard against which mutants are compared. The *Antirrhinum* mutants include genes which influence qualitative morphological traits affecting the leaves and flowers, quantitative traits such as plant height, and biochemical differences in the pigments of the flowers (anthocyanins and flavonoles) and leaves (chlorophylls and carotenoids) (see Table 1).

Gene nomenclature follows the international rules. Each locus is given a name that describes some deviation of its phenotype from the "standard" or "wild" type. The wild-type allele is given a + designation. If the heterozygote looks wild type with respect to the phenotype in question, the mutant is recessive, and the symbol is written with a small letter. If the heterozygote is clearly different from standard, even if the phenotype does not correspond completely to the homozygous mutant, then the gene is considered dominant, and its symbol is written with a capital letter. As in other species, most mutants are recessive.

Observations of deviations from expected Mendelian ratios led to the discovery of genes that influence the development of the gametophyte (Harte, 1969) and of lethals (Linnert, 1968).

Many genes exhibit a great variability of phenotypes. This may be due to variable expressivity and penetrance, to modifying genes, to multiple alleles, or to genetic instability. Variable expressivity and penetrance can be distinguished from other sources of phenotypic variability by selecting plants that show a "more normal" or an "extreme deviation from wild type" and selfing them. The phenotypic variability and relative frequency of abnormalities will remain the same in subsequent generations in both lines if the gene shows variable expressivity and penetrance (Stubbe, 1963).

Another type of variable manifestation of the mutant character is given by the alleles of the locus *transcendens*. These alleles show a tendency towards a reduced number of anthers, from 4 to 3 or 2 per flower. The variability within a family is influenced by the environment, and selection is without success (Hagemann, 1967). After crossing to other

TABLE 1.   *Phenotypes of Antirrhinum Genes*

Phenotype[a]	Symbol	Description
albostriata	*Stri*	The mutated allele is unstable; seedlings have white cotyledons; later small green spots develop as a result of backmutation to $Stri^+$
Aurea	*Aur*	Homozygotes are light yellow, lethal; heterozygotes viable with golden-yellow leaves
bipartita	*bip*	Cotyledons and first leaves with yellow-green color of the distal part and green basis; flowers slightly different from normal plants
Crispa	*Cri*	Homozygotes are lethal; in heterozygotes the leaves have necrotic areas
Cycloidea	*Cyc*	A series of multiple alleles transforming the symmetry of the flower from zygomorphic to radial
neo-hemiradialis	Cyc^{neo}	Flower not completely radial; number of anthers 4 or more
radialis	Cyc^{rad}	Flower with radial symmetry
Deficiens	*Def*	A series of multiple alleles transforming the corolla stepwise into small greenish petals and reducing male fertility
eosina	*eos*	Flowers light (eosin-) red
	eos^+	Flowers dark red
Eluta	*El*	Series of multiple alleles consisting of two stable (*El* and El^+) and several unstable alleles
	El^+	Color of corolla dark red
	El	Reduced anthocyanin content of the corolla, light red
sup-El		Suppressor of El
fimbriata	*fim*	Corolla dissected, variable expression from nearly normal to small strips
graminifolia	*gram*	Leaves very narrow, grass-like; the series consists of 2 stable alleles (*gram* and $gram^+$) and 2 mutable alleles
incolorata	*inc*	Homozygous seedlings light green, without anthocyanin; flowers ivory or light yellow
	inc^+	Required for synthesis of anthocyanins in leaves and flowers

Table 1.   Continued

Phenotype[a]	Symbol	Description
luteovirens	luv	Cotyledons and leaves yellow-green; growth reduced
marmorata	marm	Leaves with green spots on white background
nivea	niv	Corolla white; viability reduced
Pallida	Pal	Series of multiple alleles; Stepwise reduction of anthocyanin content of the corolla from dark red to ivory
	Pal^{rec}	Mutable allele
perlutea	perl	Seedlings light green, leaves yellow-green
phantastica	phan	Leaves with reduced lamina; variable expression from narrow to needlelike
rosea	ros	Light red color of corolla
transcendens	trans	Malformation of the corolla and reduction in the number of anthers
monocotyly		Only one cotyledon
polycotyly		Three or more cotyledons; characters inherited polygenically and variable in expression
albomaculatus		Cotyledons and leaves with white and green sectors; non-Mendelian inheritance; plastid mutations

[a] A more extensive list, with over 500 genes, is given by Stubbe (1966).

wild species, from the F_3 generation on, the variability in selected lines is decreased, and the development of only 2 anthers per flower is stabilized. The same reaction, i.e., stabilizing of a variable character in lines selected out of the progenies of species crosses, is observed for the increase to five anthers by the allele *neo-hemiradialis (Cyc^{neo})* of the *Cycloidea* locus. In both cases there is a selection of modifying genes (minor genes) which is made possible by the increased heterozygosity of species hybrids (Stubbe, 1952).

Multiple Alleles

In *Antirrhinum majus* several series of multiple alleles for biochemical and morphological traits are known. Examples are the series

Pallida (Pal) and *rosea (ros)*, which influence the intensity of flower pigmentation, and *deficiens (def)* and *Cycloidea (Cyc,* formerly *radialis)*,
which act on flower morphology. All. these multiple alleles are
characterized by two qualities: (1) the phenotypes of the homozygotes can
be arranged in a series of increasing quantitative differences from the standard, and (2) each series contains at least one unstable allele. This can undergo somatic and/or generative mutations to the + allele, and sometimes
also to other stable or unstable alleles of the series (see the section
Mutagenesis).

Linkage Groups

Shortly after the start of genetic investigations on *Antirrhinum*, the
first coupling of genes was detected. It concerned the 3 flower-color genes,

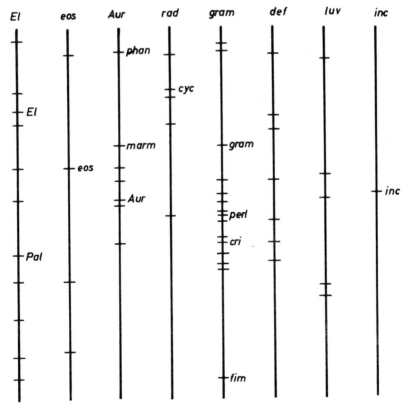

*Figure 1. Linkage maps for the 8 linkage groups of Antirrhinum majus. The
position of known loci is indicated; the symbols are included only for the genes mentioned in Table 1.*

Unicolor (now *Eluta*), *rosea*, and *Pallida* (Baur, 1911). There are 8 groups of linked genes, and each linkage group is named after a gene in it: *El* (formerly *Uni*), *eos, Aur, rad, Def* (formerly *bas*), *luv, inc,* and *gram.* The *inc* group contains only one locus; the *gram* group is the largest one with 15 loci. Only a small fraction of the known genes have been assigned to a linkage group. The reason for this is the amount of work necessary for determining the linkage relations of a gene and its map distance to other loci: crossing with at least 8 test stocks (one for each linkage group for at least 3 generations (P, F_1 and F_2), and in most cases another back-cross generation of the F_1 with the double homozygote is also required).

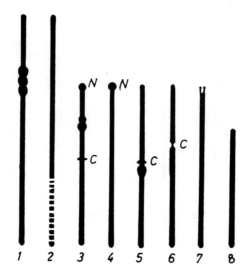

Figure 2. The pachytene chromosomes of Antir-rhinum majus Sippe 50; C = centromere, N = nucleolar organizer. Chromosome 1 is large, with a group of 3 large chromomeres; chromosome 2 is large, with a diffuse-staining end; chromosome 3 is medium-sized, with a nucleolar organizer at end of chromosome arm with 2 large chromomeres; chromosome 4 is medium-sized, with a nucleolar organizer at end of one chromosome arm; chromosome 5 is medium-sized, with a large chromomere near to cen-tromere; chromosome 6 is medium-sized, with a large, diffuse-staining centromere in heterochro-matic region; chromosome 7 is medium-sized, with two small chromomeres at one end; and chromosome 8 is small, with no special charac-teristics.

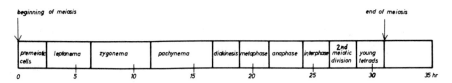

Figure 3. Timing of meiosis in pollen mother cells of Antirrhinum majus. Constructed from the data of Ernst (1938).

The main research on the localization of the genes was done between 1927 and 1938 by a single team of geneticists at the research station in Müncheberg (Figure 1).

The investigations on linkage and crossing over in *Antirrhinum* were done in several phases: (1) establishing the existence of the phenomena of linkage and crossing over in this species (1911), (2) the localization of genes (published in a series of papers during 1924–1938), (3) investigations on crossing over in species hybrids (Hoffmann, 1949), and (4) investigations on the variability of crossing over under environmental influences (Ernst, 1938; Harte, 1952; Lauritzen, 1953). To understand these problems, it is necessary to consider first the cytology of *Antirrhinum*.

Cytology

Antirrhinum has a haploid chromosome number of 8. In the pachytene stage they can be distinguished from one another by their relative sizes and some other features. There are 2 large, 5 medium, and 1 small chromosome. Chromosome 2 has a region at one end that stains only diffusely with the usual cytological stains. The other chromosomes are marked by heterochromatic knobs at one end or by one or more large chromomeres in one of the arms. The nucleolus is usually formed by 3 of the medium-sized chromosomes (No. 3, 4, and 5) (Figure 2).

The development of the 4 anthers in a flower is synchronized so that the pollen mother cells in the two anthers at a similar position at the right and left side of the flower are in the same meiotic stage. The duration of meiosis is easily determined by first investigating one pair of anthers of a flower, and then investigating the other pair 3–24 hours later. The differences of the meiotic stages observed give an exact estimate of the progress made during the time between the two fixations (see Figure 3). The whole process of meiosis takes approximately 30–34 hours. In comparison with the times known for other plants, this can be considered as very fast.

The chromosome number $n = 8$ gave rise to the suspicion that during evolution a doubling of the whole genome or of single chromosomes may have occurred. Cytological investigations of haploid plants were expected to give information on this question. In pollen mother cells of haploid plants with 8 chromosomes, all the important stages of meiosis can be identified. In zygonema, pairing between parts of the chromosomes is found. During pachynema, in many cells all the chromosomes are synapsed with a partner, but not with the same one for their entire length. In every cell the chromosomes form different, complex configurations. Foldback pairing is seen in many places. These observations suggest that illegitimate pairing is common in haploid snapdragons. In metaphase I, the chromosomes in most cells behave like univalents, but some are held together by chiasmata. Therefore, both synapsis and chiasmata formation occur between nonhomologous chromosomes in haploids (Rieger, 1957), and homologous segments cannot be detected presently, although chromosomal doubling may have occurred in the past.

Chromosomes and Linkage Groups

As in other species, the assignment of linkage groups to the chromosomes was attempted with the use of chromosome aberrations. Here it is assumed that in case of a deletion, the plant will be hemizygous for that part of the chromosome. If, in a heterozygote, the deletion occupies the place in the chromosome normally containing certain dominant alleles, the action of these chromosomes will be missing, and the plant will develop the phenotype which is correlated to the recessive alleles on the other chromosome. Such studies were done on plants whose genetic markers belonged to the *El* group. The cytological investigation of the plants with the unexpected phenotype made it possible to coordinate the *El* group to chromosome 2 (Pohlendt, 1942). The *gram* group was assigned with the help of translocations to either chromosomes 3, 4, or 6. In addition, it was concluded that the unstable alleles of the *gram* series originate, if the part of the chromosome that contains the *gram* locus is translocated into the neighborhood of one of the heterochromatic knobs, at the end of one of these chromosomes (Mechelke and Stubbe, 1954).

Trisomics

With the haploid number of 8 chromosomes, we can expect 8 different primary trisomics. They are synthesized by crossing $4n \,\female \times 2n \,\male$, and backcrossing the triploid F_1 with the diploid parent ($3n \,\female \times 2n \,\male$).

The progeny contains 5–20 percent trisomics (Sampson *et al.*, 1961). The 8 trisomics are defined by morphological characters and are confirmed cytologically. Six of the trisomics have been assigned to genetically defined linkage groups.

Chiasmata and Crossing Over

Antirrhinum is one of the few species where research has been done to analyze the variability of crossing over and the reasons for it. There are two different cytological questions: (1) Can a correlation be found between a change of crossing-over frequency and chiasma frequency? (2) Are there structural differences of the chromosomes between species of the genus *Antirrhinum* which lead to differences in the mapping of genes or to differences in the crossing-over frequency? If the chiasmata, which are seen in pollen mother cells in diakinesis or metaphase I, are the places of interchange of parts of the synapsed homologs, then they are the cytological correlate of the genetically defined crossing over. The consequence of this hypothesis is the expectation of a positive correlation between chiasma frequency and crossing-over frequency. Investigations on this problem were initiated by Oehlkers (1935, 1937).

In pollen mother cells of *Antirrhinum* it is not possible to count the chiasmata directly. If at least one chiasma is formed in a pair of chromosomes between the centromere and the end, then the partners will be held together at diakinesis by terminal chiasma. It is not possible to distinguish the chromosomes in this stage of meiosis. A high frequency of chiasmata in pachynema will lead to a high relative frequency of terminal chiasmata in diakinesis and metaphase I. In cells with a low frequency of chiasmata there will occur some chromosome arms with no chiasma at all, and, as a consequence, no terminal chiasma in later stages. On the basis of this correlation, the difference in the relative frequency of terminal chiasmata between two samples will allow a conclusion about differences in chiasma frequency in earlier stages of meiosis of these plants and of others treated during the same time in the same manner. If, in an experiment with heterozygous plants, some from each treatment are taken as a sample for the cytological observations and other plants from the same treatment are used for backcrossing of their pollen to styles of double-recessive plants, then the investigator can compare the behavior of terminal chiasmata and crossing over. In controls, approximately 99.5 percent of the chromosome arms are connected by terminal chiasmata. After experimental treatments with heat shocks or by extremely low or high water contents, the frequency of terminal chiasmata was lowered to 80 percent.

In a parallel experiment, the crossing over between the loci *Aurea* and *marmorata* decreased from 15 percent in the controls to 5 percent. Similar relations between the frequencies of terminal chiasmata and crossing over were found in experiments with plants heterozygous for genes in the linkage groups *El, bas,* and *gram.* These results are in agreement with the expectation, on the basis of the hypothesis, that chiasmata are the places of crossing over (Ernst, 1938, 1939; Harte, 1952; Lauritzen, 1953). Antirrhinum is one of the two genera (Oenothera is the other) in which experiments of this type have been done. Considering the dependence of crossing over on environmental conditions, it is not astonishing that in earlier publications authors found a great variability of the frequency of crossing over and of the calculated map distance in their experiments.

Crossing Over in Species Hybrids

In progenies of the hybrids between *A. majus* and other species of the genus (*A. molle, glutinosum, barrelieri,* and *latifolium*) it was noted that the sequence of the loci in the linkage groups *El, gram,* and *Aur* is the same for all species (see reviews by Hoffmann, 1949; Stubbe, 1966).

Phenogenetics

The genes of *Antirrhinum* have been used for a variety of investigations in both biochemical and developmental genetics.

Flower Pigments

The gene responsible for the formation of red and yellow flower pigments is niv^+. In niv/niv homozygotes the concentration of coffeic acid, ferulic acid and p-cumaric acid is increased 10-fold compared with niv^+ plants. Precursors of the flower pigments can be synthesized from the hydroxy-cinnamic acids only in the presence of niv^+. The allele inc^+ transforms a certain percentage of these precursors into anthocyanins. The alleles *eos* and eos^+ are responsible for simultaneous changes in the chemical constitution of all pigments, flavonoles as well as anthocyanins. The alleles of the *Pal* series are responsible for the synthesis of cyanidin-3-rutinoside; however, under the influence of those alleles that cause a decrease of anthocyanins, the comparable flavonole derivate (quercitin-3-rutinoside), instead of the anthocyanin, is synthesized.

The pigment pattern in different genotypes is given in Table 2. Figure 4 illustrates the interrelations between the pigments.

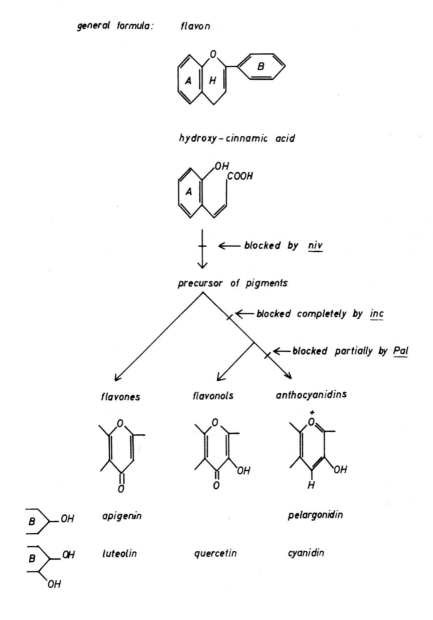

Figure 4. Biosynthesis of flower-color pigments in A. majus. All pigments have the same basic formula, i.e., derived from flavon. The differences between flavones, flavonols, and anthocyanidins are in the radicals in the positions 3 and 4 of the heterocyclic ring (H). In anthocyanidins and flavonols, different sugars can be added in position 3. Within each of these groups, of pigments, the same differences in the radicals in the positions 3 and 4 of the B ring are found. Apigenin and pelargonidin have only one hydroxy group in position 3; luteolin, quercetin, and cyanidin have a second hydroxy group in position 4.

TABLE 2.   The Pigment Pattern in Different Genotypes

	Luteolin	Apigenin	Cyanidin	Pelar-gonidin	Hydroxy-cinnamic acids
Sippe 50	+	+	+	−	+
Eluta	+	+	+	−	+
incolorata	+	+	−	−	+
eosina	−	+	−	+	+

New results show that for the formation of the flower pigments the cooperation of major genes with specific suppressor genes is of importance. In the presence of *sup-El*, the allele *El,* known as dominant, becomes recessive compared with *El*$^+$ (Linnert, 1968). Other genes influence the concentration and the distribution of the pigments in the petals (Sherrat, 1958; Sampson and Hunter, 1959; Harborne, 1963; Schmidt, 1962). The alleles *inc*$^+$, *Pal*$^+$, and *El* act quantitatively on the amount of anthocyanins. The dose effects are additive (Seyffert, 1957).

Leaf Pigments and Plastids

The synthesis of chlorophylls is disturbed by several mutants in a characteristic way. There are aberrant photostable (*meline*) and photosensitive (*albostriata, bipartita*) genotypes in which the development of the leaf color is influenced by light (Sagromsky, 1962). Baur (1910) observed white and green spotted plants (*albomaculatus*) which arose by the spontaneous mutation of the plastids. The aberrant plastids cannot form normal grana and are in most cases devoid of pigments. An experiment to increase the mutation rate of the plastids with x rays did not succeed (Maly, 1958). In spite of some promising results, an intensive study of the gene and plastom mutations to elucidate the molecular basis of the cooperation of the nuclear genes and the plastom in the development of plastids and their pigments remains to be done (Döbel, 1964; Diers, 1971).

Morphological Characters

Out of the set of genes with morphological effects, only a few have been used for investigations on phenogenetics. The first publication of this kind was on the alleles of the *Def* series (*Deficiens*), whose action leads to a reduction of corolla and anthers. The phenocritical phase in the development of the flower (the stage where the first difference from the normal can be seen) shifts to earlier stages when the reduction of the flower parts increases (Klemm, 1927).

Experiments on the development of leaves under the influence of alleles that cause narrow or small leaves have shown that there are characteristic differences in the values of the parameters of the allometric function that describes the relation of growth in length and growth in width. The mutants also show differences in the reaction of these parameters to changes of environment. All these genes have pleiotropic effects. For example, the pattern of stomata in the epidermis of the leaf and the growth pattern of excised roots are modified (Harte and Hansen, 1967, 1970, 1971).

The gene *Crispa* (*Cri;* undulated leaves, caused by disturbance in the development of the epidermis) is a good example of intergenic interaction. In plants which contain the dominant allele *Aur* (*Aurea;* yellow-green leaves) or which are homozygous for recessive alleles of other loci which act on the chlorophyll content of the plastids, the symptoms correlated to *Cri* are not found. The allele that causes a disturbance of chlorophyll synthesis or plastid development is in all these cases epistatic over *Cri*. This evidence indicates a developmental correlation between the differentiation of the epidermal cells, which normally do not contain chlorophyll, and the development of the plastids (Mainx, 1948).

Heterosis has been established for a number of loci. A model experiment about the combining ability of 14 genotypes showed that such heterosis is usually found during a short stage in development. There is no superdominance. In the heterozygotes there is a combination of genes with different dominance relations for the final size and the pattern of growth (Stern, 1958).

Mutagenesis

Investigations on spontaneous mutations were done on *A. majus* in the early period of genetic research (Baur, 1924). Spontaneous mutations are found in about 0.6 percent of the gametes (Stubbe, 1966). *Antirrhinum* has two peculiarities: (1) differences in the rate of mutations (both spontaneous and mutagen-induced) between lines and (2) mutable or unstable alleles at many loci. During the heyday of mutation research, the mutagenic action of many influences was tested: aging of seeds, radium treatment, x-ray treatment of pollen and seeds, and the influence of chemicals that were applicated to seeds or to the inflorescence before and during meiosis. Early experiments were done on chemical mutagenesis in *Antirrhinum* (Baur, 1932). It was later proved that treatment with chemicals (urethane–KCl mixture) not only induced chromosome mutations, but also mutations of single loci (Oehlkers, 1956). The first experiments on the effect of radium treatment led to the detection of radiation-induced tumors in plants. Continued investigations on this problem showed the relation between radiation damage, disturbances of cell structure, deviations in the development of the nucleus, and induced chromosomal abnormalities (Stein, 1922, 1942).

The mutable alleles are designated by the term *recurrens* and by the use of superscript *rec*. The first case of an unstable allele was the mutant *phantastica* described by Baur (1926). Following are examples of the observed reactions: In *perl^{rec}* (*perlutea recurrens*), the unstability becomes apparent in somatic cells and in generative cells or in the sporogenic tissue during the last two cell generations before meiosis, and the change always goes in one step to the stable allele *perl^+* (Döring, cited in Stubbe, 1966). In *fim* (*fimbriata*), the mutation always generates another unstable allele. The deformation of the flower varies from extreme dissection of the corolla in several narrow strips, i.e., only connected at the base, to nearly normal. Repeated selection in both directions is possible. A backmutation to the standard allele *fim^+* has never been observed (Harte, 1951).

The unstable alleles of the *gram* series are correlated with cytological peculiarities which are interpreted as translocations. Backmutation is not possible in vegetative cells, but only during meiosis. As an explanation for this behavior, it is postulated that in *gram^{rec}* a position effect occurs, and that backmutation occurs by retranslocation of the *gram* locus to its original position through crossing over (Mechelke and Stubbe, 1954; Stubbe, 1966).

The allele *Pal^{rec}* (*Pallida recurrens*) can mutate in both somatic and germ cells. The behavior of this allele resembles the unstable loci *A* and *R* of maize. The background color of the flowers is white. After mutation of *pal^{rec}* to *pal^+* or other stable alleles of the series, the cell can synthesize anthocyanins. The size of the pigmented areas varies between whole branches and small spots containing one or a few cells on the outer side of the corolla. The rate of backmutation is temperature sensitive; it is very high at 15°C, and it decreases with increasing temperature. In heterozygotes the mutation rate is influenced by the stable allele that accompanies the unstable *Pal^{rec}*. Different lines are distinguished by the rate of backmutation. The difference between "high" and "low" mutability depends on the alleles of a mutator gene, that is not linked to the major gene (Harrison and Fincham, 1964 1967, 1968).

In some stocks of "Sippe 50" there are mutator genes that influence both the spontaneous and induced x-ray mutation rate. There are also epistatic suppressor genes that determine whether or not mutations at certain other loci can be detected ($El\text{-}sup_{El}$ and $squa\text{-}sup_{squa}$ are examples) (Linnert, 1966, 1967, 1968).

Thanks to the intensive research that was done on *Antirrhinum* by many scientists using the methods of classical genetics, a genetically analyzed species is now available that can be used on modern problems of gene action and gene structure.

Literature Cited

Baur, E., 1907 Untersuchungen über die Erblichkeitsverhältnisse einer nur in Bastardform lebensfähigen Sippe von *Antirrhinum majus*. *Ber. Dtsch. Bot. Ges.* **25**:442–454.

Baur, E., 1910 Untersuchungen über die Vererbung bon Chromatophorenmerkmalen bei *Melandrium, Antirrhinum* und *Aquilegia*. *Z. Vererbungsl.* **4**:81–102.

Baur, E., 1911 Ein Fall von Faktorenkoppelung bei *Antirrhinum majus*. *Verh. Naturforsch. Vereins Brünn* **XLIX**:30–38.

Baur, E., 1924 Untersuchungen über das Wesen, die Entstehung und die Vererbung von Rassenunterschieden bei *Antirrhinum majus*. *Bibl. Genet.* **4**:1–110.

Baur, E., 1926 Untersuchungen über Faktormutationen I: *Antirrhinum majus* mut. phantastica, eine neue, dauernd zum dominanten Typ zurückmutierende rezessive Sippe. *Z. Vererbungsl.* **41**:47–53.

Baur, E., 1932 Der Einfluß von chemischen und physikalischen Reizungen auf die Mutationsrate von *Antirrhinum majus*. *Z. Vererbungsl.* **60**:467–473.

Darwin, C., 1868 *The Variation of Animals and Plants under Domestication*, John Murray, London.

Diers, L., 1971 Übertragung von Plastiden durch den Pollen von *Antirrhinum majus*. II. Der Einfluß verschiedener Temperaturen auf die Zahl der Schecken. *Mol. Gen. Genet.* **113**:150–153.

Döbel, P., 1964 Über die Plastiden einer Herkunft des Status albomaculatus von *Antirrhinum majus* L. *Z. Vererbungsl.* **95**:226–235.

Ernst, H., 1938 Meiosis und crossing-over, zytologische und zytogenetische Untersuchungen an *Antirrhinum majus*. *Z. Bot.* **33**:241–294.

Ernst, H., 1939 Zytogenetische Untersuchungen an *Antirrhinum majus* L. *Z. Bot.* **34**:81–111.

Godron, D. A., 1863 Des hybrides végétaux considérés au point de vue de leur fécondité et de la perpétuité ou non-perpétuité de leurs charactères. *Ann. Sci. Nat. IV, Série Bot. Toms* **XIX**:135–179.

Hagemann, R., 1967 Die Wirkung von Umweltfaktoren auf das Manifestierungsmuster der Mutation transcendens von *Antirrhinum majus* L. *Kulturpflanze* **15**:367–380.

Harborne, S. B., 1963 Plant polyphenols. X. Flavone and aurone glycosides of *Antirrhinum*. *Phytochemistry* **2**:327–334.

Harrison, B. J. and J. R. S. Fincham, 1964 Instability at the *Pal* locus in *Antirrhinum majus*. I. Effects of environment on frequencies of somatic and germinal mutation. *Heredity* **19**:237–258.

Harrison, B. J. and J. R. S. Fincham, 1967 Instability at the *Pal* locus in *Antirrhinum majus*. II. Multiple alleles produced by mutation of an original unstable gene. *Heredity* **22**:211–224.

Harrison, B. J. and J. R. S. Fincham, 1968   Instability at the *Pal* locus in *Antirrhinum majus*. III. A gene controlling mutation frequency. *Heredity* **23**:67–72.

Harte, C., 1951   Untersuchungen über labile Gene I. Mitteilung. Selektionsversuche an *Antirrhinum majus* L. mut. fimbriata. *Z. Vererbungsl.* **83**:392–413.

Harte, C., 1952   Untersuchungen über die Nachkommenschaft von Heterozygoten der graminifolia-Koppelungsgruppe von *Antirrhinum majus* L. *Z. Vererbungsl.* **84**:480–507.

Harte, C., 1969   Gonenkonkurrenz bei *Antirrhinum majus* L. *Theor. Appl. Genet.* **39**:339–344.

Harte, C. and H. Hansen, 1967   Untersuchungen über die Entwicklung der Blattform bei *Antirrhinum majus* L. und einigen Mutanten. *Planta (Berl.)* **73**:250–280.

Harte, C. and H. Hansen, 1970   Untersuchungen über Zellgröße und Zellanzahl in den Blättern von *Antirrhinum majus* L. und einigen ihrer Mutanten. *Biol. Zentralbl.* **89**:23–47.

Harte, C. and H Hansen, 1971   Das Spaltöffnungsmuster in der Blattepidermis einer Standardsippe und einiger Mutanten von *Antirrhinum majus* L. *Biol. Zentralbl.* **90**:1–26.

Hoffmann, W., 1949   Untersuchungen über Kopplungen bei *Antirrhinum*. IX. Mitteilung. In einigen Artkreuzungen. *Z. Vererbungsl.* **83**:165–202.

Klemm, M., 1927   Vergleichende morphologische und entwicklungsgeschichtliche Untersuchungen einer Reihe multipler Allelomorphe bei *Antirrhinum majus*. *Bot. Arch.* **20**:432–474.

Lauritzen, M., 1953   Zytogenetische Untersuchungen an den Basund El-Koppelungsgruppen von *Antirrhinum majus*. *Z. Vererbungsl.* **85**:220–237.

Linnert, G., 1966   Unterschiedliche Mutationsspektren verschiedener reiner Linien der Sippe 50 von *Antirrhinum majus* L. *Z. Vererbungsl.* **98**:25–40.

Linnert, G., 1967   Die Wirkung von Suppressorgenen bei der Induktion spezifischer Mutationen durch Röntgenstrahlen bei *Antirrhinum majus* L. *Biol. Zentralbl.* **86** (Suppl.):119–126.

Linnert, G., 1968   Genetische Analyse eines weiteren Suppressorgens bei *Antirrhinum majus*. L. *Mol. Gen. Genet.* **101**:17–28.

Mainx, F., 1948   Versuche zur Analyse der Genwirkung bei *Antirrhinum majus*. *Österr. Bot. Z.* **94**:3.

Maly, R., 1958   Die Mutabilität der Plastiden von *Antirrhinum majus* L. Sippe 50. *Z. Vererbungsl.* **89**:692–696.

Mechelke, F. and H. Stubbe, 1954   Studien an mutablen Genen. 1. *Antirrhinum majus* L. mut. graminifolia. *Z. Vererbungsl.* **86**:224–248.

Oehlkers, F., 1935   Untersuchungen zur Physiologie der Meiosis I. *Z. Bot.* **29**:1–53.

Oehlkers, F., 1937   Neue Versuche über zytologisch-genetische Probleme (Physiologie der Meiosis). *Biol. Zentralbl.* **57**:126–149.

Oehlkers, F., 1956   Die Auslösung von Mutationen durch Chemikalien bei *Antirrhinum majus*. *Z. Vererbungsl.* **87**:584–589.

Pohlendt, G., 1942   Cytologische Untersuchungen an Mutanten von *Antirrhinum majus* L. I. Deletionen im uni-Chromosom. *Z. Vererbungsl.* **80**:281–288.

Rieger, R., 1957   Inhomologenpaarung und Meioseablauf bei haploiden Formen von *Antirrhinum majus* L. *Chromosoma* **9**:1–38.

Sagromsky, H., 1962   Die Bedeutung des Lichtes für die Ausprägung eines Mutationsmerkmales der Mutante bipartita von *Antirrhinum majus*. *Kulturpflanze* **10**:158–167.

Sampson, D. R. and A. W. S. Hunter, 1959   Inheritance of shades of crimson and magenta flowers of *Antirrhinum majus. Can. J. Genet. Cytol.* **1**:173–181.

Sampson, D. R., A. W. S. Hunter and E. C. Bradley, 1961   Triploid × diploid progenies and the primary trisomics of *Antirrhinum majus. Can. J. Genet. Cytol.* **3**:184–194.

Schmidt, H., 1962   Chemische Untersuchungen über den Biosyntheseweg der Blütenfarbstoffe in Mutanten von *Antirrhinum majus. Biol. Zentralbl.* **81**:213–226.

Seyffert, W., 1957   Untersuchungen über interallele Weschselwirkungen. I. Die unvollständige Dominanz des El-Faktors von *A. majus. Z. Vererbungsl.* **88**:56–77.

Sherrat, H. S. A., 1958   The relationship between anthocyanidins and flavonoles in different genotypes of *Antirrhinum majus. J. Genet.* **56**:28–36.

Stein, E., 1922   Einfluβ von Radiumbestrahlung auf Antirrhinum. *Z. Vererbungsl.* **29**:1–15.

Stein, E., 1942   Über einige durch Radiumbestrahlung erzeugte Periklinalchimären von *Petunia* und *Antirrhinum siculum* mit Veränderung der Zellstruktur. *Biol. Zentralbl.* **62**:483–508.

Stern, K., 1958   Kombinationseignung hinsichtlich der Wachstumsergebnisse eines Modellversuches mit *Antirrhinum majus* L. *Silvae Genet.* **7**:41–57.

Stubbe, H., 1952   Über einige theoretische und praktische Fragen der Mutationsforschung. *Abh. Sächs. Akad. Wiss. Kl.* **44**:3.

Stubbe, H., 1963   Über die Stabilisierung des sich variabel manifestierenden Merkmals "Polykotylie" von *Antirrhinum majus* L. *Kulturpflanze* **11**:250–263.

Stubbe, H., 1966   *Genetik und Zytologie von Antirrhinum L. sect. Antirrhinum*, G. Fischer, Jena.

Wheldale, M., 1907   The inheritance of flower colour in *Antirrhinum majus. Proc. R. Soc. Lond. Ser. B Biol. Sci.* **79**:288–305.

14

Collinsia

EDWARD D. GARBER

Introduction

The genus *Collinsia* includes at least 22 annual herbaceous species in the family Scrophulariaceae which, with two exceptions, are found mostly in the western regions of the Pacific Coast States. Two species, *Collinsia violacea* and *Collinsia verna*, are endemic to the Midwest and have been collected in Illinois. Consequently, numerous populations of each species can be acquired without much difficulty. The genus is taxonomically isolated but related to the small genus *Tonella*. Newsom (1929) revised the taxonomy of the species in *Collinsia* and, except for conflicts in the status of several taxa, the revision has been retained in its essentials by Pennell (1951) and Munz (1959), using the usual criteria of the herbarium taxonomist. The first attempt to apply the methodology of experimental taxonomy was reported by Hiorth (1933*b*) in his study of an interspecific hybrid between *Collinsia heterophylla* (*Collinsia bicolor*) and *Collinsia bartsiaefolia*.

The genus *Collinsia* offers a number of technically advantageous characteristics for investigating the cytogenetics of interspecific hybrids at the diploid and polyploid levels: Many species yield 2–3 crops each year in the greenhouse. Three groups of species have relatively large flowers which are readily emasculated for controlled pollination. Many species produce numerous clusters of flowers over a sufficiently long period so that cytological studies can use the early buds and the later flowers are available for crosses. Seeds of most of the species germinate without delay.

EDWARD D. GARBER—Barnes Laboratory, The University of Chicago, Chicago, Illinois.

Finally, the low somatic chromosome number $(2N = 14)$ is an asset in analyzing diploid and polyploid interspecific hybrids and chromosomal aberrations in diploid species.

Culture Conditions

To insure maximum germination, seeds are harvested from capsules which have opened spontaneously. The failure to collect such seeds is usually responsible for the inviability of seeds from plants which are still flowering in the field. Although seeds germinate in soil, the seeds are usually added to moist filter paper in Petri dishes maintained at 10–15°C until the radicle emerges. Approximately 90–95 percent of the seeds germinate within 5–7 days and are planted in soil with some sand (3.5-inch unglazed pots). This procedure gives a valid count of germinated and ungerminated seeds.

To obtain vigorous seedlings, greenhouse temperatures should not exceed 25°C during the first 3 weeks. Temperatures up to 32°C have no deleterious effect after this interval. During the autumn and winter, supplemental lighting with incandescent-filament 150-watt lamps from 7 am to 7–9 pm is necessary in Chicago, Illinois (Hiorth, 1929; Gorsic, 1957). Fluorescent lamps should provide sufficient supplemental light, providing that incandescent-filament lamps are also used. The reflectors are usually maintained at least 3 feet from the tops of the plants.

Almost all of the species have a flower with a two-lobed upper and a three-lobed lower lip. The keel-shaped, middle lobe of the lower lip encloses the two pairs of anthers and the style. In the bud, the lower lateral lobes are folded within the middle one and the upper lip encloses the lower lip. When the upper lip of the flower begins to rise, but prior to its final vertical position, the flower is ready for emasculation. The style is relatively short and lies behind the anthers, which can be removed with forceps without rupturing. Approximately 5 days after emasculation, the flower is completely open and the stigma is receptive. The stigma usually remains within the middle lobe and is protected from stray pollen. The flowers on the main stem and branches do not shed pollen simultaneously; the lowest whorl sheds first, sequentially followed by the other whorls.

Pollen is collected on the tip of a rolled piece of torn paper towel, using the ragged edges as a small brush. Two pollinations within 24 hours generally yield high seed set in intraspecific hybridizations and maximum seed set in interspecific hybridization. After a successful pollination, the style loses rigidity and the corolla wilts, often falling after 48–72 hours. Unpollinated flowers generally do not wilt for at least 7–10 days after opening, and after this period the style appears to support the wilted flower, which often yields seed after pollination.

Cytological Methods

Clusters of buds are fixed in a solution (6:3:2, v/v) of methanol, chloroform, and propionic acid (Pienaar, 1955), which has been superior to the familiar ethanol–acetic fixatives. Air is evacuated from the buds by placing the vials in a sealed container and using an aspirator to produce a reduced atmosphere within the container. Buds have yielded satisfactory preparations after 1 year of storage in a freezer. Smears of pollen mother cells (PMC's) are stained with acetocarmine.

Each flower has two pairs of anthers, which differ in size and relative stage of meiosis. It is convenient to smear all of the anthers without adding a cover slip and to look for tetrads originating in the larger pair of anthers with the low-power objective. The smaller anthers generally yield PMC's at either metaphase I or anaphase I.

The pachytene chromosomes are not suited for a detailed study of their morphology. Consequently, the seven haploid chromosomes cannot be identified by their relative length, arm ratio, or such topological markers as heterochromatic segments. Furthermore, the somatic chromosomes at metaphase are unsatisfactory for identifying chromosomes because of their small size. The chromosomes are metacentric; at least six chromosomes appear to be similar in length, and the seventh chromosome somewhat shorter than the other chromosomes. At metaphase I, the bivalents display not more than one chiasma in each arm, and appear as rods or rings.

Cytotaxonomy

Newsom (1929) divided the genus into two groups. In one group, the species have "sessile" flowers congested in whorls, with pedicels shorter than to not longer than the calyces of the lower whorls of flowers, and flat mature seeds. In the other group, the species have pediceled flowers, either solitary or in whorls, with the pedicels of the solitary or lower-whorl flowers from as long as to longer than the calyces, and either flat or thick mature seeds.

A cytological study of the genus indicated that, with one exception, the species have seven bivalents at metaphase I (Hiorth, 1933b; Garber, 1956, 1958b; Ahloowalia and Garber, 1961; Garber and Dhillon, 1962b;

Garber and Unni, 1965; Hayhome and Garber, 1968). One species, *Collinsia torreyi,* with 21 bivalents (Ahloowalia and Garber, 1961) may have been erroneously identified as a member of *Collinsia.* No tetraploid species has been found in the genus.

The mean number of chiasmata per bivalent at metaphase I has cytotaxonomic value. The species included in the group with "sessile" flowers have 1.1–1.5 chiasmata per bivalent, and those in the group with pediceled flowers have 1.7–1.9 (Garber, 1956, 1958*b*). Three species, however, do not conform: *C. corymbosa, C. multicolor,* and *C. linearis.* When the species are also characterized by the variability or uniformity of their populations, it is possible to recognize five groups of species (Table 1). While Groups I and II include most of the species previously assigned to the two major groups (Newsom, 1929), Groups III, IV, and V were erected to accommodate the three exceptional species.

TABLE 1.   *Salient Characteristics of the Groups of Species of the Genus Collinsia*

Group	Characteristics[a]	Member species
I	Sessile flowers, variable populations, low chiasmata frequency	*C. heterophylla* Buist, *C. austromontana* (Newsom) Pennell, *C. concolor* Greene, *C. tinctoria* Hartwig, *C. bartsiaefolia* Bentham, *C. stricta* Greene
II	Pediceled flowers, uniform populations, high chiasmata frequency	*C. sparsiflora* Fischer and Meyer, *C. greenei* Gray, *C. bruceae* Jones, *C. grandiflora* Douglas, *C. solitaria* Kell, *C. childii* Parry, *C. parviflora* Douglas, *C. wrightii* Watson, *C. rattanii* Gray, *C. callosa* Parish, *C. Parryi* Gray, *C. violacea* Nuttall, *C. verna* Nuttall
III	Sessile flowers, uniform populations, high chiasmata frequency	*C. corymbosa* Herder
IV	Pediceled flowers, uniform populations, low chiasmata frequency	*C. multicolor* Lindley and Paxton
V	Pediceled flowers, uniform populations, low chiasmata frequency	*C. linearis* Gray

[a] Low chiasmata frequency, 1.1–1.5 per bivalent at metaphase I; high chiasmata frequency, 1.7–1.9.

TABLE 2. *Interspecific Hybridizations in the Genus Collinsita*[a]

	linearis	*multicolor*	*corymbosa*	*callosa*	*childii*	*grandiflora*	*greenei*	*parryi*	*rattanii*	*solitaria*	*bruceae*	*sparsiflora*	*stricta*	*tinctoria*	*austromontana*	*bartsiaefolia*	*concolor*
heterophylla		+	+	0	0			0	0	+	0	+	+	+	+	+	+
concolor		+	+	0				+		+	+	+	+	0	+	+	
bartsiaefolia		+	+	0				0	0	+	+	+	+	0	+		
austromontana		+	+	0						+	+	0	0	0			
tinctoria		0	+	0	0	0	0			0	0	0	0				
stricta		+	+				0		0			0					
sparsiflora		+	+					+	0	+	+						
bruceae	0	+	+			0	0	+	0	+							
solitaria		+	+			0	0	+									
rattanii	0	0															
parryi		0		0			0										
greenei		0		0	0												
grandiflora					0												
childii			0														
callosa	0																
corymbosa		+															
multicolor																	

[a] + = succeeded, 0 = failed. Sources: Garber and Gorsic, 1956; Garber, 1958a; Bell and Garber, 1961; Ahloowalia and Garber, 1961; Garber and Dhillon, 1962a,c; Garber and Unni, 1965; Hayhome and Garber, 1968.

Interspecific Hybridizations

Four species of *Collinsia* have not yet been used for interspecific hybridizations: *C. parviflora*, *C. wrightii*, *C. verna*, and *C. violacea*. Of the 153 possible interspecific hybridizations (Table 2), 45 combinations succeeded, 45 combinations failed, and 63 combinations have not yet been attempted. A number of unsuccessful interspecific hybridizations did not involve reciprocal crosses. It is possible that reciprocal crosses or crosses with plants from different populations would yield viable seed or viable seedlings. Hybrids between *C. austromontana* and *C. bruceae* and between *C. sparsiflora* and *C. multicolor* died at the seedling stage. Dwarf-lethal seedlings were obtained from reciprocal crosses between *C. bruceae* and *C. bartsiaefolia*, but viable seedlings resulted from the cross between unusual collections of *C. bartsiaefolia* as seed parent and *C. bru-ceae* (Garber and Dhillon, 1962*c*).

While no discernible pattern of successful or unsuccessful interspecific hybridizations is apparent for either intragroup or intergroup crosses, certain species yield few to no successful interspecific hybridizations: *C. tinctoria* (Group I), *C. ratanii, C. parryi, C. grandiflora, C. childii, C. callosa* (Group II), and *C. linearis* (Group V). It would be interesting to determine whether the pollen from these species germinate on the stigma or grow in the style of other species.

Interspecific hybridizations offer one means to detect cryptic or sibling species in the genus *Collinsia*. The taxonomic treatments of Newsom (1929), Pennell (1951), and Munz (1959) are replete with confused assignments of taxa. For example, *C. stricta* was viewed as a variety of *C. bartsiaefolia* or *C. tinctoria*. Garber and Unni (1965) established *C. stricta* as a valid species by interspecific hybridizations and a cytogenetic study of the interspecific hybrids.

Cytogenetics of Diploid Interspecific Hybrids

Diploid interspecific hybrids exhibit bivalents, univalents and bivalents, or 1–2 interchange complexes with 4–10 chromosomes at metaphase I and 0–3 dicentric chromatid bridges and acentric fragments at anaphase I. Garber (1960) proposed the following hypothesis to account for the probable course of speciation in the genus *Collinsia*: (1) Chromosome repatterning of the homologous genomes has been a major factor in the isolation of populations in incipient species; (2) mutations were responsible for the emergence and establishment of species; and (3) polyploidy played no significant role in speciation. There is no evidence at this time that the genus includes species with nonhomologous genomes.

Interchange Complexes

The chromosome associations at metaphase I for 39 diploid interspecific hybrids have been summarized in Table 3. Garber (1960) assigned chromosome formulas to nine species to account for the number of interchange complexes and for the number of chromosomes in each complex, using the rationale to identify the exchanged terminal segments of chromosomes in the Renner complexes of *Oenothera* (Cleland, 1936). These chromosome formulas are tentative and require supporting evidence from a larger number of interspecific hybrids and the chromosome associations in the corresponding amphiploids.

TABLE 3. Chromosome Associations in Diploid Interspecific Hybrids in the Genus Collinsia[a]

	concolor	bartsiaefolia	austromontana	stricta	sparsiflora	bruceae	solitaria	greenei	corymbosa	multicolor
heterophylla	II + I	II + I	106	II + I	204	04 + 08	04 + 08		104	106
concolor		II + I	II + I	II + I	204	04 + 08	04 + 08		II + I	106
bartsiaefolia			II + I	II + I	08	08	08		II + I	106
austromontana										108
tinctoria									106	
stricta									II + I	108?
sparsiflora							204	010	04 + 06	04 + 06
bruceae							204	010	04 + 06	04 + 08
solitaria						II			012	106
corymbosa										

[a] II + I = bivalents and univalents; 0 = interchange complex. Sources: see Table 2.

Although diploid intragroup-I interspecific hybrids have either bivalents or univalents and bivalents, the observations do not necessarily indicate that these hybrids lack an interchange complex. For example, the chromosome formulas for *C. heterophylla* and *C. austromontana* indicate that the interspecific hybrid should have an interchange complex of six chromosomes. Garber (1960) suggests that the expected interchange complex occurs as three bivalents, and has presented a diagram (see Fig. 5 in Bell and Garber, 1961) to illustrate the possible sites of chiasma formation in homologous segments of the chromosomes in the proposed interchange complex. One means to determine whether the diploid intragroup-I and diploid intragroup-II interspecific hybrids with bivalents or univalents and bivalents have an interchange complex requires amphiploids. It is possible to infer the presence of interchange complexes and the number of chromosomes in each complex for a diploid interspecific hybrid by determining the maximum number of quadrivalents in the corresponding amphiploid (Garber and Dhillon, 1962a). The diploid interspecific hybrid between *C. stricta* and *C. multicolor* gives a chain but not a ring of six chromosomes, suggesting that the interchange complex might include eight rather than six chromosomes. The corresponding amphiploid could provide information to determine the number of chromosomes in the interchange complex for this interspecific hybrid. The rationale for interpreting the chromosome associations in diploid interspecific hybrids by the maximum number of quadrivalents in the corresponding amphiploid will be presented in the section on preferential chromosome pairing.

With one exception, intragroup-II hybrids have 1–2 interchange complexes; the hybrid between *C. sparsiflora* and *C. bruceae* has 7 II. The sticky chromosomes in *C. solitaria* × *C. greenei* precluded a study of chromosome associations. The largest interchange complex to date occurs in the hybrid between *C. greenei* and *C. sparsiflora* or *C. bruceae* and includes 10 chromosomes.

Intergroup hybrids usually exhibit 1–2 interchange complexes. The only exceptions have been recorded for the hybrids between *C. corymbosa* and species in Group I. The apparent lack of interchange complexes in the exceptional hybrids may represent the failure to form the interchange complex(es), as in *C. heterophylla* × *C. austromontana*. Although Garber and Gorsic (1956) reported two interchange complexes with four chromosomes in hybrids between *C. sparsiflora* and *C. heterophylla* or *C. concolor,* one complex occured as a chain. This complex may include six rather than four chromosomes.

The orientation of the interchange complexes at metaphase I in a number of interspecific hybrids is directed, i.e., in more than 50 percent of the PMC's the alternate chromosomes proceed to the same pole at anaphase I: *C. corymbosa* × *C. multicolor*, *C. corymbosa* × *C. tinctoria*, *C. multicolor* × *C. heterophylla* (Bell and Garber, 1961), *C. solitaria* × *C. sparsiflora*, and *C. solitaria* × *C. bruceae* (Ahloowalia and Garber, 1961). The directed orientation of the interchange complexes for these hybrids was also observed in the F_2 and F_3 progenies (Ahloowalia and Garber, 1961; Garber and Bell, 1962). An explanation for the directed or nondirected orientation for interchange complexes of interspecific hybrids in the genus *Collinsia* will be deferred to the section on reciprocal translocations in *C. heterophylla*.

Heterozygous Paracentric Inversions

A crossing over within a heterozygous paracentric inversion yields a dicentric chromatid bridge and acentric fragment at anaphase I. The number of heterozygous paracentric inversions in 38 interspecific hybrids was determined by counting the number of bridges and fragments at this stage (Table 4). From one to three heterozygous paracentric inversions were observed in 18 interspecific hybrids; the intragroup-II hybrids had 0–3 detectable inversions. Bridges and fragments were not found in the hybrids between *C. solitaria* and *C. sparsiflora* or *C. bruceae*. These observations

TABLE 4.   *Number of Detectable Heterozygous Paracentric Inversions in Diploid Interspecific Hybrids in the Genus Collinsia*[a]

	multicolor	*corymbosa*	*greenei*	*solitaria*	*bruceae*	*sparsiflora*	*stricta*	*austromontana*	*bartsiaefolia*	*concolor*
heterophylla	1	2		2		2	0	0	0	0
concolor	0	0		2	1	2	0	0	0	
bartsiaefolia	0	0		2	2	2	0	0		
austromontana	1	0								
tinctoria		0								
stricta	1	0								
sparsiflora		3	2	0	0					
bruceae	1	2	3	0						
solitaria	2									
corymbosa	0									

[a] Sources: see Table 2.

coupled with data on the number and directed orientation of the interchange complexes at metaphase I clearly indicate that these three taxa in
Group II are very closely related. The number of interspecific hybrids
involving *C. corymbosa* with no detectable heterozygous paracentric hybrids was greater than that for interspecific hybrids involving *C. multicolor.*

The heterozygous paracentric inversions in the diploid interspecific
hybrids may occur in one or both bivalents or in one or two of the chromosomes in an interchange complex. The unsuitability of the pachytene
chromosomes for a detailed study requires another method to determine
the probable sites of the paracentric inversions. The chromosome associations in the corresponding amphiploids often provide evidence concerning
the location of the inversions, that is, in a bivalent or an interchange complex (see the section on preferential chromosome pairing).

Chromosomal Repatterning and Speciation

Chromosomal repatterning of the presumably homologous genomes
of the species of *Collinsia* has been detected by a cytogenetic study of
diploid interspecific hybrids which provided evidence for reciprocal
translocations and paracentric inversions. It should be noted that neither
heterozygous reciprocal translocations nor heterozygous paracentric inversions have been found during the extensive cytological surveys of a
number of populations in many species. Grant (1956) has presented a
cogent hypothesis to account for the extensive chromosome repatterning in
the species of *Collinsia*: "The difficulty of ascribing a function of linkage
to structural arrangement, which, though differing as between species, are
constant and homozygous within each species may also be resolved by the
hypothesis that the structural changes are related to former rather than
the present conditions of the populations. The existing genomes, being established in the homozygous state within each species, do not prevent gene
recombinations during regular crossbreeding in the population. Their
existence may represent instead a vestige from a previous period in the
history of the species when the formation of structural heterozygotes was
commonplace. That earlier situation may have been the ancestral hybrid
swarm or introgressive population in which the progenitors of the species
had their origin."

Fertility of Interspecific Hybrids

Approximately half of the 41 diploid interspecific hybrids are fertile
to some degree (Table 5), setting seed when self-pollinated or seed-parents

TABLE 5. Fertility (F) or Sterility (S) of Diploid Interspecific Hybrids in the Genus Collinsia[a]

	multicolor	corymbosa	greenei	solitaria	bruceae	sparsiflora	stricta	austromontana	bartsiaefolia	concolor
heterophylla	F	F		S		S	F	F	F	F
concolor	F	F		S	S	S	F	F	F	
bartsiaefolia	S	F		S	S	S	F	F		
austromontana	F	F		S						
tinctoria		F								
stricta	F	F								
sparsiflora		S	S	F	F					
bruceae	S	S	S	F						
solitaria	S	S	S							
corymbosa	F									

[a] Sources: see Table 2.

in crosses with a parental or nonparental species. Intragroup-I hybrids are fertile and, except for the high fertility of hybrids between *C. solitaria* and *C. sparsiflora* or *C. bruceae,* the intragroup-II hybrids are sterile. The hybrids between species in Groups I and II are sterile. The hybrids between species in Group I and *C. corymbosa* or *C. multicolor* are usually sterile.

The fertility of diploid interspecific hybrids is generally associated with the presence of bivalents, univalents and bivalents, or 1–2 interchange complexes with a directed orientation at metaphase I and no heterozygous paracentric inversions. The fertility of many interspecific hybrids has provided an opportunity to study the cytogenetics of progenies from crosses between these hybrids and either a parental or nonparental species (Garber and Dhillon, 1962c; Garber and Bell, 1962; Garber and Unni, 1965; Hayhome and Garber, 1968).

Preferential Chromosome Pairing in Polyploid Hybrids

Triploid interspecific hybrids and amphiploids have furnished the experimental materials (Table 6) to demonstrate preferential chromosome pairing in the genus *Collinsia.* Triploid hybrids were obtained from hybridizations between diploid species and from crosses between diploids and autotetraploids. The latter crosses are more likely to succeed when

TABLE 6. *Polyploid Interspecific Hybrids and Amphiploids in the Genus Collinsia*

Hybrid	Ploidy	Origin	Reference
Interspecific Hybrids			
C. heterophylla × C. concolor	3X	seed	
C. concolor × C. sparsiflora	3X	seed	Garber and Gorsic, 1956
C. heterophylla × C. sparsiflora	3X	seed	
C. heterophylla × C. tinctoria	3X	seed	Bell and Garber, 1961
C. austromontana × C. solitaria	3X	seed	Garber and Dhillon,
C. corymbosa × C. concolor	3X	seed	1962a
C. stricta × C. multicolor	3X	seed	Garber, 1965a
(C. heterophylla × C. stricta) × C. multicolor	3X	seed	Garber and Unni, 1965
C. concolor × (C. stricta × C. concolor)	3X, 4X	seed	Hayhome and Garber, 1968
Amphiploids			
C. sparsiflora– "C. bartsiaefolia"	4X	branch	Ahloowalia and Garber,
C. bruceae– C. solitaria	4X	branch	1961
C. corymbosa– C. concolor	4X	colchicine	Garber and Dhillon,
C. solitaria– C. heterophylla	4X	colchicine	1962a
C. heterophylla– C. corymbosa	4X	colchicine	Garber, 1965a
C. multicolor– C. heterophylla	4X	colchicine	
C. concolor– C. parryi	4X	colchicine	Hayhome and Garber, 1968

one parent is aneuploid than when both parents are euploid. Amphiploids have been obtained from seeds or fertile *gigas* branches of sterile, diploid interspecific hybrids or from treating diploid hybrid seedlings with colchicine.

The genomic contribution of each parent to a triploid hybrid can be readily determined when one species has a relatively high chiasmata frequency (1.7–1.9), and the other parental species a relatively low chiasmata frequency (1.1–1.5). For example, the spontaneous triploid hybrid between *C. sparsiflora* and *C. heterophylla* or *C. concolor* usually had 7 II + 7 I, and the bivalents had a high chiasmata frequency (1.9), indicating that *C. sparsiflora* had contributed two genomes to the triploid hybrid. The triploid hybrid between *C. heterophylla* and *C. concolor* had up to 4 III at metaphase I, simulating to some degree an autotriploid for one of these species. Garber and Gorsic (1956), Garber and Dhillon (1962c), and Garber (1956a) have reported that seven triploid hybrids between one species with a high chiasmata frequency and one hybrid with a low chiasmata frequency exhibited bivalents with a high chiasmata frequency. These observations indicate that the species with a high chias-mata frequency are likely to contribute two genomes to triploid hybrids. Finally, the triploid hybrids did not display any detectable dicentric chro-matid bridges and fragments at anaphase I even when several cor-responding diploid hybrids had such bridges and fragments. These observations indicate a preferential pairing of the chromosomes in the two genomes contributed by one parental species.

The chromosome associations at metaphase I for amphiploids and their corresponding diploid hybrids are summarized in Table 7. No amphiploid had a detectable dicentric chromatid bridge and fragment at anaphase I. The chromosome associations at metaphase I and the absence of bridges and fragments at anaphase I in the amphiploids can be explained by assuming the preferential pairing of structurally identical chromosomes. The maximum number of quadrivalents in the amphiploids presumably represents the number of bivalents with structurally identical chromosomes in the diploid hybrids. The bivalents in the amphiploids represent the chromosomes in interchange complexes or in bivalents with a heterozygous paracentric inversion in the diploid. This hypothesis can be tested by predicting the minimum and maximum number of qua-drivalents for an amphiploid; by determining the number of chromosomes in the interchange complexes and then assigning the chromosomes with a heterozygous paracentric inversion to chromosomes in interchange com-plexes or in bivalents. For example, the diploid hybrid between *C. sparsi-flora* and *C. concolor* had two interchange complexes of four chromosomes and two dicentric chromatid bridges and acentric fragments (Garber and Gorsic, 1956). Eight chromosomes are involved in the interchange com-plexes, and the chromosomes with the paracentric inversion may be in-cluded in one or two bivalents or in the interchange complexes. Con-sequently, the maximum number of quadrivalents in the corresponding amphiploid should be three, and the minimum number, one. Because the amphiploid (*C. concolor–C. sparsiflora*) had a maximum number of two quadrivalents, one chromosome with the inversion is assumed to have been in a bivalent and the other chromosome with an inversion in one of the interchange complexes for the diploid hybrid.

The amphiploid *C. corymbosa–C. concolor* was particularly interesting because the corresponding diploid hybrid did not exhibit an interchange complex of six chromosomes according to the chromosome formulas for these species (Garber, 1960) and exhibited bivalents and univalents (Bell and Garber, 1961). Garber and Dhillon (1962a) found a maximum number of 4 IV in the amphiploid, indicating that 4 II in the diploid hybrid had structurally identical chromosomes and that six chro-mosomes were included in the hypothesized interchange complex.

TABLE 7. *Chromosome Associations in Diploid Interspecific Hybrids and their Corresponding Amphiploids in the Genus Collinsia*

Hybrid	Diploid hybrids		Amphiploids	References
	Chromosome association	Heterozygous paracentric inversions	Maximum number of IV's	
C. corymbosa × C. concolor	II + I	0	4	Garber and Dhillon, 1962a
C. concolor × C. sparsiflora	204 + 2 II	2	2	Garber and Dhillon, 1962a
C. sparsiflora × C. bartsiaefolia	08 + 3 II	2	1	Garber and Dhillon, 1962a
C. solitaria × C. heterophylla	04 + 08 + 1 II	2	1	Garber and Dhillon, 1962a
C. heterophylla × C. corymbosa	04 + 5 II	2	2	Garber, 1965a
C. multicolor × C. heterophylla[a]	06 + 4 II	1	2	Garber, 1965a
C. concolor × (C. concolor × C. stricta)	II + I	0	0	Hayhome and Garber, 1968

[a] Plant of C. heterophylla had 1 04 + 5 II

If the chromosomes in a diploid interspecific hybrid in the genus *Collinsia* do not have structurally identical homologs, the corresponding amphiploid should yield only bivalents, or occasionally bivalents and two univalents, thereby simulating a genomic allotetraploid. Hayhome and Garber (1968) obtained evidence for such a situation but did not use a diploid interspecific hybrid in the usual sense. A hybridization between *C. concolor* and *C. concolor* × *C. stricta* yielded a spontaneous amphiploid which had 14 II, or occasionally 13 II + 2 I. To apply the hypothesis of preferential pairing, it is assumed that (1) the chromosomes from *C. concolor* × *C. stricta* had segments from both parental genomes and were not structurally identical with the chromosomes of either parental species (cryptic structural hybridity), and that (2) the amphiploid included two genomes from *C. concolor* and two identical "genomes" from the interspecific hybrid. The preferential pairing of the structurally identical homologs in an amphiploid with four presumably homologous genomes can simulate a genomic allotetraploid.

"New" Genomes from Fertile Diploid Interspecific Hybrids

Fertile diploid interspecific hybrids in the genus *Collinsia* presumably produce gametes whose chromosomes have segments from each of the parental chromosomes. Dhillon and Garber (1962b) used F_2 progenies with only bivalents from *C. corymbosa* × *C. tinctoria* and from *C. corymbosa* × *C. multicolor* which had one interchange complex of six chromosomes and no detectable heterozygous paracentric inversions to establish F_3 lines with bivalents. Plants from these lines were crossed with the parental species and nonparental species to determine the chromosome associations in the diploid hybrids. The chromosome associations were used to determine whether the contribution from the lines different from the interspecific hybrids would be comparable to those for hybrids involving the parental species yielding the lines.

"New" genomes were present in two lines from *C. corymbosa* × *C. tinctoria* and in one of three lines from *C. corymbosa* × *C. multicolor*. Hybrids between lines from *C. corymbosa* × *C. tinctoria* and *C. multicolor* had 7 II and were fully fertile. Hybrids involving these lines and either *C. heterophylla* or *C. corymbosa* exhibited an interchange complex of six chromosomes, as in hybrids between *C. multicolor* and each species (Bell and Garber, 1961). These cytological observations indicate that the two lines derived from *C. corymbosa* × *C. tinctoria* had a genome simulating that in *C. multicolor*. Although both *C. corymbosa* and *C. tinctoria* have short flower pedicels, the lines derived from their hybrid had a long flower pedicel, similar to that found in *C. multicolor.*

The "new" genome in one of three lines from *C. corymbosa* × *C. multicolor* simulated the genome in *C. heterophylla* in that hybrids between this line and *C. heterophylla* gave 7 II. Furthermore, plants with the "new" genome had flowers with a crossline on the upper lip, a phenotype found in *C. heterophylla* but not in either parental species.

One line derived from *C. corymbosa* × *C. multicolor* had a genome simulating the genome of *C. multicolor.* Hybrids between this line and *C. corymbosa* had one interchange complex of six chromosomes with a directed orientation at metaphase I, as was the case in hybrids between *C. corymbosa* and *C. multicolor.* Finally, this line had the long flower pedicel found in *C. multicolor.*

Although the third line from *C. corymbosa* × *C. multicolor* had a long flower pedicel, the plants and flowers combined a number of morphological characters associated with each parental species. Furthermore, the line did not yield hybrids in crosses with each parental species. If this had been found in nature, it would have probably been recognized as a distinct taxon with close relationships to *C. corymbosa* and *C. multicolor.*

The recovery of plants with "new" genomes in progeny from fertile interspecific hybrids and with 7 II in *Collinsia* opens new vistas in studying speciation at the diploid level in this genus. By appropriate crosses with the species in the genus, it should be possible to detect the "new" genomes which simulate genomes already present or represent genomes not yet found in the genus.

Autopolyploidy

Four species of *Collinsia* yielded autotetraploids when colchicine in lanolin was applied to the apical meristems of seedlings or when seeds were germinated in an aqueous solution of colchicine (Soriano, 1957; Dhillon and Garber, 1961). The autotetraploids of *C. heterophylla, C. concolor, C. tinctoria,* and *C. sparsiflora* exhibit the usual characteristics associated with autopolyploidy: larger pollen, seeds, stomata, and flowers (Dhillon and Garber, 1961). Although the autotetraploids and diploids of *C. heterophylla* do not seem to differ in their growth rate or time to flowering, the autotetraploids of the other three species grow more slowly and flower later than the corresponding diploids.

Except for *C. tinctoria,* reciprocal intraspecific hybridizations between diploids and autotetraploids do not yield seed. Crosses between diploids and autotetraploids of *C. tinctoria* gave autotriploid seed when

the diploid is used as seed parent. Intraspecific crosses between autotetraploids and self-pollinated autotetraploids yield seed. The mean number
of seeds per capsule in self-pollinated autotetraploids is greatly reduced
(by approximately 30–40 percent) when compared to that for the corresponding diploids. Some overlapping, however, occurs for the range of
seed per capsule, and seedless capsules are not unusual in the autotetraploids.

Interspecific hybridizations at the diploid level are readily accomplished for *C. concolor* and *C. heterophylla* and for *C. sparsiflora* and *C.
concolor* or *C. heterophylla* (Garber and Gorsic, 1956). Reciprocal interspecific hybridizations at the autotetraploid level, however, do not yield
seed. Two interspecific hybridizations between autotetraploids and
diploids were successful: $2N + 1$ *C. heterophylla* × $4N$ *C. sparsiflora* and
$4N$ *C. concolor* and $2N$ *C. corymbosa*.

Chomchalow and Garber (1964) investigated polyploidy in *C. tinctoria* to provide guidelines for the study of polyploidy in the genus
Collinsia. Plants with somatic chromosome numbers of 19 ($3N - 2$) to 27
($4N - 1$) were found in the progeny from a self-pollinated autotriploid.
Thirty-five trisomics were obtained from crosses between diploid seed
parents and autotriploid or $3N - 2$, pollen parents. Self-pollinated $3N - 1$
and $3N - 2$ plants gave progenies with somatic chromosome numbers
ranging from $2N$ to $3N - 1$; self-pollinated $4N - 1$ and $4N - 2$ plants
gave progenies with somatic chromosome numbers ranging from $4N - 3$
to $4N + 1$. In the former progenies, diploids constituted approximately 43
percent of the plants; a in the latter, approximately 37 percent were
autotetraploids.

Crossing autotetraploids and diploids in many species of *Collinsia*
does not yield seeds. Chomchalow and Garber (1964) proposed one
method to circumvent this problem. Hypotetraploids and hypertetraploids
are frequently obtained in the progeny from self-pollinated autotetraploids. Crosses between such aneuploids and diploids yield
autotriploids more readily than crosses between the euploids and diploids.
This method may also be used to obtain triploid interspecific hybrids.

Chromosome associations at metaphase I in autotetraploids of *C.
heterophylla, C. concolor, C. tinctoria,* and *C. sparsiflora* included
univalents, bivalents, trivalents, and quadrivalents. Bivalents and quadrivalents were the most frequent associations, and the number of
univalents usually exceeded that for trivalents. The maximum number of
7 IV was recorded in relatively few PMC's; the range was 3–7 II, with a
modal number of 5 IV per PMC. The distribution of chromosomes at
anaphase I was usually regular so that each group included 14 chromosomes.

The chromosome associations at metaphase I in autotriploids of *C. tinctoria* (Chomchalow and Garber, 1964) and *C. heterophylla* (Garber, 1964) included univalents, bivalents, and trivalents in approximately equal frequencies. The maximum number of trivalents per PMC was 5 in *C. tinctoria* and 7 in *C. heterophylla*. In *C. tinctoria*, approximately 60 percent of the PMC's at anaphase I had one group with 10 and the other group with 11 chromosomes. The mean number of chiasmata at metaphase I in the diploids and autopolyploids of four species is summarized in Table 8 (Dhillon and Garber, 1961; Garber 1964).

Chemotaxonomy

Chemotaxonomic methods have provided criteria to delineate species, to determine the parentage of interspecific hybrids, and to study the contribution of parental species to amphiploids. Paper chromatography has been used to determine whether relatively small molecules in extracts from different plants parts furnish clues to delineate taxa and to determine the composition of hybrid-swarms involving different populations of one species or different species. Electrophoresis accomplishes the same task by comparing the sites of soluble proteins or enzymes in extracts from different plant parts in starch or acrylamide gels (West and Garber, 1967*a,b*; Reddy and Garber, 1971).

Strømnaes and Garber (1963), Garber and Strømnaes (1964), and Garber (1965*b*) used root, flower, and leaf extracts for a paper chromatographic survey of 15 species of *Collinisa* and 6 interspecific hybrids and their progenies. The dimension, location (Rf) and color of "spots" in visible or ultraviolet light before and after treatment with specific reagents provided a basis for comparing and contrasting chromatograms. The results did not yield the necessary evidence to distin-

TABLE 8. *Mean Number of Chiasmata at Metaphase I in Diploids and Autopolyploids of Four Species of Collinsia[a]*

Species	2N	3N	4N
C. heterophylla	1.3–1.5	1.3	1.1–1.5
C. concolor	1.1–1.5	—	1.3–1.6
C. tinctoria	1.2	1.2–1.3	1.4–1.7
C. sparsiflora	1.8–1.9	—	1.4–1.6

[a] Chiasma frequency for 3N calculated by doubling the mean number of chiasmata per chromosome for 4N on the basis of 14 pairs of chromosomes, not necessarily bivalents.

guish the different taxa, to determine their relationship in the different groups, or to establish the parentage of interspecific hybrids and their progenies. These observations suggest that the species of *Collinsia* are so closely related that no overt differences could be detected in the extracts for the small molecules separated by paper chomatography.

Wennstrom and Garber (1965) used starch gel electrophoresis to compare zymograms for extracts from seeds, flowers, and young leaves of 12 species of *Collinsia* and one interspecific hybrid. Although the different sources of extract gave different patterns of esterase and phosphates sites, the tested species gave the same pattern for each enzyme system when one plant part and one extract were used. These observations indicate considerable homogeneity among the tested species with respect to these enzymes. Other enzyme systems and different extraction methods might yield zymograms to characterize the species of *Collinsia*. The species in Group I include variable populations which should be surveyed to detect electrophoretically variant enzymes as another measure of intraspecific variability. These species are usually cross-pollinated in nature, in contrast with the species of Group II which are usually self-pollinated. Electrophoresis could provide additional evidence to contrast the variability of species in Group I with the uniformity of species in Group II.

Garber (1965*b*) used disc polyacrylamide gel electrophoresis to separate the soluble proteins in leaf extracts from 14 species in progenies from five interspecific hybrids. The protein profiles characterized three species, but none of the remaining 11 species. The pattern of stained proteins for *C. stricta*, however, distinguished this species from the other tested species, thereby supporting the cytogenetic evidence indicating that this taxon merited species status (Garber and Unni, 1965). The parentage of the progenies from the five interspecific hybrids, however, could not be determined by their protein profiles.

Except for occasional observations, chemotaxonomic methods have not yet furnished acceptable evidence to delineate all of the species of *Collinsia* nor to determine the parentage of interspecific hybrids and their progenies. Consequently, it is still necessary to use the conventional methods of experimental taxonomy, e.g., crossability, sterility or fertility of interspecific or intertaxon hybrids, and chromosomal associations and paracentric inversions, to detect cryptic or sibling species in the genus *Collinsia*.

Supernumerary Chromosomes

Supernumerary chromosomes were found in *C. solitaria* (Garber, 1958*b*) and in interspecific hybrids between *C. sparsiflora* and *C. bruceae*

(Ahloowalia and Garber, 1961). According to Ahloowalia and Garber (1961), these three species form a closely related group, yielding fully fertile interspecific hybrids. Supernumerary chromosomes have not yet been reported for any other species in the genus.

Dhillon and Garber (1962a) investigated the cytogenetics of the supernumerary chromosomes, which are smaller and more darkly stained than the standard chromosomes at metaphase I and do not pair with the standard chromosomes. Two supernumerary chromosomes usually form a bivalent with one chiasma at metaphase I. In plants with four such chromosomes, the occasional trivalent indicates that a supernumerary chromosome can have one chiasma in each arm.

Plants with 1–4 supernumerary chromosomes are fertile; the relatively few plants with 5–8 such chromosomes have misshapen or aborted anthers and set no seed. To account for a maximum number of 8 supernumeraries in one plant, gametophytes with more than 4 and zygotes with more than 8 supernumerary chromosomes may abort or malfunction.

Individuals with more than the expected number of supernumerary chromosomes have been found in the progenies from reciprocal crosses using plants with supernumerary chromosomes as the seed parent or pollen parent. For example, plants with two supernumeraries gave progeny with 2–4 such chromosomes even though the supernumerary chromosomes proceeded normally through meiosis. Mitotic nondisjunction of supernumerary chromosomes has been reported for numerous species during the development of the male gametophyte and in relatively few species during the development of the female gametophyte to give gametic nuclei with more than the expected number of supernumerary chromosomes. Only rye and the species of *Collinsia* exhibit an increase in the number of supernumerary chromosomes in the development of the male and female gametophytes.

Intraspecific Chromosomal Aberrations

An extensive cytological study of populations of the species of *Collinsia* did not yield chromosomal aberrations other than an occasional autopolyploid. Two types of intraspecific chromosomal aberrations have been investigated: reciprocal translocations in *C. heterophylla* induced by colchicine or ionizing radiation, and primary trisomy induced in *C. heterophylla* by colchicine and detected in the progenies from autotriploids of *C. heterophylla* and *C. tinctoria*.

The Colchicine Effect in *C. heterophylla*

Progenies from 30 partially pollen-sterile plants of *C. heterophylla* surviving colchicine treatment were examined cytologically for chromosomal aberrations (Soriano, 1957). Heterozygous reciprocal translocations, primary trisomics, and autotetraploids were found in a number of these progenies. Since a treated plant gave progeny with different chromosomal aberrations, colchicine presumably produced mosaics of cells with one or the other aberration, either in one flower or in different flowers. It would be interesting to determine whether mosaics of cells with different chromosomal aberrations are present in a single anther of flower. This unusual effect of colchicine on the chromosomes of *C. heterophylla* was not observed in *C. concolor*, *C. tinctoria*, or *C. sparsiflora* (Dhillon and Garber, 1961).

Reciprocal Translocations in *C. heterophylla*

A reciprocal translocation between nonhomologous chromosomes is usually detected cytologically as an interchange complex, a ring or chain of chromosomes at diakinesis or metaphase I. The complex occurs either in an alternate or adjacent configuration at metaphase I, depending on which chromosomes proceed to the same pole at anaphase I. In some species, the frequency of PMC's with one or the other configuration is approximately equal; in other species, the number of PMC's with the alternate configuration is significantly greater than 50 percent (directed orientation). Reciprocal translocations within one species generally yield interchange complexes exhibiting either the random or the direction orientation, but not both (Burnham, 1956). A cytogenetic study of reciprocal translocations in *C. heterophylla* induced by colchicine or by ionizing radiation indicated that this species does not conform to the generality.

A sample of 18 heterozygous reciprocal translocations was assembled by selecting one plant from each of 14 progenies and two plants from each of two progenies from colchicine-treated plants (Soriano, 1957). Without exception, the reciprocal translocations were associated with an interchange complex appearing as a ring or chain of four chromosomes or 2 II. The frequency of PMC's with a ring complex was 14–34 percent, and with a chain complex, 19–45 percent. The orientation of the complexes at metaphase I was random, that is, approximately 50 percent of the PMC's had an adjacent or alternate configuration.

The percentage of stainable pollen grains in progenies from self-pollinated or outcrossed plants with a heterozygous reciprocal translocation was not a reliable criterion to distinguish individuals with bivalents or an interchange complex. The mean number of seeds per capsule (2.0–7.3) was seen to be greatly reduced for plants with an interchange complex when compared with the value (10.0–12.9) for plants with bivalents, thereby providing a reliable means to identify plants with an interchange complex.

Plants heterozygous for a reciprocal translocation were self-pollinated or outcrossed to determine the frequency of progeny with an interchange complex or bivalents. Approximately 50 percent of the progeny from 18 reciprocal translocations had an interchange complex.

The cytogenetic analysis of the colchicine-induced reciprocal translocations in *C. heterophylla* revealed several unexpected observations. First, colchicine appears to be a potent agent for breaking the chromosomes in this species, but not in other species of the genus. Second, chromosome breakage may not have involved all of the seven chromosomes in the haploid complement. Preliminary data indicate that at least four chromosomes were involved in the 18 reciprocal translocations. Third, the sites of chromosomal breakage may not be random. This assumption could account for the constant presence of a ring or chain interchange complex or bivalents in PMC's from plants with a heterozygous reciprocal translocation. In many species, heterozygous reciprocal translocations are usually associated with a high frequency of PMC's with a ring or chain interchange complex of 4 chromosomes (Burnham, 1956).

Garber and Dhillon (1962*b*) used ionizing radiation to obtain 4 interchange complexes with 4 chromosomes and 1 interchange complex with 6 chromosomes in *C. heterophylla*. These interchange complexes significantly differed from those induced by colchicine treatment: (1) The complexes occurred as a ring or chain, and rarely as bivalents, at metaphase I, and (2) the ring and chain complexes displayed a direct orientation at this stage. Although the percentages of stainable pollen grains and the mean number of seeds per capsule were reduced in plants with an interchange complex, they were not semisterile. Plants with the interchange complex of 6 chromosomes were less fertile than those with the interchange complex of 4 chromosomes.

By crossing plants with the different interchange complexes in all possible combinations, 5 of the 7 haploid chromosomes were shown to be involved in the 5 interchange complexes (Cleland, 1936). The gene for white flower, *w*, was linked with two different interchange complexes with a common chromosome, suggesting that the locus is in this chromosome.

Garber and Dhillon (1962*b*) proposed an explanation to account for the differences between the reciprocal translocations induced by colchicine and by ionizing radiation. According to Garber and Bell (1962) the chromosome associations at metaphase I in diploid interspecific hybrids can be interpreted by assuming that only one chiasma can occur in a terminal or subterminal segment in each chromosome arm. If the colchicine-induced breaks were restricted to the chiasma-forming segments and the ionizing-radiation-induced breaks were more likely to occur in the internal segment of each chromosome, then two types of cross-configurations would occur (see Fig. 1 in Garber and Dhillon, 1962*b*). At pachytene, the colchicine-induced reciprocal translocation would yield two extremely long and two extremely short "arms" for the cross-configuration; the ionizing-radiation-induced reciprocal translocation would yield less-extreme differences in the "arm" lengths. It is not clear how the differences in the lengths of the "arms" of the cross-configuration may be responsible for the random orientation of the colchicine-induced interchange complex and the directed orientation of the ionizing-radiation-induced interchange complex.

Although relatively small populations were involved, the white-flower locus, w, was completely linked to two heterozygous reciprocal translocations. If the locus were in the internal chromosome segment between the chiasam-forming segments, no recombinants would be expected. This assumption should be tested by using large populations to detect recombinants.

Interspecific hybrids in *Collinsia* have interchange complexes with a random or a directed orientation at metaphase I. These observations can be explained by assuming spontaneous breakage in the chiasma-forming segments or in the internal segment of the chromosomes.

Trisomy in *C. heterophylla*

Aneuploids in many species have provided useful tools to assign mutant genes to specific chromosomes and to correlate linkage groups with their respective chromosomes. In basic diploid species, primary trisomics ($2N + 1$) have been particularly useful for these purposes because the trisomics are frequently identifiable by their characteristically altered array of phenotypes. The primary trisomics from colchicine-treated plants and from an autotriploid of *C. heterophylla* furnished the experimental material for a cytogenetic study conducted prior to the use of these aneuploids to assign the numerous mutant genes available in this species to specific chromosomes (Dhillon and Garber, 1960, 1964).

By appropriate self-pollinations, $2N + 1$ to $2N + 3$ individuals were obtained for each of the 13 trisomics resulting from colchicine treatment. Eight of the trisomics gave $2N + 4$ and $2N + 5$ plants. In the heptasomics ($2N + 5$), one chromosome constitutes approximately 37 percent of the chromosomal material. No gross morphological changes were detected in the plants with extra chromosomes ranging from $2N + 1$ to $2N + 5$.

Without exception, the extra chromosome in each of the 13 trisomics occurred as a univalent at metaphase I and was usually included in one of the telophase-I nuclei. In $2N + 2$ plants the extra chromosomes generally formed a bivalent, but two univalents, a trivalent, and univalent or a quadrivalent was occasionally observed. Lagging chromosomes were uncommon during anaphase I. The extra chromosomes in the $2N + 3$ plants yielded a trivalent or a bivalent and univalent.

Aneuploid plants were used as seed parents or pollen parents to determine the frequency of transmission of the extra chromosomes by the pollen and eggs. The extra chromosome was found in 25–65 percent of the progeny when $2N + 1$ plants were used as pollen parent. Progenies with the $2N + 1$ plants as seed parents included individuals with $2N + 1$, $2N + 2$, or $2N + 4$. The $2N + 2$ plants produced $N + 1$ pollen grains, and aneuploids with more than two extra chromosomes gave functional pollen grains with $N + 2$ or $N + 3$ chromosomes. Trisomics were as fertile as diploid siblings; aneuploids with two or more extra chromosomes were almost sterile.

The most striking characteristics of the use of the colchicine-induced trisomics as seed parents was the occurrence of $2N + 2$ progeny. At least three explanations can be offered: (1) $N + 2$ megaspores, (2) $N + 2$ eggs from $N + 1$ megaspores, or (3) $2N + 2$ plants from $2N + 1$ zygotes. Although reasonable mechanisms can be proposed to support the first two explanations, the third explanation is considered unlikely.

Rai and Garber (1960) used 11 mutant genes and the 13 colchicine-induced trisomic lines in an unsuccessful attempt to assign at least one mutant gene to one of these trisomes. Dhillon and Garber (1962a) suggested that the trisomes obtained from colchicine treatment may be replicates of one chromosome which, as a trisome, has a number of characteristics associated with the supernumery chromosomes in this genus. Consequently, this chromosome has been termed a psuedosupernumerary chromosome. These speculations assume that the primary trisomes for the other six chromosomes of the haploid complement would behave in an orthodox fashion.

Garber (1964) obtained one trisomic from a self-pollinated plant of *C. heterophylla* with an interchange complex of four chromosomes, and nine trisomics from a spontaneous autotriploid from a self-pollinated plant with an interchange complex of six chromosomes. These trisomics provided the experimental material for a comparison with the colchicine-induced trisomics.

A trivalent occurred in 21–74 percent of the PMC's, and univalents in occasional PMC's. The frequency of trisomics from self-pollinated plants was 5–47 percent, and from trisomics as seed parents, 25–50 percent. Tetrasomic ($2N + 2$) plants were not found in progenies from self-pollinated trisomics. One double trisomic ($2N + 1 + 1$) was found in the progeny from the autotriploid. Seed set was markedly reduced in self-pollinated trisomics (3.6–6.4 seeds per capsule), compared with that for self-pollinated diploid siblings (10.9 seeds per capsule). Although trisomics were usually less vigorous, slender, and shorter than most of the diploid siblings, occasional diploids could not be distinguished from known trisomics. Attempts to associate a specific syndrome of altered morphological characteristics with trisomy or with specific trisomics were unsuccessful.

Five observations support the assumption that colchicine treatment yields trisomes which markedly differ from those obtained from the autotriploid (Table 9). All of the trisomes from colchicine-treatment shared a number of characteristics which were not found in any trisomes from the autotriploid.

Apparently unrelated observations may be explained by assuming that the unusual "seventh" chromosome (pseudosupernumerary) in *C. heterophylla* is responsible for the effect of colchicine on the chromosomes of this species. The high frequency of chromosomal breaks detected as interchange complexes in plants treated with colchicine may also be related to the presence of this chromosome. No evidence is available, however, to indicate that this chromosome is involved in the colchicine-induced reciprocal translocations.

TABLE 9.   Comparison of Trisomics Obtained by Colchicine Treatment and from an Autotriploid and a Diploid with an Interchange Complex (RT)

Observations	*Colchicine*	*3N and RT*
Trivalents, metaphase I	Absent	Present
Chromosome numbers of progeny[a]	$2N, 2N + 1, 2N + 2$	$2N, 2N + 1$
Seed set	Normal	Reduced
Phenotypic effects	Absent	Present
Maximum level of somy	$2N + 5$	$2N + 1$

[a] Trisomes either self-pollinated or seed parent.

Trisomy in *C. tinctoria*

Chomcholow and Garber (1964) investigated the cytogenetics of 35 primary trisomics from an autotriploid of *C. tinctoria*. Although the trisomics were initially assigned to seven groups by morphological characteristics, this procedure was undoubtedly influenced by the knowledge that 7 different trisomics are possible in this species. Except for the trisomics assigned to Group I, a particular array of altered phenotypes could not be associated with trisomy or with a specific trisome when the parental and progeny trisomics were compared.

The chromosome associations at metaphase I in the trisomics included univalents, bivalents, and trivalents. The frequency of PMC's with a trivalent was 0–32 percent, depending on the trisome, but the trisomics could not be characterized by this criterion.

The different primary trisomes of a species usually have a characteristic frequency of transmission through the egg and pollen. While the trisomes of *C. tinctoria* differ in their transmission through the egg, such differences were not sufficiently reliable to characterize the trisomes. Four self-pollinated trisomics gave only diploid progenies. One trisomic seed parent yielded 33 percent trisomic progeny. Another trisomic was exceptional in that the extra chromosome was transmitted with a relatively high frequency through the egg and pollen. Finally, none of the self-pollinated trisomics had tetrasomic progeny. None of the trisomes of *C. tinctoria* was comparable to the unusual "seventh" chromosome in *C. heterophylla*.

The mean number of seeds per capsule in the trisomics was 0.6–3.2, and the number of seeds per capsule was 0.5. Diploid siblings had a mean number of 5.5 seeds per capsule, and the number of seeds per capsule was 5–7. Seed set did not provide a reliable means to characterize the trisomics.

Genetics of Species in Group I

The variable populations of the species in Group I have yielded a number of contrasting phenotypes which are determined by two or more alleles at one locus (Rasmuson, 1920). Gorsic (1957) described more than 48 dominant or recessive phenotypes in *C. heterophylla* which were obtained by screening progeny from self-pollinated plants grown from field-collected seeds, and an additional 30 phenotypes have since been analyzed (Gorsic, unpublished). The phenotypes appear in seedlings with cotyledons, young seedlings, and mature plants. Numerous phenotypes can be scored within 30 days after emergence of the seedlings. Several phenotypes are determined by multiple-allelic series. Descriptions and photographs of many phenotypes appear in Gorsic's (1957) publication. Two linkage groups, one (I) with eight loci and the other (II) with nine loci, were reported by Gorsic (1957), and four additional linkage groups have since been found. Hiorth (1930, 1931, 1933a) reported three linkage groups, and these observations have been confirmed.

Mutant genes suitable for marking chromosomes for cytogenetic studies have been assembled in *C. heterophylla* and *C. tinctoria*. Similar stockpiles of mutant genes should be sought for the other four species in Group I by screening progenies from self-pollinated plants grown from field-collected seed. The fertility of intra-group-I hybrids permits an intensive study of homologous genes in the species of this group.

Summary

The genus *Collinsia* with at least 22 annual, herbaceous diploid species ($N = 7$) provides suitable experimental material to investigate speciation at the diploid level. Almost all of the species are found in the western regions of the Pacific Coast States, and they occupy diverse ecological niches. Many species yield 2–3 crops each year in the greenhouse. Progeny from self-pollinated plants grown from field-collected seeds of several species frequently include individuals with a number of contrasting phenotypes determined by alleles, therefore, genetic markers can be used in interspecific and intraspecific hybridizations. Numerous highly or partially fertile or sterile interspecific hybrids have been obtained and analyzed cytogenetically. Chromosomal repatterning in the homologous genomes by reciprocal translocation and paracentric inversion appears to have been a major factor in speciation. Spontaneous amphiploidy often results from interspecific hybridization. One species, *C. heterophylla*, is unusual in that colchicine treatment yields many reciprocal translocations and primary trisomics which have the same extra chromosome.

Acknowledgments

I am indebted to the National Science Foundation and to the Dr. Wallace C. and Clara A. Abbott Memorial Fund of The University of Chicago for their financial support of the research on the genus *Collinsia*. Furthermore, I wish to acknowledge the encouragement and support of Dr. Herman Lewis.

Literature Cited

Ahloowalia, B. S. and E. D. Garber, 1961   The genus *Collinsia*. XIII. Cytogenetic studies of interspecific hybrids involving species with pediceled flowers. *Bot. Gaz.* **122:**228–232.

Bell, S. L. and E. D. Garber, 1961   The genus *Collinsia*. XII. Cytogenetic studies of interspecific hybrids involving species with sessile flowers. *Bot. Gaz.* **122:**210–218..

Burnham, C. R., 1956   Chromosomal interchanges in plants. *Bot. Rev.* **22:**419–552.

Chomchalow, N., and E. D. Garber, 1964   The genus *Collinsia*. XXVI. Polyploidy and trisomy in *C. tinctoria. Can. J. Genet. Cytol.* **6:**488–499.

Cleland, R. E., 1936   Some aspects of the cytogenetics of *Oenothera. Bot. Rev.* **2:**316–348.

Dhillon, T. S. and E. D. Garber, 1960   The genus *Collinsia*. X. Aneuploidy in *C. heterophylla. Bot. Gaz.* **121:**125–133.

Dhillon, T. S. and E. D. Garber, 1961   The genus *Collinsia*. XIV. A cytogenetic study of four induced autotetraploids. *Indian J. Genet. Plant Breed.* **21:**206–211.

Dhillon, T. S. and E. D. Garber, 1962a   The genus *Collinsia*. XVI. Supernumerary chromosomes. *Am. J. Bot.* **49:**168–170.

Dhillon, T. S. and E. D. Garber, 1962b   The genus *Collinsia*. XIX. "New" genomes from two fertile interspecific hybrids. *Cytologia (Tokyo)* **27:**189–203.

Garber, E. D., 1956   The genus *Collinsia*. I. Chromosome number and chiasma frequency of species in the two sections. *Bot. Gaz.* **118:**71–72.

Garber, E. D., 1958a   The genus *Collinsia*. III. The significance of chiasma frequencies as a cytotaxonomic tool. *Madrono* **14:**172–176.

Garber, E. D., 1958b   The genus *Collinsia*. VII. Additional chromosome numbers and chiasmata frequencies. *Bot. Gaz.* **120:**55–56.

Garber, E. D., 1960   The genus *Collinsia* IX. Speciation and chromosome repatterning. *Cytologia (Tokyo)* **25:**233–243.

Garber, E. D., 1964   The genus *Collinsia*. XXII. Trisomy in *C. heterophylla. Bot. Gaz.* **125:**46–50.

Garber, E. D., 1965a   The genus *Collinsia*. XXV. Additional evidence for preferential pairing in two amphiploids and two triploid interspecific hybrids. *Bot. Gaz.* **126:**130–133.

Garber, E. D., 1965b   The genus *Collinsia* XXVIII. A paper-chromatographic and disc electrophoretic study of leaf extracts from fourteen species and progeny from five interspecific hybrids. *Can. J. Genet. Cytol.* **7:**551–558.

Garber, E. D. and S. L. Bell, 1962   The genus *Collinsia*. XV. A cytogenetic study of three fertile interspecific hybrids with an interchange complex of six chromosomes. *Bot Gaz.* **133:**190–197.

Garber, E. D. and T. S. Dhillon, 1962a   The genus *Collinsia*. XVII. Preferential pairing in four amphidiploids and three triploid interspecific hybrids. *Can. J. Genet. Cytol.* **4:**6–13.

Garber, E. D. and T. S. Dhillon, 1962b   The genus *Collinsia*. XVIII. A cytogenetic study of radiation-induced reciprocal translocations in *C. heterophylla. Genetics* **47:**461–467.

Garber, E. D. and T. S. Dhillon, 1962c   The genus *Collinsia*. XX. Cytogenetic studies of interspecific hybrids. *Bot. Gaz.* **123:**291–298.

Garber, E. D. and J. Gorsic, 1956   The genus *Collinsia*. II. Interspecific hybrids involving *C. heterophylla, C. concolor* and *C. sparsiflora. Bot. Gaz.* **118:**73–77.

Garber, E. D. and Ø. Strømnaes, 1964   The genus *Collinsia*. XXIII. A paper-chromoatographic study of flower extracts from fourteen species and four interspecific hybrids. *Bot. Gaz.* **125**:96–101.

Garber, E. D. and M. K. Unni, 1965   The genus *Collinsia*. XXIV. Taxonomic status of *C. stricta* Greene. *Bot. Gaz.* **126**:133–136.

Gorsic, J., 1957   The genus *Collinsia* V. Genetic studies in *C. heterophylla*. *Bot. Gaz.* **118**:208–223.

Grant, V., 1956   Chromosome repatterning and adaptation. *Adv. Genetics* **8**:89–107.

Hayhome, B. A. and E. D. Garber, 1968   The genus *Collinsia*. XXIX. Preferential pairing in diploid, triploid and tetraploid hybrids involving *C. stricta* × *C. concolor* and related species. *Cytologia (Tokyo)* **33**:246–255.

Hiorth, G., 1929   Die Anwendung elektrisches Beleuchtung fur Verebungsversuche mit Pflanzen. *Züchter* **1**:204–209.

Hiorth, G., 1930   Genetische Versuche mit *Collinsia bicolor*. I. *Z. indukt. Abstammungs.-Vererbungsl.* **55**:117–144.

Hiorth, G., 1931   Genetische Versuche mit *Collinsia*. II. Die Blatt- und Kotyledonenzeichningen von *Collinsia bicolor*. *Z. indukt. Abstammungs.-Vererbungsl.* **59**:236–269.

Hiorth, G., 1933*a*   Geneticsche Versuche mit *Collinsia*. III. Koppelungsuntersuchungen bei *Collinsia bicolor*. *Z. indukt. Abstammungs. -Vererbungsl.* **65**:253–277.

Hiorth, G., 1933*b*   Genetische Versuche mit *Collinsia*. IV. Die Analyse einer nahe zu sterilen Artbastardes. Die Diploiden Bastarde zwischen *Collinsia bicolor* und *C. bartsiaefolia*. Pts. 1, 2. *Z. indukt. Abstammungs.- Vererbungsl.* **66**:106–157, 245–274.

Munz, P. A., 1959   *A California Flora*, University of California Press, Berkeley, Calif.

Newsom, V. M., 1929   A revision of the genus *Collinsia*. *Bot. Gaz.* **87**:260–301.

Pennell, F. W., In *Illustrated Flora of the Pacific States*, Vol. 3, edited by L. Abrams, Stanford University Press, Stanford, California.

Pienaar, R. de V., 1955   Combinations and variations of techniques for improved chromosome studies in the Gramineae. *J. S. Afr. Bot.* **21**:1–8.

Rai, K. S. and E. D. Garber, 1960   The genus *Collinsia*. XI. Trisomic inheritance in *C. herophylla*. *Bot. Gaz.* **122**:109–117.

Rasmuson, H., 1920   On some hybridization experiments with varieties of *Collinsia* species. *Hereditas* **1**:178–185.

Reddy, M. M. and E. D. Garber, 1971   Genetic study of variant enzymes. III. Comparative electrophoretic studies of esterases and peroxidases for species, hybrids and amphiploids in the genus *Nicotiana*. *Bot. Gaz.* **132**:158–166.

Soriano, J. D., 1957   The genus *Collinsia*. IV. The cytogenetics of colchicine-induced reciprocal translocations in *C. heterophylla*. *Bot. Gaz.* **118**:139–145.

Strømnaes, Ø. and E. D. Garber, 1963   The genus *Collinsia*. XXI. A paper-chromatographic study of root extracts from fifteen species and six interspecific hybrids. *Bot. Gaz.* **124**:363–367.

Wennstrom, J. I. and E. D. Garber, 1965   The genus *Collinsia*. XXVII. Separation of esterases and acid phosphatases in extracts from twelve species and one interspecific hybrid by starch-gel zone electrophoresis. *Bot. Gaz.* **126**:223–225.

15

The Duration of Chromosomal DNA Synthesis, of the Mitotic Cycle, and of Meiosis of Higher Plants

JACK VAN'T HOF

Introduction

A relationship between DNA content per cell and the duration of the mitotic cycle in plants was first shown by Van't Hof and Sparrow (1963) in 1963. Since that time, work on other species has provided additional confirmatory observations, and the information in the tables which follow constitutes an introduction to the literature on the topic. The compilation is limited to the mitotic cycle of cells of higher plants as measured *in vivo* by the Quastler–Sherman method (Quastler and Sherman, 1959) which utilizes ^{3}H-thymidine and autoradiography. The 2C-DNA content was calculated according to Van't Hof (1965) when necessary and separate references are given for cases where the durations and DNA content were reported independently. Table 1 lists the 2C-DNA content, the duration

JACK VAN'T HOF—Biology Department, Brookhaven National Laboratory, Upton, New York.

TABLE 1. *The 2C-DNA Content and the Duration of the G1, S, G2, M, and CT Periods of Meristematic Cells of Several Higher Plants*

Species	2C DNA, pg	Duration, hr					Temperature, C	Reference
		G1	S	G2	M	CT		
Allium cepa var. *Excell*	33.6	2.5	10.9	—	4.0	17.4	23	Sparrow and Miksche, 1961[a]
		1.9	15.0	—	5.6	20.5	19	Van't Hof, 1965 Arcara and Ronchi-Nutti, 1967
Allium cepa var. *Stuttgart*		<0.5	9.6	—	8.2	17.8	20	Evans and Rees, 1971[b]
		3.3	12.0	3.6	4.0	23.0	21	Matagne, 1968
Allium cepa var. *Evergreen bunching*	11.7	1.5	6.5	2.4	2.3	12.8	24	Melera, 1971[a] Bryant, 1969
Allium fistulosum	25.1	2.5	10.3	—	6.0	18.8	23	Van't Hof, 1965[b]
Allium cepa × *allium fistulosum* (hybrid)	29.9		8.6	—	—	16.6	20	Evans and Rees, 1971[b]
Allium tuberosum	34.8	0-2.5	11.8	7.5-9.5	—	20.6	23.	Van't Hof, 1965[b]
		<0.5	7.1	8.0	—	15.1	20	Evans and Rees, 1971[b]
Antirrhinum majus[c] Diploid	3.7	4.0	4.5	—	1.5	10.0	~23	Bennett, 1971[a], Alfert and Das, 1969
Tetraploid		5.5	4.5	—	1.5	11.5	~23	Alfert and Das, 1969
Avena pilosa		0.	4.3	2.9	1.7	8.9	25	Alfert and Das, 1969
Avena strigosa Diploid		2.5	3.6	2.0	1.8	9.8	25	Yang and Dodson, 1970
Tetraploid		1.0	3.8	3.6	1.6	9.9	25	
Bellavalia romona		8.0	8-6	6.0	>1.0	21.0		Taylor, 1961
		6.0	7.0	5.3	2.7	21.0	25	Jona, 1966
Chrysanthemum lineare		<1.0	10.5	<4.0	1-2	15.0	22-23	Tanaka, 1966

								Reference
Chrysanthemum nipponicum	28.1	<2.5	9.0	<3.5	1–2	15.0	22–23	Baetcke et al., 1967a, Tanaka, 1966
Chrysanthemum nipponicum × Chrysanthemum lineare (hybrid)		<1.0	9.5	<3.5	1–2	15.5	22–23	Tanaka, 1966
Crepis aspera		4.3	2.4	2.3	1.3	9.8	23	Langridge et al., 1970
Crepis capillaris	2.7	4.7	3.3	2.8	—	10.8	23	Van't Hof, 1965b
Crepis neglecta		2.4	3.6	3.0	1.1	9.7	23	Langridge et al., 1970
Crepis neglecta × Crepis capillaris (hybrid)		4.5	2.9	2.2	0.7	9.8	23	
Crepis parviflora		3.0	3.1	2.5	1.1	9.3	23	
Crepis tectorum		3.3	3.4	2.1	1.3	9.6	23	Johnson, 1971b
Cucurbita pepo	5.6	5.6	3.8	2.1	1.4	12.3	30	Bennett, 1971a
Haplopappus gracilis	3.7	1.0	4.4	2.3	1.5	9.1	24	Sparvoli et al., 1966
Helianthus annuus	6.3	3.4	4.0	1.4	1.7	10.5	25	Van't Hof and Sparrow, 1963a
Hordeum vulgare var. Brage — Diploid	~13.3	1.2	4.5	1.5	0.6	7.8	25	Burholt and Van't Hof, 1971
Tetraploid		1.9	4.5	3.0	1.0	10.4	25	Bennett and Smith, 1971a, Skult, 1969
Hordeum vulgare var. Hakata		1.3	5.8	3.2	1.1	11.4	25	Skult, 1969
Hordeum vulgare var. Sultan	13.2	2.0	6.0	3.5	—	11.5	20	Kusanagi, 1966
Hordeum vulgare var. Maris Otter		2.6	3.8	4.3	1.8	12.4	20	Bennet and Finch, 1972
Hyacinthus orientalis	49.9	1.8	13.6	5.0	1.7	24.0	20	Evans and Rees, 1971b
Impatiens balsamina	2.7	0.8	3.9	9.6	—	8.8	23	Van't Hof, 1965b
Lathyrus angulatus	8.9	4.4	3.9	4.0	—	12.3	20	Evans and Rees, 1971b

TABLE 1. Continued

Species	2C DNA, pg	Duration, hr					CT	Temperature, °C	Reference
		G1	S	G2	M				
Lathyrus articulatus	12.4	3.9	4.3	—	6.1		14.3	20	Evans and Rees, 1971[b]
Lathyrus hirsutus	20.1	7.8	5.0	—	5.2		18.0	20	
Lathyrus tingitanus	18.1	6.3	5.3	—	5.2		16.8	20	
Lemna perpusilla			3.0	—	3.3		6.0	24–26	Halaban, 1972
Linum usitatissimum	1.4		4.1				11.2	20	Evans and Rees, 1971[b]
Lolium perenne	9.9	0.5	4.2	—	3.9		8.6	20	
Luzula purpura		4.5	6.5	—	6.0		17.0	18	Kusanagi, 1964
Lycopersicum esculentum	5.1	1.8	4.3	—	4.5		10.6	23	Van't Hof, 1965[b]
Melandrium		9.6	3.3	—	2		15.5	25	Choudhuri, 1969
Nicotiana plumbaginifolia		<0.6	5.7	3.4	>1.3		11.0	23	Gupta, 1969
Nicotiana tabacum (hybrid)		<3.0	3.8	1.4	>0.8		9.0	23	
Nigella damascena	21.1	1.5	10.5	—	4.0	—	16.5	20	Evans and Rees, 1971[b]
Phalaris canariensis		2.7	3.4	6.8	1.6		14.5	~23	Prasad and Godward, 1965
Phalaris minor	20.3	3.8	6.4	6.0	1.8		18.0	~23	Martin, 1966[a]
									Prasad and Godward, 1965
Picea glauca			11.1					22	Miksche and Rollins, 1971
Pinus banksiana	44.5	15.3	7.6	1.4	1.4		25.7	22	Miksche, 1967[b]
Pisum sativum var. Alaska	7.9	5.0	4.5	3.0	1.2		14.0	23	Van't Hof and Sparrow, 1963[a],
									Van't Hof, 1967
Pisum sativum var. Weitor		5.0	4.5	3.0	1.2		14.0	23	Van't Hof, 1967

									Reference
Pisum sativum var. *Witham wonder*		4.0	4.5	3.3		1.4	13.0	23	Van't Hof, 1967
Rumex thyrsiflora			7.5	—	8.0	—	15.5	20	Zuk, 1969
Scilla campanulata	55.5	7.3	10.8	12.3		2.3	32.6		Martin, 1966[a]; John and Lewis, 1969
Secale cereale	18.9	1.0	6.0	—	4.7	—	11.7	20	Evans and Rees, 1971[b]
Secale Diploid + 0B[d]		1.2	5.2	4.3		0.9	11.5	20	Kaltsikes, 1971
Diploid + 2B		2.2	6.9	2.3		1.3	12.8		Ayonoadu and Rees, 1968
Diploid + 4B		2.9	7.9	3.1		1.6	15.6		
Spiranthes sinensis		1.4	9.4	4.2		1.7	16.7	21	Tanaka, 1965
Tradescantia paludosa	38.8	4.0	9.0	6.0		3.0	22.0	21	Sparrow and Miksche, 1961[a],
		4.0	10.8	2.7		2.5	20.0		Wimber, 1960[b]
Triticale var. *Armadillo* Hexaploid		2.8	5.3	—	2.5	—	11.0	25	Kaltsikes, 1972
Triticale var. *Rosmer* Hexaploid		2.4	6.3	—	3.1	—	12.2	25	
Triticale Amphiploid (*Triticum turgidum* × *Secale cereale*)		1.2	5.1	4.8		1.0	12.0	20	Kaltsikes, 1971
Triticum aestivum var. *Indus* Hexaploid	36.1	0.8	10.0	2.0		1.2	14.0	23	Bennett, 1972[a]; Evans and Van't Hof, unpublished
T. durum var. *Aziziah*		2.3	6.5	4.3		1.0	14.0	23	Avanzi and Deli, 1969
T. durum var. *Cappelli*		1.3	9.3	4.7		0.5	16.3	23	

TABLE 1. Continued

Species	2C DNA, pg	Duration, hr					Temperature, C	Reference
		G1	S	G2	M	CT		
T. turgidum		3.3	5.5	4.3	0.8	13.8	20	Kaltsikes, 1971
Vicia faba var. *Longpod*	24.3	4.0	9.0	3.5	1.9	18.0	23	Sparrow and Miksche, 1961[a], Van't Hof, 1967[b]
primary root tip lateral root tip		2.5	6.2	3.3	2.0	14.0	21–22	Webster and Davidson, 1968
small lateral root		3.6	8.3	2.8	1.9	16.6	20	Macleod, 1971
primordia		2.7–2.8	4.4	1.9	2.4	11.4	20	
Zea mays	11.0	0.5	4.3	— 5.7 —		10.5	20	Evans and Rees, 1971[b]

[a] Determined DNA content only.
[b] Determined both CT and nuclear DNA content.
[c] Determined by Van't Hof from author's data.
[d] Diploid + OB = diploid chromosome complement plus no B chromosomes; diploid + 2B = diploid complement plus 2 B chromosomes; diploid + 4B = diploid complement plus 4 B chromosomes.

TABLE 2.   *The Influence of Temperature on the Duration of G1, S, G2, M, and CT of Root-Meristem Cells*

Temperature, °C	Duration of period, hr				
	G1	S	G2	M	Ct
Helianthus annuus var. *Russian mammoth*[a]					
10	14.8	22.3	4.9	4.4	46.4
15	6.8	11.8	2.9	1.7	23.2
20	3.8	6.1	1.6	1.0	12.5
25	1.2	4.5	1.5	0.6	7.8
30	0.4	4.3	1.1	0.5	6.3
35	0.8	4.0	1.1	0.5	6.4
38	0.8	6.2	2.5	—	—
Tradescantia paludosa[b]					
30	2.4	9.5	2.4	1.7	16.0
21	5.8	10.8	2.5	1.7	20.8
13	<15.4	22.5	8.3	5.1	<51.4

[a] Burholt and Van't Hof, 1971.
[b] Wimber, 1960.

of the mitotic cycle (CT), and its parts: the presynthetic (G1), DNA synthetic (S), postsynthetic (G2), and mitotic (M) periods. Data on diploid, tetraploid, and hyperploid species for which no quantitative DNA values are available are also given. The durations of G1, S, G2, M, and CT at different temperatures are listed in Table 2, and those of cells in different histological loci are shown in Table 3. Unless noted otherwise, the data refer to cells of root apices.

Additional Facts Worth Noting

The regression equations (see Figures 1 and 2) that express CT and S as functions of the 2C DNA content of cells have slopes that differ by a factor of 2. The mitotic cycle duration increases approximately 0.34 hour while S increases about 0.17 hr per picogram of DNA.

The duration of the mitotic cycle and its component periods in root meristem cells are similar in some diploid and tetraploid plants of related species (Alfert and Das, 1969; Skult, 1969; Yang and Dodson, 1970) and different in others (Ayonoadu and Rees, 1968; Prasad and Godward, 1965; Titu and Popovici, 1970). Colchicine-induced autotetraploid cells have a mitotic-cycle duration which is nearly the same as that of the diploid parent cells (Friedberg and Davidson, 1970; Van't Hof. 1966).

TABLE 3. The Duration of G1, S, G2, M, and CT of Cells in Different Histological Loci in Root and Shoot Apices and the Effect of Polyploidy

Species	Duration, hr					Temperature, C	Reference
	G1	S	G2	M	CT		
Allium sativum, root tip							
Cap initials	3	13	6	5	27	20	Thompson and Clowes, 1968
Quiescent center	136	17	8	4	165	20	
Stele (above quiescent center)	4	12	4	6	26	20	
Stele (200 μ above quiescent center)	4	12	4	6	26	20	
Beta vulgaris var. *Lourin*, root tip[a]							
Diploid							
Central cylinder, 50–250 μ	<6.6	5.8	<3.8	—	16.2	21	Titu and Popovici, 1970
Central cylinder, 250–450 μ	<7.5	7.5	<4.0	—	19.0	21	
Cortex, 50–250 μ	<10.3	9.2	<2.7	—	22.2	21	
Cortex, 250–450 μ	<8.3	10.8	<5.0	—	24.1	21	

Tetraploid						
Central cylinder, 50–250 μ	<5.7	8.0	<4.8	—	18.5	21 ⎫ Titu and Popovici, 1970
Central cylinder, 250–450 μ	<6.3	9.5	<5.2	—	21.0	21
Cortex, 50–250 μ	<6.1	12.2	<5.5	—	23.8	21
Cortex, 250–450 μ	<6.8	13.4	<6.0	—	26.2	21 ⎭
Zea mays var. *Golden Bantam*, root tip						
Cap initials	−0.7	4.3	5.4	1.4	10.4	23 ⎫
Quiescent center	19.3	11.7	3.3	5.1	39.6	23
Stele (just above quiescent center)	2.2	6.1	3.4	2.5	14.2	23 ⎬ Clowes, 1971
Stele (200 μ above quiescent center)	3.4	5.3	3.5	2.2	14.4	23 ⎭
Rudbeckia bicolor, shoot apical bud						
Peripheral zone	<9.0	11.6	<9.5	—	30.1	20 ⎫
Central zone	—	19.2	<14.8	—	—	20 ⎬ Jacqmard, 1970
Pith-rib meristem	<6.8	15.9	<7.8	—	30.5	20
Subapical pith	<13.1	13.0	<6.8	—	32.9	20 ⎭

a Here μ notes distance from the initial zone.

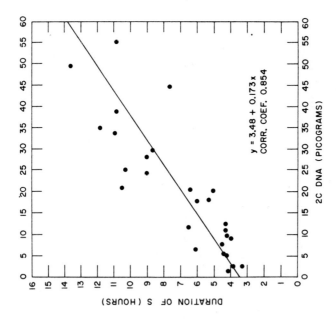

Figure 2. The duration of chromosomal DNA synthesis of root meristem cells of 27 different plant species expressed as a function of their 2C-DNA value. Also shown are the regression equation and correlation coefficient for the relationship.

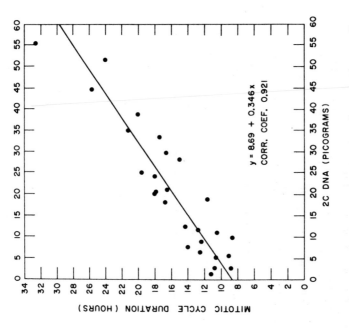

Figure 1. The mitotic-cycle duration of root meristem cells of 27 different plant species expressed as a function of their 2C-DNA value. Also shown are the regression equation and correlation coefficient for the relationship.

TABLE 4.   Duration of Meiosis, Nuclear DNA Content, and Ploidy Level of 24 Plant Species[a]

Species	DNA per cell, pg	Duration of meiosis, hr	Ploidy level [b] 2x = diploid
Agapanthus umbellulatus[c]	—	300	2x
Allium cepa	54.3	96	2x
Antirrhinum majus	5.5	24.	2x
Endymion nonscriptus	65.3	48	2x
Gasteria sp.[c]	—	96	—
Haplopappus gracilis	5.5	36	2x
Hordeum vulgare	20.7	39	2x
Lilium candidum	—	168	2x
Lilium henryi var. Baker	100.0	170	2x
Lilium longiflorum	106.0	192	2x
Oenothera sp.[c]	—	96	—
Ornithogalum virens var. Lindl [c]	19.3	96	2x
Rhoeo discolor var. Hance	23.8	48	2x
Secale cereale	28.7	51.2	2x
Secale cereale	57.4	38.0	4x
Tradescantia paludosa	58.2	126.0	2x
Tradescantia reflexa var. Raf	—	144.0	4x
Trillium erectum	120.0	274.0	2x
Triticale (hybrid)	82.7	20.8	8x
Triticum aestivum	54.3	24.0	6x
Triticum dicoccum var. Koern	38.3	30.0	4x
Triticum monococcum	20.9	42.0	2x
Tulbaghia violaceae var. Harv.	58.5	130.0	2x
Vicia faba	44.0	72.0	2x

[a] Data from Bennett (1972).
[b] 2X = diploid.
[c] Approximate duration of meiosis.

Listed in Table 4 are the duration of meiosis, nuclear DNA content, and ploidy level of 24 plant species. The data, compiled by Bennett (1972), show that meiosis lengthens with increased DNA content of diploid plant nuclei. The linear regression model for these data is, duration of meiosis = 10.7 + 1.76 (DNA content) and the correlation coefficient is 0.90. The influence of increased nuclear DNA is detected only when comparisons are made between diploid plants of unrelated species. When polyploids are compared with their diploid ancestors or relatives, the diploid plants have the longer duration. For example, the amphiploid Triticale (2n = 8x = 56 chromosomes) and its ancestors Triticum aestivum (2n = 6x = 42 chromosomes) and Secale cereale (2n = 14 chromosomes) have respective durations of meiosis of 21, 24, and 51 hours at

TABLE 5 The Duration of Stages of Meiosis of 14 Plants[a]

Species	Duration of meiotic stages, hr												Total meiotic duration	Ploidy level
	Leptotene	Zygotene	Pachytene	Diplotene	Diakinesis	Metaphase I	Anaphase I	Telophase I	Dyads	Metaphase II	Anaphase II	Telophase II		
Antirrhinum majus	6.0	3.0	3.0	—	—				9.0				24.0	
Allium cepa	72.0								24.0				96.0	
Hordeum vulgare	12.0	9.0	8.8	2.2	0.6	1.6	0.5	0.5	2.0	1.2	0.5	0.5	39.4	2x
Lilium candidum	40.0	40.0	40.0	—	—				24.0				168.0	2x
Lilium longiflorum	40.0	36.0	40.0	—	—				48.0				192.0	2x
Rhoeo discolor	24								24.0				48.0	2x
Secale cereale	20.0	11.4	8.0	1.0	0.6	2.0	1.0	1.0	2.5	1.7	1.0	1.0	51.2	2x
Secale cereale	13.0	9.0	6.4	1.0	0.6	1.8	0.7	0.7	2.0	1.4	0.7	0.7	38.0	4x
Tradescantia paludosa	48.0	24.0	24.0	—	—				13.0				126.0	2x
Tradescantia reflexa	111								33.0				144.0	4x
Trillium erectum	70.0	70.0	50.0	—	—				64.0				274.0	2x
Triticum aestivum	10.4	3.4	2.2	0.6	0.4	1.6	0.5	0.5	2.0	1.4	0.5	0.5	24.0	6x
Triticale	7.5	3.0	2.3	1.0	0.5	1.8	0.5	0.5	1.5	1.3	0.5	0.5	20.8	8x
Tulbaghia violacea	102.0								28.0				130.0	2x

[a] Data from Bennett and Smith (1972); Bennett (1971, 1972).

20°C (Bennett *et al.*, 1971). It is not known if the effect of polyploidy on the duration of meiosis is confined to the division processes from leptotene to tetrad formation or whether it also includes the premeiotic DNA synthetic period.

Neither increased nuclear DNA content nor polyploidy preferentially effect a given meiotic stage. Differences in meiotic duration in the few plants examined result from proportional changes in all stages of meiosis (Table 5).

Acknowledgment

This research was carried out at the Brookhaven National Laboratory under the auspices of the U.S. Atomic Energy Commission.

Literature Cited

Albert, M. and N. K. Das, 1969   Evidence for control of the rate of nuclear DNA synthesis by the nuclear membrane in eukaryotic cells. *Proc. Natl. Acad. Sci. USA* **63**:123–128.

Arcara, P. G. and V. N. Ronchi-Nutti, 1967   Effect of ethyl alcohol on the mitotic cycle of *Allium cepa* root meristems. *Caryologia* **20**:229–232.

Avanzi, S. and P. L. Deri, 1969   Duration of the mitotic cycle in two cultivars of *Triticum durum,* as measured by ^{3}H-thymidine labeling. *Caryologia* **22**:187–194.

Ayonoadu, U. W. and H. Rees, 1968   The regulation of mitosis by B-chromosomes in rye. *Exp. Cell Res.* **52**:284–290.

Baetche, K. P., A. H. Sparrow, C. H. Nauman and S. S. Schwemmer, 1967   The relationship of DNA content of nuclear and chromosome volumes and to radiosensitivity (LD_{50}). *Proc. Natl. Acad. Sci. USA* **58**:533–540.

Bennett, M. D., 1971   The duration of meiosis. *Proc. R. Soc. Lond. Ser. B Biol. Sci* **178**:277–299.

Bennett, M. D., 1972   Nuclear DNA content and minimum generation time in herbaceous plants. *Proc. R. Soc. Lond. Ser. B Biol. Sci* **181**:109–135.

Bennett, M. D. and R. A. Finch, 1972   The mitotic cycle time of root meristem cellls of *Hordeum vulgare. Caryologia* **25**:439–444.

Bennett, M. D. and J. B. Smith, 1971   The 4C nuclear DNA content of several *Hordeum* genotypes. *Can. J. Genet. Cytol.* **13**:607–611.

Bennett, M. D. and J. B. Smith, 1972   The effects of polyploidy on meiotic duration and pollen development in cereal anthers. *Proc. R. Soc. Lond. Ser. B Biol. Sci.* **181**:81–107.

Bennett, M. D., V. Chapman and R. Riley, 1971   The duration of meiosis in pollen mother cells of wheat, rye, and *Triticale. Proc. R. Soc. Lond. Ser. B Biol. Sci.* **178**:259–275.

Bryant, T. R., 1969   DNA synthesis and cell division in germinating onion II. Mitotic cycle and DNA content. *Caryologia* **22**:139–148.

Burholt, D. R. and J. Van't Hof, 1971   Quantitative thermal-induced changes in growth and cell population kinetics of *Helianthus* roots. *Am. J. Bot.* **58**:386–393.

Choudhuri, H. C., 1969   Late DNA replication pattern in sex chromosomes of *Meland-rium. Can. J. Genet. Cytol.* **11**:192–198.

Clowes, F. A. L., 1971   The proportion of cells that divide in root meristems of *Zea mays* L. *Ann. Bot.* **35**:249–261.

Evans, G. M. and H. Rees, 1971   Mitotic cycles in dicotyledons and monocotyledons. *Nature (Lond.)* **233**:350–351.

Friedberg, S. H. and H. Davidson, 1970   Duration of S phase and cell cycles in diploid and tetraploid cells of mixoploid meristems. *Exp. Cell Res.* **61**:216–218.

Gupta, S. B., 1969   Duration of mitotic cycle and regulation of DNA replication in *Nicotiana plumbaginifolia* and a hybrid derivative of *N. tabacum* showing chromosome instability. *Can. J. Genet. Cytol.* **11**:133–142.

Halaban, R., 1972   The mitotic index and cell cycle of *Lemna perpusilla* under different photoperiods. *Plant Physiol.* **50**:308 310.

Jacqmard, A., 1970   Duration of the mitotic cycle in the apical bud of *Rudbeckia bicolor. New Phytol.* **69**:269–271.

John, B. and K. R. Lewis, 1969   The chromosome cycle. In *Protplasmatologia,* Vol. 6B, Springer-Verlag, Vienna.

Johnson, D. L., 1971   Changes of nuclear acid and protein contents of root tip cells from intact squash plants during development of and recovery from boron deficiency and mitotic cycle changes in root-tip cells during development of boron deficiency. Ph.D. Thesis, University of Rhode Island, Kingston, R. I.

Jona, R., 1966   La durata del ciclo mitotic nella *Bellavalia romona* determinata per via autoradiografica mediante l'imprego della timidina H^3. *Caryologia* **19**:429–442.

Kaltsides, P. J., 1971   The mitotic cycle in an amphiploid (*Triticale*) and its parental species. *Can. J. Genet. Cytol.* **13**: 656–662.

Kaltsikes, P. J., 1972   Duration of the mitotic cyle in *Triticale. Caryologia* **25**:535–542.

Kusanagi, A., 1964   Cytological studies on *Luzula* chromosomes V. Duration of the mitotic cycle in root meristems of *Luzula purpurea* as measured with thymidine-H^3. *Bot. Mag. (Tokyo)* **77**:77–80.

Kusanagi, A., 1966   Rate of DNA replication in the DNA synthetic period of the barley chromosomes. *Chromosoma* **20**:125–132.

Langridge, W. H. R., T. A. O'Malley and H. Wallace, 1970   Neutral amphiplasty and regulation of the cell cycle in *Crepis* herbs. *Proc. Natl. Acad. Sci. USA* **67**:1894–1900.

Macleod, R. D., 1971   Thymidine kinase and thymidylate synthetase in meristems of roots of *Vicia faba. Protoplasma* **73**:337–348.

Martin, P. G., 1966   Variation in the amounts of nucleic acids in the cells of different species of higher plants. *Exp. Cell Res.* **44**:84–94.

Matagne, R., 1968   Chromosomal aberrations induced by dialkylating agents in *Allium cepa* root-tips and their relation to the mitotic cycle and DNA synthesis. *Radiat. Bot.* **8**:489–497.

Melera, P. W., 1971   Nucleic acid metabolism in germinating onion. I. Changes in root-tip nucleic acid during germination. *Plant Physiol.* **48**:73–82.

Miksche, J. P., 1967   Radiobotanical parameters of *Pinus banksiana. Naturwissenschaften* **12**:322.

Miksche, J. P. and J. A. Rollins, 1971   Constancy of the duration of DNA synthesis and percent G + C in white spruce from several provenances. *Can. J. Genet. Cytol.* **13**:415–421.

Prasad, A. B. and M. B. E. Godward, 1965   Comparison of the developmental responses of diploid and tetraploid *Phalaris* following irradiation of the dry seed. I. Determination of mitotic cycle time, mitotic time, and phase time. *Radiat. Bot.* **5:**465–474.

Quastler, H. and F. G. Sherman, 1959   Cell population kinetics in the intestinal epithelium of the mouse. *Exp. Cell Res.* **17:**420–438.

Ronchi, V. N. and P. G. Arcara, 1967   The chromosome breaking effect of 6-methylcoumarin in *Allium cepa* in relation to the mitotic cycle. *Mutat. Res.* **4:**791–796.

Skult, H., 1969   Growth and cell population kinetics of tritiated thymidine labeled roots of diploid and autotetraploid barley. *Acta Acad. Aboensis Ser. B* **29:**1–15.

Sparrow, A. H. and J. P. Miksche, 1961   Correlation of nuclear volume and DNA content with higher plant tolerance to chronic radiation. *Science (Wash., D.C.)* **134:**282–283.

Sparvoli, E., H. Gay and B. P. Kaufmann, 1966   Duration of the mitotic cycle in *Haplopappus gracilis. Caryologia* **19:**65–71.

Tanaka, R., 1965   ³H-thymidine autoradiographic studies on the heteropycnosis, heterochromatin and euchromatin in *Spiranthes sinensis. Bot. Mag (Tokyo)* **78:**50–62.

Tanaka, R., 1966   DNA replication in *Chrysanthemum lineare, Ch. nipponicum* and their F₁ hybrid. *Bot. Mag. (Tokyo)* **79:**447–456.

Taylor, H. J., 1961   Control of DNA synthesis. In *Conference on Molecular and Radiation Biology*, edited by R. A. Deering, pp. 12–20, National Research Council Publication 823, Washington, D.C.

Thompson, J. and F. A. L. Clowes, 1968   The quiescent centre and rates of mitosis in the root meristem of *Allium sativum. Ann. Bot.* **32:**1–13.

Titu, H. and I. Popovici, 1970   Duration of mitotic cycle phases in the radicular meristem of diploid and tetraploid sugar beet (*Beta vulgaris*). *Rev. Roum. Biol. Ser. Bot.* **15:**51–56.

Van't Hof, J., 1965   Relationships between mitotic cycle duration, S period duration, and the average rate of DNA synthesis in the root meristem cells of several plants. *Exp. Cell Res.* **39:**48–58.

Van't Hof, J., 1966   Comparative cell population kinetics of tritiated thymidine labeled diploid and colchicine-induced tetraploid cells in the same tissue of *Pisum sativum. Exp. Cell. Res.* **41:**274–288.

Van't Hof, J., 1967   Studies on the relationships between cell population and growth kinetics of root meristems. *Exp. Cell Res.* **46:**335–347.

Van't Hof, J. and A. H. Sparrow, 1963   A relationship between DNA content, nuclear volume, and minimum mitotic cycle time. *Proc. Natl. Acad. Sci. USA* **49:**897–902.

Webster, P. L. and D. Davidson, 1968   Evidence from thymidine-³H-labeled meristems of *Vicia faba* of two cell populations. *J. Cell Biol.* **39:**332–338.

Wimber, D. E., 1960   Duration of the nuclear cycle in *Tradescantia paludosa* root-tips as measured with ³H-thymidine. *Am. J. Bot.* **47:**828–834.

Wimber, D. E., 1966   Duration of the nuclear cycle in *Tradescantia* root-tips at three temperatures as measured with H³-thymidine. *Am. J. Bot.* **53:**21–24.

Yang, D.-P. and D. O. Dodson, 1970   The amounts of nuclear DNA and the duration of DNA synthetic period (S) in related diploid and autotetraploid species of oats. *Chromosoma* **31:**309–320.

Zük, J., 1969   Autoradiographic studies in *Rumex* with special reference to sex chromosomes. In *Chromosomes Today*, Vol. 2, edited by C. D. Darlington and K. R. Lewis, pp. 183–187, Oliver and Boyd, Edinburgh.

PART E
PLANT VIRUSES

16

Host-Range and Structural Data on Common Plant Viruses

AHMED F. HADIDI and HEINZ FRAENKEL-CONRAT

In the study of plant viruses, the plant is indispensible. The literature describing diseases produced by plant viruses is vast. Much of the earlier work has been summarized by Smith (1937), and more recent accounts have been given by Bos (1963) and Holmes (1964). *Descriptions of Plant Viruses,* edited by A. J. Gibbs, B. D. Harrison and A. F. Murant, and published by the Commonwealth Mycological Institute and Association of Applied Biologists (CMI/AAB)* is also of considerable usefulness. We have recently compared our independently tabulated data with these descriptions and have checked for any errors or omissions in regard to hosts. Some of our physical data are more up to date, however, than those of the CMI/AAB.

Some viruses under appropriate conditions may infect a plant without producing any obvious symptoms. Other viruses may lead to rapid death of the whole plant. Between these extremes, a wide variety of diseases can be produced. The genetic make-up of the host plant often has

* Central Sales Branch, Commonwealth Agricultural Bureau, Farnham Royal, Slough, sL2 3BN, England.

AHMED HADIDI—Department of Virology and Cell Biology, Litton Bionetics, Inc., Kensington, Maryland. HEINZ FRAENKEL-CONRAT—Department of Molecular Biology and Virus Laboratory, University of California, Berkeley, California.

TABLE 1. Structural Characteristics of 32 Common Plant Viruses

Virus	Particle				Nucleic acid[a]			Protein		Remarks
	Shape, (structure)	Dimensions, nm	Particle weight[c] $\times 10^6$	Sedimentation coefficient,[c] S_{20}	Sedimentation coefficient, S_{20}	Percent	Molecular weight $\times 10^6$	No. of subunits	Molecular weight $\times 10^3$	
Alfalfa mosaic	Bacilliform and spheroid,[b] (helical)	18 × 58(B),[b] 18 × 48(M), 18 × 36(Tb), 18 × 18(Ta)	7.3(B) to 3.7(Ta)	99(B), 89(M), 75(Tb), 68(Ta), (60, To)	27(B) 13(Ta)	18	1.3, 1.0 0.7, 0.34		25–29	Co-virus system, the three largest particles or all four RNAs necessary for infection
Barley stripe mosaic	Tubular rod, (helical)	20 × 128	26	185	20	4	1.0		21.5	Co-virus system. Particles helically constructed with 2.5 nm pitch, diameter of internal hole 3–4 nm
Bean pod mottle	Isometric, (icosahedral)	30	7.5 6.5 (5.0)	112 91 (54)		37, 30, 0	2.5 1.5	60 60	~40 ~20	Co-virus system, both particles or RNAs required for infectivity
Broad bean mottle	Isometric, (icosahedral)	26	4.8	84	17 10	22	1.1, 1.03 0.9 0.36	180	21	Co-virus system, the three larger RNAs required for infectivity, hollow 12-nm center.
Brome mosaic	Isometric (icosahedral)	26	4.7	86	27 22 14	20	1.09, 0.99 0.75 0.28	180	20	Co-virus system, the three larger RNAs required for infectivity, all RNAs have the same terminal sequences
Cauliflower mosaic	Isometric	50	28	220	20(18)	16	4.7		33	The DNA is circular and double-stranded, contains 43 percent G + C, particle center is hollow about 20-nm in diameter

Virus	Particle									Remarks
Citrus exocortis	No particle	—	—	—	11	100	0.1	—	—	Viroid, i.e., nonparticulate RNA
Cowpea chlorotic mottle	Isometric (icosahedral)	25	4.6	88	23, 18, 13	24	1.15, 1.0, 0.85, 0.32	180	19.3	Co-virus system, the three largest RNAs required for infectivity
Cowpea mosaic	Isometric	28	7.7, 6.0	115, 95, (55)	34, 26	33, 23, 0	2.55, 1.45	60, 60	42, 22	Co-virus system, both particles or RNAs required for infectivity
Cucumber mosaic	Isometric (icosahedral)	30	5.5	98	22, 19, 13	18	0.95, 0.65, 0.33	180	25	Co-virus system, particle center is hollow, unstable virus
Pea enation mosaic	Isometric	28	—	115, 95	34, 30	28, 18	1.6, 1.3		22	Co-virus system, unstable virus
Potato spindle tuber	No particle	—	—	—	10	100	~0.1	—	—	Viroid, i.e., nonparticulate RNA
Potato X	Flexuous filament (helical)	11.5 × 550	35	121		6	2.0	1300	27	Protein end groups are known, protein unstable, the width of the virus helix 3.4 nm with 10 subunits per turn
Potato Y	Flexuous filament (helical)	11 × 730				6			21.3	The width of the helix is 3.4 nm, unstable virus, characteristic inclusions
Rice dwarf	Isometric (icosahedral)	70		510		11	Range, 2.8–0.4 (12 pieces)	Many	Various	50-nm core containing double-stranded RNA of 12 distinctive sizes, RNA polymerase, many proteins, etc. RNA is not infectious and contains 44 percent G + C, the virus resembles reovirus group of mammals
Satellite of tobacco necrosis	Isometric	17	2.0	50	14	20	0.4	72	23	Smallest and deficient RNA virus, requires tobacco necrosis as helper, RNA and protein end groups (etc.) known

Table 1. Continued

Virus	Particle				Nucleic acid[a]			Protein		Remarks
	Shape, (structure)	Dimensions, nm	Particle weight[c] $\times 10^6$	Sedimentation coefficient[c] S_{20}	Percent	Sedimentation coefficient, S_{20}	Molecular weight $\times 10^6$	No. of subunits	Molecular weight $\times 10^3$	
Southern bean mosaic	Isometric	27	6.5	115	21	25	1.4	180	30	Very stable virus
Squash mosaic	Isometric, (icosahedral)	25–30	6.9 6.1 (4.5)	118 95 (57)	35 27 0		2.4 1.6	60 60	40 22	Multicomponent, probably also a co-virus system although the largest particles (1185) have been reported as being infectious
Tobacco etch	Flexuous filament, (helical)	13 × 730		154	5				27	Related to potato virus Y.
Tobacco mosaic	Rigid rods, (helical)	18 × 300	40	190	5	30	2	2130	17.5	The rod particle is a helix with 2.3 nm pitch and 130 turns, each containing 16 1/3 protein subunits, end group sequences and some internal sequences are known for the RNA, the amino acid sequences of the protein of several strains are known
Tobacco necrosis	Isometric, (icosahedral)	28	7.6	120	19	14	1.3	180	32	End group sequences are known for protein and RNA, defective strains exist
Tobacco rattle	Rigid rods: long (L) and short (S)	20.5 × 180–210(L) or 20.5 × 52–110(S)	46–50(L), 12–29(S)	300(L), 155–243(S)	5	26 12–20	2.3(L) 0.6–1.3(S)		22	Co-virus system, L particles are infective, but induce synthesis of L particle RNA only, S particles are

Common name	Particle shape									Comments
										noninfective alone but carry the gene for the viral coat protein; L and S are produced only when the inoculum contains both L and S or their RNAs; the protein C terminus is proline
Tobacco ringspot	Isometric	30	5.7, 4.9, (3.3)	127, 92, (53)	32, 24	40, 27, 0	2.2, 1.2		28	Possibly a co-virus system, the largest RNA is infectious
Tobacco streak	Isometric	28.5		113, 98, 90			1.3, 1.06, 0.78, 0.72, 0.36		28.5	Possibly a co-virus system with the largest two particles required for infectivity
Tomato bushy stunt	Isometric, (icosahedral)	31	9	132		17	1.6	180, 1	40, 80	The large protein may be a polymerase or maturation factor, very stable virus
Tomato ringspot	Isometric. (icosahedral)	28	5.6, (3.2)	127, (53)		40, 0	2.3		24	
Turnip crinkle	Isometric, (icosahedral)	29	8–9	128		17	1.5	180, 1	38, 80	Similar in all known respects to tomato bushy stunt virus, but not serologically related
Turnip yellow mosaic	Isometric, (icosahedral)	28	5.6, (3.6)	117, (54)	22	34, (0)	1.9	180	20	End group sequences of the RNA and the amino acid sequence of the protein is known, the RNA is very high in cytidine (40 percent), particles containing less RNA have also been reported
Wheat striate mosaic	Bacilliform and bullet shaped, (helical)	75 × 205, 250, 410		900		5	Many	Many	Various	Rhabdovirus resembling vesicular stomatitic virus of mammals, lipid-containing envelope, contains RNA polymerase

TABLE 1. Continued

Virus	Particle				Nucleic acid[a]			Protein		Remarks
	Shape, (structure)	Dimensions, nm	Particle weight[c] × 10^6	Sedimentation coefficient[c], S_{20}	Sedimentation coefficient, S_{20}	Percent	Molecular weight × 10^6	No. of subunits	Molecular weight × 10^3	
White clover mosaic	Filamentous (helical)	13 × 480		119		6	2.4		21	Particle axial canal about 3.5 nm in diameter, pitch of helix 3–4 nm, perhaps 11 subunits per turn.
Wild cucumber mosaic	Isometric, (icosahedral)	28	6	119 (63) (53)		35 (11) (0)	2.4	180	19	Similar and related to turnip yellow mosaic virus, RNA contains 41 percent cytidine.
Wound tumor	Isometric, (icosahedral)	70	68	514		22	(2.8, 2.3, 2.08, 1.72[d], 1.08[d], 1.01, 0.77, 0.56[d], 0.54, 0.33)	Many	4 major, 150–44	The double-stranded RNA is present as 13 strands, the virus, related to the reoviruses of mammals, contains RNA polymerase, isolated RNA is not infective; contains 38 percent G + C

[a] All plant viruses listed contain single-stranded RNA except cauliflower mosaic virus, which contains cyclic double-stranded DNA, and the rice dwarf as well as wound tumor viruses which contain double-stranded RNA of 10–18 different species.

[b] Upon density-gradient centrifugation, the bacilliform particles yield bottom (B), middle (M), and top (Tb) components, while the spheroidal particles yield the accessory top components Ta and To.

[c] Numbers between parenthesis indicate viral top component lacking nucleic acid.

[d] Two components.

TABLE 2. *Host Ranges of 32 Common Plant Viruses*

Virus	Susceptible plant species[a]		Comments	General References
	Systemic infection	Localized reaction		
Alfalfa mosaic	6, 38, 45,[b] 69, 77, 85, 110, 123, 126, 127, 130, 134,[b] 138,[b] 149, 157, 164,[b] 197, 198, 206, 207, 215	38, 45,[c] 49,[c] 158,[c] 207,[b] 211[c]	Very wide host range; many strains have been described	Bancroft et al., 1960a; Berkeley, 1947; Fezer and Ross, 1959; Kreitlow and Price, 1949; Milbrath, 1963; Oswald, 1950; Pierce, 1934; Price, 1940; Price and Spencer, 1943; Thomas, 1951; Williams et al., 1971; Zaumeyer, 1952, 1953
Barley stripe mosaic	1, 2, 3, 7, 16,[b]33, 46, 93, 94, 102, 103,[b] 146, 168, 183, 192,[b] 202,[b] 214[b]	16,[b] 18[b]44,[c] 45,[c] 49,[c] 140,[c] 143	Moderate wide host range among monacot and dicot species; appears to be the best of the grass-infecting viruses for obtaining information on the evolution of strains (genetic instability), interstrain competition, and natural selection; its many strains appear to be ideal for the study of factors influencing interstrain competition and population balances in mixed strain cultures	Atabekov and Novikov, 1971; Eslick and Afansiev, 1955; Hollings, 1957; Inouye, 1962; Kassanis and Slykhuis, 1959; McKinney, 1951a, b, 1953, 1954; McKinney and Greeley, 1960, 1965; Nitzany and Gerechter, 1962; Sill and Hansing, 1955; G. P. Singh et al., 1960; Slykhuis, 1952; Timian and Sisler, 1955; Walters, 1954

TABLE 2. *Continued*

Virus	Susceptible plant species[a]		Comments	General References
	Systemic infection	Localized reaction		
Bean pod mottle	97,[b] 158,[b] 188	155, 158[c]	Very narrow host range confined to the leguminous family, either local lesion infection or systemic infection in bean according to variety, the virus serologically related to cowpea mosaic (Shepherd, 1964) and squash mosaic (Campbell, 1964)	Campbell, 1964; Shepherd, 1964; Thomas and Zaumeyer, 1950a; Zaumeyer and Thomas, 1948
Broad bean mottle	110, 133, 158,[b] 164,[b] 197, 198, 200, 207,[b] 208	45,[b] 97	Very narrow host range, no strains reported	Bawden et al., 1951; Hollings, 1959
Brome mosaic	2, 16, 33, 69, 72, 91, 92, 98, 102, 103,[b] 143, 169, 183, 202, 214[b]	18, 44,[c] 45,[c] 47,[c] 49, 69, 77,[c] 138, 158, 211, 214[b]	Moderately wide host range, especially among grasses; very few strains are known, the virus serologically related to cowpea chlorotic mottle virus (Scott and Slack, 1971)	Chiu and Sill, 1959, 1963; Ford et al., 1970; McKinney, 1944, 1953; McKinney et al., 1942; Rochow, 1959; Scott and Slack, 1971; Sill and Chiu, 1959; Slykhuis, 1967
Cauliflower mosaic	19,[b] 20, 21, 22, 23, 24, 25,[b] 26, 27, 28, 29, 30, 31,	19,[b] 77[c]	Virus host range appears to be confined mainly to species of the cruciferous	Berkeley and Tremaine, 1954; Broadbent and Tinsley, 1953; Caldwell

Virus			Comments	References
	32, 36, 39, 43, 77,[b] 83, 112, 117, 133, 177, 178, 186, 187		family, all susceptible plants are systemically infected with the virus, inoculated leaves usually do not form local lesions, jimson weed reported as a local-lesion assay host for certain virus isolates (Lung and Pirone, 1971)	and Prentice, 1942; Hills and Campbell, 1968; Lung and Pirone, 1971; Tompkins, 1937; Walker *et al.*, 1945
Citrus exocortis	54, 55, 56, 57, 58, 59, 60, 61, 62, 100,[b] 123,[b] 149, 170		Very limited host range, hosts systemically infected with the virus	Benton *et al.*, 1949; Childs *et al.*, 1958; Garnsey and Whidden, 1970; Kapur *et al.*, 1970; Olson and Shull, 1956; Semancik and Weathers, 1972; Weathers, 1965
Cowpea chlorotic mottle	12, 40, 85, 118, 138, 149, 158, 164, 207, 209, 210, 211[b]	41, 44,[c] 47,[c] 72, 97,[c] 152	Narrow host range, the virus serologically related to brome mosaic virus (Scott and Slack, 1971)	Bancroft and Flack, 1972; Kuhn, 1964, 1968; Scott and Slack, 1971
Cowpea mosaic	69, 97, 133, 134, 138, 153, 154, 155, 156, 157, 164, 184, 210, 211, 212	18, 40, 44, 45,[c] 46, 49, 73, 77, 85, 131, 133, 134, 138, 149, 153, 155, 156, 158,[c] 184, 211,[b] 212	Moderate host range of nonleguminous plants in addition to leguminous species, the virus serologically related to bean pod mottle virus (Shepherd, 1964)	Chant, 1962; Dale, 1949; Pérez and Cortés-Monllor, 1970; Shepherd, 1964
Cucumber mosaic	7, 12, 18, 26, 38, 40, 53, 68, 69,[b] 71, 72,[b] 75, 77, 78, 87, 93, 98,	18, 45,[c] 49,[c] 53, 69,[b] 73, 77, 98, 123,[b] 133,[b] 136, 138,[b] 149, 152, 155,	Very wide host range, the virus has numerous strains	Ainsworth, 1935; Doolittle, 1920; Doolittle and Walker, 1925; Doolittle and Zaumeyer, 1953;

TABLE 2. Continued

Virus	Susceptible plant species[a]		Comments	General References
	Systemic infection	Localized reaction		
	101, 119, 123,[b] 127, 133,[b] 134,[b] 136, 138,[b] 141,[b] 149, 152, 155, 158,[b] 164, 172, 192, 197, 198, 207, 210, 211,[b] 214, 215	158,[c] 189, 207, 210, 211,[c] 214		Foster, 1972; Gibbs and Harrison, 1970; Hadidi and Hampton, 1968; Lawson, 1966; Lindberg et al., 1956; Mink et al., 1969; Orellana and Quacquarelli, 1968; Pound and Walker, 1948; Price, 1940; Sill and Walker, 1952; Tomaru and Udagawa, 1967; Tomlinson et al., 1959
Pea enation mosaic	10, 13, 52, 97, 110, 118, 119, 126, 127, 128, 133, 158, 164,[b] 197,[b] 198, 199, 200, 207,[b] 208	44,[c] 45,[b] 49,[c] 95,[c] 164,[c] 207[b]	Wide host range among leguminous species, few species in other families are susceptible	Bozarth and Chow, 1965; Gonsalves and Shephard, 1972; Hagedorn, 1957; Hagedorn et al., 1964; Izadpanah and Shepherd, 1966; McEwen and Schroeder, 1956; McWhorter and Cook, 1958; Simons, 1954
Potato spindle tuber	38, 76, 77, 98, 123,[b] 124, 130, 134, 136, 138, 149, 160, 161, 182,[b]	182[c]	Narrow host range confined mainly to plants in the potato family (Solanaceae), susceptible plants are	Diener et al., 1972; Easton and Marriam, 1963; O'Brien and Raymer, 1964; Raymer and

	190, 191[b]		systemically infected with the virus	O'Brien, 1962; R. P. Singh, 1971; R. P. Singh and O'Brien, 1970
Potato X	14, 18, 38, 69, 74, 76, 77[b] 104, 109, 113, 123, 130[b] 134[b] 136[b] 138[b] 181, 189, 190, 191[b]	7, 14, 44, 45,[c] 76,[c] 77[b] 81, 98,[c] 104, 130,[b] 134,[b] 136,[b] 138,[b] 181, 189, 205	Moderate host range, many strains differ from each other in different ways. most susceptible hosts are locally and systemically infected	Ladeburg *et al.*, 1950; Larson, 1944; Matthews, 1949; Salaman, 1939; R. P. Singh, 1969; Wilkinson and Blodgett, 1948
Potato Y	9, 38,[b] 41, 65, 74, 75, 76, 78, 104, 105, 123, 134,[b] 136, 138,[b] 139,[b] 163, 189, 190, 191,[b] 215	45,[c] 49,[c] 50, 120,[c] 121,[c] 122,[c] 130[c] 136,[c] 160[c] 162,[c] 191, 192	Moderate host range, many strains, the virus is serologically related to tobacco etch virus (Purcifull and Gooding, 1970)	Bawden and Kassanis, 1946; Delgado-Sanchez and Grogan, 1970; Dennis, 1938; Darby *et al.*, 1951; Martin, 1952; Murant *et al.*, 1970; Purcifull and Gooding, 1970; Ross, 1948; Silberschmidt and Rostom, 1955
Rice dwarf	4, 15, 16, 86,[b] 96, 103, 143,[b] 145, 147, 167, 169, 183, 202		A narrow host range confined to the grass family, all susceptible plants are systemically infected with the virus	Fukushi, 1934; Iida, 1969; Shinkai, 1962
Satellite of tobacco necrosis		133,[b] 138,[b] 158[c]	The virus multiplies only in plants infected with tobacco necrosis virus; the two viruses are not serologically related, several strains have been described which differ	Kassanis, 1970

TABLE 2. Continued

Virus	Susceptible plant species[a]		Comments	General References
	Systemic infection	Localized reaction		
			antigenically and in their ability to be activated by different strains of tobacco necrosis virus	
Southern bean mosaic	97[b], 158[b], 188, 209, 210, 211[b]	153, 155[b], 158[c], 211[c]	Narrow host range confined to the leguminous species, three distinct strains reported, local lesions only or systemic infection only in bean according to variety, however, some varieties react locally and systemically to the severe strain of the virus (Grogan and Kimble, 1964), the common strain does not infect cowpea, a good host for the cowpea strain, and the cowpea strain does not infect bean, a good host for the common strain (Kuhn, 1963)	Grogan and Kimble, 1964; Kuhn, 1963; Shepherd and Fulton, 1962; Zaumeyer and Harter, 1943
Squash mosaic	9, 53[b], 65, 67, 68, 69[b], 70, 71, 72[b],	53[b], 72[b]	Narrow host range, the virus has several strains, a host	Campbell, 1964; Demski, 1969; Freitag, 1956;

Virus			Description	References
	108, 110, 111, 129, 150, 164, 196, 201		for local lesion assay is not known, the virus is serologically related to bean pod mottle virus (Campbell, 1964)	Grogan et al., 1959; Hadidi et al., 1966; Lindberg et al., 1956; Nelson et al., 1965
Tobacco etch	38[b], 69, 77[b], 123[b], 134[b], 138[b], 149, 158, 189, 190, 191, 215	35, 42, 44[c], 45[c], 49[c], 64, 101, 138[c], 161[c], 193	Very wide host range, hosts susceptible to the virus are also susceptible to tobacco mosaic virus, the virus is serologically related to potato virus Y (Purcifull and Gooding, 1970), jimson weed is susceptible to tobacco etch virus but not to potato virus Y, several strains of the virus have been described	Holmes, 1942, 1946; Purcifull and Gooding, 1970; Shepherd and Purcifull, 1971; Stover, 1951
Tobacco mosaic	38, 42, 48, 77, 88, 93, 98, 99, 123[b], 125, 130, 133[b], 135, 136[b], 137[b], 138, 142, 149, 150, 159, 161, 176, 189, 190, 191, 192, 213, 215	18, 38, 42, 44, 45[c], 48, 49, 69[b], 77, 80, 93, 98, 121, 134[c], 136[b], 141[c], 158[c], 159, 166[b], 189, 190, 191, 193, 215	Very extensive host range, numerous viral strains have been described, among the strains described that have distinct characteristics either in host reaction, biochemical properties or both are: Cucumber strains (CV3 and CV4) (Ainsworth, 1935; Bawden and Pirie, 1937) which are confined to plants of the cucurbit family, a strain	Ainsworth, 1935; Bald and Paulus, 1963; Bawden and Pirie, 1937; Cheo, 1970; Chessin et al., 1967; Corbett, 1967; Gilmer and Kelts, 1965; Grant, 1934; Holmes, 1928, 1938, 1941; Jensen, 1933; Kirkpatrick and Lindner, 1964; Lister and Thresh, 1955; McKinney and Fulton, 1949; Nienhous and Yarwood, 1972; Price, 1930; Siegel et al., 1962;

TABLE 2. Continued

Virus	Susceptible plant species[a]		Comments	General References
	Systemic infection	Localized reaction		
			which infects common bean systemically (Lister and Thresh, 1955), ribgrass strain (Holmes, 1941) which contains histidine and methionine, *Lychnis alba* strain (Corbett, 1967), defective PM1 and PM2 strains (Siegel *et al.*, 1962), oak strain (Yarwood and Hecht-Poinar, 1970; Nienhous and Yarwood, 1972); most of the above-mentioned strains produce local lesions and systematic symptoms in common tobacco	Siegel and Wildman, 1954; Yarwood and Hecht-Poinar, 1970
Tobacco necrosis	42, 97, 98, 101, 173, 174, 175 203, 213	18. 26, 45,[c] 53, 77, 97, 98, 101, 103, 107, 133,[c] 138,[c] 149, 153,[c] 158,[c] 172, 173, 174, 175, 203, 207, 211,[c] 213, 214	Very wide host range, virus remains localized or nearly so in most susceptible hosts; in several species infection spreads and becomes systemic, many strains of	Bawden and Kassanis, 1947; Cesati and Van Regenmortel, 1969; R. W. Fulton, 1950; Hollings and Stone, 1965; Liu *et al.*, 1969; Kassanis, 1949; Kassanis and Phillips,

			the virus have been described, the virus is required for the multiplication of its defective satellite	1970; Price, 1940; Uyemoto and Gilmer, 1972
Tobacco rattle	38, 97, 110, 123, 133[b], 137, 138, 149, 191	35, 37, 45[c], 69[b], 72, 97, 98, 110, 127, 134, 136, 137, 138, 155, 158[c], 164, 191	Very wide host range, virus infection often localized in susceptible hosts	Allen, 1964; Cadman and Harrison, 1959; Noordam, 1956; Schmelzer, 1957; Semancik, 1966; Stouffer, 1965
Tobacco ringspot	6, 11, 38, 53, 63, 69[b], 72, 73, 97, 98, 101, 107, 118, 123[b], 126, 127, 133[b], 134[b], 138[b], 149, 153, 158[b], 172, 189, 191, 192, 193, 204, 207, 212[b], 213, 215	6, 11, 40,[c] 44, 45,[b] 49[b], 53, 69[b], 77, 97, 98, 101, 118, 133[c], 134[b], 138[c], 149, 153, 155[c], 158[b], 164, 189, 191, 193, 207, 211[c]	Very extensive host range, a satellite like virus associated with tobacco ringsport has been recently described (Schneider, 1971)	de Zeeuw, 1965; Gilmer et al., 1970; Hollings, 1957; Ladipo and de Zoeten, 1972; McLean, 1962a, b; Orellana, 1967; Price, 1940; Roberts and Corbett, 1961; Rush and Gooding, 1970; Schneider, 1971; Tuite, 1960; Wilkinson 1952,[b] Wingard, 1928
Tobacco streak	11, 45, 48, 49, 66, 69, 71, 73,[b] 75, 77,[b] 97, 101, 119, 123, 127, 130, 132, 134, 136,[b] 138,[b] 149, 154, 156, 157, 158,[b] 164, 165, 184, 197, 199, 208, 210, 211, 212[b]	40, 44, 45, 48, 49,[c] 73,[c] 84,[c] 85, 97, 98, 101, 118, 123, 130, 136,[b] 138,[b] 158,[c] 160, 207, 209,[c] 211	Very wide host range; bean red node is a distinct strain of the virus (Scott et al., 1961)	Brunt, 1968; Costa and Carvalho, 1961; Diachum and Valleau, 1947, 1950; Fagbenle and Ford, 1970; R. W. Fulton, 1948; 1967; Johnson, 1936; Lister and Bancroft, 1970; Mink et al., 1966; Patino and Zaumeyer, 1959; Scott et al., 1961; Thomas and Zaumeyer, 1950b

TABLE 2. Continued

Virus	Susceptible plant species[a]		Comments	General References
	Systemic infection	Localized reaction		
Tomato bushy stunt	19, 45, 49[b] 77[b] 80, 88, 123, 133[b], 148, 149[b] 195, 207	6[c], 19, 42[c] 45[c] 49[b] 69, 77[b] 80, 88, 98[b] 101, 107, 110, 123, 133[b], 134[c] 138, 148, 149[b] 158[c] 191, 195, 207, 211, 215	Moderate wide host range, it remains localized in most susceptible hosts, some hosts are infected systemically and locally with the virus	Hollings, 1959; Hollings and Stone, 1965; Kleczkowski, 1950; Martelli et al., 1971; Smith, 1935
Tomato ringspot	11, 12, 18, 45[b] 48, 49[b] 53, 68, 69[b] 71, 73, 77, 78, 98, 101, 123[b] 131, 134, 136, 138[b] 139[b] 149[b] 155, 158[b] 159, 164, 189, 190, 204, 211[b] 213, 215	11, 12, 18, 45[c] 49[c] 53, 68, 69[b] 71, 73, 77, 98, 99, 123[b] 138, 139[c] 149[b] 155, 158[b] 189, 190, 204, 211[v]	Very wide host range, most hosts are locally and systemically infected with the virus, the virus not serologically related to tobacco ringspot virus (Cadman and Lister, 1961)	Cadman and Lister, 1961; S. P. Fulton and Fulton, 1970; Kahn, 1956; McLean, 1962b; Price, 1940; Rush and Gooding, 1970; Samson and Imle, 1942; Stace-Smith, 1966; Uyemoto, 1970; Varney and Moore, 1952a,b; Wilkinson, 1952a
Turnip crinkle	17, 19, 21[b] 22, 24, 25, 26, 30[b] 31[b] 32, 36, 48[b] 90, 110, 117, 123, 148, 177, 178, 179[b] 186, 193[b] 194	6, 17, 25, 26, 30[b] 31[b] 39, 42, 43, 44, 45[c] 48[b] 49[b] 63, 66, 69, 77[b] 98, 107, 110, 123, 132, 149, 193[b]	Wide host range among cruciferous and noncruciferous species, the virus has several strains	Broadbent and Heathcote, 1958; Hollings and Stone, 1965

Virus			Description	References
Turnip yellow mosaic	17, 20, 21, 23, 24, 26, 27, 28, 29, 30,[b] 31,[b] 32, 39, 43, 83, 90, 112, 117, 177, 178, 179, 187, 194	30,[b] 31[b]	The virus has a narrow host range confined mainly in the cruciferous family, most hosts show only systemic reponse; Chinese cabbage, under special controlled conditions, was reported as a local lesion assay host for the virus (Diener and Jenifer, 1964); many strains have been described, the virus serologically related to wild cucumber mosaic virus (MacLeod and Markham, 1963)	Broadbent and Heathcote, 1958; Croxall et al., 1953; Diener and Jenifer, 1964; Lister, 1958; Markham and Smith, 1949; Sander and Schramm, 1963
Wheat striate mosaic	15, 16, 33, 34, 82, 86,[b] 89,[b] 103,[b] 115, 116, 144,[b] 145, 146, 185, 202,[b] 214[b]		Narrow host range confined mainly to plants in the grass family, susceptible hosts are systemically infected with the virus	Slykhuis, 1953, 1963
White clover mosaic	53, 66, 73, 97, 123, 126, 127, 137, 155, 156, 158,[b] 197,[b] 198,[b] 199,[b] 207,[b] 211[b]	53, 66, 69,[b] 72, 73, 77, 97, 106, 123, 126, 127, 137, 155, 156, 158,[c] 164,[b] 197,[b] 198,[b] 199,[b] 207,[b] 211[c]	Narrow host range, most susceptible hosts are locally and systemically infected	Agrawal et al., 1962; Brancroft et al., 1960b; Bos et al., 1959, 1960; Johnson, 1942; Pierce, 1935; Scott and Gold, 1959; Zaumeyer and Wade, 1935

TABLE 2. *Continued*

Virus	Susceptible plant species[a]		Comments	General References
	Systemic infection	Localized reaction		
Wild cucumber mosaic	53[b], 68, 71[b], 72[b], 73, 87, 130	53[b]	A very limited host range, cucumber is immune to virus infection, turnip yellow mosaic virus and wild cucumber mosaic virus are related serologically (MacLeod and Markham, 1963) but do not have host in common	Freitag, 1941; Lindberg *et al.*, 1956; MacLeod and Markham, 1963; Thornberry, 1966
Wound tumor	8, 39, 51, 79, 114, 127[b], 128[b], 138, 171, 179, 180[b], 193, 197[b], 198		Moderate host range, the most common symptoms are vein-enlargement and the occurrence of woody tumors on the roots of infected plants, agallian leaf hoppers are also hosts of the virus	Black, 1945, 1946, 1950, 1970; Valleau, 1947

[a] The numbers refer to plant species as listed on Table 3.
[b] Commonly used plant species.
[c] Commonly used plant species and local-lesion assay host.

TABLE 3.   Host Plants of Some Common Plant Viruses

No. in Table 2	Host plant	No. in Table 2	Host plant
1	*Agropyron intermedium* (Host) Beauv. intermediate wheat grass	28	*Brassica oleracea* L. var. *gemmifera* Zenker., brussels sprouts
2	*Agropyron* spp.	29	*Brassica oleracea* L. var. *italica* Plenk., sprouting broccoli
3	*Agrostis palustris* Huds. bent grass		
4	*Alopecurus aequalis* Sobol.	30	*Brassica pekinensis* Rupr., chinese cabbage
5	*Alopecurus japonicus* Steud.	31	*Brassica pe-tsai* Bailey., Chinese cabbage
6	*Amaranthus caudatus* L. love-lies-bleeding	32	*Brassica rapa* L., turnip
7	*Amaranthus retroflexus* L., pigweed	33	*Bromus inermis* Leyss., bromegrass
8	*Anthemis cotula* L., may weed camomile	34	*Bromus* spp.
9	*Anthriscus cerefolium* Hoffin., salad chervil	35	*Callistephus chinensis* Nees., China aster
10	*Anthyllis vulneria* L.	36	*Camelina sativa* Crantz., gold of pleasure
11	*Antirrhinum majus* L., snapdragon	37	*Campanula medium*, canterbury bells
12	*Arachis hypogea* L., peanut	38	*Capsicum frutescens* L., tabasco pepper
13	*Astragalus rubyi*.		
14	*Atropa belladona* L.	39	*Capsilla bursa-pastoris* Medic., shepherd's purse
15	*Avena fatua* L., wild oat		
16	*Avena sativa* L., oat	40	*Cassia occidentalis* L., coffee senna
17	*Barbarea vulgaris* R. Br., winter cress	41	*Cassia tora* L., sickle senna
18	*Beta vulgaris* L., sugar beet	42	*Celosea argentea* L., feather cockscomb
19	*Brassica campestris*, tendergreen mustard	43	*Cheiranthus cheiri* L., wallflower
20	*Brassica hirta* Moench., white mustard	44	*Chenopodium album* L., lamb's quarter
21	*Brassica juncea* Czern and Coss., indian mustard	45	*Chenopodium amaranticolor* Coste and Reyn., ornamental goosefoot
22	*Brassica napo-brassica* Mill., swede		
23	*Brassica napus* L., rape	46	*Chenopodium capitatum* (L) Asch., strawberry blight
24	*Brassica niger* L., black mustard	47	*Chenopodium hybridum*
25	*Brassica oleracea* L. var. *botrytis* L., cauliflower	48	*Chenopodium murale* L.
		49	*Chenopodium quinoa* Willd., quinoa
26	*Brassica oleracea* L. var. *capitata* L., cabbage	50	*Chenopodium urbicum* L., city goosefoot
27	*Brassica oleracea* L. var. *caulorapa* Pasq., kohlrabi		

TABLE 3.   Continued

No. in Table 2	Host plant	No. in Table 2	Host plant
51	*Chrysanthemum lencanthemum* L., chrysanthemum	75	*Dahlia variabilis* (Willd) Desf.
52	*Cicer arietinum* L., chick pea	76	*Datura metel* L., hindu datura
53	*Citrullus vulgaris* Schrad., watermelon	77	*Datura stramonium* L., jimson weed
54	*Citrus aurantifolia*, Mexican lime	78	*Daucus carota* var. *sativa*, carrot
55	*Citrus aurantium*, sour orange	79	*Dianthus armeria* L., deptford carnation
56	*Citrus jambhiri*, rough lemon	80	*Dianthus barbatus* L.
57	*Citrus limon*, Eureka lemon	81	*Digitalis lanata* Ehrh., gracian foxglove
58	*Citrus limonica*, rungpur lime	82	*Digitaria sanguinalis* (L). Scop., crabgrass
59	*Citrus medica*, etrog citron	83	*Diplotaxis tenuifolia* DC., perennial wall rocket
60	*Citrus paradisi* Duncan., grapefruit	84	*Dolichos biflorus* L.
61	*Citrus sinensis*, sweet orange	85	*Dolichos lablab* L., hyacinth bean
62	*Citrus sinensis* × *Poncirus trifoliata*, rusk citrange	86	*Echinochloa crusgalli* (L) Beauv., barnyard grass
63	*Cleome spinosa* L., spider flower	87	*Echinocystis lobata* (Michx) Torr and Gray., wild cucumber
64	*Collinsia bicolor* Benth., pagoda collinsia	88	*Emilia sagittata* (Vahl) DC., tassel flower
65	*Coriandrum sativum* L., coriander	89	*Eragrostics cilianensis* (All) Lutati., stink grass
66	*Crotalaria spectabilis* Roth., showy crotalaria	90	*Erysimum cheiranthoides* L., treacle mustard
67	*Cucumis anguria* L., West India or bur gherkin	91	*Euchlaena mexicana* Schrad.
68	*Cucumis melon* L., muskmelon	92	*Euchlaena perennis* Hitche.
69	*Cucumus sativus* L., cucumber	93	*Fagopyrum esculentum* Gaertn., buckwheat
70	*Cucurbita digitata* Gray., wild gourd	94	*Festuca rubra* L., red fescue
71	*Cucurbita maxima* L. Duch., winter squash	95	*Galactia* sp.
72	*Cucurbita pepo* L., pumpkin or squash	96	*Glyceria acutiflora* Torr.
73	*Cyamopsis tetragonoloba* (L) Taub., guar	97	*Glycine max* (L) Merr., soybean
74	*Cyphomandra betacea*, tree tomato	98	*Gomphrena globosa* L., globe amaranth
		99	*Gossypium hirsutum* L., cotton

TABLE 3. *Continued*

No. in Table 2	Host plant	No. in Table 2	Host plant
100	*Gynura aurantiaca* DC.	126	*Medicago sativa* L., alfalfa
101	*Helianthus annuus* L., sunflower	127	*Melilotus alba* Desr., white sweet clover
102	*Holcus sorghum* L., sorghum	128	*Melilotus officinalis* (L) Lam., yellow sweet clover
103	*Hordeum vulgare* L., barley		
104	*Hyoscyamus albus* L., henbane	129	*Momordica charantia* L., palsam pea
105	*Indigofera hirsuta*, hairy indigo	130	*Nicandra physalodes* (L) Pers., apple of Peru
106	*Ipomoea purpurea* Lam., morning glory	131	*Nicotiana alata* Link and Otto.
107	*Lactuca sativa* var. *capitata*, lettuce	132	*Nicotiana bigelovii* Wats var. *multivalvis* Gray.
108	*Lagenaria siceraria* Standl., white-flowered gourd	133	*Nicotiana clevelandii* Gray.
109	*Lamium hybridum*	134	*Nicotiana glutinosa* L.
110	*Lathyrus odoratus* L., sweet pea	135	*Nicotiana paniculata* L.
111	*Lens esculenta* Moench., lentil	136	*Nicotiana rustica* L., Aztec tobacco or Indian tobacco
112	*Lepidemium campestre* R. Br., pepperwort	137	*Nicotiana sylvestris* Speg and Comes.
113	*Linnaria sp.*	138	*Nicotiana tabacum* L., tobacco
114	*Linum grandiflorum* Desf., flowering flax	139	*Nicotiana tabacum* L. var. *Havana* 425
115	*Lolium multiflorum* Lam., Italian rye grass	140	*Nicotiana tabacum* L. var. *Samsun*
116	*Lolium perenne* L., perennial rye grass	141	*Nicotiana tabacum* var. *Xanthi* nc.
117	*Lunarria annua* L., honesty	142	*Nicotiana tomentosa* Ruiz and Pav.
118	*Lupinus albus* L., white lupine	143	*Oryza sativa* L., rice
119	*Lupinus angustifolius* L., blue lupine	144	*Panicum capillare* L., witch grass
120	*Lycium barbarum* L., barbary wolfberry	145	*Panicum miliaceum* L., broomcorn millet
121	*Lycium chinense* Mill., Chinese wolfberry	146	*Panicum spp.*
122	*Lycium rhombifolium* (Moench) Dippel.	147	*Paspalum thunbergii* Kunth.
123	*Lycopersicon esculentum* Mill., tomato	148	*Pelargonium domesticum* L., pelargonium
124	*Lycopersicon spp.*	149	*Petunia hybrida* Vilm., petunia
125	*Malus sylvestris* Mill., apple	150	*Phacelia campanularia* Gray.

TABLE 3.   Continued

No. in Table 2	Host plant	No. in Table 2	Host plant
151	*Phalaenopsis* sp., moonglow orchids	177	*Raphanus raphanistrum* L., wild radish
152	*Phaseolus acutifolius* var. *latifolius* Gray., tepary bean	178	*Raphanus sativus* L., radish
153	*Phaseolus aureus* Roxb., mungbean	179	*Reseda odorata* L., mignonette
154	*Phaseolus coccineus* L., scarlet runner bean	180	*Rumex acetosa* L., garden sorrel
155	*Phaseolus lunatus* L., sieve lima bean	181	*Salvia lanceafolia* Poir., lanceleaf sage
156	*Phaseolus mungo* L. Wolly Pyrol, urdbean	182	*Scopolia sinensis* Hemsl.
157	*Phaseolus* spp.	183	*Secale cereale* L., rye
158	*Phaseolus vulgaris* L., common bean	184	*Sesbania exaltata* (Raf.) Cory.
159	*Phlox drummondii* Hook., phlox	185	*Setaria verticillata* (L) Beauv. Bur., brittlegrass
160	*Physalis floridana* Rydberg.	186	*Sinapsis alba* L., white mustard
161	*Physalis peruviana* L., cape gooseberry	187	*Sisymbrium officinale* Scop., hedge mustard
162	*Physalis turbinata* Medic.	188	*Soja max* (L) Piper., soybean
163	*Physalis virginiana* Mill., Virginia ground cherry	189	*Solanum melongena* L., egg plant
164	*Pisum sativum* L., garden pea	190	*Solanum* spp.
165	*Plantago major* L., plantain	191	*Solanum tuberosum* L., potato
166	*Plantago* spp.	192	*Spinacia oleracea* L., spinach
167	*Poa annua* L.	193	*Tetragonia expansa* Murr., New Zealand spinach
168	*Poa compressa* L., Canada blue grass	194	*Thlaspi arvense* L., field penny-cress
169	*Poa paratensis*, Kentucky bluegrass	195	*Torenia fournieri* Lindl.
170	*Ponsirus trifoliata* (L) Raft., trifoliate orange	196	*Trichosanthes anguina* L., snake gourd
171	*Portulaca grandiflora* Hook., common portulaca	197	*Trifolium incarnatum* L., crimson clover
172	*Primula obconica*, top primrose	198	*Trifolium pratense* L., red clover
173	*Prunus armeniaca* L., apricot	199	*Trifolium repens* L., white clover
174	*Pyrus communis* L., pear	200	*Trifolium subterraneum*, subterranean clover
175	*Pyrus malus* L., apple		
176	*Quercus* spp., oak		

TABLE 3. *Continued*

No. in Table 2	Host plant	No. in Table 2	Host plant
201	*Trigonella foenum graecum* L., fenugreek	209	*Vigna cylidrica* (L) Skeels.
202	*Triticum aestivum* L., wheat	210	*Vigna sesquipedalis* (L) Fruwirth., asparagus bean
203	*Tulipa* spp., garden tulip		
204	*Ulmus americana* L., American elm	211	*Vigna sinensis Endl.,* cowpea
205	*Veronica* sp.	212	*Vinca rosea* L.
206	*Viburnum* sp.	213	*Vitis vinifera* L., grape
207	*Vicia faba* L., broad bean	214	*Zea mays* L., corn
208	*Vicia* spp.	215	*Zinnia elegans* Jacq., zinnia

a profound effect on the outcome of infection with a particular virus or any of its strains. There are many records of plant reactions to experimental infection with many viruses. Some descriptions conflict, or differ in details; this is not surprising, considering the widely differing environments in which test plants have been grown and the differences in virus strains and varieties of plant species used. It is generally believed that a certain type of resistance of a plant to virus infection results in a localized reaction, the so-called local lesion, as compared with the recognized susceptible reaction involving systemic spread of the virus. No consistant pattern of reaction is obvious among the different species of plants in many families; even different cultivars of the same plant species may react quite differently to the same virus isolates. Nevertheless, host-range studies have several uses: They may reveal new plants suitable for diagnostic or qualitative work. Sources of immunity or resistance valuable to the plant breeder may be discovered. The finding of advantageous hosts for virus maintenance and large-scale virus production may be facilitated. In addition, host-range studies can be used to differentiate similar viruses or virus strains and to suggest similiarties between other viruses.

We will list in tabular alphabetic forms (a) brief structural descriptions (Table 1) and host-range data (Table 2) for 32 common plant viruses. Neither set of data is expected to be complete. The hosts will, for brevity's sake, be listed by number, these numbers referring to the alphabetical list of these virus hosts in Table 3. The reactions of the susceptible plants are subdivided into those giving generalized systemic disease, local response, and local lesions suitable for virus quantitation. Also, a list of references pertaining to the hosts and their responses to infection by the viruses listed will be found in Table 2. For reasons of space, the data summarizing the structural characteristics of these viruses are given without references.

More information about the properties of these viruses can be obtained from the *Catalogue of Viruses,* Volume I of *Comprehensive Virology* (Fraenkel-Conrat 1974), and particularly from *Descriptions of Plant Viruses,* listed above.

Literature Cited

Agrawal, H., L. Bos and M. Chessin, 1962   Distribution of clover yellow mosaic and white clover mosaic viruses on white clover in the United States. *Phytopathology* **52:**517–519.

Ainsworth, G. S., 1935   Mosaic diseases of the cucumber. *Ann. Appl. Biol.* **22:**55–67.

Allen, T. C., Jr., 1964   Tobacco rattle virus from Oregon compared with pea early browning virus. *Phytopathology* **54:**1431.

Atabekov, J. G. and V. K. Novikov, 1971   Barley strip mosaic virus. In *Descriptions of plant viruses,* No. 68, edited by A. J. Gibbs, B. D. Harrison and A. F. Murant, Commonwealth Mycological Institute and Association of Applied Biologists, Slough, England.

Bald, J. G. and A. O. Paulus, 1963   A characteristic form of tobacco mosaic virus in tomato and *Chenopodium murale. Phytopathology* **53:**627–629.

Bancroft, J. B. and I. H. Flack, 1972   The behavior of cowpea chlorotic mottle virus in CsCl. *J. Gen. Virol.* **15:**247–251.

Bancroft, J. B., E. L. Moorhead, J. Tuite and H. P. Liu, 1960a   The antigenic characteristics and the relationship among strains of alfalfa mosaic virus. *Phytopathology* **50:**34–39.

Bancroft, J. B., J. Tuite and G. Hissong, 1960b   Properties of white clover mosaic virus from Indiana. *Phytopathology* **50:**711–717.

Bawden, F. C. and B. Kassanis, 1946   Varietal difference in susceptibility to potato virus Y. *Ann. Appl. Biol.* **33:**46–50.

Bawden, F. C. and B. Kassanis, 1947   *Primula obconica,* a carrier of tobacco necrosis viruses. *Ann. Appl. Biol.* **34:**127–135.

Bawden, F. C. and N. W. Pirie, 1937   The relationships between liquid crystalline preparations of cucumber virus 3 and 4 and strains of tobacco mosaic virus. *Brit. J. Exp. Biol.* **28:**275–290.

Bawden, F. C., R. P. Chaudhuri and B. Kassanis 1951   Some properties of broad-bean mottle virus. *Ann. Appl. Biol.* **38:**774–784.

Benton, R. J., F. T. Bowman, L. Fraser and D. G. Kebby, 1949   Selection of citrus budwood to control scaly butt in *Trifoliata* rootstock. *Agric. Gaz. N. S. W.* **60:**31–34.

Berkeley, G. H., 1947   A strain of the alfalfa mosaic virus on pepper in Ontario. *Phytopathology* **37:**781–789.

Berkeley, G. H. and J. H. Tremaine, 1954   Swedes naturally infected with two viruses. *Phytopathology* **44:**632–634.

Black, L. M., 1945   A virus tumor of plants. *Am. J. Bot.* **32:**408–415.

Black, L. M., 1946   Plant tumors induced by the combination of wounds and virus. *Nature (Lond.)* **158:**56–57.

Black, L. M., 1950   A plant virus that multiplies in its insect vector. *Nature (Lond.)* **166:**852–853.

Black, L. M., 1970 Wound tumor virus. In *Descriptions of Plant Viruses,* No. 34, edited by A. J. Gibbs, B. D. Harrison and A. F. Murants, Commonwealth Mycological Institute and Association of Applied Biologists, Slough, England.

Bos, L., 1963 *Symptoms of Virus Diseases in Plants with Indexes of Names in English, Dutch, German, French, and Italian,* Mededel · 307, Centre for Agricultural Publications and Documentation, Wageningen, Netherlands.

Bos, L., B. Delevic and J. P. H. van der Want, 1959 Investigations on white clover mosaic virus. *Tijdschr. Plantenziekten* **65:**89–106.

Bos, L., D. Z. Maat, J. B. Bancroft, A. H. Gold, M. J. Pratt, L. Quantz and H. A. Scott, 1960 Serological relationship of some European, American, and Canadian isolates of white clover mosaic virus. *Tijdschr. Plantenziekten* **66:**102–106.

Bozarth, R. F. and C. C. Chow, 1965 Pea enation mosaic virus: a new local lesion host. *Phytopathology* **55:**492–493.

Broadbent, L. and G. D. Heathcote, 1958 Properties and host range of turnip crinkle, rosette, and yellow mosaic viruses. *Ann. Appl. Biol.* **46:**585–592.

Broadbent, L. and T. W. Tinsley, 1953 Symptoms of cauliflower mosaic and cabbage black ringspot in cauliflower. *Plant Pathology* **2:**88–92.

Brunt, A. A., 1968 Tobacco streak virus in dahlia. *Plant Pathol.* **17:**119–122.

Cadman, C. H. and B. D. Harrison, 1959 Studies on the properties of soil-borne viruses of the tobacco-rattle type occurring in Scotland. *Ann. Appl. BIol.* **47:**542–556.

Cadman, C. H. and R. M. Lister, 1961 Relationship between tomato ringspot and peach yellow bud mosaic viruses. *Phytopathology* **51:**29–31.

Caldwell, J. and I. W. Prentice, 1942 A mosaic disease of broccoli. *Ann. Appl. Biol.* **29:**366–373.

Campbell, R. N., 1964 Radish mosaic virus, a crucifer virus serologically related to strains of bean pod mottle and to squash mosaic virus. *Phytopathology* **53:**1285–1291.

Cesati, R. R. and M. H. Van Regenmortel, 1969 Serological detection of a strain of tobacco necrosis virus in grapevine leafs. *Phytopathol. Z.* **64:**362–366.

Chant, S. R., 1962 Further studies on the host range and properties of Trinidad cowpea mosaic virus. *Ann. Appl. Biol.* **50:**159–162.

Cheo, P. C., 1970 Substantial infection of cotton by tobacco mosaic virus. *Phytopathology* **60:**41–46.

Chessin, M., M. Zaitlin and R. A. Solberg, 1967 A new strain of tobacco mosaic virus from *Lychnis alba. Phytopathology* **57:**452–453.

Childs, J. F. L., G. G. Norman and J. L. Eichhorn, 1958 A color test for exocortis infection in *Poncirus trifoliata. Phytopathology* **48:**426–432.

Chiu, R. J. and W. H. Sill, Jr., 1959 *Datura stramonium* L, a possible quantitative bioassay host for the bromegrass mosaic virus. *Plant Dis. Rep.* **43:**690–694.

Chiu, R. J. and W. H. Sill, Jr., 1963 Factors affecting assay of bromegrass mosaic virus on *Datura stramonium* and *Chenopodium hybridum. Phytopathology* **53:**69–78.

Corbett, M. K., 1967 Some distinguishing characteristics of the orchid strain of tobacco mosaic virus. *Phytopathology* **57:**164–172.

Costa, A. S. and A. M. B. Carvalho, 1961 Studies on Brazilian tobacco streak. *Phytopathol. Z.* **42:**113–138.

Croxall, H. E., D. C. Gwynne and L. Broadbent, 1953 Turnip yellow mosaic in broccoli. *Plant Pathol.* **2:**122–123.

Dale, W. T., 1949 Observations on a virus disease of cowpea in Trinidad. *Ann. Appl. Biol.* **36:**327–333.

Darby, J. F., R. H. Larson and J. C. Walker, 1951   Variation in virulence and properties of potato virus Y strains. *Univ. Wis. Madison Res. Bull.* **177**:1–32.

Delgado-Sanchez, S. and R. G. Grogan, 1970   Potato virus Y. In *Descriptions of Plant Viruses*, No. 37, edited by A. J. Gibbs, B. D. Harrison, and A. F. Murant, Commonwealth Mycological Institute and Association of Applied Biologists, Slough, England.

Demski, J. W., 1969   Local reaction and cross protection for strains of squash mosaic virus. *Phytopathology* **59**:251–252.

Dennis, R. W. G., 1938   A new test plant for potato virus Y. *Nature Lond.* **142**:154.

de Zeuw, D. J., 1965   Tobacco ringspot virus hosts and suscepts. *Mich. Agric. Exp. Stn. Q. Bull.* **48**:64–75.

Diachun, S. and W. D. Valleau, 1947   Reaction of 35 species of Nicotiana to tobacco streak virus. *Phytopathology* **37**:7.

Diachun, S. and W. D. Valleau, 1950   Tobacco streak virus in sweet clover. *Phytopathology* **40**:516–518.

Diener, T. O. and F. O. Jenifer, 1964   A dependable local lesion assay for turnip yellow mosaic virus. *Phytopathology* **54**:1258–1260.

Diener, T. O., D. R. Smith and M. J. O'Brien, 1972   Potato spindle tuber viroid VII. Susceptibility of several solanaceous plant species to infection with low molecular-weight RNA. *Virology* **48**:844–846.

Doolittle, S. P., 1920   The mosiac disease of cucurbits. *U.S. Dep. Agric. Bull.* **879**:1–69.

Doolittle, S. P. and M. N. Walker, 1925   Further studies on the overwintering and disseminating of cucurbit mosaic. *J. Agric. Res.* **31**:1–58.

Doolittle, S. P. and W. J. Zaumeyer, 1953   A pepper ringspot caused by strains of cucumber mosaic virus from pepper and alfalfa. *Phytopathology* **43**:333–337.

Easton, G. D. and D. C. Marriam, 1963   Mechanical inoculation of the potato spindle tuber virus in the genus *Solanum*. *Phytopathology* **53**:349.

Eslick, R. F. and M. M. Afansiev, 1955   Influence of time of infection with barley stripe mosaic on symptoms, plant yield, and seed infection of barley. *Plant Dis. Rep.* **39**:722–724.

Fagbenle, H. H. and R. E. Ford, 1970   Tobacco streak virus isolated from soybeans, *Glycine max*. *Phytopathology* **60**:814–820.

Fezer, K. D. and A. F. Ross, 1959   Assay of alfalfa mosaic virus activity. *Phytopathology* **49**:529–530.

Ford, R. E., H. H. Fagbenle and W. N. Stone, 1970   New hosts and serological identity of bromegrass mosaic virus from South Dakota. *Plant Dis. Rep.* **54**:191–195.

Foster, R. E., II, 1972   Cucumber mosaic virus buffering effects *Chenopodium amaranticolor* response. *Plant Dis. Rep.* **56**:443–445.

Fraenkel-Conrat, H., 1974   *Comprehensive Virology*, Vol. 1, (Editors H. Fraenkel-Conrat and R. Wagner) *Catalogue of Viruses*, Plenum Press, New York.

Freitag, J. H., 1941   A Comparison of the transmission of four cucurbit viruses by cucumber beetles and by aphids. *Phytopathology* **31**:8.

Freitag, J. H., 1956   Beetle transmission, host range, and properties of squash mosaic virus. *Phytopathology* **46**:73–81.

Fukushi, T., 1934   Plants susceptible to dwarf disease of rice plants. *Trans. Sapporo Nat. Hist. Soc.* **13**:162–166.

Fulton, J. P. and R. W. Fulton, 1970   A comparison of some properties of elm mosaic and tomato ringspot viruses. *Phytopathology* **60**:114–115.

Fulton, R. W., 1948   Hosts of tobacco streak virus. *Phytopathology* **38**:421–428.

Fulton, R. W., 1950   Variants of the tobacco necrosis virus in Wisconsin. *Phytopathology* **40**:298–305.

Fulton, R. W., 1967   Purification and some properties of tobacco streak and tulare apple mosaic viruses. *Virology* **32**:153–162.

Garnsey, S. M. and R. Whidden, 1970   Transmission of exocortis virus to various citrus plants by knife-cut inoculation. *Phytopathology* **60**:1292.

Gibbs, A. J. and B. D. Harrison, 1970   Cucumber mosaic virus. In *Descriptions of Plant Viruses,* No. 1, edited by A. J. Gibbs, B. D. Harrison and A. F. M. Grant, Commonwealth Mycological Institute and Association of Applied Biologists, Slough, England.

Gilmer, R. M. and L. J. Kelts, 1965   Isolation of tobacco mosaic virus from grape foliage and roots. *Phytopathology* **55**:1283.

Gilmer, R. M., J. K. Uyemoto, and L. J. Kelts, 1970   A new grapevine disease induced by tobacco ringspot virus. *Phytopathology* **60**:619–627.

Gonsalves, D. and R. J. Shepherd, 1972   Biological and physical properties of the two nucleoprotein components of pea enation mosiac virus and their associated nucleic acids. *Virology* **48**:709–723.

Grant, T. J., 1934   The host range and behavior of the ordinary tobacco mosaic virus. *Phytopathology* **24**:311–336.

Grogan, R. G. and K. A. Kimble, 1964   The relationship of severe bean mosaic virus from Mexico to southern bean mosaic virus and its related strain in cowpea. *Phytopathology* **54**:75–78.

Grogan, R. G., R. H. Hall and K. A. Kimble, 1959   Cucurbit mosaic viruses in California. *Phytopathology* **49**:366–376.

Hadidi, A. F. and R. E. Hampton, 1968   Reaction of two varieties of barley tobacco to strains of cucumber mosaic virus *Phytopathology* **58**:1049.

Hadidi, A. F., E. A. Salama, and W. H. Sill, Jr., 1966   Comparative studies of host range, biophysical and biochemical properties of pumpkin mosaic virus and squash mosaic virus. *Phytopathology* **56**:877.

Hagedorn, D. J., 1957   Host range and inoculation studies on pea enation mosaic virus. *Phytopathology* **47**:14.

Hagedorn, D. J., R. E. C. Layne and E. G. Ruppel, 1964   Host range of pea enation mosaic virus and use of *Chenopodium album* as a local lesion host. *Phytopathology* **54**:843–848.

Hills, G. J. and R. N. Campbell, 1968   Morphology of broccoli necrotic yellows virus. *J. Ultrastruct. Res.* **24**:134–144.

Hollings, M., 1957   Reactions of some additional plant viruses on *Chenopodium amaranticolor. Plant Pathol.* **6**:133–135.

Hollings, M., 1959   Host range studies with fifty two plant viruses. *Ann. Appl. Biol.* **47**:98–108.

Hollings, M. and O. M. Stone, 1965   Studies of pelargonium leaf curl virus. II. Relationships to tomato bushy stunt and other viruses. *Ann. Appl. Biol.* **56**:87–98.

Holmes, F. O., 1928   Accuracy in quantitative work with tobacco mosaic virus. *Bot. Gaz.* **86**:66–81.

Holmes, F. O., 1938   Taxonomic relationships of plants susceptible to infection by tobacco mosaic virus. *Phytopathology* **28**:58–66.

Holmes, F. O., 1941   A distinctive strain of tobacco mosaic virus from Plantago. *Phytopathology* **31**:1089–1098.

Holmes, F. O., 1942   Quantitative measurement of tobacco etch virus. *Phytopathology* **32:**1058.

Holmes, F. O., 1946   A comparison of the experimental host ranges of tobacco etch and tobacco mosaic viruses. *Phytopathology* **36:**643–659.

Holmes, F. O., 1964   Symptomology of viral diseases in plants. In *Plant Virology,* edited by M. K. Corbett and H. D. Sisler, pp. 17–38. University of Florida Press, Gainsville, Fla.

Iida, T. T., 1969   Dwarf, yellow dwarf, stripe, and black-streaked dwarf disease of rice. In *The Virus Diseases of the Rice Plant. Proceedings of a Symposium at the International Rice Research Institute, April, 1967,* John Hopkins Press, Baltimore, Md.

Inouye, T., 1962   Studies on barley stripe mosaic in Japan. *Ohara (Japan) Inst. Landw. Biol., Bd. XI, Heft* **4:**414–496.

Izadpanah, K. and R. J. Shepherd, 1966   *Galactia sp.* as a local lesion host for the pea enation mosaic virus. *Phytopathology* **56:**458–459.

Jensen, J. H., 1933   Leaf enations resulting from tobacco infection in certain species of *Nicotiana* L. *Contrib. Boyce. Thompson Inst.* **5:**129–142.

Johnson, F., 1942   The complex nature of white clover mosaic. *Phytopathology* **32:**103–110.

Johnson, J., 1936   Tobacco streak, a virus disease. *Phytopathology* **26:**285–292.

Kahn, R. P., 1956   Seed transmission of the tomato ringspot virus in the Lincoln variety of soybeans. *Phytopathology* **46:**295.

Kapur, S. P., L. G. Weathers and E. C. Calavan, 1970   Studies on the mechanical transmission, extraction, and assay of citrus exocortis virus in *Gynura aurantiaca. Phytopathology* **60:**1535.

Kassanis, B., 1949   A necrotic disease of forced tulips caused by a tobacco necrosis virus. *Ann. Appl. Biol.* **36:**14–17.

Kassanis, B., 1970   Satellite virus, In *Descriptions of Plant Viruses,* No. 15, edited by A. J. Gibbs, B. D. Harrison and A. F. Murant, Commonwealth Mycological Institute and Association of Applied Biologists, Slough, England.

Kassanis, B. and M. P. Phillips, 1970   Serological relationship of strains of tobacco necrosis virus and their ability to activate strains of satellite virus. *J. Gen. Virol.* **9:**119–126.

Kassanis, B. and J. T. Slykhuis, 1959   Some properties of barley stripe mosaic virus. *Ann. Appl. Biol.* **47:**254–263.

Kirkpatrick, H. C. and R. C. Lindner, 1964   Recovery of tobacco mosaic virus from apple. *Plant Dis. Rep.* **48:**855–857.

Kleczkowski, A., 1950   Interpreting relationships between concentrations of plant viruses and number of local lesions. *J. Gen. Microbiol.* **4:**53–69.

Kreitlow, K. W. and W. C. Price, 1949   A new virus disease of Landino clover. *Phytopathology* **39:**517–528.

Kuhn, C. W., 1963   Field occurrence and properties of the cowpea strain of southern bean mosaic virus. *Phytopathology* **53:**732–733.

Kuhn, C. W., 1964   Purification, serology and properties of a new cowpea virus. *Phytopathology* **54:**853–857.

Kuhn, C. W., 1968   Identification and specific infectivity of a soybean strain of cowpea chlorotic mottle virus. *Phytopathology* **58:**1441–1442.

Ladeburg, R. C., R. H. Larson and J. C. Walker, 1950   Origin, interrelation, and properties of ringspot strains of virus X in American potato varieties. *Univ. Wisc. Madison. Res. Bull* **165:**1–47.

Ladipo, J. L. and G. A. de Zoeten, 1972   Influence of host and seasonal variation on the components of tobacco ringspot virus. *Phytopathology* **62:**195–201.

Larson, R. H., 1944   The identity of the virus causing punctate necrosis and mottle in potatoes. *Phytopathology* **34:**1006.

Lawson, R. H., 1966   Oak leaf chlorosis symptomatic of cucumber mosaic virus infection in dahlia. *Phytopathology* **56:**343–344.

Lindberg, G. D., D. H. Hall and J. C. Walker, 1956   A study of melon and squash mosaic viruses. *Phytopathology* **46:**489–495.

Lister, R. M., 1958   Some turnip viruses in Scotland and their effect on yield. *Plant Pathol.* **7:**144–146.

Lister, R. M. and B. J. Bancroft, 1970   Alteration of tobacco streak virus component ratios as influenced by host and extraction procedure. *Phytopathology* **60:**689–694.

Lister, R. M. and J. M. Thresh, 1955   A mosaic disease of leguminous plants caused by a strain of tobacco mosaic virus. *Nature (Lond.)* **175:**1047–1048.

Liu, H. Y., F. O. Holmes and M. E. Reichmann, 1969   Satellite tobacco necrosis virus from mung beans. *Phytopathology* **59:**822–826.

Lung, M. and T. P. Pirone, 1971   *Datura stramonium* a local lesion host for certain isolates of cauliflower mosaic virus. *Phytopathology* **61:**901–902.

McEwen, F. L. and W. T. Schroeder, 1956   Host range studies in pea enation virus. *Plant Dis. Rep.* **40:**11–14.

McKinney, H. H., 1944   Studies on the virus of bromegrass moasic. *Phytopathology* **34:**973.

McKinney, H. H., 1951*a*   A seed borne virus causing false stripe symptoms of barley. *Plant Dis. Rep.* **35:**48.

McKinney, H. H., 1951*b*   A seed borne virus causing false strip in barley *Phytopathology* **41:**563–564.

McKinney, H. H., 1953   New evidence on virus diseases in barley. *Plant Dis. Rep.* **37:**292–295.

McKinney, H. H., 1954   Cultural methods for detecting seed-borne virus in Glacier barley seedlings. *Plant Dis. Rep.* **38:**152–162.

McKinney, H. H., and R. W. Fulton, 1949   Local lesion of tobacco mosaic virus on cotyledons of cucumber. *Phytopathology* **39:**806–812.

McKinney, H. H. and L. W. Greeley, 1960   Several unique strains of the barley stripe-mosaic virus. *Plant Dis. Rep.* **44:**752–753.

McKinney, H. H. and L. W. Greeley, 1965   Biological characteristics of barley stripe mosaic virus strains and their evolution. *U.S. Dept. Agric. Res. Serv. Tech. Bull.* **1324:**1–84.

McKinney, H. H., H., Fellows and C. O. Johnson, 1942   Mosaic of *Bromus inermis*. *Phytopathology* **32:**331.

McLean, D. M., 1962*a*   Common weed hosts of tobacco ringspot virus in the lower Rio Grande valley of Texas. *Plant Dis. Rep.* **46:**5–7.

McLean, D. M. 1962*b*   Differentiation of tobacco ringspot and tomato ringspot viruses by experimental host reactions. *Plant Dis. Rep.* **46:**877–881.

Macleod, R. and R. Markham, 1963   Experimental evidence of a relationship between turnip yellow mosaic virus and wild cucumber mosaic virus. *Virology* **19:**190–197.

McWhorter, F. P. and W. C. Cook, 1958   The hosts and strains of pea enation mosaic virus. *Plant Dis. Rep.* **42:**51–60.

Markham, R. and K. M. Smith, 1949   Studies on the virus of turnip yellow mosaic. *Parasitology* **39:**330–342.

Martelli, G. P., A. Quacquarelli and M. Russo, 1971 Tomato bushy stunt virus. C.M.I./A.A.B. Descriptions of Plant Viruses No. 69. 4 pp.

Martin, C., 1952 Isolement d'une souche de virus Y de la pomme de terre (*Marmor upsilon* Holmes) a partir du Dahlia et de la tomate. *Ann. Inst. Natl. Rech. Agron. Ser. C Ann. Epiphyt.* **3**:394.

Matthews, R. E. F., 1949 Reactions of *Cyphomandra betacea* to strains of potato virus X. *Parasitology* **39**:241–244.

Milbrath, J. A., 1963 An investigation of thermal inactivation of alfalfa mosaic virus. *Phytopathology* **53**:1036–1040.

Mink, G. I., K. N. Saksena and M. J. Silbernagel, 1966 Purification of the bean red node strain of tobacco streak virus. *Phytopathology* **56**:645–649.

Mink, G. I., M. J. Silbernagel and K. N. Saksena, 1969 Host range, purification, and properties of the western strain of peanut stunt virus. *Phytopathology* **59**:1625–1631.

Murant, A. F., T, Munthe and R. A. Goold, 1970 Parsnip mosaic virus, a new member of the potato virus Y group. *Ann. Appl. Biol.* **65**:127–135.

Nelson, M. R., J. C. Matejka and H. H. McDonald, 1965 Systemic infection of water-melon by a strain of squash mosaic virus. *Phytopathology* **55**:1362–1364.

Nienhous, F. and C. E. Yarwood, 1972 Transmission of virus from oak leaves fractionated with sephadex. *Phytopathology* **62**:313–315.

Nitzany, F. E. and Z. K. Gerechter, 1962 Barley stripe mosaic virus host range and seed transmission tests among Gramineae in Israel. *Phytopath. Mediterr.* **2**:11–19.

Noordam, D., 1956 Waarplanten en toetsplanten van het ratelvirus van de tabak. *Tijdschr. Plantenzickten* **62**:219–225.

O'Brien, M. J. and W. B. Raymer, 1964 Symptomless hosts of the potato spindle tuber virus. *Phytopathology* **54**:1045–1047.

Olson, E. O. and A. V. Shull, 1956 Excortis and xyloporosis-bud transmission virus diseases of rangpur and other mandarin-lime rootstocks. *Plant Dis. Rep.* **40**:939–946.

Orellana, R. G., 1967 Reactions of guar to strains of tobacco ringspot virus. *Phytopathology* **57**:791–792.

Orellana, R. G. and A. Quacquarelli, 1968 Sunflower mosaic caused by a strain of cucumber mosaic virus. *Phytopathology* **58**:1439–1440.

Oswald, J. E., 1950 A strain of the alfalfa-mosaic virus causing vine and tuber necrosis in potato. *Phytopathology* **40**:973–999.

Patino, G. and W. J. Zaumeyer, 1959 A new strain of tobacco streak virus form peas. *Phytopathology* **47**:43–48.

Pérez, J. E. and A. Cortés-Monllor, 1970 A mosaic virus of cowpea from Puerto Rico. *Plant Dis. Rep.* **54**:212–216.

Pierce, W. H., 1934 Viroses of the bean. *Phytopathology* **24**:87–115.

Pierce, W. H., 1935 Identification of certain viruses affecting legumenous plants. *J. Agric. Res.* **51**:1017–1039.

Pound, G. S. and J. C. Walker, 1948 Strains of cucumber mosaic virus pathogenic on crucifers. *Phytopathology* **38**:21.

Price, W. C., 1930 Local lesions on bean leaves inoculated with tobacco mosaic virus. *Am. J. Bot.* **17**:694–702.

Price, W. C., 1940 Comparative host ranges of six plant viruses. *Am. J. Bot.* **27**:530–541.

Price, W. C. and E. L. Spencer, 1943 Accuracy of the local lesion method for measuring virus activity. II. tobacco necrosis, alfalfa mosaic and tobacco-ringspot viruses. *Am. J. Bot.* **30**:340–346.

Purcifull, D. E. and G. V. Gooding, Jr., 1970   Immunodiffusion tests for potato Y and tobacco-etch viruses. *Phytopathology* **60**:1036–1039.

Raymer, W. B. and M. J. O'Brien, 1962   Transmission of potato spindle tuber virus to tomato. *Am. Potato J.* **39**:401–406.

Roberts, D. A. and M. K. Corbett, 1961   Bioassay of tobacco ringspot virus in *Cassia occidentalis. Phytopathology* **51**:831–833.

Rochow, W. F., 1959   *Chenopodium hybridium* as a local-lesion assay host for brome mosaic virus. *Phytopathology* **49**:126–130.

Ross, A. F., 1948   Local lesions with potato virus Y. *Phytopathology* **38**:930–932.

Rush, M. C. and G. V. Gooding, Jr., 1970   The occurrence of tobacco ringspot virus strains and tomato ringspot virus in hosts indigenous to North Carolina. *Phytopathology* **60**:1756–1760.

Salaman, R. N., 1939   The potato virus X; its strains and reactions. *Philos. Trans. R. Soc. Lond. Ser. B Biol. Sci.* **229**:137–217.

Samson, R. W. and E. P. Imle, 1942   A ringspot type of virus disease of tomato. *Phytopathology* **32**:1037–1047.

Sander, E. and G. Schramm, 1963   Die Bedeutung der Proteinhülle für die Wirtsspezifität von Pflanzenviren. *Z. Naturforsch.* **18B**:199–202.

Schmelzer, K., 1957   Untersuchungen über den Wirtspflanzenkreis des Tabakmauchevirus. *Phytopathol Z.* **30**:281–314.

Schneider, I. R., 1971   Characteristics of a satellite-like virus of tobacco ringspot virus. *Virology* **45**:108–122.

Scott, H. A. and A. H. Gold, 1959   Studies on certain viruses infecting legumes in California. *Phytopathology* **49**:525.

Scott, H. A. and S. A. Slack, 1971   Serological relationship of brome mosaic and cowpea chlorotic mottle viruses. *Virology* **46**:490–492.

Scott, H. A., M. Vincent and W. J. Zaumeyer, 1961   Serological studies of red node and pod mottle viruses. *Phytopathology* **51**:755–758.

Semancik, J. S., 1966   Purification and properties of two isolate of tobacco rattle virus from pepper in California. *Phytopathology* **56**:1190–1193.

Semancik, J. S. and L. G. Weathers, 1972   Pathogenic 10*S* RNA from exocortis disease recovered from tomato bunchy-top plants similar to potato spindle tuber virus infection. *Virology* **49**:622–625.

Shepherd, R. J., 1964   Properties of a mosaic virus of cowpea and its relationship to the bean pod mottle virus. *Phytopathology* **54**:466–473.

Shepherd, R. J. and R. W. Fulton, 1962   Identity of a seed-borne virus in cowpea. *Phytopathology* **52**:489–493.

Shepherd, R. J. and D. E. Purcifull, 1971   Tobacco etch virus. In *Description of Plant Viruses,* No. 55, edited by A. J. Gibbs, B. D. Harrison and A. F. Murant, Commonwealth Mycological Institute and Association of Applied Biologists, Slough, England.

Shinkai, A., 1962   Studies on insect transmission of rice virus diseases in Japan. *Bull. Natl. Inst. Agric. Sci. Ser. C (Plant Pathol. Entomol.)* **14**:1–112.

Siegel, A. and S. G. Wildman, 1954   Some natural relationships among strains of tobacco mosaic virus. *Phytopathology* **44**:277–282.

Siegel, A., M. Zaitlin and P. Sehgal, 1962   The isolation of defective tobacco mosaic virus strains. *Proc. Natl. Acad. Sci. USA* **48**:1845–1851.

Silberschmidt, K. and E. Rostom, 1955   A valuable indicator plant for a strain of potato virus. Y. *Am. Potato J.* **32**:222–227.

Sill, W. H., Jr. and R. J. Chiu, 1959   Kentucky bluegrass, *Poa pratensis* L. a new host of the bromegrass mosaic virus in nature. *Plant Dis. Rep.* **43**:85.

Sill, W. H., Jr. and E. D. Hansing, 1955   Some studies on barley stripe mosaic (false stripe) and its distribution in Kansas. *Plant Dis. Rep.* **39**:670–672.

Sill, W. H., Jr. and J. C. Walker, 1952   A virus inhibitor in cucumber in relation to mosaic resistance. *Phytopathology* **42**:349–352.

Simons, J. N., 1954   Vector-virus relationships of pea-enation mosaic and the pea aphid *Macrosiphum pisi* (Kalt). *Phytopathology* **44**:283–289.

Singh, G. P., D. C. Arny and G. S. Pound, 1960   Studies on the stripe mosaic of barley, including effects of temperature and age of host on disease development and seed infection. *Phytopathology* **50**:290–296.

Singh, R. P., 1969   Use of *Datura metel* L. as local lesion host for potato virus X. *Am. Potato J.* **46**:355–357.

Singh, R. P., 1971   A local lesion host for potato spindle tuber virus. *Phytopathology* **61**:1034–1035.

Singh, R. P. and M. J. O'Brien, 1970   Additional indicator plants for potato spindle tuber virus. *Am. Potato J.* **47**:367–371.

Slykhuis, J. T., 1952   Virus diseases of cereal crops in South Dakota. *S. D. Agric. Exp. Stn. Bull.* **11**:211.

Slykhuis, J. T., 1953   Striate mosaic, a new disease of wheat in South Dakota. *Phytopathology* **43**:537–540.

Slykhuis, J. T., 1963   Vector and host relations of North American wheat striate mosaic virus. *Can. J. Bot.* **41**:1171–1185.

Slykhuis, J. T., 1967   *Agropyron repens* and other perennial grasses as hosts of bromegrass mosaic virus from the U.S.S.R. and the United States. *FAO (Food Agric. Organ. U.N.) Plant Prot. Bull.* **15**:65–66.

Smith, K. M., 1935   A new virus disease of the tomato. *Ann. Appl. Biol.* **22**:731–741.

Smith, K. M., 1937   *A Text Book of Plant Virus Diseases,* p. 615, Churchill, London.

Stace-Smith, R., 1966   Purification and properties of tomato ringspot virus and an RNA-deficient component. *Virology* **29**:240–247.

Stouffer, R. F., 1965   Isolation of tobacco rattle virus from Transvaal daisy *Gerbera jamesonii. Phytopathology* **55**:501.

Stover, R. H., 1951   Tobacco etch in Ontario. *Can. J. Bot.* **29**:235–245.

Thomas, H. R., 1951   Yellow dot, a virus disease of beans. *Phytopathology* **41**:967–974.

Thomas, H. R. and W. J. Zaumeyer, 1950a   Inheritance of symptom expression of pod mottle virus. *Phytopathology* **40**:1007–1010.

Thomas, H. R. and W. J. Zaumeyer, 1950b   Red node, a virus disease of beans. *Phytopathology* **40**:832–846.

Thornberry, H. H., 1966   Index of plant virus diseases. In *Agriculture Handbook No. 307,* A. R. S., U. S. Department of Agriculture, Washington, D.C.

Timian, R. G. and W. W. Sisler, 1955   Prevalence sources of resistance, and inheritance of resistance to barley stripe mosaic (false stripe). *Plant Dis. Rep.* **39**:550–552.

Tomaru, K. and A. Udagawa, 1967   Strains of cucumber mosaic virus isolated from tabacco plants. IV. A strain causing systemic infection on legume plants. *Hatano Tobacco Exp. Stn. Bull.* **42**:63–67.

Tomlinson, J. A., R. J. Shepherd and J. C. Walker, 1959   Purification, properties, and serology of cucumber mosaic virus. *Phytopathology* **49**:293–299.

Tompkins, C. M., 1937   A transmissible disease of cauliflower. *J. Agric. Res.* **55**:33–46.

Tuite, J., 1960 The natural occurrence of tobacco ringspot virus. *Phytopathology* **50**:296–298.

Uyemoto, J. K., 1970 Symptomatically distinct strains of tomato ringspot virus isolated from grape and elderberry. *Phytopathology* **60**:1838–1841.

Uyemoto, J. K. and R. M. Gilmer, 1972 Properties of tobacco necrosis virus strains from apple. *Phytopathology* **62**:478–481.

Valleau, W. D., 1947 Club root of tobacco: a wound-tumor like graft transmitted disease. *Phytopathology* **37**:580–582.

Varney, E. H. and J. D. Moore, 1952a Mechanical transmission of a virus form mosaic-infected elm to herbaceous plants. *Phytopathology* **42**:22.

Varney, E. H. and J. D. Moore, 1952b Strain of tomato ringspot virus from American elm. *Phytopathology* **42**:477.

Walker, J. C., F. J. LeBeau and G. S. Pound, 1945 Viruses associated with cabbage mosaic. *J. Agric. Res.* **70**:379–404.

Walters, H. J., 1954 Virus diseases in small grains in Wyoming. *Plant Dis. Rep.* **38**:836–837.

Weathers, L. G., 1965 Petunia, a herbaceous host of exorcortis virus of citrus. *Phytopathology* **55**:1081.

Wilkinson, R. E., 1952a Ringspot of winter squash caused by the tomato ringspot virus. *Phytopathology* **42**:23.

Wilkinson, R. E., 1952b Woody plant hosts of the tobacco ringspot virus. *Phytopathology* **42**:478.

Wilkinson, R. E. and F. M. Blodgett, 1948 *Gomphrena globosa,* a useful plant for qualitative and quantitative work with potato virus X. *Phytopathology* **38**:28.

Williams, H. E., S. H. Smith and F. W. Schwenk, 1971 Vibur calico caused by a strain of alfalfa mosaic virus. *Phytopathology* **61**:1305.

Wingard, S. A., 1928 Hosts and symptoms of ringspot, a virus disease of plants. *J. Agric. Res.* **37**:127–153.

Yarwood, C. E. and E. Hecht-Poinar, 1970 A virus resembling tobacco mosaic virus in oak. *Phytopathology* **60**:1220.

Zaumeyer, W. J., 1952 Another strain of alfalfa mosaic virus systemically infectious on beans. *Phytopathology* **42**:344.

Zaumeyer, W. J., 1953 Alfalfa yellow mosaic virus systemically infectious to beans. *Phytopathology* **43**:38–42.

Zaumeyer, W. J. and L. L. Harter, 1943 Two new virus diseases of beans. *J. Agric. Res.* **67**:305–328.

Zaumeyer, W. J. and H. R. Thomas, 1948 Pod mottle, a virus disease of beans. *J. Agric. Res.* **77**:81–96.

Zaumeyer, W. J. and B. L. Wade, 1935 The relationship of certain legume mosaics to bean. *J. Agric. Res.* **51**:715–749.

PART F
PROTISTS OF
GENETIC INTEREST

17

Chlamydomonas reinhardi

R. P. Levine

Introduction

Only a few species of algae have been exploited for genetic studies. So far the most commonly used species is the unicellular green alga *Chlamydomonas reinhardi* (see Figure 1). Its principal virtues are ease of cultivation in the laboratory, a simple and rapid life cycle, and a rather well-known Mendelian genetics. Among the cellular phenomena that have been studied with *C. reinhardi* are the genetics of photosynthesis (Levine, 1969; Levine and Goodenough, 1970), the genetics of the flagellar apparatus (Starling and Randall, 1971), and the genetics of the cell wall. Of particluar interest is the genetics of the chloroplast (Levine and Goodenough, 1970; Surzycki *et al.*, 1970) and the presence of a non-Mendelian, uniparental system of inheritance which may govern a portion of the genetic function of chloroplasts and mitochondria (Gillham, 1969; Surzycki and Gillham, 1971; Sager, 1972).

Conditions for Growth and Life Cycle

The medium for growth of wild-type *Chlamydomonas* in the light contains a few inorganic salts, and includes certain trace elements. This medium can be modified by the addition of various nutritional supplements for the growth of auxotrophs. Vegetative cell multiplication in the light is usually accomplished with the aid of daylight fluorescent lamps at

R. P. LEVINE—The Biological Laboratories, Harvard University, Cambridge, Massachusetts.

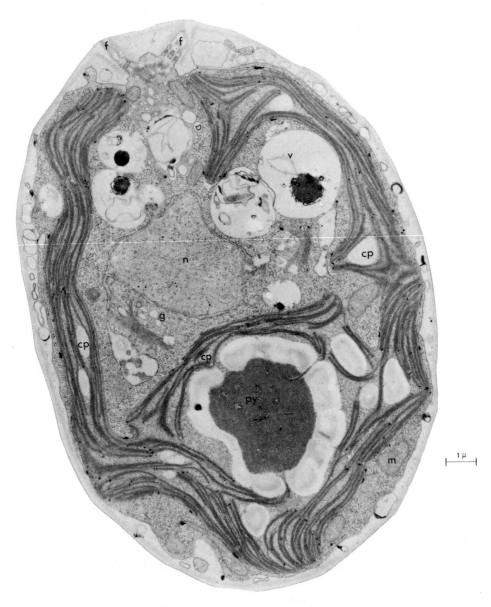

Figure 1. Electron micrograph of C. reinhardi. Portions of its two flagella (f,f) emerge through channels in the cell wall at the cell anterior. A cup-shaped chloroplast (cp) fills a large portion of the cell interior; a polygonal pyrenoid (py), surrounded by starch plates, lies at the base of the chloroplast. Other organelles in the micrograph include a central nucleus (n), mitochondria (m), stacks of Golgi membranes (g), and vacuoles (v) containing dense granules. Electron micrograph through the courtesy of Dr. Ursula Goodenough.

a range of 250–800 ft-c. Rapid vegetative growth of *C. reinhardi* with a generation time of 5 hours can be achieved in a liquid medium in which there is a tenfold increase in the mineral salts, aeration with five percent carbon dioxide in air (or air alone), and agitation in large cultures of a liter or more. Under these conditions a final cell concentration of 2×10^7/ml can be obtained.

Multiplication in the dark can be achieved with the addition of sodium acetate to the minimal medium.

Zygotes are matured on a minimal medium or, if necessary, on supplemented medium. In the case of *C. reinhardi*, maturation on minimal medium is satisfactory for most crosses. However, in crosses wherein certain auxotrophic mutants are selfed, better germination is obtained on a suitably supplemented medium.

When agar-grown haploid vegetative cells are suspended in water and illuminated, each cell becomes motile through the action of two flagella. After 2–4 hours in the light, the motile vegetative cells differentiate into gametes by depleting their available nitrogen supply. When gametes of opposite mating type are mixed, pairs or clumps containing cells of both mating types form immediately. Shortly thereafter, cells of opposite mating type begin to fuse in pairs. Then, 10–15 minutes after pairing begins, the process of cell fusion is completed by the formation of a binucleate, quadriflagellate zygote. At this time zygotes can be plated onto an agar medium, where all mating ceases, and only those cells which have started to fuse will form mature zygotes. After the resorption, or loss, of their flagella, the zygotes develop a thick wall and enlarge to approximately twice their original diameter. The zygotes require a maturation period which, at 25°C, is six days in length. The first 18–24 hours of this period is in the light (500 ft-c), while the remaining time is in the dark. Following the maturation period, zygotes can be induced to germinate by transferring them to fresh agar medium and placing them in the light. The time required for germination varies from 15 to 24 hours, depending in part on the genotype of the zygote. At germination, the zygote wall ruptures and four, or more frequently eight, haploid cells are extruded. These are the products of meiosis and are capable of vegetative growth. For further details of the biology of *Chlamydomonas* see Levine and Ebersold (1960), Sueoka (1960), and Gorman and Levine (1965).

Methods of Genetic Analysis

Two methods of genetic analysis are commonly used with *Chlamydomonas*. These are tetrad analysis and single-strand analysis from plating zygotes.

Tetrad Analysis

Various techniques have been developed for analzying *Chlamy-domonas* tetrads. All of the methods (see Levine and Ebersold, 1960) have in common the following steps: transferring zygotes singly to fresh agar medium, allowing time for germination, and separating the meiotic products. In the most commonly used method, mature zygotes together with unmated vegetative cells are scraped from the agar surface with a small metal spatula and placed on fresh agar medium. Single zygotes are then manipulated, singly, into one end of 3-mm × 25-mm rectangular lanes between parallel cuts in the agar surface. In order to kill the vegetative cells, each plate is inverted for 30 seconds over a Petri dish containing chloroform. After germination, the zygote products are separated by means of a fine glass loop with the aid of dissection microscope at 60× to form a row of cells 3 mm apart within each lane. With a little experience, several investigators have found this method practical, since it is possible to isolate the products from at least 100 zygotes in an hour. After colonies approximately 0.75 mm in diameter develop, they can be replica plated, using smooth-surfaced filter paper, to any combination of scoring media.

Zygote Plating

The zygote-plating method provides a means for analyzing a large number of single strands for recombinational events. Mature zygotes are suspended in distilled water and plated at concentrations of 100 to 2000 or more per plate. The products from each zygote are not separated but remain together in a single colony. By replica plating to the proper media, the frequency of a desired genotype can be determined. An advantage of this technique is that it is possible to test for reciprocal recombination by analyzing the original colony in which the recombinant genotype was found.

Genetic Maps

Figure 2 illustrates the linkage groups of *C. reinhardi*. Though the linkage groups have been identified by genetic analysis (Hastings *et al.*, 1965), only eight chromosomes have been detected cytologically. The basis for this discrepancy is unknown at this time.

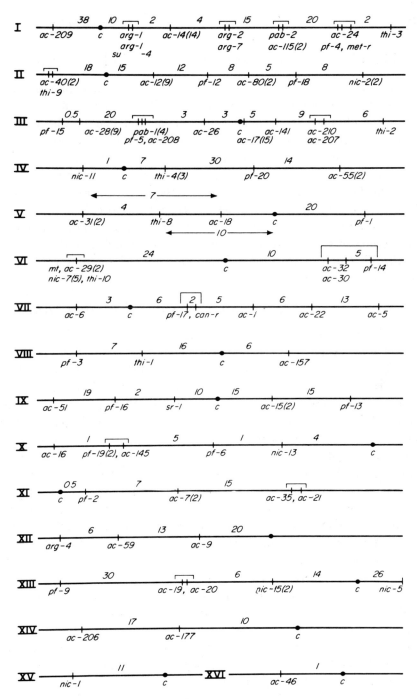

Figure 2. The 16 groups of *C. reinhardi*. Abbreviations: ac, acetate dependent; arg, arginine dependent; c, centromere; can-r, canavanine resistant; met-r, methionine sulfoxide resistant; mt, mating type; nic, nicotinic acid dependent; pab, p-amino benzoic acid dependent; pf, nonmotile (paralyzed flagella); sr, streptomycin-resistant; su^{arg-1}, suppressor of arg-1; thi, thiamin-dependent.

The Use of *C. reinhardi* in Studies of Photosynthesis

Mutant strains that are unable to carry out normal photosynthesis can be obtained with relative ease (Levine, 1971*b*). So far, mutant strains have been obtained that cannot carry out normal photosynthetic carbon metabolism (Levine, 1969), photosynthetic electron transport (Levine, 1969), and photosynthetic phosphorylation (Sato *et al.*, 1971). More recently, mutant strains of the second type cited above have proven to be of particular interest, for it has been possible to deduce the nature of major portions of the photosynthetic electron transport chain on the basis of partial photosynthetic reactions that these mutants can or cannot carry out. Studies of some of these mutant strains (Epel and Levine, 1971; Epel *et al.*, 1972) have provided information that relates to the nature of photochemical system II.

The Chloroplast Genetic System of *C. reinhardi*

The chloroplast of *C. reinhardi* possesses the genetic components of a typical prokaryotic cell (Levine and Goodenough, 1970; Levine, 1971*a*). Among them are 68S ribosomes. They are composed of 33S and 28S subunits that contain, respectively, 23S and 16S RNA (Hoober and Blobel, 1969). A 5S species of ribosomal RNA is also specifically associated with the chloroplast ribosomes (Surzycki and Hastings, 1961). They thus resemble bacterial 70S-type ribosomes. The cytoplasmic ribosomes, on the other hand, are of the 80S variety, with 43S and 30S subunits and 25S and 18S RNA's (Hoober and Blobel, 1969). They also possess a distinct 5S-RNA species (Surzycki and Hastings, 1961).

A number of discrete regions of DNA can also be distinguished within the chloroplast stroma (Ris and Plaut, 1962). The DNA appears as a meshwork of 2.5-nm fibrils. It is not known whether the several DNA regions seen in section are in fact isolated or whether they form a continuous phase within the chloroplast.

Biochemically, the chloroplast (β band) of *C. reinhardi* is one of the most extensively studied of the organelle DNA's. About 15 percent of the total DNA of *C. reinhardi* is β-band DNA (Chiang and Sueoka, 1967), and yet, to date, mutations of this DNA have not been positively identified. We have attempted to induce such mutations by providing the base analog bromodeoxyuridine to synchronous cultures at the time of β-band replication. In spite of substantial incorporation of bromodeoxyuridine into β-band DNA, these experiments have only produced some unstable strains whose growth is inhibited by certain amino acids (Surzycki, unpublished). One explanation for the apparent paucity of chloroplast-DNA mutations might be that the chloroplast genome is extensively polyploid, or redundant, in which case any single-gene mutation would be masked and undetectable for many generations. That chloroplast DNA in *C. reinhardi* is at least diploid is indicated by the fact that the gametes of at least one strain contain only half the β-band DNA of the vegetative cells (Chiang and Sueoka, 1967).

So far it has not been possible to identify in any detail the functions of chloroplast DNA beyond the observation that chloroplast ribosomal RNA appears to be coded for in chloroplast DNA. It has been suggested (Sager, 1972) that the information for the uniparental, non-Mendelian genetic system lies in chloroplast DNA, but there is also the possibility that it is in mitochondrial DNA (Surzycki and Gillham, 1971; Schimmer and Arnold, 1970a,b,c). During the sexual cycle of *C. reinhardi* certain genetic markers, most of them concerned with antibiotic sensitivity and thus presumed to have an organelle location, are transmitted only via the plus (+) mating type. Thus, in the cross *erythromycin-resistant* (mt^+) × *erythromycin-sensitive* (mt^-), all four haploid products are erythromycin resistant (*er-r*), whereas all are erythromycin sensitive (*er-s*) in the reciprocal cross. Since markers located in nuclear chromosomes exhibit the expected 2:2 segregation ratios in such crosses, one can conclude that the alleles for resistance and sensitivity are located in a genetic system that is transmitted in a non-Mendelian fashion. A number of these markers have been shown to be linked to one another, indicating that they lie in a single chromosome.

The classical explanation for non-Mendelian inheritance, in which the egg contributes the bulk of the cytoplasm to the zygote, cannot, however, be invoked in the case of *C. reinhardi*, since both mt^+ and mt^- are of equal size and contribute equally to the zygote cytoplasm when they fuse. In addition, there is no evidence for a selective degradation of the mt^- chloroplast or the mt^- mitochondria during zygote maturation. Thus, the location of the non-Mendelian genes within the cell remains unknown, as is the basis for their uniparental transmission.

Chloroplast DNA in *C. reinhardi* has been shown by denaturation–renaturation kinetics to be 26-fold redundant. Therefore, even though the organism has but a single large chloroplast, it may likely possess many copies of chloroplast genes. It also undoubtedly possesses many copies of mitochondrial chromosomes, since each cell contains at least ten mitochondria.

The Genetics of Photosynthesis

Mutant strains of *C. reinhardi* can be obtained in which photosynthetic electron transport, photosynthetic phosphorylation, or

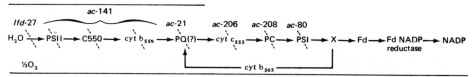

Figure 3. Sequence of components in the photosynthetic electron transport chain of C. reinhardi and the mutations (lfd. low fluorescence; ac, acetate-requiring) preventing the synthesis of active forms of the indicated components. The component(s) affected by the lfd-27 mutation are not yet identified. Pleiotropy is exhibited by the ac-141 mutation since it affects the chain at three sites. Abbreviations: PSI and II, photosystems I and II; c_{550} component with an absorbance maxima at 550 nm; cyt b_{559} and cyt b_{563} cytochrome b's with absorbance maxima at 559 and 563 nm; PQ, plastoquinone; cyt c_{553} cytochrome c with absorbance maximum at 553 nm; PC; plastocyanin; X, unidentified component; Fd, ferridoxin. From U. Goodenough and R. P. Levine, Genetics, Holt, Rinehart & Winston, 1974.

photosynthetic carbon reduction are impaired as a consequence of nuclear gene mutations (Levine, 1969). The techniques for obtaining and culturing mutant strains of this sort are described in detail elsewhere (Levine, 1971*b*).

The nuclear gene mutations affecting photosynthetic electron transport and phosphorylation have lent support to the two-light-reaction scheme for photosynthetic electron transport and have led to the pathway shown in Figure 3.

The Flagellar Apparatus

Numerous single-gene mutations have been identified at loci determining the motility of cells. The mutant strains have been used to study both the genetics and the nature of the flagellar apparatus (Ebersold *et al.*, 1962; McVittie, 1972*a,b*). These mutants all show Mendelian inheritance, and they map among the 16 linkage groups. Many of these strains have normally constituted flagella with 9 + 2 microtubules, but strains are known that lack the central pair of tubules (9 + 0 flagella) whereas others have short flagella, abnormally constituted flagellar membranes, or other disorders of flagellar morphology.

Now that it is possible to isolate and to characterize chemically the flagellar components (Witman *et al.*, 1972*a,b*), it has become highly likely that the functions of the flagellar genes will soon become known.

Miscellaneous Mutant Strains

Arginine-dependent strains are relatively easy to obtain in *C. reinhardi*, and they have been used to study arginine biosynthesis (Loppes, 1970; Loppes and Strijkert, 1972; Loppes *et al.*, 1972). Unfortunately, it has not been possible to obtain auxotrophic mutant strains for other amino acids.

Recently (Davies and Plaskitt, 1971; Hyams and Davies, 1972) mutant strains of *C. reinhardi* have been described that lack the capacity to form a normal cell wall. These strains should prove to be of some practical value in biochemical studies where the gentle lysis of the cell is an advantage.

Howell and Naliboff (1973) have described conditional mutant strains of *C. reinhardi* that are unable to complete the vegetative cell cycle. These mutant strains are of particular interest for studies of the biochemistry and genetics of cell division.

Literature Cited

Chiang, K. S. and N. Sueoka, 1967   Replication of chloroplast DNA in *Chlamydonomas reinhardi* during vegetative cell cycle, its mode and regulation. *Proc. Natl. Acad. Sci. USA* **57**:1506–1513.

Davies, D. R. and A. Plaskitt, 1971   Genetical and structural analyses of cell-wall formation in *Chlamydomonas reinhardi. Genet. Res.* **17**:33–43.

Ebersold, W. T., R. P. Levine, E. E. Levine and M. A. Olmsted, 1962   Linkage maps in *Chlamydomonas reinhardi. Genetics* **47**:531–543.

Epel, B. L. and R. P. Levine, 1971   Mutant strains of *Chlamydomonas reinhardi* with lesions on the oxidizing side of photosystem II. *Biochim. Biophys. Acta* **226**:154–170.

Epel, B. L., W. Butler and R. P. Levine, 1972   A spectroscopic analysis of low-fluorescent mutants of *Chlamydomonas reinhardi* blocked in their water-splitting oxygen-evolving apparatus. *Biochim. Biophys. Acta* **275**:395–400.

Gillham, N. W., 1969   Uniparental inheritance in *Chlamydomonas reinhardi. Am. Nat.* **103**:355–388.

Gorman, D. S. and R. P. Levine, 1965   Cytochrome *f* and plastocyanin: Their sequence in the photosynthetic electron transport chain of *Chlamydomonas reinhardi. Proc. Natl. Acad. Sci. USA* **54**:1665–1669.

Hastings, P. J., E. E. Levine, E. Cosbey, M. O. Hudlock, N. W. Gillham, S. J. Surzycki, R. Loppes and R. P. Levine, 1965 The linkage groups of *Chlamydomonas reinhardi. Microb. Genet. Bull.* **23**:17–19.

Hoober, J. K. and G. Blobel, 1969   Characterization of the chloroplastic and cytoplasmic ribosomes of *Chlamydomonas reinhardi. J. Mol. Biol.* **41**:121–138.

Howell, S. H. and J. A. Naliboff, 1973   Conditional mutants in *Chlamydomonas reinhardi* blocked in the vegetative cell cycle. I. An analysis of cell cycle block points. *J. Cell Biol.* **57**:760–772.

Hyams, J. and D. R. Davies, 1972   The induction and characterization of cell wall mutants of *Chlamydomonas reinhardi. Mutat. Res.* **14**:381–389.

Levine, R. P., 1969   The analysis of photosynthesis using mutant strains of algae and higher plants. *Annu. Rev. Plant Physiol.* **20**:523–540.

Levine, R. P., 1971*a*   Interactions between nuclear and organelle genetic systems. *Brookhaven Symp. Biol.* **23**:503–532.

Levine, R. P., 1971*b*   Preparation and properties of mutant strains of *Chlamydomonas* *reinhardi*. In *Methods in Enzymology*, Vol. 23, edited by A. San Pietro Academic Press, New York.

Levine, R. P. and W. T. Ebersold, 1960   The genetics and cytology of *Chlamydomonas. Annu. Rev. Microbiol.* **14**:197–216.

Levine, R. P. and U. W. Goodenough, 1970   The genetics of photosynthesis and of the chloroplast in *Chlamydomonas reinhardi. Annu. Rev. Genet.* **4**:397–408.

Loppes, R., 1970   Selection of arginine-requiring mutants after treatment with three mutagens. *Experientia (Basel)* **26**:660–661.

Loppes, R. and P. J. Strijkert, 1972   Arginine metabolism in *Chlamydomonas reinhardi. Mol. Gen. Genet.* **116**:248–257.

Loppes, R., R. Matagne and P. J. Strijkert, 1972   Complementation at the *arg-7* locus in *Chlamydomonas reinhardi. Heredity* **28**:239–251.

McVittie, A., 1972*a*   Genetic studies on flagellum mutants of *Chlamydomonas reinhardi. Genet. Res.* **9**:157–164.

McVittie, A., 1972*b*   Flagellum mutants of *Chlamydomonas reinhardi. J. Gen. Microbiol.* **71**:525–540.

Ris, H. and W. Plaut, 1962   Ultrastructure of DNA-containing areas in the chloroplast of *Chlamydomonas. J. Cell. Biol.* **13**:383–391.

Sager, R. 1972   *Cytoplasmic Genes and Organelles*, Academic Press, New York.

Sato, V. L., R. P. Levine and J. Neumann, 1971   Photosynthetic phosphorylation in *Chlamydomonas reinhardi*: Effects of a mutation altering an ATP-synthesizing enzyme. *Biochim. Biophys. Acta* **253**:437–448.

Schimmer, O. and C. G. Arnold, 1970*a*   Hin- und Rücksegregation eines ausserkaryotischen Gens bei *Chlamydomonas reinhardii. Mol. Gen. Genet.* **108**:33–40.

Schimmer, O. and C. G. Arnold, 1970*b*   Untersuchungen über Reversions-und Segregationsverhalten eines ausserkaryotischen Gens von *Chlamydomonas reinhardi* zur Bestimmung des Erbträgers. *Mol. Gen. Genet.* **107**:281–290.

Schimmer, O. and C. G. Arnold, 1970*c*   Über dei Zahl der Kopien eines ausserkaryotischen Gens bei *Chlamydomonas reinhardi. Mol. Gen. Genet.* **107**:366–371.

Starling, D. and J. Randall, 1971   The flagella of temporary dikaryons of *Chlamydomonas reinhardi. Genet. Res.* **18**:107–113.

Sueoka, N., 1960   Mitotic replication of deoxyribonucleic acid in *Chlamydomonas reinhardii. Proc. Natl. Acad. Sci. USA* **46**:83–91.

Surzycki, S. J. and N. W. Gillham, 1971   Organelle mutations and their expression in *Chlamydomonas reinhardi. Proc. Natl. Acad. Sci. USA* **68**:1301–1306.

Surzycki, S. J. and P. J. Hastings, 1961   Control of chloroplast RNA synthesis in *Chlamydomonas reinhardi. Nature (Lond.)* **220**:786–787.

Surzycki, S. J., U. W. Goodenough, R. P. Levine and J. J. Armstrong, 1970   Nuclear and chloroplast control of chloroplast structure and function in *Chlamydomonas reinhardi. Symp. Soc. Exp. Biol.* **24**:13–37.

Witman, G. B., K. Carlson and J. L. Rosenbaum, 1972*a*   *Chlamydomonas* flagella. II. The distribution of tubulins 1 and 2 in the outer double microtubules. *J. Cell Biol.* **54**:540–555.

Witman, G. B., K. Carlson, J. Berliner and J. L. Rosenbaum, 1972*b*   *Chlamydomonas* flagella. I. Isolation and electrophoretic analysis of microtubules, matrix, membranes, and mastigonemes. *J. Cell Biol.* **54**:507–539.

18

Cellular Slime Molds

MAURICE SUSSMAN and EDWARD F. ROSSOMANDO

Introduction

The cellular slime molds are designated as the order Acrasiales within the phylum Myxomycophyta. Although several genera and many species have been described, the bulk of current investigations have been carried out with *Dictyostelium discoideum*. This species has a particularly interesting and well-defined morphogenetic sequence and constructs fruiting bodies under a wide variety of environmental conditions. Two related species, *Dictyostelium mucoroides* and *Dictyostelium purpureum*, have also benn utilized. A member of another genus, *Polysphondylium pallidum*, has gained prominence as the first cellular slime mold to be cultivated axenically on defined media (M. Sussman, 1963; Hohl and Raper, 1963). However, as *P. pallidum* constructs fruiting bodies more slowly than *D. discoideum* and more fastidiously, the latter has been the species of choice for growth either in association with bacteria and more recently, axenically, (Sussman, 1966; Sussman and Sussman, 1967). A partial listing and classification of the more common slime molds appears below in Table 1.

Life Cycle

Growth. The growth of the organism is usually carried out in conjuction with bacteria, either in liquid (Gerisch, 1959) or on agar plates

MAURICE SUSSMAN—Department of Biology, Brandeis University, Waltham, Massachusetts. EDWARD F. ROSSOMANDO—School of Dental Medicine, The University of Connecticut Health Center, Farmington, Connecticut.

TABLE 1.   The Common Slime Molds

Genes	Species
Dictyostelium	lacteum, minutum, mucoroides, purpureum, discoideum, polycephalum
Polysphondylium	violaceum, pallidum

(M. Sussman, 1966), or axenically in defined media (R. R. Sussman and Sussman, 1967). Usually the medium is inoculated with spores and the culture allowed to grow until the food supply is depleted and development begins. At 22°C and under the standard conditions of growth on agar plates (M. Sussman, 1966), D. discoideum requires about 10 hours for spore germination, and 30–35 hours to reach the stationary phase (doubling time is 3–4 hours). In liquid culture with bacteria, the generation time is also about 3–4 hours.

Development. The cells initially appear as a homogeneous lawn; they then collect into aggregates, at first into amorphous blobs, and subsequently into conical, organized structures. These extend apically into fingers, then settle down laterally as migrating slugs or pseudoplasmodia which are sensate, i.e., responding in a purposeful manner to light, heat (Bonner et al., 1950), and pH (P. Newell and M. Sussman, unpublished) and displaying evidence of both histological and functional organization (Raper, 1940; Bonner et al., 1955). After rising apically once again, the slugs proceed through a complex sequence of morphogenetic movements in order to construct fruiting bodies. These consist of a mass of spores at the top and a cellulose-ensheathed, cellular stalk of the bottom, all resting upon a basal disc. The spores, stalk and basal disc cells are differentiated biochemically, morphologically, and functionally.

When the cells are allowed to develop on filter pads under standard conditions (M. Sussman, 1966), the developmental process is completed in 24 hours. Some species of Dictyostelium exhibit variations in the development time. For example, when compared to D. discoideum, D. mucoroides develops about three times as fast on filters while D. purpureum takes about six hours longer to complete the sequence.

Mutant Strains

Mutant strains isolated from Dictyostelium and Polysphondylium display a variety of developmental aberrations. Some mutants have arisen spontaneously, while others arise after ultraviolet treatment. More recently, nitrosoguanidine has been used successfully as the mutagen (Yanagisawa *et al.*, 1967).

The mutants can be grouped according to their performances as follows:

Aggregateless. The amoebae grow normally but fail to aggregate after reaching the stationary-growth phase. They retain their vegetative morphology and do not transform into stalk cells or spores (R. R. Sussman and Sussman, 1953; Rafaeli, 1962; Kahn, 1964).

Fruitless. Some fruitless strains develop no further than to construct amorphous or partially completed aggregates; others form tight, normal aggregates before development ceases; still others can become organized into migrating slugs but do not construct fruiting bodies (R. R. Sussman and Sussman, 1953; Rafaeli, 1962). The fruitless strains thus far encountered do not (by themselves) produce stalk cells or spores. However, certain aggregateless and fruitless strains, when mixed with one another or with the wild type, can construct fruiting bodies synergistically (Kahn, 1964; M. Sussman, 1954; M. Sussman and Lee, 1955; Ennis and Sussman, 1958).

Mutants that Construct Aberrant Fruiting Bodies. Deviations in the number and size of fruits ("fruity"), branches ("forked"), shape ("curly"), texture ("glassy"), and pigmentation ("white" and "brown") have been observed (R. R. Sussman and Sussman, 1953).

Drug-Resistant Mutants. Attempts have been made to select mutant strains resistant to drugs, temperature, ultraviolet light, etc. A few chloramphenicol-resistant mutants have been isolated (Ennis, private communication). However, in the cases where ultraviolet light was used as the mutagenic agent, the resistant strains which were isolated were unstable, reverting to the original level of sensitivity during a single passage in the absence of the lethal agent. Using nitrosoguanidine, Loomis and Ashworth (1968) have reported actidione-resistant mutants.

Temperature-Sensitive Mutants. Mutants have been isolated which have a temperature-sensitive step either in growth or development (Loomis, 1969; Loomis and Ashworth, 1968; Katz and Sussman, 1972).

Growth-Rate Mutants. The isolation of strains of *D. discoideum* with stable hereditable alterations in growth rate has been reported (Loomis and Ashworth, 1968). These strains were selected by alterations in plaque size following mutagenesis with nitrosoguanidine. Results of experiments performed by mixing small plaque formers in various combinations suggested recombination or complementation.

Cytogenetics

The development of a suitable chromosome-staining procedure (Wilson, 1953; Wilson and Ross, 1955; Ross, 1960; Sussman, 1961) has focused interest on the cytogenetics of *D. discoideum*. Observations of mitotic figures with 7 and 14 chromosomes demonstrated the existence of haploid and diploid amoebae (Wilson and Ross, 1957; Ross, 1960; R. R. Sussman, 1961), and the ploidal composition of the culture was found to be stable and clonally inherited (Ross, 1960; M. Sussman and Sussman, 1962). Three general types of strains have been studied: (1) stable haploid strains containing a negligible proportion of diploid derivatives; (2) stable diploid strains containing a negligible proportion of haploid derivatives; and (3) metastable strains containing appreciable proportions of both. The ploidal compositions of these strains can be maintained during serial subcultures and clonal reisolations. Properties of the three types of strains, including size distributions of spores and myxamoebae and certain morphogenetic capacities, have been described (M. Sussman and Sussman, 1962; R. R. Sussman and Sussman, 1963).

Stable, heterozygotic, diploid strains have been obtained from mixed cultures of haploid parents. Syngamy is rare ($\sim 10^{-5}$), but temperature-resistant, diploid heterozygotes can be selected from crosses of temperature-sensitive haploid parents. Haploid segregants are rare ($\sim 10^{-4}$) but can be selected by using a recessive drug-resistance maker (cycloheximide) and selecting for the resistant segregants among the sensitive diploid heterozygotes. Seven markers have been studied in this manner and have been found to fall into three linkage groups (Katz and Sussman, 1972).

Literature Cited

Bonner, J. T., W. Clark, C. Neely and M. Slifkin, 1950   Orientation to light and temperature gradients in the slime mold *D. discoideum*. *J. Cell. Comp. Physiol.* **36**:149–158.

Bonner, J. T., A. D. Chiquoine and M. Kolderie, 1955   A histochemical study of differentiation in the cellular slime molds. *J. Exp. Zool.* **130**:133–157.

Ennis, H. L. and M. Sussman, 1958   Synergistic morphogenesis by mixtures of *Dictyostelium discoideum* wild-type and aggregateless mutants. *J. Gen. Microbiol.* **18**:433–449.

Gerisch, G., 1959   Ein Submerskulturverfahren für Entwicklungsphysiologische Untersuchungen an *Dictyostelium discoideum*. *Naturwissenschaften* **46**:654–656.

Hohl, H. and K. B. Raper, 1963   Nutrition of cellular slime molds. Growth of *Polysphondylium pallidum* II in axenic culture. *J. Bacteriol.* **85**:199–206.

Kahn, A. J., 1964   Some aspects of cell interaction in the development of the slime mold *Dictyostelium purpureum. Dev. Biol.* **9**:1–19.

Katz, E. R. and M. Sussman, 1972   Parasexual recombination in *Dictyostelium discoideum:* Selection of stable diploid heterozygotes and stable haploid segregants. *Proc. Natl. Acad. Sci. USA* **69**:495–498.

Loomis, W. F., Jr. 1969   Temperature sensitive mutants of *Dictyostelium discoideum. J. Bacteriol.* **99**:65–69.

Loomis, W. F., Jr. and J. M. Ashworth, 1968   Plaque-size mutants of the cellular slime mold *Dictyostelium discoideum. J. Gen. Microbiol.* **53**:181–186.

Rafaeli, D. E., 1962   Studies on mixed morphological mutants of *Polysphondylium violaceum. Bull. Torrey Bot. Club* **89**:312–318.

Raper, K. B., 1940   Pseudoplasmodium formation and organization in *D. discoideum. J. Elisha Mitchell Sci. Soc.* **56**:241–281.

Ross, I. K., 1960   Studies on diploid strains of *D. discoideum. Am. J. Bot.* **47**:54–59.

Sussman, M., 1954   Synergistic and antagonistic interactions between morphogenetically deficient variants of the slime mold *Dictyostelium discoideum. J. Gen. Microbiol.* **10**:110–120.

Sussman, M., 1963   Growth of the cellular slime mold *Polysphondylium pallidum* in a simple nutrient medium. *Science (Wash. D.C.)* **139**:338.

Sussman, M., 1966   Biochemical and genetic methods in the study of cellular slime mold development. In *Methods in Cell Physiology,* Vol. II, edited by D. Prescott, pp. 397–417, Academic Press, New York.

Sussman, M. and F. Lee, 1955   Interactions among variants in wild-type strains of cellular slime molds across thin agar-membranes. *Proc. Natl. Acad. Sci. USA* **41**:70–78.

Sussman, M. and R. R. Sussman, 1962   Ploidal inheritance in *Dictyostelium discoideum. J. Gen. Microbiol.* **28**:417–429.

Sussman, R. R., 1961   A method for staining chromosomes of *D. discoideum. Exp. Cell Res.* **24**:154.

Sussman, R. R. and M. Sussman, 1953   Cellular differentiation in Dictyosteliaceae: heritable modifications of the developmental pattern. *Ann. N.Y. Acad. Sci.* **56**:949–960.

Sussman, R. R. and M. Sussman, 1963   Ploidal inheritance in the slime mold *Dictyostelium discoideum:* Haploidization and genetic segregation of diploid strains. *J. Gen. Microbiol.* **30**:349–355.

Sussman, R. R. and M. Sussman, 1967   Cultivation of *Dictyostelium discoideum* in axenic medium. *Biochem. Biophys. Res. Commun.* **29**:53–55.

Sussman, R. R., M. Sussman and H. L. Ennis, 1960   Appearance and inheritance of the I-cell phenotype in *D. discoideum. Dev. Biol.* **2**:367–392.

Wilson, C. M., 1953   Cytological study of the life cycle of *Dictyostelium. Am. J. Bot.* **40**:714–718.

Wilson, C. M. and I. K. Ross, 1955   Meiosis in the Myxomycetes. *Am. J. Bot.* **42**:743–749.

Yanigasawa, K., W. F. Loomis, Jr. and M. Sussman, 1967   Developmental regulation of the enzyme UDP Gal: polysaccharide transferase. *Exp. Cell Res.* **46**:328–334.

19

Tetrahymena pyriformis[*]

Tracy M. Sonneborn

Introduction

The study of the genetics of *Tetrahymena pyriformis* began about 20 years ago with the pioneering studies of Nanney and his co-workers. It is now developing more rapidly and extensively than ever before. Allen and Gibson (1973) have recently summarized the genetics of *T. pyriformis* in "an attempt to reinterpret" it in relation to their own current views. Allen (1967c), Nanney (1968b), and Preer (1969) have presented differently oriented reviews. Hill (1972) has summarized the physiology and bio-chemistry of *Tetrahymena,* including a chapter on genetics. The general implications of some aspects of *Tetrahymena* genetics have been discussed by Nanney (1964, 1968b, 1972). Many aspects of the biology of *Tetrahymena* are dealt with in a book edited by Elliott (1973b) which contains a remarkably extensive bibliography (65 pages in length!); the bibliography has been supplemented by Corliss (1973). Many of the methods used in work on *Tetrahymena* are the same as or similar to those used on *Paramecium* which have been summarized by Sonneborn (1950, 1970).

The present review aims only to summarize concisely the present status of knowledge and to call attention to some unresolved problems and disputed interpretations without attempting to evaluate them critically.

Genetic studies on *Tetrahymena* are almost entirely confined to *T. pyriformis*; but this "species," like *Paramecium aurelia* and other named "species" of ciliated protozoa, is not a single biological species but a group

---

[*] This is contribution No. 952 from the Zoology Department, Indiana University.

Tracy M. Sonneborn—The Zoology Department, Indiana University, Bloomington, Indiana.

of sibling species currently referred to as syngens. This term, originally proposed for the biological species of *P. aurelia,* denotes a potentially common gene pool, i.e., gene flow is possible within but not between syngens. With only one exception so far as is known, the different syngens of *T. pyriformis* do not even mate with each other. In fact, the 12 known syngens, numbered 1–12, are distinguished primarily on that basis.

These syngens are much alike in some respects. So far as known, all have the same chromosome number (Ray, 1956). They differ only slightly in respects such as the structure and pattern of the cell surface (Nanney, 1967, 1971; Nanney and Chow, 1974); but they differ markedly in isozymes (Allen and Weremiuk, 1971; Borden *et al.,* 1974), in the nucleotide sequences of their DNA (Allen and Li, 1974), and probably in their surface antigens (Loefer *et al.,* 1958; Loefer and Scherbaum, 1963; Margolin *et al.,* 1959), although the antigens have been extensively examined in only a few syngens. Elliott (1973*a*) has summarized syngenic differences in geographic distribution, temperature tolerance, and nutritional requirements. Borden *et al.* (1974) have concluded from molecular studies that the syngens of *T. pyriformis* have diverged much more than other sibling species previously studied from this point of view. On the basis of their study of divergence in DNA nucleotide sequences, Allen and Li (1974) agree with the conclusion of Borden *et al.* that "vast molecular distances" separate the syngens. This situation underscores the importance of comparative studies, especially in the field of molecular genetics, before making generalizations about *T. pyriformis.*

To make matters worse, some strains which have been extensively employed in molecular studies cannot be assigned to syngens on the basis of their mating reactions; they are amicronucleate and incapable of mating. Still worse, a recent study of strains by Borden *et al.* (1973*b*) clearly indicates that strains which go by the same designation in different laboratories and which are assumed to be the same are often different from one another but are the same as strains which go by different designations. Borden *et al.* have reclassified these strains on the basis of their characteristics. This should be kept in mind in all references here to these amicronucleate strains.

Most of our knowledge of genetics of *T. pyriformis* is restricted to syngen 1. *Unless otherwise stated, all that follows is to be understood as referring specifically to syngen 1,* although much of the account is applicable with little or no modification to other or all syngens.

The Cells

Normal, growing cells have on their surface usually 15–28 (or more under certain conditions) longitudinal rows (kineties, meridians) of cilia, an anterior oral apparatus which includes an undulating membrane and three adoral membranelles, a posterior cytoproct on the same meridian as the oral apparatus, and one or more pores of contractile vacuoles which are located near the posterior end about one-quarter of the circumference to the right of the oral meridian. Normal cells have one diploid ($2n = 10$) micronucleus and one macronucleus. High population densities, up to 10^6 cells/ml, are achieved in certain media. Growth in axenic media (e.g., proteose peptone) is easy to initiate and readily capable of yielding bulk material for biochemical studies.

The Cell Cycles

A large literature deals with various aspects of the cell division cycle, with much of the work performed on the genetically dead, i.e., amicronucleate, strains referred to above. A very high degree of division synchrony can be achieved by repeated heat shocks (for review, see Zeuthen, 1964) and by other means (e.g., see Cameron and Jeter, 1970). A somewhat lower, but still high, degree of synchrony can be achieved in genetically useful (micronucleate) strains of syngen 1 (e.g., Holz, 1960; Gavin, 1965). Individual cells can of course be isolated at the moment of division and followed through a cycle. The length of the cell cycle may be less than 2 hours under certain conditions and up to more than 14 hours under others, but it is about 3 hours under the usual conditions of culture. Another cell "cycle," in which neither nuclear nor cell division occurs, is known as oral replacement in which the existing oral apparatus is resorbed and concurrently a new one is formed. Oral replacement can be induced by amino acid starvation and by other means. Various events of both the cell division cycle and oral replacement are reviewed in several chapters (by Cameron, McDonald, and Frankel and Williams) of Elliott's (1973*b*) book.

Micronuclear and Macronuclear DNA

Micronuclear DNA synthesis begins in telophase and is completed soon after cell division. The micronuclei are thus in the G2 period through almost the whole cell cycle and the G1 period is totally lacking. Macronuclear DNA synthesis has been reported to occur at different periods in different strains (syngen often unknown) and under different conditions (for review, see McDonald's chapter in Elliott, 1973*b*); but the S period in syngen 1 probably begins after about 30 percent completion of the cycle and extends until about 60 percent completion of the cycle.

Some aspects of macronuclear DNA synthesis (see page 436), the amounts of micronuclear and macronuclear DNA, and the quantitative and qualitative relations between their "genomes" are of basic importance for the interpretation of some features of the genetics of *T. pyriformis* (see pages 441–445). The amount of DNA per haploid micronuclear genome (i.e., one-quarter of the amount in micronuclear G2) has been estimated to be either about 4×10^{10} daltons (Flavell and Jones, 1970) or $12–14 \times 10^{10}$ daltons (Gibson and Martin, 1971; Woodard *et al.*, 1972). These diverse values were arrived at using different methods and assumptions. For example, Flavell and Jones measured total cellular DNA, ascertained its kinetic complexity, equated this to the basic repeating macronuclear "genome," and tacitly assumed it to be the same as the normal micronuclear haploid genome. The similarity between the genomes of the two nuclei is supported by two independent studies. Allen and Gibson (1972) reported that genome size (kinetic complexity) of the macronuclear genome is no less than that of the micronucleus as calculated from photometric comparison with bull sperm. M-C. Yao and M. A. Gorovsky (private communication) concluded from their elegant study of hybridization of micronuclear with macronuclear DNA that 80 to 90 percent of the sequences are present in both kinds of nuclei in the same relative abundance.

The total amount of macronuclear DNA has been estimated to be about 4.3×10^{12} daltons (Flavell and Jones, 1970) or 6.1×10^{12} daltons (Gibson and Martin, 1971; Woodard *et al.*, 1972). An important question is how many copies of the macronuclear "genome" are present in The macronucleus. One method used to answer this question is to ascertain cytophotometrically the ratio of the amounts of DNA in the macronucleus and in the micronucleus. This method is of questionable validity, especially when use is made of old (senescent), infertile strains in which variable amounts of DNA have been lost from the micronuclei (see page 440); in such strains it is not known whether DNA of the macronucleus has been lost proportionately. The macronuclear:micronuclear DNA ratio needs to be ascertained for young, fertile strains with normal micronuclei. Aside from studies made on sterile strains, other studies (Woodard *et al.*, 1972; Gibson and Martin, 1971; Allen and Gibson, 1972) report that there is about 45 times as much DNA in the macronucleus as in a normal haploid micronuclear genome. Nilsson (1970), working on an amicronucleate strain, which cannot be assigned to a syngen on the basis of mating type because it does not mate, has reported evidence for about 45 units of structure in the macronucleus. Flavell and Jones (1970) have reported that the macronucleus contains about 100 copies of its basic genome as identified by kinetic complexity.

Flavell and Jones (1970) have estimated that about 5 percent of the macronuclear DNA is repetitious; Allen and Gibson (1972) have estimated that about 20 percent is repetitious; and Wille (1972) has reported that the percentage varies from 10 to 50 percent, depending on generation time.

The amount of DNA in the macronucleus varies in the course of successive cell cycles of amitotic division, but, at least in the asexual strain HSM, it is regulated toward a mean value by two mechanisms (Cleffmann, 1968). When the macronuclear DNA falls to a rather low level, two complete replications occur before the next macronuclear division. The higher the DNA level, the more DNA is extruded into the cytoplasm and disintegrated after the macronucleus divides. Among the strains of syngen 1 studied by Allen and Gibson (1972), the amount of DNA in G1 macronuclei differed much less than the corresponding macronuclear genome size. Thus, what seems to be regulated is total macronuclear DNA, not the number of copies of the genome.

Conjugation

When mature and sexually ripe cultures of any two of the seven mating types (see page 457) are mixed together, conjugation occurs after nutriment has been depleted. The necessary starvation ("initiation") takes 2 hours at 30°C and is sensitive to salinity (Bruns and Brussard, 1974). Some lines of descent, selfers, produce cultures which contain more than one mating type, so that in them conjugation does not require mixture of different cultures. On mixing together ripe cultures of different mating type, there is no mass clumping (agglutination) between cells of different mating type such as is typical of *Paramecium*. Instead, there is an ill-understood preconjugation "waiting period" of an hour or so before cells unite directly into conjugating pairs. McCoy (1972), Wolfe (1973), and Bruns and Brussard (1974) have explored aspects of the interactions between mating types occurring during the waiting period in syngen 1; Phillips (1971a) has investigated these interactions in syngen 7.

The nuclear events of conjugation (Nanney, 1953; Ray, 1956) differ only in genetically unimportant details from those described in the chapter on *P. aurelia*. The chief differences are these: (1) The prezygotic macronucleus does not fragment, but it shrinks and eventually disappears. (2) The new developing macronuclei are already present at the time the conjugating mates separate, the mates having remained united much longer (about 12–15 hours under standard conditions) than in *P. aurelia*. (3) During their long union, mates freely exchange cytoplasm (McDonald, 1966), except for larger organelles such as mitochondria (Roberts and Orias, 1973a). (4) Sometime before completion of the first postconjugation fission, one of the two initial micronuclei disappears. It is to be emphasized that, as in *P. aurelia*, the two new macronuclei which develop from products of the fertilization nucleus show no evidence, at any stage, of polyteny, chromosome fragmentation, or loss of DNA. Those remarkable events (documented particularly for certain hypotrichous ciliates) have justly attracted much attention in recent years, but have often been assumed (especially by nonspecialists) to occur universally in the development of new macronuclei in ciliates. That this is not true is made particularly apparent when the same investigator compares the development of new macronuclei in *Tetrahymena* and *Stylonychia,* one of the hypotrichous ciliates (Murti, 1973), and when micronuculear and macronuclear DNA are hybridized (see page 436).

Only during the first meiotic division can the karyotype be clearly observed (Ray, 1956). No visible difference in the karyotype has yet been reported for normal strains of syngen 1, or even for different syngens. The five chromosome pairs are each several microns long, three of them being much alike, one larger, and one smaller, and all of them possess median or submedian kinetochores.

Abnormalities of the nuclei and nuclear events at conjugation have been reported by Nanney (1953, 1957, 1959*b*, 1963), Nanney and Nagel (1964), and Allen (1963, 1967*b*). Merely observing that cells unite in pairs is no assurance that conjugation is proceeding normally. The two principal alternatives are referred to as nonconjugation and genomic exclusion.

The term nonconjugation, as commonly used, is to some extent a misnomer. It is applied to pairs in crosses between genomically diverse strains in which the two resulting clones (one from each mate) are like the two parents, respectively, both in their genic traits and in regard to sexual maturity. This merely signifies that the clones carry macronuclei descended from those of the parents instead of from the fertilization nucleus. Macronuclei from the latter would possess new combinations of genes and would establish a period of sexual immaturity. Nonconjugation thus could be either a total failure of the normal nuclear events of conjugation to occur; or these events could have occurred, including cross-fertilization, except for the development of new functional macronuclei from the fertilization nucleus. In either case, such pairs should be, and normally are, excluded from tabulations of genetic results. However, nonconjugation also occurs regularly in genomic exclusion, which has become an important tool in genetic work.

Genomic Exclusion

When normal strains are crossed to certain strains with defective micronuclei, the genome of the defective strain is excluded from the resulting genotypes (Allen, 1963, 1967*a,b*; Allen *et al.,* 1967). Such defective strains are designated by adding an asterisk to the strain designation, e.g., C* (read

"C star"); they appear to arise commonly in the course of clonal aging (see page 440). Although the micronuclei in these strains may enter into meiosis and may vary as to how far they proceed, they degenerate without having participated in fertilization.

In the normal mate, meiosis, the formation of gamete nuclei, and the migration of the male nucleus into the defective partner proceed normally. However, usually neither this haploid male nucleus nor the genically identical haploid female nucleus in the normal mate produces functional new macronuclei. Instead, the preconjugation macronucleus in each mate usually fails to disappear; it remains functional. Hence, the phenotypes of the two mates also persist; i.e., this is a form of "nonconjugation." In some unknown way at some unknown time, the identical haploid micronuclei in the two mates (or their early descendants) become diploid, as indicated by their behavior at the next conjugation.

The products of the first conjugation (round 1) in each pair are still of different mating type and sexually mature because of the persistence of the preconjugation macronuclei. Hence, if allowed to do so, they can quickly conjugate with each other again. Conveniently, the second-round mating proceeds quite normally. This time, new macronuclei arise from fertilization nuclei, they contain genes derived from only the normal strain, and the resulting clones enter an immature period.

The genetic constitution of these immature progeny depends upon the genomes of both the round-1 and the round-2 conjugants. If the normal parent was homozygous, obviously all round-2 pairs will similarly be identical homozygotes, regardless of whether pair isolation occurred in round 1 or not until round 2. If the normal parent was heterozygous, then the production of exclusively homozygous pairs can be assured by isolating round-1 pairs. The two mates of each such pair, having initially identical haploid micronuclei, and later identical homozygous diploid nuclei, can produce only the same homozygous diploid genotype at round 2. This provides the worker on *Tetrahymena* with the same advantages as those provided by autogamy for workers on *P. aurelia*. But different round-1 pairs will have different combinations of genes from the heterozygous normal parent. Hence, if the round-1 pairs are not isolated and round 2 occurs in the mass mixture, all possible combinations of matings can occur in round 2, and various homozygous and heterozygous combinations of the genes of the normal parent will result.

Genomic exclusion does not invariably follow the pattern described above. Allen *et al.* (1971) have observed that a small fraction of pairs isolated as round-1 pairs yielded immature homozygous clones. They interpreted this as due to a second round of mating without separation of the round-1 pairs. Bruns and co-workers (private communication) find that a small, variable percentage (> 1 but < 5 percent) of the first-round pairs succeed, at least in one of the mates, in forming functional new macronuclei from a (perhaps diploidized) derivative of the micronucleus of the normal mate. With appropriate genic markers and certain special techniques, these deviants from the usual pattern can be selected and used to advantage in genetic work (see page 446). Other reports (e.g., Nanney, 1963) indicate that genomic exclusion as described thus far is not the only aberrancy that occurs in crosses to strains with defective micronuclei.

The Clonal Cycle: Immaturity, Maturity, and Senescence

The generation of macronuclei from micronuclei (e.g., after conjugation or after the second round of mating in genomic exclusion) sets the clonal clock to time zero, initiating a new clonal life cycle starting with a period of sexual immaturity. This period, during which mating types are not expressed and conjugation cannot occur, is of variable duration (up to about 80 cell generations) and is not completely measured by either time or number of cell generations. Thus, Perlman (1973) found in strain D that immaturity lasted longer in time *and* extended through more cell generations (mean, 56.3) when clones were cultured at 19°C than when cultured at 30°C (mean, 45.5 cell generations).

Maturity is characterized by capacity to conjugate, i.e., to express mating type. During maturity, clones free from recessive lethals can yield a high frequency of viable sexual progeny. Sooner or later, however, a clone will show increasing frequency of nonviable progeny when its members conjugate. The beginning of this increase, which occurs at different clonal ages (from less than 100 to more than 1500 generations) in different clones, may be considered to mark the beginning of clonal senescence. Senescence ends in complete genetic death, the total inability to transmit genes by sexual reproduction; but at least some lines of descent within senescent clones continue to reproduce asexually with undiminished vigor. Apparently, the micronuclei become progressively deficient in DNA and some amicronucleate cells may be produced; these die without further fissions or after at most a few fissions (Nanney, 1957, 1959b; Allen et al., 1967). This genetic decline of clones provides the essential material for exploiting genomic exclusion (see page 438); but it also sets a term to the usefulness of genetically important stocks unless they are reconstructed by inbreeding about once a year. Formerly this was a serious impediment to genetic work, but the difficulty has been overcome by success in freeze-storing clones (Simon, 1972). Samples can be frozen when clones are young and vigorous, held at this stage as long as desired, and thawed for use as needed.

Genetics of Asexual Reproduction

The classic dictum of the genetic constancy of the clone has suffered a major impairment with regard to its applicability to *T. pyriformis,* syngen 1. The new knowledge has nothing to do with low-frequency exceptions, e.g., the occurrence of macronuclear mutations, or with genetic inconstancy of the micronuclear genome. It refers to regular and frequent diversification of phenotype within a clone. This diversification occurs regularly in heterozygotes and, in at least one special case, even in homozygotes. The basic observation in the usual case is that heterozygotic clones, after manifesting the dominant or codominant phenotype for a considerable number of cell generations, later include some subclones that become and remain pure for expression of the phenotype corresponding to one allele at the heterozygous locus and some sublines that become and remain pure for expression of the phenotype corresponding to the other allele (Allen and Nanney, 1958; and many later papers, see review in Allen and Gibson, 1973). This phenomenon has been referred to by various terms: nuclear differentiation, allelic repression, subnuclear assortment, and subnuclear determination. Because such terms involve hypotheses or inferences, I shall refer to it by a term, phenotypic assortment or simply assortment, that describes what is directly observed. In syngen 1, all known loci that can readily be and have been examined for phenotypic assortment in heterozygous clones show it, except the locus of the temperature-sensitive mutation *ts1* (Table 2).

Special aspects of assortment have been observed in two very different cases. The phosphatase locus, *P1,* also assorts a third class of constant subclones in heterozygotes. The phenotypes for genes at this locus are detected electrophoretically, and the heterozygote shows one or more bands not found in either homozygote. Assortment yields not only subclones pure for the electrophoretic pattern of each homozygote, but also subclones pure for an electrophoretic band found only in heterozygotes (Allen, 1971). Different interpretations of this special case have been put forth by Allen (1971) and Orias (1973).

The other special case, actually the one in which assortment was first discovered, concerns the mating-type locus, *mt* (Table 2). In this case, phenotypic assortment can occur even in homozygotes (Nanney and Caughey, 1955; Allen and Nanney, 1958). Homozygotes for an allele at this locus can express any one of several mating types. Sometimes a caryonide (the group of cells containing descendants of the same ancestral macronucleus) is pure for one mating type; but frequently subcaryonides (different lines of descent within a caryonide) assort, some being phenotypically pure for one mating type and others for another. Less frequently, more than two (up to four) mating types assort as phenotypically pure subcaryonides within a single caryonide. (For further details on mating types, see page 457.)

It was early noted (Allen and Nanney, 1958) that phenotypic assortment is characterized by certain mathematical regularities. In some cases at once and in others eventually, the sum of the frequencies with which phenotypically pure sublines assort from phenotypically heterozygous lines appears to stabilize at about 0.0113 per cell division. This stable value is designated the R_f value, the value per fission. Sometimes the value is stated as R_t, the value per transfer, each transfer being the isolation of one cell from among those produced by the last previous isolation made one or two days earlier. Obviously the relation between R_t and R_f is given by the mean number of cell divisions between successive transfers. R_f values are somewhat higher when reisolations are made every second day than when they are reisolated daily. This seems to be due to starvation (Nanney and Caughey, 1955) since during part of the second day of culture in slide depressions, the food is exhausted.

Subsequent studies on other loci report some differences in R_f values for different loci examined with daily reisolations (see Table 2 for descriptions of the genes): *aa*, 0.0126; *caf1*, 0.0097; *caf2*, 0.0109 [according to Carlson (1971)]; *H*, 0.0112 ± 0.0018 (Nanney and Dubert, 1960); *r1*, 0.0105 ± 0.0011 (Doerder, 1973); *TAT*, 0.0113 (Borden *et al.*, 1973a). With transfers every second day, some values are: *H*, 0.013 ± 0.0015 (Nanney and Dubert, 1960); *r1*, 0.0096 ± 0.0012 (Doerder, 1973); *r3*, 0.0087 ± 0.0008 (Doerder, 1973). As will appear, the question of whether there are significant differences in R_f values among different loci under the same conditions is of theoretical importance. For the present, this question cannot be answered with assurance because ascertainment of the values can be complicated by various factors, e.g., differences in viability or reproductive rate among assorting sublines of the same clone, uncertainties as to whether lines scored as assorted are in fact pure, and failure to identify one of the stable types when one is dominant. Currently, the value 0.0113 is generally held to apply to all assorting loci under conditions of daily reisolation.

As the number of loci examined increased, five other remarkable features of assortment became apparent. First, different loci assort at different stages of the clonal life history [see Bleyman (1971) for summary of the different assortment behaviors of several loci]. Many reports state only the stage when or by which assortment is known to have begun. This stage is "early," roughly by about the 20th cell generation, for some loci, e.g., *caf1*, *H*, *mo1*, *mt*, *r4* and *St*; there may be slight differences among some of these. The stage is "late" for *co* (McCoy, private communication), *E1*, *I*, *P1*, *r3*, *TAT*, and probably *ts2*; most of these are reported to start assorting at about 40–50 cell generations, again with indications of some differences among them. The second remarkable feature shown by comparative studies of different loci is a diversity of "output ratios," i.e., the ratio between the frequencies of the different assorting pure types within a clone. Some loci regularly assort in approximately a 1:1 ratio from the start and continue to do so. Other loci assort with variable ratios for the same locus in different clones, sometimes 1:1, but often deviating significantly from equality, sometimes reaching 25 or more :1. However, as the assortment process continues, the output ratio from residual "impure" sublines tends to approach and eventually reach 1:1. The third remarkable feature is that the constant 1:1 output ratio is characteristic of late-assorting loci while the variable ratios are characteristic of early-assorting loci. The fourth feature is that linked loci [e.g., *mt* and *E1* (Allen, 1965), *r1* and *T* (Doerder, 1973)] show independence in phenotypic assortment. Finally, the fifth feature is that for early-assorting loci with multiple assortant types (e.g., with multiple alleles, as at the *H* locus, or with multiple alternative phenotypes for the same allele, as with the *mt* alleles), certain types tend strongly to be the majority type. In the case of multiple alleles (e.g., at the *H* locus), the phenotype corresponding to a particular allele may be the majority type when this allele is present with a certain other allele, but the minority type when the same allele is present with still another allele (Nanney and Dubert, 1960; Nanney *et al.*, 1964).

The genetic implications of phenotypic assortments are in part clear and in part still unresolved. That the basis of assortment lies in events taking place in the macronucleus is clear. It is also clear that some form or forms of heterogeneity exist(s) or develop(s) within the macronucleus and that the internal diversity within a macronucleus is distributed unequally to descendant nuclei in the course of successive amitotic macronuclear divisions. This results eventually in the sorting out of homogeneous descendant macronuclei. Underlying phenotypic assortment there is thus some form of macronuclear assortment which begins when the macronuclear heterogeneity is established or when it starts to undergo unequal distribution at division (i.e., macronuclear determination). Still unclear are the nature of determination (which may be different in homozygotes, e.g., for a mating-type gene, from what it is in some—or all—heterozygotes), and the answer to the closely allied question of the nature of heterogeneity in the macronucleus.

This much knowledge, together with the assumption—as a first approximation—that the elements of heterogeneity in the macronucleus are distributed to daughter nuclei at macronuclear division equally in number but randomly in quality, permitted Schensted (1958) to develop a basic mathematics relating the kinetics of observed phenotypic assortment to the number of assorting macronuclear elements, their initial ratio ("input ratio," as inferred from output ratio), and the stage at which distributive heterogeneity arose (determination). The data fit best the mathematics for about 45 assorting elements in G1, 90 in G2. In general, it takes about a score or more cell generations for a 1:1 input to begin to yield homogeneous assortants. However, the stage at which phenotypic assortment is first found is a poor datum from which to infer the stage at which determination occurred. The extended kinetics up to or beyond the stage when 50 percent of the sublines are phenotypically pure is required for firm estimates. Even then, the mathematics makes no allowance for perturbations in macronuclear division (unequal amounts of DNA in two daughter nuclei; two rounds of doubling in some cell cycles; extrusion of DNA from nuclei in varying amounts) reported to occur (in a sterile strain) by Cleffmann (1968). Nevertheless, in spite of these difficulties, the Schensted mathematics remains the only practicable way of dealing with certain aspects of phenotypic assortment.

Various hypotheses have been proposed or considered concerning the *nature* of the heterogeneity in the macronucleus and the nature of determination. Initially, Nanney (1956) and Allen and Nanney (1958) assumed that the macronucleus consists of diploid subunits and that determination consists of interallelic repression (or intraallelic repression and/or activation in the case of the presumably complex *mt* locus). Measurements of the relative amounts of DNA in micronuclei and macronuclei (see page 436) led Woodard *et al.* (1972) to propose that, if there were 45 subunits in G1, they are haploid. Nilsson (1970) also proposed that, in strain GL, each haploid genome behaves as a unit during macronuclear DNA distribution at division and that the G1 macronucleus contains about 40 such units. Nanney (private communication) recognized, however, that the discrepancy would disappear if only about half of the DNA of a micronuclear genome was represented in the macronucleus. Then all of the macronuclear genes could still be duplex in each subunit. [See page 436 for discussion of this point.) Nanney is aware that reduction from two different alleles to one per subunit could conceivably occur at the time of determination by delayed reduction of units from diploid to haploid or by "somatic" crossing-over, yielding homozygous "twin-spots."

Allen and Gibson (1972, 1973) have suggested that the macronucleus contains only one replicative diploid genome (the "masters") and the theoretically required total number of free (i.e., discrete) nonreplicative copies ("slaves") of its genes. The masters are held normally to be exactly replicated and distributed at macronuclear division, while the slaves are distributed randomly. New slaves are assumed to be produced to compensate for halving their number at macronuclear division. On this hypothesis, determination consists of loss of one master gene, and the kinetics of assortment is due to the assortment of the nonreplicating slaves of the missing master. The idea that most of the DNA in the macronucleus consists of nonreplicating slaves seems to be at variance with reports that most, if not all, macronuclear DNA molecules replicate semiconservatively once per cell cycle in the amicronucleate strain GL (Andersen *et al.*, 1970; Andersen and Zeuthen, 1971; Andersen, 1972), unless strain GL differs from syngen 1 in this respect. Discussions of the pros and cons of various interpretations can be found in Nanney (1968*b*), Nanney and Doerder (1972), Doerder (1973), and Allen and Gibson (1973).

Whatever the correct interpretation of phenotypic assortment may turn out to be, the process has technical values for genetic work. Like autogamy in *P. aurelia,* it brings to light in the course of one generation recessives that would otherwise require further breeding to expose. Carlson (1971) and others have exploited this fortunate circumstance in screening for mutations (see page 446). Phenotypic assortment also permits identifying genotypes of hybrids without further breeding analysis. Orias' (1973) analysis of the P1 locus suggests how phenotypic assortment might be used to map intragenic sites if the independent phenotypic assortment of linked loci (Allen, 1965; Doerder, 1973) proves to be due to crossing-over. The latter might well be frequent enough to detect intragenic recombination, at least in some strains (see page 456), but convincing evidence of genetic recombination in the macronucleus is not yet available.

Inbred, Congenic and Star Strains

When attempts were first made to do genetics with *Tetrahymena,* problems were raised by the heterozygosity and low fecundity of wild stocks. Nanney (1957), therefore, carried out a long and precarious program of inbreeding and finally obtained, from a small number of wild stocks, homozygotes of high fecundity. Subsequently, he and others added more such inbred strains which are designated by symbols A, A1, . . .,F. According to Allen and Gibson (1973), these strains have now been through at least 12 and up to 22 generations of close inbreeding and selection. Strains A3, B1, and C are no longer available or usable. Doerder (1973) and Allen and Gibson (1973) list the currently available strains and the alleles of the polymorphic loci which are present in each strain. Allen and Lee (1971) constructed strains congenic with strain D by introducing, into the background genome of strain D, alleles at the *H, T, E1,*

E2, and P1 loci of strain C2. For use in genomic-exclusion crosses, in addition to C* (see page 438), two other stocks are now in use: A* III (mating-type III) and A*V (mating-type V) (F. P. Doerder, private communication).

Mutagenesis: Selective Detection and Isolation of Mutations

The favorite mutagen has been nitrosoguanidine (e.g., 10 μg/ml for 1–3 hours), but Carlson (1971) has reported success with 0.25 percent ethyl methane sulfonate. For selective detection of germinal mutation, i.e., those induced in micronuclei, the mutant micronuclei must first produce macronuclei so that the mutant gene can come to phenotypic expression. This means that conjugation must occur. If crosses are made and normal conjugation occurs, the mutant gene will almost always be in a heterozygous condition. Dominant mutations would be quickly expressed, and thus rapidly subject to selection. Even recessive mutations will soon be expressed as a result of phenotypic assortment. The method, therefore, is to expose normal clones of different mating type to mutagen, to mate them *en masse,* and then to grow the resulting clones for 40–100 cell generations (5–12 days) before selective screening. With this method, screening has been successful, e.g., for resistance to drugs or chemicals (Carlson, 1971; see also Table 1), and for serotype mutations using reaction to antisera (Doerder, 1973). Mutants identified at this stage still have micronuclei heterozygous for the mutant gene and have to be bred further to obtain the mutation in homozygotes. McCoy (1973) simplified the procedure by inducing mutations in a selfer culture. After exposure to mutagen, single-cell isolations were made and each was grown into a small mass culture which was allowed to self. Selfing pairs were then isolated, grown, and screened for mutations. Since the two members of a selfing pair come from a single ancestral cell exposed to mutagen, 25 percent of the selfing pairs from a mutated ancestor would produce synclones homozygous for the mutation.

A different strategy uses genomic exclusion (see page 438) by crossing a star strain to a mutagenized normal clone (Orias and Flacks, 1973). By isolating pairs during the first round of mating and letting the progeny of isolated pairs quickly go through a second round of mating, one obtains synclones homozygous for various combinations of genes that were present in the mutagenized parent clone. These synclones can then be screened for mutations.

A short-cut variant of this strategy developed by Bruns and co-workers (private communication) selects directly (after exposure to mutagen) the exceptional round-one pairs that form functional macronuclei from the gamete nuclei (see page 440). The procedure is as follows. A heterozygote for a dominant drug-resistance mutation, e.g., resistance to cycloheximide $(Chx1/Chx1^+)$ or to 6-methylpurine $(Mpr1/Mpr1^+)$, is allowed to multiply until phenotypic assortment occurs, yielding some lines that are sensitive to the drug but, of course, still have heterozygous micronuclei. These "functional heterokaryons" are mass-mated to a star strain. Five hours after pairs form, food (peptone) is added to inhibit a second round of mating. At 18–24 hours, a time sufficient to yield expression of the new (drug-resistant) phenotype if new macronuclei are produced, the same drug (i.e., cycloheximide or 6-methylpurine) is added. This kills cells which retain their prezygotic sensitive phenotype and thus selects for the small percentage of conjugants in which functional macronuclei, developed from products of the gamete nuclei, express the initial resistant allele now carried in homozygous condition (see page 440). Some of these nuclei could theoretically also be homozygous for mutations induced in the functional heterozygote. Preliminary breeding analysis indicates that induced (temperature-sensitive) mutants could indeed be selected in this way (see Table 2, *ts100*).

Bruns (1973) has devised an ingenious method that may prove useful in direct selection of "negative" mutants. The basis of this method is his discovery that starved cells band in a different position from growing cells in a centrifuged Ficoll gradient. This might be applied to selection of mutants that cannot grow under conditions in which wild type does grow. Selection could be attempted with mass cultures after phenotypic assortment. Table 1 gives Bruns' unpublished list of drugs, with the doses that should be used to screen for resistant or more-sensitive mutants.

Comparatively little work has been done thus far on macronuclear or mitochondrial mutations. Macronuclear mutations are suggested by observations of Nanney (1959*b*, 1962) and Bruce Byrne (unpublished). To detect macronuclear mutations, macronuclear and cell divisions are required, as for phenotypic assortment, before screening for a mutant phenotype with selective agents. Conjugation must be prevented or the macronucleus and its mutation would be lost. Roberts and Orias (1973*a*) have reported a case of chloramphenicol resistance which is probably based on a mitochondrial mutation. For such mutations, the methods used with *P. aurelia* (see chapter on Paramecium, this volume) are applicable except for those involving injections and exchange of mitochondria between mates during conjugation. *Tetrahymena* is too small for injection to be practicable and seems not to exchange organelles as large as mitochondria during mating (Roberts and Orias, 1973*a*).

TABLE 1.   *Minimal Doses of Drugs and Chemicals that Kill Strain B within 24 Hours at 30°C*[1]

Name	Dose	Solvent sterilization method	Stock solution, mg/ml
Diphenhydramine HCL (DPH)	320 μg/ml	EtOH	32
Pactamycin (Pac)	40 μg/ml	EtOH	4
Quinine (Qn)[b]	320 μg/ml	EtOH	32
Desipramine (Des)	40 μg/ml	EtOH	4
Isopropyl N-phenyl carbamate (IPC)	160 μg/ml	EtOH	16
Isopropyl N-(3-chlorophenyl) carbamate (CIPC)	40 μg/ml	EtOH	4
Quinicrine HCl (Qc)	80 μg/ml	H₂O, autoclave	8
Cycloheximide (Cy)	25 μg/ml	H₂O, autoclave	2.5
Hydroxyquinoline (Hq)	10 μg/ml	H₂O, autoclave	1
Dianisidine (Dian)[c]	320 μg/ml	H₂O, autoclave	32
p-Fluorphenylalanine (pFPA)[c]	640 μg/ml	H₂O, autoclave	10
2-Fluoroadenosine (2FA)	80 μM	H₂O, filter	2.28
6 Methylpurine (6MP)	55 μM	H₂O, filter	0.746
6-diMethylaminopurine (diMAP)	800 μM	H₂O, filter	6.52
Allyl alcohol (AA)	250 mM	use directly	
Caffeine (Caf)[d]	1.5 mg/ml	H₂O, autoclave	20
Anisomycin (Ani)	25 μg/ml	EtOH	2.5
Chlortetracycline (Chlortet)	500 μg/ml	H₂O, filter	10
Podophyllin (Podo) (keep in dark)	100 μg/ml	EtOH	10
		2% PP + 0.1% YE[e]	
6 Benzyethiopurine (6BEP)	500 μM	EtOH	10

[a] Data from Bruns (private communication).
[b] Kills by 48 hours.
[c] Does not kill, but completely inhibits growth.
[d] Kills by 96 hours.
[e] PP = proteose peptone, YE = yeast extract.

Known Genes

Table 2 lists the genes that have been reported and summarizes some useful information about each. Included are several genes which are no longer available: *aa, caf1, caf2, f,* and *tny.* The mutagen nitrosoguanidine is abbreviated as NG in the table. I am indebted to J. Frankel and L. Jenkins for revision and extension of my account of *co* and the series of *mo* genes, on which nothing is yet in print; to F. P. Doerder for the same on the genes *r1, r3,* and *r4*; to J. W. McCoy for accounts of *mp* and *r4*; to P. J. Bruns for *ts100*; and to B. Byrne for *Mpr1.* These people and also

TABLE 2.   Genes in Tetrahymena pyriformis, Syngen 1[a]

Gene locus	Characteristics
aa, aa⁺	Recessive mutant (no longer available) resistant to 250 mM allyl alcohol (which kills wild type in 12 hours) and accumulated allyl alcohol to only ¹/₉ wild-type level; isolated from cross of strains A1 and B after exposing them to 0.25 percent ethyl methane sulfonate (Carlson, 1971); assorted late
caf1, caf1⁺	Recessive mutant (no longer available) resistant to 10 mM caffeine (which kills wild type) and accumulated it to only ¹/₃ wild-type level in 30 minutes; material and methods as in *aa* above; not allelic to *caf 2*; assorted early
caf2, caf2⁺	Like *caf1* above, but caffeine accumulated to same level as in wild type; same material and methods; not allelic to *caf1*; assorted late
Chx1, Chx1⁺	Codominant mutation for resistance to 75 μg/ml cycloheximide; induced in strain D with NG by Roberts and Orias (1973*b*); selected by growth in 10 μg/ml cycloheximide; mutant requires 20 times as high a concentration of cycloheximide to duplicate the wild-type response of growth inhibition and recovery and is less sensitive than wild type to streptimidone; assorts; linked to *E1* with 25 percent recombination in certain strains; in same linkage group as *mt* and *TO;* not linked to *co, H, mo1, mo3, P1, r1, r3, r4, ts1,* or *ts2;* a similar mutation obtained with NG in strain B by Byrne (private communication) has not yet been tested for allelism with *Chxl*
co, co⁺	Recessive mutation for conical body shape obtained with NG on strain D by Doerder (private communication); body shape nearly normal during cell division, but 100 percent are conical at other stages of cell cycle; cell division unequal: posterior division product smaller (and has longer generation time) than anterior product; macro- and micronuclear divisions probably normal; assorts late; linked to *r4* with 25 percent recombination; not linked to *Chx1, H, mo1, mo3, r1, r3, ts1,* or *ts2*
*E1*ᴮ, *E1*ᶜ	Polymorphic codominant propionyl esterase alleles; phenotypes distinguished electrophoretically (Allen, 1960; see Allen and Gibson, 1973); splits α- and β-naphthyl acetate; activated by sodium taurocholate and Triton X-100; inhibited by eserine sulfate; *E1*ᶜ occurs in the inbred C strains; *E1*ᴮ in the other inbred strains; young heterozy-

[a] NG = nitrosoguanidine. See p. 441 for account of phenotypic assortment.

TABLE 2.   Continued

Gene locus	Characteristics
	gotes show both parental electrophoretic patterns, but no additional bands; multiple bands formed by homozygotes are interpreted as due to secondary structural changes; assort late and independently of *mt* (Allen, 1965) to which it is linked with 20–25 percent recombination in certain strains (Allen, 1964; Doerder, 1973), but linkage not evident in certain other crosses, e.g., strains $B \times D$ (Doerder, 1972; Borden, 1972; McCoy, 1973); in same tentative linkage group as *Chx1* and *TO*; not linked to *E2, H, I, P1, r1, r3, St, T, TAT,* or *ts1*
$E2^B$, $E2^C$	Polymorphic codominant butyrl esterase alleles; phenotypes distinguished electrophoretically (Allen, 1960, 1965; Allen and Gibson, 1973); also splits α-naphthyl valerate; insensitive to eserine; inhibited by certain concentrations of sodium taurocholate. $E2^C$ occurs in inbred strains A3, B3, C, and C2; $E2^B$, in the other inbred strains; each homozygous phenotype characterized by one distinctive band; the heterozygote has no additional bands; not linked to *E1, H, mt, P1,* or *T*
Em1, Em1$^+$	Dominant spontaneous mutation for early maturity (in less than 20 cell generations) initial low growth rate, and recessive lethal action (Bleyman and Simon, 1967); found in strain F in frequencies of origin up to 5 percent and therefore considered to be possibly a structural rearrangement (Bleyman, 1971); small-scale tests show no linkage to *Em3, H, mt, P1,* or *T*; relation to *Em2* questionable
Em2, Em2$^+$	Like *Em1* except no recessive lethal effect (Bleyman, 1971); isolated from cross of strains A × B after Nanney exposed them to NG; not closely linked to *H* or *mt*; questionable relations to *Em1* and *Em3*
Em3, Em3$^+$	Like *Em1* (Bleyman, 1971); found in strain D1; not linked to *Em1*; questionable relation to *Em2*
Em4, Em4$^+$	Like *Em1*; found in strain C3 (Simon, private communication) and analyzed by Bleyman (1972); allelism with the *Em* series has not been tested.
f, f$^+$	Spontaneous recessive lethal (Orias, 1960); no longer available; found in strain IL–12; cells got "fat" before dying
H^A, H^C, H^D, H^E	Polymorphic codominant alleles for immobilization antigens (serotypes) expressed at 20–30°C in peptone or bacterized cerophyl media (Nanney and Dubert, 1960); H^A in strains A, C1, C3; H^C in strains B2 and D1; H^D in strains A1, B,

TABLE 2.   Continued

Gene locus	Characteristics
	D, E, and F; H^E in strains A3, B3, C and C2; assort early; not linked to *Chxl, co, E1, E2, Em1, Em2, I, mol, mp, mt, P1, r1, r3, r4, St, T, TAT, TO, ts1,* or *ts2*
I^F, I^S	Polymorphic codominant alleles for fast (I^F) and slow (I^S) moving electrophoretic variants of isocitrate dehydrogenase (Borden *et al.*, 1973a); I^F in strain C2; I^S in all other inbred strains; young heterozygotes show both parental bands and an intermediate band; assort late; not linked to *E1* or *H*
M^F, M^S	Polymorphic codominant alleles for fast (M^F) and slow (M^S) moving electrophoretic variants of mitochondrial NADP-dependent malate dehydrogenase, best expressed in stationary phase cells and possibly absent during rapid growth (Borden *et al.*, 1973a); M^F in strain C2; M^S in all other inbred strains; young heterozygotes show no intermediate band; assortment probably late
mol^a, mol^b, mol^+	Temperature-sensitive recessive mutations for monster formation at 35–40°C (not expressed at 15–30°C). mol^a (Doerder, private comm.): induced with NG in strain D; fission line completely inhibited, macronuclear division severely retarded; cell growth, micronuclear division, and oral development (highly abnormal, with extra and branched membranelles) continue so that multiple oral apparatuses appear in a monstrous cell. mol^b (Jenkins, private comm.): induced with NG in strain B; at second division (40°C), fission line not completed, macronuclear division retarded, but micronuclei and oral apparatuses (normal) multiply to form monster. Assort early; not allelic to *mo2, mo6, mo8*; not linked to *Chxl, co, H, mo3, r1, r4, T,* or *ts1*; linked to *ts2* with 15 percent recombination
$mo2, mo2^+$	Recessive mutation for chain (incomplete fission) and monster formation obtained with NG in strain B (Jenkins, private communication); heterozygote shows little or no expression at 15–40°C prior to phenotypic assortment, but about 50 percent expression at 40°C and less than 5 percent at 28°C during assortment; presumed homozygous segregants show 100 percent expression at room temperature 2–3 fissions after conjugation and, being then monstrous, cannot be propagated; not allelic to *mol, mo3, mo6,* or *mo8*
$mo3^a, mo3^b, mo3^c, mo3^d, mo3^+$	Almost completely recessive mutations for incomplete cell constriction (furrowing) at fission, yielding chains of 2–8

TABLE 2. *Continued*

Gene locus	Characteristics
	cells; obtained with NG in strain B by Jenkins (private communication); in homozygotes for *mo3ᵃ*, *mo3ᶜ*, and *mo3ᵈ* (but not *mo3ᵇ*), cells connected by blocked furrows may pull apart after several hours; *mo3ᵃ* is expressed 100 percent in homozygotes at 15°C and 40°C, 10–30 percent at 20–30°C; *mo3ᵇ* and *mo3ᵈ* expressed 100 percent in homozygotes at 40°C, not at all at 15–30°C; *mo3ᶜ* expressed 10 percent or less in homozygotes at 20–40°C, about 50 percent at 15°C; expression in heterozygotes (with *mo3⁺*) 10 percent or less at 40°C, not at all at 20–30°C; (*mo3ᵇ* originally isolated together with another gene that acted as an apparent dominant partial suppressor); assort; not allelic to *mo2, mo6, mo8*; not linked to *Chx1, co, mo1, r1, r3, r4, ts1,* or *ts2*
mo6, mo6⁺	Recessive mutation for monster formation at 40°C (normal at 15–30°C); obtained with NG in strain B by Jenkins (private comm.); at 40°C, 100 percent of dividers have distorted and incomplete division furrows after one or sometimes two cell cycles; macronuclear division somewhat retarded; micronuclei and oral apparatuses multiply, leading to formation of very irregular monsters; (originally isolated together with another gene that acted as an apparent dominant suppressor); assorts; not allelic to *mo1, mo2, mo3,* or *mo8*; not linked to *r1* or *ts1*
mo8, mo8⁺	Recessive mutation for chain and monster formation at 40°C (expression increasing with time at 40°C, but never exceeding 50 percent); not expressed at all at 15–30°C; obtained with NG in strain B by Jenkins (private comm.); at 40°C, the anterior part of the new oral apparatus is frequently positioned anterior to the fission line, apparently blocking the fission furrow; in extreme cases, the whole new oral apparatus is included in the anterior fission product; assorts; not allelic to *mo1, mo2, mo3,* or *mo6*
mp, mp⁺	Recessive for abnormal membranellar pattern found by Kaczanowski (private comm.) to be characteristic of, and homozygous in, strain D; expressed in 10 percent or less of the cells during rapid growth, but up to 50–70% during prolonged stationary phase; oral membranelles show abnormal relative lengths or more than three abnormally short membranelles (usually 4 or 5) of varying relative lengths; assort, but details of assortment not yet known

TABLE 2.   Continued

Gene locus	Characteristics
Mpr1, Mpr1$^+$	Dominant mutation for resistance to 11 mM 6-methylpurine; obtained by Byrne (private comm.) in strain B; wild type (strain B), but not heterozygotes, is killed within 24 hours at 30°C by 55 μM; assorts; strain A resists higher concentrations than strain B (Masserman, private comm.)
mt^A, *mt*^{B}, *mt*^{C}	Polymorphic codominant alleles for mating-type potentialities at what may be a complex locus (Nanney and Caughey, 1953; Nanney *et al.*, 1955). *mt*^{A} occurs in strains A, A1, and A3; it permits development of mating types I, II, III, V, and VI, but not IV or VII. *mt*^{C} occurs in strains C, C1, C2, and C3; it permits development of the same mating types as *mt*^{A} but with different relative frequencies, e.g., under the same conditions (temperature being important), *mt*^{C} yields a higher frequency of type I and a lower frequency of type V than *mt*^{A}. *mt*^{B} occurs in strains B, B2, and B3; it permits development of any one of the 7 mating types except I. *mt*^{B}/*mt*^{A} and *mt*^{B}/*mt*^{C} permit development of all 7 types. See page 457 for mode of determination of types within the genetically permitted group. Literature contains references to *mt*^{D} in strains D and D1, *mt*^{E} in strain E, and *mt*^{F} in strain F, but there are no known differences between these and *mt*^{A} (Bleyman, 1971; Bleyman *et al.*, 1966; Bleyman and Simon, 1968; McCoy, 1973). Assort early; linked to *E1* in certain strains (see *E1*); linked to *TO* with 16 percent recombination in crosses that show no linkage between *mt* and *E1* (McCoy, 1973); not linked to *E2, H, P1, r1, r4, St, ts1,* or *ts2*
P1^A, *P1*^{B}	Polymorphic codominant alleles for fast (*P1*^{A}) and slow (*P1*^{B}) moving anodal electrophoretic variants (isozymes 1 and 5, respectively) of acid phosphatase (Allen *et al.*, 1963); split α-naphthyl acid phosphate and other synthetic phosphates; inhibited by d-tartaric acid and NaFl; young heterozygotes show, in addition to the parental isozymes, an intermediate (isozyme 3); assortment (late) yields both parental types and still later a stable intermediate banding at the position of isozyme 3; for different interpretations of the stable intermediate, see Allen *et al.* (1963) and Orias (1973); not linked to *Chx1, E1, E2, H, r1, r3, T,* or *TAT*
r1, r1$^+$	Recessive mutation characterized by expression of immobilization antigen (serotype) r at 15–40°C, the temperature range in which wild type expresses serotypes L, H, or T (*q.v.*); obtained with NG in strain A by Doerder (1973);

TABLE 2. *Continued*

Gene locus	Characteristics
	epistatic to *r4* under conditions for expression of serotype H; thought to be epistatic to *r3* at 40°C; assorts late; linked to *T*; about 9 percent recombination with *T* of strain D1, about 37 percent with *T* of strain D (but higher percent, probably independence, when tested by genomic exclusion); in same tentative linkage group as *ts1* and *TAT*; but *ts1* shows independent segregation; not linked to *Chx1, co, E1, H, mo1, mo3, mo6, mt, P1, r3, r4, ts1,* or *ts2*
r3, r3+	Recessive mutation like *r1* except that serotype T instead of r is expressed at high temperatures. When r is expressed (i.e., at lower temperatures), it is indistinguishable from the r serotype of *r1*; therefore, Doerder (1973), who obtained *r3* with NG in strain B, suggests that *r1* and *r3* do not code for antigen r, but merely suppress expression of the wild-type serotypes and permit another (unknown) locus for antigen r to act; epistatic to *r4* under conditions for expression of serotype H; thought to be hypostatic to *r1* at 40°C; assorts late; not linked to *Chx1, co, H, mo1, mo3, P1, r1, r4, T, ts1,* or *ts2*
r4a, r4b, r4c, r4+	Independent (identical ?) recurrences (*r4a* and *r4b* in strain B, *r4c* in strain D) of recessive mutation induced by NG (Doerder, private communication); characterized by expression of immobilization antigen (serotype) q under conditions in which wild type expresses serotypes L, H, T, or St; serotype q reacts to anti-He serum, but serotype He does not react to anti-q serum; hypostatic to *r1* and *r3* under conditions for expression of serotype H; assort early; linked to *co* with 25 percent recombination; not linked to *Chx1, H, mo1, mo3, mt, r1, r3, ts1,* or *ts2*
StA, StB	Polymorphic codominant alleles for variants of immobilization antigen (serotype) St; expressed when grown in 200 mM NaCl (Grass, 1972a,b); this concentration is lethal on direct transfer; adaptation for one day to 100 mM suffices; *StA* in strains A, A1, A3, B, D, D1, E, and F; *StB* in strains C, C1, C2, C3, and possibly B2; assort early; not linked to *E1, mt, H,* or *T*
TA, TB, TC	Polymorphic codominant alleles for immobilization antigen (serotype) T, expressed at 40–42°C in liver peptone medium (Phillips, 1967a,b); *TA* in strains A, A1, B, B3, C1, E, and F; *TB* in strains A3, C, C2, and C3; *TC* in strains B2, D, D1; assort late; linked to *ts1*, 26 percent recombination in cross B × B3 which shows no linkage of *mt* with

TABLE 2. Continued

Gene locus	Characteristics
	E1 (McCoy, 1973); linked to *TAT* with 17 percent recombination (Borden, private comm.); in same tentative linkage group as *ts1*, *TAT*, and *r1*; not linked to *E1*, *E2*, *H*, *mo1*, *mt*, *P1*, *r3*, *St*, or *TO*
TAT^F, *TAT^S*	Polymorphic codominant alleles for fast (*TAT^F*) and slow (*TAT^S*) electrophoretic variants of tyrosine aminotransferase (Borden *et al.*, 1973a); *TAT^F* in strains A1, B3, C1, C2, and LW's; *TAT^S* in strains A, B, B2, C3, D, D1, and F; assort late; linked to *T* with 17 percent recombination (Borden, private communication); in same tentative linkage groups as *ts1* and *r1*; not linked to *E1*, *H*, or *P1*
TO^F, *TO^S*	Polymorphic codominant alleles for fast (*TO^F*) and slow (*TO^S*) electrophoretic variants of tetrazolium oxidase (Borden *et al.*, 1973a); *TO^F* in strains A, B, B2, B3, C1, C2, C3, and LWB; *TO^S* in strains A1, D, D1, F, and LWA; heterozygotes show no intermediate band; linked to *mt* with 16 percent recombination in strain that shows no linkage of *mt* with *E1*; in same tentative linkage group as *Chx1* and *E1*; not linked to *H*, *T*, or *ts1*
tny, *tny$^+$*	Recessive lethal, dying as tiny cells; found in wild stock IL-12 by Orias (1960); no longer available
ts1, *ts1$^+$*	Temperature-sensitive recessive lethal (at 38°C) obtained with NG in strain Em2 after cross to strain B3 by McCoy (1973); does not assort; linked to T with 26 percent recombination; in same tentative linkage group with *TAT* and *r1*, but shows independence with *r1*; not linked to *Chx1*, *co*, *E1*, *H*, *mo1*, *mo3*, *mo6*, *mt*, *r1*, *r3*, *r4*, *TO*, or *ts2*
ts2, *ts2$^+$*	Temperature-sensitive recessive mutation induced in strain D by Orias and Flacks (1973); unable to grow above 37°C, dies at 41°C; appears to assort late; linked to *mo1* with 15 percent recombination; not linked to *Chx1*, *co*, *H*, *mo3*, *mt*, *r1*, *r3*, *r4*, or *ts1*
ts100, *ts100$^+$*	Recessive lethal induced by NG in a strain B/D heterozygous functional heterocaryon (see page 447) by Bruns (private communication); inhibited at 38°C, killed at 41°C

S. L. Allen, L. K. Bleyman, D. Borden, E. Orias and D. L. Nanney have contributed other information and valuable comments in regard to Table 2. Various details of the entries have been compiled from a number of published and unpublished sources. In general, the reader who wishes to verify or know the source of such information could probably get it from a person whose name appears in the account of the gene in question.

Linkage and Maps

In addition to the published accounts, I include here unpublished information supplied by some of the people mentioned above, particularly J. W. McCoy, who provided me with a table summarizing the results of the limited number of tests for linkage that have been made among 22 loci and with the following tentative sketch of linkage groups:

I. TO——mt——$Chx1$——$E1$
II. $ts1$——T——TAT——$r1$
III. $ts2$——$mo1$
IV. $r4$——co

Thus, 12 of the 30 loci with presently available genes have been assigned to an ordered sequence in four (of the eventually expected five) linkage groups. This formulation is still very tentative. Conceivably, groups entered separately may eventually prove to be in the same linkage group, and the order of the genes in groups I and II may have to be revised in some details.

Cross-over values have been deliberately omitted from the sketch above. Cross-over values, included in Table 2, are confusing because data on the same two loci differ greatly when tests are made on different strains. At present it seems that crosses to certain strains (B2 and D) give much higher values of recombination than crosses to other strains. Doerder (1973) has reported that the loci of $r1$ and T show about 9 percent recombination in crosses to strain D1, but 37 percent or more in crosses to strain D. McCoy (1973) and Borden (private communication) have found that the loci of mt and $E1$ show no linkage in certain crosses, although Allen (1964) and Doerder (1973) have found 20–25 percent recombination in other crosses. This situation led McCoy (1973) to state that "homologous chromosomes derived from different natural isolates appear to have very different recombination properties." He is currently investigating the basis of this difference. Whether it is related to other kinds of abnormalities in genetic results noted by Nanney (1963), particularly in strains B1 (now extinct) and D, remains to be discovered.

In view of this state of affairs, caution should also be exercised in regard to the lists of *un*linked loci entered in Table 2. Wherever the information is based on crosses involving strains that yield high frequencies of recombination, reexamination using other strains might reveal linkage. Unfortunately, the strains employed were not always known to me; when known, this information is included. I have deliberately omitted from Table 2 and from the sketch of linkage groups a few reports of possible loose linkage, especially those involving the locus of *r3*. These present difficulties too important to raise on so slender and dubious a basis.

Genetics of Mating Types

The main points on the genetics of mating types have been summarized by Nanney (1964) (see also pages 437, 441, and Table 2, locus *mt*). The most fully studied genes affecting mating types are the alleles mt^A and mt^B. Any one of the seven mating types (I–VII), except IV and VII, can be expressed by homozygotes for mt^A; any except I by homozygotes for mt^B; and the heterozygote can express any of the seven. For this locus, phenotypic assortment has two aspects: On the one hand, the heterozygote assorts as do other heterozygous loci; on the other hand, as mentioned earlier, homozygotes (and heterozygotes) can assort for the different mating types of the set producible by that genotype. The latter assortment is believed to be based on the complexity of the locus, only one of its elements—any one—being determined to be active in any subunit of the macronucleus at the time of determination of macronuclear heterogeneity for mating type.

The putative different elements of the *mt* locus differ in their probabilities of being activated (as inferred from frequencies of the mating types), and these probabilities are modifiable by the temperature prevailing until about the first cell division after conjugation, i.e., when subunit determination has been accomplished (Nanney, 1960a). A high degree of coordination appears to exist among the subunits of a macronucleus in regard to the mating type for which they are determined (Nanney and Allen, 1959): usually only one or two of the 5–7 genotypically possible mating types are found within a caryonide (see page 441). When subunits of a macronucleus are determined for different mating types, the different types later show phenotypic assortment. The two macronuclei in the same exconjugant cell show no correlation in determination. Consequently, with the possible exception of the differentiation within a caryonide, the mode of intraclonal inheritance of mating type resembles the type A or caryonidal system in *P. aurelia* (see chapter on Paramecium, this volume). It is to be emphasized that while determination occurs by the first or second cell division, there is no phenotypic expression until many cell generations later, when the clonal period of maturity is reached. There are indications that other alleles of the *mt* locus result in different probabilities of determination of the different possible mating types and/or that this result may be due to genes at other loci (Nanney, 1959*a*). Whether the *mt* locus is regulatory or whether it codes for mating-type substances remains unknown.

Genetics of Immobilization Serotypes

Most of the relevant information is given in Table 2 in the accounts of genes at the loci *H, T, St, r1, r3,* and *r4.* Here we note the serotypes for which genes have not been identified, some further information on those analyzed genetically, and some general features of the serotype system that integrate the separate gene accounts.

Margolin *et al.* (1959) recognized the existence of a set of mutually exclusive immobilization serotypes that are expressed under different conditions by cells of the same genotype, and noted the parallels to the serotype systems of *P. aurelia.* The serotypes of syngen 1 have later been named and studied in detail. One serotype, L, comes to expression at low temperature; one, H, at high temperature (20–30°C); and another, I, at these same high temperatures when the cells are grown in the presence of homologous anti-H serum. Juergensmeyer (1969) could find no serologic diversity among the L serotypes of different strains, hence no possibility of genetic analysis. She found arrays of diverse I serotypes appearing as successive transformations within single asexual lines of descent during continued growth in anti-H sera; but again strain differences susceptible to genetic analysis were not identified. Nanney (1960*b*), Nanney and Dubert (1960), and Nanney *et al.* (1963, 1964) identified and analyzed genetically a number of H serotypes dependent on a series of codominant alleles at a single locus (*H,* Table 2). In heterozygotes, there is a peck order of output ratios during phenotypic assortment. When present, H^E is the majority assortant; H^A is in the majority when combined with H^C or H^D. There is no regularity as to majority type in H^C/H^D heterozygotes. Unlike the large antigen molecules of *P. aurelia,* the H antigens are of low molecular weight (Bruns, 1971).

Phillips (1967*a,b*) reported on a "torrid" serotype expressed only at temperatures close to the maximum tolerated during growth in peptone media (see locus *T,* Table 2, for the 3 alleles at this locus). Grass

(1972*a,b*) found two alleles for a serotype expressed in media with high concentration of salt (*St*, Table 2).

A different aspect of the antigen system is shown by the r and q serotypes (Doerder, 1973, and private communication; McCoy, private communication). These are expressed under conditions which ordinarily yield some or all of the serotypes L, H, and T; but only in clones of certain genotypes. The r serotype is expressed in homozygotes for *r1* or *r3*. The q serotype is expressed in homozygotes for *r4*. (See these genes in Table 2.) It is supposed that these genes, at least *r1 and r3*, repress the expression of *L* and *H* and, in certain cases, *T* and *St*, permitting genes at some other unspecified antigen locus (not an *r* locus) to come to expression. The q serotype may have a different basis (McCoy, private communication).

Genetics of Mitochondria

Mitochondrial DNA is synthesized uniformly throughout the cell cycle (Parsons, 1965; Parsons and Rustad, 1968; Charret and André, 1968), even at times when none is synthesized in nuclei. Exposure to ^{3}H-thymidine for one complete cell cycle labels all mitochondria. The mitochondrial label is then conserved and found to be distributed randomly among all mitochondria present after two cell generations in cold medium (Parsons and Rustad, 1968). Mitochondrial DNA is double-stranded and linear (about 17.6 μm long, molecular weight about 33.8 × 10^6 daltons), *not* circular (Suyama, 1966; Suyama and Miura, 1968). There are estimated to be some 7 copies per mitochondrion. Although the mitochondrial DNA's of *T. pyriformis* and *P. aurelia* are similar in size and form, they differ in buoyant density (1.686 and 1.699 gcm^{-3}, respectively), indicating different percentages of G + C (25 and 40, respectively), and there is virtually no similarity in their base sequences as measured by coannealing in mixtures (Flavell and Jones, 1971*b*). Apparently, RNA of mitochondrial ribosomes is transcribed from mitochondria DNA (Suyama, 1967). Protein synthesis occurs *in vitro* on mitochondrial ribosomes and this is inhibited by chloramphenicol (Allen and Suyama, 1972). These studies and the well-documented evidence for mitochondrial mutations in *Paramecium* (see chapter on Paramecium, this volume) render plausible the mitochondrial localization of mutations (CA101, 102, 103) to chloramphenicol resistance accompanied by reduced growth rate in both CAM and CAM-free media reported by Roberts and Orias (1973*a*; see also page 437 on nonexchange of mitochondria during conjugation).

Genetics of the Cell Cortex

Summaries of the genetics of the cell cortex are given by Nanney (1968a, 1972), Allen and Gibson (1973), and Frankel and Williams (1973). As in *P. aurelia,* the occurrence of specifically cortical DNA is controversial. Some investigators, notably Randall and Disbrey (1965), have claimed to find DNA associated with ciliary basal bodies, but later studies (e.g., Pyne, 1968; Flavell and Jones, 1971a) have failed to confirm this and report that there is no specifically cortical DNA. The hereditary characteristics of the cortex are based partly on nuclear genes and partly on nongenic factors. Genic factors include *co* and the *mo* series of mutations (Table 2). Although the basic lesions resulting from those mutations remain unknown, they affect cell and cortical shape and cortical processes of prefission morphogenesis.

Nongenic factors have been documented chiefly by Nanney in a series of papers beginning in 1966. The chief genetic results of these studies (Nanney, 1966) concern the number of ciliary rows (kineties or meridians). The number (or corticotype) can vary greatly within a clone, from 15 or less to 35 or more; but the variations are more or less stable in asexual reproduction. The most stable numbers of kineties are in the range of 16 to 21. Change of kinety number in subclones of any one of these corticotypes occurs only 1–8 times per 1000 cell divisions. Cells with less than 16 or more than 21 rows show increasing frequencies of change (toward the more stable values) the further the number is from stable range. When crosses are made between clones with different numbers of kineties, the mates produce clones that continue to show the parental row numbers and the same rates of variation from it, i.e., the same stability patterns. Conjugation, which results in genomic identity and exchange of cytoplasm, does not alter the difference between the parents, even when they differ by only one row in the range of high stability. Moreover, the chief landmarks of the cell surface (the number of postoral kineties, the site of production of a new oral apparatus at cell division, the positions and number of contractile vacuole pores, and the position of the cytoproct) bear definite and remarkably constant geometrical relations to each other and/or to the total number of kineties. The further specifically morphogenetic analysis, of great interest in itself, is summarized in the reviews mentioned above.

Syngen 2

Elliott and Clark (1958b) have reported data suggesting that some wild stocks may be heterozygous for an incompletely recessive gene, *p,* which relieves the usual requirement for pyridoxine in defined media. According to McCoy (private communication), the immature period is usually 100–200 cell generations; at least 11 mating types are determined by alleles with serial dominance at a single locus, *mt*, which shows no phenotypic assortment in heterozygotes; at the only other known locus, *M*, for serotypes, there are similarly multiple (at least six) alleles with serial dominance and no phenotypic assortment; and the two loci (*mt* and *M*) show as little as 10 percent recombination in certain crosses and at least 45 percent (if linked at all) in other crosses. (See page 456 for variable linkage in syngen 1.)

Syngen 7

Five mating types are known in this syngen (Gruchy, 1955; Outka, 1961; Phillips, 1969). The immature period is from 70 to more than 100 cell generations (Phillips, 1969). The one stock in which mating-type I was found was sterile and thus could not be studied genetically. Phillips (1969) discovered that the other four mating types show a combination of genic and caryonidal inheritance. One allele, mt^A, permits either type II or type IV, and these alternatives are then caryonidally determined. The other allele, mt^B, permits either type III or type V, again with caryonidal determination of these alternatives. All of these 4 types can arise caryonidally in heterozygotes. Serotype genes (Phillips, 1971*b*) and phosphatase genes (Phillips, 1972) have also been reported. Two alleles, R^A and R^B, determine two intermediate-temperature (R, for room temperature), serotypes with R^A dominant and R^B recessive. These serotypes do not show phenotypic assortment. The R locus also has effects on expression of a W (warm temperature) serotype. $P1^A$, $P1^B$, and $P1^C$ are codominant alleles for electrophoretic variants of an acid phosphatase. These do show phenotypic assortment comparable to that of the $P1$ locus in syngen 1. Phillips' data show no indication of linkage among the loci *mt*, *R*, and *P1*.

Syngen 8

Orias (1963) reported for syngen 8 three mating types (I, II, III) determined by three alleles with serial dominance in the order mt^A (type I), mt^B (type III), mt^C (type II). Immature periods varied from 120 to 150 cell generations. Selfers produced lines stable for a single mating type, but this form of assortment was not further characterized. Orias' stocks are now sterile.

Syngen 9

Only one locus has been reported in this syngen (Elliott and Clark, 1958*a*). Homozygous recessives for gene *s* do not share the usual requirement for serine in defined media.

Syngen 10

Genetics of this syngen is due entirely to the work of McCoy (private communication) on stocks collected in Colorado by Doerder. Four mating types are determined by 4 alleles with serial dominance at the *mt* locus. The immature period is about 20 cell generations. Three acid phosphatases (*P1*, *P2*, and *P4*) with two alleles each and a serotype locus (*K*) with two alleles have been found. Phenotypic assortment in heterozygotes occurs for the 3 acid phosphatase loci, but not for mating types or serotypes. *P1* is linked to *mt* with about 25 percent recombination in one stock, and at least 45 percent recombination in another stock. (For variable linkage, see page 456 for syngen 1 and page 461 for syngen 2.)

Acknowledgment

Because I have not had first-hand experience with *Tetrahymena* genetics and because knowledge in this field is now increasing so rapidly, I had to call for help from a number of investigators: S. L. Allen, L. K. Bleyman, D. Borden, P. J. Bruns, B. Byrne, F. P. Doerder, J. Frankel, M. A. Gorovsky, L. Jenkins, J. W. McCoy, D. L. Nanney, and E. Orias. They responded generously by checking parts or all of the manuscript for accuracy and coverage and/or by acquainting me with unpublished work and giving permission to include it. If omissions and errors remain, the fault is not theirs.

Literature Cited

Allen, N. E. and Y. Suyama, 1972 Protein synthesis *in vitro* with *Tetrahymena* mitochondrial ribosomes. *Biochim. Biophys. Acta* **259**:369–377.

Allen, S. L., 1960 Inherited variations in the esterases of *Tetrahymena*. *Genetics* **45**:1051–1070.

Allen, S. L., 1963 Genomic exclusion in *Tetrahymena*: Genetic basis. *J. Protozool.* **10**:413–420.

Allen, S. L., 1964   Linkage studies in variety 1 of *Tetrahymena pyriformis*: A first case of linkage in the ciliated Protozoa. *Genetics* **49**:617–627.

Allen, S. L., 1965   Genetic control of enzymes in *Tetrahymena. Brookhaven Symp. Biol.* **18**:27–54.

Allen, S. L., 1967*a*   Genomic exclusion: A rapid means of inducing homozygous diploid lines in *Tetrahymena pyriformis,* syngen 1. *Science (Wash., D.C.)* **155**:575–577.

Allen, S. L., 1967*b*   Cytogenetics of genomic exclusion in *Tetrahymena. Genetics* **55**:797–822.

Allen, S. L., 1967*c*   Chemical genetics of Protozoa. In *Chemical Zoology,* edited by M. Florkin and B. T. Scheer. Vol. 1, Protozoa, edited by G. W. Kidder, pp. 617–694, Academic Press, New York.

Allen, S. L., 1971   A late-determined gene in *Tetrahymena* heterozygotes. *Genetics* **68**:415–433.

Allen, S. L. and I. Gibson, 1972   Genome amplification and gene expression in the ciliate macronucleus. *Biochem. Genet.* **6**:293–313.

Allen, S. L. and I. Gibson, 1973   Genetics of *Tetrahymena.* In *Biology of Tetrahymena,* edited by A. M. Elliott, pp. 307–373, Dowden, Hutchinson and Ross, Stroudsburg, Pa.

Allen, S. L. and P. H. T. Lee, 1971   The preparation of congenic strains of *Tetrahymena. J. Protozool.* **18**:214–218.

Allen, S. L. and C. I. Li, 1974   Nucleotide sequence divergence among DNA fractions of different syngens of *Tetrahymena pyriformis. Biochem. Genet.*: in press.

Allen, S. L. and D. L. Nanney, 1958   An analysis of nuclear differentiation in the selfers of *Tetrahymena. Am. Nat.* **92**:139–160.

Allen, S. L. and S. L. Weremiuk, 1971   Intersyngenic variations in the esterases and acid phosphatases of *Tetrahymena pyriformis. Biochem. Genet.* **5**:119–133.

Allen, S. L., M. S. Misch and B. M. Morrison, 1963   Genetic control of an acid phosphatase in *Tetrahymena*: Formation of a hybrid enzyme. *Genetics* **48**:1635–1658.

Allen, S. L., S. K. File and S. L. Koch, 1967   Genomic exclusion in Tetrahymena. *Genetics* **55**:823–837.

Allen, S. L., S. L. Weremiuk and C. A. Patrick, 1971   Is there selective mating in *Tetrahymena* during genomic exclusion? *J. Protozool.* **18**:515–517.

Andersen, H. A., 1972   Requirements for DNA replication preceding cell division in *Tetrahymena pyriformis. Expt. Cell Res.* **75**:89–94.

Andersen, H. A. and E. Zeuthen, 1971   DNA replication sequence in *Tetrahymena* is not repeated from generation to generation. *Expt. Cell Res.* **68**:309–314.

Andersen, H. A., C. F. Brunk and E. Zeuthen, 1970   Studies on the DNA replication in heat-synchronized *Tetrahymena pyriformis. C. R. Trav. Lab. Carlsberg* **38**:123–131.

Bleyman, L. K., 1971   Temporal patterns in the ciliated protozoa. In *Developmental Aspects of the Cell Cycle,* edited by I. L. Cameron, G. M. Padilla and A. M. Zimmer, pp. 67–91, Academic Press, New York.

Bleyman, L. K., 1972   A new spontaneous early mature mutation in *Tetrahymena pyriformis. Genetics* **71**:s5-s6.

Bleyman, L. K. and E. M. Simon, 1967   Genetic control of maturity in *Tetrahymena pyriformis. Genet. Res.* **10**:319–321.

Bleyman, L. K. and E. M. Simon, 1968   Clonal analysis of nuclear differentiation in *Tetrahymena. Devel. Biol.* **18**:217–231.

Bleyman, L. K., E. M. Simon and R. Brosi, 1966   Sequential nuclear differentiation in *Tetrahymena. Genetics* **54**:277–291.

Borden, D., 1972 Isozyme studies on *Tetrahymena pyriformis.* Ph.D. Thesis, Department of Zoology, University of Illinois, Champaign-Urbana, Ill.

Borden, D., E. T. Miller, D. L. Nanney and G. S. Whitt, 1973*a* The inheritance of enzyme variants for tyrosine aminotransferase, NADP-dependent malate dehydrogenase, NADP-dependent isocitrate dehydrogenase, and tetrazolium oxidase in *Tetrahymena pyriformis,* syngen 1. *Genetics* **74:**595–603.

Borden, D., G. S. Whitt and D. L. Nanney, 1973*b* Electrophoretic characterization of classical *Tetrahymena pyriformis* strains. *J. Protozool.* **20:**693–700.

Borden, D., D. L. Nanney and G. S. Whitt, 1974 Isozymic analysis of the evolutionary relationships of *Tetrahymena* syngens. *Evolution:* in press.

Bruns, P. J., 1971 Immobilization antigens of *Tetrahymena pyriformis.* I. Assay and extraction. *Expt. Cell Res.* **65:**445–453.

Bruns, P. J., 1973 Cell density as a selective parameter in *Tetrahymena. Expt. Cell Res.* **79:**120–126.

Bruns, P. J. and T. B. Brussard, 1974 Pair formation in *Tetrahymena pyriformis,* an inducible developmental system. *J. Expt. Zool.:* **188:**337–344.

Cameron, I. L. and J. R. Jeter, Jr., 1970 Synchronization of the cell cycle of *Tetrahymena* by starvation and refeeding. *J. Protozool.* **17:**429–431.

Carlson, P. S., 1971 Mutant selection in *Tetrahymena pyriformis. Genetics* **69:**261–265.

Charret, R. and J. André, 1968 La synthèse de l'ADN mitochondrial chez *Tetrahymena pyriformis.* Étude radiographique quantitative au microscope électronique. *J. Cell Biol.* **39:**369–381.

Cleffmann, A., 1968 Regulierung der DNA-Menge im makronucleus von *Tetrahymena. Expt. Cell Res.* **50:**193–207.

Corliss, J. O., 1973 Guide to the literature on *Tetrahymena*: A companion piece to Elliott's "General Bibliography." *Trans. Am. Microscop. Soc.* **92:**468–491.

Doerder, F. P., 1972 Regulatory serotype mutations in *Tetrahymena pyriformis,* syngen 1. Ph.D. Thesis, Department of Zoology, University of Illinois, Champaign-Urbana, Ill.

Doerder, F. P., 1973 Regulatory serotype mutations in *Tetrahymena pyriformis,* syngen 1. *Genetics* **74:**81–106.

Elliott, A. M., 1973*a* Life cycle and distribution of *Tetrahymena.* In *Biology of Tetrahymena,* edited by A. M. Elliott, pp. 259–286.

Elliott, A. M., editor, 1973*b* *Biology of Tetrahymena,* Dowden, Hutchinson and Ross, Stroudsburg, Pa.

Elliott, A. M. and G. M. Clark, 1958*a* Genetic studies of the serine mutant in variety 9 of *Tetrahymena pyriformis. J. Protozool.* **5:**240–246.

Elliott, A. M. and G. M. Clark, 1958*b* Genetic studies of the pyridoxine mutant in variety 2 of *Tetrahymena pyriformis. J. Protozool.* **5:**235–240.

Flavell, R. A. and I. G. Jones, 1970 Kinetic complexity of *Tetrahymena pyriformis* nuclear deoxyribonucleic acid. *Biochem. J.* **116:**155–157.

Flavell, R. A. and I. G. Jones, 1971*a* DNA from isolated pellicles of *Tetrahymena. J. Cell Sci.* **9:**719–726.

Flavell, R. A. and I. G. Jones, 1971*b* *Paramecium* mitochondrial DNA. Renaturation and hybridization studies. *Biochim. Biophys. Acta* **232:**255–260.

Frankel, J. and N. E. Williams, 1973 Cortical development in *Tetrahymena.* In *Biology of Tetrahymena,* edited by A. M. Elliott, pp. 375–409.

Gavin, R. H., 1965 The effects of heat and cold on cellular development in synchronized *Tetrahymena pyriformis* WH-6. *J. Protozool.* **12:**307–318.

Gibson, I. and N. Martin, 1971   DNA amounts in the nuclei of *Paramecium aurelia* and *Tetrahymena pyriformis. Chromosoma* **35**:374–382.

Grass, F. S., 1972*a*   An immobilization antigen in *Tetrahymena pyriformis* expressed under conditions of high salt stress. *J. Protozool.* **19**:505–511.

Grass, F. S., 1972*b*   Genetics of the St serotype system in *Tetrahymena pyriformis,* syngen 1. *Genetics* **70**:521–536.

Gruchy, D. F., 1955   The breeding system and distribution of *Tetrahymena pyriformis. J. Protozool.* **2**:178–185.

Hill, D. L., 1972   *The Biochemistry and Physiology of Tetrahymena,* Academic Press, New York.

Holz, G. G., 1960   Structural and functional changes in a generation in *Tetrahymena. Biol. Bull.* **118**:84–95.

Juergensmeyer, E. B., 1969   Serotype expression and transformation in *Tetrahymena pyriformis. J. Protozool.* **16**:344–352.

Loefer, J. B. and O. H. Scherbaum, 1963   Serological and biochemical factors relative to taxonomy of *Tetrahymena. Syst. Zool.* **12**:175–177.

Loefer, J. B., R. D. Owen and E. Christensen, 1958   Serological types among thirty-one strains of the ciliated protozoan *Tetrahymena pyriformis. J. Protozool.* **5**:209–217.

McCoy, J. W., 1972   Kinetic studies on the mating reaction of *Tetrahymena pyriformis,* syngen 1. *J. Expt. Zool.* **180**:271–278.

McCoy, J. W., 1973   A temperature-sensitive mutation in *Tetrahymena pyriformis,* syngen 1. *Genetics* **74**:107–114.

McDonald, B. B., 1966   The exchange of RNA and protein during conjugation in *Tetrahymena. J. Protozool.* **13**:277–285.

Margolin, P., J. B. Loefer and R. D. Owen, 1959   Immobilizing antigens of *Tetrahymena pyriformis. J. Protozool.* **6**:207–215.

Murti, K. G., 1973   Electron-microscopic observations on the macronuclear development of *Stylonychia mytilus* and *Tetrahymena pyriformis* (Ciliophora-Protozoa). *J. Cell Sci.* **13**:479–509.

Nanney, D. L., 1953   Nucleo-cytoplasmic interaction during conjugation in *Tetrahymena. Biol. Bull.* **105**:133–148.

Nanney, D. L., 1956   Caryonidal inheritance and nuclear differentiation. *Am. Nat.* **90**:291–307.

Nanney, D. L., 1957   Inbreeding degeneration in *Tetrahymena. Genetics* **42**:137–146.

Nanney, D. L., 1959*a*   Genetic factors affecting mating-type frequencies in variety 1 of *Tetrahymena pyriformis. Genetics* **44**:1173–1184.

Nanney, D. L., 1959*b*   Vegetative mutants and clonal senility in *Tetrahymena. J. Protozool.* **6**:171–177.

Nanney, D. L., 1960*a*   Temperature effects on nuclear differentiation in variety 1 of *Tetrahymena pyriformis. Physiol. Zool.* **33**:146–151.

Nanney, D. L., 1960*b*   The relationship between the mating type and the H serotype systems in *Tetrahymena. Genetics* **45**:1351–1358.

Nanney, D. L., 1962   Anomalous serotypes in *Tetrahymena. J. Protozool.* **9**:485–486.

Nanney, D. L., 1963   Irregular genetic transmission in *Tetrahymena* crosses. *Genetics* **48**:737–744.

Nanney, D. L., 1964   Macronuclear differentiation and subnuclear assortment in ciliates. In *The Role of Chromosomes in Development,* edited by M. M. Locke, pp. 253–273, Academic Press, New York.

Nanney, D. L., 1966   Corticotype transmission in *Tetrahymena. Genetics* **54**:955–968.

Nanney, D. L., 1967 Comparative corticotype analyses in *Tetrahymena. J. Protozool.* **14:**553–565.

Nanney, D. L., 1968*a* Cortical patterns in cellular morphogenesis. *Science (Wash., D.C.)* **160:**496–502.

Nanney, D. L., 1968*b* Ciliate genetics: Patterns and programs of gene action. *Annu. Rev. Genet.* **2:**121–140.

Nanney, D. L., 1971 Cortical characteristics of strains of syngens 10, 11 and 12 of *Tetrahymena pyriformis. J. Protozool.* **18:**33–37.

Nanney, D. L., 1972 Cytogeometric integration in the ciliate cortex. *Ann. N.Y. Acad. Sci.* **193:**14–28.

Nanney, D. L. and S. L. Allen, 1959 Intranuclear coordination in *Tetrahymena. Physiol. Zool.* **32:**221–229.

Nanney, D. L. and P. A. Caughey, 1953 Mating-type determination in *Tetrahymena pyriformis. Proc. Natl. Acad. Sci. USA* **39:**1057–1063.

Nanney, D. L. and P. A. Caughey, 1955 An unstable nuclear condition in *Tetrahymena pyriformis. Genetics* **40:**388–398.

Nanney, D. L. and M. Chow, 1974 Basal body homeostasis in *Tetrahymena. Am. Nat.* **108:**125–139.

Nanney, D. L. and F. P. Doerder, 1972 Transitory heterosis in numbers of basal bodies in *Tetrahymena pyriformis. Genetics* **72:**227–238.

Nanney, D. L. and J. M. Dubert, 1960 The genetics of the H serotype system in variety 1 of *Tetrahymena pyriformis. Genetics* **45:**1335–1349.

Nanney, D. L. and M. J. Nagel, 1964 Nuclear misbehavior in an aberrant inbred *Tetrahymena. J. Protozool.* **11:**465–473.

Nanney, D. L., P. A. Caughey and A. Tefankjian, 1955 The genetic control of mating-type potentialities in *Tetrahymena pyriformis. Genetics* **40:**668–680.

Nanney, D. L., S. J. Reeve, J. Nagel and S. De Pinto, 1963 H serotype differentiation in *Tetrahymena. Genetics* **48:**803–813.

Nanney, D. L., M. J. Nagel and R. W. Touchberry, 1964 The timing of H antigenic differentiation in *Tetrahymena. J. Expt. Zool.* **155:**25–42.

Nilsson, J. R., 1970 Suggestive structural evidence for macronuclear "subnuclei" in *Tetrahymena pyriformis* GL. *J. Protozool.* **17:**539–548.

Orias, E., 1960 The genetic control of two lethal traits in variety 1 of *Tetrahymena pyriformis. J. Protozool.* **7:**64–69.

Orias, E., 1963 Mating-type determination in variety 8, *Tetrahymena pyriformis. Genetics* **48:**1509–1518.

Orias, E., 1973 Alternative interpretation of the molecular structure and somatic genetics of acid phosphatase-1 in *Tetrahymena pyriformis. Biochem. Genet.* **9:**87–90.

Orias, E. and M. Flacks, 1973 Use of genomic exclusion to isolate heat-sensitive mutants in *Tetrahymena. Genetics* **73:**543–559.

Outka, D. E., 1961 Conditions for mating and inheritance of mating type in variety seven of *Tetrahymena pyriformis. J. Protozool.* **8:**179–184.

Parsons, J. A., 1965 Mitochondrial incorporation of tritiated thymidine in *Tetrahymena pyriformis. J. Cell Biol.* **25:**641–646.

Parsons, J. A. and R. C. Rustad, 1968 The distribution of DNA among dividing mitochondria of *Tetrahymena pyriformis. J. Cell Biol.* **37:**683–693.

Perlman, B. S., 1973 Temperature effects on maturity periods in *Tetrahymena pyriformis*, syngen 1. *J. Protozool.* **20:**106–107.

Phillips, R. B., 1967*a* Inheritance of T serotypes in *Tetrahymena. Genetics* **56:**667–681.

Phillips, R. B., 1967*b*   T serotype differentiation in *Tetrahymena. Genetics* **56**:683–692.

Phillips, R. B., 1969   Mating-type inheritance in syngen 7 of *Tetrahymena pyriformis*: Intra- and interallelic interactions. *Genetics* **63**:349–359.

Phillips, R. B., 1971*a*   Induction of competence for mating in *Tetrahymena* by cell-free fluids. *J. Protozool.* **18**:163–165.

Phillips, R. B., 1971*b*   Inheritance of immobilization antigens in syngen 7 of *Tetrahymena pyriformis*: Evidence for a regulatory gene. *Genetics* **67**:391–398.

Phillips, R. B., 1972   Similar times of differentiation of acid phosphatase heterozygotes in two syngens of *Tetrahymena pyriformis. Devel. Biol.* **29**:65–72.

Preer, J. R., Jr., 1969   Genetics of the Protozoa. In *Research in Protozoology*, Vol. 3, edited by T. T. Chen, pp. 129–278, Pergamon Press, New York.

Pyne, C., 1968   Sur l'absence d'incorporation de la thymidine tritiée dans les cinétosomes de *Tetrahymena pyriformis* (cilié holotriche). *C. R. Hebd. Seances Acad. Sci. Ser. D Sci. Nat. (Paris)* **267**:755–757.

Randall, J. T. and C. Disbrey, 1965   Evidence for the presence of DNA at basal body sites in *Tetrahymena pyriformis. Proc. R. Soc. Lond. Ser. B Biol. Sci.* **162**:473–491.

Ray, C., Jr., 1956   Meiosis and nuclear behavior in *Tetrahymena pyriformis. J. Protozool.* **3**:88–96.

Roberts, C. T. and E. Orias, 1973*a*   Cytoplasmic inheritance of chloramphenicol resistance in *Tetrahymena. Genetics* **73**:259–272.

Roberts, C. T. and E. Orias, 1973*b*   Cycloheximide-resistant mutant of *Tetrahymena pyriformis. Expt. Cell Res.* **81**:312–316.

Schensted, I. V., 1958   Model of subnuclear segregation in the macronucleus of ciliates. *Am. Nat.* **92**:161–170.

Simon, E. M., 1972   Freezing and storage in liquid nitrogen of axenically and monoxenically cultivated *Tetrahymena pyriformis. Cryobiology* **9**:75–81.

Sonneborn, T. M., 1950   Methods in the general biology and genetics of *Paramecium aurelia. J. Expt. Zool.* **113**:87–148.

Sonneborn, T. M., 1970   Methods in *Paramecium* research. In *Methods of Cell Physiology*, Vol. 4, edited by D. Prescott, pp. 241–339, Academic Press, New York.

Suyama, Y., 1966   Mitochondrial deoxyribonucleic acid of *Tetrahymena*: Its partial physical characterization. *Biochemistry* **5**:2214–2221.

Suyama, Y., 1967   The origins of mitochondrial ribonucleic acids in *Tetrahymena pyriformis. Biochemistry* **6**:2829–2839.

Suyama, Y. and K. Miura, 1968   Size and structural variations of mitochondrial DNA. *Proc. Natl. Acad. Sci. USA* **60**:235–242.

Wille, J., 1972   Physiological control of DNA nucleotide sequence redundancy in the eukaryote, *Tetrahymena pyriformis. Biochem. Biophys. Res. Commun.* **46**:677–684.

Wolfe, J., 1973   Conjugation in *Tetrahymena*: The relationship between the division cycle and cell pairing. *Devel. Biol.* **35**:221–231.

Woodard, J., E. S. Kaneshiro and M. A. Gorovsky, 1972   Cytochemical studies on the problem of macronuclear subnuclei in *Tetrahymena. Genetics* **70**:251–260.

Zeuthen, E., 1964   A temperature-induced division synchrony in *Tetrahymena*. In *Synchrony in Cell Division and Growth*, edited by E. Zeuthen, pp. 99–175, Interscience, New York.

20

*Paramecium aurelia**

Tracy M. Sonneborn

General Introduction
Early History of *Paramecium* Genetics

A year or two *before* Morgan began his classical investigations on the genetics of *Drosophila*, Jennings—looking for a favorable organism for genetic research and appreciating the potential values of microorganisms for such work—began to study the genetics of *Paramecium*. Two decades later, he (Jennings, 1929) critically summarized—with full bibliography—the genetic results obtained in research on *Paramecium* and other uni-cellular organisms, including bacteria. During this period the work on *Paramecium* was, for technical reasons, restricted to two areas; but from each, important generalizations emerged. Studies in the first area, the genetics of asexual reproduction, led Jennings to put forth the basic con-cept of the genetic uniformity of the clone, i.e., the Pure Line Theory for Asexual Reproduction. Studies in the second area, the genetics of sexual reproduction (conjugation), led Jennings to calculate and generalize the Mendelian results expected from various systems of breeding. The publications that developed these formulae were among the earliest (1912–1917) basic contributions to mathematical and population genetics. [For full bibliography of Jennings' publications, see Sonneborn (1974a).]

Jennings was driven to these theoretical explorations because he was searching for an explanation of his puzzling observation that hereditary variability continued to arise with little or no diminution in successive generations of close inbreeding of *Paramecium*. He had to conclude that Mendelian recombination may not be the whole of the matter, as also did his contemporary, Jollos. This conclusion was verified many years later,

*Contribution 970 from the Zoology Department, Indiana University.

Tracy M. Sonneborn—Indiana University, Bloomington, Indiana.

but nevertheless the obsevations were largely brought into line with knowledge of the gene by the demonstration that they were explicable in terms of nuclear differentiation and gene regulation, phenomena that underlie much of what initially appeared to be very puzzling cases of hereditary diversities between clones of identical genotype. It was not possible to discover or analyze these phenomena in the early period because it was not possible to carry out genetic analysis by crossbreeding. Studies of conjugation were restricted to matings within a uniform strain.

This stringent restriction of genetic analysis was removed by the discovery of mating types in *Paramecium* (Sonneborn, 1937*a*), which for the first time in any unicellular animal provided a simple routine procedure for crossbreeding different genotypes. Of course this quickly led to the demonstration of simple genic inheritance. Nevertheless, the old suspicion that the genetics of *Paramecium* was peculiar was strengthened by the concentration of Sonneborn and his co-workers on extranuclear heredity. Eventually, however, most of these peculiarities were traced to nuclear differentiations, symbionts, and other phenomena in the mainstream of genetics; and their analysis has provided and continues to provide—as this review will attempt to set forth—fundamental insights into major current problems of genetics such as radiation genetics and mutagenesis, molecular and developmental genetics, symbiont genetics, genetics of biological clocks, aging, organelle genetics, mitochondrial genetics, and of course nuclear differentiation and gene regulation.

The early work was summarized by Sonneborn (1947*a*, 1957) and by Beale (1954). These three full and detailed reviews, especially Beale's, are still indispensable sources of information and of references to the original literature because later reviews refer back to them instead of restating much of importance that is in them. Among the more recent reviews, J. R. Preer's (1969) review is a broad-gauged account of the genetics of the Protozoa and of course includes an account of *P. aurelia*; Gibson's (1970) review does so less fully. Methods have been assembled by Sonneborn (1950*a*, 1970*a*) and by Hanson (1974); axenic culture methods by van Wagtendonk and Soldo (1970). General implications have been discussed by Nanney (1968) and Sonneborn (1967, 1970*b*). Wichterman's (1953) book, while old, is still a useful reference for the general biology of *Paramecium*. A current review of the biology of *Paramecium*, edited by van Wagtendonk (1974), has a number of chapters on special topics of genetic interest. Other more specialized reviews will be mentioned later.

Varieties, Syngens, and Species

Like other named "species" of *Paramecium* and other Ciliates, *P. aurelia* is not a single biological species but a group of them. Some members of this group, cannot interbreed, and those that can interbreed yield either nonviable or sterile hybrids (Sonneborn, 1938a; Sonneborn and Dippell, 1946; Melvin, 1947; Levine, 1953b; Butzel, 1953a, Haggard, 1972, 1974). The genetic situation has not been changed by the discovery of chemical means of bringing about conjugation between these biological species, even between those whose mating types do not interact (Miyake, 1968), for this has not led to demonstration of gene flow between them. Although recognizing from the start, and repeatedly reiterating thereafter, the reality of these biological species, Sonneborn emphasized the futility and inexpedience of conferring species names on them because their identification was too difficult to be routinely practicable; living cultures of each of the many known mating types and considerable technical knowledge for handling them were required. Therefore, each biological species was assigned a number and referred to as a variety or, more appropriately, a syngen. The name *P. aurelia* was retained for all of them.

The situation has recently changed. Almost all of the syngens can be identified without recourse to any standard living cultures: 13 of the 14 syngens can be routinely identified by the electrophoretic patterns of a small number of cytoplasmic and mitochondrial enzymes (Tait, 1970a; Allen and Gibson, 1971; Allen *et al.*, 1971, 1973). The time has come, therefore, as others [(e.g., Hairston (1958)] have long maintained—prematurely, as it seemed to me—and as is currently being urged with good reason by others [e.g., by Adams and Allen (1974)], to recognize each syngen formally as a species, which implies conferring on it a species name. Elsewhere (Sonneborn, 1975) I have proposed to do this for at least some of the syngens. Meanwhile, in this review I shall refer to the syngens 1–14 as species 1–14.

The immediately following sections deal with general biological, cytogenetic, and genetic topics. Later sections deal with the distinctive features and genetics of each species, in numerical order. The final section deals with matters that concern two or more species such as interspecific crosses and population genetics.

General Biological Background

The cells of *P. aurelia* are cigar-shaped, heavily ciliated, 100–150 μm long, tapered bluntly at both ends, and roughly one-fourth to one-third as wide as they are long. All *P. aurelia* are confined to still or running fresh waters. No cysts or other specially resistant forms are known to occur. In the laboratory, however, both wild-type and mutant *P. aurelia* can be frozen, stored frozen for years and then thawed, and clones with apparently unaltered phenotype and genotype can be grown from thawed specimens (Simon and Schneller, 1973; Schneller, private communication). How *P. aurelia* becomes disseminated in nature is still largely unknown, but carriage externally by birds and other animals and by overflowing waters has been implicated.

Each *P. aurelia* normally contains one macronucleus, usually located centrally and close to the food-intake system, and two micronuclei closely adjacent to the macronucleus. The semirigid cortex of the cell, roughly 1–2 μm thick, is bounded externally by the cell or plasma membrane, which also bounds the approximately 5000 cilia. The cilia are arranged in some 60–80, more or less longitudinal rows (kineties) which are oriented and positioned in a virtually invariable normal pattern (Sonneborn, 1963; Gill and Hanson, 1968; Kaneda and Hanson, 1974). Each cilium (or pair of adjacent cilia in the same row) lies slightly to the right of the center of a repeating structural unit known variously as the ciliary corpuscle (Ehret and Powers, 1959), the kinetosomal territory (Pitelka, 1969), and, in papers from my laboratory since 1965, simply as the cortical unit. The cell surface is completely covered by some 4000 of these contiguous units, each with a complex, asymmetric, but regular, ultrastructure that has been described for *P. aurelia*, along with other features of the cell cortex, by Dippell (1965, 1968), Allen (1971), Hufnagel (1969a), Jurand and Selman (1969), Kaneda and Hanson (1974), and Ehret and McArdle (1974). The units, arranged in longitudinal rows like the cilia they contain, are bounded by ridges to form squares, rectangles, or hexagons. Near the middle of the transverse ridges is a specialized opening in which the tip of a trichocyst is located. The surface is interrupted near the equator of the cell by an invagination, the vestibule, from which the wall of the buccal cavity continues inward. Between the vestibule and posterior pole there is a modification of the cell surface, the cell anus or cytoproct, through which undigested remains of food vacuoles are eliminated. On the right dorsal side the cell surface bears anterior and posterior contractile vacuole pores.

Nuclei, Chromosomes, and Nuclear DNA

The two micronuclei, about 3 μm in diameter, are diploid and divide mitotically, but the nuclear membrane remains intact during mitosis. Each *wild stock,* i.e., *a stock descended from a single wild cell,* has a unique karyotype (Dippell, 1954; Jones, 1956; Kościuszko, 1965). The chromosomes are small, about 0.3–1.5 μm long during prometaphase of the first meiotic division. The haploid numbers of different stocks range from about 30 to 63. The amount of DNA in a haploid set has been calculated by different methods to be about 2.8–3.54 × 10^{-13}g, with no clear difference among the stocks and species (1, 2, 4, 7, 8) examined (Behme and Berger, 1970; Gibson and Martin, 1971). The fact that these values fall within the range of values reported for a genome of *Drosophila* is so surprising that it suggests the desirability of further investigation.

The macronucleus divides amitotically without breakdown of its membrane. The products of division may have unequal amounts of DNA and such differences may persist for some cell generations, but eventual regulation by unknown mechanisms appears to occur (Kimball, 1967). The number of chromosomes and their organization in the macronucleus have not been clearly ascertained by direct observation. Kimball (1953a) observed many fine filaments (near the limit of resolution of the optical microscope) emerging from macronuclei broken by compression and suggested that these filaments may have been free chromosomes. Wolfe (1967) examined electron microscopically the contents of broken macronuclei, and reported that bodies containing about the amount of DNA of one chromosome are attached to each other by 100-Å fibrils which are frequently paired. He interpreted his observations as supporting Sonneborn's (1940) hypothesis that the chromosomes of the macronucleus are organized into diploid subunits. This hypothesis, if correct, would reconcile amitosis with the genetic constancy of the clone and the maintenance of the genotype (even when heterozygous) through macronuclear regeneration from a small piece of macronucleus (see page 482). It conceives of amitosis as the segregation to daughter nuclei of numerous precisely replicating identical diploid genomes. An alternate hypothesis [Kimball (1953a); for the mathematics of this hypothesis, see Kimura (1957) and J. R. Preer (1969) pages 151–154] holds that free, unassociated chromosomes of the haploid set are present in such high multiples in the macronucleus that their random transmission to daughter nuclei would result in such slowly accumulating unbalance as to agree with the genetic observations and the facts (see page 476) of the clonal life cycle. The problem of the chromosomal content and organization of the macronucleus remains unsolved.

The amount of DNA in the macronucleus has been calculated to fall between 1100 and 2940 × 10^{-13}g (Gibson and Martin, 1971; Allen and Gibson, 1972; Soldo and Godoy, 1972) for various stocks of species 1, 2, 4, 7, and 8. The two extreme values were calculated for the same stock (299) of the same species (8) by different investigators (Soldo and Godoy, Gibson and Martin) using different methods and different assumptions. These values and those given above for the micronucleus set wide limits, between 155 and 525, for the ratio of the amount of DNA in the macronucleus to the amount in the micronucleus. Measures of this ratio during G1 (or corrected to G1) in species 1, 2, 4, and 8 run close to 430 (Woodard *et al.*, 1961, 1966; Behme and Berger, 1970; Gibson and Martin, 1971; Allen and Gibson, 1972; Berger, 1973*a*). This value has often been taken to show that the macronucleus contains about 430 diploid or 860 haploid sets of chromosomes. The macronucleus has consequently been considered to be highly polyploid or polygenomic. The validity of this characterization depends on at least two unsettled questions: (1) which, if either, of the two hypotheses of macronuclear organization mentioned in the preceding paragraph is correct and (2) whether all or only part of the micronuclear genome is present and replicated in the macronucleus.

The latter question has been attacked by efforts to ascertain the relation between the kinetic complexity or genome size of the macronuclear DNA and the genome size of micronuclear DNA calculated from cytophotometric comparison with a known reference material, bull sperm DNA. With this approach, Allen and Gibson (1972) reported the macronuclear genome size in species 1, 2, and 8 to be about $1.84–2.57 \times 10^{11}$ daltons, which is *not* smaller than the value they obtained photometrically for micronuclear genome size. Soldo and Godoy (1972), working on one of the same stocks (299 in species 8), reported a much smaller kinetic complexity—about 0.64×10^{11} daltons—for macronuclear DNA, but did not compare this with micronuclear genome size. Soldo and Godoy also reported more genomes per macronucleus (1680 vs. 860), a higher percentage of the genome in unique sequences (96 vs. 85), and more copies of the repeated sequences per genome (100 vs. 50–75); in addition, they found for the minor DNA component (repeated sequences) a kinetic complexity of 1.45×10^7 daltons. Both studies agreed that the main macronuclear component consists of about 29 percent G+C, but Cummings (1972) has reported 23 percent for species 1, 4, 13. Soldo and Godoy found 35 percent G + C in the repeated sequences. Until discrepancies among the studies are resolved or until hybridization studies are made between DNA from the two kinds of nuclei, the question of whether the macronucleus of *P. aurelia* contains complete or only partial micronuclear genomes remains open. In *Tetrahymena*, hybridization studies indicate that virtually entire micronuclear genomes are in the macronucleus (see page 436, this volume). In both organisms, it is possible that the unit of macronuclear structure, if there is such a unit, contains a haploid, not diploid, set of chromosomes.

The two kinds of nuclei play very different roles: the macronucleus has somatic functions; the micronucleus, at least principally, germinal functions. The phenotype of the cell is expressed, at least mainly, through the action of macronuclear genes; but only the micronucleus can undergo meiosis and fertilization. Heterocaryons carrying a dominant gene in the micronucleus and its recessive allele in the macronucleus exhibit the recessive phenotype (Sonneborn, 1946*b*, 1954*b*; Pasternak, 1967). That this may result merely from the great difference between the two types of nuclei in gene dosage, not from the total inactivity of micronuclear genes, is suggested by Pasternak's (1967) evidence for micronuclear RNA synthesis; but Nobili (1962) has presented evidence against micronuclear phenotypic activity. Amicronucleate cells can live and reproduce asexually and can undergo all but the micronuclear events associated with sexual processes. On the other hand, amacronucleate cells (produced frequently by cells of certain genotypes) quickly cease to grow, become unable to mate, and die within 2 days, nearly always without undergoing a single division (Sonneborn, 1954*a*; Nobili, 1961, 1962). Death is not due to mere starvation, for nucleated cells unable to feed or deprived of food can live for weeks. Developmental potentials of the two kinds of nuclei are strikingly different: micronuclei regularly develop into macronuclei after fertilization, but origin of a micronucleus from a macronucleus has never been reported for *Paramecium* (although it has been reported for certain other Ciliates as a rare event under extraordinary circumstances). Except as already noted, each kind of nucleus reproduces true to its kind.

The Cell Cycle

Generation time under standard conditions used in genetic work [27°C in bacterized media, see Sonneborn (1950*a*, 1970*a*)] is about 5 hours in species with small cell size, and about 6–8 hours in species with larger cell size. The periods of DNA synthesis have been reported only for species 4. Micronuclear DNA synthesis begins about halfway through the cell cycle and is completed by 0.7 of the cycle (Woodard *et al.*, 1961; Pasternak, 1967). G2 must be very short because according to Woodard *et al.* (1961), the micronuclei are already in mitotic telophase at about 0.9 of the cycle. They complete division before cell division.

There are minor differences in accounts of the macronuclear DNA cycle (Woodard *et al.*, 1961; Kimball and Barka, 1959; Kimball and Perdue, 1962; Pasternak, 1967; Berger, 1971; Smith-Sonneborn and Klass, 1974), especially as to when DNA synthesis begins: some hold that it begins at about 0.2 of the cell cycle, others that it begins 1–1.5 hours into the cycle regardless of the length of the cycle within the range of 4.5–8 hours, and others that it does not begin until about 0.5 of the cycle. All agree that it continues until the start of macronuclear division. There is thus no macronuclear G2 period.

The cytoplasmic events of the cell cycle include production of many cellular parts such as mitochondria, trichocysts, ciliary basal bodies and cilia, the repeating units of cortical structure, and the more conspicuous organelles such as the oral apparatus (which originates in the right wall of the vestibule), the cytoproct, and the contractile vacuole system. Most of these events (including the surfacing of all new basal bodies) occur during the last third of the cell cycle, actual constriction into two cells occupying only about the last 20 minutes of the cell cycle. Although these prefission events begin about the time that micronuclear DNA synthesis ends, that is surely not the signal for their initiation because they occur even in amicronucleate cells. Details of cytoplasmic events of the cell cycle are given by Sonneborn (1963, 1970b, 1974b), Dippell (1965, 1968), Gill and Hanson (1968), and Kaneda and Hanson (1974).

The Clonal Cycle

A clonal cycle begins with fertilization (autogamy, conjugation or cytogamy), passes through a series of diverse periods, and ends in clonal death. This has been demonstrated in species 1, 2, and 4 and probably occurs in all species of *P. aurelia* (Sonneborn, 1954c, 1955a, 1957). Under standard conditions of daily reisolation culture (27°C with excess bacterized organic media), clones die after, at most, about 350 cell generations. Except for the first 2 cell generations after fertilization, which are unusually long, generation time is minimal in young clones and increases as the clone ages. It is impossible to avoid ultimate extinction by selection for short generation time or any other obvious index of vigor or normality.

The first period of clonal cycles that begin with conjugation is immaturity, i.e., inability to mate and also inability to undergo autogamy. This period lasts 25–35 generations or more in many stocks of certain species, but its length is inversely proportion to the age of the parental clone (Siegel, 1961). Immaturity is totally lacking in certain species (e.g., species 4 and 8) and also in clones initiated with autogamy in all species.

The boundaries of the next period of the clonal cycle, maturity, are defined in *P. aurelia* by the relative timing of two responses to depletion of the food supply. During maturity, the cells respond by becoming capable of mating. If they lack an appropriate partner and thus do not mate, they soon cease to be capable of mating and may proceed to undergo autogamy [see Beisson and Capdeville (1966) for analysis of this sequence]. Autogamy cannot be induced in the early part of the period of maturity, only later. As the period of maturity continues, the time intervening between mating reactivity and autogamy during food depletion becomes progressively less. Then comes a time when this stimulus induces autogamy *before* the cells have become capable of mating; they do not become capable of mating until after autogamy is completed. The onset of this response to food depletion defines the end of maturity and the beginning of senescence. The duration of maturity varies among different stocks and species, from very few cell generations (as in stock 18, species 1, for clones arising at conjugation) to 75 or more generations (Sonneborn, 1957).

Senescence is the longest period of the clonal cycle and is marked by a number of progressive changes which are by no means synchronized in the various sublines of a clone. At first, autogamy is inducible in up to 100 percent of the cells within a day after the cells reach stationary phase; and it may even occur in constantly overfed, daily isolation lines. Later, autogamy slowly becomes more and more difficult to induce. Capacity to mate may appear sporadically, especially when autogamy is harder to induce. Generation time gradually increases from about 5 to about 24 hours and then is soon followed by death.

A number of other progressive changes occur during senescence. There is an increase in the frequency of abnormalities in prefission morphogenesis, in growth and division of parts of the cell, and in micronuclear mitoses (Dippell, 1955; Sonneborn and Dippell, 1960a). The frequency of cell death increases (Sonneborn, 1954c). The number of micronuclei per cell becomes more variable, with an increasing frequency of cells with no micronuclei as well as with 3 or 4 micronuclei (Mitchison, 1955; Dippell, 1955). The micronuclei become loaded with chromosomal aberrations (Dippell, 1955). The activity of DNA repair enzymes decreases (Smith-Sonneborn, 1971, 1974), as does the rate of synthesis of macronuclear DNA (Smith-Sonneborn and Klass, 1974) and the total DNA per macronucleus (Schwartz and Meister, 1973). At autogamy and conjugation, there is increasing frequency of supernumerary postzygotic micronuclei, of macronuclear regeneration (see page 482), and, during conjugation, of failure of crossfertilization (Mitchison, 1955). Great variation appears among exautogamous clones from senescent parents, fewer and fewer becoming "rejuvenated" (Sonneborn, 1954c, 1955b). The mean clonal life span of autogamous progeny is reported to decrease with age of the parent (Smith-Sonneborn et al., 1974). The frequency of production of nonviable progeny at autogamy rises from practically 0 to 100 percent (Sonneborn, 1935, 1954c; Pierson, 1938; Sonneborn and Schneller, 1955a,b, 1960b,c; Mitchison, 1955), even when the prezygotic macronucleus is retained (Mitchison, 1955). Thus, senescence is accompanied by increasing defects in the macronuclei, micronuclei and cytoplasm.

The question of whether there is a primary cause for these senescent progressions and, if so, what it is, has been much studied and discussed. Fauré-Fremiet (1953), Sonneborn (1954c, 1955a), and Siegel (1961) have all concluded that senescence results from changes in or deterioration of the macronucleus. Kimball (1953a) proposed that deterioration is based on increasing chromosomal unbalance resulting from random distribution of chromosomes to daughter nuclei during amitotic divisions. This view seemed at first to receive some experimental support (Sonneborn and

Schneller, 1955c), but further observations showed this was illusory (Sonneborn et al., 1956).

The results obtained by Sonneborn and Schneller (1960a) led them to conclude that the primary senescent changes are in the cytoplasm, not the nuclei. The cytoplasm of aged clones induces macronulcear abnormalities (Mitchison, 1955) and can quickly bring about lethal damage to normal micronuclei introduced from cells of young clones (Sonneborn and Schneller, 1960b,c). The hypothesis of a primary cytoplasmic basis of aging provides an explanation for the marked difference between *Tetrahymena pyriformis* and *P. aurelia* with respect to clonal senescence. Considerable and important parts of the cortex of *P. aurelia*, once formed, are persistent and nonrenewable during asexual reproduction (Siegel, 1970); but all areas of *T. pyriformis* grow and even the oral apparatus is replaceable when damaged. *T. pyriformis* does not show either macronuclear or cytoplasmic senescence, and clones can apparently be immortal. If macronuclear damage were primary in aging, it would be difficult to explain why the macronucleus of *P. aurelia* suffers from it, but not the macronucleus of *T. pyriformis*.

Autogamy

Autogamy, earlier mistakenly described under the name of *endomixis*, occurs in single, i.e., unpaired, cells. Both micronuclei undergo two meiotic divisions and 7 of the resulting 8 haploid nuclei disintegrate. The remaining haploid nucleus divides mitotically into genically identical "male" and "female" gamete nuclei, which then unite to form the completely homozygous, diploid fertilization nucleus, the *syncaryon*. The syncaryon divides twice, 2 of the 4 products becoming micronuclei, the other 2 growing into new macronuclei through intermediate stages called macronuclear anlagen. As in *Tetrahymena*, the macronuclear anlagen do not have stages of polytenization or massive loss of DNA; the DNA simply undergoes several successive duplications (Berger, 1973a). At the first postzygotic cell division these two anlagen do not divide but are segregated, one to each daughter cell. The asexual progeny of each of these two cells constitutes a *caryonide*, i.e., *a subclone with all macronuclei derived from a single ancestral macronucleus*. Beginning at the second cell division, the now fully developed macronucleus divides and does so regularly thereafter. The two micronuclei divide at both the first and all subsequent cell divisions. Some variations on these processes are known; for example in stock 18 of species 1, about 20 percent of the autogamous (and conjugant) cells form more than 2 (up to 11) macronuclear anlagen which segregate at the early fissions until there is only one per cell; thereafter they divide at cell divisions. In such cases there are thus more than two caryonides from a fertilized cell, and the caryonides have their origin at the second to fourth postzygotic cell division.

At the start of autogamy, the cells—like all vegetative cells—have a macronucleus. To distinguish it from the new or postzygotic macronuclei that develop from products of the syncaryon, the macronucleus present at the start is referred to as the prezygotic or old macronucleus. By the syncaryon stage of autogamy, the prezygotic macronucleus has disintegrated into many more or less spherical macronuclear fragments. Their fate depends on nutritive conditions. If the cells are provided with abundant food after autogamy, the macronuclear fragments persist until about the seventh cell generation, their mean number per cell being about halved at each fission as a result of random distribution to daughter cells. Until they disintegrate or disappear, beginning at about the seventh generation, fragments of the macronucleus continue to undergo RNA and protein synthesis, but not DNA synthesis, which rapidly declines (Berger, 1973*b*).

If the cells are starved after autogamy and so do not grow or divide, the DNA of the fragments is utilized by the growing postzygotic macronuclear anlagen, and the fragments gradually disappear until all are gone, in about 7–10 days (Berger, 1974). Under such conditions, the two postzygotic macronuclear anlagen may fuse into one, a process that is useful in attacking certain genetic problems (Nanney, 1954; page 280). In this case, on restoring food, the macronucleus divides at the first cell division, thus yielding only one instead of two caryonides from a fertilized cell.

The nuclear processes of autogamy are regularly accompanied by important extranuclear events. A small bulging of the cortex, known as the paroral cone, develops near the right wall of the vestibule. Normally one haploid product of meiosis becomes located in this cone, and there it undergoes the mitotic division that yields the "male" and "female" gamete nuclei. Also in the right vestibular wall area (as during the morphogenesis preceding fission), a new oral apparatus arises and (unlike what happens at fission) replaces the old one, which disintegrates and disappears (Roque, 1956). Whether other parts of the ventral cortex undergo structural changes (e.g., loss of cilia in certain areas) has not been reported, but this is suggested by the disappearance of ciliary mating reactivity during autogamy (as during cytogamy and conjugation when cilia *are* lost from certain ventral areas). The probable occurrence of other important extranuclear events is indicated by the directed movements of the nuclei and the differing developmental fates of those in different regions of the cell (Sonneborn, 1954*d*).

Conjugation

Conjugation is a process of fertilization that takes place while pairs of cells adhere temporarily to each other. The cells then separate, and each one gives rise to a clone by repeated fissions. The nuclear events during conjugation normally include reciprocal crossfertilization, but sometimes each conjugant fertilizes itself. The latter is referred to as *cytogamy*. In cytogamy, both mates undergo autogamy. The frequency of cytogamy is usually less than 5 percent, but its frequency is increased at both high and low temperatures and in media high in calcium concentration (Sonneborn, 1941*a*; Hallet, 1972, 1973). It is apt to occur with very high frequency in crosses involving misshapen mutants or cells differing greatly in size, and in crosses between species. In general, cytogamy occurs when the paroral cones of the two mates fail to make or to maintain contact, but it can also occur when the paroral cones seem to be in contact. Methods of selecting against cytogamous pairs are given by Sonneborn (1950*a*, pages 116–118).

The normal course of nuclear events during crossfertilizing conjugation is identical with the events of cytogamy with one exception: the fertilization nuclei are formed not by union of the male and female gamete nuclei of the *same* cell, but by union of the male nucleus of each cell with the female nucleus of the other cell. In other words, the male nuclei pass through the paroral cones into the mate and achieve crossfertilization. That is "true" conjugation. Although this is what usually happens, the only way to be sure that crossfertilization has occurred is to employ appropriate gene markers. The two clones being crossed should each be homozygous for a different recessive (or codominant) allele which is not carried by the other clone. If the two members of a conjugant pair both produce clones that are phenotypically dominant (or heterozygous for codominants) at both marker loci, then true conjugation has occurred.

The main requirement for making matings is to place together samples of cultures of complementary mating types (see page 485) under conditions in which they are sexually reactive. The chief conditions for reactivity are clonal age and nutritive state: the clones must be past the period of immaturity and preferably in the period of maturity; the culture should be neither overfed nor long-starved, but in passage from log growth into stationary phase. In addition, there are temperature optima and ranges which differ for different species, and some species have circadian rhythms of reactivity (Sonneborn, 1950*a*; pages 114–116). It is also possible to induce pairing between cells of the same mating type, even cells of the same clone, by exposing them to properly killed cells of the complementary mating type (Metz, 1947), to cell-free preparations of cilia from reactive cells of the complementary mating type (Miyake, 1964; Cronkite, 1974), to certain chemicals [see Miyake in Sonneborn (1970a) pages 287–291], or to certain killers (stocks 5 and 7). These methods have been little used in genetic work because of low frequencies of cross-fertilization and survival.

The control of cytoplasmic exchange between conjugants [for methods, see Sonneborn (1970a) pages 294–296] is of critical importance in analysis of cytoplasmic roles in heredity and in regulating genic action. Usually mates separate suddenly and quickly soon after the first division of the syncaryon, but sometimes they remain united longer at or near the paroral region. The duration of this union provides a rough measure of cytoplasmic exchange between mates (Sonneborn, 1946a). The bridge connecting the two mates may be very narrow and consist only of cortex, or it may be broad and include endoplasm, which can be seen to flow from each mate to the other. Wide bridges may become narrower and disappear or, while narrow, the two mates may pull apart. In the latter case, one mate may acquire part of the cortex of its partner, and this may have hereditary consequences (Sonneborn, 1963). Narrow bridges may become wider, and wide bridges may persist permanently. Before becoming wide and permanent, the bridge may be twisted so that the two mates are in, and remain permanently in, heteropolar orientation, or the two cells may become rigidly held in homopolar orientation. Heteropolar unions are useful in obtaining clones with inverted rows of cortical structure (Beisson and Sonneborn, 1965). Homopolar unions yield clones of homopolar doublets, much used in genetic work (Sonneborn, 1946a, 1963; Hanson, 1955, 1962; Butzel, 1968, 1973). Doublets have a common, unpartitioned endoplasm and, after several generations, a single macronucleus.

Doublets can accept two singlet mates, one on each of their two ventral surfaces. They can also accept two doublet mates, each of which can accept another doublet. In this way, up to 9 doublets in a row may conjugate with each other. If a doublet finds only one mate, conjugation occurs on its mated side, and autogamy occurs simultaneously on its un-mated side (Chun, 1969).

Various abnormal unions between cells of complementary mating types are sometimes encountered. An extra cell may be united by its anterior ventral surface to the posterior ventral surface of one member of a normally oriented pair. Both members of a pair may have such an extra partner. The free posterior ventral surface of one or both of these extra partners may have another cell adhering by its anterior ventral surface. Even longer chains of extras may be attached in this way. (Such chains are induced by strong killers of stocks 5 and 7 of species 2.) Sometimes the anterior ventral surface of a pair may have an extra cell attached in this

way. Occasionally, two normally oriented pairs are attached by their anterior ends. In all of the abnormal unions, only cells attached paroral
cone-to-paroral cone can crossfertilize; all others undergo autogamy.

Macronuclear Regeneration

Macronuclear regeneration (MR) is the development of functional
macronuclei from fragments of the prezygotic macronucleus after autogamy, cytogamy, or conjugation (Sonneborn, 1940, 1947a). MR occurs
whenever a cell contains one or more intact macronuclear fragments but
lacks a macronucleus or has a grossly defective macronucleus or macronuclear anlage. MR and normal lines of descent can arise from the same
fertilized cell when normal macronuclear anlagen or macronuclei are mis-
distributed at an early cell division so that some cells come to contain only
one or more intact fragments while others have both fragments and
postzygotic macronuclei. Cells containing only fragments undergo MR;
sister cells with postzygotic macronuclei undergo normal postzygotic
processes. Having identical micronuclei and, at least initially, identical cytoplasms, these two kinds of lines of descent provide the perfect control for
certain types of genetic analysis [Sonneborn (1954d) page 319].

In a cell lacking a macronucleus, fragments begin to grow and
continue to do so as their number per cell is reduced by random distribution to daughter cells in the course of cell divisions. When the
number is reduced to one per cell, the fragments have attained normal
macronuclear size. Thereafter, they divide at each cell division. Berger
(1973b) presents evidence that the postzygotic macronuclear anlagen and
macronuclei inhibit DNA synthesis in fragments of the prezygotic macronucleus; that whenever this inhibition is removed by the absence of the
postzygotic macronucleus, the fragments resume DNA synthesis within a
few hours; and that fragment growth is regulated to maintain a constant
total amount of DNA per cell.

Other Cytogenetic Processes

As probably happens in all organisms, almost everything that can go
wrong sometimes does. Among the aberrancies that have been more or
less clearly demonstrated are: (1) total failure of the nuclear events of
conjugation to occur in one or both members of a pair; (2) unilateral
instead of reciprocal crossfertilization during conjugation; (3) formation of
gamete nuclei from two different products instead of from one and the
same product of meiosis; (4) formation of more than two macronuclear
anlagen; and (5) formation of macronuclear anlagen from an unfertilized

and from a fertilized nucleus in the same cell. Doubtless other aberrancies occur. In order to detect these deviations from the normal processes, genic markers are particularly useful. Kimball and Gaither (1955, 1956) provide an excellent example of their use in detecting such deviations in abnormal cytological settings and after exposure to x rays.

Heredity in Asexual Reproduction

The rule of clonal constancy in hereditary traits is applicable to *P. aurelia* as it is to other organisms; but there are exceptions. A transient exception, of limited importance, is phenomic lag, i.e., the persistence of a trait for a limited number of cell generations after acquisition of a genotype for an alternative trait at fertilization. Change from the homozygous recessive to the dominant genotype is usually accompanied by phenotypic change within 1 or 2 cell generations; but the reverse change from dominant to homozygous recessive phenotype is usually slower, extending often over 4–12 generations, the number differing for different genes and for different lines of descent within the same clone. The duration of phenomic lag in generations can be greatly reduced by prolonged starvation after fertilization (Sonneborn, 1953; Berger, private communication). Phenomic lag may have two causes: (1) the production of mRNA by fragments of the prezygotic macronucleus and (2) persistence of genic products in the cytoplasm for several generations. Phenomic lag is also associated with heritable changes in genic activity that do not involve change of genotype, such as occurs with mating types (see page 485). Another special form of phenomic lag appears to occur when certain homozygous recessives conjugate with a heterozygote and both mates acquire the homozygous recessive genotype (Berger, private communication; for the case of *pwA*, see Table 5). The originally recessive mate may develop the dominant trait and retain it during several cell generations, presumably by transfer from its partner of products of the dominant gene. Berger has noted transfer of labeled leucine *during* conjugation.

Four other kinds of intraclonal hereditary diversities are more persistent and therefore more important. The first is the succession of stages in the clonal cycle from immaturity through senescence. Each stage is "inherited" for a number of cell generations (see page 476). The second is illustrated by the immobilization (i) antigens (see page 489); under certain conditions different lines of descent of the same clone produce different antigens. The third kind of intraclonal hereditary diversity occurs in killer clones (see page 490); the killer trait and its basis (kappa) can be irreversibly lost for some sublines of a clone and retained for others. The fourth kind is exemplified in the genetics of mating types (see page 485) in which clones can produce caryonides that reproduce true to different mating types.

The possibility of another kind of intraclonal hereditary diversity needs to be considered. In *T. pyriformis*, syngen 1, phenotypic assortment is a common occurrence (page 441, this volume). Heterozygotes at many loci produce three kinds of sublines: those constant for the phenotype of one allele, those constant for the phenotype of the other allele, and those that have the heterozygous phenotype (when this is identifiable) but keep producing sublines of the two constant phenotypes. Such phenotypic assortment has not been demonstrated in *P. aurelia*, but the possibility of its occurrence should be kept open. On certain hypotheses, it should take about ten times as many generations for assortment to be detectable in *P. aurelia* as in *T. pyriformis* because of the ten times greater ratio of macronuclear-to-micronuclear DNA. This would delay it until late in senescence. The few heterozygous loci that have been examined at that stage of the cycle failed to show assortment (Sonneborn *et al.*, 1956), but perhaps more loci should be tested.

Heredity in Sexual Reproduction

At autogamy in a heterozygote the meiotic product that gives rise to the two genotypically identical gamete nuclei has a 50–50 chance of carrying a particular allele. Hence, half of the autogamous cells arising in a heterozygous clone will produce clones homozygous for one allele and half for the other allele. Heterozygous clones do not normally arise at autogamy.

As a model for the genetics of crossfertilization, i.e., true conjugation, let the genotypes of the haploid male and female sister gamete nuclei in one conjugant be symbolized by A and A, those in the other conjugant by B and B. Then exchange of male nuclei and their fusion with the female nuclei will yield AB in each conjugant. Hence, crossfertilization results in identical genomes for the two clones derived from a pair of conjugants. Because of their genic identity, the two clones are together referred to as a *synclone*. Nevertheless, in the absence of cytoplasmic exchange between the mates, the two clones of a synclone are a pair of reciprocal crosses in the sense of their having the same genomes in the cytoplasms of the two different parents. The ratios in a group of synclones obtained by true conjugation are typical Mendelian ratios. For a single locus difference, all F_1 synclones are heterozygotes and the F_2 generation consists of $\frac{1}{4}$ homozygous recessive synclones: $\frac{1}{4}$ homozygous dominant synclones: $\frac{1}{2}$ heterozygous synclones.

The results obtained if cytogamy occurs are simply those expected from independent autogamies in the two mates. For example, if the clones mated are both heterozygotes for the same alleles, A and a, then each member of a pair will yield a 1:1 ratio for AA and aa. Because these results are independent in the two mates, one-fourth of the synclones will consist of two AA clones, one fourth of two aa clones, and one-half will consist of one AA and one aa clone. The latter result distinguishes cytogamy from true conjugation. One can estimate the fraction of pairs undergoing cytogamy by doubling the fraction that yields diverse clones (one AA and one aa or, more simply, one dominant when the other is recessive).

The genetic result of macronuclear regeneration is simply that the phenotype of the parent remains unchanged no matter how many heterozygous loci it carries. Another and very useful indication of the occurrence of macronuclear regeneration is that the cells are competent to undergo another conjugation or autogamy without an intervening period of immaturity or inability to undergo autogamy, respectively. Extensive searches have been made by Sonneborn and by Nobili for possible segregation of alleles when heterozygotes for codominant alleles undergo macronuclear regeneration, but heterozygosity was phenotypically maintained by all of about 1000 tested macronuclear fragments (Nobili, 1962).

Mutagenesis

Sonneborn (1970a; pages 299–304) has summarized the methods of inducing, detecting, and screening for lethals, conditional lethals, slow growers, visibles, mating-type mutations, and certain kinds of back mutations. Techniques equivalent to replica plating make large-scale screening easy and rapid. Kung (1971b) has described screening methods for swimming and behavioral mutants. Similar methods can be used to screen for backmutations in certain behavioral mutants (see, for example, gene *ssw3* in Table 5).

In general, mutagenic work is exceedingly efficient in *P. aurelia* because exposure to mutagen shortly before induction of autogamy yields homozygotes for the mutant alleles in one step and within a day or two after exposure. Currently, nitrosoguanidine is the favorite mutagen. Triethylene melamine is the best alkylating agent. Some agents, such as nitrous acid and ethyl methane sulfonate, are too toxic to be useful.

Genetics of Mating Types: Genes and Irreversible Nuclear Differentiations

J. R. Preer (1969; pages 206–219) has reviewed this topic in detail. Here we note only the main features and subsequent developments. Unlike *Tetrahymena, Paramecium* shows an immediate agglutinative sexual reaction when mature cultures of complementary mating types are brought together under appropriate conditions: cells of the two different types adhere to each other by their cilia. In this way, clumps are formed consisting of many cells. In the course of an hour or two, pairs of cells—one of each mating type—become aligned properly for conjugation, lose reactivity on exposed parts, separate from the clumps, and proceed to conjugate.

Each of the 14 species of *P. aurelia* has two mating types, designated by number, I and II in species 1, III and IV in species 2, etc. Sometimes these have been referred to as the O(odd) and E(even) mating types of each species, and I shall do so in this review. The O types of most or all of the species, and also the various E types, are believed to be analogous and homologous (Sonneborn and Dippell, 1946). There is no known exception to the rule that phenomic lag (see page 483) is conspicuous for changes from E to O, but not for the reverse [first observed in species 1 by Kimball (1939a)]. The temperature effect (described below on this page) operates on mating-type determination in the same way in different species with respect to O and E. Mating reactions occur specifically, when they occur at all, between the O of certain species and the E of others, never between two O's or between two E's (Fig. 1, page 572).

In spite of the parallels between the two mating types of different species, the genetics of mating types follows three different systems in different species. The simplest system occurs only in species 13 (Sonneborn, 1966). The O type is determined by a recessive gene, *mt*, and the E type by its dominant allele, *Mt*. Heterozygotes are type E, except that a cell destined to become type E may be type O for a few hours before it becomes E. This mode of mating-type inheritance is referred to as the C, or synclonal, system; both clones of a synclone, having the same genotype, are of the same mating type.

The second, A, or caryonidal, system (Sonneborn, 1937a; 1939; Kimball, 1937), is found in species 1, 3, 5, 9, 11, and 14. Almost all stocks are homozygous for a genome that permits both mating types. As a rule, a caryonide is phenotypically pure for a single mating type, but the two sister caryonides from a single fertilized cell are randomly either alike or different in mating type. Here random means that if p caryonides from a group of fertilized cells are type O and $1 - p$ are type E, then the two caryonides from the same fertilized cell are both O with a frequency p^2, both E with a frequency $(1 - p)^2$, and one of each type with a frequency $2p(1 - p)$. The value of p differs in different wild stocks and at least one gene affecting it has been found (see *InI*, Table 5). The value of p decreases linearly with rise of temperature during a restricted sensitive period that begins during autogamy or conjugation and ends some time before the first cell division.

The caryonidal rule of inheritance implicates the macronucleus as the seat of mating-type determination, and this has been amply confirmed by several lines of evidence. Two of these are the failure of mating type to change at macronuclear regeneration and the segregation of mating types at cell divisions 2–4 when fertilized cells form more than two macronuclei. Since the macronuclei that develop in a fertilized cell arise from products of the same micronucleus, which can be entirely homozygous (after autogamy), it is usually assumed that the macronuclei are genotypically identical in any one cell but become differentiated as to genic action. Whatever the nature of macronuclear differentiation may be, it should be understood that it is in this case irreversible in the descendent macronuclei. It may remain unexpressed phenotypically for dozens of cell generations during clonal immaturity and then come to expression when the caryonide enters the clonal period of maturity. During maturity, it is expressed only when the cells become sexually reactive, i.e., as they pass from log growth toward stationary phase under appropriate cultural conditions.

Certain wild stocks (found thus far only in species 1 among those possessing the caryonidal system) are incapable of producing mating-type E. These stocks proved to be homozygous for a recessive gene, mt^o, while so-called two-type stocks are homozygous for its dominant allele, mt^+ (or Mt^E). The caryonidal system described above depends on the presence of gene mt^+. No wild stocks incapable of producing mating-type O have been found, nor has it been possible to obtain mutants of this type in the laboratory. On the other hand, mutations from mt^+ to mt^o have been obtained in two-type stocks. Additional special features of this system of mating-type inheritance will be given in the sections devoted to species 1, 3, and 5 (pages 499, 516, 558).

The third, B, or clonal, system of mating-type genetics is found in species 2, 4, 6, 7, 8, 10, and 12. It differs in only one respect from the A, or caryonidal, system: the differentiation of postzygotic macronuclei for mating-type determination is brought about by a cytoplasmic agent which itself is under nuclear control (Sonneborn, 1954d; Nanney, 1957). Consequently, the two new macronuclei that arise in the common cytoplasm of a fertilized cell are usually determined to control the same mating type. Because the cytoplasmic factor that is present was produced under the action of the parental macronucleus, the mating type of the parental cell persists, or rather is again determined, in the fertilized cell. Thus, mating type seldom changes at autogamy, and a pair of conjugants which has not exchanged cytoplasm usually produces a clone of one type from the parent of that type and a clone of the other type from the other parent. If considerable cytoplasm is exchanged, one or both parents can produce a clone of a different mating type, or, frequently, selfers, i.e., clones that contain cells of both mating types. Some selfer clones consist of one caryonide of each mating type, but one or both caryonides of a clone may themselves be selfers. In this system, as in the caryonidal system, wild and mutant stocks may be restricted to mating-type O by virtue of their genotype. Further aspects of the clonal system will be noted in the accounts of species 2, 4, and 7 (see pages 511, 548, 560).

The three systems just described probably have a common genetic system underlying them. Butzel (1953a, 1955) proposed an interpretation for mating-type determination in species 1, which has the caryonidal, or A, system, that was later recognized as capable, with minor modifications, of accounting for the basic genetics of all three systems. I shall, therefore, present his interpretation in a slightly modified and more general form.

Assume that the O and E mating types are characterized by the presence of O and E mating-type substances, respectively, and that the O substance is a precursor of the E substance. Suppose that the mt^+ gene is involved in the conversion of O into E substance and that this gene in macronuclei is capable of existing in two stable, transmissible states, one in which it cannot be derepressed and one in which it can. Then macronuclear differentiation for mating-type determination consists in establishing one or the other of these two stable states. A caryonide would be determined for mating-type O if the non-derepressible state of mt^+ were established, and for type E if the derepressible state were established. The allele, mt^o, is held to be an inactive or inactivatable allele. In homozygotes for mt^o, the O substance could not be converted to E, and all caryonides would be type O.

The gene Mt in species 13 (page 569) with the C system could then be an mt^+ gene that exists in only one state, the state capable of derepression. The observation that cells can go through a transitory O stage before becoming type E in Mt/mt cells is consistent with the idea that O is a precursor of E. Other observations to be mentioned in accounts of various species are likewise in agreement.

The chief addition that has to be made to this hypothesis in the light of further investigations (Taub, 1963; Byrne, 1973) is that more than one locus is involved or can interact with the conversion of O to E substance. The question has also been raised as to whether the known mating-type genes, or some of them, are regulatory. Nothing is known about this. Another question to be dealt with later (see page 574) is the genetic relationship between homologous mating types and genes in different species of *P. aurelia*. On these and other questions, further advances may be delayed until the molecules involved in mating-type specificity are isolated and characterized chemically.

Genetics of the Immobilization Serotypes

All species of *P. aurelia* that have been investigated for immobilization serotypes have them, and their genetics show basic similarities [reviewed by Beale (1954, 1957a); J. R. Preer (1969; pp. 219–230), Sommerville (1970), and Finger (1967, 1974)]. Only the main points and some recent advances will be given here.

An immobilization (i) serotype is a cell type possessing on its surface (including the surface of the cilia) an antigen which, in the presence of a dilute solution of homologous (and in some cases related) antisera, leads to the immobilization of the cells as a result of the agglutination of the cell's cilia. The i antigens are large proteins, about 240,000–320,000 molecular weight, and they constitute about 30 percent of the protein content of isolated cilia. A homozygous stock of paramecia can produce cells of up to a dozen or more different i serotypes, each possessing a different i antigen on its surface. Two of the striking facts about this system are that, as a rule (exceptions later) (1), only one of the i antigens is on the surface of any cell and (2) every cell has an i antigen on its surface.

Each of the i antigens is coded by a gene at a different locus. Thus far, no two i-antigen loci have been found to be linked. The antigens coded by genes at different loci in the same wild stock are serologically distinct. The antigens coded by alleles may show all grades of serological distinction, from none to very great, and the same is true for their physical and chemical properties. The antigens coded by genes at different loci are given different letters; those coded by different alleles have the same letter together with a number designating the stock in which they were first found. Thus A and G are antigens coded by genes at different loci; 29A and 51A are the A antigens found in stocks 29 and 51, respectively; and the alleles for them are symbolized by A^{29} and A^{51}.

Even in an entirely homozygous stock, the genes at only one i-antigen locus are as a rule expressed in any one cell and, under appropriate conditions, also in its progeny. Yet it is possible by various conditions to transform the cells so that the active locus ceases to be expressed and another one comes to be expressed in the cell and its progeny. A few of the many known transforming conditions are exposure to homologous antiserum, to a different temperature, to different media, to ultraviolet irradiation, and to certain enzymes. Under identical conditions, two or more different serotypes many reproduce true to type.

When two different serotypes (e.g., 60D and 60G) of the same homozygous stock are crossed under conditions in which both serotypes are stable, each mate gives rise to a clone of its own serotype, i.e., one to 60D and one to 60G. However, if considerable cytoplasmic exchange occurs between the mates, they may both produce clones that are alike in serotype and like that of one parent (e.g., both 60D). This finding indicates that the cytoplasm contains substances that regulate expression of the i antigens and is reminiscent of the B, or clonal, mating-type system (see page 487). In the case of the serotypes, however, the regulation is not restricted to a sensitive period and is not irreversible. The diverse cytoplasms have been referred to as diverse "cytoplasmic states" by Beale (1952, 1954). When homozygous allelic serotypes (e.g., 60D × 90D) are crossed, the heterozygous hybrids ($D^{60}D^{90}$) often—but not always—have both parental antigenic specificities (e.g., those of 60D *and* of 90D) on their surface. These two examples show what is expected and what happens in a more complex cross, such as between a 60D cell with genotype $D^{60}D^{60}G^{60}G^{60}$ and a 90G cell with genotype $D^{90}D^{90}G^{90}G^{90}$; both mates acquire the genotype $D^{60}D^{90}G^{60}G^{90}$, but the 60D mate produces a clone that expresses the hybrid D character (60D + 90D) while the 90G mate produces a clone that expresses the hybrid G character (60G + 90G) (Beale, 1952). Here we have illustrated the simplest results. In later sections devoted to different species, theoretically important different results will be presented. As will appear, much is known and continues to be discovered about the genes, the molecules and the cytoplasmic states involved in the i-serotype system, but some important problems still remain unsolved.

Genetics Based on Symbiotic Bacteria

Killer paramecia and their genetics provide perhaps the most widely known episode—or series of episodes—in the whole of *Paramecium* genetics, some discussion of the topic being given in most textbooks of genetics. There is, however, far more to the story than appears in textbooks, and much of it is of considerable genetic importance as a model of phenomena that appear to be widespread among both lower and higher organisms. Killer paramecia were discovered more than a third of a century ago (Sonneborn 1939); the medium in which they were grown killed paramecia of most other stocks. The killer trait appeared at once to be inherited via the cytoplasm; some years later (Sonneborn, 1943a), the self-reproducing cytoplasmic factor involved was called kappa and its maintenance in the paramecia was shown to be gene-dependent, but no paramecium genotype was able to produce kappa *de novo*. Under certain conditions, kappa can be irreversibly lost from a clone or part of a clone (Preer, 1946; Sonneborn, 1946a).

Immediately after discovering killers, Sonneborn considered the possibility that its basis might be a symbiotic organism, but he rejected this possibility when efforts at infection failed. A decade later, he succeeded in making infections (Sonneborn, 1948*b*), but nevertheless held (Sonneborn, 1959) that better evidence was required to validate the symbiont interpretation. The evidence that kappa is indeed a bacterium came largely from the work of Preer and his associates, beginning particularly with characterization and visualization of kappa (Preer, 1946, 1948*a,b*, 1950) and going on to discoveries of its structure, composition, and properties which have continued to the present time.

Eventually kappa was found to be but one of many kinds of bacterial endosymbionts which occur in more than 50 percent of all collections of *P. aurelia* from nature (J. R. Preer *et al.*, 1974). They have thus far been found in species 1, 2, 4, 5, 6, and 8. For the many interesting and important genetic and general biological discoveries about these bacterial symbionts, the reader is referred especially to two comprehensive reviews. The earlier work was exhaustively reviewed by Sonneborn (1959) and the large body of subsequent advances is currently reviewed by J. R. Preer *et al.* (1974). Additional information on the symbionts will be given in sections below on each species in which they occur.

Mitochondrial Genetics

This is one of the most recent and promising lines of research on *P. aurelia*. It has been carried out mainly on species 1 and 4 and will be summarized below in the sections dealing with these and other species.

Genetics of the Cell Cortex

This is another area of research that has been developed mainly during the past decade and is still in full swing. It will be set forth in the section below on species 4. The principal genetic importance of this work is its demonstration that experimental alterations in the pattern of cortical structure—the number, orientation, and arrangement of cortical parts— are cytoplasmically inherited.

Biochemical, Developmental, and Population Genetics

The chief areas of biochemical genetics are those dealing with serotypes, enzymes, kappa, and the cell and clonal cycles (see page 475 and following pages). Further details as to genes with known biochemical effects are given in the lists of genes in each species. Developmental genetics is dealt with in the accounts of cell and clonal cycles and of organellar and cortical genetics, especially in the section on species 4. Population genetics will be referred to in the accounts of individual species and in the final section, which deals with studies involving more than one species.

Species 1

Introduction

Among all unicellular animals, species 1 of *P. aurelia* has several unique claims. It was the species in which were *first* discovered: (1) the existence of mating types; (2) a clearly Mendelizing gene; (3) genetic evidence that sister haploid nuclei regularly unite during autogamy; (4) genetic evidence of the occurrence of cytogamy; (5) genetic proof that the genomes of the two clones of a true conjugant synclone are identical; (6) macronuclear regeneration; and (7) the caryonidal system of mating-type determination and inheritance and its basis in macronuclear differentiation. All of these discoveries were made between 1937, when mating types were discovered, and 1942, when species 4 was discovered and began its career as the only other species of *P. aurelia* to rival and eventually surpass species 1 in the variety, intensity, and pioneering character of genetic investigation.

The details of this early period of genetic research on species 1 (and on other species up to 1954) are recounted in Beale's (1954) book, with full bibliography. Many of the most important findings have already been mentioned in this chapter. During a still earlier period (1930–1937), stock 18 (R in the literature of that period), later found to belong to species 1, had been intensively studied by Jennings and Sonneborn and their associates. It was this earlier work that laid the foundations for the delineation of the clonal cycle and for the discovery of mating types.

Characterization of Species 1

Species 1 can be distinguished from all other known species of *P. aurelia*, except species 5, by the combination of the electrophoretic patterns of its esterase C (no clear bands) (Allen and Gibson, 1971; Allen *et al.*, 1973) and of its mitochondrial beta-hydroxybutyrate dehydrogenase (three cathodal bands) (Tait, 1970*a*). Because species 5 has thus far been found only in North America and (once) in Australia, a stock collected elsewhere and conforming to the pattern for the two enzymes would almost certainly be species 1. When possibility exists for confusion with species 5, the two can be readily distinguished by the frequency of selfing caryonides arising at autogamy or conjugation: it is 50 percent or more in species 5 (see page 559) and always much less than this, usually less than 5 percent, in species 1.

Genes

The known genes in species 1 are listed in Table 1. About half of these were found in wild stocks. No linkage has yet been proved in this species, but Gibson and Beale (1961) suggest the possibility that genes *m1* and *m2* may be linked in stock 544 but not in other stocks. Not listed in the table are many lethals and detrimentals obtained in x irradiation studies by Kimball and co-workers (see page 547), but not individually characterized or retained. Because of the inevitable homozygosity arising at autogamy, it is not feasible to maintain recessive lethals in growing stocks without repeated crossing to keep them in heterozygous condition. However, it is now possible to maintain heterozygotes for recessive lethals in frozen condition. Kimball and Gaither (1956) searched for, but found no more than a very low frequency of, dominant lethals (presumably chromosomal, not point, mutations) after x irradiation; and they attributed this to the probably slight unbalance resulting from loss or gain of part of one chromosome in a diploid genome with so large a chromosome number.

Stocks

Table 2 lists the principal wild stocks of species 1. Most of those with numbers below 500 and a few in the 500's are in the Indiana University collection. In the earlier papers, stocks were given letter symbols; these were later changed to numbers corresponding to the position of the letter in the alphabet, i.e., stocks earlier assigned the letters B, P, R, S, and Z are now assigned the numbers 2, 16, 18, 19, and 26, respectively. Table 2 gives the alleles at the three main polymorphic i-antigen loci for all of the stocks on which I was able to find this information. Wild stocks carrying known alleles at other polymorphic loci are mentioned under the relevant locus in Table 1. Not included in Table 2 are stocks derived from wild stocks by introducing into their genome mutations induced in, or wild alleles derived from, other wild stocks. The principal derived stock is d1-1 which contains genes *cl1*, D^{60}, *dp*, G^{60}, and mt^o in the genome of stock 90 (Kimball and Gaither, 1955). Other derived stocks may be in Beale's collection in Edinburgh.

TABLE 1.   *Genes of Species 1*

Gene locus [a]	Characteristics
cl1	Clear, i.e., with much reduced or no crystals (Kimball, 1953*b*; Kimball and Gaither, 1955). Spontaneous in stock 90. Not allelic to *cl2*; not linked to *D, dp, G,* or *mt*.
cl2	Clear, i.e., with much reduced or no crystals (Kimball, 1953*b*). Spontaneous in stock 90. Not allelic to *cl1*.
D	Polymorphic locus for antigen D in the system of mutually exclusive immobilization (i)serotypes (Beale, 1952, 1954, 1957*a*). Serologically related to *D* of species 4. Alleles D^{33}, D^{41}, D^{60}, D^{90}, D^{143}, D^{145}, D^{175}, and D^{192} code for serologically distinguishable antigens occurring in wild stocks, designated by the superscripts, and in certain other wild stocks (see Table 2). Some serologically indistinguishable D antigens of different wild stocks are chemically different (Jones and Beale, 1963; Jones, 1965 *a,b*). At least some pairs of *D* alleles are codominant (but see *G* below). Clones, incapable of developing D antigen, presumably carrying d^0 or null alleles, segregate in the F_2 of certain stock crosses (Beale, 1958; Beale and Wilkinson, 1961). *D* is not linked to *cl1, dp, G, mt,* or *S*.
dp	Dumpy cell shape: short wide cells with blunt posterior end; dorsal surface tends to be concave (Kimball, 1953 *b*). Spontaneous mutation in stock 90. Phenotype obscured in well-fed cells, but well-expressed during starvation in mass cultures. Possibly a slight dominant effect. Not linked to *cl1, D, G,* or *mt*.
EsB	Polymorphic locus for esterase B (Cavill and Gibson, 1972). Codominant alleles *EsB*^{90} *and EsB*^{540} for electrophoretically distinct enzymes in wild stocks designated by the superscripts. Not linked to *D* or *EsC*.
EsC	Polymorphic locus for esterase C (Cavill and Gibson, 1972). *EsC*^{540} yields an electrophoretically detectable esterase and is dominant over *esC*^{90}, which yields either no detectable esterase C or one of very low activity (Allen *et al.,* 1973). Superscripts designate the wild stocks in which alleles were identified. Not linked to *EsB*.
G	Polymorphic locus for antigen G in the system of mutually exclusive immobilization (i) serotypes (Beale, 1952, 1954, 1957*a*). Serologically related to G of species 4. Alleles G^{41}, G^{60}, G^{61}, G^{90}, G^{156}, G^{168}, and G^{192} code for serologically distinguishable antigens occurring in wild stocks designated by the superscripts and in certain other wild stocks (see Table 2). Different combinations of alleles show various domi-

[a] The following system of gene nomenclature is used in tables of genes for all species. Each locus is designated by not more than three letters. Different loci with genes having similar phenotypes may be represented by the same locus symbol with the addition of a distinctive capital letter or Arabic numeral. Alleles are distinguished by a superscript, either a small letter or Arabic numeral. The dominance or recessiveness of a (nonpolymorphic) mutant gene is indicated by capital vs. small initial letter, and the wild type by a + superscript added to the mutant symbol. Codominant alleles have symbols with initial capital letters. The symbols for genes at polymorphic loci do not include a + superscript; instead each allele symbol begins with a capital or small letter to indicate its dominance (or codominance) or recessiveness, with or without a distinctive superscript. In order to create some order out of the variety of symbol systems in the *Paramecium* literature, I have sometimes had to modify published symbols, but have made the least change required for conformity. Absolute conformity, however, is not always feasible. In a few cases, the symbols here employed have had to deviate slightly from the general system.

TABLE 1. Continued

Gene locus[a]	Characteristics
	nance relations (Capdeville, 1969, 1971). Not linked to *cl1, D, dp, mt, S, ts1, ts2,* or *ts3.*
im	Polymorphic locus for length of the period of immaturity following conjugation (Siegel, 1961). *im^s* occurs in stock 18; *im^l* occurs in stocks 19 and 90. Homozygotes for *im^s* have a very short, if any, immature period. Homozygotes for *im^l* have immature periods of 15 or more cell generations. Dominance not reported.
In1	A polymorphic locus affecting the frequency of mating-type-I caryonides (Butzel, 1955). Clones possessing the dominant allele *(In1)* "increase" the frequency of type-I caryonides by 60 percent or more over the frequency in clones homozygous for *in1*. Stock 90 is homozygous for *in1*; *In1* was found following· exposure of line d40 (see *mt*) of wild stock 90 to ultraviolet.
m1	Polymorphic locus which affects maintenance of mate-killer factor, the symbiotic bacterium mu (Gibson and Beale, 1961). The allele m1 is recessive and, when *both* m1 and m2 are homozygous, results in loss of mu. Either *M1* or *M2* permits maintenance of mu. *M1* occurs in stock 540, *m1* in most other stocks (e.g., 119, 168, 217, 513, 544). Not linked to *M2* in any of these stocks except possibly stock 544.
m2	Similar to *m1* (Gibson and Beale, 1961). For maintenance of mu, either *M1* or *M2* suffices; mu is lost only in the double homozygote *m1m1, m2m2*. *M2* occurs in stock 540, *m2* in most other stocks (as with *m1*). Not linked to *M1* except possibly in stock 544.
mt	Polymorphic locus for mating type (Sonneborn, 1939). Homozygotes for the recessive allele *mt^0*(= *mt^I*) are mating-type I. Wild stocks 2, 14, 15, 16, and 21, all from the same natural source, are *mt^{0}mt^0*. Stocks from all other natural sources thus far studied are *mt^E, Mt^E* (= *mt^{I,II}, mt^{I,II}*). Alleles with effects identical to that of *mt^0* have been obtained by Nanney and by Butzel (1955) in stock 90 by heat shocks and by ultraviolet irradiation. Butzel's mutant (d40) is carried as stock d1-2. Not linked to *cl1, D, dp,* or *G* (Kimball, 1953b).
nd10	Polymorphic locus affecting discharge of trichocysts; when well-fed homozygotes for *nd10* are exposed to picric acid, the trichocysts mostly fail to discharge (Haggard, 1972). Wild type *(Nd10)* discharges trichocysts. *nd10* carried by stock d1-3. Similar phenotypes are shown by stocks 16 and 60, but genetics unknown. Stock 90 carries *Nd10*.
S	Polymorphic locus for antigen S in the system of mutually exclusive immobilization (i) serotypes (Beale, 1952, 1954, 1957a). Alleles *S^{60}*, *S^{61}*, and *S^{143}* code for serologically distinguishable antigens and occur in the wild stocks designated by the superscripts as well as in certain other wild stocks (Table 2). Not linked to *D* or *G*.
sl	A dominant mutation for asexual reproduction at about 40 percent of wild-type rate (Butzel and Vinciguerra, 1957). Cells have hooked anterior end and larger-than-normal cell size. Increased frequency of macronuclear regeneration. Obtained by ultraviolet irradiation of stock 90. Not linked to *mt*.

TABLE 1.   Continued

Gene locus[a]	Characteristics
tr	A recessive for translucent, i.e., with much reduced or no crystals (Butzel, 1953c). Cells smaller than normal. Variable expression: best expressed in young clones, most obscure in older starved clones. Isolated as an F_2 clone from cross of stock 90 to d40 (see *mt* above). More variable than *cl2*. Carried as stock d1-3. Not linked to *cl1* or *mt*.
ts1	Temperature sensitive: dies within 24 hours at 36–37°C; obtained with ultraviolet on stock 168 by Capdeville (1969, 1971); her mutant *theta-1* (Capdeville and A. M. Keller, private communication). Not linked to *G*, *ts2*, or *ts3*.
ts2	Temperature sensitive: dies in 24–36 hours at 36–37°C; obtained with ultraviolet on stock 168 by Capdeville and A. M. Keller (private communication). Not linked to *G*, *ts1*, or *ts3*.
ts3	Temperature sensitive: dies within 48 hours after several fissions at 36–37°C; obtained with ultraviolet on stock 168 by Capdeville (1969, 1971); her mutant *theta-3* (Capdeville and A. M. Keller, private communication). Not linked to *G*, *ts1*, or *ts2*.

Chromosomes and Nuclei

Haploid chromosome numbers reported by Jones (1956) are: 37 ± 1 in stock 16; 51 ± 1, stock 43; somewhat more than 30 in stock 60; 44 or 45, stock 144; and 45 ± 1, stock 516. Kościuszko (1965) reports: about 63 in stock 90, and 43–58 in 5 other European stocks not listed in Table 2 but formerly and perhaps still available in Komala's collection in Cracow. Stevenson and Lloyd (1971a) report more than 50 chromosomes in stock 540.

As might be expected from the chromosomal situation, mortality is high in the autogamous F_2 generation following a cross between two different wild stocks, as it is in all investigated species of *P. aurelia*. Up to 96 percent F_2 mortality has been reported, the percentage being correlated with the amount of difference between the stocks in chromosome number (Kościuszko, 1965). Autogamous F_2 mortality declines progressively after successive backcrosses, usually becoming very low or disappearing after the fourth generation. A standard stock for genetic work has not been selected in species 1.

Ultrastructural descriptions of micronuclear and macronuclear division are given by Stevenson and Lloyd (1971a,b). The proportions of DNA, RNA, and protein in nuclei are given by Stevenson (1967) and Cummings (1972). Cummings found two kinds of macronuclear RNA,

TABLE 2.   *Wild Stocks of Species 1*

S	G	D	Stock[b]	Stock source
60	90	90	2, 14, 15, 16, 20	Woodstock,Maryland
61	90	90	10	Baltimore, Maryland
61	90	90	18	Baltimore, Maryland
60	61	41	19	Cold Spring Harbor, New York
60	61	60	26	Stanford, California
143	61	33	33	Baltimore, Maryland
60	41	41	41	Atlanta, Georgia
60	61	60	43	Stanford, California
60	60	60	60	Burlington, Vermont
61	61	90	61	Woods Hole, Massachusetts
60	90	90	62	Woods Hole, Massachusetts
60	90	90	90	Bethayres, Pennsylvania
60	90	192	103	Philadelphia, Pennsylvania
60	90	90	119	Gwynedd, Pennsylvania
61	60	41	129	Fort Lauderdale, Florida
143	41	143	143	Falkirk, Scotland
61	41	90	144	Chantilly, France
60	60	145	145	Paris, France
60	41	60	147	Sendai, Japan
60	90	90	153	New Haven, Connecticut
60	156	60	156	New Haven, Connecticut
61	168	90	168	Sendai, Japan
143	41	60	171	Yamagata, Japan
60	41	175	175	Lake Titicaca, Puno, Peru
61	61	60	177	Santiago, Chile
61	61	60	178	Santiago, Chile
143	61	60	180	Tokyo, Japan
143	61	60	181	Tokyo, Japan
60	41	90	182	Osaka, Japan
60	41	90	13	Tokyo, Japan
—	192	192	192	Woods Hole, Massachusetts
60	60	90	513	Chantilly, France
60	60	145	514	Chantilly, France
60	60	145	515	Chantilly, France
60	60	90	516	Chantilly, France
60	41	41	520	Paris, France
—	90	60	521	Paris, France
60	61	60	523	Luzern, Switzerland
60	90	60	535	Adelaide, Australia

[a] This table is an up-dating of Beale's (1954, p. 105) Table 19. The columns headed *S*, *G*, and *D* list the superscripts of the alleles at these antigen loci (see Table 1). Dash means I was unable to find the information.

[b] Other stocks: from USA near Gulf of Mexico: 217, 241, 243, 244, 544; from Mid-USA: 74, 75, 221, 336; from California: 46, 285, 313, 548, 551, 555; from Mexico: 257, 258, 540; from Hawaii: 220; from USSR: 334, 335, 337; from Poland: 320; from Italy: 561.

molecular weights of 1.3 and 2.8 $\times 10^6$ daltons, and that the proportion of DNA was higher and the proportion of protein less in macronuclei than in micronuclei. For DNA data, see pages 472–474. Cells lacking micronuclei have been obtained by exposure to colchicine (Butzel, 1953a) and x rays (Geckler and Kimball, 1953). They also arise spontaneously in aged clones (Mitchison, 1955).

Cells and the Cell Cycle

Powelson *et al.* (1974) report, for a set of standard conditions, certain measurements and counts for cells of stocks 74, 147, 168, 175, and 221. Extreme stock means among these 5 stocks for the more important characters were: 124–137 μm cell length, 56–62 μm (and 29–34 basal bodies) between the two contractile vacuole pores; and 26–30 kineties between the posterior contractile vacuole pore and the anterior end of the cytoproct. From 4 to 5 fissions per day is the norm for vigorous, young stocks at 27°C (Sonneborn, 1954c), but some very old stocks (e.g., stock 18) reproduce more slowly (Siegel, 1958).

Hufnagel (1969a) examined the ultrastructure of isolated pellices of stock d1-1 (referred to as stock CD) and found some intracellular structural variations among the repeating units of cortical structure. The major DNA from isolated pellicles, believed to represent contaminating nuclear DNA, had a density like nuclear and whole-cell DNA; the remaining DNA (believed to be bacterial contamination) had densities corresponding to those of bacterial DNA (Hufnagel, 1969b). Selman and Jurand (1970) have described and interpreted the development of trichocysts during the cell cycle.

Clonal Cycle

The clonal cycle in stock 18 was explored in detail by Sonneborn (1937b, 1954c). This stock has a very brief or no period of immaturity, but stocks 19 and 90 become fully mature only after some 20–40 generations (Siegel, 1961). Stocks of species 1 differ considerably in the duration of immaturity and maturity (Beale, 1954; Sonneborn, 1957). Kimball and Gaither (1954) exposed cells to 80 kr of x rays and found no evidence of precocious aging. Fukushima (1974) exposed cells to fractionated doses totaling 55, 92, and 138 kr of x rays and found that the two higher doses, but not the lowest dose, did decrease clonal life span and increased the speed of decline in fission rate during senescence.

Cytogenetic Processes

See page 478 for postzygotic processes in stock 18. The calcium effect on cytogamy (see page 480) was for species 1; Miyake's chemical induction of conjugation (see page 481) is applicable even to homozygotes for mt^0 such as stock 16. Cells undergoing macronuclear regeneration can both conjugate and undergo autogamy, the regenerating fragments of the prezygotic macronucleus behaving like miniature macronuclei: the complexity of their breakdown conformation (skein) is proportional to their size, i.e., to the degree to which they have regenerated (Sonneborn and Dippell, unpublished). Amicronucleate lines of descent undergo repeated macronuclear regeneration at short intervals, and such clones are characterized by small cells with small macronuclei and other distinctive characters (Sonneborn, 1942a). Kimball and Gaither (1955) showed that in conjunction between haploids and diploids the haploid frequently fails to contribute a male gamete nucleus, and that meiotic products of haploid nuclei are unable to compete successfully with meiotic products of a diploid nucleus in cells that contain both a haploid and a diploid micronucleus. Nanney (1954) was able to cause the two macronuclear anlagen in exconjugants to fuse by prolonged starvation.

Genetics of Uniparental Reproduction

Kimball (1939a) demonstrated *transient* diversities in mating type within a caryonide as a result of phenomic lag (see page 486). Siegel (1970) correlated diversities in generation time, within a clone and at the same clonal age, with retention vs. new formation of the food-intake apparatus (retained in the anterior, and newly formed in the posterior, product of fission).

Genetics of Mating Types

Species 1 has the caryonidal system of mating-type inheritance (see page 486). The mating types of species 1 are capable of reacting over a wider range of temperatures than other species: the initial agglutination or clumping occurs at temperatures of 9–38°C; this reaction is followed by pairing and conjugation only if the cells have been above 15°C before and during the first 2 hours of mixture (Sonneborn, 1941b). A mutant "can't mate" clone was found that would give strong and prolonged clumping reactions with the complementary mating type but could not proceed to form conjugating pairs (Sonneborn, 1942c). Clones with reduced or no capacity to clump or pair were also reported by Sonneborn and Lynch (1937).

In a very extensive study never published in detail, Sonneborn (1942*b*) found that the probability for caryonidal, i.e., macronuclear, determination for a particular mating type was strongly temperature sensitive. The percentage of type-E caryonides in certain stocks of species 1 was about $18.4 + 2t$, where t is the temperature in degrees centigrade prevailing during the period from conjugation until the first postzygotic fission. This formula held over the whole range of temperature examined, i.e., from 10°C to 35°C. Later variations in temperature had no effect.

Hallet (1972, 1973) made the important discovery that exposure of stock 60 to $CaCl_2$ during the period when the cells are sensitive to temperature yields, in about 15–25 percent of the caryonides, a delay of about five cell generations in the determination of the macronuclei for mating type. Exposure to $CaCl_2$ after the first postzygotic cell division is ineffective. Delayed determination is shown by the fact that lines pure for different mating types arise not from products of the first postzygotic cell division (when macronuclear anlagen segregate), but several cell generations later. Cytological observations seem to rule out the alternative explanation that multiple macronuclear anlagen were formed in these cases. Equally remarkable is the observation that, in cytogamous pairs, delayed differentiation tends strongly to be clonal, not caryonidal, i.e., usually either both caryonides are delayed or neither one is. Delayed determination is caryonidal and somewhat less frequent after true conjugation (Hallet, 1973). Hallet's evidence for delayed determination seems to rule out hypotheses (e.g., Jones, 1956) of mating-type determination based on chromosomal segregation during the initiation of macronuclear development from micronuclei.

Although normally the overwhelming majority of caryonides are phenotypically pure for one mating type or the other, selfing caryonides containing cells of each mating type do occur in a few percent of the cases (Kimball, 1939*b*). (Possibly Seigel's (1961) high value of 27 percent in stock 90 was due to multiple macronuclear anlagen or to the Hallet phenomeon.) Usually, any cell isolated from a selfing caryonide will produce descendants of each mating type. However, sometimes sublines arise that remain constant for mating type. In the relatively few such cases studied, all the constant sublines of any one clone were alike in mating type (Kimball, 1939*b*). This raises the possibility of phenotypic assortment (see page 484) such as occurs abundantly in *Tetrahymena* (see page 441, this volume). In this connection, it is of interest that macronuclear regeneration yielded both pure and selfer sublines from a clone initiated by a fertilized cell in which the two macronuclear anlagen (one presumably determined for one mating type and the other determined for the other) had been induced by starvation to fuse into one (Nanney, 1954).

Possibly Comparable Genetic Systems

J. R. Preer (1969; pages 237–239) has reviewed papers by Génermont and Franceschi and discussed them in relation to caryonidal and cytoplasmic inheritance and to the long-lasting modifications (Dauermodifikationen) of Jollos. Génermont (1966) has pointed out that sister caryonides of the same clone show significant correlation with respect to variation in fission rate and also to variations of resistance to NaCl, but that even subcaryonides can differ in resistance to $CaCl_2$. This may, in the light of Hallet's results, be due to delayed differentiation of macronuclei for level of resistance to $CaCl_2$. In general, Génermont interprets his interesting data in terms of heritable diversities in gene action, i.e., macronuclear differentiation, analogous to that shown in the genetics of mating types and serotypes.

Genetics of Immobilization Antigens

Species 1 was the first to be examined genetically in regard to immobilization antigens or serotypes (Sonneborn, 1943*b*; Kimball, 1947, 1948). Although this early work demonstrated that both genes and nongenic factors were involved, the situation was not clarified until the general system was exposed in species 4. Soon Beale (1952, 1954) showed that the system was fundamentally the same in species 1, but some of the elements in the system were more clearly demonstrated and some new features were discovered. The various wild stocks proved to be remarkably similar in two respects: they had the same three major serotypes under standard cultural conditions—S, G, and D; in each stock, the cells were S at low temperatures, D at high temperatures, and G at intermediate temperatures. The exact temperature range for these serotypes differed for different stocks, but the *sequence* was the same. Commonly, in any one stock the temperature ranges for S and G overlapped, as did the ranges for G and D. In the overlap ranges, cells do not have both antigens, only one or the other, but both serotypes usually reproduce true to type.

These three serotypes are not the only ones possible in a stock (Beale, 1954). Sporadically, others (e.g., T) appear but are usually unstable. At least one of them, L, can be stabilized at moderate or high temperatures by adding NaCl or certain other salts to the medium. Such media also shift the temperature ranges for G and D.

Some alleles for the same serotype code for antigens so different that they cross-react little or not at all to each other's antiserum. However, there are gradations from such extreme differences down to allelic antigens that cannot be distinguished by immobilization titers or other serological tests but differ by a few spots in their fingerprints (Jones, 1965a).

The existence of serologically distinct allelic antigens has (as in species 2) led to theoretically important discoveries about allelic interrelations in hybrids. The earliest studies by Beale (1954, 1957a) showed that both alleles were expressed in a heterozygote, often one more than the other; and that antisera against the hybrid serotype contained diverse antibodies, some adsorbable by antigen of the one homozygote and the rest by that of the other homozygote. Jones (1965b) found that heterozygotes for D^{60} and D^{90} contained mainly pure 60D antigen, but they also apparently contained three classes of mixed antigen, each containing a different relative amount of 60D and 90D specificity.

Recently, Capdeville (1969, 1971) began a systematic exploration of the interrelations between pairs of alleles for antigen G as judged by immobilization responses of F_1 clones at about the twelfth cell generation after conjugation. She found that some, but not all, heterozygotic combinations (e.g., G^{33}/G^{156}) always (or nearly always) gave immobilization responses that were indistinguishable from those of one, and always the same, parent (e.g., all like 156G). In at least some cases, the gene for the missing antigen could not have been lost from the macronucleus because that antigen could be found when the clones were older. Capdeville noted the resemblance of this exclusion of an allele from expression to the exclusion of nonalleles from phenotypic expression (e.g., exclusion of S and D when G is being expressed) and (as had Finger, 1967) to the allelic exclusions of immunoglobulin production in cells of higher organisms. Nanney (1963) has discussed a comparable situation in *Tetrahymena*.

Capdeville found that other combinations of alleles behave differently. With certain pairs of alleles two kinds of clones were found; some, like those mentioned above, gave immobilization reactions indistinguishable from those of one, and always the same, parent; but other clones were immobilized by antisera against both of the two parental serotypes. In some allelic combinations, one allele was excluded from expression in the clones from *both* members of some conjugant pairs, but in the clone from only one member of other pairs. In other allelic combinations, one allele was excluded from expression in one clone of a synclone, never in both clones. The frequency and pattern of exclusion were repeatedly constant for each particular combination of alleles and consistently different for different allelic combinations.

Capdeville also noted preliminary observations that suggest some additional features of the system. First, when only one clone of a heterozygous synclone shows phenotypic exclusion of an allele, it is the clone descended from the homozygous parent that contributed the active allele. This suggests that the cytoplasmic state of a cell plays a role in determining which allele will be active. Second, some clones that had both alleles active consisted of one caryonide in which one allele was active and one caryonide in which both alleles were active. This suggests that "macronuclear differentiation" for serotypes occurs during the development of macronuclear anlagen, as in mating-type determination. Taken together with the first point, it appears that the cytoplasmic state plays a role in macronuclear differentiation.

It would be of interest to know whether different results would occur if the hybrids were produced when the parents were expressing a different serotype and were later transformed to express G; such experiments and others would indicate whether something correlated with the presence of G antigen was the decisive component of the relevant cytoplasmic state and whether the kind of macronuclear differentiation involved can be achieved only during the origin of macronuclei from products of a fertilization nucleus. Such experiments have been carried out with species 2 (see page 512). Further, in both species 2 and 4, the phenotype of a clone at the twelfth cell generation is not necessarily the same as its phenotype soon thereafter. Hence, inferences about macronuclear differentiation require more extended observations than ascertainment at the twelfth generation only, the standard stage observed by Capdeville.

Beale (1952, 1954, 1957a), exploited the great serologic differences between allelic antigens in species 1 to clarify and analyze the role of cytoplasmic states which were known to be operative in species 4. He showed, in cases where allelic exclusion did not occur, that the cytoplasmic state was locus-specific and acted on both alleles in heterozygotes, i.e., both were expressed. In one remarkable case, stock 192 seemed to lack G altogether, but when crossed to certain other stocks, the heterozygote showed two serologically diverse G's one of which must have come from the stock 192 parent. The G cytoplasmic state in stock 192 appears to be so unstable as to be virtually undetectable except in hybrids. In fact, the stability of the G states (and other states) of different stocks differ greatly and seem, at least in certain cases, to be determined by the corresponding antigen gene. That they are determined by the genome and cytoplasmically inherited was shown by finding that both clones of a synclone had the same stability for a given serotype when the synclone was heterozygous for alleles that, in the homozygous state, produced very different stabilities.

Beale (1958) discovered another phenomenon not yet found in any other species of *P. aurelia*. Among the F_2 clones obtained by crossing stock 60 to any one of three other stocks (33, 103, 145), some were characterized by a complete inability to produce antigen D. These clones appeared to have an incompetent allele, d^o, as confirmed by their failure to contribute detectable D antigen in hybrids expressing D after crossing to the other stocks. Whether these d^o alleles are intra-allelic recombinants between different alleles, whether they lack D specificity, or whether they are alleles that are regularly excluded from expression in the allelic combination tested is perhaps not yet clear.

The chemistry of the immobilization (i) antigens, first investigated in species 4, has been studied in species 1 by Bishop (1963), Jones and Beale (1963), Jones (1965a,b), Sommerville (1967, 1970), and Sinden (1973). The antigens are entirely, or almost entirely, complex proteins. Their amino acid composition is known, and they are remarkably rich in disulfide bonds. Fingerprints show 65 or more spots, and allelic D or G antigens can differ in from a few to about 20 spots. Allelic antigens are therefore not mere differences of folding of the same primary structure. Analysis suggests that there may be three basic subunit polypeptides, alpha, beta and gamma, in a subunit and two or three such subunits in the final antigen molecule of 240,000–320,000 molecular weight (but see page 552).

Population Genetics

Stocks 2, 14, 15, 16, and 20 are the only wild stocks of species 1 known to be homozygous for gene mt^o (or mt^1) and therefore genetically restricted to mating type I. They are all homozygous for genes S^{60}, G^{90} and D^{90} and were all isolated from a hillside stream near Woodstock, Maryland. Wild stocks bearing mt^o have never been found elsewhere. They were repeatedly found in the same stream from 1937 through the 1940's, coexisting with stocks of species 2 but with no other stocks of species 1 (Sonneborn, 1957, p. 213). In 1958, however, collections from this stream yielded some mating type II (Beale, private communication).

Obviously, a population composed exclusively of mt^o homozygotes cannot outbreed; it is restricted to the closest form of inbreeding (autogamy). Species 1 contains some stocks (e.g., stock 18) with virtually no immature period; these would then tend to inbreed by conjugation between close relatives. Other stocks have a considerable immature period, with more opportunity to disperse before mating and, therefore, to outbreed. Kimball *et al.* (1957) followed, in "population cages," the frequencies of alleles at three loci—*cl1*, dp, and *D*—in two nearly isogenic stocks (stock 90 and stock d1-1) which differed at these loci. Some restriction on random mating was imposed by homozygosity of stock d1-1 for gene *mt°*. In the course of 18 months, the two recessive mutants (*cl1* and *dp*) declined greatly in frequency, and one of them seemed to be entirely lost. The frequency of heterozygotes for the two codominant *D* alleles remained high and relatively constant but somewhat below expectations for random mating. The numerical results were such as would be expected if about 10 percent of the clones arose at autogamy. Hence, under conditions of high-density laboratory populations, conjugation seemed to prevail, autogamy being relatively rare. The data also indicated that deleterious recessive mutations were accumulating.

Mutagenesis and Radiation Genetics

As appears in Table 1, mutations have been routinely induced in species 1, but only Kimball (1949*a,b*, 1955) and his co-workers (Kimball and Gaither, 1951, 1953, 1956; Kimball *et al.*, 1955) have specifically investigated the genetic effects of beta radiation from ^{32}P and of x rays and ultraviolet light; their results are summarized by J. R. Preer (1969; pages 162–169). The magnitude of mutational effect of x rays is nearly an order of magnitude less when the cells are irradiated during micronuclear DNA synthesis and division than when irradiated at other stages of the cell cycle. Externally applied H_2O_2 did not produce mutations, but hypoxia decreased mutagenesis, presumably by action of H_2O_2 or HO_2 produced within the cell or nucleus. See page 493 for the dominant-lethal problem.

Killers and Bacterial Symbionts

J. R. Preer *et al.* (1974) gives an up-to-date summary of knowledge of the bacterial symbionts of all species of *P. aurelia*. Until recently, stock 540 was the only known killer stock of species 1 (Beale, 1957*b*). This and the other killer stocks (548, 551, and 555) found subsequently are mate killers (Beale and Jurand, 1966), like those first found in species 8. They contain the bacterial symbiont, mu, recently named *Caedobacter conjugatus* by J. R. Preer *et al.* (1974). Maintenance of the symbiont in stock 540 and its derivates requires the presence of gene *M1* or *M2* (Table 2) in the host (Gibson and Beale, 1961). The genetics of mu in the other stocks remains unclarified [(Beale and Jurand (1966), but see J. R. Preer *et al.* (1974)]. The metagon hypothesis (Gibson and Beale, 1962) is now in disrepute (Beale and McPhail, 1966; Byrne, 1969) and will not be discussed. Another bacterial symbiont, delta, now called *Tectobacter vulgaris* (J. R. Preer *et al.*, 1974), has been found in stock 561 which is not a killer.

Mitochondrial Genetics

The latest and one of the most promising areas of genetic research on species 1, as on species 4, is mitochondrial genetics (Beale, 1969, 1973; Beale *et al.*, 1972; Tait, 1972; Knowles, 1972, 1974; Beisson *et al.*, 1974). With or without mutagenesis, mutant mitochondria can be selected by exposure of the paramecia to growth-inhibiting concentrations of erythromycin, chloramphenicol, or mikamycin. Cell growth resumes only after mutant mitochondria have multiplied greatly. Mutants resistant to one antibiotic can be similarly selected for resistance to another antibiotic. That the altered cell character is based on mutant mitochondria was proved by injecting purified preparations of mitochondria from mutant into wild-type cells and demonstrating that antibiotic resistance had been transferred. Long culture, in nonselective media, of cells with a mixed population of mitochondria (wild-type and resistant, or resistant to two different antibiotics) reveals a system of selective advantage of various types in competition with others. Thus far, no recombination of mitochondrial genes has been reported in species 1. Erythromycin resistance seems to be correlated with altered mitochondrial ribosomes. The buoyant density of mitochondrial DNA (kinetic complexity 35×10^6 daltons, 40 percent G + C) in CsCl is 1.699 g/cm^3 as compared with 1.686 for nuclear DNA (Flavell and Jones, 1971).

Species 2

Introduction

The unique research roles played by species 2 go back so far that the earliest chapters [recounted in Jennings (1929)] have been all but forgotten. In the 19th century, Weismann contended that the distinction between the mortal soma and the potentially immortal germ line in multicellular animals did not hold for unicellular organisms because in them soma and germ were not segregated into different cell lines. Hence, argued Weismann, Ciliates are potentially immortal. On the other hand, Maupas held that the argument did not fit his observations: Ciliates were not potentially immortal, but destined to pass through a limited clonal cycle beginning with conjugation and ending in death. In the early 20th century this opposition of views was continued by Woodruff and Calkins. Working with *Paramecium*, Calkins basically confirmed Maupas, but

Woodruff provided incontrovertible evidence of potential immortality without recourse to conjugation at all. It turned out, in a surprising way, that both were right, and the resolution of the conflict (which came after Jennings' review was published) depended on new fundamental cytological and genetic discoveries made over the course of a quarter of a century (1914–1939).

Calkins studies were carried on with *P. caudatum*, Woodruff's with *P. aurelia*, in fact with strain 23 (formerly W) which he isolated from nature in 1907. Eventually, stock 23 was found to belong to species 2. It is still on hand after a record 67 years in the laboratory. Stock 23 came to be referred to as *"Paramecium Methusalah."* For many years and many thousands of generations, it retained its vigor and never conjugated; however, to everyone's surprise, in 1914 Woodruff and Erdmann reported the periodic occurrence in stock 23 of a "nuclear reorganization" which they called endomixis. According to their description, endomixis involved neither meiosis nor fertilization, but a mitotic product of the micronucleus developed into a new macronucleus and the old macronucleus disintegrated and disappeared. If the macronucleus can be considered somatic, there was thus "partial death" of the soma and replacement from the germ line (micronucleus); but not fertilization and no conjugation. Diller (1936) reinvestigated the matter in stock 23 and in a number of other stocks and confirmed macronuclear replacement, but found that the replacement was from a product of a fertilization nucleus. He called the process autogamy. Autogamy does not normally occur ·in *P. caudatum*. This was apparently the basis of the diverse results of Calkins and Woodruff: autogamy can serve as a substitute for conjugation in initiating new clonal cycles and preventing clonal senescence and death. In both *P. aurelia* and *P. caudatum*, the "germ line," but not the "soma," is immortal.

Because for many years he failed to find a single pair of conjugants in stock 23, Woodruff held that this was a "non-conjugating" stock, unable to conjugate, although over the course of many years he once or twice found a few pairs of conjugants. Much later the reasons became apparent: stock 23 consists almost exclusively of mating-type IV and mates as avidly as any other stock of species 2 under appropriate conditions, i.e., at night, at low temperatures, and when type III is available to it (Sonneborn, 1938*a*).

Characterization of Species 2

Species 2 is cosmopolitan; it has been found in North and South America, Europe, the Middle East, Asia, and New Zealand. In a number of respects, there is more diversity among wild stocks of species 2 than among those of any other known species of *P. aurelia*. However, it can apparently be uniquely defined by the combination of its electrophoretic patterns of mitochondrial beta-hydroxybutyrate dehydrogenase (one cathodal band; Tait, 1970*a*) and esterase C (one cathodal and one anodal band; Allen and Gibson, 1971). The form of conjugant pairs seems to be also unique: the longitudinal axes of the mates are parallel or almost so, and the conjugants are straight-sided; in other species, the axes diverge posteriorly and the cells are not straight-sided.

Genes and Chromosomes

Table 3 lists the known genes in species 2. See Table 3, loci *r1* and *V* for the only report of possible linkage. There are no published accounts of chromosome numbers or karyotypes.

Stocks

The wild stocks of species 2 are listed in Table 4 according to their geographical origin. The available information on the alleles they bear at

TABLE 3.   *Genes in Species 2*

Gene locus[a]	Characteristics
alf	Polymorphic locus for maintenance of symbiont alpha (Beale *et al.*, 1969). Stock 114 carries *alf* which results in loss of alpha; stock 562 carries *Alf* which permits maintenance of alpha. Stock 576 also carries a gene (allelism unknown) that will not support alpha (Gibson, 1973).
C	Polymorphic locus coding for the immobilization (i)-antigen C of serotype C. Alleles C^7, C^{30}, C^{72}, C^{83}, and C^{197} code for the serologically distinguishable C antigens inwild stocks designated by the superscripts (Finger *et al.*, 1966). Not linked to *E* or *ts1*.
E	Polymorphic locus coding for the immobilization (i)-antigen E on the surface of cells of serotype E. Alleles E^{72} and E^{197} code for the serologically distinguishable E antigens in stocks 72 and 197, respectively (Finger and Heller, 1962). Allele E^7 codes for a defective E carried by stock 7, detectable in F_1 of cross to stock 197 (Finger and Heller, 1964). Stock 30 may carry a null allele, e^{30} (Finger, 1957*b*). Not linked to *C* or *ts1*.
EsA	Polymorphic locus for esterase A (Allen and Golembiewski, 1972). Codominant alleles EsA^{50}, EsA^{93}, and EsA^{305}, occurring in stocks 50, 93, and 305, respectively, are for electrophoretically distinguishable forms of the enzyme. Not linked to EsB.
EsB	Polymorphic locus for esterase B (Allen and Golembiewski, 1972).

[a] See footnote to Table 1 for an explanation of the gene nomenclature system used here.

TABLE 3. Continued

Gene locus[a]	Characteristics
	Codominant alleles EsB^{50} and EsB^{305}, occurring in stocks 50 and 305, respectively, are for electrophoretically distinguishable forms of the enzyme. Not linked to EsA.
Idc	Polymorphic locus for cytoplasmic isocitrate dehydrogenase (Tait, 1970a). Idc^1, in stock 50 and many others, and Idc^2, in stock du41, are for electrophoretically distinguishable forms of the enzyme. Not linked to *Idm*.
Idm	Polymorphic locus for mitochondrial isocitrate dehydrogenase (Tait, 1970a). Idm^1, in stock 50 and many others, and Idm^2, in stock 583, are for electrophoretically distinguishable forms of the enzyme. Not linked to *Idc*.
k	Polymorphic locus concerned with the killer trait (Balsley, 1967). Allele *k* occurs in wild stock 1010, allele *K* in stock 7. Homozygotes for *k* lose kappa (and also a mutant kappa) and the killer trait, the maintenance of which depends on the presence of allele *K*. Not linked to *tr* or *ts1*. (Stock 1010 carries nonkiller symbiont, nu.)
r1	A hypothetical polymorphic regulatory locus which would act on and be either closely linked to, or perhaps identical with, locus *V* (Cooper, 1968). *R1* is assumed to repress *V*, while *r1* does not. If so, *R1* would be carried by wild stocks 3 and 197; *r1* by wild stocks 7, 9, and 28. Not linked to *r2* or *tr*.
r2	A polymorphic locus for regulation of activity of genes at locus *V* (Cooper, 1968). *R2* represses activity at *V*, *r2* does not. *R2* occurs in wild stocks 3 and 197; *r2* in 7 and 8. Not linked to *r1* or *V*.
s	A polymorphic locus for maintenance of killer symbiont sigma [Balsley, cited by J. R. Preer *et al.* (1974)]. Stock 11 carries *s* and is unable to maintain the sigma of stock 114. Stock 114 carries *S* which can support maintenance of sigma.
tr	A mutation of trichocyst form found in a subline of stock 197 (Preer, 1959a). Homozygotes for *tr* have undischarged trichocysts of variable abnormal shape, sometimes lacking tips, and they are slow to discharge after animals are crushed beneath a cover slip. ts^+ yields normal trichocysts. Not linked to *k, ts1,* or *V*.
ts1	A possibly polymorphic locus for temperature sensitivity (Preer, 1957). Stock 35 carries *ts1* which results, at 31°C, in no more than 6 cell generations in 4 days, most of these during the first two days; practically none on the fourth day. Stock 30 carries $ts1^+$ (or *Ts1*); like most stocks, it undergoes not less than 8, usually 10–12, generations in 4 days at 31°C. Not linked to *C, E, k,* or *tr*.
V	A polymorphic structural locus for antigen 5, not an immobilization (i) antigen (Cooper, 1968). Serological differences distinguish the V antigens coded by multiple alleles; V^3 in stocks 3 and 11; V^{30} in stock 30; V^7 in stocks 7 and 53; V^9 in stock 9; and V^{28} in stock 28. In heterozygotes, V^7 and V^9 are codominant, but V^3 is dominant over V^7, V^9, and V^{28}. Not linked to *tr* or *R2*.

TABLE 4.   Wild Stocks of Species 2[a]

Source	Stock
Eastern coastal states of the USA	1, 4, *5, 7, 8,* 9, 11, 12, 21, 23, 28, 20, *34,* 35, *36,* 49, *50,* 86, 88, 91, 93, 100, 104, 122, 149, *249,* 308-2
Intercoastal states of USA	53, 71, 72, *114,* 160, 187, 259, 260, 305, *1010*
California	*235, 318*
Chile	179, 185
Europe	
Ireland	*564*
Scotland	*511, B1-166-1, Hu35-1*
England	*576*
Norway	206, 207, 208
France	*517*
Germany	*193, 197,* 304
Italy	*526, 527, 537, 562, 563, SG*
USSR	*570, 1035, 1038, 1039, 1041*
Lebanon	291, *292*
Japan	234
New Zealand	*310*

[a] Killers and symbiont-bearers are italicized. Most of these stocks are available from either Beale in Edinburgh or the Preer-Sonneborn laboratory in Bloomington, Indiana.

polymorphic loci is included in the account of the loci in Table 3. Some derived stocks incorporating genes (especially *k, tr, ts1,* and serotype alleles) from one or more wild stocks into the genome of another wild stock have been constructed by Finger (1957*b*) and Balsley (1967).

Nuclear DNA

See pages 472–474 for references and results on this topic.

Cells and the Cell Cycle

Powelson *et al.* (1974) examined, for a set of standard conditions, measurements and counts on cells of stocks 53, 114, 149, 176, 206, and 235. Extreme stock means among these six stocks for the more important characters were: 147–169 μm cell length; 39–47 μm (and 36–41 basal bodies) between the two contractile vacuole pores: and 27–31 kineties between the posterior pore and the anterior end of the cytoproct. Generation time is about 6–8 hours at 27°C in young vigorous clones. This species cannot tolerate as high a temperature as others can. Generation time is minimal at 27–30°C and falls off sharply above that, the cells eventually dying at 34°C (Sonneborn, 1957; pages 175–176). According to Preer (1957), most stocks of species 2 live for several days at 34°C.

Clonal Cycle

Sonneborn (1937*b*, 1938*b* 1954*c*) and Pierson (1938) examined the clonal cycle of stock 23 in detail and delineated the progress of increasing mortality at autogamy with increasing clonal age. As in species 1, there are differences among stocks of species 2 with respect to the existence and duration of the period of immaturity following conjugation (Sonneborn, 1957; page 191). The maximal clonal life span observed was 303 cell generations, with marked decline in fission rate and increase in frequency of the abnormalities of senesence beginning at about 150 generations. In this species, as in others, the observed maximal life span depends largely on the number of daily reisolation lines being followed. During senescence the risks are great of termination by death or an unwanted autogamy.

Asexual Genetics

Hereditary variations arising within a clone will be mentioned in sections dealing with mating types, serotypes, and killers. The section above on the clonal cycle implies that different sublines of a clone age at different rates, as occurs in all species studied.

Mating Types

Mating types in species 2 are expressed only sporadically, if at all, at 10°C and at 25°C and above, and during the hours from 7 P.M. to 9 A.M., when cultures are exposed to the daily alternation of light and dark (Sonneborn, 1938*a*). Peak mating reactivity occurs at 17–19°C between 1 A.M. and 5 A.M. (Sonneborn, 1939). The B, or clonal, system of mating-type inheritance was found first in species 2 (Sonneborn, 1942*a*). Most of the observations (by Sonneborn in the 1940's and Rudman in the 1950's) have not been published. In general, they are similar to the observations on species 4, but there is much more diversity in detail among different wild stocks, e.g., some are extremely stable for type III, others for type IV, and stocks differ markedly in the frequency of selfing caryonides after conjugation. Rudman (Rudnyansky, 1953) observed that the frequency of selfing caryonides was decreased and the frequency of type-III caryonides increased by exposing exconjugants, prior to the first cell division, to homogenates of either type-III or type-IV cultures. This treatment reduced growth rate. In other systems (e.g., serotypes) reduced growth rate has effects similar to reduced temperature. Rudman indeed found that reduced temperature before the first fission also increased the frequency of type-III caryonides. These observations suggest that type III is the homolog of the O mating type of other species, but direct proof is impossible because neither type III nor type IV reacts with either mating type of any other species.

Serotypes

Among the reviews on serotypes, those of Finger (1967, 1974) give a particularly full account of the phenomena in species 2. In many respects, the system is essentially the same as in species 4 and 1 (Finger, 1957a,b). Preer early found 8 serotypes (A–H), three of these (A, B, and G) being serologically similar to those designated by the same letter in species 4; but the others are not. Serotype G is serologically identical in all stocks of species 2, but other serotypes differ somewhat from stock to stock. Some serotypes are lacking in certain stocks [e.g., no E in stocks 7, 30, and 83; but stock 7 can produce a defective or incomplete E protein according to Finger and Heller (1962)]. Most attention has been directed to serotypes C, E, and G.

Finger (1956) demonstrated that the serotype antigens are soluble and can be studied by gel diffusion techniques. Finger and Heller (1962), Finger *et al.* (1963), and Finger (1964, 1974) then showed, for example, that each allelic C antigen has a distinctive combination of 5–8, of a total of 11, serologically distinguishable antigenic determinants, each presumably involving a different site (or set of adjacent sites in the folded configuration) of the whole antigen molecule.

This information made possible a more penetrating and extensive analysis of the serotypes of heterozygotes than has been accomplished in any other species of *P. aurelia* (Finger and Heller, 1963, 1964; Finger *et al.*, 1966). The type of antigen produced by a heterozygote depends in part upon which one is being formed in the parents at the time they conjugate. If, for example, one parent is serotype E and the other is not, the E antigen formed in *some* of the resulting heterozygotes (for two distinguishable *E* genes) is indistinguishable from the E antigen formed by the serotype-E parent. This holds not only for the clone from the serotype-E parent, but also for the clone from non-E parent when it is induced to transform to serotype E. The *E* gene of the non-E parent thus can be excluded from expression in both clones of a synclone. Other identical heterozygotes from the same cross can, however, form a hybrid E antigen that contains some antigenic determinants distinctively characteristic of the E antigen of the one parent and some, of the other parent. The frequency of clones forming hybrid antigens is increased if the parents exchange cytoplasm during mating. Hybrid antigens are usually formed when *both* parents are of the same serotype (but of course bear different distinguishable alleles for that serotype). Detailed studies were made on these hybrid antigens in crosses between homozygous stocks expressing serotype C. Surprisingly, different clones of identical heterozygous constitution formed different hybrid C antigens. There were, in fact, four different combinations of the antigenic determinants of the C antigens of the two parents for any one parental combination. Equally surprising, any one clone seemed to form only one of the four possible combinations.

In a searching analysis of the implications of these findings, Finger *et al.* (1966) were led to conclude that the C antigen is composed of at least two subunits, each consisting of three polypeptides. They made a tentative assignment of the antigenic determinants to each of the three polypeptides. This interpretation agrees with the model of the serotype antigens arrived at by others from physical and chemical analyses of antigens in other species of *P. aurelia* (page 504 and page 552).

The work on antigens of species 2 obviously bears on Capdeville's studies on species 1 and, in general, on the regulation of genic action. Unfortunately, Finger has not yet reported a study of *caryonides*, but only clones and synclones. A caryonidal study would throw light on whether, as is suggested by the differences between the two clones of a synclone and by the cytoplasmic effect mentioned above, macronuclear differentiation underlies the various results obtained among identical heterozygotes. J. R. Preer (1969; page 224) has pointed out this theoretically important possibility.

Finger and his associates (Finger, 1968; Finger *et al.*, 1962, 1969, 1972*a,b,c*) have reported extensively on unstable clones. Although fairly stable for serotype when grown in high population densities, unstable clones give rise to sublines of varied serotypes when single cells are isolated and grown in low population densities. Some unstable clones, called Z clones, do not become immobilized when exposed to any antiserum. These clones, and presumably their cells, contain *two or more* antigens which can be extracted and shown to be serologically identical with known i-antigens. Other unstable clones have a known single serotype indistinguishable from serotypes of stable clones, yet they contain, in addition, "secondary" antigens which are serologically identical with the i surface antigens of other serotypes. These two kinds of unstable clones are apparently examples of exceptions (first found in species 4, page 551) to the usual rule of mutual exclusion, i.e., that genes at only one i-antigen locus come to expression in any one nontransforming cell. Unstable clones have been used extensively to study aspects of stabilization and transformation of serotype. The details, which include matters of considerable potential importance for understanding the control of genic actions, are given especially in the papers of Finger *et al.* (1972*a,b,c*) and are too complex and insufficiently resolved to recount here.

Mitochondrial Genetics

Preer and Preer (1959) found that the mitochondrial fraction of homogenized paramecia of stock 3 contains a distinctive antigen (their antigen 3). Finger *et al.* (1960) confirmed this and added that the specificity of the mitochondrial antigen underwent "transformations" comparable to those that characterize the surface antigens. The implied genetic possibilities have not been followed up.

Killers and Symbionts

Killers, sensitives, and stock differences in resistance of nonkillers were first found in species 2 (Sonneborn, 1939). Preer (1948*a,b*, 1950) made a series of brilliant discoveries about them. Under some conditions kappa and the paramecia reproduce at different rates. Utilizing these differential growth rates, he completely and irreversibly diluted out kappa and showed that one kappa particle sufficed to restore a full population under some conditions of slow growth. His characterization of the hypothetical particle's size by x-ray inactivation led him to discover that it could be seen microscopically and was present in the previously inferred numbers per cell. He (Preer and Stark, 1953; Preer *et al.*, 1953) then proceeded (1) to discover that there were two kinds of kappa per cell: N, or nonbright, particles and B, or bright, particles containing a refractile, or R, body; and (2) to show that killing activity was associated with the B particles. Preer and Preer (1964) isolated the R bodies of B particles and found killing activity to be associated with them. Preer and Preer (1967) then discovered that there is a viruslike particle associated with the R body. J. R. Preer *et al.* (1971) isolated and characterized these particles which appear to be defective DNA phages. Kappa is thus a lysogenic bacterium. L. B. Preer *et al.* (1974) succeeded in inducing the prophage to mature by exposure to ultraviolet. J. R. Preer *et al.* (1974) list and name the following classes of bacterial symbionts found in stocks of species 2: kappa (*Caedobacter taeniospiralis*); sigma (*Lyticum flagellatum*), the rapid-lysis killer agent (Sonneborn *et al.*, 1959); mu (*Caedobacter conjugatus*), the mate-killer symbiont; nu (*Caedobacter falsus*), a non-killer symbiont; delta (*Tectobacter vulgaris*), also a non-killer symbiont (Beale *et al.*, 1969); and alpha (*Cytophaga caryophila*), a non-killer and readily infectible symbiont that lives in the macronucleus (L. B. Preer, 1969). No other species of *P. aurelia* can boast such a variety of known symbionts.

Species 3

Introduction

Although relatively little work has been done on species 3, the first homology between mating types of different species was based on reactions between mating-type II of species 1 and type V of species 3 (Sonneborn, 1938*a*) and the effect of temperature on mating-type determination during a sensitive period was first discovered in species 3 (Sonneborn, 1938*a*, 1939). Work on species 3 has centered almost exclusively around enzymatic characters, mating types and serotypes.

Characterization of Species 3

Species 3 is distinguished from all other species of *P. aurelia* by the combination of electrophoretic patterns for two or three enzymes: one cathodal band for mitochondrial beta-hydroxybutyrate dehydrogenase (Tait, 1970*a*) and one anodal band for esterase D; but because of somewhat erratic results with the latter enzyme, the absence of a definite band for esterase C is a helpful additional criterion (Allen *et al.*, 1973).

Genes, Chromosomes, and Stocks

No gene or chromosome studies have been reported for species 3. This species has been found only in North America. Of the nearly 40 natural sources in which it has been found, only one (Florida, stock 324) is located south of the latitudes of Palo Alto (one stock, 286), Kentucky (one stock, 186), and Maryland (three stocks, M, Y, and 279). One of the few collections from Alaska yielded stock 275. Other stocks from the mid-Atlantic states are 42, 77, 78, 79, 92, 95, 98, 152, 154, and 331; from south Canada and northern states of the USA are stocks 37, 112, 121, 125, 161, 188, 191, 231, 254, 261, and 283; and from Indiana are stocks 52, 55, 58, 70, 133, and 136.

Mating Types

The caryonidal system of mating-type inheritance in species 3 seems to be obscured when *both* conjugation and the period preceding the first postzygotic fission take place at 10°C: the type-V mate regularly produces a type-V clone, and the type-VI mate produces a type-VI clone (Sonneborn, 1939). This is the result that would be expected if macronuclear regeneration occurred in all exconjugants. At that time, macronuclear regeneration had not been discovered and the matter has subsequently not been investigated. At other temperatures, typical caryonidal determination of mating type and typical phenomic lag occur (Sonneborn, 1939).

Like species 2, species 3 has a rhythm of mating reactivity. It becomes reactive only at 27°C or lower, best at 24°C or lower (Sonneborn, 1938a). Reactivity, when cultures are exposed to the normal daily alternation of light and dark, is restricted to the period between about 1 A.M. and 1 P.M. but is optimal only between 4 A.M. and 11 A.M. Sonneborn claimed that the period of reactivity was gradually extended by continuous culture in the dark, until the cells could react at all hours after 5 days in the dark. Conversely reactivity was restricted progressively by continuous culture in light, until it totally disappeared after 5 days. Karakashian (1968) made a more-extended and precise study of the rhythm of mating reactivity. She showed that it was a typical circadian rhythm with a free running period of about 22.2 hours, that the rhythm persists at least 4 days in continuous darkness, and that it disappears quickly in continuous illumination. However, Karakashian found some stock differences in certain aspects of the rhythm. Breeding analysis of these differences has not been reported.

Serotypes

Melechen (1954) described the serotype system of species 3. His results agree, as far as they went, with the previously obtained results of others on species 1, 2, and 4. He identified altogether 8 serotypes. Some of these were serologically homologous with serotypes of other species: there were homologs of the A and B of species 2 and 4; of the D of species 1 and 4; of the G of species 1, 2, and 4; and of the S of species 1. Other than this, little or no work of genetic interest was carried out.

Species 4

Introduction

Although species 4 was not discovered until several years after species 1, 2, and 3 (Sonneborn and Dippell, 1943), it soon took the lead in genetic work on *P. aurelia* and has been more extensively and intensively studied than any other species.

Characterization

From the account of Allen *et al.* (1973) species 4 is the only one that forms two anodal electrophoretic bands of esterase B(using axenic cultures and electrostarch gel).

Genes

Far more genes are known in species 4 (Table 5) than in any of the other species. Almost all of the genes are maintained in the genome of stock 51. This stock has long been the standard stock of species 4 and is by far the genetically best-known stock of any unicellular animal. Because of the karyotypic differences among stocks and the incompatible chromosomal combinations produced by recombination after stock crosses, selection of a single, standard, wild stock for genetic work appeared to be a practical necessity. At the time the choice was made by Sonneborn, so much work had already been done on stock 51 that it clearly seemed the material of choice, although its haploid genome does not have the smallest number of chromosomes.

In order to make Table 5 as useful as possible, I have included a large amount of unpublished information, partly because so much of our knowledge of the genes in species 4 has been recently acquired. J. R. Preer's (1969) list contained less than 20 loci; Table 5 includes about 120 loci. Although I have tried to find, assemble, and check as much information as possible in the time available, the task has not been done to my satisfaction. There are doubtless errors of omission and commission, and failures to find available pertinent information. Table 5 should therefore be considered as a preliminary attempt, but one which may begin to satisfy the needs of researchers and hopefully lead to a better production in the near future. As indicated in Table 5, many people have contributed unpublished information. I am heavily indebted to all of them, but especially to Mary L. Austin and Myrtle V. Schneller who have, in addition, retrieved and summarized for me not only their own data, but also the data of others in our laboratory, and have helped check the whole table. Table 6 classifies the genes by their phenotypic effects.

Linkage and Chromosomes

Table 7 lists the 15 genes for which linkage has been found. These tentatively mark 7 linkage groups in stock 51. Dippell (1954) reported

TABLE 5. *Genes of Species 4*

Gene locus[a]	Characteristics[b]
A	Immobilization (i)-antigen (serotype) A. Serologically distinguishable forms are coded by codominant polymorphic alleles, e.g., A^{29}, A^{32}, A^{51} (TS, 1950*b*). (The superscripts designate the stock in which each allele was found.) A^{29} is carried as stock d4-1; A^{32} (with E^{32}) as stock d4-45. Serotypes 29A, 32A, and 128A are difficult to distinguish. Serologically different from these, but difficult to distinguish from each other, are serotypes 47A, 51A, 116A, and 277A. Probably different from both of these groups are serotypes 126A and 139A. Not closely linked to *big4* (MI), *E* (MS), *ts1001* (MS), *ts1012* (MS), *ts2108* (MS). Not linked to *am* (148; MS), *big1* (77; MA), *cl2* (80; JD, MS), *cyr* (77; MA), *D* (228; IT, TS, MA), F (186; TS *et al.*, 1953*b*), *fna* (77; MA), *H* (115; TS *et al.*, 1953*b*), *hr* (204; MA), *k* (305; TS *et al.*, 1953*b*), *mtA* (600; MS, MA), *ndB* (77; MA), *nd3* (180; MS), *nd6* (134; MS), *PaA* (204; MA), *ptA* (132; MS), *scl* (77; MA), *sn* (803; AS, MS, MA), *tam8* (114; MS, MI), *th* (TS *et al.*, 1953*b*), *ts111* (283; MA), *ts401* (100; MS), *tsm21* (100; MS), or *tw1* (607; MS, MA)
am	Amacronucleate (TS, 1954*a,b*; RN, 1959, 1960, 1961). Obtained with UV on stock 51. Frequent (20–25 percent) grossly unequal distribution of macronucleus at cell division yielding

[a] See footnote to Table 1 for the system of gene nomenclature. With a few obvious exceptions, the loci listed and the information on their independence or linkage refer to the genome of standard stock 51 or approximately isogenic derivatives of it, symbolized by d4 followed by a dash and another number. Some mutations are listed separately, even when tests for allelism with some or all other mutations having similar phenotypes are lacking, thus calling attention to gaps that need to be filled in. Genes are listed as nonallelic if they complement in F_1 but were not followed through the F_2. They are listed as not *closely* linked if the segregation ratio in a *small* F_2 (about 50 clones or less) was not significantly different from expectation for independent assortment. They are listed as not linked if the F_2 consisted of more than 50 clones, usually very many more. Large F_2's do not necessarily include all available data. No F_2 showing more than 40 percent recombination (up to 60 percent!) differed significantly from independent assortment. More than 90 percent of the F_2's with more than 40 percent recombination actually showed between 44 and 56 percent. Obviously, further data may show loose linkage between some of the loci listed as not linked. Information about pairs of loci is listed under both loci of the pair. The first entry, under the locus coming first in the alphabetical list, includes, in parentheses after the locus symbol, the size of the segregating F_2 and the source of the information whenever I was able to obtain it. The second entry, under the locus symbol coming second in the alphabetical list, does not repeat this information; all repeated entries as to independence are grouped together in alphabetical order between parentheses.

[b] ABBREVIATIONS: q.v., which see; ts, temperature-sensitive; NG, nitrosoguanidine; UV, ultraviolet. Sources of information are usually indicated by initials, as follows: *AS*, Anne Smith (Wells College); *BaB*, Barbara Byrne (Wells College); *BB*, Bruce Byrne (Wells College); *B & R*, J. Beisson and M. Rossignol (Laboratoire de Génétique Moléculaire, Gif-sur-Yvette, France); *CK*, Ching Kung (University of California, Santa Barbara); *C & K*, Chang and Kung; *EB*, Elias Balbinder (University of Rochester); *IT*, Irwin Tallan (University of Toronto); *JB*, Janine Beisson (see B & R); *JD*, Judy Dilts; *JDB*, James D. Berger (University of British Columbia); *MA*, Mary L. Austin; *MI*, Mary Ingle; *MS*, Myrtle V. Schneller; *RK*, Richard Kimball (Oak Ridge National Laboratory, Tenn.); *RN*, Renzo Nobili (University of Pisa, Italy); *RW*, Robert Whittle (University of Sussex, England); *SI*, Satomi Igarashi (University of Alberta); *SP*, Sidney Pollack; *TS*, Tracy Sonneborn; *VH & K*, Van Houten and Kung (see CK); *W & S*, R. Whittle and Lily Chen-Shan (University of West Virginia). Where no address is given above, it is Indiana University. Initials plus a date refer to a publication; where no date is given, the information is a private communication. Most private communications from SP are in his Ph.D. Thesis (Pollack, 1970).

TABLE 5. *Continued*

Gene locus[a]	Characteristics[b]
	some cells with no macronucleus. At the first cell division after fertilization, both macronuclear anlagen often (70 percent of the time) pass to the posterior product of division, the anterior product then undergoing macronuclear regeneration (see page 482); at later cell divisions, as long as one or more intact fragments of the prezygotic macronucleus are present, the same result can follow (20–25 percent). Completely amacronucleate cells lacking macronuclear fragments cease to feed, quickly become thin and unable to mate, 'nd die within 24–48 hours. Clones homozygous for *am* have a clonal life cycle of about 130 cell generations, little more than one-third of the normal (RN, 1960). Slightly reduced growth rate (5–8 instead of 9–10 generations in two days) at 35°C. For uses as a tool in genetic experiments, see RN (1962, 1964), RK and Prescott (1964), Smith-Sonneborn and Plaut (1969), and JDB (1973*b*). Carried as stock d4-39. Not closely linked to *cl2* (MS), *E* (MS), or *ndB* (MI). Not linked to (*A*), *D* (54; MS), *H* (269; MS), *k* (53; MS), *nd3* (312; MI), *nd6* (312; MS, B & R, MI), *ptA* (84; MS), *sml* (156; MI), *sn* (225; AS), *tam8* (148; MS), *ts111* (B & R), *ts401* (156; MI), or *ts2109* (157; MI).
ata	Atalanta (C & K). Obtained with NG on stock 51. Named after the Grecian runner who stopped to pick up golden apples! Responds to Ba solution (8 mM Ba, 1 mM Ca) by making frequent stops during forward locomotion, but not backing. In culture medium, behavior nearly normal; but incapable of sustained backward movement. According to C & K, not linked to *bdB*, *fna*, *PaA*, *pwA*, *pwB*, *pwC*, or *sp*.
B	Immobilization (i)-antigen (serotype) B. Different wild stocks (e.g., stocks 32 and 51) have serologically distinguishable B antigens presumably coded by different alleles; but no breeding analysis.
bdA	Body deformation (CK). Obtained with NG in stock 51. Variable expression: exaggerated oral groove to extremely twisted body. Cells may spiral to left or to right while swimming (JD). Carried as stock d4-92. Not linked to *bdB* (CK), *k* (161; JD), *pwB* (CK), *stA* (89; JD), *ts111* (CK), or *ts2112* (JD).
bdB	Body deformation (CK, 1971*b*; mutant C68). Obtained with NG on stock 51. Variable expression: normal, to bent anterior end, to monstrous distortion (C & K, 1973*b*). Carried as stock d4-93. Not linked to (*bdA*), *fB*, *fna*, or *ts111* (CK, 1971*b*); to *ata*, *jk*, *PaA*, *pwA*, *pwB*, *pwC*, *slg*, or *sp* (C & K); to *na*[ts] (VH & K).
big1	Big cells. Obtained by CK with NG on stock 51; found and characterized by MA. Cells have bent anterior ends, swim

TABLE 5. Continued

Gene locus[a]	Characteristics[b]
	slowly and in wide spirals, divide normally, but have a fission rate about 4 fissions per day at 27°C; back and whirl in Dryl's solution (see *fA*); some cells die during moderate starvation. In combination with *fna*, *sc*, and *tw1*, the double mutant shows effects of both genes and is distinguishable from both single mutants. Carried as stock d4-128. Linked to *big4* with 28.5 percent recombination (MA). Not closely linked to (*am*), *nd6* (MI), *tam8* (MI). Not linked to (*A*), *cl2* (467), *cyr* (535), *d* (646), *fna* (1394), *hr* (516), *mtA* (184), *ndB* (>1500), *nd3* (161), *PaA* (200), *scl* (1125), *sn* (196), *ts111* (197), *ts401* (157), or *tw1* (199)—all by MA.
big4	Big cells. Obtained by CK with NG on stock 51; found and characterized by MA. Differs from *big1* as follows: anterior end less blunt; cells longer and narrower; unequal cell (and, occasionally, macronuclear) divisions, the posterior cell being smaller than and not tandem to the anterior cell; swims fast; normal fission rate; no backing in Dryl's solution (see *fA*); doesn't die during moderate starvation. Carried with *big1* and *ndB* as stock d4-130. The double mutant (*big1*, *big4*) is distinguished from both single mutants by slower fission rate (2–3 per day), by greater morphological abnormality and more variability in form (ventral surface sometimes flat), by greater size difference between the two products of fission, by high death rate in depression cultures, and by settling of cells to bottom of slide depressions. Linked to *big1* (q.v.). Not closely linked to (*A* or *am*). Not linked to *ndB* (321; MA).
big38	Big cells (B & R). Obtained with NG on stock d4-2. Grows slowly and is somewhat sensitive to 35–36°C. Not linked to *ts401* (B & R).
C	Immobilization (i)-antigen (serotype) C. Serologically distinguishable forms in stocks 29, 47, and 51; but the C's of stocks 32, 128, and 173 are only slightly different from 47C. No breeding analysis, but different C's are presumed to be due to multiple alleles.
ch1	Chain-former (BB). Obtained with NG on stock 51. Fails to complete cell constriction during fission, more frequently in shallow than in deep culture. Sporadic expression in heterozygotes; perhaps more so at 35°C than at 27°C (MS). Unlike *sn* (q.v.), *ch1* does not cease to form chains when iron is added to the medium (JDB). Carried as stock d4-78 which dies in 2 days at 36°C. Not yet clear whether the ts effect is due to *ch1* or to a linked gene. Not allelic to *mtB* (JD) or *sn* (JDB). Not linked to *cl2* (406; SP), *ftA* (506; SP), *k* (457; SP, MS), *scl* (350; MS, SP), or *tsm21* (179; MS).
cil2	Some cells lose their cilia and parasomal sacs and then die (JB). Clones perpetuated by cells which retain cilia but continue to

TABLE 5. *Continued*

Gene locus[a]	Characteristics[b]
	produce some deciliated nonviable progeny. Obtained with NG on stock d4-2. Not linked to *ts111* (JB).
cl1	Clear: absence or great reduction in size of cytoplasmic crystals. Obtained by RK (1953*b*) with x rays on stock 51. Carried as stock d4-49. Not closely linked to *tw2* (JD). Not linked to *cl2* (57; MS). Another gene, designated *cl1* by Sainsard *et al.* (1974), is listed here as *sg1* (q.v.); it is not English "clear" but French "slow growth" (croissance lente).
cl2	Clear: like *cl1*, but clearer and less difficult to score than *cl1*. Variable expressivity makes scoring problems. When present, crystals form a circular mass (SP). Obtained by RK (1953*b*) with x rays on stock 51. Carried as stock d4-50. Linked to *hr* with about 33 percent recombination (MA); in same linkage group as *pwB*, but assorts independently (403; MA). Not closely linked to (*am*). Not linked to (*A, big1, ch1, cl1*), *cyr* (161; MA), *D*(126; MA), *fna* (328; MA), *ftA* (316; SP), *k* (468; SP), *mtA* (90; JD, MS), *mtB* (247; JD), *ndB* (552; MA), *nd3* (244; JD), *PaA* (90; JD, MS), *ptA* (145; MA, MS), *scl* (840; SP, MS, JD), *sn* (90; JD, MS), *ssw1* (308; MA), *ts111* (90; JD, MS), *ts401* (337; JD, MA), *ts1001* (271; MA), *ts2108* (271; MA), *tw1* (180; MA, MS and JD), or *tw2* (302; MA).
cl3	Clear (B & R). Obtained with NG on stock d4-2. Not linked to *ts401* (B & R).
cl4	Clear (B & R). Obtained with NG on stock d4-2. Not linked to *ts401* (B & R).
cro1	Crochu (hooked): bent and long anterior (i.e., anterior to vestibule); sporadic patches of misplaced and misoriented kinetosomes (B & R). Obtained with UV on stock d4-2. Not linked to *kin241, sg1, tam8, ts401*—all by B & R.
cu^r, Cu^s	Polymorphic locus for resistance (*r*) or sensitivity (*s*) to cupric ions (D. Nyberg). *Cu^s* confers sensitivity (death) to concentrations above about 10 μM; homozygotes for *cu^r* resist 3–4 times as high a concentration. Tolerance level varies with the culture medium; values given are for 0.125-percent cerophyl medium. See Table 9 for the allele present in different stocks. There may be minor differences in level of resistance among stocks listed as having the same allele.
cyr	Cycloheximide resistance. Obtained by CK and Cronkite with NG on stock 51; found and characterized by MA. In 1 mg/ml cycloheximide, wild type dies but homozygous mutant lives with retarded growth. Accurate scoring requires use of more than one concentration near 1 mg/ml, and testing cultures 8–12 generations after fertilization. Carried as stock d4-135. Not linked to (*A, big1, cl2*), *D* (546), *fna* (136), *hr* (607), *mtA* (177), *ndB* (634), *nd3* (124), *PaA* (194), *scl* (802), *sn* (187), *ts111* (188), *ts401* (120), or *tw1* (187)—all by MA.

TABLE 5. *Continued*

Gene locus[a]	Characteristics[b]
D	Immobilization (i)-antigen (serotype) D. Serologically distinguishable forms coded by codominant alleles: D^{29}, D^{32}, D^{47}, D^{51}, D^{111}, D^{172} (superscripts designate the stock in which the allele occurs). The D alleles in stocks 127, 128, and 170 are the same or very similar to D^{32}; that of stock 146 to D^{47}; and that of stock 139 to D^{51}. Stock d4-20 carries D^{32}. Not closely linked to *ptA* (MS). Not linked to (*A, am, big1, cl2, cyr*), E (164; AS), F (IT), *fna* (831; MA), H (TS *et al.*, 1953*b*), *hr* (209; MA), *k* (238; TS *et al.*, 1953*b*), M (Margolin, 1956*a*), *mtA* (93; MA), *ndB* (310; MA), *nd3* (124; MA), *PaA* (102; MA), *sc1* (434; MA), *sn* (260; AS, MA), *th* (TS *et al.*, 1953*b*), *ts111* (99; MA), *ts401* (122; MA), *tw1* (99; MA).
dc	Defective constriction of dividing cells (Maly, 1958; his gene *ds*, q.v.). Found in stock 51. Results in chains and monsters (compare *ch1, sn*). Interpreted to act by inducing mutations of a cytoplasmic genetic factor [see J. R. Preer (1969) page 198, for review]. Not allelic with *sn* (Maly, 1960*a,b*).
dk1	Dark cells as a result of excess cytoplasmic crystals (B & R). Obtained with NG on stock d4-2. Not linked to *ts401* (B & R).
ds	Delayed separation of conjugants (Wood, 1953). Polymorphic locus: stock 32 has *ds* and stocks 29 and 51 have *Ds*. The *ds* allele prevents and the *Ds* allele permits the formation of cytoplasmic bridges between conjugants. Time bridge persists is roughly proportional to the amount of cytoplasmic transfer between mates (TS, 1946*a*). Not allelic with *k* (Wood, 1953).
E	Immobilization (i)-antigen (serotype) E. Serologically distinguishable forms coded by codominant alleles: E^{32}, E^{51}, E^{172}. The superscripts designate the stock in which the alleles occur. Stocks 47 and 239 carry alleles which are the same as or very similar to E^{32}. E^{32} carried as stock d4-46. Not closely linked to (*A* or *am*). Not linked to (*D*), H (178; AS), or *k* (109; EB, MS).
F	Immobilization (i)-antigen (serotype) F. Serologically distinguishable forms coded by codominant alleles: F^{29}, F^{148}, and F^{169} (superscripts designate the stocks in which the alleles occur). Stocks 239, 316, and 329 carry an allele for an F antigen which is serologically very similar to that produced by F^{29} or F^{148} and these two differ only slightly. F^{29} carried in stock d4-23. Stocks 146 and 174 have alleles which are the same or very similar to F^{169}. Stock 51 carries recessive allele f^{51}, apparently a null allele, for this stock fails to produce an F antigen (TS *et al.*, 1953*b*). Not linked to (*A, D*); to H, *k*, or *th* (TS *et al.*, 1953*b*); or to *sn* (183; AS).

TABLE 5. Continued

Gene locus[a]	Characteristics[b]
fA	Fast swimmer in ordinary culture medium when disturbed; unlike *fna* (q.v.), sensitive to sodium, i.e., gives avoiding reactions like wild type when introduced into Dryl's solution (8 mM Na, 0.3 mM Ca); unlike *fB* (q.v.), cells are small (CK, 1971a,b; his *fast 1*, line 11-3-38). Obtained by CK with NG on stock 51. Carried in stock d4-98. In combination with *PaA*, both phenotypes expressed. Difficult to score in combination with other small mutants. Not closely linked to *sm1* (MI), *ts2109* (MI). Not linked to *fB, fna, PaA,* or *pwA*—all by CK.
fB	Fast swimmer: like *fA* except not quite so fast and cells of normal size (CK, 1971a,b). Obtained by CK with NG on stock 51; his *fast 1*, line 11-4-25. Carried in stock d4-97. Not linked to (*bdB, fA*), *fna, PaA,* or *pwA*—all by CK.
fl	Furrowless: at 35°C, but not at 27°C, variable degrees of suppression of cleavage furrow at fission, resulting in chains and abnormal kinety patterns (W & S, 1972; their mutant *211*). Obtained with NG on stock 51. Not linked to *sc1* (161; RW), *sm1* (92; RW), *ts111* (RW), *tw1* (103; RW), *tw2* (80; RW), or *tw56* (55; RW).
fna	Fast swimmer: like *fA* except insensitive to sodium (e.g., swims rapidly forward instead of backing when introduced into Dryl's solution; see *fA*); but avoids when put into solution of 8 mM Ba, 1 mM Ca or of 8 mM KCl, 0.3 mM CaCl$_2$. Obtained by CK (1971a,b; his line 1-24) with NG on stock 51. Carried as stock d4-91. Another allele at this locus, *fna^P*, is paranoiac—a strikingly different phenotype (like *PaA*, q.v.). Behaviorly, *fna* is epistatic to *PaA, fA, fB,* and *jk*; hypostatic to *pwA* and *pwB*. In combination with *fAA**, *cells are smaller and swim faster than fA* cells. Difficult to score by size and speed in combination with *hr^a*, but sodium test is decisive (MA). Not linked to (*A, ata, bdB, big1, cl2, cyr, D, fA, fB*); to *PaA, pwA, pwB, ts111* (CK, 1971b); to *jk, PaB, pwC, slg, sp* (C & K); to *na^{ts}* (VH & K); to *hr* (555), *mtA* (180), *ndB* (>1200), *nd3* (162), *sc1* (>1300), *sn* (192), *ts401* (159), *tw1* (194)—all by MA.
ftA	Football-shaped trichocysts (SP, 1970, 1974). Three alleles, *ftA^1*, *ftA^2*, *ftA^3*, differ slightly in the fission rates of homozygotes at 27°C. All reproduce less than 3.2 fission per day at 27°C, have fewer trichocysts than normal, have trichocysts mostly lacking terminal tips and failing to reach or attach to the cell surface or to discharge when picric acid is added to whole cells. Some trichocysts may be found near the surface. Cells die at 35°C. Tendency to unequal distribution of macronucleus at cell division, resulting in a low frequency of amacronucleate cells (MS). Up to 15 generation phenomic

TABLE 5. Continued

Gene locus[a]	Characteristics[b]
	lag in going from *ftA*$^+$ to *ftA*^{3} (SP, 1970). Epistatic to *ndA, ndB, stA,* and the cigar character of *scl^c*. *ftA*^{1} and *ftA*^{2} are SI's (1966a) *ts1194* and *1189* which were found by SP to be nondischarge. *ftA*^{3} was obtained by SP with x rays on stock d4-186. Carried as stocks d4-116 (*ftA*^{1}), d4-117 (*ftA*^{2}), and d4-102 (*ftA*^{3}). Not linked to (*chl*), *ftB* (147; SP, 1970), *k* (487; SP), *ndA* (257; SP, 1970), *ndB* (160; SP, 1970), *ptA* (631; SP, 1970), *scl* (536; SP, 1970), *stA* (160; SP, 1970; some crosses gave deficiency of *stA* in F$_2$), *tam8* (B & R), *tl* (82; SP, 1970), or *ts111* (B & R).
ftB	Football-shaped trichocysts. Like *ftA*; grows at about 2.8 fissions per day at 27°C. Obtained with NG on stock 51 by CK (his line F92); characterized by SP (1970, 1974). Epistatic to *ndB, stA,* and to cigar character of *scl^c*. Carried as stock d4-103. Not linked to (*ftA*), *ndB* (71; SP, 1970), *scl* (71; SP, 1970), or *stA* (71; SP, 1970).
G	Immobilization (i)-antigen (serotype) G. Serologically distinguishable forms probably coded by codominant alleles: G^{51}, G^{126}, G^{172}. The superscripts designate the stocks in which the alleles occur. The antigens in stocks 29 and 32 are identical or very similar to 51G; those in stocks 127, 146, and 148 to 126G.
H	Immobilization (i)-antigen (serotype) H. Serologically distinguishable forms coded by codominant alleles: H^{29}, H^{32}, H^{51}, H^{169} (TS and EB, 1953). The superscripts designate the stocks in which the alleles occur. The allele in stock 146 is identical or very similar to H^{169}; those in stock 127, 128, 172, and 173 to H^{51}. H^{29} carried in stock d4-31, H^{32} in d4-38; d4-41 carries H^{51} in stock 32 genome; d4-55 carries H^{29} in stock 169 genome. Reisner (1955) obtained with x rays on stock 169 three independent recessive mutations (*H*) at the *H* locus which resulted in loss of the capacity to produce the H antigen; these are carried (in stock 169 genome) as stock d4-52, 53, and 54. The H locus is not closely linked to *ptA* (MS) or linked to (*A, am, D, E, F*), *k* (TS et al., 1953b), *sn* (187; AS), or *th* (TS et al., 1953b).
Hbd	Beta-hydroxybutyrate dehydrogenase of mitochondria (Tait, 1970a). This enzyme from stock 51 and four other stocks yields a single electrophoretic band while the one from stock 174 yields five bands. No breeding analysis reported.
hr	High reactor: mating reactivity retained during up to 9 days of starvation; wild-type cells lose reactivity in hours (MA). Two alleles, *hr^a* and *hr^b*, found and characterized by MA among lines of stock 51 mutagenized with NG by CK. These carried as stocks d4-109 and d4-110, respectively. *hr^a* characteristics: 4 fissions per day at 27°C; high population density in

TABLE 5.   *Continued*

Gene locus[a]	Characteristics[b]
	stationary phase; 24–40 generations after fertilization before autogamy can be induced (as compared with 12–15 in wild type); small cells; swim fast with wobbly wide spiral, rarely resting on bottom of depression slide; posterior opaque spot when moderately thin; survives many months of starvation. hr^b cells differ from hr^a cells in being less small, in suddenly dropping to the bottom, remaining motionless, and being less responsive to jarring; usually swim less fast and often change direction greatly. Linked to both *pwB* (34 percent recombination) and *cl2* (33 percent recombination) (MA). Not linked to (*A, big1, cyr, D, fna*), *mtA* (378), *ndB* (383), *PaA* (399), *ptA* (145), *sc1* (389), *sn* (398), *ts111* (395), or *tw1* (577)—all by MA.
J	Immobilization (i)-antigen (serotype) J. Two serologically distinguishable forms in stocks 32 and 51 probably determined by alleles J^{32} and J^{51}.
jk	Jerker: in ordinary culture medium, spontaneously dash backward some five body lengths before resumption of normal forward movement (C & K). Obtained with NG on stock 51. Not linked to (*bdB, fna*), *PaA, pwA*, or *pwB*—all by C & K.
k	Polymorphic locus for killer trait (TS; 1943a). The basis of the killer trait, the bacterial symbiont kappa, is lost from *kk* clones but maintained in *K-* clones. Table 9 lists stocks carrying each allele. Stock d4-186 carries the k from stock 29. Not allelic to (*ds*), *ts2148* (JD), or closely linked to *sm1* (MI), *ts1001* (MS), or *ts2109* (JD). Not linked to (*A, am, bdA, chl, cl2, D, E, F, ftA, H*), *mtA* (94; MS), *nd3* (58; JD, MS), *PaA* (94; MS) *ptA* (96; SP), *S1* (EB; 1959), *S2* (EB; 1959), *sc1* (532; MS, SP), *sn* (190; AS), *ssw216* (B & R), *stA* (161; JD), *th* (TS et al., 1953b), *ts72* (Hipke and Hanson, 1974), *ts111* (450; BaB, MS), *ts401* (>480; BaB, B & R), *ts2112* (153; JD), *ts2108* (94; MS), *ts2149* (50; Igarashi, 1969), *tsm21* (100; MS), or *tw1* (94; MS).
kin241	Kinetosomes misplaced and misoriented in patches on some cells; grows slowly at 27°C; dies within one day at 36°C (B & R). Obtained with UV on stock d4-2. Not linked to (*cro1*), *mit34, ndB, sg1, tam8*, or *ts401*—all by B & R.
lp	Lethal with pleiotropic effects: before dying, cells become shorter, develop large amounts of cytoplasmic crystals, and sometimes extra contractile vacuoles (Hunter and Hanson, 1973; their line A4-2). Maintained in combination with its suppressor, *su-lp* (q.v.), with which it is not linked.
M	Immobilization (i)-antigen (serotype) M. Alleles M^{32} and M^{172} in stocks 32 and 172 are codominant for serologically distinguishable forms of antigen M (Margolin, 1956a). Not allelic to (*D*).

TABLE 5.   Continued

Gene locus[a]	Characteristics[b]
mic44	Four micronuclei per cell usually; forms four macronuclear anlagen and four micronuclei at autogamy; grows slowly; occasional abnormal cell divisions; some cells with forked or extra row of basal bodies in quadrulus (Adoutte and Ruiz). Obtained with UV on branch (E^R_{102}) of stock d4-2 resistant to erythromycin.
mit34	Mitochondrial morphology abnormal and slow growth (B & R). Obtained with NG on stock d4-2. Not linked to (*kin241*), *ts111*, or *ts401*—all by B & R.
mtA	Mating type (BB, 1973). Mutant allele *mtA^o* restricts to mating-type O (= VII); permits either presence or absence of E cytoplasmic factor. Obtained with NG on stock 51. Carried as stock d4-76. The wild-type allele, *mtA$^+$*, permits either O or E mating type and cytoplasmic factor (see page 549). Not closely linked to *tw2* (JD) or linked to (*A, big1, cl2, cyr, D, fna, hr, k*), *mtB* (BB, 1973), *mtC* (BB, 1973), *mtD, ndB* (179; MA), *nd3* (90; JD, MS), *PaA* (482; MA, MS), *pwA* (BB, 1973), *sc1* (296; JD, MS), *sn* (323; MS), *ts111* (BB, 1973; MS, JD), *ts401* (90; JD, MS), *ts2108* (94; MS), or *tw1* (133; MS).
mtB	Mating type (BB, 1973). Wild-type allele, *mtB$^+$*, and mutant allele, *mtB^o*, as for *mtA*. Another mutant allele, *mtB^S*, results in selfing. Stock d4-82 carries *mtB^o* and stock d4-80 carries *mtB^S* with *ts111*. Both *mtB^o* and *mtB^S* obtained with NG on stock 51. Not closely linked to (*ch1*) or linked to (*cl2, mtA*), *mtC* (BB, 1973), *mtD, nd3* (244; JD), *sc1* (269; JD), *ts111* (BB, 1973), or *ts401* (49; JD).
mtC	Mating type (BB, 1973). Wild type *mtC$^+$* and mutant *mtC^o* like *mtA$^+$* and *mtA^o*. Obtained with NG on stock 51. Carried as stock d4-83. Not linked to (*mtA, mtB*), *pwA* (BB, 1973), *ssw1* (59; JD), *ts111* (BB, 1973), *ts1001* (91; JD), or *tw2* (88; JD).
mtD	Mating type (Brygoo, 1973). Alleles *mtD^{51}* and *MtD^{32}* in stocks 51 and 32, respectively, are involved in mating-type determination as set forth on page 549. Not linked to (*mtA, mtB*), *nd9* or *ts401*.
na^{ts}	Temperature-sensitive reaction to sodium (VH & K). Little or no avoiding reaction to Dryl's solution (see *fA*) when first grown to starvation at 23°C, but avoids when grown first at 35°C. Obtained with NG on stock 51. Not linked to (*bdB* or *fna*).
ndA	Non-discharge of trichocysts when picric acid added to intact cells (SP, 1970, 1974). Obtained with NG on stock d4-186. Reduced number of trichocysts; they fail to migrate to the cell surface; in cell squashes, some trichocysts discharge abnormally; fission rate about 2.8 per day (SP) or more (MS) at 27°C; cells usually die after 4 fissions or less at 35°–36°C. Some unequal macronuclear division at fission, resulting in a

TABLE 5.   Continued

Gene locus[a]	Characteristics[b]
	low frequency of amacronucleate cells (MS). Epistatic to *ndB* and *nd3*; hypostatic to *ftA*, *ptA*, and *tl.* Carried as stock d4-100. Not linked to *(ftA)*, *ndB* (132; SP, 1970), *nd3* (113; MI), *nd6* (JDB), *ptA* (272; SP, 1970), *sc1* (86; SP, 1970), *stA* (349; SP, 1970), or *tl* (73; SP).
ndB	Non-discharge of trichocysts (SP, 1970, 1974). Like *ndA* except: trichocysts migrate to cell cortex and align normally at cell surface, but nevertheless fail to discharge when picric acid added; in cell squashes, discharge internally; fission rate about 4.3 per day at 27°C; usually grows well at 35–36°C, but sometimes dies. Scoring after origin of homozygote from heterozygote should be delayed until after generation 15 because of long phenomic lag. Hypostatic to *ftA, ftB, ndA, ptA*, and *tl.* Carried as stock d4-101. Not closely linked to *(am)*, *nd6* (MI), *tam8* (MI). Not linked to *(A, big1, big4, cl2, cyr, D, fna, ftA, ftB, hr, kin241, mtA, ndA)*, *nd3* (148; MA), *PaA* (190; MA), *ptA* (>150; MA, SP), *pwB* (235; MA), *sc1* (>1209; MA, SP), *sn* (187; MA), *stA* (71; SP, 1970), *tl* (79; SP), *ts111* (190; MA), *ts401* (156; MA), or *tw1* (190; MA, SP).
nd3	Non-discharge of trichocysts. Two mutant alleles, *nd3ᵃ* and *nd3ᵇ*; *nd3ᵃ* carried in stock d4-84 along with *ts401* (q.v.), *nd3ᵇ* carried as stock d4-111 along with *ts1001* (q.v.). The phenotypes of these two alleles are identical, with one exception: in both, the trichocysts are aligned normally at the cell surface and both fail to discharge trichocysts when picric acid is added to well-fed cells, but starved *nd3ᵃ* cells discharge a low percentage of their trichocysts while starved *nd3ᵇ* cells discharge none, when picric acid is added to the cells. JD found *nd3ᵃ* in stock d4-84 and *nd3ᵇ* in stock d4-111; origin of mutations unknown. Not linked to *(A, am, big1, cl2, cyr, D, fna, k, mtA, mtB, ndA, ndB)*, *nd6* (318; MI), *PaA* (90; JD, MS), *pwA* (97; JD), *sc1* (244; JD), *sm1* (192; MI), *sn* (90; JD, MS), *tam8* (142; MI), *ts111* (90; JD, MS), *ts401* (244; JD, MS), *ts1001* (109; MI), *ts2108* (54; JD), or *tsm21* (362; MS).
nd6	Non-discharge of trichocysts: like *nd3ᵇ.* Found by JDB in stock d4-43 which also carries *am* and *k*; origin of mutation unknown. Linked to *sm1* with about 15 percent recombination (MI). Not closely linked to *(big1* or *ndB).* Not linked to *(A, am, nd3)*, *tam8* (>150; MS, B & R), *ts111* (B & R), *ts401* (156; MI), or *ts2109* (161; MS).
nd7	Non-discharge of trichocysts (Sainsard). Found in stock carrying *sg1* (q.v.); origin of *nd7* unknown. Not linked to *sg1*, *ts111*, or *ts401.*
nd9	Non-discharge of trichocysts (Brygoo): wild type when grown at 18°C, non-discharge when grown at 27°C. Not linked to *(mtD)*, *ts111*, or *ts401.*

TABLE 5. Continued

Gene locus[a]	Characteristics[b]
nd146	Non-discharge of trichocysts. Found by D. Nyberg in stock culture of stock 146. Not allelic to *nd203*.
nd163	Non-discharge of trichocysts. Found by D. Nyberg in stock culture of stock 163. Not allelic to *nd203*.
nd203	Non-discharge of trichocysts. Found by D. Nyberg in stock culture of stock 203. Not allelic to *nd146* or *nd163*.
nd1183	Non-discharge of trichocysts. Found by JD in SI's x-ray induced temperature-sensitive line 1183, which shows slightly reduced growth rate (8 instead of 10 generations in two days) and dies in 3 days at 35°C. Carried as stock d4-114.
PaA	Paranoiac (CK, 1971*b*). When put into solutions rich in sodium, cells repeatedly swim backward; spontaneous long backing occurs occasionally when grown in certain sodium phosphate-buffered culture media, e.g., baked lettuce and grass, but not cerophyl. Heterozygote *PaA*/*PaA*[+] is partially paranoiac. Obtained with NG in stock d4-85 (A^{29}, *k*, *ts111*) and carried as stock d4-90. Linked to *PaC* with about 4 percent recombination (C & K). Not linked to (*A*, *ata*, *bdB*, *big1*, *cl2*, *cyr*, *D*, *fA*, *fB*, *fna*, *hr*, *jk*, *k*, *mtA*, *ndB*, *nd3*), *PaB* (C & K), *pwA* (CK), *pwB* (CK), *pwC* (C & K), *sc1* (290; JD, MS, MA), *slg* (C & K), *sn* (323; MS), *sp* (C & K), *ts111* (>400; MA, CK), *ts401* (90; JD, MS), *ts2108* (94; MS) or *tw1* (133; MS).
PaB	Paranoiac (Chang, VH & K): like *PaA*. Obtained with NG on stock 51. Not linked to (*bdB*, *fna*, or *PaA*).
PaC	Paranoiac (C & K): like *PaA*. Obtained NG on stock 51. Linked to *PaA* with about 4 percent recombination.
ptA	Pointless trichocysts: trichocysts lack tips (SP, 1970, 1974). The tip material is displaced to the side of the trichocyst in the form of a keel. Only a small percentage of the trichocysts reach the surface and align correctly, but—like *nd* mutants—the trichocysts do not discharge when picric acid is added to intact cells, although some discharge in squashes. Two mutant alleles. *ptA*[1] grows about 3.7 fissions per day at 27°C and dies at 36°C, but survives at slightly lower temperature (SP, 1970, 1974). About 50 percent of dividing cells have grossly unequal distribution of macronucleus to the two daughter cells and about 20 percent of cells are amacronucleate (MS). Long lag in development of *nd* phenotype when parent was *nd*[+]; test after 15 generations. Obtained with x rays on stock d4-186 by SP. Carried as stock d4-104. *ptA*[2], obtained with NG on stock d4-2 by B & R, has a similar phenotype except that it grows at about 3.1 fissions per day at 27°C, that its macronuclear behavior is unreported, and that, according to SP, it is slightly leaky, some few trichocysts having normal tips. Not linked to (*A*, *am*, *cl2*, *ftA*, *hr*, *ndA*,

TABLE 5. Continued

Gene locus[a]	Characteristics[b]
	ndB), pt2 (B & R), scl (>400; SP, B & R), stA (258; SP, 1970), tam8 (B & R; 75, MS), tl (79; SP), ts111 (B & R); ts401 (B & R).
pt2	Pointless trichocysts: no tips, but only during rapid cell growth, not in stationary phase (B & R). Obtained with NG on stock d4-2. Not linked to (ptA), ts111 (B & R), or ts401 (B & R).
pwA	Pawn: cells unable to swim backward; therefore, do not give avoiding reaction when introduced into solutions high in concentration of Na$^+$, K$^+$, or Ba^{++} ions (CK, 1971b). Twenty-two independent mutations at this locus have been obtained with NG on stock 51 by CK (1971b), C & K (1973a,b), and Chang et al. (1974). Their phenotypes range from typical pawn, to various degrees of leakiness, to heat-sensitive pawns that are nearly normal when grown at 27°C but pawn when grown at 35°C; some grow slower than wild type and have truncated cell form. Mutation pwA^1 is a typical pawn and is one of the best gene markers; it is identifiable in seconds by introducing cell into a test solution without sacrifice of the cell. Carried as stock d4-94. Mutation pwA^2 avoids at 27°C, but not at 35°C (C & K, 1973a). Carried by CK as stock d4-132. This locus is linked to ssw1 with about 4 percent recombination (JD and MS). Not closely linked to (ptA) or linked to (ata, bdB, fA, fB, fna, jk, mtA, mtC, nd3, PaA), pwB (CK), pwC (C & K), sn (135; TS, MS), sp (CK), ts111 (CK), ts401 (88; JD), ts1001 (137; JD), or tw2 (138; JD).
pwB	Pawn; like pwA (CK, 1971b). CK (1971b) and Chang et al. (1974) obtained 27 mutations at this locus with NG on stock 51; most not leaky and none temperature sensitive for pawn behavior. Mutant pwB^1 (carried as stock d4-95) is a typical pawn but sometimes presents technical difficulties; pwB^2 (carried as stock d4-96) is a slightly leaky pawn giving transient slowing before swimming forward when introduced into solutions high in concentration of Na$^+$ or K$^+$ ions (CK, 1971a,b). Linked to hr with about 34 percent recombination (MA); in same linkage group at cl2, but shows independent assortment (MA). Not linked to (ata, bdA, bdB, fna, jk, ndB, PaA, pwA), pwC (C & K, 1973b), sp (C & K), stA (89; JD), ts2112 (154; JD), or tw1 (77; MA).
pwC	Pawn: like pwA except wild type when grown at room temperature and pawn when grown at 35°C (C & K, 1973b). Carried by CK as stock d4-131. Not linked to (ata, bdB, fna, PaA, pwA, pwB), or sp (C & K).
S1	Polymorphic locus affecting maintenance of the bacterial symbiont, kappa (EB, 1959, 1961). Allele S1, in stock 29, induces loss of kappa under certain conditions. Allele s1, in stock 51,

TABLE 5.   Continued

Gene locus[a]	Characteristics[b]
	permits maintenance of kappa. The actions of *S1* and *S2* (q.v.) are similar and cumulative, but not equal. Not linked to (*k*) or *S2* (EB).
S2	Like *S1*, also occurs in stock 29, *s2* occurring in stock 51 (EB, 1959, 1961). Not linked to (*k* or *S1*).
sc1	Screwy cell shape and locomotion (TS). Variable body shape: round or pear-shaped and extremely twisted, like a corkscrew, when grown in depression slides with excess food; when starved, shape approaches normal, but is somewhat bent like cashew nut. Corkscrew-shaped cells rotate rapidly on their long axis during locomotion. Alleles *sc1*[a,b and d] yield normal trichocysts; alleles *sc1*[c1-7] yield cigar-shaped trichocysts (SP, 1970). *sc1*[a]: grows at 3 fissions per day at 27°C; obtained with NG on stock 51 by Cronkite; carried as stock d4-136. *sc1*[b]: cells twisted to either right or left; obtained with NG on stock d4-85 by CK; genetics and cytology in W & S (1972; their mutant *65*). *sc1*[d]: cells twisted strongly to right; form chains by incomplete fissions; obtained with NG on stock 51 in genetics class (Indiana University); genetics and cytology in W & S (1972; their mutant *80*). *sc1*[c1]: cigar trichocysts; obtained by TS with UV on stock 51; carried as stock d4-75. *sc1*[c2]: cigar trichocysts; anterior end of round cells protrudes; cells large, spin less fast, and form more monsters than *sc1*[c6]; obtained with x rays by RK (his line 1961-38-8); carried in stock d4-137. *sc1*[c3]: cigar trichocysts; cells very twisted; posterior products of fission are more abnormal than anterior products; undergoes incomplete fissions and produces monsters; grows at variable rates, 1.8–3.0 fissions per day; obtained with NG on stock d4-2 by B & R (1969; their *m1*). *sc1*[c4] (B & R, 1969; their *m2*): like *sc1*[c3]; grows at 1.6 fissions per day at 27°C; obtained with NG on stock d4-2. *sc1*[c5]: obtained by RK (his line 1961-45-701); not yet tested for allelism. *sc1*[c6] (SP, 1970, 1974): cigar-shaped trichocysts; well-fed cells apt to develop into monsters in overfed slide depression cultures; grows at about 2.8 fissions per day at 27°C; obtained with NG on stock 51 by Cronkite; carried as stock d4-99. *sc1*[c7] (SP, 1970; his line d4-186ml): same growth rate as *sc1*[c6]; obtained with x rays on stock d4-186. Not linked to (*A, big1, ch1, cl2, cyr, D, fl, fna, ftA, ftB, hr, k, mtA, mtB, ndA, ndB, nd3, PaA, ptA*), *sc66* (179; RW), *sm1* (W & S, 1972), *sn* (286; MS, JD, MA), *stA* (264; SP, 1970), *tam8* (B & R), *tl* (217; SP), *ts111* (>360; JD, MS, MA, RW), *ts401* (B & R; 49, JD), *ts2108* (54; JD), *tsm21* (B & R); *tw1* (168; MS, RW), *tw2* (71; RW), or *tw63* (172; RW).
sc64	Like *sc1* in body shape and swimming (MA); normal trichocysts (SP). Obtained by RW with NG on stock 51 (W & S, 1972;

TABLE 5.   *Continued*

Gene locus[a]	Characteristics[b]
	their mutant 64). Not closely linked to *ts111* (RW) or linked to *tw63* (103; RW).
sc66	Like *sc64* (MA, SP) except that it rotates clockwise during forward locomotion (MA). Backward locomotion is normal (Tamm). Obtained by RW with NG on stock 51 (W & S, 1972; their mutant *66*). Not linked to (*sc1*), *ts111* (RW), *tw1* (143; RW), or *tw63* (68; RW).
sg1	Slow growth: 1–2 fissions per day at 27°C; wild-type mitochondria become abnormal (Sainsard *et al.*, 1974; who symbolize it as *cll* for "croissance lente," a symbol used otherwise, q.v.). Obtained with UV on stock d4-2. Not linked to (*cro1, kin241, nd7*), *tam8, ts111*, or *ts401*—all by Sainsard and B & R.
sg12	Slow growth (B & R). Obtained with NG on stock d4-2.
slg	Sluggish locomotion: swimming much slower than *ssw* mutants, the cells barely creeping and often stopping on bottom of culture depressions; growth at about 2 fissions per day at 27°C (C & K). Obtained with NG on stock 51. Not linked to (*bdB, fna*, or *PaA*).
sm1	Small cells; tear-drop shape (W & S, 1972; their mutant *213*). At 35°C, only 4–5 generations in 2 days (i.e., half as many as normal). To score cell shape and size, grow at 27°C in depression slides with excess food. Difficult to score in combination with *fA* or *fB*; difficult to cross to *ssw* mutants. Obtained with NG on stock 51. Linked to *nd6* with about 15 percent recombination (MI). Not closely linked to (*fA*), *ts111* (RW), or *tw63* (RW). Not linked to (*am, fl, k, nd3, sc1*), *ts401* (145; MI), *ts2109* (145; MI), or *tw1* (143; RW).
sm29	Small cells (B & R). Obtained with NG on stock d4-2. Not linked to *ts401*.
sn	Snaky (RK): incomplete fissions yield chains of 2–6 cells. Extensively studied by Maly (1960a,b; 1961, 1962): penetrance increased by EDTA and iodoacetate; normalized by low O_2 tension, CO, 2,4-dinitrophenol, $FeSO_4$, cobalt, triose phosphates; Maly concluded that *sn* yields an altered aldolase. Chains more frequent at 19°C than at 27°C and when provided growth-limiting amounts of food (TS, MS). Phenomic lag in transition from wild to mutant phenotype longer than with *ch1* (JDB). Heterozygotes (*sn/sn*[+]) occasionally produce chains, but of only two cells. In combination with *am*, chain members that lack macronuclei do not mate (TS). In combination with *tw1*, members of chain apt to deviate from tandem orientation and to develop into multiple monsters (MA). Obtained with x rays on stock 51 by RK (his line 1955-73-258); current stock is from fourth BC generation of *sn* to stock 51 (TS and AS). Carried as stock d4-48. Not allelic to (*ch1* or *dc*) or closely linked to *ts2108* (MS). Not

TABLE 5.   Continued

Gene locus[a]	Characteristics[b]
	linked to (A, am, big1, cl2, cyr, D, F, fna, H, hr, k, mtA, ndB, nd3, PaA, pwA, sc1), ts111 (246; MA, MS), ts401 (90; JD, MS), or tw1 (206; MA).
sp	Spinner: cells spin in place on their longitudinal axis for a few seconds when they come into contact with the edge of the container; spinning prolonged when put into Na solution (20 mM Na, 0.3 mM Ca) (C & K). Obtained with NG on stock 51. Not linked to (ata, bdB, fna, PaA, pwA, pwB, or pwC).
ssw1	Slow swimming (RW, his line 2250). The first 4 generations from a cell isolated in a slide depression usually remain on the bottom of the depression but thereafter move more and are less easily scored (MS). Obtained by RW with NG on stock 51. Epistatic to ssw2 (JD). Linked to pwA with about 4 percent recombination (JD and MS). Not closely linked to ts111 (RW). Not linked to (cl2, mtC), ssw2 (100; RW, JD), ssw3 (69; RW), ts1001 (203; JD, MA), or tw2 (341; JD, MA).
ssw2	Slow swimming (RW; his line 1442). Swims slower and normalizes less than ssw1 during growth of depression slide culture; grows at 3 fissions per day at 27°C. Scoring most reliable before isolated cell has produced six generations of progeny (JD). Obtained by RW with NG on stock 51. Not linked to (ssw1), ssw3 (69; RW), or ts111 (RW).
ssw3	Extremely slow swimming at about one-third the normal rate; usually lies on bottom of depression; fission rate 2–3 per day at 27°C in shallow cultures; grows poorly or not at all in deep cultures unless aerated (RW, his line 1266). Obtained by RW with NG on stock 51. Whittle selected apparently true backmutations by stopping aeration in deep cultures aftr mutagenization, autogamy, and growth for several generations with aeration; reverse mutants survive and swim to the top of the culture. Not linked to (ssw1, ssw2), or ts111 (54; RW).
ssw216	Slow swimming; doesn't swim while dividing (B & R). Obtained by B & R with UV on stock d4-2. Not linked to (k) or ts401 (B & R).
stA	Stubby trichocysts (SP, 1970, 1974): short tips (partly shifted to side) and short trichocyst body of variable form (plum, hourglass, etc.); few attach to cell membrane or discharge when picric acid added, most remaining deeper in cell; during starvation, trichocysts are more numerous and more like wild type, many discharging in picric acid; fission rate about 3.4 per day at 27°C; well-fed cells shorter and squatter than wild type, but normalize when starved; about 15 generations until 100 percent stubby trichocysts, when genotype changes from stA/stA$^+$ to stA/stA. Low percentage of amacronucleate cells and of unequal distribution of macronucleus at fission (MS). Obtained with NG on stock 51 by CK

TABLE 5.   Continued

Gene locus[a]	Characteristics[b]
	(his line 847). Partially epistatic to cigar character of *scl*^c Carried as stock d4-106, which dies at 36°C; not yet known whether the temperature sensitivity is due to *stA* or another gene. Not closely linked to *ts2112* (JD). Not linked to (*bdA*, *ftA*, *ftB*, *k*, *ndA*, *ndB*, *ptA*, *pwB*, *scl*), *stB* (238; SP, 1970), or *tl* (62; SP).
stB	Stubby trichocysts: like *stA* (including temperature sensitivity) except that normalization of trichocysts does not occur during starvation and that fission rate is about 2.6 per day at 27°C (SP, 1970, 1974). Low percentage of amacronucleate cells (MS). Carried as stock d4-107. Not linked to (*stA*).
su-lp	Suppressor of *lp* (q.v.) (Hunter and Hanson, 1973). Not linked to (*lp*).
tam6	Trichocysts mostly not aligned at cell surface; unequal macronuclear distribution at fission yielding, in extreme cases, amacronucleate cells (JB). Obtained by JB with NG on stock d4-2. Not linked to *tam8* (B & R), *ts111* (JB).
tam8	Trichocysts not aligned at surface and fail to discharge in picric acid (JB); frequent misdivisions of macronucleus, yielding amacronucleate cells (MS, JB). Obtained by B & R with NG on stock d4-2. Epistatic to *nd3*. Carried in stock d4-108. Not closely linked to (*big1* or *ndB*). Not linked to (*A*, *am*, *cro1*, *ftA*, *kin241*, *nd3*, *nd6*, *ptA*, *scl*, *sg1*, *tam6*), *ts111* (B & R), or *ts401* (181; MI).
tam38	Abnormalities of trichocysts and of division of micronuclei (0–20 per cell), macronuclei (vegetative and especially–about 30 percent–postzygotic amacronucleates), and cells (Adoutte and Ruiz). Grows slowly at 27°C; dies at 36°C. Obtained with UV on an erythromycin-resistant branch (E_{102}^{R}) of stock d4-2.
th	Thin cells when underfed; fast swimmer; subnormal fission rate; greatly reduced sexual reactivity and rarely mates; low frequency of autogamy; supports large populations of kappa and is a very strong killer (TS *et al.*, 1953*b*). Obtained by TS and MS with UV on stock 51. Carried in stock d4-36. Not linked to (*A*, *D*, *F*, *H*, or *k*). The *th* mutant resembles the fast-swimming "mutant" that arose spontaneously in stock 51 (J. R. Preer) and was studied by Cooper (1965). Cooper's "mutant" could not be mated at all; hence, genic basis unknown.
tl	Trichocystless (SP, 1970, 1974): total lack of trichocysts, except perhaps their earliest detectable developmental stages (Jurand and Saxena, 1974); fission rate about 2.1 per day at 27°C; dies at 35–36°C. Low percentage of macronuclear misdivision and amacronucleate cells (MS). Epistatic to *ftA*, *ptA*, *stA*, and to the trichocyst character of *scl*^c. Carried as stock d4-105. Not linked to (*ftA*, *ndA*, *ndB*, *ptA*, *scl*, or *stA*).

TABLE 5.   *Continued*

Gene locus[a]	Characteristics[b]
ts51	Temperature sensitive: dies in less than 2 days at 36°C (B & R). Obtained by B & R with NG on stock d4-2. Not linked to *ts401* (B & R).
ts72	Temperature sensitive (Hipke and Hanson, 1974): dies at 31°C, not at 19° or 25°C; reproduces at same rate at 19° and 25°C. Obtained with NG. Not linked to (*k*).
ts83	Temperature sensitive (B & R); dies within 2 days at 36°; reduced fission rate at 27°C. Obtained by B & R with NG on stock d4-2. Not linked to *ts401* (B & R).
ts111	Temperature sensitive (B & R, 1969): forms round brown corpses before lysing; dies within one day after 1–2 fissions at 35–36°C; normal at 31°C and lower. Obtained by B & R with UV on stock d4-2. Epistatic to *ts401* and to *ts1012* (JD, MS). Carried, along with A^{29} and *k*, in stock d4-85. Not allelic to *ts1001* or closely linked to (*ssw1*) or *tw2* (RW). Not linked to (*A, am, bdA, bdB, big1, cil2, cl2, cyr, D, fl, fna, ftA, hr, k, mit34, mtA, mtB, mtC, ndB, nd3, nd6, nd7, nd9, PaA, ptA, pt2, pwA, sc1, sc66, sg1, sm1, sn, ssw2, ssw3, tam6, tam8*), *ts401* (>500; BaB, JD, MS, B & R), *ts2108* (92; MS), *tsm21* (B & R), *tw1* (222; RW; JD, MS), or *tw63* (228; RW).
ts401	Temperature sensitive (B & R, 1969): swollen and yellowish before dying within 2 days after four fissions or less at 35–36°C; grows normally at 30°C and lower. The *ts* effect is stronger in cerophyl than in grass or lettuce medium. Obtained by B & R with UV on stock d4-2. Carried, along with *nd3*, as stock d4-84. Hypostatic and linked to *tsm21* with perhaps 3 (probably less than 10) percent recombination (B & R, 1969). Not linked to (*A, am, big1, big38, cl2, cl3, cl4, cro1, cyr, D, dk1, fna, k, kin241, mit34, mtA, mtB, mtD, ndB, nd3, nd6, nd7, nd9, PaA, ptA, pt2, pwA, sc1, sg1, sm29, sn, ssw216, tam8, ts51, ts83, ts111*), *ts2108* (54; JD), *ts2109* (112; MI), or *tw1* (90; JD, MS).
ts1001–ts2149	According to SI (1966a), these are not linked to each other, with one possible exception which was not identified. All of these obtained with x rays on stock 51 or on nearly isogenic derivates of it.
ts1001	Temperature sensitive (SI, 1966a): dies within one day, usually within about 6 hours at 35°C (MA). Carried, along with *nd3*[b], in stock d4-111. Not closely linked to (*A, k,* or *ts111*). Not linked to (*cl2, mtC, nd3, pwA, ssw1*), or *tw2* (91; JD).
ts1012	Temperature sensitive (SI, 1966a): only 6 instead of 10 generations in 2 days at 35°C. Partial to complete failure to discharge trichocysts but the trichocyst trait is genetically aberrant. Obtained with x rays on stock d4-8 (A^{29}, F^{29}, *k*). Carried, with A^{29}, F^{29}, and *k*, as stock d4-112. The *ts* effect not closely linked to (*A*).

TABLE 5. *Continued*

Gene locus[a]	Characteristics[b]
ts1187	Temperature sensitive (SI, 1966*a*): 3–6 instead of wild type 10 generations in 2 days at 35°C. Carried in stock d4-115.
ts2053	Temperature sensitive (SI, 1966*a*): 6 instead of wild type 10 generations in 2 days at 35°C. Carried, along with *k*, in stock d4-118.
ts2102	Temperature sensitive (SI, 1966*a*): dies after one fission at 35°C. Carried, with *k*, in stock d4-119.
ts2108	Temperature sensitive (SI, 1966*a*): dies within one day at 35°C. Epistatic to *ts111* and *ts401* (MS). Carried, with *k*, in stock d4-120. Not closely linked to (*A* or *sn*). Not linked to (*cl2, k, mtA, nd3, PaA, sc1, ts111, ts401*), or *tw1* (94; MS).
ts2109	Temperature sensitive (SI, 1966*a*): dies within one day at 35°C. Epistatic to *ts401* (MS). Carried, with *k*, in stock d4-121. Not closely linked to (*fA* or *k*). Not linked to (*am, nd6, sm1*, or *ts401*).
ts2110	Temperature sensitive (SI, 1966*a*): dies within one day at 35°C. Carried, with *k*, in stock d4-122.
ts2112	Temperature sensitive (SI, 1966*a*): dies within 1 day at 35°C. Carried, with *k*, as stock d4-123. Not closely linked to (*stA*). Not linked to (*bdA, k*, or *pwB*).
ts2148	Temperature sensitive (SI, 1966*a*): dies within 1 day without fission at 35°C. Carried, with *k*, as stock d4-124. Not allelic to (*k*).
ts2149	Temperature sensitive (SI, 1966*a*): dies within 1 day at 35°C. Wild-type catalase activity at 27°C, not at 35.5°C (SI, 1969). Carried, with *k*, as stock d4-125. Not linked to (*k*).
tsm21	Temperature sensitive and morphological: dies within about 36 hours at 35–36°C; fission rate 2–3 per day at 28°C; cells larger than normal and swim slower (B & R, 1969). Obtained with NG on stock d4-2. Linked to *ts401* with less than 10, perhaps about 3, percent recombination (B & R, 1969). Not linked to (*A, ch1, k, nd3, sc1*, or *ts111*).
tso1–8	All 8 mutations obtained by Baumann *et al.* (1973) with NG on stock 51. They are neither allelic nor closely linked. All information on these is from Baumann *et al.* (1973) and JDB.
tso1	Temperature sensitive with zero fissions at 35.7°C if cells growing at 27°C are transferred to 35.7°C before going through 0.75 of the cell cycle since the last fission; macronuclear DNA synthesis blocked at 35.7°C, not at 27°C.
tso2	Like *tso1* (q.v.).
tso3	Like *tso1* (q.v.) except that DNA synthesis does take place in the macronucleus at 35.7°C.
tso4	Like *tso1* (q.v.) except that there is also a defect in food vacuole formation.
tso5	Like *tso1* (q.v.).
tso6	Like *tso1* (q.v.).

TABLE 5.   Continued

Gene locus[a]	Characteristics[b]
tso7	Like tso1 (q.v.).
tso8	Like tso3 (q.v.).
tw1	Twisted body, distinguished from screwy (sc1, 64, and 66, q.v.) by failure to form round or pear-shaped cells that twirl rapidly and look like a corkscrew. tw1[a] is W & S (1972) mutant 515, obtained by them with NG on stock 51; tw1[b] is another, but unidentified, mutant of RW (JD and MA). tw1[a]: posterior bent to the left; swims with a wobble, sometimes in circles; when crossed to tw56, tends to form doublets; in combination with sn, forms many multiple monsters. tw1[b]: shorter, less bent, tear-drop shaped; swims more normally; the two parts of dividing cells often not tandem (MA). Both tw1[a] and tw1[b] tend to be less abnormal as they approach starvation. Linked to tw2 with about 33 percent recombination; data by RW (involving 552 F_2), analysis by MA. Not linked to (A, big1, cl2, cyr, D, fl, fna, hr, k, mtA, ndB, PaA, pwB, sc1, sc66, sm1, sn, ts111, ts401, ts2108), tw56 (RW), or tw63 (69, RW).
tw2	Twisted body (W & S, 1972; their mutant 1924; obtained with NG on stock 51): like tw1 except, according to MA, that tw2 is hardier, less distorted, and a better marker than tw1 because more readily scored when starved, the cell then being longer than a tw1 cell and having a sharply bent anterior part while the rest of the cell is straight; the cells do not swim far forward without stopping to swim in circles. Linked to tw1 (q.v.) with about 33 percent recombination. Not closely linked to (cl1 or mtA). Not linked to (cl2, fl, mtC, pwA sc1, ssw1, ts1001), tw56 (W & S, 1972), or tw63 (57; RW).
tw56	Twisted body: cells differ in twist direction, some showing partial twists in both directions (W & S, 1972; their mutant 56, obtained with NG on stock 51). Tends to form stable doublets when crossed to tw1[a]. Not linked to (fl, tw1, or tw2).
tw63	Twisted body, sometimes swimming in circles (W & S, 1972; their mutant 63). Not linked to (sc1, sc66, sm1, ts111, tw1, or tw2).
V	Antigen 5, not resulting in immobilization when antiserum added (Jones, 1964). Wild-type antigen, due to gene V^{51}, occurs in stocks 32, 47, 51, 111, 126, 127, 139, 146, and 169. Their antigen 5 appears to be serologically alike, but the amount of antigen 5 is variable in stock 47. Allele V^{29} in stock 29 results in absence of antigen 5. Whether these alleles are codominant or whether one (V^{29}?) is recessive could not be ascertained because of inability to grow enough F_1 for antigen study before autogamy.

TABLE 6.   *Classification of Genes (and Presumed Genes) of Species 4 by Their Phenotypic Effects*

Phenotypic effect	Gene loci
Molecular, biochemical	
Antigens (protein)	*A, B, C, D, E, F, G, H, J, M, V*
Enzymes	*Hbd, ts2149, sn*
DNA synthesis	*tso1, tso2, tso3, tso4, tso5, tso6, tso7, tso8*
Crystals (cytoplasmic)	*cl1, cl2, cl3, cl4, dk, lp*
Resistance	*cu^r, cyr*
Behavior	
Reactions to ions (some of these believed to involve molecular changes in cell membrane)	*ata, fA, fB, fna, na^{ts}, PaA, PaB, PaC, pwA, pwB, pwC, sp.*
Altered speed or pattern of swimming	*ata, bdA, big1, big4, fA, fB, fna, hr, jk, PaA, PaB, PaC, pwA, pwB, pwC, scl, sc64, sc66, slg, sp, ssw1, ssw2, ssw3, ssw216, th, tsm21, tw1, tw2, tw63*
Reactions to starvation (increased resistance or normalization of mutant traits)	*big1, hr, nd3^a, pt2, scl, ssw1, ssw2, stA, tw1*
Reactions to other cells (in mating, choice of partner, intensity of reaction, eventual separation)	*ds, hr, mtA, mtB, mtC, th*
Abnormal cell division	*big4, ch1, dc, fl* (at high temperature), *mic44, scl^d, scl^c, sn, tam38, tso1–8* (at high temperatures), *tw1, tw2*
Abnormal macronuclear division	*am, big4, ftA, ndA, ptA, stA, stB, tam6, tam8, tam38, tl*
Cell shape abnormalities	*bdA, bdB, big1, big4, cro1, kin241, mic44, pwA, scl, sc64, sc66, sm1, stA, stB, tam38, th, tw1, tw2, tw56, tw63*
Cell size	*big1, big4, big38, cro1, fA, hr, kin241, lp, scl^{c2}, sm1, sm29, stA, stB, tsm21, tw1, tw2*
Organelle abnormalities in number, size, structure, development, movement, arrangement, or functioning	
Nuclei	*mic44, tam38* (see also abnormal macronuclear division, above)
Mitochondria	*mit34, sg1*
Trichocysts	*ftA, ftB, ndA, ndB, nd3, nd6, nd7, nd9, nd146, nd163, nd203, nd1183, ptA, pt2, scl^c, stA, stB, tam6, tam8, tam38, tl, ts1012*
Cilia, ciliary basal bodies, parasomal sacs, or kineties	*cil2, cro1, kin241,* (see also the *sc* and *tw* genes)
Oral apparatus	*mic44*
Food vacuoles	*tso4*
Contractile vacuoles	*lp*

TABLE 6.   *Continued*

Phenotypic effect	Gene loci
Symbiont maintenance	*k, S1, S2*
Reduced growth rate	
Under standard conditions	*big1, ftA, ftB, kin241, mic44, mit34, ndA, ptA, pwA* (certain alleles), *sc1, sg1, sg12, slg, ssw2, ssw3, stA, stB, tam38, th, tl, ts83, tsm21*
Only at high temperature	*big38, nd1183, sm1, ts1012, ts1187, ts2053*
Conditional lethals (at high tempera-ture)	*ch1, ftA, kin241, ndA, ndB, ptA, tam38, ts51, ts72, ts83, ts111, ts401, ts1001, ts2102, ts2108, ts2109, ts2110, ts2112, ts2148, ts2149, tsm21,* and (presumably) *tso1–8*

haploid counts, plus or minus 2, of 43 for stock 51; and 35, 39, 45, and 49 for stocks 29, 173, 32, and 47, respectively. Several minute, dotlike chromosomes are present in stock 51, none in stocks 47, 172, 173. Other karyotypic differences between stocks were also noted.

Multiple-Marker Stocks

Table 8 lists multiple-gene-marker stocks in the Indiana University collection. These were constructed for Sonneborn by M. L. Austin, J. Dilts, M. Ingle, and M. V. Schneller by introducing the marker genes into the genome of wild stock 51. This work is still in progress; some pairs of stocks with only three of four markers are to be combined into a single stock. Note that pairs of linked genes (*tsm21* and *ts401*; *big1* and *big4*; and *pwA* and *ssw1*) are included in some of the stocks.

Wild Stocks

The Indiana University collection contains wild stocks of species 4 collected in the temperate and subtropical zones of the three Americas, Australia, Japan, and Europe. Most of them are maintained in frozen as well as growing conditions. Nearly twenty polymorphic loci are known (or implied) and many more may exist. Table 9 lists the stocks, where they were collected, the immobilization serotypes that have been found in

TABLE 7. *Linkage in Species 4[a]*

Group	Linked markers	Approximate % recombination	Reference[b]	Stock Designation	Unlinked markers in stock
3	big1–big4	28.5	Austin, p.c.	d4-130	ndB
2	cl2–hr–pwB	34, 33	Austin, p.c.	d4-142	
5	nd6–sm1	15	M. Ingle, p.c.	d4-141	
6	PaA–PaC	4	Chang and Kung, p.c.	d4-144	
4	pwA–ssw1	4	Dilts and Schneller, p.c.	d4-140	ts1001, tw2
1	ts401–tsm21	3	Beisson and Rossignol (1969)	d4-139	
7	tw1–tw2	33	Whittle and Austin, p.c.	d4-143	

[a] Each stock listed carries its linked markers in the genetic background of stock 51. The figures in the table are roughly the mean percentages of recombination (which varies somewhat in repeated tests). Tests for linkage between genes in different groups are still incomplete, so some groups listed separately may eventually prove to be in the same chromosome. In group 2, the percentage double recombination is about 9–10, indicating little or no interference (Austin, private communication).

[b] p.c. signifies private communication.

TABLE 8.   Multiple-Gene-Marker Stocks with Genome
of Stock 51, Species 4

Stock	Gene markers
d4-6	A^{29}, D^{32}, F^{29}, H^{29}, k
d4-72	D^{32}, E^{32}, H^{29}, sn, am
d4-88	A^{29}, mtA^{o}, PaA, sn, ts111, tw1, (killer)
d4-89	cl2, mtB^{o}, $nd3^{a}$, scl^{c2}, ts401, (killer)
d4-126	A^{29}, hr^{a}, mtA^{o}, PaA, sn, ts111, tw1
d4-127	big1, cyr, D^{32}, fB, hr^{b}, ndB, scl^{c1}
72-M-6	bdA, k, pwB^{1}, stA, ts2112
72-M-20	pwA, ssw1, ts1001, tw2, (killer)
73-I-5	$nd3^{b}$, sml, ts2109, (killer)
73-M-3	ssw1, ts1001, tw2
73M4-75	A^{29}, tsm21, ts401
73M4-141	A^{29}, $nd3^{a}$, tsm21, ts401
73M4-383	k, tsm21, ts401
d4-130	big1, big4, ndB

each stock, and a few other polymorphic characters. Each serotype implies
the presence of a gene that codes for the corresponding protein antigen.
Although corresponding serotypes of different stocks are symbolized by the
same letter (e.g., serotype A is known to be producible in 11 of the
stocks), the antigen involved is known to be coded by different alleles in
some of the stocks (see A, D, E, F, etc. in Table 5). However, most of the
stocks have not been investigated sufficiently to discover the full array of
alleles at each serotype locus. Therefore, only the serotypes, not the alleles
are listed in Table 9. This information was obtained in Sonneborn's laboratory with the help of many co-workers, especially M. L. Austin, E.
Balbinder, B. Bartel, A. LeSuer, P. Margolin (1956b, for stock 172), F.
Ogasawara (for stock 32), A. Reisner (1957, for stock 169), M. V.
Schneller, and P. D. Skaar (1956, for stock 47). The information on
sensitivity and resistance to cupric ions and on the reaction of trichocysts
of wild stocks to picric acid was provided by D. Nyberg. M. V. Schneller
provided most of the information about the stock distribution of genes k
and K and on the probable distribution of S genes, e.g., S1 and S2 (Table
5). Table 5 gives information on other polymorphic loci (ds, Hbd, mtD)
not listed in Table 9.

Other Genetically Useful Stocks

A number of stocks have been constructed so as to contain one or
more genes from one wild stock in the genome of another wild stock. The

TABLE 9. *Wild Stocks of Species 4*

Stock[a]	Source	Serotypes	Cupric ions[b]	Trichocyst Discharge[c]	Kappa maintenance[d]
29	Maryland	A,B,C,D,F,G,H,J	S	+	k, S1, S2
32	Maryland	A,B,C,D,E,G,H,M,N	S	+	k
47	California	A,B,C,D,E,G,J,N	S	±	K
51	Indiana	A,B,C,D,E,G,H,I,J,N,Q,U	S	+	K, s1, s2
111	Maryland	B,C,D,E	S	+	K?
116	Indiana		S	±	K
126	Florida	A,B,C,D,F,G,Q,T	S	−	k
127	Florida	A,C,D,E,G,N,R	S	−	k
139	Florida	A,B,C,D,G,H,J	S	−	K
146	Lake Kogawara, Japan	B,C,D,F,G,H	S	−	k
148	Lake Kogawara, Japan	A,B,C,D,E,F,G,J,M	S	+	K, S?
163	Pennsylvania	C,D,F	S	−	k
169	Morioka City, Japan	A,B,C,D,E,F,H,M,N	S	±	K
170	Yamagata, Japan	C,D,F	S	+	k
172	Macchu Picchu, Peru	A,B,C,D,G,H,I,J,M,N,P	S	+	K?
173	El Tabo, Chile	C,D,E,F,H	S	+	K
174	Valparaiso, Chile	C,D,E,F	R	+	K, S?
203	Sydney, Australia	D,J	R	−	K,S?
230	Florida		S	+	K, S?
(239)	Florida	C,D,E,F,H,J	S	+	k
242	Florida	C,D,E	S	−	k
277	Florida	A	S	+	K
278	Florida	B,C,D	S	+	K, S?
280	Virginia	D	S	+	K, S?
298	Panama	D	S	+	K
315	Carrerra, Italy	C,D	S	+	
316	Amsterdam, Holland	C,D,F	S	+	
329	Cracow, Poland		S	+	

[a] Stocks that are in italics carry kappa and are killers; stock 239 carries lambda and is a killer.
[b] S = sensitive; R = resistant.
[c] + = discharge; ± = partial discharge; − = nondischarge in picric acid.
[d] k, K, S1, s1, S2, and s2 are explained in Table 5. S? means probably contains genes that are like S1 and S2; K? means probably contains K.

foreign gene has been introduced by a succession of usually eight back-crosses with an autogamy intervening between each pair of successive backcrosses. This assures a high degree of isogenicity with the backcross parent. The autogamous F_2 mortality typical of crosses between wild stocks decreases rapidly during this breeding program and is usually absent or almost so, after the fourth or fifth backcross and autogamy, indicating that karyotypic identity with the backcross parent has been achieved. An autogamous line obtained after the last backcross then becomes the retained constructed stock. Since it is necessarily homozygous, the genotype persists regardless of subsequent autogamies or selfing conjugation. All derived stocks in species 4 are given the symbol d4 followed by a number. Table 10 lists the principal available (at Indiana University) derived stocks containing foreign genes.

Nuclei and Nuclear DNA

Much of what was set forth on this topic on pages 472–475 was based on studies of species 4: isolation and purification of nuclei and analysis of their macromolecular composition (Isaacs *et al.*, 1969; Cummings, 1972); estimates of the amount of DNA in the macronucleus and the ratio of the amounts in macronuclei and micronuclei (Behme and Berger, 1970; Gibson and Martin; 1971; Allen and Gibson, 1972); im-

TABLE 10.   *Species 4 Derived Stocks Containing Foreign Genes in Genome of Wild Stocks*

Stock	Stock genome	Foreign genes[a]
d4-1	51	A^{29}
d4-2	51	$A^{29}k$
d4-20	51	D^{32}
d4-23	51	F^{29}
d4-31	51	H^{29}
d4-38	51	H^{32}
d4-45	51	$A^{32}E^{32}$
d4-46	51	E^{32}
d4-186	51	k
d4-40	32	D^{51}
d4-41	32	H^{51}
d4-73	172	D^{32}
d4-74	172	E^{51}

[a] See Table 5 for genes.

precision of distribution of macronuclear DNA to daughter macronuclei at cell division and its eventual regulation (Kimball, 1967); the possible subunit structure of the macronucleus (Wolfe, 1967); and the use of amacronucleate, amicronucleate, and heterocaryotic cells to infer the roles of the two kinds of nuclei in determination of the cell phenotype (Sonneborn, 1946a, 1954a,b; Nobili, 1961, 1962; Pasternak, 1967).

Kimball (1964c) has noted that variation occurs in the number of micronuclei, from 0 to 3 or more instead of the normal two, and that absence of micronuclei is correlated with reduced growth rate. Amicronucleate lines have been produced by exposure to colchicine (Butzel, 1953b) and to ultraviolet irradiation (Sonneborn *et al.*, 1953a). Amacronucleate cells are produced spontaneously in clones of certain genotypes (see Table 5: *am, big4, ftA, ftB, ndA, ptA, stA, stB, tam6, 8, 38*) and can be obtained by selecting singlet cells derived from doublets (Margolin, 1954). Some characteristics of amacronucleate cells were described on page 475; they are also unable to initiate autogamy or conjugation (Sonneborn, 1955b). Dippell and Sinton (1963) examined the distribution of DNA and RNA in the macronucleus and identified the nucleoli and chromosomal material at the ultrastructural level.

The Cell and Cell Cycle

Powelson *et al.* (1974) obtained the following data on stocks 51, 139, 172, 174, 230, and 242. Extreme stock means were 105–125 μm for cell length, 43–54 μm (and 26–33 basal bodies) between the two contractile vacuole pores, and 23–27 kineties between the posterior vacuole pore and the anterior end of the cytoproct. Generation time is usually about 5 hours in young normal clones at 27°C in standard bacterized media. For DNA synthetic periods in the two kinds of nuclei see page 475. Just before cell constriction begins, the macronucleus migrates from its position close to the cytopharynx, proceeding to a position close to the cortex, far forward on the dorsal side. Then it stretches posteriorly and assumes a narrow elongated form, still close to the dorsal cortex, and usually half anterior and half posterior to the cleavage plane at the equator of the cell. Beisson and Rossignol (1974) find that these prefission movements can be disturbed by exposure to vinblastine and cytochalasin B. The cytoplasmic events of the cell cycle in species 4 have been described by Sonneborn (1963), 1970b, 1974b), Dippell (1965, 1968), Gill and Hanson (1968) and Kaneda and Hanson (1974). In general, they are similar to those described on page 475. Experimental analyses of the cytoplasmic events will be recounted in the section on the genetics of the cell cortex.

The Clonal Cycle

The clonal cycle has been studied more fully in species 4 than in any other species of *P. aurelia*. An immature period is lacking or virtually lacking. Maturity and senesence arc in general characterized as set forth on pages 476–478. During the first 12–15 cell generations after the beginning of a clonal cycle at autogamy or conjugation (longer in certain mutants, e.g., *hr*, Table 5), autogamy cannot as a rule be induced by starvation, but the cells will become reactive for mating. During the remainder of the mature period, cells will undergo autogamy when starved if they are prevented from conjugating. Beisson and Capdeville (1966) showed that blocking protein synthesis at the time log-phase cells are washed free of food does not prevent them from becoming sexually reactive and undergoing normal conjugation, but it does prevent them from proceeding into autogamy. Instead, they remain sexually reactive for days; but they can, at any time protein synthesis is permitted to occur, proceed to lose sexual reactivity and go on into autogamy. According to Beisson and Capdeville the mRNA for the protein required for autogamy is probably not transcribed until after all the protein requirements for mating reactivity and conjugation have been synthesized.

During senescence, cell divisions show more frequent and greater abnormalities (Dippell, 1955; Sonneborn and Dippell, 1960a): failures of growth during prefission morphogenesis, mistiming of the prefission events (e.g., micronuclear division and development of the new oral apparatus occur after instead of before macronuclear division), missegregation of micronuclei resulting in cells with 1–4 micronuclei (already evident by age 150 generations), and anaphase bridges and chromosomal clumping on the spindle (evident by age 180 generations). These and other abnormalities lead to increasing frequency of long generation times and cell death; but by selection, some lines of descent can be maintained for up to 325 generations before total clonal extinction.

The frequency of death at autogamy averaged only 0.7 percent in 50 successive autogamies induced each time in parent clones which were about 25 generations old; but at about 80 generations of age the frequency began to rise sharply, reaching 100 percent by parental ages of about 220 generations (Sonneborn, 1955a; Sonneborn and Schneller, 1955a,b; 1960a,b,c). Some of this death is due to the destructive action of old cytoplasm on the nuclear processes of autogamy and conjugation, as evidenced by death of the cells before the zygotic genome can begin to operate and by death of the descendants of only the old mate in true crosses between old and young clones. Much of the death is, however, also due to accumulated chromosomal and genic mutations with age. This is unambiguously shown in cases of normal, vigorous, F_1 clones from matings of old by young. They segregate many reduced-vigor and lethal clones at the next autogamy, even when they are young. Nevertheless, almost up to the time that an old clone yields 100 percent death at autogamy, i.e., up to ages of about 200 generations, it is still possible to select some autogamous progeny that are free of micronuclear damage.

A still unexplained and puzzling result is the effect of macronuclear regeneration (Sonneborn and Schneller, 1955c; Nobili, 1960). In young clones, little or no variability is found among lines of descent whose macronuclei are derived from different fragments of the same prezygotic macronucleus; practically all such lines are vigorous. In old clones, on the contrary, such a group of lines shows much variation in vigor and viability: some are as vigorous as the parent or more so while others die. At first it was believed that this was due to aneuploidy resulting from random segregation of macronuclear chromosomes to the macronuclear fragments because heterozygous pairs of alleles seemed to segregate among the fragments; but Sonneborn *et al.* (1956) showed this was not so. I also have unpublished data showing that very old clones, which yielded 100 percent death at autogamy, yielded some viable lines at each of four successive macronuclear regenerations. Even more remarkable is Nobili's (1960) report of survival through ten successive macronuclear regenerations over a span of fissions greater than the life span of the parental clone reproducing in the absence of macronuclear regeneration. These data seem to imply (1) that random chromosomal distribution to fragments does not occur and probably has nothing to do with clonal aging; and (2) that genome integrity exists in the macronucleus and during its divisions (see subunit hypothesis, pages 473–474).

For other results on clonal aging in species 4, see the accounts of the studies by Smith-Sonneborn and co-workers (Smith-Sonneborn, 1971, 1974; Smith-Sonneborn and Klass, 1974; Smith-Sonneborn *et al.*, 1974) and Schwartz and Meister (1973) on page 477. For a stimulating general discussion in terms of nuclear control, see Siegel (1967).

Autogamy, Conjugation, and Macronuclear Regeneration

Woodard *et al.* (1966) reported spectacular macronuclear changes during the first few minutes of conjugation, i.e., during the time that the two mating types were agglutinating and before intimate cell adhesion took place. In the first 5 minutes of agglutination, the amount of macronuclear DNA appeared to increase 44 percent, and this was followed by loss of nearly half of the DNA and subsequent rapid synthesis. Further, 70–80 percent of cytoplasmic RNA appeared to be lost in the first few minutes, the residue almost immediately doubled, and then the amount rose steadily but gradually. There were no dramatic changes in the amount of cytoplasmic protein in the early stages. The rapidity and extent of the DNA and RNA changes are so surprising as to call for reexamination.

Although the regularity and precision of the nuclear events of conjugation are impressive, variants have been observed. For example, Sonneborn (1954d) found a clone in which the nuclear events appeared to be out of phase with the cytoplasmic events. Macronuclear breakdown was delayed and destruction of the products of meiosis occurred before the progenitor of the gamete nuclei could reach or attach to the cell membrane of the paroral cone and thus be protected from destruction. Consequently, all micronuclei were destroyed. The mistiming of events was evident by comparison with those in normal mates.

The nuclear events of conjugation and autogamy in doublet cells, which are important in genetic work, have been studied by Chun (1969). She observed that when both halves of a doublet possess a macronucleus, both go synchronously through their normal maneuvers at autogamy and conjugation even if only half of the doublet has a mate, and that two sets of macronuclear anlagen and micronuclei arise after fertilization, one set in each half of the doublet. Chun tested the cytogenetic processes in doublets by gene markers and found that meiosis and fertilization occur independently in the two halves of a doublet.

During autogamy and conjugation, a new oral apparatus arises in the same way as during fission, but it does not move. It replaces the prezygotic oral apparatus which disintegrates and disappears (Roque, 1956). Here, as in other well-known examples in Ciliates, identical complex structures may be simultaneously arising and disappearing side-by-side within a single cell. During conjugation, a temporary or permanent fusion may occur spontaneously between mates (Sonneborn, 1946a) and can be induced by exposure to immobilizing antiserum (Sonneborn, 1950a). Wood (1953) identified a gene (ds, Table 5) that prevents mate fusion and identified various physiological conditions that favor fusion. Permanent fusion in heteropolar or homopolar position (Sonneborn, 1942d) is important in the study of cortical genetics (page 557).

Some postzygotic events can be altered by centrifugation during the stage of their occurrence (Sonneborn, 1954d). Instead of half of the products of the fertilization nucleus forming micronuclei and half forming macronuclei, there is an excess of micronuclei. Sometimes all form micronuclei and none macronuclei. There is also a tendency to form more than four products of the syncaryon before nuclear type is established. Haploid nuclei (hemicarya) likewise tend to divide three times, instead of twice, before establishing nuclear type. For other aspects of postzygotic nuclear events in species 4, see the accounts of the studies of Berger (1971, 1973*a,b*,1974) on pages 478, 479, and 482 on the fate of prezygotic macronuclear fragments, their macromolecular syntheses after conjugation and during macronuclear regeneration, and the course of DNA synthesis during the development of macronuclear anlagen.

Mutagenesis

About 1949, Powers and his co-workers began a program of study of mutagenesis in species 4 and continued it until 1955 [reviewed by Powers (1955)]. At about the same time, Kimball began a comparable program on species 1 and continued it until about 1957 (see page 505). Then he shifted to the study of species 4. This species is technically much more favorable because of the young clonal age at which autogamy can be induced. The fraction of nonviable and reduced-growth-rate clones arising at the first autogamy after exposure to a mutagenic agent was used as an index of mutation rate, and the validity of this index was experimentally confirmed. Between 1959 and 1965 Kimball and his co-workers published a series of studies on mutagenesis in species 4 that were pioneer investigations of the nature of the mutational process; these include studies of repair and fixation of premutational damage and their relation to stage of the cell cycle, to metabolic activity, and to the kind of mutagenic agent [for reviews and bibliography, see Kimball (1962, 1963, 1964*a,b*, 1965)].

A wide variety of mutagenic agents (x rays, alpha particles, nitrogen mustard, ultraviolet light, triethylene melamine and other alkylating agents, etc.) were employed. In general, few or no mutations were produced by exposures during the micronuclear S phage or prophase, but mutagen effectiveness rose sharply thereafter (during G2 and the following G1 phases), reaching a peak immediately before S phase. Kimball showed that the cell-cycle effect was due to fixation of premutational damage during DNA synthesis and to loss or repair of this damage during the preceding G1 and G2. All metabolic inhibitors tested (caffeine, iodoacetate, chloramphenicol, and streptomycin) decreased the *rate* of repair but also prolonged the time available for repair (i.e., the time until the next S phase); the latter results in fewer lesions left to be fixed during the S phase. Starvation had the same effect as the inhibitors. Kimball also showed that both one-hit (point) and two-hit (chromosomal) mutations were subject to repair and that ionizing radiations gave mainly mutations of both exposed DNA strands, while alkylating agents gave many fractionals.

Igarashi (1966*a,b*) reported that the frequency of x-ray-induced temperature-sensitive mutations decreased, while the frequency of lethals increased, with clonal age at the time of irradiation. In this connection, see the results with species 4 by Smith-Sonneborn and co-workers (see page 477) on changes in the cell cycle and DNA repair of ultraviolet-induced lesions as clones age; and the results on species 1 by Fukushima (see page 498) on effects of x-rays on the clonal cycle. Fukushima applied the radiation repeatedly during clonal aging. Its effectiveness may have been due in part to the greater sensitivity of the clones to irradiation during their older phases.

Mating Types

The B, or clonal, system of mating-type determination and inheritance was first analyzed in species 4. Mating type usually fails to change at autogamy and at conjugation; but it does change when a broad cytoplasmic bridge forms between conjugants (Sonneborn, 1942*d*). The possibility that the usual failure to change is due to cytogamy was excluded by showing that a gene marker passed from one mate to the other (Sonneborn, 1943*a*). These results and the absence of mating type segregation in the F_2 left no doubt that the two mating types existed in identical genomes, as in the A, or caryonidal, system and that there was a decisive cytoplasmic difference between the two mating types, the cytoplasmic factor not being kappa, the killer factor (Sonneborn, 1946*c*).

Further delineation of the system came rapidly [summary in Beale (1954) pages 131–141]. The chief features are that the cytoplasmic factor is a product of the prezygotic nucleus, that this factor then acts on postzygotic macronuclear anlagen to differentiate them for mating type, and that as a rule all descendants of the differentiated macronucleus determine the same mating-type character and the corresponding cytoplasmic factor or state (Sonneborn, 1954*d*; Nanney, 1954). Selfer clones, selfer caryonides, and sister caryonides of different mating type occur with relatively low frequency. Nanney (1957) presents a full discussion of the system and most of the basic data on which its characterization depends. Usually, exchange of cytoplasm at conjugation results in change of the type-VII mate to type VIII or to selfing. The frequency of type VIII increases somewhat with temperature at the time of macronuclear anlagen origin and development. However, conjugation between selfers usually has the opposite result: the type-VIII mate becomes type VII with high frequency. As Nanney points out, this implies a degree of separation and distinction between the two macronuclear activities, one determining mating type and the other determining the cytoplasmic factor that influences the developing postzygotic macronuclear anlagen. The nature of the difference between macronuclei that determine the three caryonidally diverse conditions—type VII, type VIII, and selfers—has been the subject of speculation (Nanney, 1957), but remains obscure.

Butzel (1968) obtained with axenic media results similar to those mentioned above for bacterized media except for the temperature effect; but he used only 19°C and 27°C, while Nanney (1957) sampled the range 13–32°C and observed the clearest difference between the two extreme temperatures. Koizumi (1971) injected cytoplasm from type-VIII cells into autogamous and conjugant type-VII cells. The frequency of change to mating-type VIII by the recipient was three times as great as in uninjected controls when the donor cytoplasm came from vegetative cells, and six times as great when it came from autogamous or conjugant cells. Further work with this approach is much needed.

Mating-type genetics in fused conjugants and their singlet and doublet progeny have been reported briefly by Sonneborn (1942d) and Chun (1969) and in detail by Nanney (1957) and Butzel (1973). Nanney found that nearly all doublet clones are type VIII; but some are selfers and some type VII. Butzel reported that differences in mating type may exist among lines from the early singlets produced from doublets of the same clone. He has interpreted this, and the selfers also found, as evidence of prolonged instability of mating type or delayed determination, as in the work of Hallet on the calcium effect in species 1 (see page 500).

Byrne (1973) obtained six genic mutations restricting homozygotes to mating-type VII. Three of these (*mtA°*, *mtB°*, and *mtC°*, Table 5) were fully studied and found to be neither allelic nor closely linked. With these mutants, restriction to type VII is apparently not accompanied by restriction of the corresponding cytoplasmic state; reintroduction of the wild-type allele can yield either type VII or type VIII. These genes are not completely recessive: double heterozygotes are often selfers. A seventh mutation, *mtB^s* (an allele of *mtB°*, Table 5), yields selfers when homozygous, but only when the homozygote is derived from a type-VIII parent. Byrne points out that his mutational analysis can be accommodated in Butzel's interpretation of mating-type genetics (see page 488).

A new aspect of the system has been discovered by Brygoo (1973). This concerns the interaction between genes and the cytoplasmic state. Within each of the two stocks 32 and 51 and in the F_1 of crosses between them, the system is as already set forth. However, after the cross of stock 51 type VII by stock 32 type VIII, the F_1 clone from the 32t· VIII parent often fails to yield at autogamy the usual result, i.e., all (or nearly all) type VIII. Instead, the autogamous F_2 segregates types VII and VIII in a 1:1 ratio. This indicates segregation of a pair of alleles, designed at mtD^{51} and MtD^{32}, derived from the parent stocks. The F_2 type-VII clones are designated 7* and are held to be mtD^{51}/mtD^{51} in stock 32 VIII cytoplasm. Further analysis, by crossing 7* to the two parental stocks, yields results that conform to the following interpretation: The gene MtD^{32}, but not gene mtD^{51}, can respond to the type VIII cytoplasmic factor of 7* clones and the relevant property of this cytoplasmic factor is attributable to the activity of gene MtD^{32}. The unique feature of this analysis is its demonstration of a difference between the type-VIII cytoplasmic factors of two wild stocks.

Serotypes

The main features of the genetics of serotypes in *P. aurelia* (see page 489) were first discovered in species 4 (Sonneborn, 1947*b*, 1948*a*, 1950*c*; Sonneborn and LeSuer, 1948) and are recounted in detail by Beale (1954). In brief, the important points are: (1) Loss of the immobilizing reaction to one antiserum is accompanied by gain of reaction to another antiserum, i.e., there is a system of mutual exclusion, but one serotype can transform into another and cells always have one or another serotype. (2) Each homozygous wild stock can express a number of different serotypes, each in different cells. (3) Under certain conditions several different serotypes of the same stock can reproduce true to different serotypes. (4) Crosses between different serotypes of the same homozygous stock follow clonal inheritance comparable to that of the B mating-type system, including the fact that inheritance can become synclonal when mates exchange cytoplasm. Hence, there is no genotypic difference between serotypes within a stock; the differences follow the cytoplasm in inheritance. (5) Interstock crosses and breeding analysis showed that the capacity to become a certain serotype, the specificity of a serotype, and the response to transforming conditions are all gene dependent. Hence, the cytoplasmic component of the system also is under genic control. (6) Transformations can be reliably and specifically directed, and they are all reversible. Each stock thus has a persistent gene-dependent array of serotypes and transformation patterns; transformations are reversible and involve no loss of serotypic capabilities.

Developments since 1954 have not been fully summarized. Austin (1957, 1959, 1963*a,b*) and Austin *et al.* (1956) discovered how to induce, stabilize, and direct the transformation of virtually all of the 12 serotypes of stock 51 and of some serotypes of other stocks. Increasing growth rate, temperature, and pH individually and together had similar effects. In a few cases, particular ions (e.g., Ca^{++}) or increased total salt concentration were the keys to obtaining and stabilizing certain otherwise rare or unstable types; and certain transformations could be specifically directed by patulin and acetamide. Austin also obtained evidence by gene substitutions that the stability and transformation pattern of a particular serotype depend partly on which allele for that serotype is present and partly on the residual genome. Sonneborn *et al.* (1953*c*) showed that the residual genome effect was mainly or exclusively localized at the loci of other serotype genes. Austin *et al.* (1967*a,b*) made a study of the effects of metabolic inhibitors on various transformations. The results supported the view that transformation involves inhibition of synthesis of the pre-existing antigen and repression of the activity of the gene that codes for it, and also derepression of the gene for another antigen and synthesis of its corresponding antigen.

Skaar (1956) described the transformation pattern of stock 47 and found that the same serotype would transform to different serotypes depending on the growth rate during a considerable number of generations prior to exposure to transforming conditions. He suggested that such shifts from one to another transformation pattern might be expressions of caryonidal differences established when a caryonide arises and persisting throughout its history. Dryl's (1965) evidence supported this suggestion, as did Preer *et al.* (1963) in finding a high frequency of transformation of B to A in normal exconjugants, but very little in sister lines undergoing macronuclear regeneration. The latter result is not obviously accommodated by Finger's (1974) different interpretation of Skaar's data.

Margolin (1956*a,b*) delineated the serotype pattern of stock 172 and found the first and most striking exception to mutual exclusion of serotypes: this stock could regularly express together in the same cell both serotypes D and M. By gene substitutions, he showed that the double type could still be expressed when D^{172} is replaced by D^{32}, but not when M^{172} is replaced by M^{32}. Reisner (1955, 1957) delineated the serotype pattern of stock 179 and induced mutations that lost their capacity to form antigen 169H.

Sonneborn and Ogasawara (1953) and Sonneborn *et al.* (1953*c*) delineated the serotype pattern of stock 32 and examined the temperature optima and ranges for expression and stability of the serotypes of various wild stocks and of F_2 recombinants after interstock crosses. In general, the stocks showed the following sequence of serotypes from low to high temperature: F, H, G, E, B, A, D, C. However, exceptions to this sequence were observed in certain wild stocks and in selected recombinants, unlike the more regular situation in species 1 (see page 501).

The expression of serotype alleles in heterozygotes has been less studied in species 4 than in species 1 and 2. At first, the heterozygote A^{29}/A^{51} was believed to express only the A^{51} allele, but Dippell (1953) showed that with appropriate antisera, reactions characteristic of both parents appear in the heterozygotes, although the relative intensities of the two reactions (indicating relative amounts of the parental specificities) vary. Using doublets, Chun (1969) was able to vary the dosages of the alleles and concluded that excess of one allele usually led to failure to detect any phenotypic effect of the other allele. Sonneborn and Balbinder (1953) examined the heterozygote H^{29}/H^{51} and found the hybrid serotype at 19°C, but only 29H at 27°C. In stock 29, H is stable over a wide range of temperature; but in stock 51, H is ordinarily expressed only at 19°C or lower . Thus, the H cytoplasmic state appeared to be unable to bring 51H to expression in the heterozygote at temperatures well above its normal temperature for expression. Sonneborn *et al.* (1956) reported that the temperature conditions for expression of hybrid serotypes changed with clonal age.

Serotype antigens of *P. aurelia* were first isolated and characterized in species 4 (Preer, 1959*b,c,d*). Antigen A of stock 51 was found to be a protein with about 16 percent nitrogen and of molecular weight about a quarter of a million. Other stock 51 antigens were also shown to be proteins and to fall into certain similarity groups based on several molecular criteria: A, B, G, and Q form one group; D, J, and M another; and C, E, and H differ from both groups and from each other. The isoelectric point of 51A is 4.0 (Preer, 1959*b*; Steers, 1961), of 51B and C, 3.9 and 4.3, respectively. Steers (1962) obtained distinctive fingerprints for 51A, B, and D, but no difference could be detected between A of stocks 51 and 32. These results showed that antigens due to genes at different loci are in fact different proteins, not different foldings of the same protein. Steers (1965) then made a detailed study of antigen A of stock 51. He ascertained the amino acid composition which included 272 half-cystines and 136 disulfide bonds per mole of native antigen of molecular weight 310,000. Claiming that reduction yields monomers of 35,000 molecular weight, Steers concluded that the native antigen contains 9 such monomers of three different primary structures. On the other hand, Reisner *et al.* (1969*a,b*) claim that each of the antigens 51A, B, and D is a single polypeptide chain. Hansma (1975) confirms this for 51A and B and attributes reports of subunits to protease activity of the antigen preparations. It is still not obvious how to reconcile all of the biochemical and genetic data mentioned on pages 501–504 and 512–514.

Symbionts and Killers

Killer paramecia, first found in species 2, were first analyzed genetically in species 4. The work on this species, together with that on other species, has been repeatedly reviewed (Beale, 1954; Sonneborn, 1959; Preer, 1971; J. R. Preer *et al.*, 1974). Therefore, we shall merely summarize it briefly. The initial and key discoveries (Sonneborn, 1943a, 1946a) were: (1) The killer, resistant stock 51, and the non-killer, sensitive stocks 29 and 32, differ in one major relevant gene. The killer is KK, the sensitives, kk (Table 5). (2) Some non-killer, sensitive stocks are also KK. (3) In crosses between KK killers and KK non-killers, the killer character follows the cytoplasm in heredity. In the absence of cytoplasmic exchange between mates, each mate produces a clone of its own type. When mates exchange cytoplasm, the result depends on the amount of exchange, as indicated by the time the connecting cytoplasmic bridge persists. With persistence of about 30 minutes, both mates produce killer clones. With persistence of less than about 3 minutes, the sensitive never produces killer progeny. With intermediate persistences, the sensitive mate·may yield progeny that are sensitive, resistant, or various strengths of killer. These differences are due to differences in the quantity of a cytoplasmic factor, kappa, that passes from the killer to the sensitive mate across the connecting bridge. (4) Crosses between KK killers and kk sensitives yield, in the F_2 by autogamy in the F_1 killer clone, a $1:1$ segregation of killers (KK) and sensitives (kk). The latter soon lose kappa irreversibly and do not regain it when K is reintroduced. K is therefore essential for the killer trait by permitting kappa to be maintained, but it cannot initiate production of kappa. Kappa is self-reproducing. (5) By exposure to 34°C, kappa is progressively destroyed, and in this way one can extrapolate back to the number of kappa in a killer and obtain irreversible conversion of KK killers into KK sensitives by total destruction of kappa. (6) By slowing the growth rate at 27°C, cells with as little as one kappa produce progeny with more kappa, up to the usual maximum of several hundred (see page 514 for the prior demonstration of this in species 2 by Preer).

The chief later advances were: (1) Kappa is infectious (Sonneborn, 1948b; Tallan, 1959, 1961; Mueller, 1963, 1964). (2) Kappa is mutable (Dippell, 1948, 1950; Preer, 1948a; Ehret and Powers, 1952; Hanson, 1956; Widmayer, 1965). (3) Kappa can be seen (Preer, 1948b, 1950; Preer and Stark, 1953) and exists in two forms, N and B particles; the N particles are reproductive and the B particles (containing R bodies) are nonreproductive killing particles (Preer et al., 1953; Mueller, 1963). [The killing particles were formerly called paramecin, of which one suffices to kill a sensitive cell (Austin, 1948a,b, 1951).] (4) Kappa is, in short, a symbiotic bacterium, judging by its structure (Dippell, 1958, 1959a,b; Hamilton and Gettner, 1958) and by its chemistry (Smith-Sonneborn and van Wagtendonk, 1964; Kung, 1970, 1971c). The kappa in species 4 has been christened *Caedobacter taeniospiralis* by J. R. Preer et al. (1974). (5) The N form of this bacterium in stock 51 is lysogenic, carrying a defective DNA bacteriophage which, when induced, converts the N form into the B form (Preer and Preer, 1967; L. B. Preer *et al.*, 1974; Dilts, 1974). (6) At one time, it was maintained that an RNA product of the *K* gene, called the metagon, was a stable, transmissible, but nonreproducing particle, of which one was sufficient to maintain a full intracellular population of kappa; but later studies (Beale and McPhail, 1966; Byrne, 1969) have failed to confirm this. Nevertheless, both Chao (1955) and Byrne (1969) found that kappa persists and multiplies for long periods in some *kk* clones. The basis of this phenomenon is still obscure. For further details of recent developments see the review of J. K. Preer *et al.* (1974), in which are also given accounts of other bacterial symbionts (lambda and delta) that occur in species 4.

Organellar Genetics

Mitochondria.   Mitochondrial genetics is one of the most recent and promising lines of research with species 4. As in other organisms, the mitochondria have their own DNA. Suyama and Preer (1965) found its buoyant density to be 1.702 g/cm^3 (as compared with 1.689 for nuclear DNA) and estimated that there were $3.7 \times 10^{-16}g$ per mitochondrion.

Adoutte and Beisson (1970, 1972) obtained spontaneous and ultraviolet-induced mitochondrial mutations to antibiotic resistance by selection in media containing erythromycin (E), chloramphenicol (C), or spiramycin (S). Wild type is sensitive (E^s, C^s, S^s) to all of these agents. The mutants differ in levels of resistance to the selective antibiotic, in levels of cross-resistance to other antibiotics, in growth rate of the cells at 27°C in media containing the antibiotic agent, and in growth rate and/or survival of the cells at 36°C in normal medium. Stocks have been prepared carrying various mutant mitochondria and various nuclear gene markers, e.g., *mit34, sg1, ts111, ts401,* or *kin241* (Table 5).

Mixed populations of two kinds of mitochondria can be obtained by various amounts of cytoplasmic transfer between conjugants carrying different kinds of mitochondria. Some populations contain mixtures of wild-type and erythromycin-resistant (E^R) or chloramphenicol-resistant (C^R) mitochondria; others contain mixtures of different E^R mutants or of E^R and C^R mutants. Under different culture conditions, each type in a mixture can be obtained in pure populations among the progeny cells. In growing populations of two kinds of mutants, usually one type of mitochondrion sooner or later overgrows the other. In general, wild type eventually overgrows resistant mutants under nonselective conditions, but mixtures of C^R and C^S mitochondria may remain mixed for more than 80 generations. The speed of overgrowth is correlated with the degree of thermosensitivity of the mutant. In cells with mixed mitochondrial populations, at 27–28°C and in the absence of antibiotics, the type of mitochondrion that eventually takes over is the one that is correlated with the higher cell growth rate in cellular popuations composed of cells pure for each of the two mitochondrial types. The time for overgrowth in a mixture (from 10 to more than 90 generations) is inversely related to the degree of difference in growth rates between the two pure types of cells (Adoutte *et al.*, 1973). Cells with two kinds of mutant mitochondria, e.g., $E^R C^S$ and $E^S C^R$, give no evidence of complementation when exposed to both antibiotics at the same time. A few cases of recombination have been observed (Adoutte, 1974).

Adoutte *et al.* (1972) showed that erythromycin and chloramphenicol progressively alter the structure of wild-type mitochondria: the number of cristae decreases, some abnormal lamellar cristae and long, rigid plates appear in the mitochondria, and the organelle becomes markedly elongated. Resistant mitochondria, on the contrary, show little or no effect of the corresponding antibiotic. Using this means of identifying wild-type and resistant-mutant mitochondria, Perasso and Adoutte (1974) studied the changing proportions of the two kinds of mitochondria in cells with mixed populations exposed to antibiotic and observed that loss of cellular sensitivity to antibiotic was correlated with loss of wild-type mitochondria. A cell with a minority of erythromycin-resistant mitochondria ($E^R{}_{102}$) can become pure for resistant mitochondria within three or probably fewer cell generations. This finding implies (1) an uncoupling between cell and mitochondrial multiplication; (2) an actual disappearance of wild-type mitochondria, not mere dilution by failure to keep pace with cell division; and (3) that cell growth occurs only when a population of functional mitochondria (more than 90 percent of the normal complement) is acquired.

Sainsard *et al.* (1974) have studied the interrelations between a nuclear gene (*sg1*, Table 5; *cl1* in the paper quoted) and the mitochondria. The original homozygous mutant multiplies slowly (3 fissions per day at 27°C) but has mitochondria of normal structure. When the mutant is crossed to wild type (without cytoplasmic exchange) and an F_2 generation is obtained by autogamy in each of the two F_1 clones, four classes of F_2 clones are obtained (all homozygous): *sg1* with mitochondria m_{sg1} (i.e., mitochondria from the mutant parent), $sg1^+$ with m_{sg1} mitochondria, $sg1^+$ with m_+ mitochondria (i.e., those from the nonmutant parent), and *sg1* with m_+ mitochondria. The first class is like the mutant parent; the last class grows even more poorly (1–2 fissions per day) than the mutant parent and develops abnormal mitochondria which are very poor or lacking in mitochondrial cristae. Crosses, with cytoplasmic bridges, were made between *sg1* clones bearing erythromycin-resistant m_{sg1} mitochondria and $sg1^+$ clones bearing erythromycin-sensitive m_+ mitochondria; also the reciprocal cross involving erythromycin-resistant mitochondria in the $sg1^+$ parent and sensitive mitochondria in the $sg1$ parent was made. After both crosses autogamous F_2 were obtained, and it was found that invariably the m_+ mitochondria quickly disappeared from $sg1$ segregants and the m_{sg1} mitochondria from the $sg1^+$ segregants. Clearly each kind of mitochondrion is strongly selected under the influence of the gene with which it was originally associated. The nature of the difference between the two kinds of mitochondria is not yet entirely clarified. Observations made thus far suggest that they differ not in the mitochondrial genome, but in some other feature—perhaps membrane organization—which tends to be reproduced but is not absolutely stable.

Trichocysts. These organelles, of uncertain function, provide a model of developmental genetic analysis in species 4. Large numbers of trichocysts—thousands per cell—are precisely located in the cross-ridges between cortical units, where they are oriented perpendicular to the cell surface. On stimulation with picric acid, trichocysts are fired out of the cell and into the medium in the form of tipped and banded shafts, apparently lateral aggregations of filamentous structures. Steers *et al.* (1969) have found that the trichocyst is composed mainly of a 36,000-molecular weight protein consisting of two 17,000-molecular weight monomers. Normal development of the trichocyst has been little studied in species 4, but has been described for related species by Ehret and de Haller (1963), Yusa (1963), and Selman and Jurand (1970). A trichocyst arises deep in the cell as a membrane-bound vesicle which develops internal crystalline structure, then develops the typical unfired structure (body, tip, hood), and finally migrates to the cell surface. The possibility of DNA in trichocysts has been suggested but not clearly proved. Single gene mutations affecting trichocysts have been found in species 4 by Pollack (1970, 1974) and by Beisson and Rossignol (1969, and private communication). These are listed in Tables 5 and 6. De Haller and ten Heggeler (1969) have described a genetically unanalyzed trichocyst variant. Thus far, no differences have been found in the trichocyst protein of any mutant. A mutation at the *tl* locus blocks development at a very early stage (Jurand and Saxena, 1974). At other loci (*ft, st, sc*^{c}, and *pt*), mutations alter the form and structure. Mutations at a number of loci (*nd*) block the firing of trichocysts in response to picric acid. Some mutations block movement toward the surface or attachment to it (e.g., *tam6, tam8*); interestingly, these also affect movement of the macronucleus during cell division. Pollack (1970) has studied the interactions between mutant trichocyst genes in clones homozygous for two mutations. The epistatic relations or independence of effects indicate the relationship and relative timing of the processes affected by the mutations. Progressively later blocks are brought about in the sequence *tl, ft, pt-sc^c-st, ndA, ndB*.

The Organelles and Organelle Complexes of the Cell Cortex. The cell cortex, from the plasma membrane on the outside to the fluid endoplasm on the inside, is only about 1–2 μm thick, yet it contains most of the visible structures of the cell except the nuclei. Effects on the cortex are surely or probably produced by the genes listed in Table 6 under antigens, behavior, reactions to other cells, cell shape and size, organelle abnormalities (except nuclei and mitochondria), and by some of those listed under abnormal cell division—in other words, by a considerable portion of all the known genes. It has even been concluded (Hanson and Kaneda, 1968) that the sequence of steps in the cell cycle, marked by an elaborate sequence of complex morphogenetic events in the cortex, is controlled by a built-in fixed sequence of transcriptions of genes and that blocking this sequence by a short exposure to actinomycin D causes a fission delay of one cell cycle because the whole cycle of genic transcriptions has to go on and around to the previously blocked point before the cortical events can proceed. On the other hand, Kaczanowska *et al.* (1974) do not find a regular one-cycle block by exposure to actinomycin D; the duration of delay in fission varies with the stage exposed to the antibiotic. Nevertheless, there is no doubt, in the light of the effects of the mutations mentioned above, that every feature of the cortex can be altered by genic mutation, as has been emphasized by de Haller (1965, 1969).

The surprising and theoretically important finding is that hereditary alterations in the cortex occur in the absence of genic changes. I refer here not to changes in genic action, such as underlie hereditary differences between mating types or serotypes in cells of the same genotype, but to hereditary diversities that exist even in the absence of activity differences in the genes. This area of research has been pursued almost exclusively on species 4. The first example was the inheritance of singlet and doublet cortical structure (Sonneborn, 1942*d*), i.e., differences in the number of copies of cortical organelles per cell. This difference was subjected to exhaustive genetic analysis (Sonneborn and Dippell, 1960*b*; Sonneborn, 1963) and shown not to be based on any aspect of nuclear constitution or activity, but on the self-perpetuation of existing systems of cortical organelles. Transmission follows the cortical cytoplasm and is keyed to a particular part of the cortex, the region of the vestibule and cytopharynx. This was demonstrated by losses, or destructions, and implantations of the region (Hanson, 1955, 1962; Hanson and Ungerleider, 1973; Sonneborn, 1963). Other hereditary cortical variations were likewise shown to be cortically inherited: various sorts of incomplete doublets (Sonneborn and Dippell, 1961), multiple contractile vacuole pores (Sonneborn and Dippell, 1962), inverted kineties (Beisson and Sonneborn, 1965), and others not yet reported.

The basis of this kind of inheritance lies in the organizing and developmental function of existing cortical structure, both visible and molecular, and is referred to as cytotaxis. The products of genic action cannot initiate this structure in the absence of an existing, guiding pattern of structure and cannot correct certain "errors" in the structure. The errors perpetuate themselves through cell cycles and autogamy as cytoplasmic "mutations." Accounts of the evidence and discussion of its genetic significance are given by Sonneborn (1963, 1970b,c, 1974b) and by Beisson (1972).

Species 5

Species 5 is not distinguishable from species 1 by the electrophoretic patterns of the nine enzymes thus far examined; but the two species can be distinguished by the mating character of caryonides (see page 492). Of the 24 wild stocks that have been collected, 23 come from the USA: 12 from New Jersey and Pennsylvania (stocks 76, 80, 81, 82, 87, 94, 96, 102, 106, 107, 117, and 118); 6 from Ohio, Michigan, and Indiana (stocks 63, 132, 134, 135, 190, 210); 2 from Tennessee and Mississippi (stocks 109 and 314); and 3 from South Dakota and Nevada (stocks 120, 123, 236). Surprisingly, the only other wild stock was collected at Bullock's Hole in western Australia (stock 311). Jones (1956) reported 37 ± 1 chromosomes as the haploid number for stock 63. Stocks 87 and 314 bear the non-killer symbiont, nu (Holtzman, 1959a; Beale *et al.*, 1969), now called *Caedobacter falsus* (J. R. Preer *et al.*, 1974). No genes have been identified by breeding. Nevertheless genetic studies have been carried out on mating types and mitochondria.

Mating-Type Genetics

Sonneborn and Dippell (1946) found that species 5 has the caryonidal mating-type system, including the temperature effect and one-directional phenomic lag. Bleyman (1967a) discovered a unique feature of the system in species 5: caryonides pure for mating-type E (i.e., type X in this species) do not exist. This is not because the stocks are genetically restricted to type O; no such stocks are known in species 5. All stocks consist of three kinds of caryonides: at 19°C, about 20 percent are pure type O (= IX), about 5 percent are selfers that eventually yield some pure O sublines, and about 75 percent are selfers that do not. Increasing temperature during the first cell cycle after fertilization increases the frequency of selfers, i.e., the frequency of caryonides that can produce type-E cells, as is typical for mating-type determination.

The predominant kind of selfing caryonide was studied in detail by Bleyman (1967*b*). As such a culture approaches stationary phase, the first cells to become sexually reactive are mating-type E. Later, type-O cells arise, but they are not transformations of E to O. They are type O when they first become reactive. On the other hand, cells that are initially reactive as type O can later become reactive as type E. This change occurs in the course of about 1 hour and does not require either cell division or contact with an E cell. These differentiations of mating type are not permanent: any isolated sexually reactive cell of either mating type, if fed and allowed to multiply, again produces a selfing culture containing cells of both mating types. Bleyman concluded that the basis of these shifts in cell type reflected changes in gene activity, in accord with the Butzel hypothesis (see page 488). However, it is important to emphasize that both the distinction between pure-O and selfer caryonides and the temperature effect on their frequencies imply also inherited macronuclear differentiation with respect to whether the gene (or genes) for type E can be activated. In species 5, the distinctive feature is that the level of genic activity is insufficient to achieve complete exclusion of type-O reactivity.

Mitochondrial Genetics

Beale (1973) has reported successful transfer of mitochondria of species 1 into species 5, but attempts to transfer them to species 7 failed. The donor mitochondria were marked by bearing a mutation to mikamycin resistance. Since the recipients acquired the resistance, presumably the species 5 mitochondria were effectively replaced by those from species 1. No details on growth rate were given. The results further indicate the close relationship between species 1 and 5. However, species 7 can accept mitochondria from species 1: Knowles, using erythromycin-resistant mitochondria of species 1, achieved the transfer, but with much lower frequency of success than with transfer to species 5 (Beisson *et al.*, 1974).

Species 6

Nothing is known about the genetics of this species other than that it exhibits the B, or clonal, system of mating-type determination and may undergo somewhat more frequent changes of mating type at autogamy than do other species possessing this system. The few stocks of this species that have been examined by Tait (1970*a*) differ from all other stocks he examined by the combination of electrophoretic patterns for mitochondrial (two cathodal bands) and cytoplasmic (one anodal band) isocitrate dehydrogenase. Stocks 101, 225, 226, 284, 301, 302, and 303 were found along the east coast of the United States as far north as Pennsylvania, but they were found mostly in the south; stock 159 is from Puerto Rico; stock 326 from Kenya; and all the rest of the stocks (164, 165, 166, 167, 265, 266, 267, 268, and 269) are from India, mostly or all near Bangalore. Stock 225 was originally a paralysis killer, but has lost this character and probably the symbiont that caused it. After the loss, it was found to contain delta particles (now christened *Tectobacter vulgaris*), which are not known to confer on their hosts a killer trait (J. R. Preer *et al.*, 1974). Powelson *et al.* (1974) give the following information about extreme stock means based on a study of stocks 159, 165, 167, 225, and 226: cell length, 143–160 μm; 64–71 μm (and 32–38 basal bodies) between the two contractile vacuole pores; 29–32 kineties between the posterior pore and the anterior end of the cytoproct.

Species 7

Identification and Characterization

Species 7 was identified and characterized (by its mating reactions) by Sonneborn and Dippell (1946). It is of very local distribution; the only five known stocks (38, 227, 228, 253, and 325) come from Florida and Alabama. Species 7 can be identified by its combination of one cathodal band for mitochondrial beta-hydroxybutyrate dehydrogenase (Tait, 1970a) and one cathodal band for esterase C (Allen *et al.*, 1973). No chromosome studies have been reported for species 7. Studies on nuclear DNA are mentioned on pages 472–474. Three genic loci are known: *mt* and *n*, both in the mating-type system (see below), and *nd7* (nondischarge of trichocysts in picric acid; Haggard, 1972), all obtained in stock 227. A serotype closely related to 90G of species 1 has been found in stock 325.

Mating-Type Genetics

Wild stocks 38 and 228 are genetically restricted to mating-type O (= XIII); the other three wild stocks are two-type stocks, i.e., although entirely homozygous, their genotype permits either mating-type O or E (= XIV). Taub (1963) showed that species 7 has the clonal mating-type system. His studies (Taub, 1959, 1963, 1966a,b) were pioneering in several important respects: computer analysis of the population genetics of mating types; development of methods for obtaining mating-type mutations; uncoupling expressed mating type from the cytoplasmic determiner of macronuclei for mating type; and a model analysis of selfers.

Taub's analysis of crosses of a pure O stock to a two-type stock showed clearly a single gene difference, here symbolized as mt^o and mt^+ (Taub's mt^{XIII} and $mt^{XIII\text{-}XIV}$). The mt^+ allele permits development of either mating type; homozygotes for mt^o are restricted to type O. (Isogenic stocks possessing each of these alleles in the genetic background of the other kind of stock were produced by Taub and are maintained in the Indiana University collection.) Surprisingly, whenever a cross introduces mt^+ into a cell that was $mt^o mt^o$ (and therefore, mating-type O) at the time of the cross, the resulting heterozygote regularly produces a *synclone* of type E. Thus, the type-O mate has the cytoplasmic factor for type E, but is genically incompetent to respond to it. Thorough analysis showed that the mt^o gene (or one that was closely linked to it and was never seen to separate from it) determined the presence of the E cytoplasmic factor and that it was dominant for this effect, although recessive for its restriction of mating-type to O. Thus, the expressed mating type and the kind of cytoplasmic factor were uncoupled; they need not be correlated.

Taub's (1959) computer analysis of the population genetics of mating types, when both mt^o and mt$^+$ are present, showed that a population approaches a stable equilibrium in which the mating-type ratio is 1:1. The frequencies of matings among the various genotypes are, of course, influenced by a number of variables.

Taub induced a mutation, *n*, which also restricted homozygotes for it to mating-type O, but *n* and *mt* are neither allelic nor linked. Mating-type E can arise only when caryonides are both $n^+/-$ and $mt^+/-$. Gene *n*, unlike mt^o, has no effect on the cytoplasmic factor. Homozygotes and heterozygotes for *n* can have either O or E cytoplasm.

A considerable proportion of caryonides in species 7 are selfers. Taub (1966*a*) made a detailed study of the changes with time in subcultures as they approached stationary phase (i.e., approached sexual reactivity) and as they passed through the phase of sexual reactivity. He also examined changes with time in a sequence of subcultures of a caryonide as it grew older. Both studies showed the same direction of shift in the mating-type composition of the subcultures: from O toward E. He further showed (Taub, 1966*b*) that individual cells of selfing cultures shifted from O to E, sometimes passing through a brief "hermaphrotypic" stage in which they could react to both O and E cells. Taub's results obviously can be fitted to the Butzel hypothesis (see page 488). The basis of the changes in selfing caryonides with time is unknown. Taub (1966*a,b*) considers the possibilities of delayed determination of some macronuclear subunits, early determination followed by assortment, and other possibilities. The question remains open.

Species 8

Species 8 is the species in which both mate-killers (Siegel, 1954) and rapid-lysis killers (Schneller, 1958) were first found and genetically analyzed (Siegel, 1953; Levine, 1953a; Schneller *et al.*, 1959) and in which genetic differences in the chemical inducibility of mating were first found (Cronkite, 1972, 1974). For studies of DNA in species 8, see pages 472–474. Species 8 can be identified by its unique electrophoretic pattern for a single enzyme: it is the only species that forms two anodal bands (in addition to one or two cathodal bands) for esterase C (Allen and Gibson, 1971). Levine (1955) identified some 20 enzymes in stock 131 (and found the same enzymes in stock 51 of species 4).

Table 11 lists the wild stocks in Sonneborn's laboratory, their places

TABLE 11.   Wild Stocks of Species 8

Stock	Source	Serotypes
31	Maryland	A,B,D,E,F,G,H,O,T
130	Florida	A,B,D,E,F,G
131	Florida	A,B,C,D,E,G,H,M,N
137	Florida	A,B,D,E,G,H
138	Florida	A,D,F,H,T
140	Florida	A,B,D,G,H
141	Florida	C
142	Florida	A,B,D,E,G,H
150	Florida	A,B,F,G,H
151	Florida	A,B,C,D,E,N
202	New Mexico	
213	Florida	
214	Florida	B,D,E,H,N,O
216	Florida	A,B,E,H,O
218	Florida	
224	Florida	
229	Florida	F
252	Florida	
276	Texas	
281	South Carolina	
299	Panama	B,E,F,H,N,O
300	Panama	
307	Panama	
322	Panama	B,C,D,M
327	Florida	A
330	Georgia	
565	Uganda	

of origin, and their known serotypes. Species 8 seems to be restricted to warm or hot climates: 17 of the 27 stocks are from Florida, only one has been found north of South Carolina, and only one (from Uganda) is outside the Americas. Chromosomes have been counted only in stock 138; its haploid number is 60 ± 1 (Jones, 1956). The known genes of species 8 are listed in Table 12.

Mating Types and the Genetics of Chemically Induced Mating

Species 8 shows the typical B, or clonal, system of mating-type determination and inheritance. Cronkite (1974) found that cells of the same mating type of 20 different stocks could be induced to conjugate by Miyake's (1968) chemical method, but that the method failed with stocks 151 and 276. He then analyzed the genetic basis of these stock differences. The main results are stated in Table 12 in the descriptions of loci *kau1*, *kau2*, and *Su(kau2)*. The two recessive genes block different steps in the response of the cells to chemical inducers of conjugation.

Serotypes

Corresponding serotypes (designated by the same letter) in different stocks (Table 11) often show minor serologic differences. All of the letter symbols employed have the same meaning as in species 4, i.e., serotypes designated by the same letter in the two species are homologous in the sense of being immobilizable by antiserum against the same serotype in the other species. Species 8 stocks have various combinations of the serotypes A–H, M, N, O, and T. Genetic analyses of stock serotype differences gave results entirely comparable to the results in species 4 (Sonneborn and co-workers, unpublished). Although few have been studied, differences between the same serotype in different stocks may be assumed to be due to multiple alleles at a polymorphic locus.

Killers and Symbionts

Siegel (1953, 1954) discovered that stock 138 had a new type of killing action, which he characterized and analyzed genetically. The stock is a mate-killer, i.e., when crossed to non-mate-killer sensitive stocks, such as 151 or 176, the sensitives were killed. The killing action is not mediated by anything in the culture medium, but is accomplished only by prolonged contact. Fertilization is not necessary, however; pairs in the initial phase of conjugation can be mechanically separated and isolated

TABLE 12.   Genes of Species 8

Gene locus[a]	Characteristics
A	Polymorphic locus for immobilization (i)-antigen A. A^{138} and A^{299} in stocks 138 and 299 (Cavill and Gibson, 1972). (NB: serotype A has not been detected in stock 299 in Sonneborn's laboratory.) Presumably not linked to ac or EsA.
ac	Polymorphic locus affecting activity of esterase A (Cavill and Gibson, 1972). Alleles ac^{138} and ac^{299} yield different activities of the enzyme, but do not code for the enzyme. Presumably not linked to A or EsA.
B, C, D, E	Polymorphic loci for immobilization (i)-antigens B, C, D, and E, respectively. See text, page 563.
EsA	Polymorphic locus for esterase A with codominant alleles EsA^{138} and EsA^{299} in stocks 138 and 299, respectively (Cavill and Gibson, 1972). Presumably not linked to A or ac.
F, G, H	Polymorphic loci for immobilization (i)-antigens F, G, and H, respectively. See text page 563.
kau1	Polymorphic locus for one step (not the same step as that affected by kau2) in the response to induction of conjugation by exposure to chemicals (Cronkite, 1974). kau1 (K-acriflavine-uninducible) is in stock 151; its allele Kau1 is in stock 31. Probably not linked to kau2, Su(kau2), ts1, or ts8.
kau2	Polymorphic locus of one step (not the same step as that affected by kau1) in the response to induction of conjugation by exposure to chemicals (Cronkite, 1974). kau2 is in wild stock 276; its allele Kau2 is in stock 31. Probably not linked to kau1, Su(kau2), ts1, or ts8.
M, N, O, T	Polymorphic loci for immobilization (i)-antigens M, N, O, and T. See text, page 563.

[a] For system of symbols, etc., see footnote to Table 1.

before meiosis and the sensitive members of these "split pairs" also are killed sometimes. When much cytoplasm is exchanged durng conjugation between mate-killers and sensitives, both mates sometimes produce killer clones. Siegel showed that mate-killing is due to a visible cytoplasmic particle which he called mu [later recognized as a symbiotic bacterium and now called *Caedobacter conjugatus* by J. R. Preer *et al.* (1974)], and that its maintenance depends on the presence of gene M in the host paramecia (now called Mk, Table 12). Mu in stock 138 can be diluted out by rapid growth of the paramecia, and sensitive, mu-free progeny can be propagated. When the number of mu per paramecium is merely reduced by dilution, and the cells are then crossed to a sensitive, the sensitive mate is not immediately killed, but can reproduce up to 100 cell generations or more. However, these clones are "doomed;" so, often, are the recessive autogamous progeny of heterozygous mate-killers: while

TABLE 12. *Continued*

Gene locus[a]	Characteristics
mk	Polymorphic locus for genes affecting maintenance of the mate-killer symbiont (mu) (Siegel, 1953). Stock 138 (and probably stocks 130 and 131; Levine, 1953a) carry *Mk*, which is essential for maintenance of mu; stock 31 carries *mk*, which leads to loss of mu. *Mk* is also carried in the genome of stock 31 as stock d8-2.
nd8	Locus affecting discharge of trichocysts in picric acid (Haggard, 1974). The nitrosoguanidine-induced mutation, non-discharge (*nd8*), is carried as stock d8-6 (see *ts1* below). The wild-type allele, *nd8*⁺, which permits discharge, occurs in stock 31.
rl	Polymorphic locus affecting maintenance of the rapid-lysis killer symbiont, lambda (Schneller *et al.*, 1959). Stock 214 is homozygous for *rl* (referred to as *l* in paper cited), which results in loss of lambda. Stock 299 (and presumably other rapid-lysis killer stocks 216, 229, and 327) is homozygous for the dominant allele *Rl*, which is essential for the maintenance of lambda.
Su(kau2)	A dominant partial suppressor of the action of *kau2* which permits induction of conjugant pairing by Mg⁺⁺-acriflavine, but not by K⁺-acriflavine (Cronkite, 1972). Carried by stock 276. Probably not linked to *kau1*, *kau2*, *ts1*, or *ts8*.
ts1	Temperature sensitive: dies at 35°C in less than 2 days (Haggard, 1974). Induced by nitrosoguanidine. Carried as stock d8-5 and, with *nd8*, as stock d8-7. Probably not linked to *kau1*, *kau2*, *Su(kau2)*, or *ts8*.
ts8	Temperature sensitive: not more than 3 fissions in 2 days at 35°C (Haggard, 1974). Induced by nitrosoguanidine. Carried as stock d8-4. Probably not linked to *kau1*, *kau2*, *Su(kau2)*, or *ts1*.

their mu population is being reduced on the way to being lost, they are fatally injured. Doomed clones are often amicronucleate or have visibly abnormal micronuclei which seem to be deficient in DNA. This interesting action of the mu of stock 138 is worthy of further study. Doomed amicronucleate clones can be revived and saved by introduction of normal nuclei by conjugation.

Levine (1953a) found two other stocks (130 and 131) of mu-bearing mate-killers that showed similar phenomena. However, mate-killers of these two stocks are killed by mate-killers of stock 138. Also, mate-killer 130 killed mate-killers of stock 131. This "pecking-order" was shown not to be due to differences among the stocks in their genes, but in the kind of mu they bore. Neither stock 130 nor stock 131 produced any detectable micronuclear damage on sensitives when the amount of mu per killer was reduced by dilution. Instead, they produced only a transient reduction of growth rate in the sensitive mate. Consequently, no doomed clones arose.

Two other kinds of bacterial symbionts have been found in species 8. An extremely fast-acting killer that carried a large symbiont was first reported by Schneller (1958) in stock 299; later, it was also found in stocks 229 and 216 and was called lambda (Sonneborn *et al.*, 1959). It is now recognized as a bacterium, christened *Lyticum flagellatum* (J. R. Preer *et al.*, 1974). Schneller *et al.* (1959) showed that it depends for its maintenance on a gene (see *rl* in Table 12) and that stock 214, a different kind of killer, carries an allele which can not maintain lambda. Sonneborn (1956) found that sensitives killed by stock 214 became enlarged and spherical before dying, and Beale *et al.* (1969) found the same kind of killing and the same kind of small symbiont in stock 565. They christened the symbiont gamma, and J. R. Preer *et al.* now call it *Caedobacter minutus*. Schneller *et al.* (1959) found that stock 299 (rapid-lysis killer) carries one or more genes that cannot maintain the *C. minutus* of stock 214. These observations indicate that each kind of symbiont depends for its maintenance on a different gene or genes of the host, and that genes which support one kind of symbiont do not support another. Van Wagtendonk *et al.* (1963) report axenic *in vitro* culture of lambda; Williams (1971) confirmed this and also reported the culturing of mu.

Species 9

Characterization

Species 9, first recognized and characterized by Beale and Schneller (1954), has been found thus far only in Europe. It has the caryonidal system of mating-type determination and inheritance and requires cultivation at 25°C or lower in order for the mating reaction to occur. The four stocks (from widely scattered sources) thus far examined are alike but differ from all other species of *P. aurelia* in their electrophoretic pattern of esterase A; this enzyme of species 9 yields only two anodal bands, while at least three are formed by the esterase A of all other species (Allen *et al.*, 1973). Tait (1968, 1970a) found, among stocks of species 9, electrophoretic variants of two other enzymes. Mitochondrial beta-hydroxybutyrate dehydrogenase has a slow-moving anodal band in most stocks (e.g., stock 317), and a fast-moving band in stock hu43. These bands were shown to be due to a pair of alleles, Hbd^2 and Hbd^1, respectively. The heterozygote forms five bands, one like each parental type and three of intermediate mobility, indicating that the enzyme is a tetramer. Mitochondrial NADP isocitrate dehydrogenase, probably a dimer, shows a single cathodal band in most stocks (e.g., 317), but the same band and two slower bands in a few stocks (P7a from Pisa and he2 from Norwich). Most cultivated stocks of species 9 are in Beale's collection in Edinburgh. Some are referred to above and below.

Serotypes

Aside from the preceding enzyme studies, the only genetic work reported on species 9 concerns the immobilization (i) antigens or serotypes (Pringle, 1956; Pringle and Beale, 1960). A number of unreported serotypes occur, but there are just two principal serotypes: G, homologous to the G of other species, and X, not identified as homologous to any other known serotype of *P. aurelia*. Serotype G is expressed up to temperatures of 28°C, X at 30–33°C. Although X in species 9 has the temperature character of D in species 1, they are not serologically related. Five different G alleles were identified by serological distinctions among their antigens: G^1 in stock 510, G^2 in stock 506, G^3 in stock 509, G^4 in stock 503, and G^5 in stock 505; likewise, there are 7 different X alleles: X^1 in stock 510, X^2 in stock 530, X^3 in stock 531, and different X's in stocks 503, 504, 505, and 532. The G and X loci are not linked.

Pringle and Beale (1960) sampled the frequencies of various G and X alleles in cells taken from the same pond over a period of 6 years. The frequencies of the alleles were not significantly different at different times or in different parts of the pond. Population density was low. The frequencies of homozygotes and heterozygotes indicated that there were significantly fewer heterozygotes than expected from random mating. Autogamy was presumed to account for the excess of homozygotes. However, the heterozygotes were apparently identified by phenotype only, on the assumption of codominance. In view of later studies on species 1 and 2, indicating that one allele in a heterozygote may not be expressed (see pages 502 and 512), the matter should perhaps be re-examined.

Holzman (1959b) made a study of a genetically marked population introduced into a pond near Bloomington, Indiana. He used two alleles at each of the two serotype loci and introduced into the pond known relative frequencies (far from frequencies expected from random mating) of the various genotypes, a total of millions of paramecia. Samples taken periodically showed some shifts in the gene frequencies and the gene combinations, but did not yield ratios expected from random mating. Further study (not published) showed that during most of the year there was little or no change or growth in the population, but that for a short period in spring and summer there was considerable reproduction and change in gene and genome frequencies. After a few years, the population disappeared by a flood washout of the pond.

Species 10

Only one stock (223) of species 10 has been found; it came from Florida. This stock differs from all others examined by Tait (1970a) and Allen *et al.* (1973) in the combination of the electrophoretic patterns of mitochondrial beta-hydroxybutyrate dehydrogenase (three cathodal bands) and esterase D (three anodal bands), but esterase D tends to be erratic. It has the B, or clonal, mating-type system, has no immature period after conjugation, and remains mature for about 15 cell generations. No genes have been identified by breeding analysis in species 10.

Species 11

Only three stocks (247, 248, and 306) have been found, the first two in Mississippi and the third in Texas. In spite of its southern distribution, it conjugates well at 19°C and poorly or not at all at 24°C or above. It becomes sexually reactive in late afternoon and remains so until mid-morning. It has the A, or caryonidal, mating-type system. Tait (1970a) examined two of the stocks and found they differed from all other stocks examined by him in that they had one cathodal band for mitochondrial and two anodal bands for cytoplasmic isocitrate dehydrogenase. No genes have been identified by breeding analysis in species 11.

Species 12

All six known stocks (246, 251, 270, 271, and 273, and 274) of this species came from southern states of the United States bordering on the Gulf of Mexico. [This is listed as *P. aurelia*, syngen 14, in Sonneborn (1957); the organisms referred to there as syngen 12 are not *P. aurelia* but *P. jenningsi*.] The only two stocks (270, 273) examined are unique in having 2 instead of 3 anodal bands for the erratic esterase D (Allen *et al.*, 1973). Species 12 has the B, or clonal, mating-type system, lacks a period of immaturity after conjugation, and remains mature for only about 15 cell generations (Sonneborn, 1957). No genes have been identified by breeding analysis in species 12.

Species 13

Characterization

First recognized by Rafalko and Sonneborn (1959), species 13 has been collected from only three natural sources: the Seine River in Paris (stock 209); Benenitra, Madagascar (stock 238); and the Cuernavaca-Taxco Valley, Mexico (stock 321). Intercrosses among these three stocks yield very high mortality in the autogamous F_2, indicating considerable difference in chromosome numbers. The three stocks are alike and differ from stocks of all other species of *P. aurelia* in the combination of electrophoretic patterns of two enzymes: two cathodal bands for mitochondrial NADP isocitrate dehydrogeneriase and two anodal bands for cytoplasmic NADP isocitrate dehydrogenase (Tait, 1970*a*). See page 474 for percentage G+C in DNA. Table 13 lists the known genes of species 13. [Sonneborn (1957) page 197, lists as syngen 13 what was corrected in the addendum, page 314, to *P. multimicronucleatum*; throughout that review note that syngen 13 is *not* the present species 13.]

Mating-Type Genetics

Sonneborn (1966) gives mating-type information, but more is known than is published. Cultures grown at 27–31°C give good mating reactions, but at lower temperatures they react only erratically if at all. Species 13 is the only known species of *P. aurelia* that has the C, or synclonal, system of mating-type determination and inheritance. The two mating types, O (= XXV) and E (=.XXVI), have different genotypes in this system: homozygotes for *mt* are type O, homozygotes for *Mt* are type E. The heterozygote produces cells of both mating type, but the type-O state is transient; unfed O cells become nonreactive or E in an hour or so. These changes occur as a culture passes into stationary phase. E cells remain E until they lose mating reactivity. Stock 209 is pure type O; stocks 238 and 321 are pure type E.

Species 14

Only one stock (328 from Emily Gap, Australia) has been found in species 14. No genes have been identified by breeding analysis; genetic work is limited to showing that it has the A, or caryonidal, system of mating-type determination and inheritance. It is distinguished from all other electrophoretically examined stocks of *P. aurelia* by two cathodal and no anodal bands for esterase C (Allen *et al.*, 1973).

Interspecific Genetics

Readers of the preceding pages cannot fail to be impressed by the many similarities among the 14 species in their form and structure, their cell and clonal cycles, their cytogenetic processes, their mating types and

TABLE 13.   *Known Genes in Species 13*[a]

Gene locus	Characteristics
am	Unequal macronuclear division at fission, yielding about 7 percent amacronucleate cells in isolation lines at 23°C. Growth rate about 2 fissions per day. Partially characterized by Leslie Harrold. Not linked to *mt* or *ts5*.
ch	Chain-former; frequent failure of cell constriction at fission. Carried as stock d13-3 and, in the genome of stock 321, as stock d13-4. Not linked to *mt, tr, ts5,* or *ts11*.
mt	Polymorphic locus for mating types (see text, page 569). Stock 209 carries *mt* (for type 0); stocks 238 and 321 carry *Mt* (for type E); stock d13-2 carries *Mt* from stock 321 in genome of stock 209. Probably not linked to any of the other listed loci.
sp	Aberrant swimming, but normal body shape. Cells spin rapidly on their longitudinal axis. Carried as stock d13-13 and, in the genome of stock 321, as d13-14. Not linked to *mt*.
tr	Excess trichocysts and other variable abnormalities: long cells, narrow anterior to the vestibule; vestibule shifted posterior of cell center; etc. Carried as stock d13-29 and, in the genome of stock 321, as d13-30. Not linked to *ch, mt,* or *ts5*.
ts1	Temperature sensitive: 2–7 fissions in 2 days at 35°C. Carried as stock d13-5 and, in the genome of stock 321, as stock d13-6. Not linked to *mt* or *ts5*.
ts2	Temperature sensitive: no more than one fission in one day at 37°C. Carried as stock d13-9 and, in the genome of stock 321, as stock d13-10. Not linked to *mt* or *ts5*.
ts3	Temperature sensitive: not more than 2 fissions in one day at 37°C. Carried as stock d13-11 and, in the genome of stock 321, as stock d13-12. Not linked to *mt* or *ts5*.

[a] See footnote of Table 1 for system of symbols. All mutations were induced with nitrosoguanidine and were selected and bred by Sonneborn and Schneller. Mutations *am, tr,* and *ts11* were induced in stock d13-2 which carries gene *Mt* in the genome of stock 209; all other mutations were induced in stock 209. In regard to the deviation of the *ts* mutants from controls, note that controls reproduce at 4 to 5 fissions per day at 35° and 37°C.

serotypes, and, fundamentally, even in their genetics of these characteristics. Indeed, for over a century *P. aurelia* was considered to be one species. Yet, as has been pointed out, the 14 species of *P. aurelia* are genetically isolated and can be distinguished biochemically. However, some of them are not sexually isolated and some are often found living together in the same pond or stream. Their competitive coexistence in nature and their ecology have engaged the attention of Hairston (1958, 1967), Hairston and Kellerman (1965), Tait *et al.* (1971), Gill (1972), and Gill and Hairston (1972). Their sexual relations have alerted geneticists to the possibilities of attacking interspecific genetic problems. Interspecific mating

TABLE 13.   Continued

Gene locus	Characteristics
ts4	Temperature sensitive: not more than 7 fissions in 2 days at 35°C. Carried as stock d13-15 and, in the genome of stock 321, as stock d13-16. Not linked to *mt* or *ts5*.
ts5	Temperature sensitive: usually dies, never more than 4 fissions, in 2 days at 35°C. Carried as stock d13-17 and, with *ch,* as stock d13-31; in the genome of stock 321, as stock d13-18; and with *ch* in the genome of stock 321, as stock d13-32. Not linked to *am, ch, mt, tr, ts1, ts2, ts3, ts4, ts6, ts9, ts10,* or *ts11.*
ts6	Temperature sensitive: usually less than 5 fissions, often only 2, in 1 day at 35°C. Carried as stock d13-19 and, in the genome of stock 321, as stock d13-20. Not linked to *mt* or *ts5*.
ts7	Temperature sensitive: dies or undivided in one day at 37°C. Carried as stock d13-21 and, in the genome of stock 321, as d13-22. Not linked to *mt*.
ts8	Temperature sensitive: dies in one day at 37°C. Carried as stock d13-23 and, in the genome of stock 321, as d13-24. Not linked to *mt*.
ts9	Temperature sensitive: has no more than 3 fissions in one day at 35°C. Carried as stock d13-25 and, in the genome of stock 321, as d13-26. Not linked to *mt* or *ts5*.
ts10	Temperature sensitive: some die, none has more than 4 fissions, in one day at 35°C. Carried as stock d13-27 and, in the genome of stock 321, as stock d13-28. Not linked to *mt* or *ts5*.
ts11	Temperature sensitive: dies at 27°C, but lives at 19°C. Carried as stock d13-29. Not linked to *ch, mt,* or *ts5*.

between any two species of *P. aurelia* can be induced chemically (Miyake, 1968), but this possibility has not yet been exploited genetically. It is not known yet whether cross-fertilization and survival can be achieved by this method. On the other hand, interspecific matings consummated by normal mating reactions have been obtained and analyzed.

This work is based on the system of interspecific mating reactions summarized in Figure 1. These were among the chief criteria for identifying species before the electrophoretic patterns of enzymes were known. Species 2, 6, 9, 11, 13, and 14 are omitted from the table because they have not yet been observed to give interspecific mating reactions. For the eight species entered in Figure 1, the interspecific mating reactions are always limited to type O of one species with type E of another; and there are enough of these to show that all of these O's are homologous, and that all of the E's are (Sonneborn and Dippell, 1946). The homologies in the other six species are based on the temperature effect (see page 486), the

Figure 1. System of Mating between Species of P. aurelia

Species	Mating types	1 O	1 E	3 O	3 E	4 O	4 E	5 O	5 E	7 O	7 E	8 O	8 E	10 O	10 E	12 O	12 E
1	O	−	+++	+	−			++	++	++	++		±				
	E	C3	−					++	++	++	++						
3	O			C1	+++		−			±			++				
	E			−	C3	C3	−										
4	O					−	+++		−	+++	+++	+++	+++	+		+	
	E					−	C3										
5	O					−		C2	−				±	+			
	E						C3	+++	C2								

		C1			C1		C1		C3		
7	O									−	+++
	E								C3	−	
8	O				C3				−	+++	
	E		C1		C1			C3	−	+	
10	O							−	+++	?	
	E						C3	−			
12	O								−	+++	
	E					C3	−				

Symbols above the diagonal give degrees of mating reaction: ? = dubious; ± = brief and weak contacts one pair at a time; + = only a few percent of cells stick together at any one time; ++ = about 40 percent of cells make more or less persistent contacts when most reactive; +++ = about 90–95 percent of cells in durable contact under optimal reactive conditions.

Symbols below the diagonal: C = true conjugant pairs form. The number following C indicates the proportion of clumped cells that form conjugating pairs. C1 = very small proportion, several percent or less; C2 = of the order of 40 to 50 percent; C3 = of the order of 90 percent.

unidirectional phenomic lag (see page 486), and in the case of species 13, on dominance relations (all known O-restricted mutants being recessive). As appears in Figure 1, there are many more interspecific combinations that react sexually to each other than go on to complete conjugation. Genetic analysis is therefore restricted to crosses of species 1×3, 1×5, 1×7, 3×8, and 4×8.

All of these crosses were initially explored by Sonneborn and Dippell [unpublished, but some results given by Sonneborn (1950*a*, page 118)] and by Butzel, Levine, and Haggard in papers to be cited. Briefly, Sonneborn and Dippell obtained viable hybrids from all of the crosses except species 4×8, the one that mates best of all. The hybrids were in general slow-growing and hyperreactive sexually, i.e., their mating types reacted even in the presence of excess food; and, most important, they seemed to be completely unable to breed fruitfully with either parent species or among themselves. One major evidence of hybridity, and a point of genetic interest, was the dual character of mating types in the hybrids. Most of them behaved as if they were double mating types. Some reacted *strongly* with the E types of *both* parent species; most (by far) reacted *strongly* with the O types of *both* parent species. This happened even when one parent was stock 16 of species 1 or stock 38 of species 7, stocks that were genetically restricted to type O; the type E hybrid, nevertheless, acted as if it had both E specificities, i.e., even the specificity genetically missing in the species 1 or 7 parent. Moreover, the O hybrids from the cross of species 1×7 acquired a third mating reactivity, not possessed by either parent species, the capacity to mate with type E of species 3.

Species 4×8

This was studied further by Melvin (1947), Levine (1953*b*), and Haggard (1974). Melvin (see summaries in the papers of Levine and Haggard) used a serotype marker to validate hybridity. All of her hundreds of F_1 were abnormal and died after at most several fissions; the clones from the species 4 member were more quickly and extremely affected prior to death. She presented some evidence for a cytoplasmic role in the F_1 mortality. Levine got 5 percent survival in "F_1" but showed that most of the survivors were not hybrids. The 19 true hybrids that survived were sterile. He presented evidence that hybrid death was due to incompatibility between the cytoplasm of each species and the nuclei of the other and that brief exposure to 36°C during the first cell cycle alleviated this incompatibility. Haggard was unable to confirm the observations on which these conclusions were based. Levine's most important result was derived from crosses of amicronucleate cells of one species to normals of the other, in both combinations. In each combination he got two surviving clones from the originally amicronucleate member of the cross and found that these clones were indistinguishable from the species that contributed the nucleus. Whether cytoplasm had also been contributed was not known.

Haggard, unlike his predecessors, used good gene markers in the cross. He obtained 32 percent survival in "F_1," but, like Levine, found that most survivors were nonhybrid. All of the 9 percent (19 clones) that were hybrid and viable were produced from the species-8 members of the pairs. The hybrids were not hyperreactive; they were sterile and no segregation of markers could be found before F_2 and backcross clones died. He attempted to prove that the gene K, possessed by the species-4 parent, was in the hybrids. The method was to infect with kappa, but infections could not be produced. This result, however, is indecisive because it is known that genes other than K may be involved in kappa maintenance (see *S1* and *S2*, Table 5).

Species 1 × 7

This is the most interesting cross because it is between species with different systems of mating-type determination. It has been studied by Butzel (1953a, 1955) and Haggard (1972); both confirmed the early results of Sonneborn and Dippell. Butzel and Haggard both described more fully the character of the hybrids, especially their abnormalities and fragility. Butzel crossed a heterozygote (mt^o/mt^+) of type E, species 1, with type O [mt^o/mt^o, as proved later by Taub (1963)] of species 7 and obtained indirect but suggestive evidence that the restriction to type O in the two species is by allelic genes. Butzel also presented an hypothesis to explain the dual-mating-type character of interspecific hybrids. He proposed that there were not only similar genes in all of the species for presumed O and E mating-type substances (or their precursors), but that, in addition, each species had specificity genes that uniquely modified these substances (or precursors) in each species.

Between the time of Butzel's work and that of Haggard, Taub (1963) discovered that species 7 has the clonal system of mating-type determination. Haggard therefore undertook, with the aid of good gene markers, to repeat the species 1 × 7 cross and not only retest the question of gene flow, but also seek insight into the genetic basis of the difference between the clonal and caryonidal mating-type systems. He reported no gene flow; the hybrids were sterile. The outstanding fact concerning mating types was that all of the hybrids were type E; but the number observed was insufficient to rule out the low percentage of type O found by earlier investigators. Nevertheless, Haggard suggested that the caryonidal system may include the cytoplasmic component characteristic of the clonal system, but that this component is effective only on genes in species having the clonal system. In hybrids between clonal-system species 7 and caryonidal-system species 1, responsive genes would be present. Haggard recognized the need for more decisive tests.

Interspecific Transfers of Cytoplasmic Genetic Elements

Koizumi and Preer (1966) have developed a technique of sucking out cytoplasm of paramecia into a micropipette and injecting it (or other preparations) into other paramecia. This technique has been applied to transfers of kappa (Koizumi and Kobayashi, 1974), including transfers from one species to another. The technique has also been used for various symbionts by Gibson (1973) and for mitochondria by Beale (1973) and by Knowles (1974). This method, which is in early stages of exploitation, has already proved its value and may open the door to the isolation and characterization of cytoplasmic factors that differentiate macronuclei for mating-type and serotype determinations (Koizumi, 1971). Another more limited and still unexploited method has been suggested by Haggard (1972): In crosses between species, even when all true hybrids die, many viable nonhybrid clones are obtained; the conjugants that give rise to some of these could acquire a piece of the cortex of their mate [as reported for intraspecific conjugants by Sonneborn (1963)]. If so, this would add to the means of studying cortical genetics.

In the future, genetic analysis of *Paramecium* may be expected to capitalize on these and other approaches in attacking the fundamental problems of molecular and developmental genetics which have already been exposed in genetic work on *P. aurelia*.

Literature Cited

Adams, J. P., and S. L. Allen, 1974   Polymorphism and differentiation in *Paramecium*. In *Proceedings of the Third International Isozyme Conference*, Academic Press, New York.

Adoutte, A., 1974   Mitochondrial mutations in *Paramecium*: Phenotypical characterization and recombination. In *The Biogenesis of Mitochondria* (edited by A. Kroon and C. Saccone), pp. 263–271 Academic Press, New York.

Adoutte, A. and J. Beisson, 1970   Cytoplasmic inheritance of erythromycin-resistant mutations in *Paramecium aurelia*. Mol. Gen. Genet. **108,** 70–77.

Adoutte, A. and J. Beisson, 1972   Evolution of mixed populations of genetically different mitochondria in *Paramecium aurelia*. Nature (*Lond.*) **235:**393–395.

Adoutte, A., M. Balmefrezol, J. Beisson, and J. André, 1972   The effects of erythromycin and chloramphenicol on the ultrastructure of mitochondria in sensitive and resistant strains of *Paramecium. J. Cell Biol.* **54:**8–19.

Adoutte, A., A. Sainsard, M. Rossignol, and J. Beisson, 1973 Aspects génétiques de la biogenèse des mitochondries chez la *Paramécie. Biochimie* **55:**793–799.

Allen, R. D., 1971 Fine structure of membranous and microfibrillar systems in the cortex of *Paramecium caudatum. J. Cell. Biol.* **49:**1–20.

Allen, S. L., and I. Gibson, 1971 Intersyngenic variations in the esterases of axenic stocks of *Paramecium aurelia. Biochem. Genet.* **5:**161–181.

Allen, S. and I. Gibson, 1972 Genome amplification and gene expression in the Ciliate macronucleus. *Biochem. Genet.* **6:**293–313.

Allen, S. L., and P. A. Golembiewski, 1972 Inheritance of esterases A and B in syngen 2 of *Paramecium aurelia. Genetics* **71:**469–475.

Allen, S. L., B. C. Byrne and D. L. Cronkite, 1971 Intersyngenic variations in the esterases of bacterized *Paramecium aurelia. Biochem. Genet.* **5:**135–150.

Allen, S. L., S. W. Farrow and P. A. Golembiewski, 1973 Esterase variations between the 14 syngens of *Paramecium aurelia* under axenic growth. *Genetics* **73:**561–573.

Austin, M. L., 1948a The killing action and rate of production of single particles of paramecin 51. *Physiol. Zool.* **21:**69–86.

Austin, M. L., 1948b The killing substance, paramecin: activity of single particles. *Am. Nat.* **82:**51–59.

Austin, M. L., 1951 Sensitivity to paramecin in *Paramecium aurelia* in relation to stock, serotype and mating type. *Physiol. Zool.* **24:**196–204.

Austin, M. L., 1957 The action of patulin and zinc chloride on various serotypes of stocks 51, 29 and 32 of variety 4, of stock 31 of variety 8, and on one serotype of stock 90, variety 1, of *Paramecium aurelia. Microb. Genet. Bull.* **15:**5.

Austin, M. L., 1959 The effect of high and of low *p*H and of gelating and liquefying agents on antigenic transformation in *Paramecium aurelia. Science (Wash., D.C.)* **130:**1412.

Austin, M. L., 1963a Progress in control of the emergence and the maintenance of serotypes of stock 51, syngen 4, of *Paramecium aurelia. J. Protozool.* **10** (Suppl.): 21.

Austin, M. L., 1963b The influence of an exchange of genes determining the D antigens between stocks 51 and 32, syngen 4, *Paramecium aurelia*, on the transformation of D to B and A in patulin. *Genetics* **48:**881.

Austin, M. L., D. Widmayer and L. M. Walker, 1956 Antigenic transformation as adaptive response of *Paramecium aurelia* to patulin; relation to cell division. *Physiol. Zool.* **29:**261–287.

Austin, M. L., J. Pasternak and B. M. Rudman, 1967a Studies on the mechanism of serotype transformation in *Paramecium aurelia*. I. The effects of actinomycin D, puromycin, and chloramphenicol on an antiserum-induced transformation. *Exp. Cell Res.* **45:**289–305.

Austin, M. L., J. Pasternak and B. M. Rudman, 1967b Studies on the mechanism of serotype transformation in *Paramecium aurelia*. II. The effects of actinomycin D, puromycin, and chloramphenicol on transformations induced by patulin, acetamide and antiserum. *Exp. Cell Res.* **45:**306–322.

Balbinder, E., 1959 The genotypic control of kappa in *Paramecium aurelia*, syngen 4, stock 51. *Genetics* **44:**1227–1241.

Balbinder, E., 1961 Intraclonal variation in genotypically determined mixed (killer-sensitive) clones of *Paramecium aurelia*, syngen 4. *Physiol. Zool.* **34:**184–201.

Balsley, M., 1967 Dependence of the kappa particles of stock 7 of *Paramecium aurelia* on a single gene. *Genetics* **56:**125–131.

Baumann, P. J., E. L. Peterson and J. D. Berger, 1973   Mass selection of temperature-sensitive DNA synthesis deficient mutants in *Paramecium aurelia*. *J. Protozool.* **20**:507.

Beale, G., 1952   Antigen variation in *Paramecium aurelia*, variety 1. *Genetics* **37**:62–74.

Beale, G. H., 1954   *The Genetics of Paramecium aurelia*, Cambridge University Press, London.

Beale, G. H., 1957*a*   The antigen system of *Paramecium aurelia*. *Interntl. Rev. Cytol.* **6**:1–23.

Beale, G. H., 1957*b*   A mate-killing strain of *Paramecium aurelia*, variety 1, from Mexico. *Proc. R. Phys. Soc. Edinburgh* **26**:11–14.

Beale, G. H., 1958   The role of the cytoplasm in antigen determination in *Paramecium aurelia*. *Proc. R. Soc. Lond. Ser. B Biol. Sci.* **148**:308–314.

Beale, G. H., 1969   A note on the inheritance of erythromycin-resistance in *Paramecium aurelia*. *Genet. Res.* **14**:341–342.

Beale, G. H., 1973   Genetic studies on mitochondrially inherited mikamycin-resistance in *Paramecium aurelia*. *Mol. Gen. Genet.* **127**:241–248.

Beale, G. H. and A. Jurand, 1966   Three different types of mate-killer (mu) particle in *Paramecium aurelia* (syngen 1). *J. Cell Sci.* **1**:31–34.

Beale, G. H. and S. McPhail, 1966   Some additional results on the maintenance of kappa particles in *Paramecium aurelia* (stock 51) after loss of the gene *K*. *Genet. Res.* **9**:369–373.

Beale, G. H. and M. V. Schneller, 1954   A ninth variety of *Paramecium aurelia*. *J. Gen. Microbiol.* **11**:57–58.

Beale, G. H. and J. F. Wilkinson, 1961   Antigenic variation in unicellular organisms. *Annu. Rev. Microbiol.* **15**:263–296.

Beale, G. H., A. Jurand and J. R. Preer, Jr., 1969   The classes of endosymbiont of *Paramecium aurelia*. *J. Cell Sci.* **5**:65–91.

Beale, G. H., J. K. C. Knowles and A. Tait, 1972   Mitochondrial genetics in *Paramecium*. *Nature (Lond.)* **235**:396–397.

Behme, R. J. and J. Berger, 1970   The DNA content of *Paramecium aurelia*, stock 51. *J. Protozool.* **17** (**Suppl.**):20.

Beisson, J., 1972   Déterminants nucléaires et cytoplasmiques dans la biogenèse des structures chez les Protozoaires. *Ann. Biol.* **11**:401–411.

Beisson, J. and Y. Capdeville, 1966   Sur la nature possible des étapes de differentiation conduisant à l'autogamie chez *Paramecium aurelia*. *C. R. Hebd. Seances Acad. Sci. Ser. D. Sci. Nat.* **263**:1258–1261.

Beisson, J. and M. Rossignol, 1969   The first case of linkage in *Paramecium aurelia*. *Genet. Res.* **13**:85–90.

Beisson, J. and M. Rossignol, 1974 unpublished. Mouvements et positionnement des organelles chez la Paramécie.

Beisson, J. and T. M. Sonneborn, 1965   Cytoplasmic inheritance of the organization of the cell cortex in *Paramecium aurelia*. *Proc. Natl. Acad. Sci. USA* **53**:275–282.

Beisson, J., G. H. Beale, A. Sainsard, J. K. C. Knowles, A. Adoutte and A. Tait,1974   Genetic control of mitochondria in Paramecium. *Proc. 13th Interntl. Congr. Genet.*, unpublished.

Berger, J. D., 1971   Kinetics of incorporation of DNA precursors from ingested bacteria into macronuclear DNA of *Paramecium aurelia*. *J. Protozool.* **18**:419–429.

Berger, J. D., 1973*a*   Nuclear differentiation and nucleic acid synthesis in well-fed exconjugants of *Paramecium aurelia*. *Chromosoma (Berl.)* **42**:247–268.

Berger, J. D., 1973*b*   Selective inhibition of DNA synthesis in macronuclear fragments in

Paramecium aurelia exconjugants and its reversal during macronuclear regeneration. *Chromosoma (Berl.)* **44**:33–48.

Berger, J. D., 1974 Selective autolysis of nuclei as a source of DNA precursors in *Paramecium aurelia* exconjugants. *J. Protozool.* **21**:145–152.

Bishop, J. O., 1963 Immunological assay of some immobilization antigens of *Paramecium aurelia*, variety 1. *J. Gen. Microbiol.* **30**:271–280.

Bishop, J. O. and G. Beale, 1960 Genetical and biochemical studies of the immobilization antigens of *Paramecium aurelia. Nature (Lond.)* **186**:734.

Bleyman, L., 1967a Determination and inheritance of mating type in *Paramecium aurelia*, syngen 5. *Genetics* **56**:49–59.

Bleyman, L, 1967b Selfing in *Paramecium aurelia,* syngen 5: persistent instability of mating type expression. *J. Exp. Zool.* **165**:139–146.

Brygoo, Y., 1973 Mise en évidence d'un gène intervenant dans la détermination du type sexuel chez *Paramecium aurelia* (varieté 4). *Progr. Protozool. Proc. 4th Internatl. Congress Protozoology,* page 69.

Butzel, H. M. Jr., 1953a Genetic studies on *Paramecium aurelia*. Ph.D. Dissertation, Indiana University, Bloomington, Ind.

Butzel, H. M., Jr., 1953b Production of amicronucleate lines of *Paramecium aurelia* by colchicine treatment. *Microb. Gen. Bull.* **7**:32.

Butzel, H. M., Jr., 1953c A morphological mutant of stock 90, variety 1, of *Paramecium aurelia. Microb. Genet. Bull.* **8**:5–6.

Butzel, H. M., Jr., 1955 Mating type mutations in variety 1 of *Paramecium aurelia*, and their bearing upon the problem of mating type determination. *Genetics* **40**:321–330.

Butzel, H., Jr., 1968 Mating type determination in stock 51, syngen 4 of *Paramecium aurelia* grown in axenic culture. *J. Protozool.* **15**:284–290.

Butzel, H. M., Jr., 1973 Abnormalities in nuclear behavior and mating type determination in cytoplasmically bridged exconjugants of doublet *P. aurelia. J. Protozool.* **20**:140–143.

Butzel, H. M., Jr. and B. Vinciguerra, 1957 A fisson-rate mutant of stock 90, variety 1, of *Paramecium aurelia. Microb. Genet. Bull.* **15**:7.

Byrne, B. C., 1973 Mutational analysis of mating type inheritance in syngen 4 of *Paramecium aurelia. Genetics* **74**:63–80.

Bryne, B. J., 1969 Kappa, mu, and the metagon hypothesis in *Paramecium aurelia. Genet. Res.* **13**:197–211.

Capdeville, Y., 1969 Sur les interactions entre allèles contrôlant le type antigénique G chez *Paramecium aurelia. C. R. Hebd. Seances Acad. Sci. Ser. D. Sci. Nat.* **269**:1213–1215.

Capdeville, Y., 1971 Allelic modulation in *Paramecium aurelia* heterozygotes. Study of G serotypes in syngen 1. *Mol. Gen. Genet.* **112**:306–316.

Cavill, A. and I. Gibson, 1972 Genetic determination of esterases of syngen 1 and 8 in *Paramecium aurelia. Heredity* **28**:31–37.

Chang, S-Y. and C. Kung, 1973a Temperature-sensitive pawns: conditional behavioral mutants of *Paramecium aurelia. Science (Wash., D.C.)* **180**:1197–1199.

Chang, S-Y. and C. Kung, 1973b Genetic analyses of heat-sensitive pawn mutants of *Paramecium aurelia. Genetics* **75**:49–55.

Chang, S-Y., J. van Houten, L. J. Robles, S. S. Lui and C. Kung, 1974 An extensive behavioral and genetic analysis of the pawn mutants in *Paramecium aurelia. Genet. Res.* **23**:165–173.

Chao, P. K., 1955 Hereditary modification of the action of macronuclear *k* genes in the maintenance of the killer trait in variety 4 of *Paramecium aurelia. Rec. Genet. Soc. Am.* **24**:547.

Chun, A. W. S., 1969 Genetic studies of doublets of *Paramecium aurelia*. M. A. Thesis, University of Hong Kong, Hong Kong.

Cooper, J. E., 1965 A fast swimming "mutant" in stock 51 of *Paramecium aurelia*, variety 4. *J. Protozool.* **12**:381–384.

Cooper, J. E., 1968 An immunological and genetic analysis of antigen 5 in *Paramecium aurelia*, syngen 2. *Genetics* **60**:59–72.

Cronkite, D. L., 1972 Genetics of chemical induction of conjugation in *Paramecium aurelia*. Ph.D. Thesis , Department of Zoology Indiana University, Bloomington, Ind.

Cronkite, D., 1974 Genetics of chemical induction of conjugation in *Paramecium aurelia*. *Genetics* **76**:703–714.

Cummings, D. J., 1972 Isolation and partial characterization of macro- and micronuclei from *Paramecium aurelia*. *J. Cell. Biol.* **53**:105–115.

de Haller, G. 1965 Sur l'hérédité de charactéristiques morphologiques du cortex chez *Paramecium aurelia*. *Arch. Zool. Exp. Gén.* **105**:169–178.

de Haller, G., 1969 Interaction nucléocytoplasmique dans la morphogénèse des Ciliés. *Ann. Biol.* **8**:115–138.

de Haller, G. and B. ten Heggeler, 1969 Morphogénèse expérimentale chez les Ciliés: III. Effet d'une irradiation UV sur la génèse des trichocystes chez *Paramecium aurelia*. *Protistologica* **1**:115–120.

Diller, W. F., 1936 Nuclear reorganization processes in *Paramecium aurelia*, with descriptions of autogamy and "hemixis." *J. Morphol.* **59**:11–52.

Dilts, J. A., 1974 Evidence for covalently closed circular DNA in the kappa endosymbiont of *Paramecium aurelia*. *Genetics*, **77 (Suppl.)**:s17.

Dippell, R. V., 1948 Mutations of the killer plasmagene, kappa, in variety 4 of *Paramecium aurelia*. *Am. Nat.* **82**:43–50.

Dippell, R. V., 1950 Mutation of the killer cytoplasmic factor in *Paramecium aurelia*. *Heredity* **4**:165–187.

Dippell, R. V., 1953 Serotypic expression in heterozygotes of variety 4, *Paramecium aurelia*. *Microb. Genet. Bull.* **7**:12.

Dippell, R. V., 1954 A preliminary report on the chromosomal constitution of certain variety 4 races of *Paramecium aurelia*. *Caryologia* **Suppl.**: 1109–1111.

Dippell, R. V., 1955 Some cytological aspects of aging in variety 4 of *Paramecium aurelia*. *J. Protozool.* **2 (Suppl.)**: 7.

Dippell, R. V., 1958 The fine structure of kappa in killer stock 51 of *Paramecium aurelia*. *J. Biophys. Biochem. Cytol.* **4**:125–128.

Dippell, R. V., 1959a Cytological observations on kappa in killer stock 51 of *Paramecium aurelia*. *Anat. Rec.* **134**:554.

Dippell, R. V., 1959b Distribution of DNA in kappa particles of Paramecium in relation to the problem of their bacterial affinities. *Science (Wash., D.C.)* **130**:1415.

Dippell, R. V., 1965 Reproduction of surface structure in *Paramecium*. *Progr. Protozool., Excerpt. Med. Interntl. Congr. Ser.* **91**:65.

Dippell, R. V., 1968 The development of basal bodies in Paramecium. *Proc. Natl. Acad. Sci. USA* **61**:461–468.

Dippell, R. V. and S. E. Sinton, 1963 Localization of macronuclear DNA and RNA in *Paramecium aurelia*. *J. Protozool.* **10 (Suppl.)**: 23.

Dryl, S., 1965 Antigenic transformation in relation to nutritional conditions and the interautogamous cycle in *Paramecium aurelia*. *Exp. Cell Res.* **37**:569–581.

Ehret, C. F. and G. de Haller, 1963 Origin, development and maturation of organelles and organelle systems of the cell surface of *Paramecium*. *J. Ultrastruct. Res. (Suppl.)* **6**:1–42.

Ehret, C. F. and E. W. McArdle, 1974   The structure of *Paramecium* as viewed from its constituent levels of organization. In *Paramecium: A Current Survey*, edited by W. J. van Wagtendonk, Elsevier, Amsterdam.

Ehret, C. F. and E. L. Powers, 1952   Antibiotic sensitivity of kappa and the influence of X irradiation on the production of aureomycin-resistant "killers" in *Paramecium aurelia*, 51 VIII K. *J. Protozool.* **3 (Suppl.):** 14.

Ehret, C. F. and E. L. Powers, 1959   The cell surface of *Paramecium*. *Interntl. Rev. Cytol.* **8:**97–133.

Fauré-Fremiet, E., 1953   L'hypothèse de la sénescence et les cycles de réorganization nucléaire chez Ciliés. *Rev. Suisse Zool.* **60:**426–438.

Finger, I., 1956   Immobilizing and precipitating antigens of *Paramecium*. *Biol. Bull.* **111:**358–363.

Finger, I., 1957a   Immunologic studies of the immobilization antigens of *Paramecium aurelia*, variety 2. *J. Gen. Microbiol.* **16:**350–359.

Finger, I., 1957b   The inheritance of the immobilization antigens of *Paramecium aurelia*, variety 2. *J. Genet.* **55:**361–374.

Finger, I., 1964   Use of simple gel-diffusion techniques to assign antigenic markers to native proteins. *Nature (Lond.)* **203:**1035–1039.

Finger, I., 1967   The control of antigenic type in Paramecium. In *The Control of Nuclear Activity,* edited by L. Goldstein., Prentice-Hall, Inc., Engelwood Cliffs, N.J., pages 377–411.

Finger, I., 1968   Gene activation by cell products. *Trans. N.Y. Acad. Sci. Ser. II* **30:**968–976.

Finger, I., 1974   Surface antigens of *Paramecium aurelia*. In *Paramecium: a Current Survey*, edited by W. J. van Wagtendonk, Elsevier, Amsterdam.

Finger, I. and C. Heller, 1962   Immunogenetic analysis of proteins of Paramecium. I. Comparison of specificities controlled by alleles and by different loci. *Genetics* **47:**223–239.

Finger, I. and C. Heller, 1963   Immunogenetic analysis of proteins of Paramecium. IV. Evidence for the presence of hybrid antigens in heterozygotes. *J. Mol. Biol.* **6:**190–202.

Finger, I. and C. Heller, 1964   Cytoplasmic control of gene expression in Paramecium. I. Preferential expression of a single allele in heterozygotes. *Genetics* **49:**485–498.

Finger, I., M. Kaback, P. Kittner and C. Heller, 1960   Immunologic studies of isolated particulates of *Paramecium aurelia*. I. Antigenic relationships between cytoplasmic organelles and evidence for mitochondrial variations as demonstrated by gel diffusion. *J. Biophys. Biochem. Cytol.* **8:**591–601.

Finger, I., C. Heller and A. Green, 1962   Immunogenetic analysis of proteins of Paramecium. II. Coexistence of two immobilizing antigens within animals of a single serotype. *Genetics* **47:**241–253.

Finger, I., C. Heller and J. P. Smith, 1963   Immunogenetic analysis of proteins of *Paramecium*. III. A method for determining relations among antigenic proteins. *J. Mol. Biol.* **6:**182–189.

Finger, I., F. Onorato, C. Heller, and H. B. Wilcox, 1966   Biosynthesis and structure of Paramecium hybrid antigen. *J. Mol. Biol.* **17:**86–101.

Finger, I., F. Onorato, C. Heller, and L. Dilworth, 1969   Role of non-surface antigens in controlling *Paramecium* surface antigen synthesis. *J. Protozool.* **16:**18–25.

Finger, I., C. Heller, L. Dilworth, and C. von Allmen, 1972a   Clonal variation in *Paramecium*. I. Resistant unstable clones. *Genetics* **72:**17–33.

Finger, I., F. Onorato, C. Heller, and L. Dilworth, 1972b   Clonal variation in *Parame-*

cium. II. A comparison of stable and unstable clones of the same serotype. *Genetics* **72**:35–46.

Finger, I., C. Heller and S. Magers, 1972c   Clonal variation in *Paramecium*. III. Heterogeneity within clones of identical serotypes. *Genetics* **72**:47–62.

Flavell, R. A. and I. G. Jones, 1971   Paramecium mitochondrial DNA. Renaturation and hybridization studies. *Biochim. Biophys. Acta* **232**:255–260.

Fukushima, S., 1974   Effect of X-irradiations on the clonal life-span and fission rate in *Paramecium aurelia*. *Exp. Cell Res.* **84**:267–270.

Geckler, R. P. and R. F. Kimball, 1953   Effect of X rays on micronuclear number in *Paramecium aurelia*. *Science (Wash., D.C.)* **117**:80–81.

Génermont, J., 1966   Le déterminisme génétique de la vitesse de multiplication chez *Paramecium aurelia* (syng. 1). *Protistologica* **2**:45–51.

Gibson, I., 1970   Interacting genetic systems in Paramecium. In *Advances in Morphogenesis*, Vol. 8, edited by M. Abercrombie and J. Brachet, pp. 159–208, Academic Press, New York.

Gibson, I., 1973   Transplantation of killer endosymbionts in *Paramecium*. *Nature (Lond.)* **241**:127–129.

Gibson, I. and G. H. Beale, 1961   Genic basis of the mate-killer trait in *Paramecium aurelia*, stock 540. *Genet. Res.* **2**:82–91.

Gibson, I. and G. H. Beale, 1962   The mechanism whereby the genes M1 and M2 in *Paramecium aurelia*, stock 540, control growth of the mate-killer (mu) particles. *Genet. Res.* **3**:24–50.

Gibson, I. and N. Martin, 1971   DNA amounts in the nuclei of *Paramecium aurelia* and *Tetrahymena pyriformis*. *Chromosoma (Berl.)* **35**:374–382.

Gill, D. E., 1972   Intrinsic rates of increase, saturation densities, and competitive ability. I. An experiment with *Paramecium*. *Am. Nat.* **106**:461–471.

Gill, D. E. and N. G. Hairston, 1972   The dynamics of a natural population of Paramecium and the role of interspecific competition in community structure. *J. Animal Ecol.* **41**:81–96.

Gill, K. S. and E. D. Hanson, 1968   Analysis of prefission morphogenesis in *Paramecium aurelia*. *J. Exp. Zool.* **167**:219–236.

Haggard, B., 1972   Interspecies crosses in *Paramecium aurelia:* syngen 4 by syngen 8 and syngen 1 by syngen 7. Ph.D. Thesis, Department of Zoology Indiana University, Bloomington, Ind.

Haggard, B. W., 1974   Interspecies crosses in *Paramecium aurelia* (syngen 4 by syngen 8). *J. Protozool.* **21**:152–159.

Hairston, N. G., 1958   Observations on the ecology of *Paramecium* with comments on the species problem. *Evolution* **12**:440–450.

Hairston, N. G., 1967   Studies on the limitation of a natural population of *Paramecium aurelia*. *Ecology* **48**:904–910.

Hairston, N. G. and S. L. Kellerman, 1965   Competition between varieties 2 and 3 of *Paramecium aurelia:* the influence of temperature in a food-limited system. *Ecology* **46**:134–139.

Hallet, M. M., 1972   Effet du chlorure de calcium sur le déterminisme génetique de type sexual chez *Paramecium aurelia* (Souche 60, syng. 1). *Protistologica* **8**:387–396.

Hallet, M. M., 1973   Perturbation du rôle du macronucleus dans le déterminisme génetique du type sexuel chez *Paramecium aurelia*, syngen 1, sous l'effet du chlorure de calcium: facteurs génetique nucléaires et extranucléaires. *Progr. Protozool. Proc. 4th Internatl. Congress Protozoology* 170.

Hamilton, L. D. and M. E. Gettner, 1958   Fine structure of kappa in *Paramecium aurelia*. *J. Biophys. Biochem. Cytol.* **4**:122–123.

Hansma, H. G., 1975   Proteins in the ciliary membrane of Paramecium. unpublished.

Hanson, E. D., 1955   Inheritance and regeneration of cytoplasmic damage in *Paramecium aurelia. Proc. Natl. Acad. Sci. USA* **41**:783–786.

Hanson, E. D., 1956   Spontaneous mutations affecting the killer character in *Paramecium aurelia*, variety 4. *Genetics* **41**:21–30.

Hanson, E. D., 1962   Morphogenesis and regeneration of oral structures in *Paramecium aurelia:* an analysis of intracellular development. *J. Exp. Zool.* **150**:45–68.

Hanson, E. D., 1974   Methods in the cellular and molecular biology of Paramecium. *Meth. Cell Biol.* (edited by D. M. Prescott) **8**:319–363. Academic Press, New York.

Hanson, E. D. and M. Kaneda, 1968   Evidence for sequential gene action within the cell cycle of *Paramecium. Genetics* **60**:793–805.

Hanson, E. D. and R. M. Ungerleider, 1973   The formation of the feeding organelle in *Paramecium aurelia. J. Exp. Zool.* **185**:175–188.

Hipke, H. and E. D. Hanson, 1974   Induction of temperature sensitivity in *Paramecium aurelia* by nitrosoguanidine. *J. Protozool.* **21**:349–352.

Holzman, H. E., 1959*a*   A kappa-like particle in a non-killer stock of *Paramecium aurelia*, syngen 5. *J. Protozool.* **6 (Suppl)**: 26.

Holzman, H., 1959*b*   Some consequences of introducing a genetically defined population of a European syngen of *Paramecium aurelia* into an American pond. *Science* (Wash., D.C.) **130**:1418.

Hufnagel, L. A., 1969*a*   Cortical ultrastructure of *Paramecium aurelia*. Studies on isolated pellicles. *J. Cell Biol.* **40**:779–801.

Hufnagel, L., 1969*b*   Properties of DNA associated with raffinose-isolated pellicles of *Paramecium aurelia. J. Cell. Sci.* **5**:561–573.

Hunter, J. D. and E. D. Hanson, 1973   A suppressor gene in *Paramecium aurelia. J. Protozool.* **20(4)**:505.

Igarashi, S., 1966*a*   Temperature-sensitive mutation in *Paramecium aurelia*. I. Induction and inheritance. *Mutat. Res.* **3**:13–24.

Igarashi, S., 1966*b*   Temperature-sensitive mutation in *Paramecium aurelia*. II. Modification of the mutation frequency by pre- and postirradiation conditions. *Mutat. Res.* **3**:25–33.

Igarashi, S., 1969   Temperature-sensitive mutation. III. Temperature-sensitive catalase reaction of a *ts* mutant in *P. aurelia. Can. J. Microbiol.* **15**:1415–1418.

Isaacs, R. E., B. G. Santos and W. J. van Wagtendonk, 1969   Fractionation of nuclei from *Paramecium aurelia* by density gradient centrifugation. *Biochim. Biophys. Acta* **195**:268–270.

Jennings, H. S., 1929   Genetics of the Protozoa. *Bibliogr. Genet.* **5**:106–330.

Jones, A., 1964   Description and genetic analysis of an antigen of *Paramecium aurelia*. Dissertation, Department of Biology University of Pennsylvania, Philadelphia, Pa.

Jones, I. G., 1965*a*   Studies on the characterization and structure of the immobilization antigens of *Paramecium aurelia. Biochem J.* **96**:17–23.

Jones, I. G., 1965*b*   Immobilization antigen in heterozygous clones of *Paramecium aurelia. Nature (Lond.)* **207**:769.

Jones, I. G. and G. H. Beale, 1963   Chemical and immunological comparisons of allelic immobilization antigens in *Paramecium aurelia. Nature (Lond.)* **197**:205–206.

Jones, K. W., 1956   Nuclear differentiation in Paramecium. Ph.D. Thesis, Department of Biology Aberystwyth University, Wales.

Jurand, A. and D. M. Saxena, 1974   Ultrastructural features of the trichless mutant of *Paramecium aurelia. Acta Protozool.* **12**:307–311.

Jurand, A., and G. G. Selman, 1969   *The Anatomy of Paramecium aurelia*. Macmillan, London.

Kaczanowska, J., N. Hyvert and G. de Haller, 1974. Actinomycin D experiments on *Paramecium aurelia*. I. Effects on generation time and morphogenesis during the cell cycle. unpublished.

Kaneda, M. and E. D. Hanson, 1974   Growth patterns of morphogenetic events in the cell cycle of *Paramecium aurelia*. In *Paramecium: A Current Survey*, edited by W. J. van Wagtendonk, pp. 221–265, Elsevier, Amsterdam.

Karakashian, M., 1968   The rhythm of mating in *Paramecium aurelia*, syngen 3. *J. Cell Physiol.* **71**:197–210.

Kimball, R. F., 1937   The inheritance of sex at endomixis in *Paramecium aurelia*. *Proc. Natl. Acad. Sci. USA* **23**:469–474.

Kimball, R. F., 1939a   A delayed change of phenotype following a change of genotype in *Paramecium aurelia*. *Genetics* **24**:49–58.

Kimball, R. F., 1939b   Change of mating type during vegetative reproduction in *Paramecium aurelia*. *J. Exp. Zool.* **81**:165–179.

Kimball, R. F., 1947   The induction of inheritable modification in reaction to antiserum in *Paramecium aurelia*. *Genetics* **32**:486–499.

Kimball, R. F., 1948   The role of cytoplasm in inheritance in variety 1 of *Paramecium aurelia*. *Am. Nat.* **82**:79–84.

Kimball, R. F., 1949a   The induction of mutations in *Paramecium aurelia* by beta radiation. *Genetics* **34**:210–222.

Kimball, R. F., 1949b   Inheritance of mutational changes induced by radiation in *Paramecium aurelia*. *Genetics* **34**:412–424.

Kimball, R. F., 1953a   The structure of the macronucleus of *Paramecium aurelia*. *Proc. Natl. Acad. Sci. USA* **39**:345–347.

Kimball, R. F., 1953b   Three new mutants and the independent assortment of five genes in variety 1 of *Paramecium aurelia*. *Microb. Genet. Bull.* **8**:10–11.

Kimball, R. F., 1955   The role of oxygen and peroxide in the production of radiation damage in Paramecium. *Ann. N.Y. Acad. Sci.* **59**:638–647.

Kimball, R. F., 1962   Chromosome duplication and mutation. In *Strahlenwirkung und Milieu*, edited by H. Fritz-Niggli, special issue of *Strahlentherapie* **51**:116–122.

Kimball, R. F., 1963   The relation of repair to differential radiosensitivity in the production of mutations in *Paramecium*. In *Repair from Genetic Radiation*, edited by F. H. Sobels, pp. 167–178. Pergamon Press, Oxford.

Kimball, R. F., 1964a   Relation of the studies on mutagenesis in *Paramecium* to the dose-rate problem. In *Mechanisms of the Dose Rate Effect of Radiation at the Genetic and Cellular Levels. Jap. J. Genet.* **40 (Special Suppl.)**:264–274.

Kimball, R. F., 1964b   Studies on radiation mutagenesis in microorganisms. *Genet. Today, Proc. XI Internatl. Congr. Genet.*:227–234.

Kimball, R. F., 1964c   The distribution of X-ray-induced mutations to chromosomal strands in *Paramecium aurelia*. *Mutat. Res.* **1**:129–138.

Kimball, R. F., 1965   The induction of reparable premutational damage in *Paramecium aurelia* by the alkylating agent triethylene melamine. *Mutat. Res.* **2**:413–425.

Kimball, R. F., 1967   Persistent intraclonal variations in cell dry mass and DNA content in *Paramecium aurelia*. *Exp. Cell Res.* **48**:378–394.

Kimball, R. F. and T. Barka, 1959   Quantitative cytochemical studies on *Paramecium aurelia*. II. Feulgen microspectrophotometry of the macronucleus during exponential growth. *Exp. Cell Res.* **17**:173–182.

Kimball, R. F. and N. Gaither, 1951   The influence of light upon the action of ultraviolet on *Paramecium aurelia*. *J. Cell Comp. Physiol.* **37**:211–231.

Kimball, R. F. and N. Gaither, 1953   Influence of oxygen upon genetic and nongenetic effects of ionizing radiation on *Paramecium aurelia*. *Proc. Soc. Exp. Biol. Med.* **82**:471–477.

Kimball, R. F. and N. Gaither, 1954   Lack of an effect of a high dose of X-rays on aging in *Paramecium aurelia*, variety 1. *Genetics* **39**:977.

Kimball, R. F. and N. Gaither, 1955   Behavior of nuclei at conjugation in *Paramecium aurelia*. I. Effect of incomplete chromosome sets and competition between complete and incomplete nuclei. *Genetics* **40**:878–889.

Kimball, R. F. and N. Gaither, 1956   Behavior of nuclei at conjugation in *Paramecium aurelia*. II. The effects of X-rays on diploid and haploid clones, with a discussion of dominant lethals. *Genetics* **41**:715–728.

Kimball, R. F. and S. W. Perdue, 1962   Quantitative cytochemical studies on *Paramecium*. V. Autoradiographic studies of nucleic acid syntheses. *Exp. Cell Res.* **27**:405–415.

Kimball, R. F. and D. M. Prescott, 1964   RNA and protein synthesis in amacronucleate *Paramecium aurelia*. *J. Cell Biol.* **21**:496–497.

Kimball, R. F., J. Z. Hearon and N. Gaither, 1955   Tests for a role of H_2O_2 in X-ray mutagenesis. II. Attempts to induce mutations by peroxide. *Radiat. Res.* **3**:435–443.

Kimball, R. F., N. Gaither and S. Wilson, 1957   Genetic studies of a laboratory population of *Paramecium aurelia*, variety 1. *Evolution* **11**:461–465.

Kimura, M., 1957   Some problems of stochastic processes in genetics. *Ann. Math. Stat.* **28**:882–901.

Knowles, J. K. C., 1972   Observations on two mitochondrial phenotypes in single paramecium cells. *Exp. Cell Res.* **70**:223–226.

Knowles, J. K. C., 1974   A study of cytoplasmic erythromycin resistance in *Paramecium aurelia* using an improved microinjection technique. *Exp. Cell Res.* unpublished.

Koizumi, S., 1971   The cytoplasmic factor that fixes macronuclear mating type determination in *Paramecium aurelia*, syngen 4. *Genetics* **68**:34.

Koizumi, S. and S. Kobayashi, 1974   Microinjection and transfer of cytoplasm in Paramecium: I. Experiments on the transfer of kappa particles into cells at different stages. *Exp. Cell. Res.*, unpublished.

Koizumi, S. and J. R. Preer, Jr., 1966   Transfer to cytoplasm by microinjection in *Paramecium aurelia*. *J. Protozool.* **13 (Suppl.):** 27.

Kościuszko, H., 1965   Karyologic and genetic investigations in syngen 1 of *Paramecium aurelia*. *Fol. Biol.* **13**:339–368.

Kung, C., 1970   The electron transport system of kappa particles from *Paramecium aurelia*, stock 51. *J. Gen. Microbiol.* **61**:371–378.

Kung, C., 1971*a*   Genic mutations with altered system of excitation in *Paramecium aurelia*. I. Phenotypes of the behavioral mutants. *Z. Vergl. Physiol.* **71**:142–164.

Kung, C., 1971*b*   Genic mutants with altered system of excitation in *Paramecium aurelia*. II. Mutagenesis, screening and genetic analysis of the mutants. *Genetics* **69**:29–45.

Kung, C., 1971*c*   Aerobic respiration of kappa particles from *Paramecium aurelia*, stock 51. *J. Protozool.* **18**:328–332.

Levine, M., 1953*a*   The diverse mate-killers of *Paramecium aurelia*, variety 8; their interrelations and genetic basis. *Genetics* **38**:511–578.

Levine, M., 1953*b*   The interaction of nucleus and cytoplasm in the isolation and evolution of species of *Paramecium*. *Evolution* **7**:366–385.

Levine, M., 1955   An enzymatic comparison of two interbreeding but genetically isolated varieties of *Paramecium aurelia*. *J. Cell. Comp. Physiol.* **45**:409–420.

Maly, R., 1958   Eine genetisch bedingte Störung der Zelltrennung bei *Parmaecium aurelia*, var. 4. Ein Beitrag zum Problem der Mutabilität plasmatischer Systeme. *Z. Vererbungsl.* **89**:397–421.

Maly, R., 1960a   Die Normalisierung genetisch bedingter Defecte der Zelltrennung bei *Paramecium aurelia* durch Sauerstoffmangel und Kohlenmonoxyd. *Z. Vererbungsl.* **91**:226–236.

Maly, R., 1960b   Die Wirkung eines Komplexbildners und von Metallionen auf die Ausprägung des Snaky-und Monstra-Charakters bei *Paramecium aurelia*. *Z. Vererbungsl.* **91**:333–337.

Maly, R., 1961   Die Aufhebung eines Defectes der Zelltrennung beim Snaky-Stamm von *Paramecium aurelia* in Salzlösungen. *Z. Vererbungsl.* **92**:462–464.

Maly, R., 1962   Die Analyse eines Teilprozesses der Zytokinese mit Hilfe einer Mutante von *Paramecium aurelia*. *Verhandl. Deut. Zool. Ges. Wien.* (**1962**):84–86.

Margolin, P., 1954   A method for obtaining amacronucleated animals in *Paramecium aurelia*. *J. Protozool.* **6**:174–177.

Margolin, P., 1956a   An exception to mutual exclusion of the ciliary antigens in *Paramecium aurelia*. *Genetics* **41**:685–699.

Margolin, P., 1956b   The ciliary antigens of stock 172, *Paramecium aurelia*, variety 4. *J. Exp. Zool.* **133**:345–388.

Melechen, N. E., 1954   Antigenic stability and variability in *Paramecium aurelia*, variety 3. Ph.D. Thesis, Department of Biology University of Pennsylvania, Philadelphia, Pa.

Melvin, J. B., 1947   Hybrid mortality in intervarietal crosses in *Paramecium aurelia*. M.A. Thesis, Department of Zoology and Physiology Wellesley College, Wellesley, Mass.

Metz, C. B., 1947   Induction of "pseudoselfing" and meiosis in *Paramecium aurelia* by formalin-killed animals of opposite mating type. *J. Exp. Zool.* **105**:115–140.

Mitchison, N. A., 1955   Evidence against micronuclear mutations as the sole basis for death at fertilization in aged, and in the progeny of ultra-violet-irradiated, *Paramecium aurelia*. *Genetics* **40**:61–75.

Miyake, A., 1964   Induction of conjugation by cell-free preparations in *Paramecium multimicronucleatum*. *Science* (*Wash., D.C.*) **146**:1583–1585.

Miyake, A., 1968   Induction of conjugation by chemical agents in Paramecium. *J. Exp. Zool.* **167**:359–380.

Mueller, J. A., 1963   Separation of kappa particles with infective activity from those with killing activity and identification of the infective particles in *Paramecium aurelia*. *Exp. Cell Res.* **30**:492–508.

Mueller, J. A., 1964   Cofactor for infection by kappa in *Paramecium aurelia*. *Exp. Cell Res.* **35**:464–476.

Nanney, D. L., 1954   Mating type determination in *Paramecium aurelia*. A study in cellular heredity. In *Sex in Microorganisms,* edited by D. H. Wenrich, American Association for Advancement of Science, Washington, D.C.

Nanney, D. L., 1957   Mating type inheritance at conjugation in variety 4 of *Paramecium aurelia*. *J. Protozool.* **4**:89–95.

Nanney, D. L., 1963   Aspects of mutual exclusion in *Tetrahymena*. In *Biological Organization at Cellular and Supracellular Levels*, edited by R. J. C. Harris, pp. 91–109, Academic Press, New York.

Nanney, D. L., 1968   Ciliate genetics: patterns and programs of gene action. *Annu. Rev. Genetics* **2**:121–140.

Nobili, R., 1959   The effects of aging and temperature on the expression of the gene *am* in variety 4, stock 51, of *Paramecium aurelia*. *J. Protozool.* **6(Suppl.)**: 29.

Nobili, R., 1960   The effect of macronuclear regeneration on vitality in *Paramecium aurelia*, syngen 4. *J. Protozool.* **7(Suppl.):** 15.

Nobili, R., 1961   L'azione del gene *am* sull'apparato nucleare di *Paramecium aurelia* durante la riproduzione vegetativa e sessuale in relazione all'eta del clone ed alla temperatura di allevamento degli animali. *Caryologia* **14:**43–58.

Nobili, R., 1962   Il dimorfismo nucleare dei Ciliati: inerzia vegetativa del micronucleo. *Accad. Nazl. Lincei Rend. Classe Sci. Fisiche Mat. Nat. Ser. 8* **32:**392–396.

Nobili, R., 1964   Perdita del "kappa" associata a rigenerazione del macronucleo e all'autogomia, rispettivamente nel syngen 4, stock d4-47 e nel syngen 2, stock SG di *Paramecium aurelia. Accad. Nazl. Lincei Rend. Classe Sci. Fisiche, Mat. Nat. Ser. 8* **36:**1–10.

Pasternak, J., 1967   Differential genic activity in *Paramecium aurelia. J. Exp. Zool.* **165:**395–417.

Perasso, R. and A. Adoutte, 1974   The process of selection of erythromycin-resistant mitochondria by erythromycin in *Paramecium. J. Cell Sci.* **14:**475–497.

Pierson, B. F., 1938   The relation of mortality after endomixis to the prior interendomictic interval in *Paramecium aurelia. Biol. Bull.* **74:**235–243.

Pitelka, D. R., 1969   Fibrillar structures of the ciliate cortex: the organization of kinetosomal territories. *Progr. Protozool. Proc. 3rd Internl. Congr. Protozool.:* 44–46.

Pollack, S., 1970   Studies in the genetics and development of trichocysts in *Paramecium aurelia,* syngen 4. Ph.D. Thesis, Department of Biology University of Pennsylvania, Philadelphia, Pa.

Pollack, S., 1974   Mutations affecting the trichocysts in *Paramecium aurelia*. I. Morphology and description of the mutants. *J. Protozool.* **21:**352–362.

Powelson, B. E., M. A. Gates and J. Berger, 1974   A biometrical analysis of 22 stocks of 4 syngens of *Paramecium aurelia. Can. J. Zool.,* unpublished.

Powers, E. L., 1955   Radiation effects in Paramecium. *Ann. N.Y. Acad. Sci.* **59:**619–636.

Preer, J. R., Jr., 1946   Some properties of a genetic cytoplasmic factor in Paramecium. *Proc. Natl. Acad. Sci. USA* **32:**247–253.

Preer, J. R., Jr., 1948*a*   A study of some properties of the cytoplasmic factor "kappa," in *Paramecium aurelia,* variety 2. *Genetics* **33:**349–404.

Preer, J. R., Jr., 1948*b*   The killer cytoplasmic factor kappa: its rate of reproduction, the number of particles per cell, and its size. *Am. Nat.* **82:**35–42.

Preer, J. R., Jr., 1950   Microscopically visible bodies in the cytoplasm of the "killer" strains of *Paramecium aurelia. Genetics* **35:**344–362.

Preer, J. R., Jr., 1957   A gene determining temperature sensitivity in *Paramecium. J. Genet.* **55:**375–378.

Preer, J. R., Jr., 1959*a*   Nuclear and cytoplasmic differentiation in the Protozoa. In *Developmental Cytology,* edited by D. Rudnick, pp. 3–20, Ronald Press, New York.

Preer, J. R., Jr., 1959*b*   Studies on the immobilization antigens of *Paramecium*. II. Isolation. *J. Immunol.* **83:**378–384.

Preer, J. R., Jr., 1959*c*   Studies on the immobilization antigens of *Paramecium*. III. Properties. *J. Immunol.* **83:**385–391.

Preer, J. R., Jr., 1959*d*   Studies on the immobilization antigens of *Paramecium*. IV. Properties of different antigens. *Genetics* **44:**803–814.

Preer, J. R., Jr., 1969   Genetics of the Protozoa. In *Research in Protozoology* Vol. 3, edited by T. T. Chen, pp. 130–278, Pergamon Press, Oxford.

Preer, J. R., Jr., 1971   Extrachromosomal inheritance: hereditary symbionts, mitochondria, chloroplasts. *Annu. Rev. Genet.* **5:**361–406.

Preer, J. R., Jr. and L. B. Preer, 1959   Gel diffusion studies on the antigens of isolated cellular components of Paramecium. *J. Protozool.* **6**:88–100.

Preer, J. R., Jr. and L. B. Preer, 1967   Virus-like bodies in killer paramecia. *Proc. Natl. Acad. Sci. USA* **58**:1774–1781.

Preer, J. R., Jr. and P. Stark, 1953   Cytological observations on the cytoplasmic factor "kappa" in *Paramecium aurelia. Exp. Cell Res.* **5**:478–491.

Preer, J. R., Jr., R. W. Siegel and P. S. Stark, 1953   The relation between kappa and paramecin in *Paramecium aurelia. Proc. Natl. Acad. Sci. USA* **39**:1228–1233.

Preer, J. R., Jr., M. Bray and S. Koizumi, 1963   The role of cytoplasm and nucleus in the determination of serotype in *Paramecium. Proc. XI Interntl. Congr. Genet.:* 189.

Preer, J. R., Jr., L. B. Preer, B. Rudman and A. Jurand, 1971   Isolation and composition of bacteriophage-like particles from kappa of killer paramecia. *Mol. Gen. Genet.* **111**:202–208.

Preer, J. R., Jr., L. B. Preer and A. Jurand, 1974   Kappa and other endosymbionts in *Paramecium aurelia. Bacteriol Rev.* **38**:113–163.

Preer, L. B., 1969   Alpha, an infectious macronuclear symbiont of *Paramecium aurelia. J. Protozool.* **16**:570–578.

Preer, L. B. and J. R. Preer, Jr., 1964   Killing activity from lysed particles of *Paramecium. Genet. Res.* **5**:230–239.

Preer, L. B., B. Rudman and J. R. Preer, Jr., 1974   Induction of R bodies by ultraviolet light in killer paramecia. *J. Gen. Microbiol.,* **80**:209–215.

Pringle, C. R., 1956   Antigenic variation in *Paramecium aurelia,* variety 9. *Z. Indukt. Abstammungs.-Vererbungsl.* **87**:421–430.

Pringle, C. R. and G. H. Beale, 1960   Antigenic polymorphism in a wild population of *Paramecium aurelia. Genet Res.* **1**:62–68.

Rafalko, M. and T. M. Sonneborn, 1959   A new syngen (13) of *Paramecium aurelia* consisting of stocks from Mexico, France and Madagascar. *J. Protozool.* **6 (Suppl.)**:30.

Reisner, A., 1955   A method of obtaining specific serotype mutants in *Paramecium aurelia,* stock 169, var. 4. *Genetics* **40**:591.

Reisner, A. H., 1957   Studies on the surface antigens of stocks 169 and 51 of *Paramecium aurelia,* variety 4. Ph.D. Thesis, Department of Zoology Indiana University, Bloomington, Ind.

Reisner, A. H., J. Rowe and H. M. Macindoe, 1969a   The largest known monomeric globular proteins. *Biochim. Biophys. Acta* **188**:196–206.

Reisner, A. H., J. Rowe and R. W. Sleigh, 1969b   Concerning the tertiary structure of the soluble surface proteins of Paramecium. *Biochemistry* **8**:4637–4644.

Roque, M., 1956   L'évolution de la ciliature buccale pendant l'autogamie et la conjugaison chez *Paramecium aurelia. C. R. Hebd. Seances Acad. Sci. Ser. D. Sci. Nat.* **242**:2592–2595.

Rudnyansky, B. M., 1953   The effect of breis made from sexually reactive animals on mating type determination in *Paramecium aurelia,* variety 2. *Microb. Genet. Bull.* **7**:19.

Sainsard, A., M. Claisse and M. Balmefrezol, 1974   A nuclear mutation affecting structure and function of mitochondria in *Paramecium. Mol. Gen. Genet.,* **130**:113–125.

Schneller, M. V., 1958   A new type of killing action in a stock of *Paramecium aurelia* from Panama. *Proc. Ind. Acad. Sci.* **67**:302.

Schneller, M. V., T. M. Sonneborn and J. A. Mueller, 1959   The genetic control of kappa-like particles in *Paramecium aurelia. Genetics* **44**:533–534.

Schwartz, V. and H. Meister, 1973 Eine Alterveränderung des Macronucleus von *Paramecium. Z. Naturforsh.* Sect. B **28c:**232.

Selman, G. G. and A. Jurand 1970 Trichocyst development during the fission cycle of Paramecium. *J. Gen. Microb.* **60:**365–372.

Siegel, R. W., 1953 A genetic analysis of the mate-killer trait in *Paramecium aurelia*, variety 8. *Genetics* **38:**550–560.

Siegel, R. W., 1954 Mate-killing in *Paramecium aurelia*, variety 8. *Physiol. Zool.* **27:**89–100.

Siegel, R. W., 1958 Hybrid vigor, heterosis and evolution in *Paramecium aurelia. Evolution* **12:**402–416.

Siegel, R. W., 1961 Nuclear differentiation and transitional cellular phenotypes in the life cycle of *Paramecium. Exp. Cell Res.* **24:**6–20.

Siegel, R. W., 1967 Genetics of ageing and the life cycle in Ciliates. *Symp. Soc. Exp. Biol.* **21:**127–148.

Siegel, R. W., 1970 Organellear damage and revision as a possible basis for intraclonal variation in *Paramecium. Genetics* **66:**305–314.

Simon, E. M. and M. V. Schneller, 1973 The preservation of ciliated Protozoa at low temperature. *Cryobiology* **10:**421–426.

Sinden, R. E., 1973 The synthesis of immobilization antigen in *Paramecium aurelia*: in *vitro* localization of antigen in ribosomal cell fractions. *J. Protozool.* **20(2):**307–315.

Skaar, P. D., 1956 Past history and pattern of serotype transformation in *Paramecium aurelia. Exp. Cell. Res.* **10:**646–656.

Smith-Sonneborn, J., 1971 Age-correlated sensitivity to ultraviolet radiation in *Paramecium. Radiat. Res.* **46:**64–69.

Smith-Sonneborn, J., 1974 Age-correlated effects of caffeine on non-irradiated and UV-irradiated *Paramecium aurelia*. unpublished.

Smith-Sonneborn, J. and M. Klass, 1974 Changes in the DNA synthesis pattern of Paramecium with increased clonal age and interfission time. *J. Cell Biol.* **61:**591–611.

Smith-Sonneborn, J. and W. Plaut, 1969 Studies on the autonomy of pellicular DNA in *Paramecium. J. Cell Sci.* **5:**365–372.

Smith-Sonneborn, J. E. and W. J. van Wagtendonk, 1964 Purification and chemical characterization of kappa of stock 51, *Paramecium aurelia. Exp. Cell Res.* **33:**50–59.

Smith-Sonneborn, J., M. Klass and D. Cotton, 1974 Parental age and life span versus progeny life span in *Paramecium. J. Cell. Sci.* **14:**691–699.

Soldo. A. T. and G. A. Godoy, 1972 The kinetic complexity of Paramecium macronuclear deoxyribonucleic acid. *J. Protozool.* **19:**673–678.

Sommerville, J., 1967 Immobilization antigen synthesis in *Paramecium aurelia*: The detection of labelled antigen in a cell-free amino acid incorporating system. *Biochim. Biophys. Acta* **149:**625–627.

Sommerville, J., 1970 Serotype expression in *Paramecium. Adv. Microb. Physiol.* **4:**131–178.

Sonneborn, T. M., 1935 The relation of endomixis to vitality in *Paramecium aurelia. Anat. Rec.* **64:**103.

Sonneborn, T. M., 1937a Sex, sex inheritance and sex determination in *Paramecium aurelia. Proc. Natl. Acad. Sci. USA* **23:**378–385.

Sonneborn, T. M., 1937b The extent of the interendomictic interval in *Paramecium aurelia* and some factors determining its variability. *J. Exp. Zool.* **75:**471–502.

Sonneborn, T. M., 1938a Mating types in *Paramecium aurelia*: Diverse conditions for mating in different stocks; occurrence, number and interrelations of the types. *Proc. Am. Philos. Soc.* **79**:411–434.

Sonneborn, T. M., 1938*b* The delayed occurrence and total omission of endomixis in selected lines of *Paramecium aurelia. Biol. Bull.* **74**:76–82.

Sonneborn, T. M., 1939 *Paramecium aurelia*: Mating types and groups; lethal interactions; determination and inheritance. *Am. Nat.* **73**:390–413.

Sonneborn, T. M., 1940 The relation of macronuclear regeneration in *Paramecium aurelia* to macronuclear structure, amitosis and genetic determination. *Anat. Rec.* **78(4)(Suppl.):** 53–54.

Sonneborn, T. M., 1941*a* The occurrence, frequency and causes of failure to undergo reciprocal cross-fertilization during mating in *Paramecium aurelia*, variety 1. *Anat. Rec.* **81 (Suppl.):** 66.

Sonneborn, T. M., 1941*b* The effect of temperature on mating reactivity in *Paramecium aurelia. Anat. Rec.* **81 (Suppl.):** 131.

Sonneborn, T. M., 1942*a* Inheritance in ciliate Protozoa. *Am. Nat.* **76**:46–62.

Sonneborn, T. M., 1942*b* Inheritance of an environmental effect in *Paramecium aurelia*, variety 1, and its significance. *Genetics* **27**:169.

Sonneborn, T. M., 1942*c* Evidence for two distinct mechanisms in the mating reaction of *Paramecium aurelia. Anat. Rec.* **84(4):**92–93.

Sonneborn, T. M., 1942*d* Double animals and multiple simultaneous mating in variety 4 of *Paramecium aurelia* in relation to mating types. *Anat. Rec.* **84:**29–30.

Sonneborn, T. M., 1943*a* Gene and cytoplasm. I. The determination and inheritance of the killer character in variety 4 of *Paramecium aurelia. Proc. Natl. Acad. Sci. USA* **29:**329–338.

Sonneborn, T. M., 1943*b* Acquired immunity to a specific antibody and its inheritance in *Paramecium aurelia. Proc. Ind. Acad. Sci.* **52:**190–191.

Sonneborn, T. M., 1946*a* Experimental control of the concentration of cytoplasmic genetic factors in *Paramecium. Cold Spring Harbor Symp. Quant. Biol.* **11:** 236–255.

Sonneborn, T. M., 1946*b* Inert nuclei: inactivity of micronuclear genes in variety 4 of *Paramecium aurelia. Genetics* **31:**231.

Sonneborn, T. M., 1946*c* A system of separable genetic determiners in the cytoplasm of *Paramecium aurelia*, variety 4. *Anat. Rec.* **94:**346.

Sonneborn, T. M., 1947*a* Recent advances in the genetics of *Paramecium* and *Euplotes. Adv. Genet.* **1:**263–358.

Sonneborn, T. M., 1947*b* Developmental mechanisms in *Paramecium. Growth Symp.* **11:**291–307.

Sonneborn, T. M., 1948*a* The determination of hereditary antigenic differences in genically identical *Paramecium* cells. Proc. Natl. Acad. Sci. USA **34:**413–418.

Sonneborn, T. M., 1948*b* Symposium on plasmagenes, genes and characters in *Paramecium aurelia*. Introduction. *Am. Nat.* **82:**26–34.

Sonneborn, T. M., 1950*a* Methods in the general biology and genetics of *Paramecium aurelia. J. Exp. Zool.* **113:**87–148.

Sonneborn, T. M., 1950*b* The cytoplasm in heredity. *Heredity* **4:**11–36.

Sonneborn, T. M., 1950*c* Cellular transformations. *Harvey Lect.* **44:**145–164.

Sonneborn, T. M., 1953 Environmental control of the duration of phenomic or cytoplasmic lag in *Paramecium aurelia. Microb. Genet. Bull.* **7:**23.

Sonneborn, T. M., 1954*a* Gene-controlled, aberrant nuclear behavior in *Paramecium aurelia. Microb. Genet. Bull.* **11:**24–25.

Sonneborn, T. M., 1954*b*   Is gene *K* active in the micronucleus of *Paramecium aurelia?* *Microb. Genet. Bull.* **11**:25–26.

Sonneborn, T. M., 1954*c*   The relation of autogamy to senescence and rejuvenescence in *Paramecium aurelia. J. Protozool.* **7**:38–53.

Sonneborn, T. M., 1954*d*   Patterns of nucleocytoplasmic integration in *Paramecium. Caryologia* **Suppl.**: 307–325.

Sonneborn, T. M., 1955*a*   Heredity, development and evolution in *Paramecium. Nature (Lond.)* **175:** 1100.

Sonneborn, T. M., 1955*b*   Macronuclear control of the initiation of meiosis and conjugation in *Paramecium aurelia. J. Protozool.* **2 (Suppl.):** 12.

Sonneborn, T. M., 1956   The distribution of killers among the varieties of *Paramecium aurelia. Anat. Rec.* **125**:567–568.

Sonneborn, T. M., 1957   Breeding systems, reproductive methods, and species problems in Protozoa. In *The Species Problem*, edited by E. Mayr, pp. 155–324, American Association for Advancement of Science, Washington, D.C.

Sonneborn, T. M., 1959   Kappa and related particles in *Paramecium. Adv. Virus Res.* **6**:229–356.

Sonneborn, T. M., 1963   Does preformed structure play an essential role in cell heredity? In *The Nature of Biological Diversity*, edited by J. M. Allen, pp. 165–221, McGraw Hill, New York.

Sonneborn, T. M., 1966   A non-conformist genetic system in *Paramecium aurelia. Am. Zool.* **6**:589.

Sonneborn, T. M., 1967   The evolutionary integration of the genetic material into genetic systems. In *Heritage from Mendel* edited by R. A. Brink, Chapt. 19., pp. 375–401, University of Wisconsin Press, Madison, Wisc.

Sonneborn, T. M., 1970*a*   Methods in *Paramecium* research. *Meth. Cell Physiol.* **4**:241–339.

Sonneborn, T. M., 1970*b*   Gene action in development. *Proc. R. Soc. Lond. Ser. B Biol. Sci.* **176**:347–366.

Sonneborn, T. M., 1970*c*   Determination, development and inheritance of the structure of the cell cortex. *Symp. Interntl. Soc. Cell. Biol.* **9**:1–13.

Sonneborn, T. M., 1974*a*   Herbert Spencer Jennings, *Biograph. Mem. Natl. Acad. Sci.* **47,** unpublished.

Sonneborn, T. M., 1974*b*   Ciliate morphogenesis and its bearings on general cellular morphogenesis. *Actualités Protozoologiques (4th Interntl. Congr. Protozool.)* **1**:327–355.

Sonneborn, T. M., 1975   Species designations and descriptions of the 14 syngens of *Paramecium aurelia. Trans. Am. Microscop. Soc.*, unpublished.

Sonneborn, T. M. and E. Balbinder, 1953   The effect of temperature on the expression of allelic genes for serotypes in a heterozygote of *Paramecium aurelia. Microb. Genet. Bull.* **7**:24–25.

Sonneborn, T. M. and R. V. Dippell, 1943   Sexual isolation, mating, and sexual responses to diverse conditions in variety 4, *Paramecium aurelia. Biol. Bull.* **85**:36–43.

Sonneborn, T. M. and R. V. Dippell, 1946   Mating reactions and conjugation between varieties of *Paramecium aurelia* in relation to conceptions of mating type and variety. *Physiol. Zool.* **19**:1–18.

Sonneborn, T. M. and R. V. Dippell, 1960*a*   Cellular changes with age in Paramecium. In *The Biology of Aging, American Institute of Biological Science Symposium 6* edited by B. L. Strehler, p. 285, Waverly Press, Baltimore, Md.

Sonneborn, T. M. and R. V. Dippell, 1960*b*   The genetic basis of the difference between single and double *Paramecium aurelia*. *J. Protozool.* **7 (Suppl.):** 26.

Sonneborn, T. M. and R. V. Dippell, 1961   Self-reproducing differences in the cortical organization in *Paramecium aurelia*, syngen 4. *Genetics* **46:** 900.

Sonneborn, T. M. and R. V. Dippell, 1962   Two new evidences of cortical autonomy in syngen 4 of *Paramecium aurelia*. *J. Protozool.* **9 (Suppl.):** 28.

Sonneborn, T. M. and A. LeSuer, 1948   Antigenic characters in *Paramecium aurelia* (variety 4): determination, inheritance, and induced mutations. *Am. Nat.* **82:** 69–78.

Sonneborn, T. M. and R. S. Lynch, 1937   Factors determining conjugation in *Paramecium aurelia*. III. A genetic factor: the origin at endomixis of genetic diversities. *Genetics* **22:** 284–296.

Sonneborn, T. M. and F. R. Ogasawara, 1953   The ciliary antigenic system of stock 32, variety 4, of *Paramecium aurelia*. *Proc. Ind. Acad. Sci.* **63:** 270.

Sonneborn, T. M. and M. V. Schneller, 1955*a*   Are there cumulative effects of parental age transmissible through sexual reproduction in variety 4 of *Paramecium aurelia? J. Protozool* **2 (Suppl.):** 6.

Sonneborn, T. M. and M. V. Schneller, 1955*b*   Genetic consequences of aging in variety 4 of *Paramecium aurelia*. *Rec. Genet. Soc. Am.* **24:** 596.

Sonneborn, T. M. and M. V. Schneller, 1955*c*   The basis of aging in variety 4 of *Paramecium aurelia*. *J. Protozool.* **2 (Suppl.):** 6.

Sonneborn, T. M. and M. V. Schneller, 1960*a*   Physiological basis of aging in Paramecium. In *The Biology of Aging, Amer. Institute of Biological Science Symposium 6* edited by B. L. Strehler, p. 283–284, Waverly Press, Baltimore Md.

Sonneborn, T. M. and M. V. Schneller, 1960*b*   Age-induced mutations in *Paramecium*. In *The Biology of Aging, American Institute of Biological Science Symposium 6* edited by B. L. Strehler, p. 286–287, Waverly Press, Baltimore, Md.

Sonneborn, T. M. and M. V. Schneller, 1960*c*   Measures of the rate and amount of aging on the cellular level. In *The Biology of Aging, American Institute of Biological Science Symposium 6* edited by B. L. Strehler, p. 290–291, Waverly Press, Baltimore, Md.

Sonneborn, T. M., R. V. Dippell, M. V. Schneller and I. Tallan, 1953*a*   The explanation of "anomalous" inheritance following exposure to ultraviolet in variety 4 of *Paramecium aurelia*. *Microb. Genet. Bull.* **7:** 25–26.

Sonneborn, T. M., I. Tallan, E. Balbinder, F. Ogasawara and B. Rudnyansky, 1953*b*   The independent inheritance of 5 genes in variety 4 of *Paramecium aurelia*. *Microb. Genet. Bull.* **7:** 27–28.

Sonneborn, T. M., F. Ogasawara and E. Balbinder, 1953*c*   The temperature sequence of the antigenic types in variety 4 of *Paramecium aurelia* in relation to the stability and transformation of antigenic types. *Microb. Genet. Bull.* **7:** 27.

Sonneborn, T. M., M. V. Schneller and M. F. Craig, 1956   The basis of variation in phenotype of gene-controlled traits in heterozygotes of *Paramecium aurelia*. *J. Protozool* **3 (Suppl.):** 8.

Sonneborn, T. M., J. A. Mueller and M. V. Schneller, 1959   The classes of kappa-like particles in *Paramecium aurelia*. *Anat. Rec.* **134:** 642.

Steers, E., Jr., 1961   Electrophoretic analysis of immobilization antigens of *Paramecium aurelia*. *Science (Wash., D.C.)* **133:** 2010–2011.

Steers, E., Jr., 1962   A comparison of the tryptic peptides obtained from immobilization antigens of *Paramecium aurelia*. *Proc. Natl. Acad. Sci. USA* **48:** 867–874.

Steers, E., Jr., 1965   Amino acid composition and quaternary structure of an immobilizing antigen from *Paramecium aurelia*. *Biochem.* **4:** 1896–1901.

Steers, E., Jr., J. Beisson and V. T. Marchesi, 1969   A structural protein extracted from the trichocyst of *Paramecium aurelia. Exp. Cell Res.* **57**:392–396.

Stevenson, I., 1967   A method for the isolation of macronuclei from *Paramecium aurelia. J. Protozool.* **14**:412–414.

Stevenson, I. and F. P. Lloyd, 1971*a*   Ultrastructure of nuclear division in *Paramecium aurelia.* I. Mitosis in the micronucleus. *Austr. J. Biol. Sci.* **24**:963–975.

Stevenson, I. and F. P. Lloyd, 1971*b*   Ultrastructure of nuclear division in *Paramecium aurelia.* II. Amitosis of the macronucleus. *Austr. J. Biol. Sci.* **24**:977–987.

Suyama, Y. and J. R. Preer, Jr., 1965   Mitochondrial DNA from Protozoa. *Genetics* **52**:1051–1058.

Tait, A., 1968   Genetic control of β-hydroxybutyric dehydrogenase in *Paramecium aurelia. Nature (Lond.)* **219**:941.

Tait, A., 1970*a*   Enzyme variation between syngens in *Paramecium aurelia. Biochem. Gen.* **4**:461–470.

Tait, A., 1970*b*   Genetics of NADP-dependent isocitrate dehydrogenase in *Paramecium aurelia. Nature (Lond.)* **225**:181–182.

Tait, A., 1972   Altered mitochondrial ribosomes in an erythromycin-resistant mutant of *Paramecium. FEBS (Fed. Eur. Biochem. Soc.) Lett.* **24**:117–120.

Tait, A., G. H. Beale and A. R. Oxbrow, 1971   Enzyme polymorphism in S. E. England. *J. Protozool.* **18 (Suppl.)**: 26.

Tallan, I., 1959   Factors involved in infection by the kappa particles in *Paramecium aurelia,* syngen 4. *Physiol. Zool.* **32**:78–89.

Tallan, I., 1961   A cofactor required by kappa in the infection of *Paramecium aurelia* and its possible action. *Physiol. Zool.* **34**:1–13.

Taub, S. R., 1959   Mathematical analysis of the mating type determining system in *Paramecium aurelia,* syngen 7. *Science (Wash., D.C.)* **130**:1424.

Taub, S. R., 1963   The genic control of mating type differentiation in *Paramecium. Genetics* **48**:815–834.

Taub, S. R., 1966*a*   Regular changes in mating type composition in selfing cultures and in mating type potentiality in selfing caryonides of *Paramecium aurelia. Genetics* **54**:173–189.

Taub, S. R., 1966*b*   Unidirectional mating type changes in individual cells from selfing cultures of *Paramecium aurelia. J. Exp. Zool.* **163**:141–150.

van Wagtendonk, W. J., editor, 1974   *Paramecium: A Current Survey,* Elsevier, Amsterdam.

van Wagtendonk, W. J. and A. T. Soldo, 1970   Methods used in the axenic cultivation of *Paramecium aurelia. Meth. Cell Physiol.* **4**:117–130.

van Wagtendonk, W. J., J. A. D. Clark and G. A. Godoy, 1963   The biological status of lambda and related particles in *Paramecium aurelia. Proc. Natl. Acad. Sci. USA* **50**:835–838.

Whittle, J. R. S. and L. Chen-Shan, 1972   Cortical morphogenesis in *Paramecium aurelia*: mutants affecting cell shape. *Genet. Res.* **19**:271–279.

Wichterman, R., 1953   *The Biology of Paramecium,* Blakiston, New York.

Widmayer, D. J., 1965   A non-killer resistant kappa and its bearing on the interpretation of kappa in *Paramecium aurelia. Genetics* **51**:613–623.

Williams, J., 1971   The growth *in vitro* of killer particles from *Paramecium aurelia* and the axenic culture of this Protozoan. *J. Gen. Microbiol.* **68**:253–262.

Wolfe, J., 1967   Structural aspects of amitosis: a light and electron microscope study of the isolated macronuclei of *Paramecium aurelia* and *Tetrahymena pyriformis. Chromosoma* **23**:59–79.

Wood, H. K., 1953   Some factors affecting delayed separation of conjugants in *Parame-cium aurelia*. Ph.D. Thesis, Department of Zoology Indiana University, Bloomington, Ind.

Woodard, J., B. Gelber, and H. Swift, 1961   Nucleoprotein changes during the mitotic cycle in *Paramecium aurelia*. *Exp. Cell Res.* **23**:258–264.

Woodard, J., M. Woodard, B. Gelber and H. Swift, 1966   Cytochemical studies of conjugation in *Paramecium aurelia*. *Exp. Cell Res.* **41**:55–63.

Woodruff, L. L. and R. Erdmann, 1914   A normal periodic reorganization process without cell fusion in Paramecium. *J. Exp. Zool.* **17**:425–518.

Yusa, A., 1963   An electron microscope study on regeneration of trichocysts on *Parame-cium caudatum*. *J. Protozool.* **10**:253–262.

Author Index*

Aalders, L. E., 149
Abashian, D. V., 313
Abdullaev, H. A., 165
Abercrombie, M., 582
Abraham, A., 306
Abrams, L., 361
Achmedov, A. Y., 178
Adams, J. P., 576
Adoutte, A., 576, 577, 578, 587
Afansiev, M. M., 406
Agrawal, H., 404
Ahloowalia, B. S., 360
Ahnström, G., 165, 179
Ahuja, M. R., 306
Ainsworth, G. S., 404
Äkerberg, E., 218
Albert, M., 375
Alexander, D. E., 27
Allard, R. W., 109
Allen, J. M., 591
Allen, N. E., 462
Allen, R. D., 577
Allen, S. L., 462, 463, 466, 576, 577
Allen, T. C., Jr., 404
Allison, D. C., 127
Allmen, C. von, 581
Almgärd, G., 211
Ambler, J. E., 275
Ames, I. H., 306

Anand, R., 165
Anders, F., 314
Andersen, H. A., 463
Andersen, W. R., 274
Anderson, E. G., 26, 30
Anderson, L. M., 86
Andeweg, J. M., 149
André, J., 464, 576
Andrus, C. F., 280
Appa Rao, K., 306, 310
Apple, J. L., 310
Aravindan, K. V., 220
Arcara, P. G., 375, 377
Armstrong, J. J., 426
Arnold, C. G., 165, 426
Arny, D. C., 412
Ar-Rushdi, A. H., 306
Ashman, R. B., 26
Ashraf, J., 165
Ashworth, J. M., 431
Atabekov, J. G., 404
Athwal, D. S., 53
Athwal, R. S., 86, 88
Atkins, I. M., 57
Atteck, O. M., 129
Aurand, K., 177
Austin, M. L., 577
Avanzi, S., 375
Avery, P., 309
Avetisov, V. A., 167
Axtell, J. D., 26

Ayonoadu, U. W., 375
Ayyangar, G. N. R., 57

Babkishev, A. G., 168
Bacalangco, E. H., 53
Baetche, K. P., 375
Bailey, L. H., 143
Balbinder, E., 577, 591, 592
Bald, J. G., 404
Balkema, G. H., 211
Balmefrezol, M., 576, 588
Balsley, M., 577
Bancroft, B. J., 409
Bancroft, J. B., 404, 405
Baquar, S. R., 212
Barabás, Z., 165
Barber, H. N., 211
Barka, T., 584
Barthelmess, I., 165
Barton, D. W., 53, 211, 220, 278
Baskin, C. C., 165
Baskin, J. M., 165
Bates, L. S., 28
Bathelt, H., 245
Batra, S., 149
Baum, J. W., 169
Bauman, L. F., 27, 29
Baumann, P. J., 578
Baur, E., 329

* This lists authors *only* where they appear in the bibliographies following the chapters.

595

Bawden, F. C., 404
Bayer, M. H., 306
Beachell, H. M., 53, 55, 57
Beadle, G. W., 26, 27
Beale, G. H., 578, 579, 582, 583, 588, 593
Beasley, J. O., 127
Beckett, J. B., 26
Behme, R. J., 578
Beisson, J., 576, 577, 578, 592
Belfield, A. M., 90
Bell, S. L., 360
Belling, J., 242
Bemis, W. P., 143, 144
Benepal, P. S., 143
Bennett, M. D., 375
Benton, R. J., 404
Berger, B., 165
Berger, G., 165
Berger, J. D., 578, 579, 587
Bergman, L., 306
Berkeley, G. H., 404
Berliner, J., 426
Bernier, G., 165
Berset, D., 174
Besnard-Wibaut, C., 165
Bhaduri, P. N., 53
Bhatia, C. R., 166, 167, 307
Bhowal, J. G., 86
Bianchi, A., 274
Bielig, L. M., 86
Bird, J., 127
Bishop, J. O., 579
Black, L. M., 404
Blakeslee, A. F., 242, 243
Blank, L. M., 127
Bleyman, L. K., 463, 579
Blixt, S., 211, 212
Blobel, G., 425
Blodgett, F. M., 413
Bluethner, W. D., 89
Boardman, N. K., 110
Böhme, H., 274
Bohn, G. W., 143, 144, 149, 150
Bonner, J. T., 430
Bonzi, G., 166, 174
Bora, K. C., 53

Borden, D., 464
Bos, L., 404, 405
Bouharmont, J., 53, 166
Bourgin, J. P., 307
Bowden, W. M., 87
Bowman, F. T., 404
Boynton, J. E., 110, 275, 278
Boza Barducci, T., 127
Bozarth, R. F., 405
Bozzini, A., 87
Brabec, F., 275
Brachet, J., 582
Bradley, E. C., 331
Braun, A. C., 307
Bray, M., 588
Bremer, A. H., 211
Brewbaker, J. L., 28
Brewer, G., 87
Brink, R. A., 26, 591
Broadbent, L., 405
Brock, R. D., 166, 275, 277
Brodführer, U., 166
Brodführer-Franzgrote, U., 172
Bronckers, F., 166
Brosi, R., 463
Brotherton, J. W., 212
Brown, J. A. M., 166, 167, 173
Brown, J. C., 275
Brown, M. S., 127, 128, 129, 130
Brown, S. W., 53
Brunk, C. F., 463
Bruns, P. J., 464
Brunt, A. A., 405
Brussard, T. B., 464
Bryant, T. R., 375
Brygoo, Y., 579
Buiatti, M., 275, 307
Burger, D., 171
Burholt, D. R., 375
Burk, L. G., 307, 308, 312, 313
Burnham, C. R., 26, 110, 360
Burns, J. A., 307, 309
Butler, L., 275, 278

Butler, W., 425
Butzel, H. M., Jr., 579
Byrne, B. C., 577, 579
Byrne, B. J., 579
Byrne, O. R., 169

Cadman, C. H., 405
Calavan, E. C., 408
Caldwell, J., 405
Cameron, D. R., 306, 307, 311
Cameron, I. L., 463, 464
Campell, R. N., 405, 407
Canas, A. P., 88
Capdeville, Y., 578, 579
Carlson, K., 426
Carlson, P. S., 307, 464
Carmer, S. G., 143
Caruso, J. L., 275
Carvalho, A. M. B., 405
Castetter, E. F., 143
Catcheside, D. G., 242
Cauderon, Y., 87
Caughey, P. A., 466
Cavill, A., 579
Cesati, R. R., 405
Cetl, I., 167
Chandler, R. F., 53
Chandrasekharan, P., 54
Chaney, R. L., 275
Chang, T. D., 54, 56
Chang, T. M., 54
Chang, T. T., 54
Chang, S-Y., 579
Chang, W. T., 54
Chant, S. R., 405
Chao, L. F., 54
Chao, P. K., 579
Chaplin, J. F., 307, 310
Chapman, V., 87, 89, 90, 375
Charret, R., 464
Chatterjee, D., 54
Chaudhuri, R. P., 404
Chen, T. T., 467, 587
Cheng, T.-Y., 307
Chen-Shan, L., 593

Cheo, P. C., 405
Cherry, J. P., 127
Chessin, M., 404, 405
Chevalier, A., 54
Chiang, K. S., 425
Childs, J. F. L., 405
Chilton, S. J. P., 55
Chiquoine, A. D., 430
Chiu, R. J., 405, 412
Chmielewski, T., 275
Chomchalow, N., 360
Choudhuri, H. C., 376
Chow, C. C., 405
Chow, M., 466
Christensen, E., 465
Christoff, G. T., 312
Chu, Y. E., 54
Chun, A. W. S. 580
Chung, S. C., 167, 176
Claisse, M., 588
Clark, F. J., 26
Clark, G. M., 464
Clark, J. A. D., 593
Clark, W., 430
Clary, G. B., 28
Clausen, R. E., 307, 308, 309
Clayberg, C. D., 275
Clayton, E. E., 308
Cleffmann, A., 464
Cleland, R. E., 242, 243, 245, 360
Clement, D., 131
Clowes, F. A. L., 376, 377
Cochran, F. D., 149
Cockerham, C. C., 311
Cocking, E. C., 311
Coe, E. H., 26, 28, 29
Cole, R. D., 277
Collins, G. B., 308, 310
Conklin, M. E., 308
Comstock, R. E., 312
Cook, W. C., 409
Cooper, J. E., 580
Corbett, M. K., 405, 408, 411
Corcos, A., 167
Corliss, J. O., 464

Cortés-Monllor, A., 410
Cosbey, E., 425
Costa, A. S., 405
Cotton, D., 589
Coucoli, H., 90
Craig, M. F., 592
Creech, R. G., 26
Creighton, H. B., 27
Cronkite, D. L., 577, 580
Crowden, R. K., 213
Croxall, H. E., 405
Cruse, D., 165
Cua, L. D., 54
Cummings, D. J., 580
Cummings, M. B., 143
Cummings, S. E., 311
Curtis, L. C., 143
Cutler, H. C., 143
Cutter, E. G., 275

Da Costa, C. P., 150
Dale, W. T., 405
Daly, K., 167, 308, 313
D'Amato, F., 87
Darby, J. F., 406
Darlington, C. D., 27, 309, 377
Darvey, N. L., 87
Darwin, C., 329
Das, N. K., 375
Dawson, R. F., 308
Davidson, D., 377
Davidson, H., 376
Davies, D. R., 425
Davis, G. N., 150
Deakin, J. R., 150
Dearing, R. D., 307
Debergh, P., 275
Deering, R. A., 377
de Haller, G., 580, 584
de la Roche, I. A., 27
Delevic, B., 405
Delgado, L., 168
Delgado-Sanchez, S., 406
Delwiche, E. H., 213
Demchenko, S. I., 167
Demerec, M., 27
Dempsey, E., 29

Dempsey, W. H., 278
Demski, J. W., 406
de Nettancourt, D., 308
Denna, D. W., 143
Dennis, R. W. G., 406
Deri, P. L., 375
Dermen, H., 307, 308
De Pinto, S., 466
Devreux, M., 219
De Vries, H., 243
de Zeuw, D. J., 406
de Zoeten, G. A., 409
Dhillon, T. S., 360
Diachun, S., 406
Dickerson, L. M., 212
Diener, T. O., 406
Dierks, W., 167
Diers, L., 243, 329
Dieterman, L. J., 314
Dietz, R. S., 127
Diller, W. F., 580
Dilts, J. A., 580
Dilworth, L., 581
Dippell, R. V., 580, 591, 592
Disbrey, C., 467
Döbel, P., 329
Dobrovolná, J., 167
Dodds, K. S., 212
Dodson, D. O., 377
Doerder, F. P., 464, 466
Donini, B., 167
Doolittle, S. P., 406
Dougall, D. K., 278, 309
Douglas, C. R., 127
Douwes, H., 127, 128
Dow, K. P., 219
Doy, C. H., 167, 168, 275
Doyle, G. G., 27
Dragavtsev, V. A., 168
Drijfhout, E., 212
Driscoll, C. J., 86, 87, 89
Dryl, S., 580
Dubert, J. M., 466
Duggar, B. M., 309
Dulieu, H., 308
Durham, G. B., 144
Duvick, D. N., 27, 29

East, E. M., 308
Easton, G. D., 406
Ebersold, W. T., 425, 426
Ecochard, R., 275
Edwardson, J. R., 308
Ehrenberg, L., 168, 179, 211
Ehret, C. F., 580, 581
Eichhorn, J. L., 405
Eig, A., 87
Eisa, H. M., 143
Ekelund, R., 212
Elliott, A. M., 463, 464
Elstad, V. B., 178
Emerson, R. A., 27
Emerson, S. H., 243
Emery, G., 275
Emmerling, M. H., 29
Endrizzi, J. E., 87, 127, 128, 132, 133
Ennis, H. L., 430, 431
Enomoto, N., 54
Epel, B. L., 425
Erdmann, R., 594
Ernst, H., 329
Erwin, A. T., 143
Eslick, R. F., 406
Espiritu, L. G., 220
Evans, G. M., 376
Evans, L. E., 87, 88

Fagbenle, H. H., 406
Farrow, S. W., 577
Fauré-Fremiet, E., 581
Favret, E. A., 110
Feaster, C. V., 132
Fedak, G., 109
Fedotov, V. S., 212
Feenstra, W. J., 168, 174
Feldman, M., 87, 91
Fellows, H., 409
Fezer, K. D., 406
File, S. K., 463
Finch, R. A., 375
Fincham, J. R. S., 329, 330
Finger, I., 581, 582
Fischbeck, G., 91
Flack, I. H., 404

Flacks, M., 466
Flavell, R. A., 464, 582
Florkin, M., 463
Ford, R. E., 406
Fork, D. C., 243
Foster, R. E., II, 406
Fraenkel-Conrat, H., 406
Francis, R. R., 143
Frankel, J., 464
Franklin, I. R., 275
Fraser, A. C., 27
Fraser, L., 404
Freitag, J. H., 406
Friedberg, S. H., 376
Fritz-Niggli, H., 584
Fry, P. R., 221
Frydenberg, O., 109
Fryxell, P. A., 128
Fujii, T., 168
Fujishita, N., 150
Fukushi, T., 406
Fukushima, E., 55
Fukushima, S., 582
Fulton, J. P., 406
Fulton, R. W., 406, 407, 409, 411

Gaither, N., 584, 585
Galinat, W. C., 27
Garber, E. D., 311, 360, 361
Gardella, C., 243
Garnsey, S. M., 407
Garrecht, M., 276
Gates, M. A., 587
Gavin, R. H., 464
Gay, H., 377
Geckler, R. P., 582
Gelber, B., 594
Gelin, O., 211, 212
Gelinas, D. A., 27
Génermont, J., 582
Gerechter, Z. K., 410
Gerish, G., 430
Gerlack-Cruse, D., 168
Gerstel, D. U., 128, 131, 307, 308, 309, 310

Gettner, M. E., 582
Ghatge, M. B., 54
Ghose, R. L. M., 54
Gibbs, A. J., 404, 405, 406, 407, 408, 411
Gibbs, J. L., 309
Gibson, I., 463, 465, 577, 579, 582
Gichner, T., 168, 173, 178, 179
Giesbrecht, J., 27
Giles, J. A., 128
Gill, D. E., 582
Gill, K. S., 582
Gillham, N. W., 425, 426
Gilmer, R. M., 407, 413
Ginter, E. K., 178
Globerson, D., 142
Glotov, N. V., 178
Glover, D. V., 27
Glubrecht, H., 168
Godoy, G. A., 589, 593
Godron, D. A., 329
Godward, M. B. E., 376
Gold, A. H., 405, 411
Goldenberg, J. B., 212
Goldstein, L., 581
Golembiewski, P. A., 577
Gómez-Campo, C., 168
Gonsalves, D., 407
Good, R., 128
Goodenough, U. W., 426
Gooding, G. V., Jr., 411
Goodspeed, T. H., 307, 308, 309
Goold, R. A., 410
Gorman, D. S., 425
Gorovsky, M. A., 467
Gorsic, J., 360, 361
Gottschalk, W., 212
Govindaswami, S., 57
Govorov, L. I., 212
Gowans, C. S., 172
Gowen, J., 312
Grant, A. F. M., 407
Grant, T. J., 407
Grant, V., 361
Grass, F. S., 465
Grebenscikov, I., 143

Greeley, L. W., 29, 409
Green, A., 581
Green, J. M., 128
Gresshoff, P. M., 167, 168, 275
Griffing, B., 168, 169, 171
Gritton, E. T., 212, 213
Gröber, K., 275, 276
Grogan, R. G., 406, 407
Grosso, J. J., 307, 308
Grout, A. J., 212
Grover, N. S., 169
Gruchy, D. F., 465
Gwynne, D. C., 405
Guha, S., 309
Gunckel, J. E., 309, 313
Günther, E., 276
Gupta, P., 309
Gupta, S. B., 309, 376

Haan, H. De, 213
Haccius, B., 276
Hadacova, V., 178
Hadidi, A. F., 407
Hagberg, A., 109, 110
Hagedorn, D. J., 407
Hagemann, R., 276, 329
Hagen, G. L., 306, 309, 313
Haggard, B. W., 582
Hairston, N. G., 582
Håkansson, A., 213
Halaban, R., 376
Hall, C. V., 143
Hall, D. H., 409
Hall, J. L., 310
Hall, R. H., 407
Hallet, M. M., 582
Hamill, D. E., 313
Hamilton, L. D., 582
Hampton, R. E., 407
Hansen, H., 330
Hansing, E. D., 412
Hanson, E. D., 582, 583, 584
Hara, S., 56
Harborne, S. B., 329
Hardon, J. J., 276
Hare, W. W. 213

Härer, L., 169
Harland, S. C., 128, 129, 213
Harle, J. R., 169
Harris, R. J. C., 586
Harrison, B. D., 404, 405, 406, 407, 408, 411
Harrison, B. J., 329, 330
Härstedt. E;. 213
Hart, G. E., 309
Harte, C., 243, 330
Harter, L. L., 413
Harwood, R. R., 150
Hastings, P. J., 425, 426
Haupt, W., 213
Hawthorn, L. R., 150
Hayes, H. K., 150
Hayhome, B. A., 361
Hayman, B. I., 309
Hearon, J. Z., 585
Heathcote, G. D., 405
Heber, U., 243
Hecht-Poinar, E., 413
Heggestad, H. E., 307, 308
Heimlich, L. F., 150
Helgason, S. B., 109
Heller, C., 581, 582
Hemberg, T., 212
Henderson, M. T., 54, 58
Henningsen, K. W., 110
Henry, T. A., 309
Heringa, R. J., 213
Herrmann, F., 276
Highkin, H. R., 213
Hildebrandt, A. C., 309, 314
Hildering, G. J., 276
Hill, D. L., 465
Hills, G. J., 407
Hiorth, G., 361
Hipke, H., 583
Hirono, Y., 169, 176
Hissong, G., 404
Ho, K. M., 54
Hoeppener, E., 243
Hoffmann, W., 330
Hogenboom, N. G., 277
Hohl, H., 430
Hollaender, A., 168
Hollings, M., 407

Holmes, F. O., 309, 407, 408, 409
Holmes, H. L., 311
Holz, G. G., 465
Holzman, H. E., 583
Hong, M. S., 56
Hoober, J. K., 425
Hooker, A. L., 30
Hopp, R., 143
Hoshino, Y., 213
Howell, S. H., 425
Howland, G., 243
Hsieh, S. C., 54, 56
Hu, C. H., 54
Huang, C. S., 54
Huang, T. S., 56
Huart, R., 171
Hudlock, M. O., 425
Hufnagel, L. A., 583
Hunter, A. W. S., 331
Hunter, J. D., 583
Hurd, P. D., Jr., 143
Hussein, H. A. S., 169
Hutchins, A. E., 143
Hutchinson, J. B., 129
Hyams, J., 425
Hyde, B. B., 243
Hylander, N., 169
Hyvert, N., 584

Igarashi, S., 583
Iida, T. T., 408
Ikenaga, M., 168
Imai, S., 309
Imber, D., 279
Imle, E. P., 411
Immer, F. R., 110, 169
Inouye, T., 408
International Rice Research Institute, 54
Isaacs, R. E., 583
Israfilova, U., 178
Itai, C., 279
Ivanov, M. A., 310
Ivanov, V. I., 169, 174, 178
Iwata, N., 54
Iyengar, N. K., 129
Izadpanah, K., 408
Izard, C., 310

Jachuck, P. J., 54
Jacobs, M., 169, 170, 171
Jacqmard, A., 376
Jahn, A., 212
Jain, H. K., 277
James, T., 170
Japha, B., 243
Jeffrey, R. N., 307, 310
Jenifer, F. O., 406
Jenkins, B. C., 87
Jenkins, E. W., 143
Jenkins, J. A., 277, 278
Jenkins, J. M., Jr., 150
Jenkins, M. T., 27
Jennings, H. S., 583
Jennings, P. R., 55
Jensen, J. H., 408
Jeter, J. R., Jr.,
Jobitharaj, S., 57
Jodon, N. E., 55, 58
John, B., 376
Johnson, B. L., 87, 91, 129
Johnson, C. O., 409
Johnson, D. L., 376
Johnson, F., 408
Johnson, J., 408
Johnson, R., 87, 90
Johnston, M. J., 213
Jona, R., 376
Jones, A., 583
Jones, A., 583
Jones, C. M., 150
Jones, D. R., 150
Jones, I. G., 464, 582, 583
Jones, J. W., 53, 55
Jones, K. W., 583
Jones, L. M., 28
Jones, M. E., 170
Jones, W. E., 275
Jones, W. H., 180
Josephson, L. M., 27
Juergensmeyer, E. B., 465
Jurand, A., 578, 583, 588, 589
Justus, N., 129

Kaback, M., 581
Kaczanowska, J., 584

Kadam, B. S., 55
Kahler, A. L., 109
Kahn, A. J., 431
Kahn, R. P., 408
Kajanus, B., 213
Kaltsikes, P. J., 376
Kamieńska, A., 172
Kammacher, P., 129
Kammacher, P. A., 129
Kaneda, M., 583, 584
Kaneshiro, E. S., 467
Kannangara, G. C., 110
Kao, K. N., 109
Kapoor, B. M., 213
Kapur, S. P., 408
Karakashian, M., 584
Karibasappa, B. K., 55
Karlsson, K-E., 109
Kasha, K. J., 109
Kasperbauer, M. J., 308, 310
Kassanis, B., 404, 408
Kasyanenko, A. G., 170
Katayama, T., 55
Kato, T. A., 27
Katterman, F. R. H., 127
Katz, E. R., 431
Kauffman, H. E., 54
Kaufmann, B. P., 377
Kaznowski, L., 213
Kebby, D. G., 404
Keeble, F. W., 213
Kehr, A. E., 310, 313
Kellenbarger, Sh., 213
Kellerman, S. L., 582
Kelts, L. J., 407
Kempanna, C., 87, 89
Kermicle, J. L. 27
Kerr, E. A., 275
Keyser, J. S., 221
Khafaji, S. D. Al
Khamankar, Y. G., 277
Khan, S. H., 55
Khangildin, V. V., 213
Khanna, K. R., 310
Khush, G. S., 53, 55, 57, 277, 278
Kidder, G. W., 463
Kihara, H., 55, 88

Kimball, R. F., 582, 584, 585
Kimber, G., 86, 87, 88, 90, 91, 129, 310
Kimble, K. A., 407
Kimura, M., 585
Kinoshita, T., 56
Kirkpatrick, H. C., 408
Kishi, Y., 150
Kittner, P., 581
Kiyosawa, S., 55
Klas, J., 213
Klasova, A., 213
Klass, M., 589
Kleczkowski, A., 408
Klein, H. D., 213
Klein, W. H., 178
Klemm, M., 330
Knight, R. L., 129
Knott, D. R., 88
Knowles, J. K. C., 578, 585
Kobayashi, S., 585
Koch, S. L., 463
Koga, Y., 56, 58
Kohel, R. J., 87, 128, 130, 133
Koizumi, S., 585, 588
Kolderie, M., 430
Kolek, J., 213
Kondo, A., 55
Koo, F. K. S., 170
Kooistra, E., 213
Kosciuszko, H., 585
Kostoff, D., 310
Kovacs, E. I., 310
Kowles, R. V., 26
Kranz, A. R., 170
Kreitlow, K. W., 408
Kribben, F. J., 170
Krishnamurty, K. V., 306, 310
Kroon, A., 576
Kruse, J., 277
Kučera, J., 170
Kugler, I., 170
Kuhn, C. W., 408
Kung, C., 579, 585
Kurakami, H., 56
Kuriyama, H., 55

Kusanagi, A., 376
Kuwada, Y., 55

Labib, G., 313
Ladeburg, R. C., 408
Ladipo, J. L., 409
Laibach, F., 170, 171
Lamberts, H., 109
Lamm, R., 213, 214
Lamprecht, H., 214, 215, 216, 217, 218
Land, J. B., 171
Lang, A., 213
Langridge, J., 168, 171, 277
Langridge, W. H. R., 376
Larson, L. A., 218
Larson, R. H., 406, 408, 409
Larter, E. N., 88
Laughnan, J. R., 27, 29
Lauritzen, M., 330
Lavrinetskaya, T. E., 171
Law, C. N., 88, 90
Lawrence, C. W., 171
Lawrence, M. J., 178, 179
Lawson, R. H., 409
Layne, R. E. C., 407
LeBeau, F. J., 413
Ledoux, L., 171
Lee, F., 431
Lee, P. H. T., 463
Lee, R. E., 310
Lee, Y. H., 88
Lee-Chen, S., 171, 172, 178
Legg, P. D., 308, 310
LeSeur, A., 592
Levin, D., 243
Levine, E. E., 425
Levine, M., 585
Levine, R. P., 425, 426
Levings, C. S., 27
Lewis, C. F., 127, 130, 133
Lewis, D., 243
Lewis, K. R., 27, 57, 89, 90, 309, 376
Li, C. H., 88
Li, C. I., 463
Li, H. W., 88

Li, S. L., 172, 176
Lindberg, G. D., 409
Linder, R. C., 408
Lindqvist, K., 219
Ling, K. C., 54
Linnane, A. W., 110
Linnert, G., 244, 330
Linsley, E. G., 143
Linsmaier, E. M., 310
Lister, R. M., 405, 409
Liu, H. P., 404
Liu, H. Y., 409
Lloyd, F. P., 593
Lock, R. H., 219
Locke, M. M., 465
Lockhart, J., 172
Loden, H. D., 130
Loefer, J. B., 465
Loegering, W. Q., 90
Loewenberg, J. R., 172
Loginov, M. A., 178
Longley, A. E., 27
Longley, A. F., 55
Loomis, W. F., Jr., 431
Loppes, R., 425, 426
Löve, Å., 172
Lui, S. S., 579
Lundqvist, U., 109
Lung, M., 409
Luthra, J. K., 310
Lutz, A., 244
Lwanga, K., 218
Lyall, L. H., 219
Lyman, J. T., 168, 169
Lynch, R. S., 592

Maat, D. Z., 405
McArdle, E. W., 581
McClintock, B., 27, 28
McCoy, J. W., 465
McCray, F. A., 310
McCullogh, J. M., 172
McDonald, B. B., 465
McDonald, H. H., 410
MacDonald, T., 28
Mace, F., 166
Macer, R. C. F., 90
McEwen, F. L., 409

McFadden, E. S., 88
McFarland, E. F., 130
MacGillivray, J. M., 26
Machold, O., 277
Macindoe, H. M., 588
McIntosh, R. A., 88
McKelvie, A. D., 172
Mac Key, J., 88
McKinney, H. H., 29, 409
McLean, D. M., 409
Macleod, R., 409
Macleod, R. D., 376
McMichael, S. C., 128, 130
MacNeish, R. S., 132
McPhail, S., 578
McVittie, A., 426
McWhorter, F. P., 409
Madoo, R. M., 127
Magers, S., 582
Magoon, M. L., 310
Maheshwari, S. C., 309
Mahler, J., 127
Mains, E. B., 143
Mainx, F., 330
Majid, R., 277
Maly, R., 330, 585, 586
Mann, M. C., 308
Mann, T. J., 307, 309, 310, 311, 312
Manske, R. H. F., 311
Manzinger, D. F., 310, 311
Marchesi, G., 274
Marchesi, V. T., 592
Margolin, P., 465, 586
Marion, L., 311
Markarian, D., 150
Markham, R., 409, 410
Marks, G. E., 219
Marquardt, H., 244
Marriam, D. C., 406
Martelli, G. P., 410
Martin, C., 410
Martin, D. W., 130
Martin, N., 465, 582
Martin, P. G., 376
Marx, G. A., 219
Masima, I., 55
Masuyama, Y., 58
Matagne, R., 376, 426

Matejka, J. C., 410
Mathan, D. S., 277, 278
Mather, K., 311
Matsumura, S., 88
Matsuura, H., 219
Matthews, P., 212, 219
Matthews, R. E. F., 410
Matzinger, D. F., 174, 314
Mayo, O., 244
Mayr, E., 591
Mechelke, F., 330
Meister, H., 588
Melchers, G., 172, 311, 312, 313
Melechen, N. E., 586
Melera, P. W., 376
Mello-Sampayo, T., 87, 88
Melvin, J. B., 586
Mendel, G., 219
Menzel, M. Y., 127, 130, 278
Merkx, G., 275
Merritt, J. F., 131
Mertz, E. T., 28
Měsíček, J., 172
Mesken, M., 172
Mettin, D., 89
Metz, C. B., 586
Meyer, J. R., 129, 130
Meyer, V. G., 130
Michniewicz, M., 172
Miksche, J. P., 167, 173, 376, 377
Milbrath, J. A., 410
Miller, C. O., 312
Miller, E. T., 464
Mink, G. I., 410
Miravalle, R. J., 214
Misch, M. S., 463
Mishanec, W., 211, 219
Misra, R. C., 173
Misra, R. N., 58
Mitchison, N. A., 586
Miura, K., 467
Miyake, A., 586
Moav, R., 307, 311
Mochizuki, A., 89
Moens, P., 278
Mohan Rao, P. K., 58

Mohanty, R. N., 53
Monti, L. M., 219
Moore, J. D., 413
Moorhead, E. L., 404
Morinaga, T., 55, 56
Morris, R., 89
Morrison, B. M., 463
Morrison, J. W., 219
Mosely, M. E., 132
Moseman, J. G., 109
Mothes, K., 311
Mrkos, H., 218
Mudaliar, S. D., 57
Mueller, J. A., 586, 588, 592
Mukai, Y., 58
Müller, A. J., 173, 178
Munck, L., 109
Munger, H. M., 143
Munthe, T., 410
Munz, P. A., 244, 361
Muramatsu, M., 89
Murant, A. F., 404, 405, 406, 408, 410, 411
Murfet, I. C., 219
Murray, J. C., 130
Murti, K. G., 465
Murty, G. S., 310

Näf, U., 311
Nagai, I., 56
Nagamatsu, T., 56
Nagao, S., 56
Nagata, T., 311
Nagel, J., 466
Nakagahara, M., 54
Nakamori, E., 56
Nakata, K., 313
Naliboff, J. A., 425
Nandi, H. K., 56
Nanney, D. L., 463, 464, 465, 466, 586
Napp-Zinn, K., 166, 173, 174
Narayanan, K. R., 166
Nasyrov, Y. S., 165, 170, 178
Natarajan, A. T., 53, 89, 165

Nath, P., 143
Nauman, C. H., 375
Neales, T. F., 174
Neatby, K. W., 89
Neely, C., 430
Nelson, M. R., 410
Nelson, O. E., Jr., 28, 30
Neuffer, M. G., 28
Neumann, J., 426
Nevo, Y., 279
Newman, J. S., 129
Newsom, V. M., 361
Nielsen, G., 109
Nielsen, O. F., 110
Niemann, E. G., 174
Nienhous, F., 410
Niethammer, A., 174
Niizeki, H., 56
Nikolov, C. V., 174
Nilan, R. A., 26, 109, 110
Nilsson, E., 219
Nilsson, J. R., 466
Nishimura, Y., 56
Nitsch, C., 275
Nitsch, J. P., 174, 307, 311
Nitzany, F. E., 410
Nobili, R., 586, 587
Noordam, D., 410
Norman, G. G., 405
Norton, G., 171
Novikov, V. K., 404
Nuffer, M. G., 28, 29
Nuttal, V. W., 219

Oakes, A. J., 130
O'Brien, M. J., 406, 410, 411, 412
Oehlkers, F., 243, 244, 330
Ogasawara, F. R., 592
Ohlrogge, A. J., 28
Öhlund, K., 211
Oka, H. I., 56
Oka, M., 311
Okamoto, M., 89, 90
Okura, E., 57
Olby, R. C., 311
Olmsted, M. A., 425
Olson, E. O., 410

O'Malley, T. A., 376
Omar, A., 220
O'Mara, J. G., 89
Omura, T., 54, 56
Ono, K., 56
Ono, S., 58
Onorato, F., 581
Oostindier-Braaksma, F.,
    174
Orellana, R. G., 410
Orias, E., 466, 467
Osterman-Golkar, S., 179
Oswald, J. E., 410
Ou, S. H., 54
Outka, D. E., 466
Owen, R. D., 465
Owings, A., 130
Oxbrow, A. R., 593

Paddock, E. F., 278
Padilla, G. M., 463
Pao, W. K., 88
Parnell, F. R., 57
Parsons, J. A., 466
Parthasarathy, N., 57
Pasternak, J., 577, 587
Pathak, M. D., 53, 54
Patino, G., 410
Paton, D. M., 211
Patrick, C. A., 463
Paulus, A. O., 404
Pearson, O. H., 143
Pederson, D. G., 174
Pellew, C., 213, 220
Pennell, F. W., 361
Perasso, R., 587
Perdue, S. W., 585
Pérez, J. E., 410
Perkus, M. P., 312
Perlman, B. S., 466
Perry, C. M., 176
Person, C. L., 89
Persson, G., 110
Peterson, E. L., 578
Peterson, P. A., 28
Pettigrew, R., 89
Phillips, L. L., 128, 130,
    131, 132

Phillips, M. P., 408
Phillips, R. B., 466, 467
Phillips, R. L., 26
Pienaar, R. de V., 361
Pierce, W. H., 410
Pierson, B. F., 587
Pirie, N. W., 404
Pirone, T. P., 409
Pitelka, D. R., 587
Plaskitt, A., 425
Plaut, W., 426, 589
Plurad, S. B., 176
Pohjakallio, O., 220
Pohlendt, G., 330
Poisson, C., 131
Pokorný, V., 179
Pollack, S., 587
Pollard, J. K., 313
Polyakova, T. F., 174
Popovici, I., 377
Porter, R. H., 150
Postlethwait, S. N., 27
Pound, G. S., 410, 412, 413
Powelson, B. E., 587
Power, J. B., 311
Powers, E. L., 581, 587
Prabhakara Rao, M. V., 89
Prasad, A. B., 376
Pratt, M. J., 405
Preer, J. R., Jr., 467, 578,
    585, 587, 588, 593
Preer, L. B., 587, 588
Prentice, I. W., 405
Prescott, D. M., 431, 467,
    583, 585
Price, H. J., 178
Price, J. M., 278
Price, W. C., 408, 410
Pringle, C. R., 588
Privalov, G. F., 168
Provvidenti, R., 211
Pura, C. D., 53
Purcifull, D. E., 411
Pyne, C., 467

Quacquarelli, A., 410
Quantz, L., 405
Quastler, H., 377
Quisenberry, K. S., 89

Rafaeli, D. E., 431
Rafalko, M., 588
Ragazzini, R., 275
Rai, K. S., 361
Rajewsky, B., 175
Ramage, R. T., 110
Ramanna, M. S., 278
Ramanujam, S., 57
Ramiah, K., 55, 57, 58
Randall, J., 426, 467
Randolph, L. R., 28
Ranga Rao, D. R., 58
Rao, M. B. V. N., 57
Rao, N. S., 53
Raper, K. B., 430, 431
Raskin, R. S., 278
Rasmuson, H., 361
Rasmusson, J., 220
Ratcliffe, D., 175
Rau, N. S., 57
Raut, R. N., 277
Raven, P. H., 244
Ray, C., Jr., 467
Raymer, W. B., 410, 411
Reddy, G. M., 28
Reddy, M. M., 311, 361
Rédei, G. P., 165, 167, 169,
    172, 175, 176
Rees, H., 375, 376
Reeve, R. M., 220
Reeve, S. J., 466
Rehwaldt, C. A., 176
Reichmann, M. E., 409
Reinholz, E., 176, 177
Reisner, A. H., 588
Reitz, L. P., 89
Renner, O., 243, 244, 245
Rhoades, M. M., 28, 29
Rhodes, A. M., 143, 144
Rhyne, C. L., 131
Richmond, T. R., 130, 133
Rick, C. M., 57, 275, 277,
    278
Rieger, R., 330
Rienits, K. G., 89
Rijven, A. H. G. C., 177
Riker, A. J., 309
Riley, R., 57, 87, 89, 90,
    127, 310, 375

Riner, M. E., 143
Ris, H., 426
Röbbelen, G., 165, 166, 167, 168, 169, 170, 172, 174, 175, 176, 177, 178, 179
Roberts, C. T., 467
Roberts, D. A., 411
Robertson, D. S., 29
Robertson, D. W., 110
Robinson, H. F., 311, 312
Robinson, R. W., 275, 277
Robles, L. J., 579
Robson, D. S., 308, 313
Rochow, W. F., 411
Rodenhiser, H. A., 90
Rolfe, B. G., 167, 168
Rollins, J. A., 376
Roman, H., 29
Ronchi, V. N., 377
Ronchi-Nutti, V. N., 375
Roque, M., 588
Roschevicz, R. J., 57
Rosen, G. von, 220
Rosenbaum, J. L., 426
Ross, A. F., 406, 411
Ross, H. F., 311
Ross, I. K., 431
Rossignol, M., 577, 578
Rostom, E., 412
Roux, J. B., 129
Rowe, J., 588
Rudloff, C. F., 245
Rudman, B., 588
Rudman, B. M., 577
Rudnick, D., 587
Rudnyansky, B., 588, 592
Ruhland, W., 172
Ruppel, E. G., 407
Rush, M. C., 411
Russo, M., 410
Rustad, R. C., 466
Rutger, J. N., 55
Ryker, T. C., 55

Saccardo, F., 219
Saccone, C., 576
Sachs, T., 220
Sacristan, M. D., 312

Sadasivaiah, R. S., 308
Sager, R., 426
Sagromsky, H., 330
Sainsard, A., 577, 578, 588
Sakaguchi, S., 56
Sakai, K., 57
Saksena, K. N., 410
Salama, E. A., 407
Salaman, R. N., 411
Sampath, S., 57
Sampson, D. R., 331
Samson, R. W., 411
Sand, S. A., 312, 313
Sander, E., 411
Sanina, A. V., 169
San Pietro, A., 426
Santos, B. G., 583
Sarkar, K. R., 29
Sarkar, P., 90
Sarvella, P., 128, 130, 131
Sastrosumarjo, S., 91
Sato, V. L., 426
Saunders, J. H., 131
Saxena, D. M., 583
Scandalios, J. G., 29, 220
Scarascia-Mugnozza, G. T., 87, 219
Scarchuk, J., 143
Schaeffer, G. W., 312
Schairer, L. A., 167, 313
Scheer, B. T., 463
Schensted, I. V., 467
Scherbaum, O. H., 465
Schimmer, O., 426
Schlegel, G., 89
Schlehuber, A. M., 90
Schmelzer, K., 411
Schmidt, H., 331
Schneider, I. R., 411
Schneller, M. V., 578, 588, 589, 592
Scholz, G., 274, 278
Schötz, F., 243, 245
Schramm, G., 411
Schroeder, W. T., 211, 220, 409
Schwartz, D., 29
Schwartz, V., 588
Schwemmer, S. S., 375

Schwendiman, S., 129
Schwenk, F. W., 413
Schwerin, K. H., 131
Schwind, F., 170
Scossiroli, R. E., 314
Scott, D. H., 143
Scott, H. A., 405, 411
Sears, E. R., 87, 88, 89, 90, 91
Sears, L. M. S., 171, 178
Sebesta, E. E., 90
Seetharaman, R., 57
Sehgal, P., 411
Seitz, F. W., 245
Selman, G. G., 583, 589
Semancik, J. S., 411
Sen, S. K., 57
Senser, F., 245
Serova, R. Y., 167
Sethi, B., 57
Seyffert, W., 178, 331
Shah, H. M., 57
Shama Rao, H. K., 91
Shands, H., 91
Shands, R. G., 110
Sharma, B., 220
Sharma, S. D., 57
Sharp, W. R., 178, 278
Shastry, S. V. S., 57, 58
Shaver, D. L., 29
Shaw, H. F., 166
Shebalov, V. V., 178
Sheen, S. J., 91, 312
Shen-Miller, J., 178
Shepherd, R. J., 407, 408, 411, 412
Sherman, F. G., 377
Sherrat, H. S. A., 331
Shifriss, O., 143, 144, 150
Shimotsuma, M., 150
Shinkai, A., 411
Shropshire, W., Jr., 172, 178
Shu-Chang, L., 219
Shull, A. V., 410
Shull, G. H., 29
Shumway, L. K., 29
Siddiqui, K. A., 91

Siegel, A., 411
Siegel, R. W., 588, 589
Sikka, S. M., 89
Silbernagel, M. J., 410
Silberschmidt, K., 412
Sill, W. H., Jr., 405, 407, 412
Silow, R. A., 129, 131
Simon, E. M., 463, 467, 589
Simons, J. N., 412
Sinden, R. E., 589
Sinha, S. K., 29
Sing, C. F., 87
Singh, A., 29
Singh, D., 144
Singh, G. P., 412
Singh, R. P., 412
Singh, S. P., 144
Singleton, W. R., 29
Sinnott, E. W., 144
Sinton, S. E., 580
Sisler, H. D., 408
Sisler, W. W., 412
Skaar, R. D., 589
Skoog, F., 310, 312
Skovsted, A., 131, 132
Skult, H., 377
Slack, S. A., 411
Sleigh, R. W., 588
Slifkin, M., 430
Slykhuis, J. T., 408, 412
Smillie, R. M., 110
Smith, C. E., Jr., 132
Smith, C. R., 313
Smith, D. R., 406
Smith, H. H., 167, 169, 306, 307, 308, 310, 312, 313
Smith, J. B., 375
Smith, J. P., 581
Smith, K. M., 410, 412
Smith, L., 91, 110
Smith, S. H., 413
Smith-Sonneborn, J., 589
Smolina, E. A., 170
Snape, J. W., 178
Snoad, B., 220, 279
Snyder, R. J., 29
Sobels, F. H., 584

Sober, H. A., 28
Solari, R. M., 110
Solberg, R. A., 405
Soldo, A. T., 589, 593
Solovjeva, V. K., 220
Solt, M. L., 308
Sommer, H. E., 278
Sommerville, J., 589
Sonneborn, T. M., 467, 578, 588, 589, 590, 591, 592
Soressi, G. P., 274
Soriano, J. D., 361
Sosna, M., 178
Sparrow, A. H., 167, 178, 180, 309, 312, 313, 375, 377
Sparrow, R. C., 167
Sparvoli, E., 377
Spencer, E. L., 410
Sprague, G. F., 29
Sproll, W. P., 127
Stace-Smith, R., 412
Stadler, L. J., 29, 30
Stark, P., 588
Starling, D., 426
Startsev, G. A., 178
Stebbins, G. L., 58, 90
Steers, E., Jr., 592
Steffensen, D. M., 30
Stein, E., 331
Steiner, E., 243, 245
Steinitz-Sears, L. M., 172, 178
Steitz, E., 313
Stephens, S. G., 129, 131, 132
Stern, K., 331
Stettler, R. F., 278, 279
Stevens, M. A., 279
Stevenson, H. Q., 313
Stevenson, I., 593
Steward, F. C., 313
Stewart, R. N., 307
Stokes, G. W., 308
Stone, O. M., 407
Stone, W. N., 406
Stouffer, R. F., 412
Stout, J. T., 26
Stover, R. H., 412

Strehler, B. L., 591, 592
Strekhalov, A. A., 171
Strickland, M. A., 131
Strijkert, P. J., 426
Strømnaes, Ø., 361
Stubbe, H., 245, 279, 330, 331
Stubbe, W., 245
Stuber, C. W., 27
Sturtevant, A. H., 243
Styles, E. D., 26, 30
Subrahmanyam, V., 54
Sueoka, N., 425, 426
Sunderland, N., 313
Surzycki, S. J., 425, 426
Sussman, M., 430, 431
Sussman, R. R., 431
Suyama, Y., 462, 467, 593
Svachulová, J., 178
Sverdrup, A., 220
Swaminathan, M. S., 89
Swift, H., 594
Synder, L. A., 91

Tageeva, S. V., 165
Tait, A., 578, 593
Takahashi, M. E., 56, 58
Takahashi, N., 58
Takahashi, R., 110
Takebe, I., 311, 313
Takenaka, Y., 313
Tal, M., 279
Tallan, I., 592, 593
Tanaka, M., 313
Tanaka, R., 377
Tateoka, T., 58
Taub, S. R., 593
Tavcar, A., 220
Taylor, H. J., 377
Taylor, T., 128
Tazelaar, M. F., 213
Teas, H. J., 30
Tedin, H., 220, 221
Tedin, O., 220, 221
Tefankjian, A., 466
ten Heggeler, B., 580
Thein, M. M., 129
Thomas, H. R., 412, 413

Thompson, J., 377
Thompson, J. F., 313
Thompson, K. H., 169, 313
Thompson, M. C., 309
Thompson, P. J., 172
Thornberry, H. H., 412
Thresh, J. M., 409
Tiedjens, V. A., 150
Timian, R. G., 412
Timofeeff-Ressovsky, H. A.,
    169
Timofeeff-Ressovsky, N. V.,
    170, 178
Tinsley, T. W., 405
Tiranti, I. N., 132
Titu, H., 377
Tomaru, K., 412
Tomes, M. L., 280
Tomlinson, J. A., 412
Tompkins, C. M., 413
Touchberry, R. W., 466
Tremaine, J. H., 404
Tsai, C.-Y., 30
Tschermak, E., 221
Tso, T. C., 312, 313
Tsuboi, M., 178
Tsuchiya, T., 109, 110
Tsunewaki, K., 91
Tuite, J., 404, 413
Tulpule, S. H., 30
Turcotte, E. L., 132

Uber, F. M., 30
Uchiyamada, H., 55
Udagawa, A., 412
Underbrink, A. G., 178
Ungerleider, R. M., 583
Unni, M. K., 361
Unrau, J., 91
Usmanov, P. D., 165, 178
Uyemoto, J. K., 407, 413

Valleau, W. D., 406, 413
Varney, E. H., 413
Van Brederode, G. H. M.,
    179
Vandendries, R., 178

Van der Veen, J. H., 166,
    172, 179, 276
van der Want, J. P. H., 405
van Houten, J., 579
Van Norel, A., 213
Van Regenmortel, M. H.,
    405
Van't Hof, J., 375, 377
van Wagtendonk, W. J.,
    581, 583, 584, 589, 593
Vasil, V., 314
Vaughn, J. G., 179
Vavilov, N. L., 212
Veleminský, J., 165, 168,
    178, 179
Venkataswamy, T., 57
Verhalen, L. M., 130
Verkerk, K., 276, 280
Vester, F., 314
Vincent, M., 411
Vinciguerra, B., 579
Vines, A., 311
Vis, M. F., 179
Viviand-Morel, M., 179

Wachtmeister, C. A., 168
Wade, B. L., 221, 413
Wagenaar, E. B., 91
Wagner, R., 406
Waines, J. G., 91
Wakakuwa, S., 88
Walker, J. C., 213, 406,
    408, 409, 410, 412, 413
Walker, L. M., 577
Walker, M. N., 406
Wall, J. R., 144, 280
Wallace, H., 376
Walles, S., 179
Walters, H. J., 413
Ware, J. C., 132
Washington, W. J., 89, 91
Watanabe, Y., 58
Weathers, L. G., 408, 411,
    413
Weaver, E. A., 313
Weaver, J. B., Jr., 132, 133
Webber, J. M., 133
Weber, E. J., 27

Webster, P. L., 377
Wehrmeyer, W., 178, 179
Weiling, F., 144
Weinheimer, W. H., 26
Weir, B. S., 109
Weiseth, G., 211
Wellensiek, S. J., 221
Wellington, R., 150
Wells, D. G., 221
Wender, S. H., 314
Wennstrom, J. I., 361
Wenrich, D. H., 586
Weremiuk, S. L., 463
Wernsman, E. A., 311, 314
Westerman, J. M., 179
Wettstein, D. von, 109, 110
Wettstein-Knowles, P. von,
    109, 110, 280
Weybrew, J. A., 310
Wheldale, M., 331
Whidden, R., 407
Whitaker, T. W., 143, 144,
    149, 150
White, J. A., 167
White, O., 221
White, P. R., 314
White, T. G., 133
Whitt, G. S., 464
Whittle, J. R. S., 593
Whyte, R. O., 179
Wibaut, C., 179
Wichterman, R., 593
Widmayer, D. J., 577, 593
Wiebe, G. A., 110
Wienhues-Ohlendorf, A., 91
Wilcox, H. B., 581
Wildman, S. G., 411
Wilkes, H. G., 30
Wilkinson, D. R., 30
Wilkinson, J. F., 578
Wilkinson, R. E., 413
Wille, J., 467
Williams, H. E., 413
Williams, J., 593
Williams, N. E., 464
Wilson, C. M., 431
Wilson, F. D., 133
Wilson, S., 585
Wimber, D. E., 377

Wingard, S. A., 413
Winge, Ö., 221
Winkler, H., 245
Wirtz, P., 179
Witman, G. B., 426
Wittmer, G., 314
Wolf, F. A., 314
Wolfe, J., 467, 593
Wood, H. K., 593
Woodard, J., 467, 594
Woodard, M., 594
Woodruff, L. L., 594
Wricke, G., 180
Wulff, E. V., 212

Yamaguchi, Y., 58

Yamakawa, K., 180
Yanagashima, M., 178
Yang, D.-P., 377
Yanigasawa, K., 431
Yao, S. Y., 58
Yarwood, E. E., 410, 413
Yasuda, S., 110
Yasui, K., 58
Yeager, A. F., 280
Yeh, B., 58
Yen, D. E., 221
Yokoyams, K., 180
Yoneda, Y., 313
Young, P. A., 275
Young, R. M., 90
Yu, S., 280
Yunoki, T., 58

Yunosov, S. Y., 178
Yusa, A., 594

Zaitlin, M., 405, 411
Zaumeyer, W. J., 406, 410, 411, 412, 413
Zeller, F. J., 91
Zenker, A. M., 171, 180
Zeuthen, E., 463, 467
Ziebur, M. K., 180
Zimmer, A. M., 463
Zobel, R. W., 278, 280
Zohary, D., 91
Zsiros, E., 169
Zuber, M., 28
Zük, J., 377

Subject Index

Acrasiales, 427–431
Acridine orange, 250
Aegilops, 59
  sitopsis group, 62, 63
Agropyron, 59, 62, 65, 66
Alfalfa mosaic virus, 382
  host range, 387
  structural characteristics, 382
Anabasine, 296
Anthocyanidine, 325
Anthocyanin, 45
  in tomato fruit, 272
Antigen
  C, 513
  immobilization, 458–459, 483, 489, 504,
      538, 540, 567
  Tetrahymena, 434
Antirrhinum majus, 315–331
  advantages for genetic study, 315–316
  allele, multiple, 318–319
  anther reduction, 317
  Baur's work (1907), 315
  chiasmata, 323–324
    frequency, 323
  chlorophyll synthesis in mutant, 326
  chromosome
    aberration, 322
    number, 322
    pachytene, 320
    pairing, illegitimate, 322
  corolla reduction, 317
  crossing-over, 321, 323–324

Antirrhinum majus (cont.)
  frequency, 323
  in species hybrids, 324
  variability, 323
  cycloidea locus, 318
  cytology, 321–322
  Darwin's work, 315
  description of plant, 315–316
  De Vries' work, 315
  expressivity, 316
  flower pigment, 324–326
    biosynthesis of, 325
  gene
    coupling, 319
    of gametophyte, 316
    lethal, 316
    localization, 321
    minor, 318
    nomenclature, 316
  genetics, Mendelian, 316–318
  Godron's work, 315
  heterosis, 327
  history, 315
  leaf
    development, 317
    pigment, 326
  linkage group, 319–322
  linkage map, 319
  meiosis, timing of, 321
  Mendel's work, 315
  Mendelian genetics, 316–318
  model for genetic study, 315–316

Antirrhinum majus (*cont.*)
  morphology, 326–327
  mutagenesis, 327–329
  mutant, 316, 327–329
  mutation, spontaneous, 327
  nucleolus, 321
  penetrance, 316
  phenogenetics, 324–327
  phenotypes of various genes, 317–318
  pigment gene, 324
  plastid, 326
  radiation effect, 328
  Sippe *50,* 316, 320, 328
  species hybrids, crossing-over, 324
  transcendens locus, 316
  trisomics, 322–323
  tumor, induced, 328
  Wheldale's work, 315
Apigenin, 325, 326
Arabidopsis thaliana, 151–180
  auxotrophy, 158
  chlorophyll mutant, 159–160
  chloroplast differentiation, 160
  chromatid segregation, 154
  chromosome number, 153
  chromosome rearrangement, 153
  culture technique, 152
  cytology, 153
  distribution, geographic, 164
  DNA uptake, foreign, 159
  embryo, isolation of, 152
  embryogenesis, 161
  flowering, 162
  gametophyte factor, 154
  gene localization, method for, 154
  genes in natural populations, 164
  genetics, developmental, 160–162
  genetics, physiological, 158–160
  germination, 162
  haploidy, 153
  heterosis, mechanism of, 163
  history, 151
  hybrids, 164
  inheritance, extrachromosomal, 163
  inheritance, quantitative, 163
  isozymes, 159
  leucine mutant, 158
  linkage, 154

Arabidopsis thaliana (*cont.*)
  megasporogenesis, 161
  model for genetic study, 152
  mutagenesis by chemical mutagens,
    157–158
  mutant, chlorophyll-, 159–160
  mutation, 154–156
    forward, 155
    plastome, 163
    rate
      calculation of, 155
      reverse, 156
      spontaneous, 155–156
  nucleolus, 153
  nucleus volume, 153
  photoreactivation, 157
  pollen morphology, 161
  polyploidy, 153
  population genetics, 163–164
  prochromosome, 153
  radiation effects, 156–158
  radiation tolerance, 156
    modification of, 157
  radiosensitivity, 156
  recombination, 154
  response, photoperiodic, 162
  seed disinfection, 152
  taxonomy, 164
  thiamine auxotrophy, 156
  thiamine pathway, 158
  trisomics, primary, 153
Arabidopsis Information Service, 151
Auxin, 300

Barley, 93–110
  character, genetic, 99–100, 107–109
    chlorophyll, 107
    commercial, 108
    DDT resistance, 108
    eceriferum, 107
    enzyme mutant, 108
    erectoides, 107
    growth habit, 108
    insect susceptibility, 109
    meiotic mutant, 109
    sterility mutant, 109
  chromosome, 94–95

Barley (*cont.*)
  arm-length ratio, 95
  at diakinesis, 96
  haploid, 96
  length, 95
  map, 100
  number, 95–98
  pachytene, 95
  structure, 95–98
  telocentric, 97
genes, list of, 101–105
gene loci, 99
gene symbols, 101–105
haploid karyotype, 94
interchange, 98
inversion, spontaneous, 98
karyotype, haploid, 94
linkage map, 99–100
as model for induced mutation, 93, 99
mutation, induced, 98–109
tetraploid, 97
translocation, reciprocal, 98
  breakpoint, 106
triploid, 96
trisomics, primary, 97
  tertiary, 97
Barley Genetics Newsletter, 94, 106
Barley stripe mosaic virus
  host range, 387
  structural characteristics, 382
Bean pod mottle virus
  host range, 388
  structural characteristics, 382
Boron in tomato fruit, 271
Broad bean mottle virus
  structural characteristics, 382
Brome mosaic virus
  host range, 388
  structural characteristics, 382
Brown plant hopper, *see Nilaparvata lugens*
Brown spot of rice, 48

Caedobacter conjugatus (mu particle), 505, 515, 564–565
Caedobacter falsus (nu particle), 515, 558

Caedobacter minutus (gamma particle), 566
Caedobacter taeniospiralis (kappa particle), 514, 553
Cauliflower mosaic virus
  host range, 388
  structural characteristics, 382
Cowpea mosaic virus
  host range, 389
  structural characteristics, 383
Cellular slime molds, 427–431 *see* Slime mold, cellular
Cercospora leaf spot, 48
Cercospora oryzae, see Cercospora leaf spot
Chlamydomonas reinhardi, 417–426
  arginine-dependent strain, 424
  cell fusion, 419
  chloroplast, 422–423
  DNA, 422–423
  electron micrograph, 418
  flagella, 424
  genetic analysis, method of, 419–420
  genetic map, 420–421
  genetics of photosynthesis, 423–424
  growth condition, 417–419
  life cycle, 417–419
  linkage group, 420–421
  medium for, 417
  motility, 419
  multiplication, 419
  mutant, 424–425
  photosynthesis, 422
    genetics of, 423–424
  replica plating, 420
  RNA, 423
  tetrad analysis, 420
  zygote, 419
    plating, 420
Chromomere, 5–6
Citrate in tomato fruit, 271
Citrus exocortis virus
  host range, 389
  structural characteristics, 383
Colchicine, 119, 121, 123, 348, 353–357, 369, 498
Collinsia, 333–361
  amphiploid, 343–347
  anaphase, 341

Collinsia (cont.)
  aneuploid, 355–356
  anther, 335
  autopolyploid, 348–350, 352
  autotetraploid, 348, 353
  bud, 335
  chemotaxonomy, 350–351
  chiasma, 336
  chiasmata number, 350
  chromatid bridge, 341
  chromosome, 335
    aberration, intraspecific, 352–358
    association, 339, 346–350
    pairing in polyploid hybrids, 343–347
    pseudonumerary, 357
    repatterning, 342
    supernumerary, 351–352
  colchicine effect, 348, 353–357
  culture condition, 334
  cytogenetics of diploid, 338–343
  cytology methods, 335
  cytotaxonomy, 335
  diploid, 338–343
  flower, 334
  Garber's hypothesis, 338
  genetics of species in group I, 358–359
  germination, 334
  Grant's hypothesis, 342
  hybrid
    diploid, interspecific, 339–342,
      346–349, 351–352
    fertility, 342–343
    polyploid, 343–347
    sterility, 343
  hybridization, interspecific, 337–343,
    347–348
  interchange complex, 338–341, 354–357
  intragroups, 340, 343
  inversion, heterozygous paracentric,
    341–342
  linkage group, 355, 359
  as model in genetics, 333
  mutant gene, 359
  phenotype, contrasting, 358
  pollen, 334
  polyploidy, 349
  radiation, ionizing, 355
  species groups, 336

Collinsia (cont.)
  taxonomy, 333
  translocation, 353, 355
  triploid hybrid, 343–344
  trisomy, 355–358
Collinsia heterophylla, 353–355
  colchicine effect, 353
  translocation, reciprocal, 353–355
  trisomy, 355–357
Collinsia tinctoria trisomy, 358
Committee on Nomenclature of Barley
      Genes, 106
Corn, 3–30
  centromere, 7
  chromomere, 5–6
  chromosomes, cytological map, 5–6
  description of plant, 3
  genetic markers (list of), 9–25
  linkage map, 7–8
  markers, genetic (list of), 9–25
  as a model in genetics, 3–4, 8
  phenotypes, 9–25
  pollination, 3
Corn and Corn Improvement by G. F.
      Sprague, 8
Cotton, 111–133
  allohexaploid, 125
  amphidiploid, 116
  aneuploidy, 125–127
    monosomy, 125–126
    origin, 125
    telocentrics, 126–127
    trisomy, 126
  apomixis, 123
  chimera, 123
  chromosome interchange, 124–125
  cytodifferentiation, 118–119
  cytotaxonomy, 111–115
  diploids, 111–115
  embryogenesis, arrested, 124
  evolution, early, 114
  genes, list of, 122
  genetics
    comparative, 119–121
    developmental, 121
    qualitative, 121
  genome group, 113
  haploidy, 121, 123

Cotton (*cont.*)
  hexaploidy, 120
  hybrid, 115, 120
    diploid × diploid, 118–119
    diploid × tetraploid, 119
  incompatibility, interspecific, 123–124
    lethals, 124
    semilethals, 124
  linkage study, 120–121
  in the New World, 115–118
  polyembryony, 121
  semigamy, 123
  species, list of, 112–113
    A genome, 113, 116, 117, 120
    B genome, 113, 120
    D genome, 114, 116, 117, 120
    E genome, 113, 120
  tetraploidy, 115–118
  tetrasomics, 126
  tetrasomy, 126
  translocation, 125
  triploidy, 120
Cowpea chlorotic mottle virus
  host range, 389
  structural characteristic, 383
Crithodium, 59
Cucumber mosaic virus
  host range, 389
  structural characteristic, 383
Cucumis, 145–150
  cross-sterile groups, 146
  crossing, interspecific, 146
  cucumber, 145
  cytology, 145–149
  description of plant, 145
  genetics, 145–149
    of *C. melo,* 147–149
    of *C. sativum,* 147–149
  gherkin, 145
  heterosis, 146
  inbreeding, 146
  inheritance of economically important
      characters in
    C. melo, 149
    C. sativum, 148
  monoploidy, 146
  muskmelon, 145
  parthenocarpy, 147

Cucumis (*cont.*)
  polyploidy, 146
  tetraploidy, 145, 146
Cucurbita, 135–144
  crossing, interspecific, 137–139
  cytology, 135–142
  description of plant, 135
  genetics, 135–142
    of *C. maxima,* 139–142
    of *C. pepo,* 139–142
  heterosis, 136–137
  inbreeding, 136–137
  inheritance of several characters in
    C. maxima, 140–142
    C. pepo, 140–142
  isozyme studies, 138
  polyploidy, 138–139
  species, 136–137
  taxonomy, numerical, 138
Cucurbitacin, 147
Cyanidin, 325–326
Cytophaga caryophila (symbiont), 515

Deoxyribonucleic acid, *see* DNA
DNA
  chromosomal synthesis in higher plants,
      363–377
  2C-DNA content in higher plants,
      364–368, 372
Dictyostelium discoideum, 427–431
Dictyostelium mucoroides, 427
Dictyostelium purpureum, 427
Die Mutationstheorie by H. De Vries, 223
Diethyl sulfate, 99
Dimethyl sulfoxide, 157
Duration
  of meiosis in higher plants, 373–374
    relation to DNA content, 373
  of mitotic cycle in plants, 363–377
    in meristematic cells, 364–368
    in meristematic root cells, 369
    in root apices, 370–371
    in shoot apices, 370–371
  of S phase in relation to 2C-DNA
      content, 372

Effectiveness of radiation, 157
Emmer, *see* Wheat, tetraploid

Ethyl methane sulfonate, 99, 250, 485
Eugenol in tomato fruit, 271
European Wheat Aneuploids Cooperative,
    69

Flavon, 325
Flavone, 325
Flavonol, 325

Germanium in tomato fruit, 271
Gossypium, see Cotton
Grassy stunt virus, 48
Green leafhopper, *see Nephotetix virescens*

Harzflora by J. Thal, 151
Haynaldia, 59, 62
Helminthosporum oryzae, see Brownspot
    of rice
Hordeum vulgare, see Barley
Hydrazine, 250
Hydroxycinnamic acid, 325, 326

International Barley Genetics Symposium,
    94, 106
International Rice Research Institute, 51,
    53
IRRI, *see* International Rice Research
    Institute
2-Isobutylthiazole in tomato fruit, 271

Jenning's pure line theory of asexual
    reproduction, 469

Kappa particle, *see Caedobacter
    taeniospiralis*
Kinetin, 300

Leaf blight, bacterial, 48
Luteolin, 325–326
Lycopersicon esculentum, see Tomato
Lycticum flagellatum (lambda particle),
    514, 566

Maize, *see* Corn
Maize Genetics Cooperation Newsletter, 8
Malate in tomato fruit, 271
Methane sulfonic ester, 157
Methyl salicylate in tomato fruit, 271
Mildew, powdery, 147, 149
Milo maize, *see* Sorghum
Mitotic cycle
    duration in higher plants, 363–377
    Quastler–Sherman method (1959), 363
    temperature effect on, 369
Mouse ear cress, *see Arabidopsis*
Muskmelon and powdery mildew, 147, 149
Myxomycophyta, 427–431

Nephotetix virescens, 48
Nicotiana, 281–334
    alkaloids
        anabasine, 296
        composition, 296
        inheritance, 295–297
        nicotine, 296
        nornicotine, 296
        synthesis, 296
    allopolyploidy, 284–286
    amphiploidy, 285–286
    aneupolyploidy, 285–293
    biometrical studies, 297–298
    chromosome H, 294
    classification, 281–284
    enzymes, 284
    evolution, 284
    gene effect, additive, 297–298
    genome, multiple, 285–293
    haploid, 304
    hybrid, 303
        instability, 294–295
        interspecific, 284–285
    hybridization, 286, 293–295
        parasexual, 305
    inheritance
        of continuous variation, 297
        cytoplasmic, 298–299
        extrachromosomal, 298–299
    instability, genetic, 293–295
    interaction, allelic, 297
    megachromosome, 295

Nicotiana (cont.)
  model system for somatic cell genetics,
    303–306
  monosomics, 285, 287–292
  phylogeny, 281–284
  protoplast, culture of, 304, 306
  recombination, interspecific, 293–295
  sterility
    barrier, 293
    male-, 298
  subgenera, 282–283
  substitution, alien race, 293
  tissue culture, 303
  tumor
    environmental factors, 302
    gene regulation, 302
    genetic, 299–302
    hybrids forming tumors, 299–300
    spontaneous, 300–301
Nicotine, 296
Nilaparvata lugens, 48
Nitrogen mustard, 546
Nitrosamine, 157
Nitrosoguanidine, 157, 485
Nitroso-imidazolidine, 158
Nitrosourea, 157
Nitrous acid, 158, 485
Nornicotine, 296
North Central Regional Plant Introduction
    Station, Ames, Iowa, 251
nu particle, *see Caedobacter falsus*

Oenothera, 223–245
  Belling's hypothesis (1927), 228, 230
  character:cruciate, 234
  character:missing petal, 234
  chlorosis, 237
  chromosome
    behavior, 230–231
    circular, 228
    configuration, 239
    segment, 230
    size, 230
    structure, 230–231
  Cleland's work, 227, 239
  complex heterozygote, 226, 232,
    240–241

Oenothera (cont.)
  complex homozygote, 228–229
  complexes, 226, 228–230, 232, 240
  crossing-over, 232–233
  De Vries' work, 223, 227, 235
  gene
    conversion, 234
    identification, 232
  genetic analysis, 231–232
  genetic mapping, 231
  genome–plastid interaction, 237–238
  half-mutant, 235
  history of research on, 223–230
  hybrid, 231
  incompatibility allele, 236–237
  megaspore composition, 231
  meiosis, 231
  mutagen, 236
  mutation, 235–236
    induced, 235
  in North America, 224
  petal, cruciate, 234
  petal, missing, 234
  phenotypes, 226–227
  plastid–genome interaction, 237–238
  plastid mutation, 238
  pollen tube development, 231
  population evolution, 238–242
  position effect, 233
  Renner's work, 227, 237
  self incompatibility, 236–237
  species, 239–240
  spore types, 225, 226
  Stubbe's work, 238
  taxonomy, 223–227
  translocation, 225, 226, 233
  triploidy, 235
  trisomics, 235
  Winkler's hypothesis of gene conversion,
    234
Oenothera—Cytogenetics and Evolution by
    R. E. Cleland, 227
Oryza, see Rice

Paramecium aurelia, 469–594
  agglutination, 545
  alpha particle, 515

Paramecium aurelia (*cont.*)

antibiotic resistance, 506

antigen, *see also* serotype
  3, 514
  genetics of, 550–552

assortment, phenotypic, 484, 500

autogamy, 476–479, 484, 505, 507, 544, 546
  frequency of death at, 544

bacteria as symbionts, 490–491

biochemical genetics, 491–492

biology, 471

C antigen, 513

caryonide, 478, 484, 500, 559

cell cycle, 475–476, 498

chloramphenicol resistance, 506

chromosome, 472–475, 496
  number, 473, 496

cilia, 472
  ciliary corpuscle, 472

clone
  constancy, 483
  cycle, 476–478, 483, 498
  unstable, 514

conjugation, 469, 476, 480–482, 484, 486, 546
  chemical induction of, 499

constancy, clonal, 483

corpuscle, ciliary, 472

cortex, 472, 557–558
  genetics of, 491
  unit, 472

cytogamy, 476, 480
  calcium effect on, 499

cytogenetics, 499

cytological aberrancies, 482–483

cytoplasmic states, diverse, 490

death, partial, 507

delta particle, 560

deoxyribonucleic acid (DNA)
  nuclear, 472–475
  synthesis, 482

description, 472

determination, delayed, 500

developmental genetics, 491–492

differentiation, nuclear, irreversible, 484–489

doublet, 481

Paramecium aurelia (*cont.*)

endomixis, 507

erythromycin resistance, 506

esterase
  A, 566
  B, 517
  C, 492, 508, 560, 562
  D, 492, 515, 560, 568

exchange, cytoplasmic, 481

exclusion, allelic, 502–503
  phenotypic, 502–503

generation time, 475, 543
  of species *2,* 510

genes, 485–489, 493–497
  of species *2,* 508–510
  of species *4,* 518–536
  markers, multiple, 540
  and phenotypic effects, 537–538

genetics
  of cell cortex, 491
  history of, 469–470
  of immobilization serotypes, 489–490
  interspecific, 569–576
  of mating types, 485–489
    of species *1,* 499–501
  of mitochondria, 491, 506
  and symbiont bacteria, 490–491
  of uniparental reproduction, 499

heredity
  in asexual reproduction, 483–484
  extranuclear, 470
  in sexual reproduction, 484–485

history of *Paramecium* genetics, 469–470

hybrid, sterile, 471

β-hydroxybutyrate dehydrogenase, mitochondrial, 492, 508, 515, 560, 566, 568

immaturity, 476

immobilization antigen, 483, 489, 504, 538, 540, 567
  genetics of, 489–490

inheritance, cytoplasmic, 501

Jenning's work, 469

K gene, 554

kappa, 514
  particle, 491
  properties, 553
  transfer, 576

Paramecium aurelia (*cont.*)
  killer paramecia, 505–506
    clone, 483
    discovery, 490, 552–553
    stock, 505
  lag, phenomic, 499
  linkage in species *4,* 539
  macronucleus, 472–475
    regeneration of, 482, 545
  mating reactivity, rhythm of, 516
  mating type
    discovery by Sonneborn, 470, 492
    genetics of, 485–489
    A (caryonidal) system, 486, 488,
      499–501, 568
    B (clonal) system, 487, 511, 548, 559,
      563, 568
    C (synclonal) system, 486, 488
    E (even), 486, 488, 559, 561, 569,
      571–573
    O (odd), 486, 488, 559–561, 569,
      571–573
    VII, 548, 549
    VIII, 548, 549
  maturity, 476, 544
  Maupas' theory of mortality of ciliates,
    506
  metagon hypothesis, 505
  micronucleus, 472–475, 543
  mikamycin resistance, 506
  mitochondria
    genetics of, 491, 506, 554–556
    mutation, 506, 554–556
  MR, *see* regeneration, macronuclear
  mu particle, 514
  mutagen 485, 547
  mutagenesis, 485, 505, 547
  mutant, mitochondrial, 506
  NADP isocitrate dehydrogenase, 566,
    568, 569
  nu particle, 514, 558, 564
  nucleus
    DNA in, 472–475
    types of, 472–475, 496
  phenomic lag, 499
  plasma membrane, 472
  population
    cage, 505

Paramecium aurelia (*cont.*)
  genetics, 491–492, 504–505
  pure line theory of asexual reproduction,
    469
  radiation effect, 505
  regeneration, macronuclear, 482, 485,
    545
  resistance to antibiotics, 506
  rhythm
    circadian, 516
    of mating reactivity, 516
  rule, of caryonidal inheritance, 487
  selfing caryonide, 559
  senescence, 476–477, 544
  serotype, *see also* antigen
    3, 516
    D, 501, 504
    G, 501–503, 567
    S, 501
    X, 567
    genetics of, 550–552
  sigma particle, 514
  species, 470–471
    1, 492–506
    2, 506–515
    3, 515–516
    4, 516–558
    5, 558–559
    6, 559–560
    7, 560–561
    8, 562–566
    9, 566–567
    10, 568
    11, 568
    12, 568
    13, 568
    14, 569
  species *1* × *7,* 575–576
  species *4* × *8,* 574–575
  symbiont, bacterial, 505–506
    Caedobacter conjugatus, 505, 515
    C. falsus, 515, 558
    C. taeniospiralis, 514, 553
    Cytophaga caryophila, 515
    Lycticum flagellatum, 514
    Tectobacter vulgaris, 506, 515
  syncaryon, 478
  synclone, 484

Paramecium aurelia (cont.)
  syngen, 470–471
    defined, 471
  transfer, interspecific
    of cytoplasmic genetic elements, 576
  trichocyst, 556–557
    function, 556
    mutation, 556
  union
    heteropolar, 481
    homopolar, 481
  varieties, 470–471
  Weismann's theory of immortality of
      ciliates, 506
  wild stock
    of species *2*, 508, 510
    of species *4*, 541–542
  Z clone, 513
Paramecium methusalah, 507
Pea, 181–221
  centromere position on genetic map, 185
  chromosome
    number, 184–185
    rearrangement, 187
  cytology, 184–187
  embryo, 182
  flower, 183
  gene loci, 188–209
  genetic map, 186–210
  germination, 182
  inflorescence, 183
  Lamprecht's work, 188
  leaf, 182
  marker genes, 189–209
  Mendel's work, 187–188
  mutants,
    grouped, 210
    lethal, 182
    list of, 189–209
  mutation
    case, 210
    group, 210
    induced, 183–184
    spontaneous, 183–184
    type, 210
  ontogeny, 181–183
  ovary, 181–183
  phenotypes, 189–209

Pea *(cont.)*
  radiation sensitivity, 184
  seed, 182
  translocation, 184, 187
  trisomics, 187
  type line, 210
  Vilmorin and Bateson's work, 188
Pea enation mosaic virus
  host range, 390
  structural characteristics, 383
Pelargonidin, 325, 326
Petanin, in tomato fruit, 272
Pisum, see Pea
Pisum Genetics Association, 210
Plant viruses, 381–413
  host plants, 399–403
  host range, 387–398
  structure of 32 common plant viruses,
      382–386
  see individual viruses
Polysphondylium pallidum, 427–428
Potato spindle tuber virus
  host range, 390
  structural characteristics, 383
Potato X virus
  host range, 391
  structural characteristics, 383
Potato Y virus
  host range, 391
  structural characteristics, 383
Protists, 417–594
  cellular slime molds, 427–431
  Chlamydomonas reinhardi, 417–426
  Paramecium aurelia, 469–594
  Tetrahymena pyriformis, 433–467

Quercetin, 325

RBE, *see* Effectiveness, relative biological
Rice, 31–58
  allotetraploid, 36
  aneuploid, 43
  autotriploid, 41
  biosystematics, 32–36
  breeding
    for wider adaptability, 50–53
    for higher productivity, 50–53

Rice (*cont.*)
  chromosome aberration, 39
  chromosome number, 33, 41
  chromosomes at pachynema, 37–38
  chromosome translocation, 39
  color
    at flowering, 49
    at ripening, 49
  complementary alleles, 45
  cytogenetics, 36–39
  disease resistance, 48
  distribution, geographic, 33
  dwarfing gene, 52
  genes, list of, 46–48
  genome classification, 33
  haploid, 39–40
  IR8 variety, 51–52
  karyotype, 36–39
  linkage group, 48–49
  linkage map, 50
  marker genes, 44–48
  monosomy, 43
  pachytene chromosomes, classification,
        38
  phenotypes of mutants, 46–48
  pigmentation, 48
  species, 31–36
  subgroups, 32–36
    angustifoliae, 34
    granulatae, 34
    latifoliae, 32
    sativae, 32, 34–36
  tetraploid, 42–43
  triploid, 40–42
  trisomics, 43
  types, 31–32
    indica, 31, 36
    japonica, 32, 36, 51
    javanica, 32, 36
  yield, low, 50
Rice blast disease, 48
Rice dwarf virus
  host range, 391
  structural characteristics, 383
*Rules for Standardization of Gene
        Symbols*, 45
Rye, 59, 62, 65
  -wheat amphiploid, 62

Satellite of tobacco necrosis virus
  host range, 391
  structural characteristics, 383
Scrophulariaceae, 333
Secale, see Rye
Slime mold, cellular, 427–431
  cytogenetics, 430
  development, 428
  diploid, 430
  fruiting body, 428
  growth, 427
  haploid, 430
  life cycle, 427–428
  mutants, list of, 428–429
  pseudoplasmodia, 428
  slugs, 428
  spore germination, 428
  syngamy, 430
Snapdragon, *see Antirrhinum majus*
Sodium azide, 99
Sorghum, 3
Southern bean mosaic virus
  host range, 392
  structural characteristics, 384
Squash mosaic virus
  host range, 392
  structural characteristics, 384

Tectobacter vulgaris (delta particle), 506,
        514, 553, 560
Tetrahymena pyriformis, 433–467
  antigen, 434
  assortment, phenotypic, 441–446
  cell, 434–435
    cycle, 435
  cilia, 434–435
  clonal cycle, 440
  conjugation, 437–438
  cross-over value, 456
  DNA
    of ciliary body, 460
    cortical, 460
    macronuclear, 435–437
    micronuclear, 435–437
    mitochondrial, 459
    synthesis, 435–437
  exclusion, genomic, 438–440, 446

Tetrahymena pyriformis (*cont.*)
    of defective characters, 438–439
    genes
      list of, 448–456
    genetics
      of asexual reproduction, 441–446
      of cell cortex, 460
      of immobilization serotypes, 458–459
      of mating types, 457–458
      of mitochondria, 459
    immaturity, 440
    immobilization serotypes, 458–459
    inbreeding, 445
    linkage groups, 456–457
    linkage maps, 456–457
    macronucleus, 435
    mating types, 457–458
    maturity, 440
    micronucleus, 435
    mitochondria, genetics of, 459
    mutagenesis, 446–448
    Nanney's work, 433
    phenotype, diversification of, 441
    reproduction, asexual, 441–446
    senescence, 440
    strain
      congenic, 445–446
      inbred, 445–447
      star, 445–446
    syngen *1,* 434–460
    syngen *2,* 460–461
    syngen *7,* 461
    syngen *8,* 461
    syngen *9,* 462
    syngen *10,* 462
Tobacco, *see Nicotiana*
Tobacco etch virus
    host range, 393
    structural characteristics, 384
Tobacco mosaic virus
    host range, 393
    structural characteristics, 384
Tobacco necrosis virus
    host range, 394
    structural characteristics, 384
Tobacco rattle virus
    host range, 395

Tobacco rattle virus (*cont.*)
    structural characteristics, 384
Tobacco ringspot virus
    host range, 395–396
    structural characteristics, 385
Tobacco streak virus
    host range, 395
    structural characteristics, 385
Tomato, 247–280
    aneuploidy, 251
    anthocyanin mutant, 272
    boron deficiency, 271
    carotenoid, 271
    centromere, positioning on genetic
      linkage map, 266
    chimera, 248
    chlorophyll variegation, 270
    chromosome map, 268
    cultivation, 247
    deficiency, interstitial, 266
    disomics, 266
    fruit, 271
    gene
      distribution, 269–270
      number per chromosome, 269
    genetics
      developmental, 271–273
      physiological, 271–273
    hybrid, 248
    linkage, 251
      map, 251, 267, 269
      markers, list of, 252–265
      test, 268–269
    mineral transport, 271–272
    monosomics, 266
    mutants, list of, 252–265
      anthocyanin, 272
      by chemicals, 249–250
      chlorophyll, 273
      chlorotic, 273
      gene, 251
      induced, 249–250
      laceolate leaved, 272
      by neutrons, 249
      temperature sensitive, 272
      thiamine-requiring, 273
      unstable, 270–271

Tomato (*cont.*)
   wilty, 272
   by x-rays, 249
   phenotypes, list of mutant, 252–265
   pigmentation of fruit, 271
   plasticity, regenerative, 248–249
   propagation method, 247–249
   relationship, interspecific, 273–274
   seed yield, 247
   trisomics, 251, 266
   trisomic ratio method, 251
Tomato bushy stunt virus
   host range, 396
   structural characteristics, 385
Tomato Genetics Cooperative, 251, 269
Tonella, 333
Triethylene melamine, 485, 547
Triticale, see Wheat-rye amphiploid
Triticum, see Wheat
Turnip crinkle virus
   host range, 396
   structural characteristics, 385
Turnip yellow mosaic virus
   host range, 397
   structural characteristics, 385

Viruses of plants, 381–413
   see individual viruses

Wall cress, *see Arabidopsis*
Wheat, 59–91
   addition of alien chromosomes, 77–80
   aestivum-type common wheat, evolution
      of, 63
   allopolyploid, 60–61
   amphiploid, 62
   aneuploid, 66–73
      definition of, 71
      monosomics, 66–70
      nullisomics, 66–70
      secondary, 71–73
      tetrasomics, 70–71
      trisomics, 70–71
   Bowden's classification, 59
   centromere, 71, 73

Wheat (*cont.*)
   Chinese spring variety, 66–67
   chromosomes, 78–79
      alien, 65, 81
      pairing, 64
         genetic control of, 81–84
      telocentric, 64, 71–73
   classification, 59–62
   diploid, 60, 62
   disease resistance, 73
   ditelosomics, 72
   durum, 64
   emmer, 62–64
   evolution of aestivum-type common, 63
   genes, 73–77
      and chromosome location, 74–76
      list of, 74–76
      maps, 73–77
      *p*H, 81–83
      symbols, 74–76
   gigas type, 70
   hexaploid, 62, 66, 79, 84
   isochromosome, 72
   linkage data, 79
   monotelosomics, 72
   mutants, 85–86
   mutation, 85–86
   nullisomy, 80
   origin, 62–66
   recessive, 67, 68
   relationships, 62–66
   species, list of, 60–61
   substitution
      alien, 80–81
      lines, 68–69, 80
   tetraploid, 62–63, 85
   tetrasomics, 71
   transfer of characters from alien species
      to wheat, 83–84
   translocation, 84–85
   trisome transmission, 70
   univalent shift, 69
Wheat grass, *see Agropyron*
Wheat striate mosaic virus
   host range, 397
   structural characteristics, 385
White clover mosaic virus

White clover mosaic virus *(cont.)*
    host range, 397
    structural characteristics, 386
Wild cucumber mosaic virus
    host range, 397
    structural characteristics, 386
Wound tumor virus
    host range, 398
    structural characteristics, 386

Xanthomonas oryzae, see Leaf blight

Zea mays, see Corn
Zentralinstitut für
           Kulturpflanzenforschung in
           Gatersleben, East Germany,
           270

Contents of Other Volumes

Volume 1: Bacteria, Bacteriophages, and Fungi

The Classification and Evolution of
Prokaryotes and Eukaryotes
    LYNN MARGULIS
    Biological Science Center
    Boston University
    Boston, Massachusetts

The Bacteria

Diplococcus pneumoniae
    ELENA OTTOLENGHI-NIGHTINGALE
    University Affiliated Center for Child
      Development
    Georgetown University Hospital
    Washington, D.C.

Pseudomonas
    BRUCE W. HOLLOWAY
    Monash University
    Clayton, Victoria, Australia

Bacillus subtilus
    FRANK E. YOUNG AND
    GARY A. WILSON
    University of Rochester
    School of Medicine and Dentistry
    Rochester, New York

Escherichia coli—An Overview
    ROY CURTISS III
    FRANCIS L. MACRINA, AND
    JOSEPH O. FALKINHAM III
    Department of Microbiology
    Institute of Dental Research
    Cancer Research and Training
      Center
    University of Alabama
    Birmingham, Alabama

A Linkage Map and Gene Catalogue for
Escherichia coli
    AUSTIN L. TAYLOR AND
    CAROL DUNHAM TROTTER
    University of Colorado Medical
      Center
    Denver, Colorado

F-prime Factors of Escherichia coli
    K. BROOKS LOW
    School of Medicine
    Yale University
    New Haven, Connecticut

The Lactose Operon of Escherichia coli
    RICHARD E. SANDERS
    Massachusetts Institute of
      Technology
    Cambridge, Massachusetts

The Lac Z Gene of Escherichia coli
  IRVING ZABIN
  School of Medicine and Molecular
    Biology Institute
  University of California
  Los Angeles, California

*Tryptophan Synthetase of Escherichia coli
and Its Structural Gene*
  CHARLES YANOFSKY
  Stanford University
  Stanford, California

*The Alkaline Phosphatase of
Escherichia coli*
  ANNAMARIA TORRIANI
  Massachusetts Institute of
    Technology
  Cambridge, Massachusetts

Bacterial Ribosomes
  JULIAN DAVIES
  University of Wisconsin
  Madison, Wisconsin

Salmonella
  MARK LEVINTHAL
  Purdue University
  West Lafayette, Indiana

  KENNETH E. SANDERSON
  University of Calgary
  Calgary, Alberta, Canada

*The Histidine Operon of
Salmonella typhimurium*
  BRUCE N. AMES
  University of California
  Berkeley, California

  PHILIP E. HARTMAN
  Johns Hopkins University
  Baltimore, Maryland

Streptomyces coelicolor
  DAVID A. HOPWOOD AND
  KEITH F. CHATER
  John Innes Institute
  Norwich, England

The Bacteriophages

Bacterial Viruses of Genetic Interest
  ELLEN G. STRAUSS AND
  JAMES H. STRAUSS
  Division of Biology
  California Institute of Technology,
  Pasadena, California

RNA Bacteriophages
  WALTER FIERS
  State University of Ghent
  Ghent, Belgium

Episomes
  ALLAN CAMPBELL
  Stanford University
  Stanford, California

Bacteriophage Lambda
  WACLAW SZYBALSKI
  University of Wisconsin Medical
    Center
  Madison, Wisconsin

Bacteriophage $\phi X174$
  ROBERT L. SINSHEIMER
  California Institute of Technology
  Pasadena, California

Bacteriophage T_4
  WILLIAM B. WOOD
  California Institute of Technology
  Pasadena, California

The Lysozyme Cistron of T_4 Phage
  YOSHIMI OKADA
  Faculty of Science,
  University of Tokyo
  Tokyo, Japan

The Fungi

Phycomyces
  ENRIQUE CERDÁ-OLMEDO
  Universidad de Sevilla
  Sevilla, Spain

Saccharomyces
FRED SHERMAN AND
CHRISTOPHER W. LAWRENCE
University of Rochester School of
Medicine and Dentistry
Rochester, New York

Schizosaccharomyces pombe
HERBERT GUTZ
University of Texas at Dallas
Richardson, Texas

HENRI HESLOT
Institut National Agronomique
Paris-Grignon, France

URS LEUPOLD
Institut für allgemeine Mikrobiologie
der Universität
Bern, Switzerland

NICOLA LOPRIENO
Instituto di Genetica dell'Università e
Laboratorio di Mutagenesi e Dif-
ferenziamento del Consiglio Na-
tionale della Ricerche
Pisa, Italy

Aspergillus nidulans
A. JOHN CLUTTERBUCK
University of Glasgow
Glasgow, Scotland

Neurospora crassa
RAYMOND W. BARRATT
School of Science
Humboldt State College
Arcata, California

Podospora anserina
KARL ESSER
Leh Ruhr-Universität Bochum
Bochum, West Germany

Sordaria
LINDSAY S. OLIVE
University of North Carolina
Chapel Hill, North Carolina

Ascobolus
BERNARD DECARIS
JACQUELINE GIRARD, AND
GÉRARD LEBLON
Université de Paris-Sud
Orsay, France

Ustilago maydis
ROBIN HOLLIDAY
National Institute for Medical
Research
Mill Hill
London, England

Schizophyllum commune
JOHN R. RAPER
Harvard University
Cambridge, Massachusetts

ROBERT M. HOFFMAN
Genetics Unit
Massachusetts General Hospital
Boston, Massachusetts

Coprinus
JEAN LOUIS GUERDOUX
Centre National de la Recherche
Scientifique
Gif-sur-Yvette, France

Volume 3: Invertebrates of Genetic Interest

Molluscs of Genetic Interest
JAMES J. MURRAY, JR.
University of Virginia
Charlottesville, Virginia

Insects of Genetic Interest

The German Cockroach, Blattella germanica
MARY H. ROSS AND
DONALD G. COCHRAN
Virginia Polytechnic Institute
Blacksburg, Virginia

The Domesticated Silkmoth, Bombyx mori
YATARO TAZIMA
National Institute of Genetics
Misima, Japan

HIROSHI DOIRA
Kyushu University
Fukuoka, Japan

HIROMU AKAI
National Sericultural Experiment Station
Tokyo, Japan

The Mediterranean Meal Moth, Ephestia kühniella
ERNST W. CASPARI
University of Rochester
Rochester, New York

FREDERICK J. GOTTLIEB
University of Pittsburgh
Pittsburgh, Pennsylvania

The Flour Beetles Tribolium castaneum and T. confusum
ALEXANDER SOKOLOFF
California State College
San Bernadino, California

The Honey Bee, Apis mellifera
WALTER C. ROTHENBUHLER
Ohio State University
Columbus, Ohio

The Parasitoid Hymenoptera, Habrobracon and Mormoniella
JOSEPH D. CASSIDY, O.P.
Northwestern University
Evanston, Illinois

Lower Diptera with Giant Chromosomes

Rhynchosciara
CRODOWALDO PAVAN AND
PATRICIA FERN SANDERS
University of Texas
Austin, Texas

ANTONIO BRITO DA CUNHA
Universidade de São Paulo
São Paulo, Brazil

Sciara
NATALIA GABRUSEWYCZ-GARCIA
State University of New York
Syracuse, New York

Chironomus
KLAUS HAGELE
Ruhr-Universität Bochum
Bochum, West Germany

Glyptotendipes
LUDWIG WALTER
Ruhr-Universität Bochum
Bochum, West Germany

Mosquitoes and Flies of Genetic Interest

Anopheline Mosquitoes
MARIO COLUZZI
University of Rome
Rome, Italy

JAMES B. KITZMILLER
University of Illinois
Urbana, Illinois

Aedes
    KARAMJIT S. RAI
    University of Notre Dame
    Notre Dame, Indiana

    W. KEITH HARTBERG
    Georgia Southern College
    Statesboro, Georgia

Culex
    A. RALPH BARR
    School of Public Health
    University of California
    Los Angeles, California

The House Fly, Musca domestica
    RICCARDO MILANI
    University of Pavia
    Pavia, Italy

The Australian Sheep Blow Fly,
Lucilia cuprina
    MAXWELL J. WHITTEN
    GEOFFREY G. FOSTER
    JAMES T. ARNOLD, AND
    CHRISTINE KONOWALOW
    Division of Entomology
    Commonwealth Scientific and
       Industrial Research Organization
    Canberra, Australia

The Genus *Drosophila*

The Phylogeny, Ecology, and Geography
of Drosophila
    LYNN H. THROCKMORTON
    University of Chicago
    Chicago, Illinois

A Catalogue of Drosophila Metaphase
Chromosome Configurations
    FRANCES E. CLAYTON
    University of Arkansas
    Fayetteville, Arkansas

    MARSHALL R. WHEELER
    University of Texas
    Austin, Texas

Drosophila ananasae
    DAIGORO MORIWAKI
    National Institute of Genetics
    Misima, Japan

    YOSHIKO N. TOBARI
    Tokyo Metropolitan University
    Tokyo, Japan

Drosophila pseudoobscura and Its
American Relatives, D. persimilis and
D. miranda
    THEODOSIUS DOBZHANSKY
    University of California
    Davis, California

    JEFFREY R. POWELL
    Yale University
    New Haven, Connecticut

The willistoni Group of Sibling Species of
Drosophila
    THEODOSIUS DOBZHANSKY
    University of California
    Davis, California

    JEFFREY R. POWELL
    Yale University
    New Haven, Connecticut

Drosophila melanogaster

Drosophila melanogaster: An Introduction
    ROBERT C. KING
    Northwestern University
    Evanston, Illinois

Temperature-Sensitive Mutations in
Drosophila melanogaster
    DAVID T. SUZUKI
    University of British Columbia
    Vancouver, B. C., Canada

The Biochemical Mutations in
Drosophila melanogaster
    STEPHEN J. O'BRIEN
    National Institutes of Health
    Bethesda, Maryland

Behavioral Mutants of Drosophila
  JOSEPH GROSSFIELD
  The City College of the City
    University of New York
  New York, New York

Mutants Affecting the Vision of
Drosophila melanogaster
  WILLIAM L. PAK
  Purdue University
  Lafayette, Indiana

Mutations Influencing Male Fertility in
Drosophila melanogaster
  LYNN J. ROMRELL
  Harvard Medical School
  Boston, Massachusetts

Y-Linked Factors Affecting Male Fertility
in Drosophila melanogaster and D. hydei
  OSWALD HESS
  University of Düsseldorf
  Düsseldorf, Germany

The Genetic Control of Oogenesis in
Drosophila melanogaster
  ROBERT C. KING
  Northwestern University
  Evanston, Illinois

  J. DAWSON MOHLER
  University of Iowa
  Iowa City, Iowa

The Puffing Activities of the Salivary
Gland Chromosomes
  MICHAEL ASHBURNER
  University of Cambridge
  Cambridge, England

The Drosophila viruses
  PHILIPPE L'HÉRITIER
  University of Clermont-Ferrand
  Clermont-Ferrand, France

Drosophila Cell Culture
  IMOGENE SCHNEIDER
  Walter Reed Army Institute of
    Research,
  Washington, D.C.

Volume 4: Vertebrates of Genetic Interest

Amphibia

The Axolotl, Ambystoma mexicanum
  RUFUS R. HUMPHREY
  University of Indiana
  Bloomington, Indiana

Frogs of the Genus Rana
  LEON W. BROWDER
  University of Calgary
  Calgary, Alberta, Canada

Xenopus laevis
  JOHN B. GURDON AND
  HUGH R. WOODLAND
  University of Oxford
  Oxford, England

The RNA of the Mature Xenopus Oocyte
  ERIC H. DAVIDSON AND
  BARBARA R. HOUGH
  California Institute of Technology
  Pasadena, California

Working Maps of Amphibian Lampbrush
Chromosomes
  HAROLD G. CALLAN AND
  LYDIA LLOYD
  The University, St. Andrews
  Fife, Scotland

Fishes and Birds

The Platyfish, Xiphophorus maculatus
  KLAUS KALLMAN
  New York Aquarium
  Seaside Park
  Coney Island, New York

The Medaka, Oryzias latipes and the
Guppy, Lebistes reticularis
  Toki-o Yamamoto
  Meijo University
  Nagoya, Japan

Avian Genetics
  Ursula K. Abbott
  University of California
  Davis, California

Mammals

The Laboratory Mouse, Mus musculus
  Margaret C. Green
  The Jackson Laboratory
  Bar Harbor, Maine

The Laboratory Rat, Rattus norvegicus
  Joy Palm
  The Wistar Institute
  Philadelphia, Pennsylvania

The Genus Peromyscus
  David I. Rasmussen
  Arizona State University
  Tempe, Arizona

The Golden Hamster,
Mesocricetus auratus
  Roy Robinson
  St. Stephen's Road Nursery
  London, England

The Guinea Pig, Cavia porcellus
  Roy Robinson
  St. Stephen's Road Nursery
  London, England

The Rabbit, Oryctolagus cuniculus
  Richard R. Fox
  The Jackson Laboratory
  Bar Harbor, Maine

The Chinchilla, Chinchilla lanigera
  Roy Robinson
  St. Stephen's Road Nursery
  London, England

The Horse, Equus caballus, and the
Donkey, E. asinus
  Ralph Bogart
  Oregon State University
  Corvalis, Oregon

The Domestic Cat, Felis catus
  Roy Robinson
  St. Stephen's Road Nursery
  London, England

The American Mink, Mustela vison
  Roy Robinson
  St. Stephen's Road Nursery
  London, England

The Red Fox, Vulpes vulpes
  Roy Robinson
  St. Stephen's Road Nursery
  London, England

The Dog, Canis familiaris
  John M. Stewart
  The Jackson Laboratory
  Bar Harbor, Maine

  J. Paul Scott
  Bowling Green State University
  Bowling Green, Ohio

Blood-Group Alleles of Domesticated
Animals
  Benjamin A. Rasmusen
  University of Illinois at Urbana
  Urbana, Illinois

Distribution and Probable Homologies of
Some Genes Determining Coat Color in
Mammals
  Anthony G. Searle
  Medical Research Council
    Radiobiology Unit
  Harwell, Oxfordshire, England

A Bibliography of Reference Works
Bearing on Human Genetics—1963–1973
  Robert C. King
  Northwestern University
  Evanston, Illinois

Data on Human Biochemical Genetics
   DAVID W. HOLLISTER AND
   DAVID L. RIMOIN
   Harbor General Hospital
   Torrance, California

Identification of Human Mitotic
Chromosomes
   KARIN E. BUCKTON
   Western General Hospital
   Edinburgh, Scotland

Gene Localizations and Other Properties of
Human Chromosomes
   ROBERT C. KING
   Northwestern University
   Evanston, Illinois

The Structure of Immunoglobulins and
Immunoglobin Genes
   JOSEPH A. GALLY
   Meharry Medical College
   Nashville, Tennessee

Immunoglobin Allotypes
   GEORGE A. GUTMAN
   Stanford University
   Stanford, California

Animal   Viruses   and   Somatic   Cell
Lines

Animal Viruses of Genetic Interest
   FRANK FENNER
   Australian National University
   Canberra, Australia

Reference Animal Cell Lines
   JOHN E. SHANNON AND
   MARVIN L. MACY
   American Type Culture Collection
   Rockville, Maryland

Somatic Cell Hybrids Formed Between
Lines of Cultured Cells
   ELIZABETH K. SELL AND
   R. S. KROOTH
   Columbia University
   College of Physicians and Surgeons
   New York, New York

Volume 5: Molecular Genetics (Tentative Listing)

The Repeated DNA in Eukaryotes
   NEIL A. STRAUSS
   University of Toronto,
   Toronto, Ontario, Canada

The Cytological Localization of Repetitive
DNA
   RONALD A. ECKHARDT
   Brooklyn College of CUNY
   Brooklyn, New York

The Chromosomal Proteins
   RU CHIH C. HUANG AND
   REX HJELM
   Johns Hopkins University
   Baltimore, Maryland

Organization of Genetic Material in the
Macronucleus of Hypotrichous Ciliates
   K. G. MURTI
   University of Colorado
   Boulder, Colorado

The Proteins that Interact with Nucleic
Acids
   LOREN A. DAY
   Public Health Research Institute of
      the City of New York
   New York, New York

The Histones of Sperm
   DAVID BLOCK
   University of Texas
   Austin, Texas

The Organization and Size of Replicons
   Robert B. Painter
   University of California Medical
     Center
   San Francisco, California

RNAs of Nuclei
   Bertil Daneholt
   Karolinska Institutet
   Stockholm, Sweden

Cytoplasmic Messenger RNA
   Anton J. M. Berns and
   Hans Bloemendal
   University of Nijmegen
   Nijmegen, Holland

The Transfer RNAs
   James T. Madison
   U.S. Plant, Soil, and Nutrition
     Laboratory
   U.S. Department of Agriculture
   Ithaca, New York

Eukaryotic Ribosomes
   Mary G. Hamilton
   Sloan-Kettering Institute
   New York, New York

Chloroplast DNA
   Ruth Sager
   Hunter College
   New York, New York

Chloroplast Ribosomes
   Erhardt Stutz
   University of Neuenberg
   Neuenberg, Switzerland

   Arminio Boschetti
   University of Bern
   Bern, Switzerland

*Nucleocytoplasmic Interactions in
Acetabularia*
   Hans G. Schweiger
   Max Planck Institute for Cell Biology
   Wilhelmshaven, Germany

Mitochondrial DNA
   Margit M. K. Nass
   University of Pennsylvania School of
     Medicine
   Philadelphia, Pennsylvania

Mitochondrial Ribosomes
   Thomas W. O'Brien and
   David E. Matthews
   University of Florida
   Gainesville, Flordia

*Genetic Variants of Enzymes Detected by
Zone Electrophoresis*
   Charles R. Shaw and
   Rupi Prasad
   M. D. Anderson Hospital and Tumor
     Institute
   Houston, Texas